Intermediate
Algebra
An Interactive Approach

Intermediate
Algebra

An Interactive Approach

SECOND EDITION

Linda Pulsinelli
Patricia Hooper

Department of Mathematics
Western Kentucky University

MACMILLAN PUBLISHING COMPANY
NEW YORK

Collier Macmillan Publishers
London

For two of our teachers who sparked our early interest in mathematics: Ann Hancock Collins and Norman E. Cromack

Copyright © 1987, Macmillan Publishing Company, a division of Macmillan, Inc.

Printed in the United States of America

Earlier edition copyright © 1983 by Macmillan Publishing Company.

Macmillan Publishing Company
866 Third Avenue, New York, New York 10022

Collier Macmillan Canada, Inc.

Library of Congress Cataloging in Publication Data

Pulsinelli, Linda Ritter.
  Intermediate algebra.

  Includes index.
    1. Algebra.   I. Hooper, Patricia (Patricia I.)
II. Title.
QA154.2.P85   1987   512.9   86-8497
ISBN 0-02-396990-3

Printing: 2 3 4 5 6 7 8   Year: 7 8 9 0 1 2 3 4 5

ISBN   0-02-396990-3

Preface

Intermediate Algebra: An Interactive Approach is designed to be useful to students in bridging the gap between high school and college algebra. It is generally assumed that a student using this textbook has been exposed to an introductory algebra course but is not ready to tackle traditional college algebra.

In order to make this mathematics textbook readable, we have worded explanations in clear, concise language understandable to students at this level. Our general approach is to start from a fundamental idea with which students are familiar and proceed to a related concept in the most straightforward, intuitive way possible.

Our experience with intermediate algebra students leads us to believe that they must practice each new skill as soon as it has been presented. For this reason the book includes several unique features that are designed to provide maximum reinforcement. Each chapter in the book follows the same basic structure.

Motivational Applied Problem

At the beginning of each chapter we have presented an applied problem that can be solved after the student has mastered the skills in that chapter. Its solution appears within the chapter.

Explanations

We have tried to avoid the "cookbook" approach to algebra by including a straightforward and readable explanation of each new concept. Realizing that students at the intermediate level become easily bogged down in reading lengthy explanations, we have attempted to make our explanations as brief as possible without sacrificing rigor.

Highlighting

Definitions, properties, theorems, and formulas are highlighted in boxes throughout the book for easy student reference. In most cases a rephrasing of a generalization in words accompanies the symbolic statement, and it is also highlighted in a box.

Examples

Immediately following the presentation of a new idea, several completely worked-out examples appear along with several partially worked-out examples with blanks to be filled in by the student. These examples are completed correctly at the end of each section and the student is advised to check his or her work immediately.

Trial Runs

Sprinkled throughout each section are several short Trial Runs, a list of six or eight problems to check on the student's grasp of a new skill. The answers appear at the end of the section.

Exercise Sets

Each section concludes with an extensive Exercise Set in which each odd-numbered problem corresponds closely to the following even-numbered problem.

Stretching the Topics

At the end of each Exercise Set there are several problems designed to challenge the better students by extending the skills learned in the chapter to the next level of difficulty.

Checkups

Following each Exercise Set, a list of about 10 problems checks on the student's mastery of the most important concepts in the section. Each Checkup problem is keyed to comparable examples in the section for restudy if necessary.

Problem Solving

One section of almost every chapter involves switching from words to algebra. By including such a section in each chapter, we are attempting to treat problem solving as a natural outgrowth of acquiring algebraic skills.

Chapter Summaries

Each chapter concludes with a summary in which the important ideas are again highlighted, in tables when possible. New concepts are presented in symbolic form and verbal form, accompanied by a typical example.

Speaking the Language of Algebra

Following the summary, we have included a group of sentences to be completed *with words* by the student. Algebra students (especially those in self-paced programs) often lack the opportunity to ''speak mathematics.'' We hope that these short sections will help them develop a better mathematics vocabulary.

Review Exercises

A list of exercises reviewing all the chapter's important concepts serves to give the student an overview of the content. Each problem is keyed to the appropriate section and examples.

Practice Test

A Practice Test is included to help the student prepare for a test over the material in the chapter. Once again, each problem is keyed to the appropriate chapter sections and examples.

Sharpening Your Skills

Finally, we have included a short list of exercises that will provide a cumulative review of concepts and skills from earlier chapters. Retention seems to be a very real problem with students at this level, and we hope that these exercises will serve to minimize that problem. Each cumulative review exercise is keyed to the appropriate chapter and section.

Throughout the book we have adhered to a rather standard order of topics, making an attempt to connect new concepts to old ones whenever appropriate. This modified spiraling technique is designed to help students maintain and overview of the content. Success in future courses seems to us to hinge on students' seeing that algebra is a logical progression of ideas rather than a set of unrelated skills to be memorized and forgotten.

The answers to the odd-numbered exercises in the Exercise Sets appear in the back of the book together with answers for *all* items in Stretching the Topics, Checkups, Speaking the Language of Algebra, Review Exercises, Practice Tests, and Sharpening Your Skills.

More assistance for students and instructors can be found among the supplementary materials that accompany this book.

Instructor's Manual with Test Bank

The Instructor's Manual contains the answers for all exercises in the Exercise Sets and Stretching the Topics. In addition there are six Chapter Tests (four open-ended and two multiple choice) for each chapter and three Final Examinations (two open-ended and one multiple choice). Answers to these tests and examinations also appear in the Instructor's Manual.

Student's Solutions Manual

The Student's Solutions Manual, written by Rebecca Stamper, contains step-by-step solutions for the even-numbered exercises in the Exercise Sets and for *all* items in the Review Exercises, Practice Tests, Sharpening Your Skills, and exercises involving word problems. Using the same style as appears in the text, these solutions emphasize the procedure as well as the answer.

Video Tapes

A series of 10 video tapes (each 20 to 30 minutes in length) provides explanations for some of the more difficult topics in the course.

Audio Tapes

A series of 10 audio cassettes (each 20 to 30 minutes in length) also offer explanations for the more difficult topics. Keyed to examples in the text, these cassettes encourage students to work along.

Computerized Test Generator

A set of computer-generated tests is available for producing either a 10-item test for each chapter *or* tests of any length from objective-referenced items. Cumulative tests and final exams may also be constructed using the objective-referenced items.

Acknowledgments

The writing of this book would not have been possible without the assistance of many people. We express our appreciation to our indefatigable typist Maxine Worthington, to Becky Stamper for carefully working all our problems, to our families for tolerating our obsessive work schedules, to our Mathematics Editors Gary Ostedt and Bob Clark for their enthusiastic support, and to our Production Supervisor Elaine Wetterau for her efficiency and expertise.

We also thank our reviewers: Mary Jean Brod, University of Montana; Donald R. Johnson, Scottsdale Community College; Adele Le Gere, Oakton Community College; Gerald J. LePage, Bristol Community College; Lois Miller, El Camino College; Harold M. Nerr, University of Wisconsin; Dan Streeter, Portland State University; and Jack W. Rotman, Lansing Community College; for their careful scrutiny and helpful comments.

L.R.P.
P.I.H.

Contents

Working with Numbers and Variables

1

Charlie is a part-time college student. He must pay $50 per course credit hour and he has saved $800. Write an algebraic expression for the amount Charlie will have left after enrolling for *h* credit hours.

We review operations with real numbers here at the start, so that we can use them with confidence throughout the remainder of the text. In this chapter we recall how to:

1. Recognize different kinds of numbers.
2. Perform operations with real numbers.
3. Use the properties of the real numbers.
4. Work with variables and constants in algebraic expressions.
5. Find the absolute value of a quantity.
6. Switch from word expressions to algebraic expressions.

1.1 Working with Sets of Numbers

In the same way that your knowledge of different kinds of numbers has changed throughout your life, mathematicians have introduced new sets of numbers as they were needed. We shall use standard set notation to represent our number sets, naming each with a capital letter and listing the members of each set with commas between them and enclosed in braces.

Recognizing Sets of Numbers

The first set of numbers ever used for measuring and counting was the set of **natural numbers**, represented by N.

Natural Numbers

$N = \{1, 2, 3, 4, \ldots\}$

The dots following the number 4 indicate that this set continues forever; it is an **infinite set** containing no last element. However, N is a **well-defined** set, which means that we can determine exactly what numbers belong to the set without listing all its members.

The need for a symbol to represent *none* of some measure led to the introduction of **zero**, and the set of **whole numbers** was invented. We shall use W to represent this set.

Whole Numbers

$W = \{0, 1, 2, 3, 4, \ldots\}$

Notice that the set of whole numbers is also a well-defined, infinite set. Notice, too, that every natural number is a whole number.

In order to measure losses, mathematicians expanded their set of numbers to include the opposites (or **negatives**) of the natural numbers, and the set of **integers** came into existence. We represent the set of integers with the letter J.

Integers

$J = \{\ldots, -3, -2, -1, 0, 1, 2, 3, \ldots\}$

This set is also well defined and infinite, and it contains all the whole numbers.

Let's take a moment here to introduce some mathematical shorthand. To say that a number belongs to a set (or is an **element** of a set) we use the symbol "ϵ."

Symbol	Meaning	Examples
ϵ	Is an element of	$5 \in N$
		$-6 \in J$
$\notin$	Is not an element of	$-2 \notin W$
		$\dfrac{1}{2} \notin J$
$=$	Is equal to	$2 + 3 = 5$
		$7 \cdot 4 = 28$
$\neq$	Is not equal to	$-6 \neq 0$
		$\dfrac{2}{3} \neq \dfrac{3}{2}$

A need to represent *parts* of integers led mathematicians to define the set of **rational numbers** containing every number that can be expressed as a fraction with an integer as the numerator and a nonzero integer as the denominator. Using a to represent the numerator, b to represent the denominator, and Q to represent the set of rational numbers, we define this new set.

> **Rational Numbers**
>
> $$Q = \left\{ \frac{a}{b} : a \in J, \, b \in J, \, b \neq 0 \right\}$$

To be a member of Q a number must be expressible as a quotient of integers. You should agree that

$$\frac{2}{3}, \quad \frac{1}{2}, \quad \frac{7}{93}, \quad \frac{-5}{8}, \quad \frac{17}{11}, \quad \frac{-1}{10}, \quad \frac{-9}{5}$$

are all rational numbers.

Since every integer can be expressed as a quotient of integers, every integer is also a rational number. Some decimal numbers are also rational numbers; in particular, any decimal number that *terminates* or *repeats* in a fixed block of digits is a rational number.

> $6 \in Q$ because $6 = \dfrac{6}{1}$
>
> $-3 \in Q$ because $-3 = \dfrac{-3}{1}$
>
> $0 \in Q$ because $0 = \dfrac{0}{1}$
>
> $0.37 \in Q$ because $0.37 = \dfrac{37}{100}$
>
> $0.333 \ldots \in Q$ because $0.333 \ldots = \dfrac{1}{3}$
>
> $-0.2727 \ldots \in Q$ because $-0.2727 \ldots = \dfrac{-3}{11}$

> To recognize members of the set of rational numbers, we must look for numbers that are integers or common fractions or repeating decimals or terminating decimals.

Although you may think that we have considered all the possible sets of numbers, there are many numbers, called **irrational numbers**, which are *not* expressible as quotients of integers. Some examples might jog your memory:

$$\sqrt{2}, \quad \sqrt{3}, \quad -\sqrt{5}, \quad \pi, \quad 3 + \sqrt{7}$$

are members of the set of irrational numbers, H.

If we consider all the rational numbers together with all the irrational numbers, we can form the very important set of **real numbers**, represented by R.

Real Numbers

$R = \{$numbers that are rational *or* irrational$\}$

Every number that we have discussed so far belongs to the set of real numbers. Until further notice, all our work in algebra will deal with real numbers.

Example 1. To which number sets does 3 belong?

Solution

$$3 \in N$$
$$3 \in W$$
$$3 \in J$$
$$3 \in Q$$
$$3 \in R$$

Now try Example 2.

Example 2. To which number sets does -0.5 belong?

Solution

$$-0.5 \in \text{\_\_\_\_\_}$$
$$-0.5 \in \text{\_\_\_\_\_}$$

Check your work on page 9. ▶

You complete Example 3.

Example 3. For the set $\{-5, 0, 0.1, \sqrt{5}, \pi, 4, \frac{24}{5}\}$, complete the following list.

Natural numbers: 4

Whole numbers: 0, 4

Integers: ____, ____, ____

Rational Numbers: ____, ____, ____, ____, ____

Irrational numbers: ____, ____

Check your work on page 9. ▶

Using the Number Line

To better understand the different sets of numbers, we can use a **real number line**. To construct a number line, we draw a straight line and then choose a zero point and a length to represent

1 unit. All points spaced 1 unit apart are labeled to correspond to the integers in order, with positive integers to the right of zero and negative integers to the left of zero.

The arrows at either end of the number line show that the line extends indefinitely in both directions, so **we can locate a point corresponding to any rational or irrational number on this number line**. Although our location of irrational numbers will not be completely accurate, we shall be satisfied with a good approximation. On the number line, we locate points corresponding to real numbers by placing a solid dot in the appropriate position. This is called **graphing a number**.

Always be sure to label your points so that a reader will not be confused about your meaning. See if you can graph the numbers in Example 4.

Example 4. On a real number line, graph the numbers 3, 1.1, $-\sqrt{2}$, $\dfrac{9}{2}$, -2.5, and 0.

Solution

Check your work on page 9. ▶

The number line provides us with a handy means of visualizing the set of real numbers and allows us to make observations about *order* within that set. Suppose that we let a and b represent any real numbers.

> If a lies to the *right* of b on the number line, a must be **greater than** b.
> If a lies to the *left* of b on the number line, a must be **less than** b.
> If a and b occupy the *same* position on the number line, a must **equal** b.

Mathematicians have invented symbols to represent these three situations.

Symbols	Meaning	Number Line
$a > b$	a is greater than b	
$a < b$	a is less than b	
$a = b$	a is equal to b	

Because of the orderly nature of the real numbers, it is always possible to compare two real numbers using exactly one of these three statements.

Trichotomy Principle. Given any two real numbers, a and b, exactly *one* of the following statements must be true:

$$a > b \quad \text{or} \quad a < b \quad \text{or} \quad a = b$$

The statements $a > b$ and $a < b$ are called **inequalities**, and the number line helps us see that the following property for inequalities makes sense.

Transitive Property. Let a, b, and c be real numbers. If $a < b$ and $b < c$, then $a < c$.

For instance, if we know that $\sqrt{2} < 1.5$ and $1.5 < \sqrt{3}$, we may safely conclude that $\sqrt{2} < \sqrt{3}$.

Example 5. Compare the numbers using the symbols $<$, $>$, or $=$.

$$-2 < 4$$
$$0 > -1$$
$$3 < \pi$$
$$\sqrt{25} = 5$$

Now you try Example 6.

Example 6. Compare the numbers using the symbols $<$, $>$, or $=$.

$$-6 \underline{\quad} -5.5$$
$$2 \underline{\quad} \sqrt{3}$$
$$\sqrt{9} \underline{\quad} 3$$

Check your work on page 9. ▶

If a real number lies to the *right* of zero on the number line, we know that it is *positive*. If a real number lies to the *left* of zero on the number line, we know that it is *negative*. Therefore, we can state whether a number is positive or negative using our symbols of inequality.

$a > 0$	means	"a is a positive real number."
$a < 0$	means	"a is a negative real number."

Example 7. Use an inequality to state whether each number is positive or negative.

$$5, \quad 0.01, \quad -\pi$$

Solution

$$5 > 0$$
$$0.01 > 0$$
$$-\pi < 0$$

Now you try Example 8.

Example 8. Use an inequality to state whether each number is positive or negative.

$$\frac{-9}{2}, \quad \sqrt{3}, \quad -\sqrt{3}$$

Solution

$$\frac{-9}{2} \underline{\quad} 0$$
$$\sqrt{3} \underline{\quad} 0$$
$$-\sqrt{3} \underline{\quad} 0$$

Check your work on page 9. ▶

Suppose that we look at the real numbers 5 and -5. On the number line, each of these numbers corresponds to a point 5 units from zero. The point corresponding to 5 lies 5 units to the *right* of zero; the point corresponding to -5 lies 5 units to the *left* of zero. Such numbers are called **opposites** of each other and we say that

$$-5 \text{ is the opposite of } 5$$

$$5 \text{ is the opposite of } -5$$

Every real number has an opposite.
Let's complete Example 9.

Example 9. Complete each statement.

-2 is the opposite of _____ .   _____ is the opposite of π.

1.6 is the opposite of _____ .   _____ is the opposite of $-\sqrt{2}$.

_____ is the opposite of $\dfrac{2}{3}$.   _____ is the opposite of 0.

Check your work on page 9. ▶

IIIⅢ➡ Trial Run

1. On the number line plot the points corresponding to the numbers $\dfrac{7}{2}$, $-\sqrt{3}$, 2.2, -3.1, 0 and 1.

<------+----+----+----+----+----+----+----+----+----+------>

Compare the numbers using $<$, $>$, or $=$.

2. 4 _____ $\sqrt{10}$ **3.** $\dfrac{5}{2}$ _____ 3 **4.** -2 _____ -2.5

Use an inequality to state whether each number is positive or negative.

_____ **5.** 0.53 _____ **6.** $-\sqrt{5}$ _____ **7.** $\dfrac{2}{3}$

Complete each statement.

8. 3 is the opposite of _____ .

9. _____ is the opposite of $\dfrac{-4}{5}$.

10. -0.25 is the opposite of _____ .

Answers are on page 10.

Finding Absolute Values of Numbers

The number line is a useful tool for understanding the important mathematical concept of **absolute value**. Stated very simply, the **absolute value of a real number is its distance from zero on the number line**.

The absolute value of 3 is 3 because its distance from zero is 3 units. The absolute value of -3 is also 3 because its distance from zero is 3 units.

> To find the absolute value of a real number, count the units between that number and zero on the number line.

To denote the absolute value of a number, we use a pair of vertical bars.

Symbols	Meaning	Evaluation
$|3|$	Absolute value of 3	$|3| = 3$
$|-3|$	Absolute value of -3	$|-3| = 3$

Example 10. Evaluate $|2|$, $|0|$, and $|-\sqrt{2}|$.

Solution

$|2| = 2$ because there are 2 units between 0 and 2.

$|0| = 0$ because there are 0 units between 0 and 0.

$|-\sqrt{2}| = \sqrt{2}$ because there are $\sqrt{2}$ units between 0 and $-\sqrt{2}$.

You complete Example 10.

Example 11. Evaluate $|-16|$, $|0.25|$, and $\left|\dfrac{-2}{3}\right|$.

Solution

$|-16| = $ _____ because there are _____ units between 0 and _____ .

$|0.25| = $ _____ because there are _____ units between 0 and _____ .

$\left|\dfrac{-2}{3}\right| = $ _____ because there are _____ units between 0 and _____ .

Check your work on page 9. ▶

From these examples, we make an important observation about absolute value.

> The absolute value of a real number is never negative.

⫸ Trial Run

Evaluate.

_____ **1.** $|5|$

_____ **2.** $|-3.2|$

_____ **3.** $|0|$

_____ **4.** $|-\sqrt{3}|$

_____ **5.** $|7| + |-2|$ _____ **6.** $|-8| - |3|$

_____ **7.** $|7 - 2|$ _____ **8.** $|-8| + |-3|$

Answers are on page 10.

▶ Examples You Completed

Example 2. To which number sets does -0.5 belong?

Solution

$$-0.5 \in Q$$
$$-0.5 \in R$$

Example 3

Solution

Natural numbers: 4

Whole numbers: 0, 4

Integers: $-5, 0, 4$

Rational numbers: $-5, 0, 4, 0.1, \dfrac{24}{5}$

Irrational numbers: $\sqrt{5}, \pi$

Example 4. On a real number line, plot the points corresponding to the numbers 3, 1.1, $-\sqrt{2}$, $\dfrac{9}{2}$, -2.5, and 0.

Solution

Example 6. Compare the numbers using the symbols $<$, $>$, or $=$.

$$-6 < -5.5$$
$$2 > \sqrt{3}$$
$$\sqrt{9} = 3$$

Example 8. Use an inequality to state whether each number is positive or negative.

$$\frac{-9}{2}, \quad \sqrt{3}, \quad -\sqrt{3}$$

Solution

$$\frac{-9}{2} < 0$$
$$\sqrt{3} > 0$$
$$-\sqrt{3} < 0$$

Example 9

Solution

-2 is the opposite of 2.

1.6 is the opposite of -1.6.

$-\dfrac{2}{3}$ is the opposite of $\dfrac{2}{3}$.

$-\pi$ is the opposite of π.

$\sqrt{2}$ is the opposite of $-\sqrt{2}$.

0 is the opposite of 0.

Example 11

$|-16| = 16$ because there are 16 units between 0 and -16.

$|0.25| = 0.25$ because there are 0.25 units between 0 and 0.25.

$\left|-\dfrac{2}{3}\right| = \dfrac{2}{3}$ because there are $\dfrac{2}{3}$ units between 0 and $-\dfrac{2}{3}$.

Answers to Trial Runs

page 7   **1.**

2. $4 > \sqrt{10}$ **3.** $\dfrac{5}{2} < 3$ **4.** $-2 > -2.5$ **5.** $0.53 > 0$ **6.** $-\sqrt{5} < 0$ **7.** $\dfrac{2}{3} > 0$

8. -3 **9.** $\dfrac{4}{5}$ **10.** 0.25

page 8   **1.** 5   **2.** 3.2   **3.** 0   **4.** $\sqrt{3}$ **5.** 9   **6.** 5   **7.** 5   **8.** 11

EXERCISE SET 1.1

Name the number sets to which each of the following numbers belong.

_____ 1. $-\dfrac{2}{3}$ _____ 2. $\dfrac{4}{5}$

_____ 3. 0 _____ 4. 1

_____ 5. $\sqrt{2}$ _____ 6. $\sqrt{5}$

_____ 7. 3.52 _____ 8. -1.39

From the set $\{-8, \ -5.2 \ -\sqrt{3}, \ -\dfrac{1}{2}, \ 0, \ \sqrt{2}, \ \dfrac{7}{2}, \ \pi, \ 5.666. \ . \ . \ ., \ 9\}$, list the members that belong to the following sets.

_____ 9. Whole numbers   _____ 10. Natural numbers

_____ 11. Rational numbers   _____ 12. Irrational numbers

_____ 13. Negative integers   _____ 14. Nonpositive integers

Decide whether each statement is true or false.

_____ 15. Every natural number is a whole number.

_____ 16. Every real number is a rational number.

_____ 17. Every rational number can be written as a quotient integers.

_____ 18. Every integer is a rational number.

_____ 19. Every irrational number is a real number.

_____ 20. Every real number is rational or irrational.

21. Use a real number line to graph the numbers $\dfrac{8}{3}, \ -\sqrt{7}, \ 3.4, \ -5.12, \ 0, \ -2, \ 0.333 \ . \ . \ .$

22. Use a real number line to graph the numbers $\dfrac{9}{2}, \ 5.2, \ -3.6, \ 1, \ -3, \ 0.66 \ . \ . \ ., \ -\sqrt{17}$.

Compare the following numbers using $<$, $>$, or $=$.

_____ 23. 3 and $\sqrt{7}$ _____ 24. $\sqrt{10}$ and 4

_____ 25. $-\dfrac{9}{2}$ and -3 _____ 26. $-\dfrac{7}{3}$ and -5

_____ 27. $\dfrac{10}{2}$ and $\sqrt{25}$ _____ 28. $\dfrac{18}{3}$ and $\sqrt{36}$

_____ 29. -1.6 and -0.7 _____ 30. -3.9 and -2.5

_____ 31. $-\dfrac{8}{3}$ and $-\sqrt{16}$ _____ 32. $-\dfrac{12}{5}$ and $-\sqrt{4}$

Use inequalities to state whether each of the following numbers is positive or negative.

_____ 33. 0.29 _____ 34. 0.74

_____ 35. $-\sqrt{6}$ _____ 36. $-\sqrt{11}$

_____ 37. $\dfrac{3}{4}$ _____ 38. $\dfrac{2}{3}$

_____ 39. -9 _____ 40. 7

Find the missing numbers in each statement.

41. 5 is the opposite of _____ .   42. -4 is the opposite of _____ .

43. $-\dfrac{7}{8}$ is the opposite of _____ .   44. $\dfrac{4}{5}$ is the opposite of _____ .

45. -3.2 is the opposite of _____ .   46. 5.6 is the opposite of _____ .

47. $\sqrt{5}$ is the opposite of _____ .   48. $-\sqrt{6}$ is the opposite of _____ .

Evaluate each of the following.

_____ 49. $|6|$ _____ 50. $|13|$

_____ 51. $|-6.8|$ _____ 52. $|-1.3|$

_____ 53. $|-\sqrt{3}|$ _____ 54. $|-\sqrt{5}|$

_____ 55. $|3| + |-2|$ _____ 56. $|7| + |-8|$

_____ 57. $|-10| - |2|$ _____ 58. $|-6| - |4|$

☆ Stretching the Topics _____

_____ 1. Evaluate $|a + 8|$ for $a = -15$.

_____ 2. Complete $-4(7 + 3)$ _____ $- |-41|$ using $<$, $>$, or $=$.

_____ 3. Evaluate $|-40.39| - (|-3.72| + |18.02|) - |-18.65|$.

Check your answers in the back of your book.

If you can complete **Checkup 1.1**, you are ready to go on to Section 1.2.

CHECKUP 1.1

_____ **1.** To which number sets does -9 belong?

_____ **2.** From the set $\left\{-3,\ -1.2,\ -\dfrac{3}{4},\ 0,\ \sqrt{2},\ \dfrac{5}{4},\ 4\right\}$ list the members that belong to the set of integers.

Compare the following numbers using $<$, $>$, or $=$.

_____ **3.** $\dfrac{8}{2}$ and $\sqrt{16}$ _____ **4.** $\dfrac{-7}{2}$ and -2.5

Use inequalities to state whether the following numbers are positive or negative.

_____ **5.** $-\dfrac{9}{5}$ _____ **6.** $\sqrt{36}$

Find the missing number in each statement.

7. $-\dfrac{5}{12}$ is the opposite of _____ .

8. 9 is the opposite of _____ .

Evaluate each of the following.

_____ **9.** $|-3|$ _____ **10.** $|7| + |-2|$

Check your answers in the back of your book.

If You Missed Problems:	You Should Review Examples:
1, 2	1–3
3, 4	5, 6
5, 6	7, 8
7, 8	9
9, 10	10, 11

1.2 Operating with Real Numbers

Although you learned to add, subtract, multiply, and divide real numbers in earlier algebra courses, a brief review is provided here to refresh your memory.

Adding and Subtracting Real Numbers

Recall that the result of performing **addition** is called a **sum** and the numbers being added are called **terms**. Some students find the number line helpful in finding sums of real numbers.

Most of us would agree that $6 + 7 = 13$. Let's see how we could find that sum using the number line. We start at the point corresponding to the number 6 and then move 7 more units to the *right*.

Indeed, we end up at the point corresponding to positive 13.

Let's try the sum $-3 + (-5)$. Here we must start at the point corresponding to -3. Then we must move 5 more units to the *left*, since we are adding -5.

Our final point is located 8 units to the *left* of zero, so our sum is negative 8.

$$-3 + (-5) = -8$$

When we added two positive numbers, our sum was positive. When we added two negative numbers, our sum was negative. The following rule describes the results of our additions.

> To add two numbers having the *same* sign, add the units (absolute values) together and give your answer that sign.

Example 1. Find $0.5 + 0.1$.

Solution: Both numbers are *positive* and the sum of the units is 0.6, so

$$0.5 + 0.1 = 0.6$$

You complete Example 2.

Example 2. Find $-1 + (-3)$.

Solution: Both numbers are _____ and the sum of the units is _____ , so

$$-1 + (-3) = \underline{}$$

Check your work on page 22. ▶

Now let's consider the sum $-2 + 8$. On the number line we start at the point corresponding to -2. Since we are adding a *positive* 8, we now move 8 units to the *right*.

Our final point lies 6 units to the right of zero, so

$$-2 + 8 = 6$$

Example 3. Use the number line to find the sum $10 + (-14)$.

Solution: We start at positive 10 and move 14 units to the *left*.

So $10 + (-14) = -4$.

You complete Example 4.

Example 4. Use the number line to find the sum $-7 + 6$.

Solution: We start at _____ 7 and move 6 units to the _____ .

So $-7 + 6 =$ _____ .

Check your work on page 22. ▶

From these examples, can you see a pattern for finding the sum of two numbers with *different* signs? Doesn't the following rule seem reasonable?

> To add two numbers with different signs, find the *difference* between the units (absolute values) and give your answer the sign of the original number having greater absolute value.

Example 5. Find the sum $5 + (-17)$.

Solution: The signs are different. The answer will have a *negative* sign. The difference between 5 units and 17 units is 12 units.

$$5 + (-17) = -12$$

You complete Example 6.

Example 6. Find the sum $-3 + 9$.

Solution: The signs are different. Here the answer will have a _____ sign. The difference between 3 units and 9 units is _____ units.

$$-3 + 9 =$$ _____

Check your work on page 22. ▶

In performing these addition problems, it was assumed that you could find the difference between numbers of units. Now, however, we must consider a more detailed discussion of **subtraction**. Remember that the result of performing subtraction is called a **difference**.

Some subtraction problems cause us no difficulty. We know that

$$27 - 16 = 11 \quad \text{because} \quad 11 + 16 = 27$$
$$8 - 3 = 5 \quad \text{because} \quad 5 + 3 = 8$$

> If a, b, and c are real numbers, then
>
> $a - b = c$ provided that $c + b = a$

Unfortunately, this fact does not help us to see quickly answers to more complicated subtraction problems. Let's try an approach that will change each subtraction problem into an *addition* problem.

Subtraction Statement	Addition Statement
$27 - 16 = 11$	$27 + (-16) = 11$
$8 - 3 = 5$	$8 + (-3) = 5$
$22 - 7 = 15$	$22 + (-7) = 15$

In fact, every subtraction statement can be written as an addition statement.

> To subtract two real numbers, *add* the *opposite* of the number being subtracted.

This is the mathematical definition for subtraction, and letting a and b represent any real numbers, we may write:

> **Definition of Subtraction**
>
> $$a - b = a + (-b)$$

Subtraction Statement	Addition Statement	Answer
$3 - 10$	$3 + (-10)$	-7
$-3 - 4$	$-3 + (-4)$	-7
$5 - (-6)$	$5 + 6$	11
$-8 - (-2)$	$-8 + 2$	-6

Example 7. Find $-22 - 14$.

Solution

$$-22 - 14$$
$$= -22 + (-14)$$
$$= -36$$

You try Example 8.

Example 8. Find $13 - (-7)$.

Solution

$$13 - (-7)$$
$$= 13 + \underline{\qquad}$$
$$= \underline{\qquad}$$

Check your work on page 22. ▶

Example 9. Find $\dfrac{-2}{3} - \dfrac{-1}{5}$.

Solution

$$\frac{-2}{3} - \frac{-1}{5}$$
$$= \frac{-2}{3} + \frac{1}{5}$$
$$= \frac{-10}{15} + \frac{3}{15}$$
$$= \frac{-7}{15}$$

Now complete Example 10.

Example 10. Find $-18 - (-18)$.

Solution

$$-18 - (-18)$$
$$= -18 + \underline{\qquad}$$
$$= \underline{\qquad}$$

Check your work on page 23. ▶

⇒ Trial Run

Simplify.

_____ **1.** $-3 + (-8)$

_____ **2.** $-11 + 23$

_____ **3.** $\dfrac{15}{4} + \dfrac{-9}{2}$

_____ **4.** $-8 + 8$

_____ **5.** $23 - 50$

_____ **6.** $-9 - 13$

_____ **7.** $-25 - (-38)$

_____ **8.** $3.5 - 5.0$

Answers are on page 23.

Multiplying and Dividing Real Numbers

To review the operation of **multiplication**, we should recall that quantities being multiplied are called **factors** and the result of multiplication is called a **product**.

It is assumed that you know your basic multiplication tables but that you may need a short refresher on the rules for *signs* of products. Recall that the sign of a product is determined by the signs of the factor being multiplied. Let's review the basic rules for multiplying real numbers.

The product of two positive numbers is a *positive* number.

$$\text{Example:} \quad 5 \cdot 7 = 35$$

To multiply two negative numbers, multiply the units (absolute values) and give your answer a *positive* sign.

$$\text{Example:} \quad (-2)(-6) = 12$$

To multiply a positive number and a negative number, multiply the units (absolute values) and give your answer a *negative* sign.

$$\text{Example:} \quad -3(6) = -18$$
$$2(-7) = -14$$

Example 11. Find the product $6(-1.2)$.

Solution

$$6(-1.2) = -7.2$$

Complete Example 12.

Example 12. Find the product $-11(-3)$.

Solution

$$-11(-3) = \underline{\quad\quad}$$

Check your work on page 23. ▶

In a **division** problem, the number being divided is called the **dividend** and the number doing the dividing is called the **divisor**. The result of division is called the **quotient**, and the quotient of two real numbers, a and b, can be written as $\dfrac{a}{b}$ or $a \div b$. To investigate the rules for the sign of a quotient, recall that there is a multiplication statement that corresponds to every division statement.

Division Statement		*Multiplication Statement*
$\dfrac{15}{3} = 5$	because	$15 = 3 \cdot 5$
$28 \div 7 = 4$	because	$28 = 7 \cdot 4$
$\dfrac{-15}{3} = -5$	because	$-15 = 3(-5)$
$\dfrac{28}{-7} = -4$	because	$28 = -7(-4)$
$\dfrac{-18}{-6} = 3$	because	$-18 = -6(3)$

From these examples, we may conclude that:

The quotient of two positive numbers is positive.
The quotient of two negative numbers is positive.
The quotient of a positive number and a negative number is negative.

Example 13. Find the quotient $\dfrac{49}{-7}$.

Solution

$$\frac{49}{-7} = -7$$

You complete Example 14.

Example 14. Find the quotient $\dfrac{-125}{-5}$.

Solution

$$\frac{-125}{-5} = \underline{\hspace{1.5cm}}$$

Check your work on page 23. ▶

Example 15. Find the quotient $\dfrac{-3}{4} \div \dfrac{3}{5}$.

Solution

$$\frac{-3}{4} \div \frac{3}{5}$$

$$= \frac{-3}{4} \cdot \frac{5}{3}$$

$$= \frac{-5}{4}$$

Now complete Example 16.

Example 16. Find the quotient $\dfrac{17}{-1}$.

Solution

$$\frac{17}{-1} = \underline{\hspace{1.5cm}}$$

Check your work on page 23. ▶

⫸ Trial Run

Simplify.

_____ **1.** $-11(5)$

_____ **2.** $(-8)(-7)$

_____ **3.** $9(-6)$

_____ **4.** $-1(-1.9)$

_____ 5. $\dfrac{72}{-8}$ _____ 6. $\dfrac{-110}{-10}$

_____ 7. $\dfrac{5}{8} \div \dfrac{-3}{4}$ _____ 8. $\dfrac{-8.3}{-1}$

Answers are on page 23.

Working with Zero

Having discussed addition, subtraction, multiplication, and division of real numbers, we turn our attention to performing those operations with the real number **zero**. Let's review the properties of zero. We shall let a represent any real number.

Property Name	Statement	Examples
Addition property of zero	$a + 0 = a$ $0 + a = a$	$-6 + 0 = -6$ $0 - \dfrac{1}{2} = 0 + \left(\dfrac{-1}{2}\right) = \dfrac{-1}{2}$ $\pi + 0 = \pi$
Subtraction property of zero	$a - 0 = a$	$5.3 - 0 = 5.3$ $\dfrac{-3}{2} - 0 = \dfrac{-3}{2}$
Multiplication property of zero	$a \cdot 0 = 0$ $0 \cdot a = 0$	$0(-2.5) = 0$ $3 \cdot 0 = 0$

You complete Examples 17, 18, and 19.

Example 17. Find each sum.

$11 + 0 = $ _____

$0 - 8 = 0 + $ _____

$= $ _____

$0 - (-1) = 0 + $ _____

$= $ _____

$0 + 0 = $ _____

Check your work on page 23. ▶

Example 18. Find each difference.

$20 - 0 = $ _____

$-1 - 0 = $ _____

$\dfrac{2}{3} - 0 = $ _____

$0 - 0 = $ _____

Check your work on page 23. ▶

Example 19. Find each product.

$10 \cdot 0 = $ _____

$0(-1) = $ _____

$\dfrac{1}{2}(0) = $ _____

$0 \cdot 0 = $ _____

Check your work on page 23. ▶

Let's investigate the use of zero in _division_ problems. What happens when we divide zero by a real number other than zero?

$$\frac{0}{6} = 0 \qquad \text{because} \qquad 0 = 6 \cdot 0$$

$$\frac{0}{-5} = 0 \qquad \text{because} \qquad 0 = -5 \cdot 0$$

If we divide zero by a nonzero number, the quotient is zero.

What about dividing a nonzero number by zero?

$$\frac{13}{0} = \underline{\qquad} \qquad \text{because} \qquad 13 = 0(\underline{\qquad})$$

$$\frac{-25}{0} = \underline{\qquad} \qquad \text{because} \qquad -25 = 0(\underline{\qquad})$$

We can find *no* number to use in these blanks. We conclude that division *by* zero is impossible or **undefined**.

Suppose that we try to divide zero by zero. Will we get an answer this time?

$$\frac{0}{0} = \underline{\qquad} \qquad \text{because} \qquad 0 = 0(\underline{\qquad})$$

Aa a matter of fact, we can use *any* number in the blanks.

$$\frac{0}{0} = 0 \qquad \text{because} \qquad 0 = 0(0)$$

$$\frac{0}{0} = 1 \qquad \text{because} \qquad 0 = 0(1)$$

$$\frac{0}{0} = 837 \qquad \text{because} \qquad 0 = 0(837)$$

In this case there are *too many* answers! We conclude that division of zero by zero is **not uniquely defined**.

Let's summarize our observations about zero in division.

Property Name	Statement	Examples
Division of zero	$\dfrac{0}{a} = 0$ $(a \neq 0)$	$\dfrac{0}{11} = 0$ $\dfrac{0}{-1} = 0$
Division by zero	$\dfrac{a}{0}$ is undefined $(a \neq 0)$	$\dfrac{3 \cdot 9}{0}$ is undefined
Division of zero by zero	$\dfrac{0}{0}$ is not uniquely defined	$\dfrac{5 - 5}{0} = \dfrac{0}{0}$ is not uniquely defined

Now complete Examples 20 and 21.

Example 20. Complete each statement.

$$\frac{0}{11} = \underline{\hspace{1cm}} \quad \text{because} \quad 0 = 11(\underline{\hspace{0.6cm}})$$

$$\frac{0}{-1} = \underline{\hspace{1cm}} \quad \text{because} \quad 0 = -1(\underline{\hspace{0.6cm}})$$

Check your work on page 23. ▶

Example 21. Find each quotient.

$$\frac{0}{12} = \underline{\hspace{1cm}}$$

$$\frac{-1}{0} = \underline{\hspace{1cm}}$$

$$\frac{0}{-1} = \underline{\hspace{1cm}}$$

$$\frac{0}{0} = \underline{\hspace{1cm}}$$

Check your work on page 23. ▶

⇒ Trial Run

Simplify.

_____ **1.** $7 + 0$

_____ **2.** $-9 + 0$

_____ **3.** $0 - 3.2$

_____ **4.** $16 \cdot 0$

_____ **5.** $0(-4)$

_____ **6.** $\frac{0}{-2}$

_____ **7.** $0\left(\frac{2}{3}\right)$

_____ **8.** $\frac{8}{0}$

_____ **9.** $\frac{0}{0}$

_____ **10.** $0 - (-1)$

Answers are on page 23.

▶ Examples You Completed

Example 2. Find $-1 + (-3)$.

Solution: Both numbers are negative and the sum of the units is 4, so

$$-1 + (-3) = -4$$

Example 4. Use the number line to find the sum $-7 + 6$.

Solution: We start at negative 7 and move 6 units to the right.

So $-7 + 6 = -1$.

Example 6. Find the sum $-3 + 9$.

Solution: The signs are different. Here the answer will have a positive sign. The difference between 3 units and 9 units is 6 units.

$$-3 + 9 = 6$$

Example 8. Find $13 - (-7)$.

Solution

$$13 - (-7)$$

$$= 13 + 7$$

$$= 20$$

Example 10. Find $-18 - (-18)$.

Solution

$$-18 - (-18)$$
$$= -18 + 18$$
$$= 0$$

Example 12. Find the product $-11(-3)$.

Solution

$$-11(-3) = 33$$

Example 14. Find the quotient $\dfrac{-125}{-5}$.

Solution

$$\frac{-125}{-5} = 25$$

Example 16. Find the quotient $\dfrac{17}{-1}$.

Solution

$$\frac{17}{-1} = -17$$

Example 17. Find each sum.

$$11 + 0 = 11$$
$$0 - 8 = 0 + (-8) = -8$$
$$0 - (-1) = 0 + 1 = 1$$
$$0 + 0 = 0$$

Example 18. Find each difference.

$$20 - 0 = 20$$
$$-1 - 0 = -1$$
$$\frac{2}{3} - 0 = \frac{2}{3}$$
$$0 - 0 = 0$$

Example 19. Find each product.

$$10 \cdot 0 = 0$$
$$0(-1) = 0$$
$$\frac{1}{2}(0) = 0$$
$$0(0) = 0$$

Example 20. Complete each statement.

$$\frac{0}{11} = 0 \quad \text{because} \quad 0 = 11(0)$$
$$\frac{0}{-1} = 0 \quad \text{because} \quad 0 = -1(0)$$

Example 21. Find each quotient.

$$\frac{0}{12} = 0$$
$$\frac{-1}{0} = \text{undefined}$$
$$\frac{0}{-1} = 0$$
$$\frac{0}{0} = \text{not uniquely defined}$$

Answers to Trial Runs

page 18 **1.** -11 **2.** 12 **3.** $-\dfrac{3}{4}$ **4.** 0 **5.** -17 **6.** -22 **7.** 13 **8.** -1.5

page 19 **1.** -55 **2.** 56 **3.** -54 **4.** 1.9 **5.** -9 **6.** 11 **7.** $-\dfrac{5}{6}$ **8.** 8.3

page 22 **1.** 7 **2.** -9 **3.** -3.2 **4.** 0 **5.** 0 **6.** 0 **7.** 0 **8.** Undefined
9. Not uniquely defined **10.** 1

EXERCISE SET 1.2

Simplify.

_____ **1.** $8 + 6$ _____ **2.** $9 + 7$

_____ **3.** $(-3) + (-7)$ _____ **4.** $(-9) + (-3)$

_____ **5.** $(-4) + 8$ _____ **6.** $(-5) + 9$

_____ **7.** $4 + (-13)$ _____ **8.** $7 + (-15)$

_____ **9.** $\dfrac{-2}{9} + \dfrac{-5}{9}$ _____ **10.** $\dfrac{-3}{20} + \dfrac{-11}{20}$

_____ **11.** $5 - 12$ _____ **12.** $3 - 8$

_____ **13.** $(-6) - (-4)$ _____ **14.** $(-8) - (-3)$

_____ **15.** $(-1.7) - (-3.2)$ _____ **16.** $(-1.9) - (-4.1)$

_____ **17.** $(-27) - 13$ _____ **18.** $(-25) - 19$

_____ **19.** $32 - (-42)$ _____ **20.** $47 - (-22)$

_____ **21.** $(-8)(0.5)$ _____ **22.** $(-6)(1.5)$

_____ **23.** $(-9)(-8)$ _____ **24.** $(-3)(-21)$

_____ **25.** $7(-15)$ _____ **26.** $5(-13)$

_____ **27.** $(-9)(0)$ _____ **28.** $15(0)$

_____ **29.** $(-2)(-8)$ _____ **30.** $(5)(-2)$

_____ **31.** $\dfrac{-24}{6}$ _____ **32.** $\dfrac{-42}{7}$

_____ **33.** $\dfrac{-21}{-3}$ _____ **34.** $\dfrac{-25}{-5}$

_____ **35.** $\dfrac{-5}{6} \div \dfrac{35}{36}$ _____ **36.** $\dfrac{-6}{7} \div \dfrac{12}{49}$

_____ **37.** $\dfrac{-144}{-12}$ _____ **38.** $\dfrac{-169}{-13}$

_____ **39.** $\dfrac{0}{-3}$ _____ **40.** $\dfrac{0}{8}$

_____ **41.** $0 - (-23)$ _____ **42.** $0 - (-37)$

_____ **43.** $\left(\dfrac{4}{5}\right)(0)$ _____ **44.** $\left(\dfrac{7}{8}\right)(0)$

_____ **45.** $\dfrac{0}{-3 - (-3)}$ _____ **46.** $\dfrac{0}{8 + (-8)}$

_____ **47.** $0 + (-13)$ _____ **48.** $0 + (-51)$

_____ **49.** $\dfrac{-9}{0}$ _____ **50.** $\dfrac{12}{0}$

☆ Stretching the Topics

_____ **1.** Evaluate $-4a + (-7b) + 6c$ for $a = -1$, $b = -3$, and $c = -5$.

_____ **2.** Simplify $\dfrac{-9 - 5 + 6 - 2}{-5 - 7 + 8 + 4}$.

_____ **3.** The temperature is dropping $3°$ per hour. The temperature right now is $10°$. Find the temperature 3 hours from now and 4 hours ago.

Check your answers in the back of your book.

If you can complete **Checkup 1.2**, you are ready to go to Section 1.3.

Name _____ Date _____

 CHECKUP 1.2

Simplify.

_____ **1.** $(-31) + (-21)$ _____ **2.** $14 + (-52)$

_____ **3.** $24 - (-9)$ _____ **4.** $(-19) - (-26)$

_____ **5.** $0 - (-7)$ _____ **6.** $(-4)(-7)$

_____ **7.** $(-3)(9)$ _____ **8.** $\dfrac{-81}{9}$

_____ **9.** $\dfrac{-75}{-5}$ _____ **10.** $\dfrac{7 - (-2)}{4 + (-4)}$

Check your answers in the back of your book.

If You Missed Problems:	You Should Review Examples:
1, 2	1–6
3, 4	7–10
5	17, 18
6, 7	11, 12
8, 9	13–16
10	21

1.3 Working with Real Number Properties

Number properties are the rules involved in playing the game of algebra. They tell us what we can do and cannot do in performing operations with real numbers.

We have observed that sums and products of real numbers are always themselves real numbers. Add two real numbers and you get a real number; multiply two real numbers and you get a real number. Moreover, there is one and only one answer when a given operation is performed with two numbers. Every time we add $2 + 3$, the answer is 5; every time we multiply $-6 \cdot 7$, the answer is -42. Think how confusing mathematics would be if this were not so; we depend on our answers' being **unique**.

We summarize these first facts as the *closure properties* for addition and multiplication of real numbers.

Closure Properties. If a and b are real numbers, then:	*Illustrations*
(1) $a + b$ is a unique real number	(1) $3 + 2 = 5$ and $5 \in R$
(2) $a \cdot b$ is a unique real number	(2) $5 \cdot 2 = 10$ and $10 \in R$

The order in which we add or multiply two real numbers does not change the sum or product. We can summarize these facts as the **commutative properties** for addition and multiplication of real numbers.

Commutative Properties. If a and b are real numbers, then:	*Illustrations*
(1) $a + b = b + a$	(1) $-5 + 6 = 6 + (-5)$
(2) $a \cdot b = b \cdot a$	(2) $(-7)(-2) = (-2)(-7)$

How do we add or multiply *three* numbers? To find

$$2 + 6 + 11 \quad \text{or} \quad 7 \cdot 2 \cdot 3$$

requires some mental grouping, since our minds can handle only two numbers at once. We shall use **parentheses** to indicate such grouping; parentheses always tell us to "do this first."

There are two approaches to grouping the terms in our sum.

Approach 1	*Approach 2*
$2 + 6 + 11$	$2 + 6 + 11$
$= (2 + 6) + 11$	$= 2 + (6 + 11)$
$= 8 + 11$	$= 2 + 17$
$= 19$	$= 19$

With either approach, the sum is 19. The way in which we group the terms does not change the sum.

Once again, there are two approaches to grouping the factors in our product.

Approach 1	*Approach 2*
$7 \cdot 2 \cdot 3$	$7 \cdot 2 \cdot 3$
$= (7 \cdot 2) \cdot 3$	$= 7 \cdot (2 \cdot 3)$
$= 14 \cdot 3$	$= 7 \cdot 6$
$= 42$	$= 42$

Notice again that the answer is the same in each case.

These examples lead us to the **associative properties** for addition and multiplication of real numbers.

Associative Properties. If a, b, and c are real numbers, then:

(1) $a + (b + c) = (a + b) + c$

(2) $a(bc) = (ab)c$

Illustrations

(1) $10 + (7 + 1) = (10 + 7) + 1$

(2) $-1(3 \cdot 5) = (-1 \cdot 3) \cdot 5$

Notice how the associative properties can help you cut down on complicated arithmetic in the following examples.

Example 1. Find $-99 + (99 + 76)$.

Solution

$$-99 + (99 + 76)$$
$$= (-99 + 99) + 76$$
$$= 0 + 76$$
$$= 76$$

Example 2. Find $\left(37 \cdot \dfrac{1}{2} \right) 20$.

Solution

$$\left(37 \cdot \frac{1}{2} \right) 20.$$
$$= 37 \left(\frac{1}{2} \cdot 20 \right)$$
$$= 37(\underline{\hspace{1cm}})$$
$$= \underline{\hspace{1cm}}$$

Check your work on page 37. ▶

The commutative and associative properties allow us to change the positions of numbers in a sum or product to make our computation more convenient. In an addition problem, you may wish to group all the positive numbers together and all the negative numbers together before adding.

Example 3. Simplify $-13 + 17 + 4 - 18 - 1$.

Solution

$$-13 + 17 + 4 - 18 - 1$$
$$= -13 + (-18) + (-1) + 17 + 4$$
$$= -32 + 21$$
$$= -11$$

You try Example 4.

Example 4. Simplify $2 - 5 - 7 + 3 - (-9)$.

Solution

$$2 - 5 - 7 + 3 - (-9)$$
$$= 2 + 3 + 9 + (-5) + (-7)$$

Check your work on page 37. ▶

Using the commutative and associative properties in a multiplication problem, the "easy" products can be found first.

Example 5. Simplify $6(-3)(2)(5)(-1)$.

Solution

$$6(-3)(2)(5)(-1)$$
$$= 6(-6)(-5)$$
$$= 6(30)$$
$$= 180$$

You try Example 6.

Example 6. Simplify $-2(3)(8)(10)$.

Solution

$$-2(3)(8)(10)$$

Check your work on page 37. ▶

When a problem contains several different operations, some guidelines are needed. Recall that grouping symbols (parentheses, brackets, and braces) say "do this first" and should be dealt with by working from the innermost pairs outward. Fraction bars also act as symbols of grouping, separating the numerator and the denominator.

Order of Operations

1. Deal with any symbols of grouping.
2. Perform multiplications and/or divisions from left to right.
3. Perform additions and/or subtractions from left to right.

Try completing Example 7.

Example 7. Simplify

$$5[3 - 2(7 - 10)] - 2$$

Solution

$$5[3 - 2(7 - 10)] - 2$$
$$= 5[3 - 2(-3)] - 2$$
$$= 5[3 + \underline{\qquad}] - 2$$

Check your work on page 37. ▶

Example 8. Simplify

$$\frac{5(9 - 4) - 2(7 - 3)}{-2\left(\frac{1}{2}\right)}.$$

Solution

$$\frac{5(9 - 4) - 2(7 - 3)}{-2\left(\frac{1}{2}\right)}$$

$$= \frac{5(5) - 2(4)}{-2\left(\frac{1}{2}\right)}$$

$$= \frac{25 - 8}{-1}$$

$$= \frac{17}{-1}$$

$$= -17$$

Suppose that we try two different approaches to find the product $-2(5 + 1)$.

Approach 1	*Approach 2*
$-2(5 + 1)$	$-2(5 + 1)$
$= -2(6)$	$= -2(5) + (-2)(1)$
$= -12$	$= -10 + (-2)$
	$= -12$

The second approach illustrates the **distributive property** for multiplication over addition of real numbers.

Distributive Property. If a, b, and c are real numbers, then

$$a(b + c) = a \cdot b + a \cdot c$$

Illustration

$$3(2 + 5) = 3 \cdot 2 + 3 \cdot 5$$

Example 9. Use the distributive property to find $-6(-2 - 8)$.

Solution

$$-6(-2 - 8)$$
$$= -6(-2) + (-6)(-8)$$
$$= 12 + 48$$
$$= 60$$

You try Example 10.

Example 10. Use the distributive property to find $7(-4 + 3)$.

Solution

$$7(-4 + 3)$$
$$= 7(-4) + 7(3)$$

Check your work on page 37. ▶

You may be thinking that the distributive property is more complicated than our earlier method for finding these products, and for problems involving only rational numbers, you are probably right. However, we shall soon encounter algebraic expressions that can be simplified only by using the distributive property. As a matter of fact, *the distributive property is one of the most useful of all the properties of the real numbers.*

⫸ Trial Run

Simplify.

_____ **1.** $(19 + 15) - 15$

_____ **2.** $\frac{1}{3}(12 \cdot 11)$

_____ **3.** $4(5 + 2) - 3 \cdot 10$

_____ **4.** $9 - 2[7(3 - 9) + 2]$

_____ **5.** $\dfrac{7(10 - 5) - 3(8 - 7)}{-16\left(\dfrac{1}{2}\right)}$

Simplify using the distributive property for real numbers.

_____ **6.** $-3(6 + 2)$

_____ **7.** $-7(3 - 11)$

_____ **8.** $2(\pi - 9)$

Answers are on page 38.

We have already discussed the very special nature of the real number zero. We should emphasize that zero is unique in that it is the *only* real number that leaves every real number unchanged under the operation of addition. Because zero added to any real number gives an answer that is *identical* to that original number, we call zero the **identity for addition**.

Do you suppose that the set of real numbers also contains an identity for multiplication? Can you think of a real number that leaves every real number unchanged under multiplication? You should realize that 1 is the only such number. We conclude that 1 is the unique **identity for multiplication**.

Identity for Addition Is 0	Identity for Multiplication Is 1
$3 + 0 = 3$	$3 \cdot 1 = 3$
$-7 + 0 = -7$	$-7 \cdot 1 = -7$
$\dfrac{1}{3} + 0 = \dfrac{1}{3}$	$\dfrac{1}{3} \cdot 1 = \dfrac{1}{3}$
$\pi + 0 = \pi$	$\pi \cdot 1 = \pi$

We summarize the identity properties for the real numbers as follows:

> **Identity Properties**
>
> **1.** The unique identity for addition is 0. For every $a \in R$, $a + 0 = a$.
> **2.** The unique identity for multiplication is 1. For every $a \in R$, $a \cdot 1 = a$.

In our discussion of the number line, we discovered that every real number has an *opposite* that is also a real number. The opposite of a number is also called its **additive inverse**. Every time we add a number to its additive inverse (or opposite), the sum is 0 (the additive identity).

Number	Additive Inverse	Reason
5	-5	$5 + (-5) = 0$
0.6	-0.6	$0.6 + (-0.6) = 0$
-8	$-(-8) = 8$	$-8 + 8 = 0$
a	$-a$	$a + (-a) = 0$

We may conclude that the additive inverse (or negative) of a number can be found by multiplying that number by -1.

Example 11. Find the additive inverse of $\pi + 2$.

Solution: The additive inverse of $\pi + 2$ is

$$-(\pi + 2)$$
$$= -1(\pi + 2)$$
$$= -1 \cdot \pi + (-1)(2)$$
$$= -\pi - 2$$

Notice that we used the distributive property to multiply our quantity by -1.

You try completing Example 12.

Example 12. Find the additive inverse of $5 - \sqrt{3}$.

Solution: The additive inverse of $5 - \sqrt{3}$ is

$$-(5 - \sqrt{3})$$
$$= -1(5 - \sqrt{3})$$
$$= (-1)(\underline{}) + (-1)(\underline{})$$
$$=$$

Check your work on page 37. ▶

Now that we have discussed additive inverses, we might also consider the existence of **multiplicative inverses**. The multiplicative inverse for every real number is its **reciprocal**. Every real number (except 0) has its own unique multiplicative inverse. When we multiply a number by its multiplicative inverse, the product is 1 (the multiplicative identity).

Number	Multiplicative Inverse	Reason
5	$\dfrac{1}{5}$	$5 \cdot \dfrac{1}{5} = 1$
-7	$\dfrac{-1}{7}$	$-7\left(\dfrac{-1}{7}\right) = 1$
$\dfrac{5}{8}$	$\dfrac{8}{5}$	$\dfrac{5}{8}\left(\dfrac{8}{5}\right) = 1$
-0.3	$\dfrac{-1}{0.3}$	$-0.3\left(\dfrac{-1}{0.3}\right) = 1$
π	$\dfrac{1}{\pi}$	$\pi\left(\dfrac{1}{\pi}\right) = 1$

As you can see, we find the multiplicative inverse of a real number by inverting it, or turning it upside down.

Inverse Properties

1. Every real number a has a unique *additive inverse* (or opposite) denoted by $-a$, and

$$a + (-a) = 0$$

2. Every real number a (except 0) has a unique *multiplicative inverse* (or reciprocal) denoted by $\dfrac{1}{a}$, and

$$a \cdot \frac{1}{a} = 1$$

Example 13. Find the additive inverse and the multiplicative inverse of $\frac{1}{7}$.

Solution: The additive inverse of $\frac{1}{7}$ is $\frac{-1}{7}$ because

$$\frac{1}{7} + \left(\frac{-1}{7}\right) = 0$$

The multiplicative inverse of $\frac{1}{7}$ is 7 because

$$\frac{1}{7} \cdot 7 = 1$$

You complete Example 14.

Example 14. Find the additive inverse and the multiplicative inverse of -4.

Solution: The additive inverse of -4 is _____ because

$$-4 + \underline{\hspace{1cm}} = 0$$

The multiplicative inverse of -4 is _____ because

$$-4(\underline{\hspace{1cm}}) = 1$$

Check your work on page 37. ▶

The existence of a multiplicative inverse corresponding to every nonzero real number allows us to write every division expression as a multiplication expression.

Division Expression	Multiplication Expression	Answer
$\frac{6}{3}$	$6 \cdot \frac{1}{3}$	2
$\frac{-10}{5}$	$-10 \cdot \frac{1}{5}$	-2
$\frac{24}{-8}$	$24\left(\frac{-1}{8}\right)$	-3

To divide two numbers, find the reciprocal of the divisor and multiply.

Indeed, this is the formal definition of division in terms of multiplication. Letting a and b represent real numbers, we may write:

Definition of Division

$$\frac{a}{b} = a \cdot \frac{1}{b} \qquad \text{where } b \neq 0$$

To be sure that we understand how to deal with expressions involving several operations, let's try a few more examples.

Example 15. Simplify

$$3 - 2\left[\frac{5(-4) - 2(-1)}{10(3) - 6(4)}\right].$$

Solution

$$3 - 2\left[\frac{5(-4) - 2(-1)}{10(3) - 6(4)}\right]$$

$$= 3 - 2\left[\frac{-20 + 2}{30 - 24}\right]$$

$$= 3 - 2\left[\frac{-18}{6}\right]$$

$$= 3 - 2(-3)$$

$$= 3 + 6$$

$$= 9$$

Now carefully complete Example 16.

Example 16. Simplify

$$-7\left(8 + \frac{4 - 1}{2 - 3}\right) - 10.$$

Solution

$$-7\left(8 + \frac{4 - 1}{2 - 3}\right) - 10$$

$$= -7\left(8 + \frac{3}{-1}\right) - 10$$

$$= -7(8 - \underline{\hspace{1cm}}) - 10$$

Check your work on page 37. ▶

⫸ **Trial Run**

Name the property of real numbers illustrated by each statement.

_____ **1.** $-6 + 0 = -6$ _____ **2.** $8\left(\frac{1}{8}\right) = 1$

_____ **3.** $-3(1) = -3$ _____ **4.** $5 + (-5) = 0$

5. The multiplicative inverse of -5 is _____ .

6. The additive inverse of -10 is _____ .

Simplify.

_____ **7.** $-(3 - \sqrt{2})$ _____ **8.** $5 + 2\left[\frac{7 + (-2)(-10)}{6 - 3(5)}\right]$

Answers are on page 38.

We mention here that although both addition and multiplication are commutative and associative, subtraction and division are *not*. For example,

$$3 - 5 \neq 5 - 3 \qquad 8 - (5 - 3) \neq (8 - 5) - 3$$

$$6 \div 2 \neq 2 \div 6 \qquad 24 \div (6 \div 2) \neq (24 \div 6) \div 2$$

EXERCISE SET 1.3

Name the property of real numbers that is illustrated by each of the following statements.

_____ **1.** $5(3 \cdot 9) = (5 \cdot 3)9$

_____ **2.** $2 + 0 = 2$

_____ **3.** $\frac{4}{5} + (-5)$ is a real number

_____ **4.** $(-3 + 5) + 10 = -3 + (5 + 10)$

_____ **5.** $(-12)(1) = -12$

_____ **6.** $23 + (-23) = 0$

_____ **7.** $-8(7 + 5) = (-8)7 + (-8)5$

_____ **8.** $9 + (-13) = (-13) + 9$

_____ **9.** $9\left(\frac{1}{9}\right) = 1$

_____ **10.** $5 + 2$ is a real number

Complete the following statements using the indicated property.

11. $-4 + 8 = $ _____ (commutative)

12. $-2(3 \cdot 5) = $ _____ (associative)

13. $(-9 + 3) + 4 = $ _____ (associative)

14. $(-0.9)(1) = $ _____ (identity)

15. $(-15) + 15 = $ _____ (inverse)

16. $-7(6 + 4) = $ _____ (distributive)

17. $(-19) + 0 = $ _____ (identity)

18. $\frac{1}{4}(4) = $ _____ (inverse)

19. $2(3) + 2(-5) = $ _____ (distributive)

20. $5(7) + 5(8) = $ _____ (distributive)

Simplify.

_____ **21.** $-55 + (55 + 23)$

_____ **22.** $15 + (-15 + 6)$

_____ **23.** $\left(29 \cdot \frac{1}{3}\right) 30$

_____ **24.** $15\left(\frac{1}{5} \cdot 17\right)$

_____ **25.** $-8 + 12 + 7 - 13 - 1$

_____ **26.** $-9 + 15 - 18 + 3 - (-6)$

_____ **27.** $4(-3)(2)(-7)$

_____ **28.** $-3(6)(2)(-5)$

_____ **29.** $6(-1)(9)(0)(21)$

_____ **30.** $9(-13)(7)(0)(15)$

_____ **31.** $-9(5 - 7)$

_____ **32.** $-10(5 - 12)$

_____ **33.** $(-6)(11) - 21$

_____ **34.** $9(-5) - 15$

_____ **35.** $5\left(\frac{1}{5}\right) - 3(7)$

_____ **36.** $9\left(\frac{1}{9}\right) - 4(5)$

_____ **37.** $7 - 9[3 - 5(-1)]$

_____ **38.** $6 - 5[8 - 3(-2)]$

_____ **39.** $(7 - 9)[4 - 7(0)]$

_____ **40.** $(6 - 5)[13 - 9(0)]$

_____ 41. $\dfrac{5(-2) - \frac{1}{2}(-20)}{3(8 - 6)}$

_____ 42. $\dfrac{6(-3) - \frac{1}{2}(-36)}{7(9 - 5)}$

_____ 43. $\dfrac{20(-4)}{10} + \dfrac{21 - 6}{-3}$

_____ 44. $\dfrac{36(-2)}{12} + \dfrac{27 - 9}{-6}$

_____ 45. $\dfrac{-9(7 - 1) - 8(-3)}{-5[(-8) + 6]}$

_____ 46. $\dfrac{-8(10 - 3) - 9(-4)}{-2[(-9) + 4]}$

_____ 47. $\dfrac{-6\left(\frac{1}{2}\right) + 3[2 - (-3)]}{9[12 - 2(6)]}$

_____ 48. $\dfrac{-8\left(\frac{1}{4}\right) + 6[8 - (-1)]}{11[15 - 3(5)]}$

_____ 49. $\dfrac{\frac{-15}{5} + \frac{0}{7}}{\frac{35}{-5} + 11}$

_____ 50. $\dfrac{\frac{18}{6} + \frac{0}{9}}{\frac{-30}{5} - 9}$

_____ 51. $8 - (-6)\left[\dfrac{2(-3) - 5(4)}{-8(6) - 4}\right]$

_____ 52. $-9 - 6\left[\dfrac{11(-1) + 9}{4(-3) - 2(-5)}\right]$

_____ 53. $8 - (-2)\left[\dfrac{3 - 5(7 - 1)}{6 - 1(-3)}\right]$

_____ 54. $9 - (-5)\left[\dfrac{8 - 3(6 - 12)}{7 - 3(-2)}\right]$

_____ 55. $4\left(-9 + \dfrac{7 - 3}{6 - 8}\right) - 3$

_____ 56. $5\left(-10 + \dfrac{8 - 3}{15 - 20}\right) - 6$

_____ 57. $\dfrac{8(-5 + 7) - (-9)}{(-2)4 + 13(-4)}$

_____ 58. $\dfrac{12(-6 + 4) - (-4)}{-3(4) + 7(-4)}$

☆ Stretching the Topics

_____ 1. Evaluate $a + bc - d$ when $a = 2.1$, $b = -3.2$, $c = 4.5$, and $d = -5.6$.

_____ 2. Evaluate $\dfrac{m + (2n - 3) - (n - 4)}{3(n - 1) - 2n}$ when $n = 2$ and $m = -3$.

_____ 3. The population of a city was 495,000. Over the next 5 years the urban planning chairman recorded these changes: a gain of 2365, a loss of 3729, a gain of 5728, a loss of 1796, and a loss of 768. Find the population of the city after 5 years. Find the average gain or loss.

Check your answers in the back of your book.

If you can complete **Checkup 1.3**, you are ready to go on to Section 1.4.

CHECKUP 1.3

Name the property of real numbers that is illustrated by each of the following statements.

_____ **1.** $5(-2) = -2(5)$

_____ **2.** $\dfrac{-1}{3}(-3) = 1$

Simplify.

_____ **3.** $-23 + (23 + 17)$

_____ **4.** $-4(5 - 12)$

_____ **5.** $-5 + 7 - 10 - (-13)$

_____ **6.** $5(-3)(6)(2)$

_____ **7.** $4 - 8[6 - 2(-3)]$

_____ **8.** $(5 - 10)[7 - 2(5)]$

_____ **9.** $\dfrac{5(-6) - \dfrac{1}{5}(15)}{-6(8 - 19)}$

_____ **10.** $\dfrac{-12(5) - 3[8 - 2(4)]}{-9[17 - 3(5)]}$

Check your answers in the back of your book.

If You Missed Problems:	You Should Review Examples:
1, 2	(Real number properties)
3	1
4	9, 10
5	3, 4
6	5, 6
7, 8	7
9, 10	8, 15, 16

1.4 Working with Variables

Our work with the properties of the real numbers is intended to lay the groundwork for the study of algebra, which requires the use of constants and variables. A **constant** is a number such as 3 or -4 whose value is fixed and never changes. A **variable** is a symbol that stands for a number. We generally use letters to represent variables. For example,

17 is a constant

x is a variable

y is a variable

0 is a constant

Combinations of constants and variables involving the operations of addition, subtraction, multiplication, and division are called **algebraic expressions**. We continue to use familiar symbols to show what operation is to be performed.

Expression	Meaning
$10 + x$	Add 10 and the variable x
$y - 3$	Subtract 3 from the variable y
$\dfrac{a}{2}$	Divide the variable a by 2
$9x$	Multiply 9 times the variable x
$-3y$	Multiply -3 times the variable y

Example 1. Write an algebraic expression that means "divide the sum of x and 3 by 7."

Solution

$$\frac{x + 3}{7}$$

Now try Example 2.

Example 2. Write an algebraic expression that means "multiply 6 times the sum of -1 and x."

Solution

 $6(\underline{\quad} + \underline{\quad})$

Check your work on page 51. ▶

Evaluating Algebraic Expressions

Algebraic expressions will have different numerical values depending on what value the variable represents. Finding the numerical value of an algebraic expression for a particular numerical value of the variable is called **evaluating** an expression. We evaluate an expression by substituting the given number wherever the variable appears in the expression and then performing the necessary arithmetic.

Example 3. Evaluate $3(x + 6) - 7$ when $x = -1$.

Solution: We must substitute -1 for x in the expression.

$$3(x + 6) - 7$$
$$= 3(-1 + 6) - 7$$
$$= 3(5) - 7$$
$$= 15 - 7$$
$$= 8$$

You complete Example 4.

Example 4. Evaluate $9y - 37z$ when $y = 0$ and $z = 1$.

Solution: We must substitute _____ for y and 1 for _____ in the expression.

$$9y - 37z$$
$$= 9(\underline{\quad}) - 37(\underline{\quad})$$

Check your work on page 51. ▶

Example 5. Evaluate $x(x + 3) - 4$ when $x = -6$.

Solution

$$x(x + 3) - 4$$
$$= -6(-6 + 3) - 4$$
$$= -6(-3) - 4$$
$$= 18 - 4$$
$$= 14$$

Now complete Example 6.

Example 6. Evaluate $(x + 1)(1 - x)$ when $x = -2$.

Solution

$$(x + 1)(1 - x)$$
$$= (-2 + 1)[1 - (-2)]$$

Check your work on page 52. ▶

Algebraic expressions often contain quantities with **exponents**. In general, we say that if a is a real number (called the **base**) and n is a natural number (called the exponent):

Definition of a Power

$$a^n = \underbrace{a \cdot a \cdot a \cdot \ldots \cdot a}_{n \text{ factors}}$$

This definition allows us to give meaning to more complicated algebraic expressions.

Exponent Form	Factor Form
5^2	$5 \cdot 5$
x^2	$x \cdot x$
$(-2)^4$	$(-2)(-2)(-2)(-2)$
-2^4	$-1 \cdot 2 \cdot 2 \cdot 2 \cdot 2$
$-2y^3$	$-2 \cdot y \cdot y \cdot y$
$(x + 4)^2$	$(x + 4)(x + 4)$

Use the definition of a power to evaluate the expression in Example 7.

Example 7. Evaluate $-x^2 - 3x + 2$ when $x = -4$.

Solution

$$-x^2 - 3x + 2$$
$$= -(-4)^2 - 3(-4) + 2$$
$$= -1(-4)(-4) - 3(-4) + 2$$

Check your work on page 52. ▶

Example 8. Evaluate $4(x - 3)^2 - 2xy - y^3$ when $x = 3$ and $y = -1$.

Solution

$$4(3 - 3)^2 - 2(3)(-1) - (-1)^3$$
$$= 4(0)^2 - 6(-1) - (-1)(-1)(-1)$$
$$= 4(0)(0) + 6 - (-1)$$
$$= 4(0) + 6 + 1$$
$$= 0 + 6 + 1$$
$$= 7$$

Example 9. The surface area of a box with a square base and no top is found by the formula $S = 4sh + s^2$, where s is the length of one side of the base and h is the height. Evaluate this formula when $s = 12$ inches and $h = 20$ inches.

Solution

$$S = 4sh + s^2$$
$$= 4(12)(20) + (12)^2 \qquad \text{Substitute 12 for } s, 20 \text{ for } h.$$
$$= 960 + 144 \qquad \text{Remove parentheses.}$$
$$= 1100 \text{ square inches} \qquad \text{Simplify.}$$

Combining Like Terms

In an algebraic expression the parts being added are called the **terms** of that expression. Each term consists of a constant alone *or* a constant times one or more variables. The constant part of a variable term is called the **numerical coefficient** for that term.

Expression	Constant Term	Variable Terms	Numerical Coefficients
$x^2 - 7x + 2$	2	x^2	1
		$-7x$	-7
$3 - x$	3	$-x$	-1
$5x^2 - xy + y^2$	None	$5x^2$	5
		$-xy$	-1
		y^2	1

In addition and subtraction of algebraic expressions, we may combine only **like terms**, which are terms whose variable parts are *exactly* the same. Most of us would agree that

$$3x + 5x - 2x$$

is an expression containing three like terms and that the sum of these terms is $6x$.

> To combine like terms, we combine the numerical coefficients and keep the same variable part.

The commutative and associative properties of real numbers allow us to rearrange the terms in an expression with the like terms together. We then use our addition and subtraction rules to combine the coefficients of the like terms.

Example 10. Simplify $3x + 2y - 4x - y + 3$.

Solution

$$3x + 2y - 4x - y + 3$$
$$= 3x - 4x + 2y - 1y + 3$$
$$= -1x + 1y + 3$$
$$= -x + y + 3$$

You complete Example 11.

Example 11. Simplify $x^2 - 6x + 3x^2 - x$.

Solution

$$x^2 - 6x + 3x^2 - x$$

Check your work on page 52. ▶

From Example 10, you should notice that numerical coefficients of 1 or -1 are not usually written before a variable; they are simply "understood" to be there.

⟾ Trial Run

Evaluate.

_____ **1.** $(x - 6)(x + 7)$ when $x = -7$

_____ **2.** $x^2 - 5x + 6$ when $x = -1$

_____ **3.** $4(x - 2)^2 - 12x$ when $x = 0$

_____ **4.** $9x^2 - xy + y^2$ when $x = 1$ and $y = -2$

Simplify by combining like terms.

_____ **5.** $3x - 5 - 7x + 10$

_____ **6.** $x^2 + 3x - 5 - 2x^2 + 7x - 1$

_____ **7.** $2x^2 - 5xy + 3y^2 - 4x^2 + 7xy - y^2$

_____ **8.** $a^2 - 2a + 3 - a^2 - 2a - 5$

Answers are on page 53.

Multiplying Variable Terms by Constants

To multiply a constant times a variable term, we recall the associative property for multiplication.

$$2(5x) \quad = \quad (2 \cdot 5)x \quad = \quad 10x$$

$$-2\left(\frac{-1}{3}x\right) = \left[\left(-2\right)\left(\frac{-1}{3}\right)\right]x = \frac{2}{3}x$$

$$7(-x^2) \quad = \quad [7(-1)]x^2 \quad = \quad -7x^2$$

$$-(3x) \quad = \quad (-1 \cdot 3)x \quad = \quad -3x$$

$$-(-4y^2) \quad = \quad [-1(-4)]y^2 \quad = \quad 4y^2$$

> To multiply a constant times a variable term, we multiply the constant times the numerical coefficient and keep the same variable part.

Example 12. Simplify $3(2x) - 7(4x)$.

Solution

$$3(2x) - 7(4x)$$
$$= 6x - 28x$$
$$= -22x$$

You complete Example 13.

Example 13. Simplify $-3(2y) - 5(-y)$.

Solution

$$-3(2y) - 5(-y)$$

Check your work on page 52. ▶

Since we agreed earlier that division can be expressed in terms of multiplication, the following example of dividing a variable term by a constant should make sense. Let's simplify $\frac{12x}{3}$.

$$\frac{12x}{3} \qquad \text{Original problem.}$$

$$= (12x)\left(\frac{1}{3}\right) \qquad \text{Definition of division.}$$

$$= \left(\frac{1}{3}\right)(12x) \qquad \text{Commutative property for multiplication.}$$

$$= \left(\frac{1}{3} \cdot 12\right)x \qquad \text{Associative property for multiplication.}$$

$$= 4x \qquad \text{Simplification.}$$

We conclude that

$$\frac{12x}{3} = \left(\frac{12}{3}\right)x$$
$$= 4x$$

To divide a variable term by a constant, we divide the numerical coefficient by the constant and keep the same variable part.

$$\frac{10y}{2} = \left(\frac{10}{2}\right)y = 5y$$

$$\frac{-2x^2}{2} = \left(\frac{-2}{2}\right)x^2 = -x^2$$

$$\frac{16ab}{-4} = \left(\frac{16}{-4}\right)ab = -4ab$$

Simplify the expression in Example 14.

Example 14. Simplify $\dfrac{10y^2}{5} - \dfrac{14x^2}{7}$.

Solution

$$\frac{10y^2}{5} - \frac{14x^2}{7}$$

Check your work on page 52. ▶

Example 15. Simplify

$$\frac{27x}{3} - (-2x) + \frac{5x}{-5}.$$

Solution

$$\frac{27x}{3} - (-2x) + \frac{5x}{-5}$$

$$= \left(\frac{27}{3}\right)x - (-2x) + \left(\frac{5}{-5}\right)x$$

$$= 9x + 2x + (-1x)$$

$$= 11x - x$$

$$= 10x$$

A product such as $3(x + 1)$ should remind you of the **distributive property** that we used earlier. According to that property,

$$3(x + 1)$$
$$= 3 \cdot x + 3 \cdot 1$$
$$= 3x + 3$$

Example 16. Simplify $-5(x^2 + x - 2)$.

Solution

$$-5(x^2 + x - 2)$$
$$= -5 \cdot x^2 + (-5 \cdot x) + (-5)(-2)$$
$$= -5x^2 + (-5x) + 10$$
$$= -5x^2 - 5x + 10$$

You complete Example 17.

Example 17. Simplify $0.2(4x - 11y)$.

Solution

$$0.2(4x - 11y)$$
$$= 0.2(4x) + 0.2(-11y)$$

Check your work on page 52. ▶

As soon as you feel more comfortable with the distributive property, you will need fewer steps to reach your answer. Be careful, however, to avoid careless errors.

$$-3(x - 1) = -3x + 3$$

but

$$-3(x - 1) \neq -3x - 1$$

$$-3(x - 1) \neq -3x - 3$$

Let's practice simplifying some expressions involving several operations. Recall the rules for the order in which we preform operations.

Order of Operations

1. Simplify powers.
2. Remove symbols of inclusion (from the inside out).
3. Perform multiplications and/or divisions from left to right.
4. Perform additions and/or subtractions from left to right.

You try simplifying the expression in Example 18.

Example 18. Simplify

$$\frac{1}{2}(4a - 10) - 3(4 - a).$$

Solution

$$\frac{1}{2}(4a - 10) - 3(4 - a)$$

$$= \frac{1}{2} \cdot 4a - \frac{1}{2} \cdot 10 - 3 \cdot 4 - 3(-a)$$

Check your work on page 52. ▶

Example 19. Simplify

$$7 + 2[x - 8(1 + x)].$$

Solution

$$7 + 2[x - 8(1 + x)]$$

$$= 7 + 2[x - 8 - 8x]$$

$$= 7 + 2[-8 - 7x]$$

$$= 7 + 2(-8) + 2(-7x)$$

$$= 7 - 16 - 14x$$

$$= -9 - 14x$$

In Example 19 you may have been tempted to combine $7 + 2$ right at first. This would be incorrect because the 2 must be multiplied by the quantity within the brackets *before* adding the result to 7. Remember the rules governing the order of operations!

▥➡ **Trial Run**

Simplify.

_____ **1.** $5(3x) - 4(-x)$ _____ **2.** $-(2x) + 3(4y) + \dfrac{16x}{2}$

_____ **3.** $\dfrac{45x^2}{-9} - (-3x^2) + 5(3x) + \dfrac{28x}{-14}$

_____ **4.** $3(a - 1) - 2(5 + a)$

_____ **5.** $8 + 3[x - 7(1 - x)]$

_____ **6.** $\dfrac{-1}{3}(3x - 6y) + 5[x - (x - 2y)]$

Answers are on page 53.

Finding the Absolute Value of an Expression

In Section 1.1 we learned that the absolute value of a constant is its distance from zero on the number line. This definition works very well as long as we are dealing with constants because the numerical value of a constant is *fixed*; it never changes.

At this time, however, we are ready to consider a definition for absolute value that will work for constants *and* for variables. We must find a defintion for the absolute value of any real number x, which we write as $|x|$.

Suppose that x represents a positive number or zero ($x \geq 0$):

$$\text{if} \quad x = 3 \quad \text{then} \quad |x| = |3| = 3$$

$$\text{if} \quad x = 0 \quad \text{then} \quad |x| = |0| = 0$$

$$\text{if} \quad x = \frac{2}{3} \quad \text{then} \quad |x| = \left|\frac{2}{3}\right| = \frac{2}{3}$$

> If x is a *positive* number or *zero*, the absolute value of x is x itself.

Suppose that x represents a negative number ($x < 0$):

$$\text{if} \quad x = -2 \quad \text{then} \quad |x| = |-2| = 2$$

$$\text{if} \quad x = \frac{-1}{5} \quad \text{then} \quad |x| = \left|\frac{-1}{5}\right| = \frac{1}{5}$$

$$\text{if} \quad x = -183 \quad \text{then} \quad |x| = |-183| = 183$$

> If x is a *negative* number, the absolute value of x is the *opposite* of x (or $-x$).

The definition for the absolute value of x is a *two-part* definition. If we *know* that x represents a *positive number or zero*, we use the first part of the definition. If we *know* that x represents a *negative number*, we use the second part of the definition. However, if we *don't know* whether x represents a positive number or a negative number or zero, we must consider both parts of the definition.

> **Definition of Absolute Value.** If x represents a real number, then
>
> $$|x| = \begin{cases} x & \text{if } x \geq 0 \\ -x & \text{if } x < 0 \end{cases}$$

Use the definition of absolute value to complete Example 20.

Example 20. Find $|\pi|$ and $|-\sqrt{2}|$.

Solution: Since $\pi > 0$,

$$|\pi| = \underline{\quad\quad}$$

Check your work on page 52. ▶

Solution: Since $-\sqrt{2} < 0$,

$$|-\sqrt{2}| = -(-\sqrt{2})$$

$$= \underline{\quad\quad}$$

Check your work on page 52. ▶

Example 21. Find $|x + 2|$.

Solution: Since we do not know the nature of $x + 2$, we must say that

$$|x + 2| = \begin{cases} x + 2 & \text{if } x + 2 \geq 0 \\ -(x + 2) & \text{if } x + 2 < 0 \end{cases}$$

Example 22. Find $|1 - \sqrt{2}|$.

Solution: Since $1 < \sqrt{2}$, we know that $1 - \sqrt{2}$ is *negative*. Therefore,

$$|1 - \sqrt{2}|$$

$$= -(1 - \sqrt{2})$$

$$= -1 + \sqrt{2}$$

Example 23. Find $-|3 - 7|$.

Solution

$$-|3 - 7|$$

$$= -|-4|$$

$$= -(4)$$

$$= -4$$

Complete Example 24.

Example 24. Find $-|3| - |-7|$.

Solution

$$-|3| - |-7|$$

$$= -(3) - (\underline{\quad\quad})$$

Check your work on page 52. ▶

⫸ **Trial Run** ▬▬▬▬▬▬▬▬▬▬▬▬▬▬▬▬▬▬▬▬▬▬▬▬▬▬▬▬▬▬▬▬▬▬▬▬▬▬▬

Evaluate each expression.

_____ **1.** $|-3|$

_____ **3.** $|x - 3|$

_____ **5.** $|2 - \sqrt{11}|$

_____ **2.** $|-\sqrt{5}|$

_____ **4.** $|\pi - 2|$

_____ **6.** $|-3| - |5|$

Answers are on page 53.

▶ **Examples You Completed** ▬▬▬▬▬▬▬▬▬▬▬▬▬▬▬▬▬▬▬▬▬▬▬▬▬▬▬▬▬▬▬

Example 2. Write an algebraic expression that means "multiply six times the sum of -1 and x."

Solution

$$6(-1 + x)$$

Example 4. Evaluate $9y - 37z$ when $y = 0$ and $z = 1$.

Solution: We must substitute 0 for y and 1 for z in the expression.

$$9y - 37z$$

$$= 9(0) - 37(1)$$

$$= 0 - 37$$

$$= -37$$

Example 6. Evaluate $(x + 1)(1 - x)$ when $x = -2$.

Solution

$$(x + 1)(1 - x)$$
$$= (-2 + 1)[1 - (-2)]$$
$$= (-1)(1 + 2)$$
$$= (-1)(3)$$
$$= -3$$

Example 7. Evaluate $-x^2 - 3x + 2$ when $x = -4$.

Solution

$$-x^2 - 3x + 2$$
$$= -(-4)^2 - 3(-4) + 2$$
$$= -1(-4)(-4) - 3(-4) + 2$$
$$= -16 + 12 + 2$$
$$= -2$$

Example 11. Simplify $x^2 - 6x + 3x^2 - x$.

Solution

$$x^2 - 6x + 3x^2 - x$$
$$= x^2 + 3x^2 - 6x - x$$
$$= 4x^2 - 7x$$

Example 13. Simplify $-3(2y) - 5(-y)$.

Solution

$$-3(2y) - 5(-y)$$
$$= -6y + 5y$$
$$= -y$$

Example 14. Simplify $\dfrac{10y^2}{5} - \dfrac{14x^2}{7}$.

Solution

$$\frac{10y^2}{5} - \frac{14x^2}{7}$$
$$= (2)y^2 - (2)x^2$$
$$= 2y^2 - 2x^2$$

Example 17. Simplify $0.2(4x - 11y)$.

Solution

$$0.2(4x - 11y)$$
$$= 0.2(4x) + 0.2(-11y)$$
$$= 0.8x + -2.2y$$
$$= 0.8x - 2.2y$$

Example 18. Simplify $\dfrac{1}{2}(4a - 10) - 3(4 - a)$.

Solution

$$\frac{1}{2}(4a - 10) - 3(4 - a)$$
$$= \frac{1}{2} \cdot 4a - \frac{1}{2} \cdot 10 - 3 \cdot 4 - 3(-a)$$
$$= 2a - 5 - 12 + 3a$$
$$= 2a + 3a - 5 - 12$$
$$= 5a - 17$$

Example 20. Find $|\pi|$ and $|-\sqrt{2}|$.

Solution: Since $\pi > 0$, $|\pi| = \pi$.

Solution: Since $-\sqrt{2} < 0$,
$$|-\sqrt{2}| = -(-\sqrt{2})$$
$$= \sqrt{2}$$

Example 24. Find $-|3| - |-7|$.

Solution

$$-|3| - |-7|$$
$$= -(3) - (7)$$
$$= -10$$

Answers to Trial Runs

page 46   **1.** 0   **2.** 12   **3.** 16   **4.** 15   **5.** $-4x + 5$ **6.** $-x^2 + 10x - 6$
          7. $-2x^2 + 2xy + 2y^2$ **8.** $-4a - 2$

page 49   **1.** $19x$ **2.** $6x + 12y$ **3.** $-2x^2 + 13x$ **4.** $a - 13$ **5.** $24x - 13$ **6.** $-x + 12y$

page 51   **1.** 3   **2.** $\sqrt{5}$ **3.** $x - 3$ if $x - 3 \geq 0$; $-(x - 3)$ if $x - 3 < 0$ **4.** $\pi - 2$
          5. $-2 + \sqrt{11}$ **6.** -2

EXERCISE SET 1.4

Evaluate each of the following expressions.

_____ **1.** $x(2x - 3)$ when $x = 3$ _____ **2.** $2x(x - 9)$ when $x = 5$

_____ **3.** $(x - 8)(x + 1)$ when $x = -2$ _____ **4.** $(x - 1)(x + 5)$ when $x = -9$

_____ **5.** $-x^2 - 4x - 3$ when $x = -3$ _____ **6.** $-x^2 - 7x - 9$ when $x = -1$

_____ **7.** $5(x - 3)^2 - 9x$ when $x = 0$ _____ **8.** $9(x - 2)^2 - 8x$ when $x = 2$

_____ **9.** $7x^2 - 3xy - 4y^2$ when $x = 2$ and $y = -1$

_____ **10.** $10x^2 - 3xy + y^2$ when $x = 1$ and $y = -4$

_____ **11.** $P = 2l + 2w$ when $l = 12$ and $w = 7$.

_____ **12.** $K = \dfrac{1}{2} h(a + b)$ when $h = 16$, $a = 21$, and $b = 13$.

_____ **13.** $V = lwh$ when $l = 8.2$, $w = 5.6$, and $h = 10$.

_____ **14.** $i = prt$ when $p = \$4000$, $r = 0.09$, and $t = 4$.

_____ **15.** $M = \dfrac{c(1 + p)}{1 - d}$ when $c = 7.5$, $p = 0.30$, and $d = 0.25$.

_____ **16.** $s = \dfrac{rl - a}{r - 1}$ when $r = 5$, $l = 18.4$, and $a = 0.4$.

_____ **17.** $S = 100t - 16t^2$ when $t = 5$.

_____ **18.** $A = p(1 + r)^2$ when $p = \$1000$ and $r = 0.05$.

Simplify.

_____ **19.** $3a - 9 - 7a + 2$ _____ **20.** $5a - 6 - 9a + 3$

_____ **21.** $9x^2 - 4x - 7x^2 - 7x$ _____ **22.** $13x^2 - 5x - 9x^2 - 4x$

_____ **23.** $-7y^2 + 3y - 8 - 2y^2 - 3y + 7$ _____ **24.** $-15y^2 + 9y - 1 - 5y + 4y^2 - 6$

_____ **25.** $17x - 8y + 3 - 6x + 2y - 5$ _____ **26.** $21x - 9y + 5 - 8x + 3y - 9$

_____ **27.** $8x^2 - 3xy - 7y^2 - 5x^2 - 2xy + 8y^2$ _____ **28.** $21x^2 - 7xy - 9y^2 - 3xy + 5y^2$

_____ **29.** $-17a^2 - 9b^2 + c^2 + 5a^2 - 2b^2 - c^2$

_____ **30.** $8a^2 - 3b^2 - 2c^2 - 11a^2 + 3b^2 - 2c^2$

_____ **31.** $3(7a) - 5(4a)$ _____ **32.** $-3(2a) + 7(-a)$

_____ **33.** $-(3x) + 5(-2x) + \dfrac{18}{3}x$ _____ **34.** $-(4x) + 6(-3x) + \dfrac{21}{3}x$

_____ **35.** $\dfrac{28}{4}y^2 + 3(-y^2) - 2(-4y^2)$ _____ **36.** $\dfrac{32}{8}y^2 + 5(-y^2) - 3(-5y^2)$

_____ **37.** $\dfrac{63}{-9} + 4(2x) - 2(-3x^2) - \dfrac{72}{9}x$

_____ **38.** $\dfrac{56}{-8}x^2 + 3(5x) + \dfrac{14}{2}x^2 - 2(-10x)$

_____ **39.** $-8(2y) + \dfrac{91}{7}y^2 - 2(-7)$ _____ **40.** $5(-2y) + \dfrac{38}{2}y^2 - 5(-3)$

_____ **41.** $\dfrac{1}{2}(16x) - 5(-3y) - 4(13)$ _____ **42.** $\dfrac{1}{3}(-15x) - 7(-2y) - 2(15)$

_____ **43.** $(-9x - 2y) - (5x - 2y)$ _____ **44.** $(-5m - 2n) - (m - 2n)$

_____ **45.** $3(5m - 4) + 4(m + 2)$ _____ **46.** $5(2m - 1) - 2(m + 5)$

_____ **47.** $-3(x - 2y) + 4(-x + y)$ _____ **48.** $-2(2x - y) + 3(-x + 2y)$

_____ **49.** $-4[3(2x - 1) + 10x] - \dfrac{1}{5}(15x)$ _____ **50.** $-8[2(3x + 2) + 7x] - \dfrac{1}{8}(16x)$

_____ **51.** $7 - 5[x - 3(2 - x)]$ _____ **52.** $8 - 3[x - 6(5 - x)]$

_____ **53.** $2x - 4[x - 6(3x + 5)]$ _____ **54.** $7x - 3[x - 8(5x - 7)]$

_____ **55.** $\dfrac{-1}{2}(4x - 6y) + 7[x - (x - 5y)]$ _____ **56.** $\dfrac{-1}{3}(9x - 12y) + 8[x - (x - y)]$

_____ **57.** $-2[5(a - 2b) + 7(2a + b)]$ _____ **58.** $-3[4(2a - b) + 5(a + 3b)]$

_____ **59.** $7[3x - (x + 2y)] - [5(x - 2y) - 9y]$

_____ **60.** $3[4x - (2x + y)] - [5(x - 2y) - 7y]$

Evaluate each expression.

_____ **61.** $|-11|$ _____ **62.** $|-8|$

_____ **63.** $|3x|$ _____ **64.** $|5x|$

_____ **65.** $|5 - \pi|$ _____ **66.** $|9 - \pi|$

_____ **67.** $|x - 2|$ _____ **68.** $|x + 1|$

_____ **69.** $|\sqrt{7} - 3|$ _____ **70.** $|\sqrt{6} - 1|$

_____ **71.** $|-4| - |9|$ _____ **72.** $|-5| - |11|$

_____ **73.** $|-4 - 9|$ _____ **74.** $|-5 - 11|$

☆ Stretching the Topics _____

_____ **1.** Simplify $-2[5x - 2(2x - 5y)] + 7[4(x - 2y) - (3x + y)]$, then evaluate when $x = -8$ and $y = -1$.

_____ **2.** Subtract the opposite of $[3 - (4 - x)]$ from $8x - (4 - 2x)$.

_____ **3.** Evaluate $2|x + 3| - |2x + 5| - 4(3x - 1)$ when $x = -2$.

Check your answers in the back of your book.

If you can complete **Checkup 1.4**, you are ready to do Section 1.5.

✓ CHECKUP 1.4

Evaluate.

_____ **1.** $(x - 3)(2x - 1)$ when $x = 5$ _____ **2.** $s = \frac{1}{2} gt^2$ when $g = 32$ and $t = 7$

Simplify.

_____ **3.** $-5x^2 + 3x - 7 + x^2 - 2x + 3$

_____ **4.** $18x^2 - 4xy - 5y^2 - 15x^2 + 4xy - 4y^2$

_____ **5.** $7(-2a) - 3(4a)$ _____ **6.** $-2(-3y) + \dfrac{34y^2}{2} - 4(2)$

_____ **7.** $-3(2n - 7) + 8[n - (2n - 3)]$ _____ **8.** $9y - 7[x - 3(2y - x)]$

Evaluate.

_____ **9.** $|4x|$ _____ **10.** $|\sqrt{2} - 9|$

Check your answers in the back of your book.

If You Missed Problems:	You Should Review Examples:
1	6
2	9
3, 4	10, 11
5	12, 13
6	14, 15
7, 8	18, 19
9, 10	20–23

1.5 **Switching from Words to Algebraic Expressions**

Now that we have learned how to deal with simple algebraic expressions, we must try our hand at switching some everyday word expressions into expressions involving variables and constants.

To switch a word expression to a variable expression, we must first decide what the variable is to represent. Then we translate the words into a variable expression. The key to success here is in writing down each word phrase as a variable pharse. (Skipping steps will only confuse the matter.) Finally, we write a sentence that summarizes our findings.

Example 1. Martina has a part-time job paying $3.50 per hour. Write an algebraic expression for her total earnings after she has worked for x hours.

Solution

We Think	*We Write*
The variable x represents the number of hours worked.	Let x = number of hours
Earnings are found by *multiplying* rate per hour times number of hours worked	$\$3.50x$ = earnings

The expression $\$3.50x$ represents Martina's earnings.

Example 2. If a rectangular garden is 15 feet longer than it is wide, write an algebraic expression for the number of feet of fencing needed to enclose this garden.

Solution

We Think	*We Write*
The number of feet of fencing needed is just the *perimeter* of the rectangle.	Perimeter = 2 lengths + 2 widths
We need expressions for length and width. We know less about the width than we know about the length.	Let w = width (feet)
We know that the length is 15 feet *more than* the width.	Let $w + 15$ = length (feet)
We know that the perimeter is the *sum* of 2 lengths and 2 widths.	$2(w + 15) + 2w$ = perimeter (feet)
We can simplify this expression.	$2w + 30 + 2w$ $= 4w + 30$

The expression $4w + 30$ represents the number of feet of fencing needed to enclose this garden.

You try completing the missing parts of Example 3.

Example 3. Charlie is a part-time college student who has saved $800 for his tuition. He must pay $50 per course credit hour. Write an algebraic expression for the amount Charlie will have left after enrolling for h credit hours.

 Solution

We Think	*We Write*
The variable h represents the number of credit hours.	Let _____ = number of credit hours

The total cost is found by *multiplying* the rate per credit hour times the number of credit hours.

_____ = total cost (dollars)

The money left after enrolling is found by *subtracting* the total cost from $800.

800 − _____ = dollars left

After enrolling, Charlie will have _____ dollars left. What happens if $h > 16$? _____ What does that mean to Charlie? _____

Check your work on page 66. ▶

Example 4. On a recent trip, Francine drove for 8 hours, part through towns and part on the highway. Her rates were 35 miles per hour in town and 55 miles per hour on the highway. Write an expression for the total distance Francine drove.

Solution

We Think	We Write
Francine's total distance will be the *sum* of her distance through towns and her distance on the highway.	Distance = town distance + highway distance
Each distance is found by *multiplying* rate times time.	Distance = rate × time
Town distance is found by *multiplying* rate in town times hours in towns.	Town distance = 35 · hours in town
Highway distance is found by *multiplying* rate on highway times hours on highway.	Highway distance = 55 · hours on highway
All we know about the two times is that they *sum* to 8 hours. We can choose the variable t to represent the time spent driving in town.	Let t = hours in town
The time remaining for highway driving is the total time *minus* the time spent in town.	$8 - t$ = hours on the highway
Now we can write expressions for town distance and highway distance.	Town distance = $35t$ miles Highway distance = $55(8 - t)$ miles
Finally, we can *sum* the two distances to write an expression for total distance.	$35t + 55(8 - t)$ miles
We can simplify this expression.	$35t + 55(8 - t)$ $= 35t + 440 - 55t$ $= 440 - 20t$

The expression $440 - 20t$ represents Francine's total distance in miles.

To switch word expressions to variable expressions, we have discovered that it is helpful to:

1. Read the problem carefully and sort out the given information.
2. Identify the variable.
3. Switch each word expression in the problem to a variable expression.
4. Be sure that your expression represents the quantity you are asked to find.
5. Simplify the variable expression if possible.

Suppose that you try completing Example 5 using these steps.

Example 5. Each month Mark's mother earns twice as much as he does, and his father earns $900 more than Mark. Write an algebraic expression for the total monthly earnings of Mark's family.

Solution

We Think	*We Write*
The family's total earnings will be the *sum* of the members' separate earnings.	Total earnings = Mark's earnings + mother's earnings + father's earnings
We know the least about Mark's earnings.	Let x = Mark's earnings
Mark's mother earns *twice* as much as Mark earns.	_____ = mother's earnings
Mark's father earns $900 *more than* Mark earns.	$x +$ _____ = father's earnings
Now we find the *sum* of all the earnings.	Total earnings = $x +$ ____ $+ (x +$ ____$)$
We can simplify this expression.	$x +$ _____ $+ x +$ _____ = _____ $+$ _____

The expression _____ $+$ _____ represents the total monthly earnings for Mark's family

Check your work on page 66. ▶

When a company manufactures and sells a product, the money received from selling the product is called the company's **revenue**. The dollars spent in manufacturing the product are called the company's **costs**. The *difference* between the company's revenue and costs represents the company's **profit**.

> Revenue = dollars received
>
> Costs = dollars spent
>
> Profit = revenue − costs

Usually, the *revenue* from the sale of a product is found by multiplying the selling price times the number of units sold. For instance, if a shoe company sells 500 pairs of shoes at $20 a pair, the company's revenue is

$$\text{Revenue} = (\text{selling price})(\text{number of units})$$
$$= (\$20)(500)$$
$$\text{Revenue} = \$10,000$$

Example 6. If Supercal Calculator Company sells x calculators at $23.95 each, write an algebraic expression for the company's revenue.

Solution

We Think	*We Write*
We know the selling price.	$23.95 = selling price
The number of units (calculators) is x.	x = number of units

Revenue is found by multiplying the selling price times the number of units sold.

$$\text{Revenue} = 23.95x$$

The company's revenue is represented by the expression $23.95x$ dollars.

In manufacturing a product (say, shoes), a company must consider two kinds of *costs*. **Variable costs** are the costs that depend on the number of pairs of shoes produced. Variable costs include wages, raw materials, and utilities used in the manfuacturing process. **Fixed costs**, on the other hand, are expenses that the company must pay even if it does not manufacture any shoes for a period of time. Fixed costs include rent, depreciation of equipment, and payments on company loans.

Total costs are the *sum* of variable costs and fixed costs.

$$\text{Costs} = \text{variable costs} + \text{fixed costs}$$

For instance, suppose the shoe company knows that its *fixed costs* are $1000 per week and its *variable costs* are $9 for each pair of shoes. If the company produces 500 pairs of shoes in a week, what are the company's costs?

$$\text{Fixed costs} = \$1000$$
$$\text{Variable costs} = \$9(500)$$
$$= \$4500$$
$$\text{Costs} = \text{variable costs} + \text{fixed costs}$$
$$= \$4500 + \$1000$$
$$= \$5500$$

Example 7. The Supercal Calculator Company knows that its fixed costs are $7300 per month and its variable costs are $8.45 for each calculator. Write an algebraic expression for the company's costs when x calculators are produced in a month.

Solution

We Think	*We Write*
We know that fixed costs are $7300.	$\$7300 = \text{fixed costs}$
The variable costs are found by *multiplying* the number of calculators times the cost of producing each calculator.	$\$8.45 = \text{cost to produce each calculator}$ $x = \text{number of calculators produced}$ $\$8.45x = \text{variable costs}$
Total costs are found by *adding* variable costs and fixed costs.	$\text{Costs} = \$8.45x + \$7300$

The expression $\$8.45x + \$7300$ represents the company's cost for producing x calculators in one month.

To calculate *profit*, a company must *subtract* its costs from its revenue. For instance, when the shoe company produces and sells 500 pairs of shoes in one week, we found

$$\text{Revenue} = \$20(500) \quad \text{and} \quad \text{Costs} = \$9(500) + \$1000$$
$$= \$10{,}000 \qquad\qquad = \$5500$$

Now we can find the shoe company's profit:

$$\text{Profit} = \text{revenue} - \text{costs}$$
$$= \$10,000 - \$5500$$
$$= \$4500$$

Example 8. The Supercal Company manufactures calculators at a cost of $8.45 each and sells them for $23.95 each. The company's fixed costs are $7300 per month. Write an expression for the company's *profit* if x calculators are produced and sold in one month.

    Solution

We Think	*We Write*
Profit is found by *subtracting* costs from revenue.	Profit = revenue − costs
Revenue is found by *multiplying* selling price times number of units sold.	Revenue = $23.95x$
Costs are found by *adding* variable costs and fixed costs.	Variable costs = $8.45x$ Fixed costs = 7300 Costs = $8.45x + 7300$
Profit = revenue − costs.	Profit = $23.95x - (8.45x + 7300)$ $= 23.95x - 8.45x - 7300$ $= 15.5x - 7300$

The expression $15.5x - 7300$ dollars represents the Supercal Company's profit for one month.

    Now you try Example 9.

Example 9. The Hamburger Heaven makes hamburgers for $0.81 each and sells them for $1.59 each. The restaurant's fixed costs are $2100 per week. Write an expression for the restaurant's *profit* if x hamburgers are made and sold in one week.

    Solution

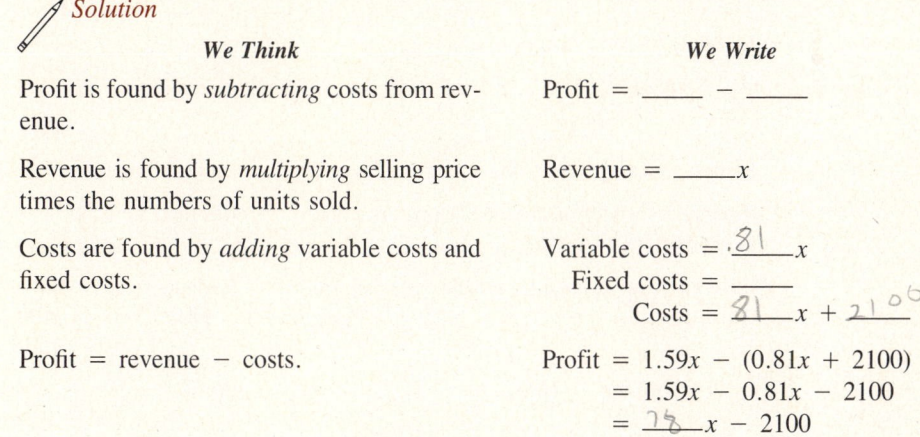

We Think	*We Write*
Profit is found by *subtracting* costs from revenue.	Profit = _____ − _____
Revenue is found by *multiplying* selling price times the numbers of units sold.	Revenue = _____ x
Costs are found by *adding* variable costs and fixed costs.	Variable costs = $.81\,x$ Fixed costs = _____ Costs = $81\,x + 21^{00}$
Profit = revenue − costs.	Profit = $1.59x - (0.81x + 2100)$ $= 1.59x - 0.81x - 2100$ $= 78\,x - 2100$

The expression $78\,x - 2100$ dollars represents the Hamburger Heaven's profit when x hamburgers are sold in one week.

Check your work on page 66. ▶

⇒ Trial Run

Change each indicated word expression to a variable expression. Then simplify if possible.

_____ **1.** If Julio drives for 6 hours at a constant rate of r miles per hour, write an algebraic expression for his total distance.

_____ **2.** Tony bought 10 pounds of chicken parts for his barbecue, some thighs and some legs. If thighs cost $1.19 per pound and legs cost $0.99 per pound, write an algebraic expression for the total cost of the chicken parts.

_____ **3.** A recent concert was attended by 200 children, 500 adults, and 300 senior citizens. A child's ticket costs half as much as an adult's ticket and a senior citizen's costs $1 less than an adult's ticket. Write an algebraic expression for the total dollars from concert ticket sales.

Answers are on page 66.

▶ Examples You Completed

Example 3

Solution

We Write

Let h = number of credit hours
$50h$ = total cost
$800 - 50h$ = dollars left

After enrolling, Charlie will have $800 - 50h$ dollars left. What happens if $h > 16$? $800 - 50h$ will be negative. What does that mean to Charlie? He cannot enroll for more than 16 hours.

Example 5

Solution

We Write

Total earnings = Mark's earnings
+ mother's earnings
+ father's earnings
Let x = Mark's earnings
$2x$ = mother's earnings
$x + 900$ = father's earnings
Total earnings = $x + 2x + (x + 900)$
$x + 2x + x + 900$
= $4x + 900$

The expression $4x + 900$ represents the total monthly earnings for Mark's family.

Example 9

Solution

We Think

Profit is found by *subtracting* costs from revenue.

Revenue is found by *multiplying* selling price times the number of units sold.

Costs are found by *adding* variable costs and fixed costs.

Profit = revenue − costs.

We Write

Profit = revenue − costs

Revenue = $1.59x$

Variable costs = $0.81x$
Fixed costs = 2100
Costs = $0.81x + 2100$

Profit = $1.59x - (0.81x + 2100)$
= $1.59x - 0.81x - 2100$
= $0.78x - 2100$

The expression $0.78x - 2100$ dollars represents Hamburger Heaven's profit when x hamburgers are sold in one week.

Answers to Trial Run

page 65   **1.** $6r$ miles   **2.** $0.2x + 9.9$ dollars   **3.** $900x - 300$

EXERCISE SET 1.5

Change the following word expressions to variable expressions and simplify.

_____ 1. Lydia is depositing $27.50 a month from her paycheck in a savings account. Write an algebraic expression for the total savings after she has saved for x months.

_____ 2. The width of a rectangle is 5 feet less than the length. Write an algebraic expression for the perimeter of the rectangle.

_____ 3. A car rental agency charges a fee of $18 a day plus 30 cents per mile driven. If Lee rents a car for one day and drives n miles, write an algebraic expression for the amount he owes at the end of one day.

_____ 4. Every morning Abdul jogs and walks for 3 hours. If he jogs at 5 miles per hour and walks at 3 miles per hour, write an algebraic expression for the total distance he covers.

_____ 5. If Otho wishes to build a pen for his dog that is twice as long as it is wide, write an algebraic expression for the number of feet of fencing needed.

_____ 6. Tillie is buying framing for a rectangular picture that is three times as long as it is wide. Write an algebraic expression for the number of inches of framing she should buy.

_____ 7. In a physical fitness class Carlos did 7 more than twice as many pushups as his father. If his father did x pushups, write an algebraic expression for the number of pushups Carlos did.

_____ 8. At Weight Watchers Gretchen discovered that she had lost 5 pounds more than one-half as many pounds as Jane. If Jane lost p pounds, write an algebraic expression for the number of pounds Gretchen lost.

_____ 9. Hector earns a base salary of $940 per month plus a $10 bonus for each suit he sells that costs $100 or more. Write an algebraic expression for Hector's salary in December if he sold n suits costing $100 or more that month.

_____ 10. Tony is saving money for a vacation during spring break. If he deposits $50 and then saves $12 a week for x weeks, write an algebraic expression for the amount that he will have at the end of x weeks.

_____ 11. When Mardell went to see her sister she made part of the trip by bus and the other part by taxi. The time she traveled by bus was twice the time she rode in a taxi. If she averaged 40 miles per hour on the bus and 30 miles per hour in the taxi, write an algebraic expression for the total distance.

_____ 12. When Gary went to college, he drove 6 hours, part on the interstate and part on two-lane highways. If his rate was 55 miles per hour on the interstate and 45 miles per hour on two-lane highways, write an expression for the total distance Gary drove.

_____ 13. Sidney spends twice as much on food as he does on clothing and six times as much for rent as for clothing. If he spends x dollars per month on clothing, write an algebraic expression for the total of these three expenditures.

_____ **14.** When baking a cake, Pat notices that the batter requires twice as much butter as the icing and the filling requires half as much butter as the icing. If the recipe uses n tablespoons of butter for the icing, write an algebraic expression for the total number of tablespoons of butter needed to bake the cake.

_____ **15.** A person burns up about three times as many calories while walking as while sitting. Suppose that Imogene spends 12 hours sitting and 4 hours walking during the day. Write an algebraic expression for the total number of calories burned that day.

_____ **16.** For her flower bed, Callie needs four times as many boxes of petunias as boxes of marigolds. If petunias cost $1 a box and marigolds cost $0.75 a box, write an algebraic expression for the total cost of planting the flower bed.

_____ **17.** Marcus is making 20 pounds of party mix, part peanuts and part cashews. If cashews cost $2.19 a pound and peanuts cost $1.39 a pound, write an algebraic expression for the total cost of the party mix.

_____ **18.** Hagar is buying 30 pounds of apples for her Halloween party, some winesaps and some golden delicious. If the winesaps cost 79 cents a pound and the golden delicious apples cost 99 cents a pound, write an algebraic expression for the total cost of the apples.

_____ **19.** The cashier at French's market has a stack of bills, some five-dollar bills and some ten-dollar bills. There are 7 fewer five-dollar bills than ten-dollar bills. Write an algebraic expression for the total value of the stack of bills.

_____ **20.** John sold a box of books at a garage sale. The number of paperbacks was 3 less than twice the number of hardback books. If he sold the paperbacks for $1 each and the hardbacks sold for $3 each, write an algebraic expression for the amount he received from the sale of his books.

_____ **21.** The Basket Place knows that its fixed costs are $2500 per month and its variable costs are $3.85 for each basket produced. Write an algebraic expression for the company's costs when x baskets are produced in a month.

_____ **22.** Movac manufactures small motors at a cost of $53.50 each. The company's fixed costs are $8500 per month. If the motors sell for $123.95 each, write an expression for the company's profit when x motors are produced and sold in one month.

_____ **23.** The Brunson Battery Company can manufacture a battery for $18.65. The company's fixed costs each week are $2475. If the batteries sell for $59.95 each, write an expression for the company's profit when x batteries are produced and sold in one week.

_____ **24.** The Pizza Palace makes pizzas for $2.30 each and sells them for $9.75 each. The fixed costs each week are $2800. Write an expression for the Pizza Palace's profit if x pizzas are made and sold in one week.

_____ **25.** Mr. Waggoner has a small farm operation. The fixed costs for a year are $10,250. If he can produce a bushel of corn for $2.35 and sell it for $4.10, write an expression for Mr. Waggoner's profit if x bushels are produced that year.

☆ Stretching the Topics

_____ **1.** If $n - 2$ represents an integer, write an expression for the sum of the next higher consecutive integer and the next lower consecutive integer.

_____ **2.** Doug earns $1\frac{1}{2}$ times as much per hour for overtime as he does regular salary. Write an expression for Doug's earnings if he works a 40-hour week plus 12 hours of overtime.

_____ **3.** Drennan has $12,000 invested, part at $7\frac{1}{2}\%$ interest and the remainder at $8\frac{3}{4}\%$ interest. Write an expression for the total amount of money he will have at the end of one year.

Check your answers in the back of your book.

If you can complete **Checkup 1.5**, you are ready to do the Review Exercises.

✓ CHECKUP 1.5

_____ **1.** Carlos earns $420 a week and $12 per hour for each hour of overtime. If he worked x hours of overtime last week, write an algebraic expression for his earnings for that week.

_____ **2.** The width of a room is 5 feet less than the length. Write an algebraic expression for the number of feet of ceiling border the paperhanger needs to purchase.

_____ **3.** The caterer tells the mother of the bride that he will cater the reception for a fee of $50, plus $3 per person for each guest expected. Write an algebraic expression for the total cost of the reception if n guests are expected to attend.

_____ **4.** On a recent trip to see her grandmother, Lola drove 4 hours, part through towns and part on the highway. If her rate in town averaged 30 miles per hour and on the highway averaged 55 miles per hour, write an algebraic expression for the total distance she traveled.

_____ **5.** When Sue, Ellen, and Sarah went shopping, Sue spent twice as much as Sarah and Ellen spent $50 more than Sarah. Write an algebraic expression for the total amount the girls spent while shopping.

Check your answers in the back of your book.

If You Missed Problem:	You Should Review Example:
1	1
2	2
3	3
4	4
5	5

Summary

In this chapter we have discussed the basic ideas that form the foundation for all future work in algebra. After reviewing the rules for operations with real numbers, we introduced algebraic expressions and discovered how those same rules can be applied to variables that represent real numbers.

The most important real number properties and definitions are summarized here for easy reference, with examples for each. We assume that all variables used represent real numbers.

Property	Statement	Examples
1. Closure	$a + b \in R$ and unique $a \cdot b \in R$ and unique	$x + 3 \in R$ $5x \in R$
2. Commutative	$a + b = b + a$ $a \cdot b = b \cdot a$	$x + 2 = 2 + x$ $2(x + y) = (x + y)2$
3. Associative	$a + (b + c) = (a + b) + c$ $a(bc) = (ab)c$	$3 + (5 + x) = (3 + 5) + x$ $2(7x) = (2 \cdot 7)x$
4. Distributive	$a(b + c) = ab + ac$	$5(x - y) = 5x - 5y$
5. Identities	$a + 0 = a$ $a \cdot 1 = a$	$2x + 0 = 2x$ $(xy)(1) = xy$
6. Inverses	$a + (-a) = 0$ $a \cdot \dfrac{1}{a} = 1 \quad (a \neq 0)$	$6y + (-6y) = 0$ $5 \cdot \dfrac{1}{5} = 1$

Some special properties of zero were also noted.

Property	Statement	Examples
1. Addition	$a + 0 = a$	$-5x + 0 = -5x$
2. Subtraction	$a - 0 = a$	$y - 0 = y$
3. Multiplication	$a \cdot 0 = 0$	$(15x)(0) = 0$
4. Division	$\dfrac{0}{a} = 0 \quad (a \neq 0)$	$\dfrac{0}{23} = 0$
	$\dfrac{a}{0}$ is undefined	$\dfrac{x}{0}$ is undefined
	$\dfrac{0}{0}$ is not uniquely defined	$\dfrac{3 - 3}{x + (-x)}$ is not uniquely defined

Finally, we made several definitions that allowed us to write one expression in terms of another.

Operation	Definition	Example
1. Subtraction	$a - b = a + (-b)$	$2x - 3x = 2x + (-3x)$
2. Division	$\dfrac{a}{b} = a \cdot \dfrac{1}{b}$	$\dfrac{5}{3} = 5 \cdot \dfrac{1}{3}$
3. Exponentiation	$\underbrace{a^n = a \cdot a \cdot \ldots \cdot a}_{n \text{ factors}}$	$2^3 = 2 \cdot 2 \cdot 2$ $x^4 = x \cdot x \cdot x \cdot x$
4. Absolute value	$\lvert x \rvert = \begin{cases} x & \text{if } x \geq 0 \\ -x & \text{if } x < 0 \end{cases}$	$\lvert 17 \rvert = 17$ $\lvert -5 \rvert = 5$

After much practice using these properties to simplify algebraic expressions, we learned to translate word expressions into variable expressions.

❏ Speaking the Language of Algebra

Complete each sentence with the appropriate word or phrase.

1. In this chapter we learned to perform operations with _____ (numbers whose values are fixed) and _____ (symbols that represent numbers).

2. All numbers that we dealt with were _____ numbers.

3. In an addition (or subtraction) problem, the quantities being added (or subtracted) are called the _____ . The answer in an addition problem is called the _____ and the answer in a subtraction problem is called the _____ .

4. To add (or subtract) algebraic expressions, we must _____ _____ _____ .

5. In a multiplication problem, the quantities being multiplied are called the _____ and the answer is called the _____ .

6. In a division problem, the quantity being divided is called the _____ . The quantity doing the dividing is called the _____ , and the answer is called the _____ .

7. When we divide zero by any nonzero number, the answer is _____ . When we divide any nonzero number by zero, the answer is _____ .

8. The absolute value of a real number is never _____ .

REVIEW EXERCISES for Chapter 1: Working with Numbers and Variables

From the set $\{-4, -2.3, -\sqrt{2}, \frac{-5}{7}, 0, \sqrt{3}, 2.333 \ldots, \frac{1}{2}, 5\}$ *list the members that belong to each of the following sets.*

_____ **1.** Irrational numbers

_____ **2.** Integers

Compare the following numbers using $<, >,$ *or* $=$.

_____ **3.** $\frac{15}{3}$ and $\sqrt{25}$

_____ **4.** $\frac{-9}{4}$ and 13

_____ **5.** -19 and $\frac{-3}{4}$

_____ **6.** 5 and $\sqrt{7}$

Use inequalities to state whether the following numbers are positive or negative.

_____ **7.** $-\sqrt{6}$

_____ **8.** 8.75

Simplify.

_____ **9.** $(-36) + (-19)$

_____ **10.** $7 + (-21)$

_____ **11.** $3.9\,(-1.1)$

_____ **12.** $(-17) - (-13)$

_____ **13.** $9 - (-8)$

_____ **14.** $(-5)(-9)$

_____ **15.** $(-7)(0.8)$

_____ **16.** $-91 \div \frac{7}{2}$

_____ **17.** $\frac{-55}{-11}$

_____ **18.** $\frac{9 - (-9)}{3 + -3}$

_____ **19.** $-12 + 18 + 9 - 6 - 5$

_____ **20.** $19 - 3 - 21 + 6 - 1$

_____ **21.** $4(-2)(7)(-5)(3)$

_____ **22.** $-6(7)(10)(8)(-1)$

_____ **23.** $-6(15 - 23)$

_____ **24.** $5 - 7[9 - 2(-1)]$

_____ **25.** $(8 - 12)[9 - 3(11)]$

_____ **26.** $\dfrac{4(-3) - \frac{1}{4}(16)}{-8(9 - 5)}$

_____ **27.** $\dfrac{-11(4) - 5[6 - 2(9)]}{-2[13 - 3(4)]}$

_____ **28.** $\dfrac{\frac{1}{5}(35) - [13 + 2(-3)]}{\frac{44}{-11} - 5(-3)}$

Evaluate.

_____ **29.** $(2x - 3)(x - 5)$ when $x = 2$

_____ **30.** $2x^2 - 7x + 5$ when $x = -1$

_____ **31.** $A = p(1 + rt)$ when $p = \$2000$, $r = 0.09$, and $t = 4$.

_____ **32.** $S = a + (n - 1)d$ when $a = -5$, $n = 10$, and $d = 3$.

Simplify.

_____ **33.** $-6x^2 + 4x - 9 + 4x^2 - 3x + 2$

_____ **34.** $17x^2 - 3xy - 4y^2 - 14x^2 + 3xy + y^2$

_____ **35.** $9(-3a) - 5(2a)$

_____ **36.** $-3(7y) + \dfrac{32y^2}{8} - 5(3)$

_____ **37.** $-7(n - 5) + 9(n + 2)$

_____ **38.** $(11 - 8)[9x - 4(2x + 3)]$

_____ **39.** $8 - 3[x - 5(1 - 3x)]$

_____ **40.** $\dfrac{-1}{5}(10x + 15y) + 3[x - (x + 2y)]$

_____ **41.** $-4[3(2x - y) + 5(3x + y)]$

_____ **42.** $9[7x - (x - 3y)] - [3(2x - 5y) - 11y]$

Evaluate each expression.

_____ **43.** $|-21|$ _____ **44.** $|a|$

_____ **45.** $|\sqrt{10} - 6|$ _____ **46.** $|9 - 13|$

_____ **47.** $|9| - |13|$ _____ **48.** $|9| + |-13|$

Change the following word expressions to variable expressions and simplify.

_____ **49.** If the width of a rectangular tablecloth is 7 feet less than twice the length, write an algebraic expression for the total number of feet of lace needed to make a border.

_____ **50.** The Bookrack is having a spring sale and is selling all its paperback books, some for $0.95 and the rest for $1.75 each. Carlos bought 5 more of the $1.75 books than he did of the cheaper ones. Write an algebraic expression for the total amount he spent at the sale.

_____ **51.** In the last election for city attorney, the Democratic candidate received twice as many votes as the Independent candidate and the Republican candidate received 15 less than three times as many as the Independent. Write an algebraic expression for the total number of votes cast in the election.

_____ **52.** Charlene is using her savings to attend an exercise salon twice a week to improve her figure. If she must pay $6 each time she attends and she has saved $500, write an algebraic expression for the amount Charlene will have left after attending for x weeks.

Check your answers in the back of your book.

If You Missed Exercises:	You Should Review Examples:	
1, 2	SECTION 1.1	1–3
3–6		5, 6
7, 8		7, 8
9, 10	SECTION 1.2	5, 6
11–13		7–10
14, 15		11, 12
16, 17		13–16
18		21
19, 20	SECTION 1.3	3, 4
21, 22		5, 6
23–28		7, 8, 15, 16
29, 30	SECTION 1.4	3–8
31, 32		9
33, 34		10, 11
35–37		12–16
38–42		17, 19
43–46		20–22
47, 48		24, 25
49–52	SECTION 1.5	1–6

If you have completed the Review Exercises and corrected your errors, you are ready to take the Practice Test.

PRACTICE TEST for Chapter 1: Working with Numbers and Variables

		Section	Examples
_____	**1.** Compare $\sqrt{10}$ and 3 using $<$, $>$, or $=$.	1.1	5, 6
_____	**2.** Use an inequality to state that $-\sqrt{5}$ is a negative number.	1.1	7, 8

Simplify.

		Section	Examples
_____	**3.** $-35 + (-12)$	1.2	2
_____	**4.** $14 + (-17)$	1.2	5
_____	**5.** $-8 - (-23)$	1.2	9
_____	**6.** $-11(-4)$	1.2	12
_____	**7.** $\dfrac{-28}{7}$	1.2	15
_____	**8.** $4(-2)(11)(0)$	1.3	5, 6
_____	**9.** $10 - 6[3 - (-4)]$	1.3	7
_____	**10.** $(7 - 11)[5 - 3(4)]$	1.3	7
_____	**11.** $\dfrac{\dfrac{-18}{6} - 4[-5 - 2(-7)]}{-[8(-2) + 3]}$	1.3	8
_____	**12.** Evaluate $A = 2lw + 2wh + 2lh$ when $l = 5$, $w = 3$, and $h = 7$.	1.4	9

Simplify.

		Section	Examples
_____	**13.** $-2x^2 - 5x - 7 - x^2 + 3x - 4$	1.4	10, 11
_____	**14.** $6a^2 - 2ab + 3b^2 - 4b^2 + ab - 6a^2$	1.4	10, 11
_____	**15.** $3(-m) - 6(-2m)$	1.4	12, 13
_____	**16.** $3x(-3x) + \dfrac{16x}{-8} - 5(4)$	1.4	15
_____	**17.** $-5(2 - a) + 7(a - 2)$	1.4	18
_____	**18.** $1 - 4[2y - (6y - 5)]$	1.4	18
_____	**19.** $(1 - 4)[3x - 4(1 - x)]$	1.4	17, 19
_____	**20.** $\dfrac{-1}{5}(10x - 5y) + 2[3x - (y - x)]$	1.4	17, 19

_____ **21.** $-|12 - 19|$ 1.4 22

_____ **22.** $|-8| - |-14|$ 1.4 24

_____ **23.** $|\sqrt{5} - 3|$ 1.4 22

_____ **24.** On a recent trip, Martina drove for 3 hours, part of the way on 1.5 4
interstate highway and part of the way on two-lane roads. If her rate
on the interstate was 55 mph and her rate on the two-lane roads was
45 mph, write an algebraic expression for the total distance she
traveled.

_____ **25.** Harriet's rectangular bedspread is three times as long as it is wide. If 1.5 2
she sews a 4-inch ruffle around the spread, write an algebraic
expression for the perimeter of the ruffled bedspread.

Solving First-Degree Equations and Inequalities

2

Mary left Hometown at 6 P.M., driving due east on Route 22 at 55 miles per hour. At the same time, Todd left Anytown driving at 60 miles per hour due west on Route 22. If Hometown is 368 miles due west of Anytown, at what time will Mary and Todd meet?

Now that we have learned to simplify algebraic expressions, it is only natural to move on to solving equations and inequalities. In Chapter 1 we learned that numbers can be compared using symbols of equality (=) or inequality ($<$, $>$, $\neq$). In this chapter we learn how to find those values of the variable that make statements of equality or inequality true. In this chapter we learn how to:

1. Solve first-degree equations by addition, subtraction, multiplication, and division.

2. Solve first-degree inequalities by addition, subtraction, multiplication, and divison.

3. Graph solutions of equations and inequalities on the real number line.

4. Solve equations and inequalities that contain absolute-value expressions.

5. Switch from word statements to first-degree equations and inequalities.

2.1 Solving First-Degree Equations

An **equation** is an algebraic statement which says that one quantity is equal to another quantity. You can probably solve some equations "on sight" just by substituting a likely number for the variable to see if it "works."

$$\text{If} \quad x + 3 = 7 \quad \text{then} \quad x = 4 \quad \text{because} \quad 4 + 3 = 7$$

$$\text{If} \quad x - 9 = -2 \quad \text{then} \quad x = 7 \quad \text{because} \quad 7 - 9 = -2$$

$$\text{If} \quad 2x = 16 \quad \text{then} \quad x = 8 \quad \text{because} \quad 2 \cdot 8 = 16$$

$$\text{If} \quad \frac{x}{3} = 4 \quad \text{then} \quad x = 12 \quad \text{because} \quad \frac{12}{3} = 4$$

Remember, to *solve an equation*, we must find the value of the variable that makes the statement true. Such a value is called a **solution** for the equation.

A number is a **solution** for an equation if substituting that number for the variable makes the equation a true statement.

We solved the first-degree equations above on sight, using trial and error, but you should remember that there are many equations for which trial and error would be a very inefficient way to arrive at a solution. Let's recall what *first-degree* equations look like:

$$x - 6 = -7 \quad \text{is a first-degree equation}$$

$$3y + 2 = 17 \quad \text{is a first-degree equation}$$

$$\frac{4y - 3}{5} = 5 \quad \text{is a first-degree equation}$$

$$x^2 + x = 3 \quad \text{is } not \text{ a first-degree equation}$$

$$y^3 = 27 \quad \text{is } not \text{ a first-degree equation}$$

First-Degree Equation. An equation of the form

$$ax + b = c$$

where a, b, and c are real numbers (and $a \neq 0$), is called a first-degree equation in x.

In solving first-degree equations, our main goal is to *isolate the variable* on one side of the equation. To accomplish this, we must recall some very important properties of equality.

We may add (or subtract) the same quantity to (or from) both sides of an equation.

In other words for a, b, and c real numbers, we have:

Addition Property of Equality

If $\qquad a = b$

then $\quad a + c = b + c$

and $\quad a - c = b - c$

Let's see how this property helps us isolate the variable in an equation.

$x - 6 = -7$	We wish to isolate x.
$x - 6 + 6 = -7 + 6$	Add 6 to both sides of the equation.
$x + 0 = -1$	Simplify the sums.
$x = -1$	

We can **check** our solution by substituting -1 for x in the *original equation*.

$$x - 6 = -7$$
$$-1 - 6 \overset{?}{=} -7$$
$$-7 = -7$$

Our solution checks.

Example 1. Solve $-3 = 5 + y$ and check.

Solution: We must isolate y.

$-3 = 5 + y$	
$-5 - 3 = -5 + 5 + y$	Subtract 5.
$-8 = y$	Simplify.

CHECK

$-3 = 5 + y$	
$-3 \overset{?}{=} 5 + (-8)$	Substitute -8 for y.
$-3 = -3$	Simplify.

Now you try Example 2.

Example 2. Solve $x - 12 = 0$ and check.

Solution: We must isolate _____ .

$$x - 12 = 0$$

CHECK

Check your work on page 90. ▶

To solve equations where the numerical coefficient of the variable is different from 1, we'll need another property of equality.

We may multiply or divide both sides of an equation by the same nonzero quantity.

In other words, for the real numbers a, b, and c ($c \neq 0$), we have

---

Multiplication Property of Equality

If $\qquad a = b$

then $\quad a \cdot c = b \cdot c$

and $\qquad \dfrac{a}{c} = \dfrac{b}{c}$

Let's use this property to solve an equation.

$$\frac{x}{3} = 11 \qquad \text{We must isolate } x.$$

$$\frac{3}{1} \cdot \frac{x}{3} = 3 \cdot 11 \qquad \text{Multiply both sides of the equation by 3.}$$

$$\overset{1}{\underset{1}{\cancel{3}}} \cdot \frac{x}{\underset{1}{\cancel{3}}} = 33 \qquad \text{Simplify.}$$

$$x = 33$$

CHECK

$$\frac{x}{3} = 11$$

$$\frac{33}{3} \overset{?}{=} 11$$

$$11 = 11$$

Let's try another equation.

$$-5y = 13 \qquad \text{We must isolate } x.$$

$$\frac{-5y}{-5} = \frac{13}{-5} \qquad \text{Divide both sides of the equation by } -5.$$

$$y = \frac{-13}{5} \qquad \text{Simplify.}$$

CHECK

$$-5y = 13$$

$$-5\left(\frac{-13}{5}\right) \overset{?}{=} 13$$

$$13 = 13$$

Now you try Example 3.

Example 3. Solve $6 = -12x$.

Solution

$$6 = -12x$$

Check your work on p. 90. ▶

Example 4. Solve $\dfrac{-2}{5}x = 10$.

Solution

$$\frac{-2}{5}x = 10$$

$$\frac{-5}{2}\left(\frac{-2}{5}x\right) = \frac{-5}{2} \cdot \frac{10}{1}$$

$$x = -25$$

⫸ Trial Run

Solve and check.

_____ **1.** $x - 6 = 1$ _____ **2.** $a - 5 = 0$

_____ **3.** $x + 6 = 6$ _____ **4.** $-2 + y = -3$

_____ **5.** $-3 = x + 7$ _____ **6.** $6x = 42$

_____ **7.** $-12x = 4$ _____ **8.** $5 = \dfrac{x}{7}$

_____ **9.** $\dfrac{-3x}{4} = 5$ _____ **10.** $0 = \dfrac{x}{-3}$

Answers are on page 91.

Many first-degree equations require that more than one operation be performed on both sides before the variable can be isolated. Most of the time it is best to use the *addition* property of equality *before* the *multiplication* property of equality when both are needed to solve an equation.

$$3y + 2 = 17 \qquad \text{We must isolate } y. \qquad\qquad \text{CHECK}$$
$$3y + 2 - 2 = 17 - 2 \qquad \text{Subtract 2 from both sides.} \qquad 3y + 2 = 17$$
$$3y = 15 \qquad \text{Simplify.} \qquad 3(5) + 2 \overset{?}{=} 17$$
$$\frac{3y}{3} = \frac{15}{3} \qquad \text{Divide both sides by 3.} \qquad 15 + 2 \overset{?}{=} 17$$
$$17 = 17$$
$$y = 5 \qquad \text{Simplify.}$$

Example 5. Solve $7 - x = 39$.

Solution

$$7 - x = 39$$
$$-7 + 7 - x = -7 + 39 \qquad \text{Subtract 7.}$$
$$-x = 32 \qquad \text{Simplify.}$$
$$\frac{-x}{-1} = \frac{32}{-1} \qquad \text{Divide by } -1.$$
$$x = -32 \qquad \text{Simplify.}$$

Now try Example 6.

Example 6. Solve $\dfrac{x}{2} - 8 = 0$.

Solution

$$\frac{x}{2} - 8 = 0$$

Check your work on page 91. ▶

 Trial Run

Solve and check.

_____ **1.** $5x + 3 = 13$ _____ **2.** $9 - 2x = 6$

_____ **3.** $\dfrac{x}{3} + 6 = 0$ _____ **4.** $5 - \dfrac{a}{2} = -1$

_____ **5.** $8 = -3m + 7$ _____ **6.** $-3y + 9 = 27$

Answers are on page 92.

When an equation contains variable terms on *both sides*, we must first perform the additions and/or subtractions needed to get all the variable terms on the *same side* of the equation. For instance, consider the equation $5x = 3x + 10$. To get all the x terms on the left-hand side, we must get rid of the $3x$ on the right. How? By *subtracting* $3x$ from both sides of the equation.

$$5x = 3x + 10$$

$$5x - 3x = 3x + 10 - 3x \qquad \text{Subtract } 3x.$$

$$2x = 10 \qquad \text{Simplify.}$$

$$\frac{2x}{2} = \frac{10}{2} \qquad \text{Divide by 2.}$$

$$x = 5 \qquad \text{Simplify.}$$

CHECK

$$5x = 3x + 10$$

$$5(5) \stackrel{?}{=} 3(5) + 10$$

$$25 \stackrel{?}{=} 15 + 10$$

$$25 = 25$$

Example 7. Solve $13 + 7x = 10x - 2$ and check.

Solution

$$13 + 7x = 10x - 2$$

$$13 + 7x - 7x = 10x - 2 - 7x$$

$$13 = 3x - 2$$

$$13 + 2 = 3x - 2 + 2$$

$$15 = 3x$$

$$\frac{15}{3} = \frac{3x}{3}$$

$$5 = x$$

CHECK

$$13 + 7x = 10x - 2$$

$$13 + 7(5) \stackrel{?}{=} 10(5) - 2$$

$$13 + 35 \stackrel{?}{=} 50 - 2$$

$$48 = 48$$

Now try Example 8.

Example 8. Solve $10 - 3x = 16 - x$ and check.

Solution

$$10 - 3x = 16 - x$$

$$10 - 3x + x = 16 - x + x$$

$$10 - 2x = 16$$

$-2x = 6$

$-x = 3$

$x = -3$

CHECK

$$10 - 3x = 16 - x$$

$$10 - 3(\underline{\quad}) \stackrel{?}{=} 16 - (\underline{\quad})$$

Check your work on page 91. ▶

These examples illustrate that it does *not* matter on which side of the equation you decide to put the variable. However, once you make your decision, you must do what is necessary to get the constant terms to the other side.

First-degree equations can also contain *parentheses* or other grouping symbols. First we should simplify such an equation by removing the grouping symbols. Then we will be able to solve by the usual methods.

Example 9. Solve and check

$$3(y - 7) + 7y = 5 - 2(5y + 3).$$

Solution

$$3(y - 7) + 7y = 5 - 2(5y + 3)$$

$$3y - 21 + 7y = 5 - 10y - 6$$

$$10y - 21 = -10y - 1$$

$$10y + 10y - 21 = 10y - 10y - 1$$

$$20y - 21 = -1$$

$$20y - 21 + 21 = -1 + 21$$

$$20y = 20$$

$$\frac{20y}{20} = \frac{20}{20}$$

$$y = 1$$

CHECK

$$3(y - 7) + 7y = 5 - 2(5y + 3)$$

$$3(1 - 7) + 7(1) \overset{?}{=} 5 - 2[5(1) + 3]$$

$$3(-6) + 7 \overset{?}{=} 5 - 2[5 + 3]$$

$$-18 + 7 \overset{?}{=} 5 - 2 \cdot 8$$

$$-11 \overset{?}{=} 5 - 16$$

$$-11 = -11$$

Now you complete Example 10.

Example 10. Solve and check

$$-3(x - 1) = 2(-3 - x).$$

Solution

$$-3(x - 1) = 2(-3 - x)$$

$$-3x + 3 = -6 - 2x$$

CHECK

$$-3(x - 1) = 2(-3 - x)$$

$$-3(\underline{\hspace{1cm}} - 1) \overset{?}{=} 2(-3 - \underline{\hspace{1cm}})$$

Check your work on page 91. ▶

Example 11. Solve $5x - 2(x + 1) = 3(x + 2)$.

Solution

$$5x - 2(x + 1) = 3(x + 2)$$

$$5x - 2x - 2 = 3x + 6 \qquad \text{Remove parentheses.}$$

$$3x - 2 = 3x + 6 \qquad \text{Combine like terms.}$$

$$3x - 3x - 2 = 3x - 3x + 6 \qquad \text{Subtract } 3x.$$

$$-2 = 6 \qquad \text{A false statement.}$$

There is no solution for this equation.

Example 12. Solve $5x - 2(x - 3) = 3(x + 2)$.

Solution

$$5x - 2(x - 3) = 3(x + 2)$$

$$5x - 2x + 6 = 3x + 6$$

$$3x + 6 = 3x + 6$$

$$6 = 6$$

Any real number will satisfy this equation.

These examples illustrate two important facts about equations.

> If, in solving an equation, we arrive at a statement that is *never* true, the equation has *no solution*.
>
> If, in solving an equation, we arrive at a statement that is *always* true, the equation is called an **identity**. It is satisfied by *any real number*.

Whenever fractions appear in an equation, we can eliminate them by multiplying both sides of the equation by the **lowest common denominator** (LCD) for all the fractions in the equation. Then we use our usual steps to solve for the variable.

Example 13. Solve $\dfrac{x + 3}{7} = 11$.

Solution: The LCD is 7.

$$\frac{7}{1}\left(\frac{x + 3}{7}\right) = 7 \cdot 11 \qquad \text{Multiply by 7.}$$

$$x + 3 = 77 \qquad \text{Simplify.}$$

$$x = 74 \qquad \text{Subtract 3.}$$

Now you complete Example 14.

Example 14. Solve $\dfrac{x}{7} + 3 = 11$.

Solution: The LCD is _____ .

$$\frac{7}{1}\left(\frac{x}{7}\right) + 7 \cdot 3 = 7 \cdot 11$$

Check your work on page 91. ▶

Example 15. Solve $\dfrac{5x - 1}{4} = \dfrac{7}{2}$.

Solution: The LCD is 4.

$$\frac{4}{1}\left(\frac{5x - 1}{4}\right) = \frac{4}{1}\left(\frac{7}{2}\right) \qquad \text{Multiply by 4.}$$

$$5x - 1 = 14 \qquad \text{Simplify.}$$

$$5x = 15 \qquad \text{Add 1.}$$

$$x = 3 \qquad \text{Divide by 5.}$$

Example 16. Solve $\dfrac{2x}{3} + \dfrac{4}{3} = 4x$.

Solution: The LCD is 3.

$$\frac{3}{1}\left(\frac{2x}{3}\right) + \frac{3}{1}\left(\frac{4}{3}\right) = 3(4x)$$

$$2x + 4 = 12x$$

$$4 = 10x$$

$$\frac{4}{10} = x$$

$$\frac{2}{5} = x$$

Let's summarize the steps we have developed for solving first-degree equations.

> **1.** Remove fractions by multiplying both sides by the LCD.
> **2.** Remove parentheses and combine like terms on each side of the equation.
> **3.** Get all variable terms on one side of the equation and get all constant terms on the other side by addition and/or subtraction.
> **4.** Solve for the variable by multiplication or division.
> **5.** Check your solution in the original equation.

⟩ Trial Run

Solve and check.

_____ **1.** $4m - 3 = m + 7$ _____ **2.** $8x + 5 = -23 - 6x$

_____ **3.** $3y - 9 = 18y - 6$ _____ **4.** $-3x + 7 = 2x - 18$

_____ **5.** $3(2a - 7) - 1 = 5(3a + 1)$ _____ **6.** $\dfrac{2y - 5}{3} + 5 = 8$

_____ **7.** $2(b + 1) - 3b = 3(3 + 2b)$

_____ **8.** $2x - (5 - 3x) = 2[x - (1 - x)]$

Answers are on page 92.

Solving Equations for a Specified Variable

Sometimes algebraic equations (or formulas) contain more than one variable. If we wish to isolate a particular variable in such an equation, we may do so by performing the usual operations on both sides of the equation.

Example 17. Solve $I = Prt$ for P.

Solution: This is the formula for simple interest. We are asked to isolate P.

$$I = Prt$$

$$\frac{I}{rt} = \frac{Prt}{rt} \qquad \text{Divide by } rt.$$

$$\frac{I}{rt} = P \qquad \text{Simplify.}$$

Example 18. Solve $P = 2l + 2w$ for l.

Solution: This is the formula for the perimeter of a _____ . We are asked to isolate _____ .

$$P = 2l + 2w$$

$$P - 2w = 2l + 2w - 2w$$

$$P - 2w = 2l$$

$$\frac{P - 2w}{2} = l$$

This procedure is called **solving an equation for a specified variable**. If the equation contains symbols of grouping, we continue to follow our usual order of operations for isolating the variable.

Example 19. Solve $C = \frac{5}{9}(F - 32)$ for F.

Solution: This is the formula for changing Fahrenheit temperatures to Celsius temperatures. We must isolate F.

$$C = \frac{5}{9}(F - 32)$$

$$\frac{9}{5}C = \frac{9}{5} \cdot \frac{5}{9}(F - 32)$$

$$\frac{9}{5}C = F - 32$$

$$\frac{9}{5}C + 32 = F$$

You try Example 20.

Example 20. Solve $A = \frac{h(a + b)}{2}$ for a.

Solution: This is the formula for the area of a trapezoid. We must isolate _____ .

$$A = \frac{h(a + b)}{2}$$

$$2 \cdot A = \frac{2}{1}\left[\frac{h(a + b)}{2}\right]$$

$$2A = h(a + b)$$

$$2A = ha + hb$$

Check your work on page 91. ▶

⮊ **Trial Run**

Solve for the specified variable.

_____ **1.** $C = 2\pi r$ for r

_____ **2.** $A = \frac{bh}{2}$ for h

_____ **3.** $A = P(1 + r)$ for r

_____ **4.** $y = mx + b$ for x

Answers are on page 92.

▶ **Examples You Completed**

Example 2. Solve $x - 12 = 0$ and check.

Solution: We must isolate x.

$$x - 12 = 0$$

$$x - 12 + 12 = 0 + 12$$

$$x = 12$$

CHECK

$$x - 12 = 0$$

$$12 - 12 \stackrel{?}{=} 0$$

$$0 = 0$$

Example 3. Solve $6 = -12x$.

Solution

$$6 = -12x$$

$$\frac{6}{-12} = \frac{-12x}{-12}$$

$$\frac{-1}{2} = x$$

Example 6. Solve $\dfrac{x}{2} - 8 = 0$.

Solution

$$\frac{x}{2} - 8 = 0$$

$$\frac{x}{2} - 8 + 8 = 0 + 8$$

$$\frac{x}{2} = 8$$

$$\frac{2}{1}\left(\frac{x}{2}\right) = 2(8)$$

$$x = 16$$

Example 8. Solve $10 - 3x = 16 - x$ and check.

Solution

$$10 - 3x = 16 - x$$

$$10 - 3x + x = 16 - x + x$$

$$10 - 2x = 16$$

$$10 - 2x - 10 = 16 - 10$$

$$-2x = 6$$

$$\frac{-2x}{-2} = \frac{6}{-2}$$

$$x = -3$$

CHECK

$$10 - 3x = 16 - x$$

$$10 - 3(-3) \overset{?}{=} 16 - (-3)$$

$$10 + 9 \overset{?}{=} 16 + 3$$

$$19 = 19$$

Example 10. Solve and check

$$-3(x - 1) = 2(-3 - x).$$

Solution

$$-3(x - 1) = 2(-3 - x)$$

$$-3x + 3 = -6 - 2x$$

$$-3x + 3 + 2x = -6 - 2x + 2x$$

$$-x + 3 = -6$$

$$-x + 3 - 3 = -6 - 3$$

$$-x = -9$$

$$\frac{-1x}{-1} = \frac{-9}{-1}$$

$$x = 9$$

CHECK

$$-3(x - 1) = 2(-3 - x)$$

$$-3(9 - 1) \overset{?}{=} 2(-3 - 9)$$

$$-3(8) \overset{?}{=} 2(-12)$$

$$-24 = -24$$

Example 14. Solve $\dfrac{x}{7} + 3 = 11$

Solution: The LCD is 7.

$$\frac{7}{1}\left(\frac{x}{7}\right) + 7 \cdot 3 = 7 \cdot 11$$

$$x + 21 = 77$$

$$x = 56$$

Example 20. Solve $A = \dfrac{h(a + b)}{2}$ for a.

Solution: This is the formula for the area of a trapezoid. We must isolate a.

$$A = \frac{h(a + b)}{2}$$

$$2 \cdot A = \frac{2}{1}\left[\frac{h(a + b)}{2}\right]$$

$$2A = h(a + b)$$

$$2A = ha + hb$$

$$2A - hb = ha$$

$$\frac{2A - hb}{h} = a$$

Answers to Trial Runs

page 84   **1.** $x = 7$ **2.** $a = 5$ **3.** $x = 0$ **4.** $y = -1$ **5.** $x = -10$ **6.** $x = 7$
7. $x = \dfrac{-1}{3}$ **8.** $x = 35$ **9.** $x = \dfrac{-20}{3}$ **10.** $x = 0$

page 85   **1.** $x = 2$ **2.** $x = \dfrac{3}{2}$ **3.** $x = -18$ **4.** $a = 12$ **5.** $m = \dfrac{-1}{3}$ **6.** $y = -6$

page 89   **1.** $m = \dfrac{10}{3}$ **2.** $x = -2$ **3.** $y = \dfrac{-1}{5}$ **4.** $x = 5$ **5.** $a = -3$ **6.** $y = 7$
7. $b = -1$ **8.** $x = 3$

page 90   **1.** $r = \dfrac{C}{2\pi}$ **2.** $h = \dfrac{2A}{b}$ **3.** $r = \dfrac{A - P}{P}$ **4.** $x = \dfrac{y - b}{m}$

EXERCISE SET 2.1

Solve and check.

_____ 1. $x - 11 = -6$ _____ 2. $x - 9 = -8$

_____ 3. $4 = y + 6$ _____ 4. $5 = 9 + y$

_____ 5. $-3 + m = -9$ _____ 6. $-5 + m = -15$

_____ 7. $-15y = -10$ _____ 8. $-18y = -9$

_____ 9. $7a = 0$ _____ 10. $9a = 0$

_____ 11. $-5 = 9x$ _____ 12. $-3 = 7x$

_____ 13. $\dfrac{x}{2} = 6$ _____ 14. $\dfrac{x}{4} = 11$

_____ 15. $\dfrac{-x}{2} = 13$ _____ 16. $\dfrac{-x}{3} = 5$

_____ 17. $3x + 5 = 26$ _____ 18. $4x + 6 = 26$

_____ 19. $7 - 4x = 9$ _____ 20. $6 - 15x = 9$

_____ 21. $21 = 12y - 15$ _____ 22. $17 = 6y + 7$

_____ 23. $\dfrac{x}{6} + 3 = 5$ _____ 24. $\dfrac{x}{7} + 4 = 9$

_____ 25. $6 - \dfrac{x}{4} = 7$ _____ 26. $9 - \dfrac{x}{2} = 12$

_____ 27. $-9 = \dfrac{a}{5} - 9$ _____ 28. $-7 = \dfrac{a}{8} - 7$

_____ 29. $3x + 2 = x + 10$ _____ 30. $5x + 3 = 3x + 15$

_____ 31. $-4y + 2 = y - 13$ _____ 32. $-5y + 3 = 2y - 11$

_____ 33. $3x - 8 = 5x$ _____ 34. $7x - 15 = 12x$

_____ 35. $-13 + 4m = -2m - 3$ _____ 36. $-18 + 3m = -3m - 8$

_____ 37. $7 - 2a = -3 + 8a$ _____ 38. $9 - 3a = -6 + 2a$

_____ 39. $10 = 2(x - 5) - 5(x + 2)$ _____ 40. $-16 = 3(x - 6) - (2x + 5)$

_____ 41. $-2[3(x - 2) - 5] = 22$ _____ 42. $4[2(x - 2) - (x + 4)] = 0$

_____ 43. $4(2x - 3) + 2 = 2(4x - 5)$ _____ 44. $3(6x - 5) + 6 = 9(2x - 1)$

———— **45.** $\dfrac{11y - 3}{4} + 6 = 3y$ ———— **46.** $\dfrac{7y - 5}{2} + 7 = 4y$

———— **47.** $2(x + 1) + 3(x - 2) = 5x - 37$ ———— **48.** $4(x + 1) + (x - 5) = 5x - 19$

———— **49.** $15 = \dfrac{2x - 1}{3} - \dfrac{1}{2}$ ———— **50.** $19 = \dfrac{4x - 6}{2} - \dfrac{2}{3}$

———— **51.** $2[x + 3(x - 2)] = 5(x - 4) - 1$ ———— **52.** $3[2x + (x - 5)] = 2(x - 7) + 6$

———— **53.** $3[2x - 3(2x - 4)] = 5[2(x - 4) + 2]$

———— **54.** $2[3x - (x + 4)] = 3[2(5 - x) - 6]$

Solve for the specified variable.

———— **55.** $A = \dfrac{1}{2} bh$ for h.   ———— **56.** $V = \dfrac{1}{3} bh$ for b.

———— **57.** $d = rt$ for t.   ———— **58.** $V = lwh$ for h.

———— **59.** $A = P + Prt$ for t.   ———— **60.** $A = \pi R^2 + 2\pi rh$ for h.

———— **61.** $3x - 2y = 1$ for x.   ———— **62.** $2x - 5y = 3$ for x.

———— **63.** $ax + by = 5$ for x.   ———— **64.** $cx + dy = 9$ for y.

———— **65.** $3x - 8a = 5(x - 4a)$ for x.   ———— **66.** $2y - 2b = -5(b - y)$ for y.

———— **67.** $l = a + (n - 1)d$ for n.   ———— **68.** $A = p(1 + rt)$ for r.

———— **69.** $d = \dfrac{V}{2r}$ for r.   ———— **70.** $W = \dfrac{wl}{L}$ for L.

☆ Stretching the Topics ————————————————

———— **1.** Solve $3ax + 7b = ax + 10b$ for x.

———— **2.** Solve $n = \dfrac{l - a}{d} + 1$ for a.

———— **3.** Solve $\dfrac{2}{3}(x - 4) = 1 + \dfrac{x + 5}{3}$ for x.

Check your answers in the back of your book.

If you can complete **Checkup 2.1**, you are ready to go on to Section 2.2.

✔ CHECKUP 2.1

Solve and check.

_____ **1.** $x - 35 = -8$ _____ **2.** $11x = -36$

_____ **3.** $\dfrac{-a}{4} = 6$ _____ **4.** $17 - 6x = -1$

_____ **5.** $-5x + 1 = 7x - 3$ _____ **6.** $2(5y - 1) - 5 = 3(7 + y)$

_____ **7.** $8x - 3(2x + 4) = 2(x - 5)$ _____ **8.** $\dfrac{9x - 8}{4} - 2 = 5x$

_____ **9.** $\dfrac{4x - 3}{6} = \dfrac{9}{2}$ _____ **10.** $A = P(1 + rt)$ for r

Check your answers in the back of your book.

If You Missed Problems:	You Should Review Examples:
1	1, 2
2	3
3, 4	4, 5
5	7, 8
6	9, 10
7	11
8, 9	14, 15
10	19, 20

2.2 Graphing First-Degree Inequalities

A thorough understanding of first-degree equations will help us learn to solve first-degree inequalities. Before we attempt such solutions, however, let's take a look at some inequalities in which the variable has already been isolated.

In each case there will not be just one answer; in fact, there will be an infinite number of solutions for each inequality. Listing all the solutions is an impossible task. The *number line* allows us to picture our answers without listing each one. Using a number line to show solutions is called **graphing** the solutions.

Example 1. Graph the solutions for $x > 2$.

Solution: To satisfy $x > 2$, x must be a real number greater than 2. All numbers to the right of 2 on the number line are solutions. The number 2 itself is *not* a solution, but it acts as a boundary.

The *open dot* at 2 indicates that the number corresponding to that point is *not* a solution but is a boundary. The arrow at the end of the graph indicates that it extends forever to the right.

Example 2. Graph the solutions for $x \leq 0$.

Solution: To satisfy $x \leq 0$, x must be a real number less than 0 *or* equal to 0. All numbers to the left of 0 and 0 itself are solutions.

The *closed dot* at 0 indicates that the number corresponding to that point *is* a solution as well as a boundary. The arrow indicates that the graph extends forever to the left.

Example 3. Graph the solutions for $x \geq -3$.

Solution: There will be a closed dot at -3 and a line extending to the right.

You try Example 4.

Example 4. Graph the solutions for $x < \dfrac{1}{2}$.

Solution: There will be a(n) _____ dot at _____ and a line extending to the _____ .

Check your work on page 103. ▶

We make note of the pattern we have observed. Suppose that *a* represents any real number.

Inequality	Words	Graph
$x < a$	x is less than a	
$x > a$	x is greater than a	
$x \leq a$	x is less than or equal to a	
$x \geq a$	x is greater than or equal to a	

⫸ Trial Run ━━━

Graph the solutions on a number line.

1. $x > 3$

2. $a \leq -2$

3. $y \geq 0$

4. $m < \dfrac{5}{2}$

5. $a \geq \dfrac{-1}{3}$

Answers are on page 103.

Finding Intersections of Graphs

Sometimes it is necessary to satisfy more than one inequality at the same time. In each of these examples, there are two inequalities that must be satisfied.

Example 5. A citizen of this state must be at least 16 years old to obtain a driver's license. To vote, a citizen must be at least 18 years old. At what ages will a citizen be able to vote *and* obtain a driver's license?

　Solution: Both statements make reference to a citizen's age, so we can let *a* represent a citizen's age.

The first statement says:	The second statement says:
$a \geq 16$	$a \geq 18$

The question is: When will *both* statements be true? For what ages will *both* inequalities be satisfied? Let's graph both inequalities on the same number line.

To decide where both inequalities are satisfied, we must find where *both* lines appear in our graph. It should be clear that both lines appear at 18 and to the right of 18. In other words, at all ages greater than or equal to 18, a citizen will be able to drive *and* vote.

　We can write our solution as the single inequality

$$a \geq 18$$

Example 6. To be used in tournament play, a tennis ball must bounce more than 53 inches but fewer than 58 inches (when dropped 100 inches onto a concrete surface). Write a single inequality describing the acceptable bounce heights.

Solution: We can let *b* represent the bounce of the tennis ball.

The bounce must be more than 53 inches: The bounce must be less than 58 inches:

$$b > 53$$ $$b < 58$$

Again we can graph both inequalities on the same number line to see where *both* conditions are met.

We see both lines between 53 and 58 but *not* at the points 53 and 58 themselves. When a variable is located *between* two numbers, we write that statement as an inequality in the following way:

  $53 < b < 58$

This inequality says:

b is greater than 53 *and* less than 58

or

53 is less than *b*, and *b* is less than 58

> The inequality $a < b < c$ means that *b* is *between* *a* and *c*.

In Examples 5 and 6 we were asked to satisfy one inequality *and* another inequality at the same time. In each case we looked for all points *common* to the set of real numbers satisfying the first inequality *and* the set of real numbers satisfying the second inequality. This set of common points is called the **intersection** of the two sets being discussed.

> **Intersection of Sets *A* and *B*.** The intersection of sets *A* and *B* is the set of elements that set *A* and set *B* have in common.

Remember that in seeking an intersection, we are demanding that *both* conditions be met at the same time. The best way to find the intersection of two inequalities is to graph both on the same number line and look for numbers that are included in both graphs.

Example 7. Solve $x < -3$ and $x \leq 2$.

Solution: Graph both inequalities on the same number line.

Both lines appear to the left of (and not including) the point -3. Therefore, we describe the solutions by

$$x < -3$$

Example 8. Solve $x \geq -1$ and $x < 4$.

Solution: Graph both inequalities on the same number line.

Both lines appear between -1 and 4 (including -1 but not including 4). Therefore, we describe the solutions by

$$-1 \leq x < 4$$

Example 9. Solve $x > 2$ and $x \leq 0$.

Solution: Graph both inequalities on the same number line.

We see that *both* lines appear *nowhere* at the same time on the graph. There are *no* numbers that will satisfy both inequalities at the same time. Therefore, our intersection contains no numbers at all; it is an **empty set**. Mathematicians sometimes use the symbol $\emptyset$ to stand for the empty set. Here we say that the intersection of $x > 2$ and $x \leq 0$ is $\emptyset$.

� Trial Run

Find the solution for each pair of inequalities by graphing on a number line.

_____ **1.** $x < -4$ and $x \leq 1$

_____ **2.** $x > 0$ and $x \geq 2$

_____ **3.** $x < -3$ and $x > 1$

_____ **4.** $x \leq 3$ and $x > 0$

_____ **5.** $x < 5$ and $x \geq \dfrac{-1}{2}$

_____ **6.** $x \leq 1$ and $x \geq 1$

Answers are on page 103.

Finding Unions of Graphs

Let's consider some more examples in which two inequalities are discussed.

Example 10. A local movie theater offers tickets at half-price to children (age 12 or below) and to senior citizens (age 62 or above). Describe the set of people eligible for half-price tickets using inequality notation.

Solution: We can let a represent the age of a moviegoer.

Children: Senior citizens:

$a \leq 12$ $a \geq 62$

To qualify for a half-price ticket, must a moviegoer meet *both* age conditions? Surely not. No one could qualify as a child and as a senior citizen at the same time. We are *not* seeking an *intersection* here.

Instead, we agree that a moviegoer will qualify for a half-price ticket by satisfying the first inequality ($a \leq 12$) *or* the second inequality ($a \geq 62$). Let's graph our inequalities and find the numbers that satisfy one *or* the other of them.

In a situation such as this we are *not* concerned with regions where *both* lines appear at the same time. *This is not an intersection problem.* Instead, we are concerned with regions where *either* of the lines appears. The solution here is

$$a \leq 12 \quad \text{or} \quad a \geq 62$$

Example 11. A student is eligible for a federal educational grant if the family's income is below $13,000. A student is eligible for university financial aid if the family's income is below $10,000. For what incomes is a student eligible for federal aid *or* university aid?

Solution: We can let I represent family income.

Federal grant:	University aid:
$I < 13,000$	$I < 10,000$

We graph our two inequalities.

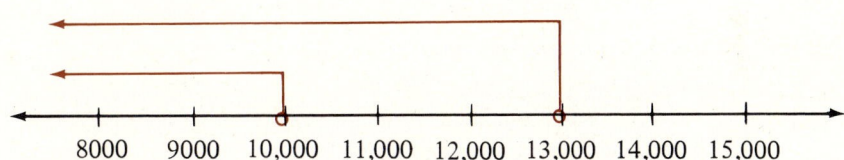

Since we are looking for incomes that will make a student eligible for one kind of aid *or* the other, we look for the numbers appearing in either graph. Our solution is

$$I < \$13,000$$

In Examples 10 and 11 we found a set called the **union** of our original sets.

> **Union of Sets A and B.** The union of sets A and B is the set containing all elements in set A or in set B (or possibly in both).

To find the union of two inequalities we graph them on the same number line and look for numbers appearing in either graph (or possibly in both).

You try to find the union described in Example 12.

Example 12. Solve $x \geq 7$ or $x > 3$.

Solution: We graph both inequalities on the number line.

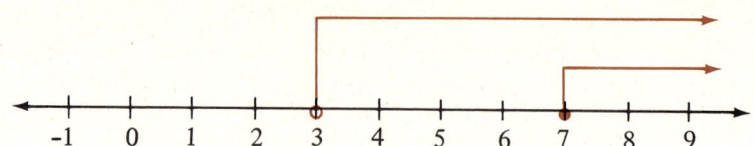

A line appears everywhere to the right of (but not including) the point _____ . Therefore, our solutions are described by the inequality

$$x > \underline{3}$$

Check your work on page 103. ▶

Example 13. Solve $x > 1$ or $x < 1$.

Solution: We graph both inequalities on the number line.

Lines appear everywhere except at the point 1. Therefore, we describe the solutions by the original inequality

$$x > 1 \quad \text{or} \quad x < 1$$

which may also be written as

$$x \in R \quad \text{and} \quad x \neq 1$$

Example 14. Solve $x > -2$ or $x \leq 3$.

Solution: We graph both inequalities on the number line.

All real numbers are included in one line or the other. Therefore, solutions that satisfy

$$x > -2 \quad \text{or} \quad x \leq 3$$

may be described by

$$x \in R$$

Let's summarize our work with pairs of inequalities.

To satisfy one inequality **and** another, a number must satisfy *both* inequalities. On the number line we look for numbers appearing in *both* graphs.

To satisfy one inequality **or** another, a number may satisfy one of the inequalities *or* the other (or possibly both). On the number line we look for numbers appearing in either one graph *or* the other.

▶ Trial Run

Find the solution for each pair of inequalities by graphing on the number line.

_____ **1.** $x \geq 3$ or $x > 5$ _____ **2.** $x < 0$ or $x \leq -2$

_____ **3.** $x > 2$ or $x < 2$ _____ **4.** $x > 3$ or $x < -1$

_____ **5.** $x > -4$ or $x < 2$ _____ **6.** $x \geq 1$ or $x < 1$

Answers are on page 103.

▶ Examples You Completed

Example 4. Graph the solutions for $x < \frac{1}{2}$.

Solution: There will be a(n) open dot at $\frac{1}{2}$ and a line extending to the left.

Example 12. Solve $x \geq 7$ or $x > 3$.

Solution: We graph both inequalities on the number line.

A line appears everywhere to the right of (but not including) the point 3. Therefore, our solutions are described by the inequality

$$x > 3$$

Answers to Trial Runs

page 98   **1.**

2.

3.

4.

5.

page 100   **1.** $x < -4$ **2.** $x \geq 2$ **3.** $\emptyset$ **4.** $0 < x \leq 3$ **5.** $\frac{-1}{2} \leq x < 5$ **6.** $x = 1$

page 102   **1.** $x \geq 3$ **2.** $x < 0$ **3.** $x \in R, x \neq 2$ **4.** $x < -1$ or $x > 3$ **5.** $x \in R$ **6.** $x \in R$

EXERCISE SET 2.2

Graph each of the following inequalities on a number line.

1. $x > -5$ **2.** $x > 1$

3. $a \leq 3$ **4.** $a \leq -2$

5. $y \geq 0$ **6.** $y \leq 0$

7. $m > \dfrac{7}{2}$ **8.** $m > \dfrac{5}{3}$

9. $x \leq -\dfrac{3}{2}$ **10.** $x \leq -\dfrac{5}{2}$

Find the solution of each of the following by graphing on a number line.

_____ **11.** $x < -3$ and $x \leq 4$ _____ **12.** $x < -5$ and $x < 1$

_____ **13.** $x > -1$ and $x \geq 3$ _____ **14.** $x > -2$ and $x \geq 0$

_____ **15.** $x < -1$ and $x > 2$ _____ **16.** $x < 0$ and $x > 2$

_____ **17.** $x \leq 5$ and $x > 2$ _____ **18.** $x \leq 1$ and $x > -2$

_____ **19.** $x \geq -1$ and $x \leq 4$ _____ **20.** $x \geq -3$ and $x \leq 1$

_____ **21.** $x < 0$ and $x \geq 0$ _____ **22.** $x < 2$ and $x \geq 2$

_____ **23.** $x \geq -3$ and $x \leq -3$ _____ **24.** $x \geq 4$ and $x \geq 4$

_____ **25.** $x \geq 2$ or $x > 5$ _____ **26.** $x \geq -1$ or $x > 1$

_____ **27.** $x < 4$ or $x \leq -1$ _____ **28.** $x \leq 3$ or $x < 7$

_____ **29.** $x > 3$ or $x < 3$ _____ **30.** $x > -4$ or $x < -4$

_____ **31.** $x < 2$ or $x \geq -1$ _____ **32.** $x < 5$ or $x \geq 1$

_____ **33.** $x < -\dfrac{1}{2}$ or $x \geq 0$ _____ **34.** $x \leq 1$ or $x > \dfrac{3}{2}$

_____ **35.** $x > -2$ or $x < 3$ _____ **36.** $x > -2$ or $x < 3$

_____ **37.** $x < -3$ and $x \leq 0$ _____ **38.** $x < 4$ and $x \leq \dfrac{1}{2}$

_____ **39.** $x > -4$ and $x < \dfrac{7}{2}$ _____ **40.** $x > -\dfrac{5}{2}$ and $x < 1$

_____ **41.** $x < -1$ or $x > 0$ _____ **42.** $x < -3$ or $x > 2$

_____ **43.** $x < 5$ or $x \le 1$ _____ **44.** $x \le 10$ or $x < 5$

_____ **45.** $x > 0$ and $x \ge \dfrac{3}{5}$ _____ **46.** $x > -1$ and $x \ge 0$

_____ **47.** $x < 5$ or $x > -1$ _____ **48.** $x < 12$ or $x \ge 2$

_____ **49.** $x \ge \dfrac{1}{2}$ and $x < \dfrac{1}{2}$ _____ **50.** $x > 9$ and $x \le 9$

☆ Stretching the Topics

Find the solution by graphing on a number line.

_____ **1.** $x \ge 3.5$ and $x \le \dfrac{7}{2}$

_____ **2.** $x > -0.75$ or $x \ge \dfrac{-3}{4}$

_____ **3.** $x \le \dfrac{22}{7}$ and $x \ge \pi$

Check your answers in the back of your book.

If you can complete **Checkup 2.2**, you are ready to go on to Section 2.3.

✓ CHECKUP 2.2

Graph each of the following inequalities.

1. $x \geq 3$

2. $a \leq -0.5$

3. $y > \dfrac{7}{2}$

Find the solution for each of the following by graphing.

————— 4. $x > 0$ and $x \geq 2$

————— 5. $x < 3$ or $x \leq -1$

————— 6. $x < -2$ and $x > \dfrac{1}{2}$

————— 7. $x > -2$ or $x < 4$

————— 8. $x < 5$ and $x \geq 0$

————— 9. $x < -3$ or $x > 2$

————— 10. $x < \dfrac{1}{2}$ or $x > \dfrac{1}{2}$

Check your answers in the back of your book.

If You Missed Problems:	You Should Review Examples:
1–3	1–4
4	5, 7
5	11, 12
6	9
7	14
8	8
9	10
10	13

2.3 Solving First-Degree Inequalities

Having learned to solve first-degree equations in Section 2.1, you will be relieved to know that solving first-degree inequalities requires similar techniques.

Solving Inequalities by Addition or Multiplication

In solving equations, we learned that it was permissible to add (or subtract) the same quantity to (or from) both sides of an equation. Is this also permissible with inequalities?

Let see what happens when we add and subtract some numbers to both sides of the inequality $3 < 7$.

Add 4	Subtract 10	Add -6	Subtract -5
$3 + 4 \overset{?}{<} 7 + 4$	$3 - 10 \overset{?}{<} 7 - 10$	$3 + (-6) \overset{?}{<} 7 + (-6)$	$3 - (-5) \overset{?}{<} 7 - (-5)$
$7 \overset{?}{<} 11$	$-7 \overset{?}{<} -3$	$-3 \overset{?}{<} 1$	$8 \overset{?}{<} 12$
True	True	True	True
$7 < 11$	$-7 < -3$	$-3 < 1$	$8 < 12$

> We may add (or subtract) the same quantity to (or from) both sides of an inequality. The inequality is preserved in the same direction.

The statement "The inequality is preserved in the same direction" means that if we start with a less-than inequality ($<$ or $\leq$), we end up with a less-than inequality. If we start with a greater-than inequality ($>$ or $\geq$), we end up with a greater-than inequality.

We can state this fact using a, b, and c to represent real numbers.

Addition Property of Inequality

If $a < b$ If $a > b$

then $a + c < b + c$ then $a + c > b + c$

and $a - c < b - c$ and $a - c > b - c$

This property allows us to use addition or subtraction to isolate the variable in solving some inequalities.

Example 1. Solve $x - 9 \leq 1$ and graph the solutions.

Solution

$$x - 9 \leq 1$$
$$x - 9 + 9 \leq 1 + 9 \qquad \text{Add 9 to both sides.}$$
$$x \leq 10 \qquad \text{Simplify.}$$

We graph our solutions on a number line.

You complete Example 2.

Example 2. Solve $y + 3 > -2$ and graph the solutions.

Solution

$$y + 3 > -2$$
$$y + 3 - \underline{\quad} > -2 - \underline{\quad} \qquad \text{Subtract } \underline{\quad} \text{ from both sides.}$$
$$y > \underline{\quad} \qquad \text{Simplify.}$$

Check your work on page 117. ▶

In solving first-degree equations we were also allowed to multiply or divide both sides by the same nonzero quantity. Let's see if a similar rule can be developed for inequalities by noticing what happens when we multiply and divide both sides of the inequality $8 < 12$.

Multiply by 3	Divide by 2	Multiply by -5	Divide by -4
$3(8) \overset{?}{<} 3(12)$	$\dfrac{8}{2} \overset{?}{<} \dfrac{12}{2}$	$-5(8) \overset{?}{<} -5(12)$	$\dfrac{8}{-4} \overset{?}{<} \dfrac{12}{-4}$
$24 \overset{?}{<} 36$	$4 \overset{?}{<} 6$	$-40 \overset{?}{<} -60$	$-2 \overset{?}{<} -3$
True	True	False	False
$24 < 36$	$4 < 6$	$-40 > -60$	$-2 > -3$

We may multiply (or divide) both sides of an inequality by the same *positive* quantity, and the direction of the inequality will be *preserved*.

We may multiply (or divide) both sides of an inequality by the same *negative* quantity, but the direction of the inequality will be *reversed*.

Using a, b, and c to represent real numbers, we can state:

Multiplication Property of Inequality

If $a < b$ and c is positive ($c > 0$), then $a \cdot c < b \cdot c$ and $\dfrac{a}{c} < \dfrac{b}{c}$.

If $a > b$ and c is positive ($c > 0$), then $a \cdot c > b \cdot c$ and $\dfrac{a}{c} > \dfrac{b}{c}$.

If $a < b$ and c is negative ($c < 0$), then $a \cdot c > b \cdot c$ and $\dfrac{a}{c} > \dfrac{b}{c}$.

If $a > b$ and c is negative ($c < 0$), then $a \cdot c < b \cdot c$ and $\dfrac{a}{c} < \dfrac{b}{c}$.

These properties allow us to solve first-degree inequalities in which the variable appears with a coefficient different from 1. See if you can complete Example 3.

Example 3. Solve and graph $\dfrac{x}{3} \geq -2$.

Solution

$$\frac{x}{3} \geq -2$$

$$3\left(\frac{x}{3}\right) \geq 3(-2)$$

$$x \geq \underline{\hspace{1cm}}$$

Check your work on page 118. ▶

Example 4. Solve and graph $-2x > -3$.

Solution

$$-2x > -3$$

$$\frac{-2x}{-2} < \frac{-3}{-2}$$

$$x < \frac{3}{2}$$

Notice that the direction of the inequality was reversed when we divided by -2.

⫸ Trial Run

Solve and graph each inequality.

_____ **1.** $-3 + y < 0$

_____ **2.** $x + 7 \geq 2$

_____ **3.** $-2x \leq 6$

_____ **4.** $3x > 15$

_____ **5.** $\dfrac{x}{-5} < 2$

Answers are on page 118.

Now it is time to solve some first-degree inequalities requiring addition and/or subtraction, as well as multiplication and/or division. Such problems should be solved with the same procedure as that used for equations. But remember to be careful whenever you multiply or divide both sides of an inequality by a *negative* number.

Example 5. Solve $-4x + 10 \leq 34$.

Solution

$$-4x + 10 \leq 34$$

$$-4x \leq 24 \qquad \text{Subtract 10.}$$

$$\frac{-4x}{-4} \geq \frac{24}{-4} \qquad \text{Divide by } -4.$$

$$x \geq -6 \qquad \text{Simplify.}$$

Now you try Example 6.

Example 6. Solve $\dfrac{x}{2} - 1 > 1$.

Solution

$$\frac{x}{2} - 1 > 1$$

$$\frac{x}{2} > 2 \qquad \text{Add \_\_\_\_ .}$$

$$2\left(\frac{x}{2}\right) > 2(2) \qquad \text{Multiply by \_\_\_\_ .}$$

$$x > \text{\_\_\_\_} \qquad \text{Simplify.}$$

Check your work on page 118. ▶

If a variable appears on both sides of an inequality, we proceed as with equations.

Example 7. Solve and graph $5x + 1 \geq 3x + 7$.

Solution

$$5x + 1 \geq 3x + 7$$

$$2x + 1 \geq 7 \qquad \text{Subtract } 3x \text{ from both sides to isolate variable term.}$$

$$2x \geq 6 \qquad \text{Subtract 1 from both sides.}$$

$$\frac{2x}{2} \geq \frac{6}{2} \qquad \text{Divide both sides by 2.}$$

$$x \geq 3 \qquad \text{Simplify.}$$

Now you complete Example 8.

Example 8. Solve and graph

$$\frac{3 - 2x}{4} \le -1.$$

Solution

$\frac{3 - 2x}{4} \le -1$	LCD: 4.
$4\left(\dfrac{3 - 2x}{4}\right) \le 4(-1)$	Multiply by LCD.
$3 - 2x \le \underline{\quad}$	Simplify.
$-2x \le \underline{\quad}$	Subtract 3.
$\dfrac{-2x}{-2} \ge \dfrac{\underline{\quad}}{-2}$	Divide by -2.
$x \ge \underline{\quad}$	Simplify.

```
  ┝━━┿━━┿━━┿━━┿━━┿━━┿━━┿━━┿━━┥
 -3  -2  -1   0   1   2   3   4   5   6
```

Check your work on page 118. ▶

Example 9. Solve and graph

$$5(x + 1) - x \le 1 - 2x.$$

Solution

$5(x + 1) - x \le 1 - 2x$	
$5x + 5 - x \le 1 - 2x$	Remove parentheses.
$4x + 5 \le 1 - 2x$	Combine like terms.
$6x + 5 \le 1$	Add $2x$.
$6x \le -4$	Subtract 5.
$\dfrac{6x}{6} \le \dfrac{-4}{6}$	Divide by 6.
$x \le \dfrac{-2}{3}$	Simplify.

```
  ←━━━━━━┓
  ┝━━┿━━┿●┿━━┿━━┿━━┿━
 -2  -1  ↑  0   1   2   3
        -2/3
```

▥▶ **Trial Run**

Solve.

_____ **1.** $5a - 3 \ge 12$

_____ **2.** $\dfrac{y}{7} - 1 < -2$

_____ **3.** $\dfrac{2x}{3} + 4 \le 7$

_____ **4.** $2a - 4 < 5a - 1$

_____ **5.** $5x - (x - 4) \le 2x + 5$

_____ **6.** $2(z - 2) - 3 \le 5(z + 5) - 23$

Answers are on page 119.

Solving Pairs of Inequalities

In Section 2.2 we learned how to find solutions satisfying one inequality *and/or* another. Now we can extend those techniques to solving pairs of inequalities whose solutions are not immediately obvious.

Example 10. Solve $3x - 1 < 2$ and $2(5 - x) \leq 16$.

Solution: We must find the solutions for each inequality and graph them on the same number line.

$$3x - 1 < 2 \qquad \text{and} \qquad 2(5 - x) \leq 16$$
$$3x - 1 + 1 < 2 + 1 \qquad\qquad 10 - 2x \leq 16$$
$$3x < 3 \qquad\qquad 10 - 2x - 10 \leq 16 - 10$$
$$\frac{3x}{3} < \frac{3}{3} \qquad\qquad -2x \leq 6$$
$$\qquad\qquad\qquad \frac{-2x}{-2} \geq \frac{6}{-2}$$
$$x < 1 \qquad \text{and} \qquad x \geq -3$$

Since we are seeking the *intersection* of our sets of solutions, we must find where both lines appear at the same time. We write the corresponding inequality as

$$-3 \leq x < 1$$

Example 11. Solve $7 - x > 3$ and $4(x + 1) - x < 4$.

Solution

$$7 - x > 3 \qquad \text{and} \qquad 4(x + 1) - x < 4$$
$$-x > -4 \qquad\qquad 4x + 4 - x < 4$$
$$\frac{-x}{-1} < \frac{-4}{-1} \qquad\qquad 3x + 4 < 4$$
$$\qquad\qquad\qquad 3x < 0$$
$$\qquad\qquad\qquad \frac{3x}{3} < \frac{0}{3}$$
$$x < 4 \qquad \text{and} \qquad x < 0$$

We look for solutions that satisfy $x < 4$ and $x < 0$. The intersection is: $x < 0$.

Finding the union of two inequalities involves a similar procedure. After solving and graphing the original inequalities, however, we look for those solutions that appear in one graph *or* the other.

Example 12. Solve $3(2x - 4) + 1 > 7$ or $7x - 5 \geq 9$.

Solution

$$3(2x - 4) + 1 > 7 \qquad \text{or} \qquad 7x - 5 \geq 9$$
$$6x - 12 + 1 > 7 \qquad\qquad\qquad 7x \geq 14$$
$$6x - 11 > 7 \qquad\qquad\qquad \frac{7x}{7} \geq \frac{14}{7}$$
$$6x > 18$$
$$\frac{6x}{6} > \frac{18}{6}$$
$$x > 3 \qquad \text{or} \qquad\qquad x \geq 2$$

We know that $x \geq 2$ or $x > 3$ must be satisfied. Any number to the right of (and including) 2 will satisfy one inequality or the other. Our union is described by $x \geq 2$.

Example 13. Solve $4x - 7 > 9$ or $1 - 2(x - 3) \geq 3$.

Solution

$$4x - 7 > 9 \qquad \text{or} \qquad 1 - 2(x - 3) \geq 3$$
$$4x > 16 \qquad\qquad\qquad 1 - 2x + 6 \geq 3$$
$$\frac{4x}{4} > \frac{16}{4} \qquad\qquad\qquad -2x + 7 \geq 3$$
$$-2x \geq -4$$
$$\frac{-2x}{-2} \leq \frac{-4}{-2}$$
$$x > 4 \qquad \text{or} \qquad\qquad x \leq -2$$

Our union is all solutions satisfying one inequality *or* the other. Here the union must be written $x \leq -2$ or $x > 4$.

As the examples illustrate, sometimes an intersection or union of inequalities will give us an answer that is a single inequality, sometimes it will be a three-part inequality, and sometimes it will be two separate inequalities. There is no sure way to predict which kind of solution will occur. We must solve and graph the original inequalities and then look for the intersection or union.

Finding the values of the variable that satisfy a **three-part inequality** such as

$$-3 < 2x + 1 < 7$$

is not terribly difficult if we stop and think about what three-part inequalities really mean. Recall that

$$a < x < b \qquad \text{means} \qquad a < x \qquad \text{and} \qquad x < b$$

So

$$-3 < 2x + 1 < 7 \quad \text{means} \quad -3 < 2x + 1 \ \text{and} \ 2x + 1 < 7$$

We must solve both inequalities and find the *intersection*. Since both inequalities require the same steps for solving, let's try to deal with all three parts of the inequality at once. Remember, we would like to end up with x by itself in the middle.

$$-3 < \quad 2x + 1 \quad < 7$$

$$-3 - 1 < 2x + 1 - 1 < 7 - 1 \qquad \text{Subtract 1 from all three parts.}$$

$$-4 < \quad 2x \quad < 6 \qquad \text{Simplify.}$$

$$\frac{-4}{2} < \quad \frac{2x}{2} \quad < \frac{6}{2} \qquad \text{Divide all three parts by 2.}$$

$$-2 < \quad x \quad < 3 \qquad \text{Simplify.}$$

The graph of our solutions is

Now you try completing Example 14.

Example 14. Solve and graph

$$-14 \leq 5(x - 3) + 1 \leq 11.$$

Solution

$$-14 \leq 5(x - 3) + 1 \leq 11$$

$$-14 \leq 5x - 15 + 1 \leq 11$$

$$-14 \leq \ 5x - 14 \ \leq 11$$

Check your work on page 118. ▶

Example 15. Solve $4 < \dfrac{1 - 3x}{4} \leq 7$.

Solution

$$4 < \quad \frac{1 - 3x}{4} \quad \leq 7$$

$$4 \cdot 4 < \quad 1 - 3x \quad \leq 7 \cdot 4$$

$$16 < \quad 1 - 3x \quad \leq 28$$

$$16 - 1 < 1 - 3x - 1 \leq 28 - 1$$

$$15 < \quad -3x \quad \leq 27$$

$$\frac{15}{-3} > \quad \frac{-3x}{-3} \quad \geq \frac{27}{-3}$$

$$-5 > \quad x \quad \geq -9$$

Notice that we reversed directions when we divided by -3. This inequality is correct, but let's write it in better order, starting with the -9.

$$-9 \leq x < -5$$

From our work with inequalities you should notice that a solution set whose graph is *one continuous interval* on the number line can always be written using a *single inequality*.

But solutions whose graph is a *split interval* on the number line must be written using *two separate inequalities*.

This pair of inequalities should *never* be written $a \geq x > b$ because this statement implies that $x > b$ and $x \leq a$ at the same time, which is clearly impossible.

⫸ Trial Run

Solve by graphing.

_____ **1.** $5x - 4 > 6$ and $2(3 - x) \leq 7$

_____ **2.** $9 - x > 5$ and $3(x - 2) - 2x > -6$

_____ **3.** $\dfrac{x + 7}{2} \geq 4$ or $3(2x - 5) - 4x \geq -9$

_____ **4.** $3(x - 4) - 2(x - 5) \leq -2$ or $\dfrac{5x}{2} + 1 \geq 6$

Solve and graph.

_____ **5.** $5 < 1 - 2x \leq 9$

_____ **6.** $0 \leq 2(x - 3) < 10$

Answers are on page 119.

▶ Examples You Completed

Example 2. Solve $y + 3 > -2$ and graph the solutions.

Solution

$$y + 3 > -2$$
$$y + 3 - 3 > -2 - 3 \qquad \text{Subtract 3 from both sides.}$$
$$y > -5 \qquad \text{Simplify.}$$

Example 3. Solve and graph $\frac{x}{3} \geq -2$.

Solution

$$\frac{x}{3} \geq -2$$

$$3\left(\frac{x}{3}\right) \geq 3(-2)$$

$$x \geq -6$$

Example 6. Solve $\frac{x}{2} - 1 > 1$.

Solution

$$\frac{x}{2} - 1 > 1$$

$$\frac{x}{2} - 1 + 1 > 1 + 1$$

$$\frac{x}{2} > 2$$

$$2\left(\frac{x}{2}\right) > 2(2)$$

$$x > 4$$

Example 8. Solve and graph $\frac{3 - 2x}{4} \leq -1$.

Solution

$$\frac{3 - 2x}{4} \leq -1$$

$$4\left(\frac{3 - 2x}{4}\right) \leq 4(-1)$$

$$3 - 2x \leq -4$$

$$3 - 2x - 3 \leq -4 - 3$$

$$-2x \leq -7$$

$$\frac{-2x}{-2} \geq \frac{-7}{-2}$$

$$x \geq \frac{7}{2}$$

Example 14. Solve and graph $-14 \leq 5(x - 3) + 1 \leq 11$.

Solution

$$-14 \leq 5(x - 3) + 1 \leq 11$$
$$-14 \leq 5x - 15 + 1 \leq 11$$
$$-14 \leq 5x - 14 \leq 11$$
$$-14 + 14 \leq 5x - 14 + 14 \leq 11 + 14$$
$$0 \leq 5x \leq 25$$
$$\frac{0}{5} \leq \frac{5x}{5} \leq \frac{25}{5}$$
$$0 \leq x \leq 5$$

Answers to Trial Runs

page 111   **1.** $y < 3$

                                                      2. $x \geq -5$

3. $x \geq -3$

4. $x > 5$

5. $x > -10$

page 113   **1.** $a \geq 3$ **2.** $y < -7$ **3.** $x \leq \dfrac{9}{2}$ **4.** $a > -1$ **5.** $x \leq \dfrac{1}{2}$ **6.** $z \geq -3$

page 117   **1.** $x > 2$ **2.** $0 < x < 4$ **3.** $x \geq 1$ **4.** $x \leq 0$ or $x \geq 2$

5. $-4 \leq x < -2$

-5 -4 -3 -2 -1 0

6. $3 \leq x < 8$

2 3 4 5 6 7 8 9

EXERCISE SET 2.3

Solve and graph.

_____ **1.** $x - 5 \geq -3$

_____ **2.** $x - 7 \geq -2$

_____ **3.** $3 - y > 2$

_____ **4.** $7 - y > 5$

_____ **5.** $10a \leq -5$

_____ **6.** $9a \leq 3$

_____ **7.** $\dfrac{x}{2} > -1$

_____ **8.** $\dfrac{x}{3} > -2$

_____ **9.** $3a + 9 \geq 15$

_____ **10.** $7a - 6 \geq -13$

_____ **11.** $-\dfrac{2x}{3} + 5 < 4$

_____ **12.** $-\dfrac{5x}{2} - 7 < 8$

_____ **13.** $8(m - 3) \leq 8$

_____ **14.** $3(m - 6) \leq -3$

_____ **15.** $\dfrac{2x - 9}{3} > 5$

_____ **16.** $\dfrac{5x + 1}{4} > 9$

_____ **17.** $7x - (x + 5) \leq 3x + 2$

_____ **18.** $9x - (2x + 3) \leq 5x + 8$

_____ **19.** $3(y - 4) - 10 \leq 9(y - 1) + 11$

_____ **20.** $5(y - 3) - 7 \leq 8(y + 1) - 3$

_____ **21.** $2x - 5 > 1 \quad \text{and} \quad 3(x - 1) < 6$

_____ **22.** $5x - 7 > 3 \quad \text{and} \quad 4(x - 2) < 0$

_____ **23.** $\dfrac{x - 7}{5} \leq -2 \quad \text{and} \quad 4(1 - x) < -6$

_____ **24.** $\dfrac{x - 6}{4} \leq -2 \quad \text{and} \quad 5(2 - x) < -15$

_____ **25.** $4x - 7 < 3 - x \quad \text{and} \quad 7x - 1 \geq x - 1$

_____ **26.** $3x + 5 \geq x - 1 \quad \text{and} \quad 4x - 3 < 2x + 1$

_____ **27.** $\dfrac{x - 8}{5} < 2 - x \quad \text{or} \quad \dfrac{x}{4} + 3 < 7$

_____ **28.** $\dfrac{2x - 6}{4} < 3 - x \quad \text{or} \quad \dfrac{2x}{9} + 5 < 7$

_____ **29.** $x + 4 \geq 2x - 2 \quad \text{or} \quad 3x + 1 < x + 11$

_____ **30.** $2x - 1 \geq 2 - x \quad \text{or} \quad 3x + 4 \leq 2x + 1$

_____ **31.** $3(2x - 1) \geq x + 7 \quad \text{or} \quad 2(4 - x) > x - 1$

_____ **32.** $5(3x - 4) \leq 9x + 4$ or $4(5x - 1) > 12x + 6$

_____ **33.** $2 \leq m - 5 \leq 7$ _____ **34.** $-3 < m + 1 \leq 5$

_____ **35.** $5 \leq -2a - 7 < 9$ _____ **36.** $1 < -3a - 1 \leq 5$

_____ **37.** $-5 \leq \dfrac{2y - 1}{3} < 3$ _____ **38.** $-2 \leq \dfrac{3y - 2}{4} < 1$

_____ **39.** $0 \leq 3(-x + 5) - 9 < 6$ _____ **40.** $0 \leq 4(-x - 3) - 8 \leq 4$

☆ Stretching the Topics _____

Solve and graph.

_____ **1.** $3b - [7 - (2 - b)] \geq b + 8$

_____ **2.** $5 + 7x \geq -(3 - 2x) - 3x$ and $-7 + 5x - (x + 3) \geq x + 2$

_____ **3.** $-4 < 3x + 5 < 14$ and $-16 \leq 4(x - 2) < 13$

Check your answers in the back of your book.

If you can complete **Checkup 2.3**, you are ready to go on to Section 2.4.

✔ CHECKUP 2.3

Solve and graph.

_____ **1.** $3x - 5 \geq 4$ _____ **2.** $-\dfrac{a}{4} + 2 < 1$

_____ **3.** $3(y - 2) > 4(y + 1)$ _____ **4.** $6x - (x + 3) \leq 2x - 9$

_____ **5.** $\dfrac{x - 8}{3} < -2$ and $4(1 - x) < 2$

_____ **6.** $3x - 5 \leq 5x + 3$ and $2x - 1 > x + 2$

_____ **7.** $\dfrac{2x - 3}{4} < 1 + x$ or $2(3x - 4) < 7x - 6$

_____ **8.** $3(2 - x) \geq 0$ or $5(x - 1) > 2x + 4$

_____ **9.** $-3 \leq 2a + 5 \leq 1$ _____ **10.** $0 \leq \dfrac{-y + 2}{6} < 1$

Check your answers in the back of your book.

If You Missed Problems:	You Should Review Examples:
1, 2	5, 6
3, 4	7–9
5, 6	10, 11
7, 8	12, 13
9, 10	14, 15

2.4 Solving Absolute-Value Equations and Inequalities

Solving absolute-value equations and inequalities requires little more than a thorough understanding of the number line and an ability to solve first-degree equations and inequalities.

Absolute-Value Equations

Let's look at some simple absolute-value equations and identify the solutions.

Equation	Solutions	Reasons
$\lvert x \rvert = 3$	$x = 3$ or $x = -3$	$\lvert 3 \rvert = 3$ $\lvert -3 \rvert = 3$
$\lvert x \rvert = 0$	$x = 0$	$\lvert 0 \rvert = 0$
$\lvert x \rvert = -2$	No solution	Absolute values are *never* negative

We see that an equation of the type $\lvert x \rvert = a$ will have solutions only when a is not negative.

If a is *positive* and	If a is *zero* and	If a is *negative* and
$\lvert x \rvert = a$	$\lvert x \rvert = a$	$\lvert x \rvert = a$
the solutions are	the solution is	there are *no* solutions.
$x = a$ or $x = -a$	$x = 0$	

You try completing Example 1.

Example 1. Solve $\lvert x \rvert = \dfrac{1}{3}$ and graph the solutions.

Solution

$$\lvert x \rvert = \frac{1}{3}$$

$x = \underline{\hspace{1cm}}$ or $x = \underline{\hspace{1cm}}$

Check your work on page 133. ▶

It should be clear to you that if the absolute value of some *quantity* equals some positive number a, then either that quantity equals a *or* that quantity equals $-a$. Let's use this idea to solve an equation in which the quantity within the absolute value bars contains more terms than x by itself.

Example 2. Solve $|x + 1| = 3$. Check the solutions and graph.

Solution: Since the absolute value of a quantity is 3, we know that the quantity must equal 3 or -3.

$$x + 1 = 3 \qquad \text{or} \qquad x + 1 = -3$$
$$x + 1 - 1 = 3 - 1 \qquad\qquad x + 1 - 1 = -3 - 1$$
$$x = 2 \qquad \text{or} \qquad x = -4$$

CHECK $\qquad\qquad\qquad\qquad\qquad$ CHECK

$$x = 2 \qquad\qquad\qquad\qquad x = -4$$
$$|x + 1| = 3 \qquad\qquad\qquad |x + 1| = 3$$
$$|2 + 1| \overset{?}{=} 3 \qquad\qquad\qquad |-4 + 1| \overset{?}{=} 3$$
$$|3| \overset{?}{=} 3 \qquad\qquad\qquad\qquad |-3| \overset{?}{=} 3$$
$$3 = 3 \qquad\qquad\qquad\qquad\quad 3 = 3$$

The graph of our solutions is

$$\begin{array}{ccccccccc} & -5 & -4 & -3 & -2 & -1 & 0 & 1 & 2 & 3 \end{array}$$

Notice that the two solutions are *not* opposites of each other; usually, they will *not* be.

Now you complete Examples 3 and 4.

Example 3. Solve $|2x - 3| = 7$.

Solution: $|2x - 3| = 7$ means

$$2x - 3 = 7 \qquad \text{or} \quad 2x - 3 = -7$$
$$2x = \underline{\quad\quad} \qquad\qquad 2x = \underline{\quad\quad}$$
$$x = \underline{\quad\quad} \quad \text{or} \qquad x = \underline{\quad\quad}$$

Check your work on page 133. ▶

Example 4. Solve $|6 - 9x| = 0$.

Solution:

$$|6 - 9x| = 0 \text{ means}$$
$$6 - 9x = 0$$

Check your work on page 133. ▶

Example 5. Solve $\dfrac{|7 - 2x|}{4} = 2$.

Solution

$$\frac{|7 - 2x|}{4} = 2$$

$$\frac{4}{1} \cdot \frac{|7 - 2x|}{4} = 4(2)$$

$$|7 - 2x| = 8$$

$$7 - 2x = 8 \qquad \text{or} \qquad 7 - 2x = -8$$
$$-2x = 1 \qquad\qquad\qquad -2x = -15$$
$$\frac{-2x}{-2} = \frac{1}{-2} \qquad\qquad\qquad \frac{-2x}{-2} = \frac{-15}{-2}$$
$$x = \frac{-1}{2} \qquad \text{or} \qquad x = \frac{15}{2}$$

▥➡ **Trial Run**

Solve.

_____ **1.** $|x| = 5.1$ _____ **2.** $|a| = \dfrac{3}{4}$

_____ **3.** $|2x - 5| = 7$ _____ **4.** $|3 - 12x| = 0$

_____ **5.** $\left|\dfrac{a}{2} + 3\right| = 1$ _____ **6.** $|5x - 1| = -4$

_____ **7.** $\dfrac{|5 - 10x|}{3} = 15$ _____ **8.** $|21 - 7x| = 21$

Answers are on page 134.

Absolute-Value Inequalities (Less-Than Type)

Suppose that we wish to determine the location on the number line of all the numbers that satisfy the inequality $|x| < 3$. We know the location of all the numbers for which $|x| = 3$.

But what about all the other numbers on this line? Where are all those numbers whose absolute values are *less than* 3? Where are the solutions for $|x| < 3$?

Numbers to the Left of -3	Numbers Between -3 and 3	Numbers to the Right of 3
$|-4| \overset{?}{<} 3$ No	$|-2| \overset{?}{<} 3$ Yes	$|3.1| \overset{?}{<} 3$ No
$|-3.5| \overset{?}{<} 3$ No	$|0| \overset{?}{<} 2$ Yes	$|5| \overset{?}{<} 3$ No
	$|2.95| \overset{?}{<} 3$ Yes	

It seems that only those numbers *between* -3 and 3 will satisfy our inequality. Therefore, we write the solution for $|x| < 3$ as

$$-3 < x < 3$$

The graph of our solutions is an interval on the number line.

In general, we may say:

If a is *positive*, then $|x| < a$ means that x lies between $-a$ and a on the number line.

Try Example 6.

Example 6. Solve $|x| \leq 5$ and graph.

Solution: $|x| \leq 5$ means _____ $\leq x \leq$ _____

Check your work on page 133. ▶

Example 7. Solve $|x| < -1$.

Solution: $|x| < -1$ has *no solution* because -1 is a negative number and the absolute value of a quantity must be positive or zero.

If a is positive, then	If a is positive or zero then	If a is negative, then
$|x| < a$	$|x| \leq a$	$|x| < a$
means	means	has no solution.
$-a < x < a$	$-a \leq x \leq a$	

Example 8. Solve $|x + 1| < 6$ and graph.

Solution: We realize that the quantity within the absolute-value bars must lie between -6 and 6.

$$-6 < \quad x + 1 \quad < 6 \qquad \text{We must isolate } x \text{ in the middle.}$$
$$-6 - 1 < x + 1 - 1 < 6 - 1 \qquad \text{Subtract 1 from all parts.}$$
$$-7 < \quad x \quad\quad < 5 \qquad \text{Simplify.}$$

Now you try Example 9.

Example 9. Solve $|2x - 3| \leq 7$ and graph.

Solution: The quantity within the absolute-value bars must lie between (and including) _____ and _____ on the number line.

$$-7 \leq 2x - 3 \leq 7$$

Check your work on page 133. ▶

Example 10. Solve $|5 - x| \leq 3$ and graph.

Solution: $|5 - x| \leq 3$ means

$$-3 \leq \quad 5 - x \quad \leq 3$$
$$-3 - 5 \leq 5 - x - 5 \leq 3 - 5$$
$$-8 \leq \quad -x \quad \leq -2$$
$$\frac{-8}{-1} \geq \quad \frac{-x}{-1} \quad \geq \frac{-2}{-1}$$

Notice that we reversed the direction of our inequalities. Why?

$$8 \geq x \geq 2$$

Rewriting the inequalities in better order (from left to right on the number line), we have

$$2 \leq x \leq 8$$

⇒ Trial Run

Solve.

_____ **1.** $|a| \leq 3.5$

_____ **2.** $|y| < \dfrac{4}{5}$

_____ **3.** $|x| < 0$

_____ **4.** $|3x| < 9$

_____ **5.** $|4x + 1| \leq 7$

_____ **6.** $|7 - 2x| < 11$

_____ **7.** $|5x - 3| \leq -2$

_____ **8.** $\left|\dfrac{2x - 1}{3}\right| \leq 5$

Answers are on page 134.

Absolute-Value Inequalities (Greater-Than Type)

Now let's determine the location on the number line of all the numbers that satisfy the inequality $|x| > 3$.

Numbers to the Left of -3	Numbers Between -3 and 3	Numbers to the Right of 3
$\lvert -5 \rvert \overset{?}{>} 3$ Yes	$\lvert -2 \rvert \overset{?}{>} 3$ No	$\lvert 3.5 \rvert \overset{?}{>} 3$ Yes
$\lvert -4 \rvert \overset{?}{>} 3$ Yes	$\lvert 0 \rvert \overset{?}{>} 3$ No	$\lvert 4 \rvert \overset{?}{>} 3$ Yes
$\lvert -3.1 \rvert \overset{?}{>} 3$ Yes	$\lvert 2.5 \rvert \overset{?}{>} 3$ No	$\lvert 10 \rvert \overset{?}{>} 3$ Yes

As we expected, numbers to the *left* of -3 or to the *right* of 3 will satisfy the inequality $\lvert x \rvert > 3$. We write those solutions as

$$x < -3 \quad or \quad x > 3$$

and the graph is two separate intervals on the number line.

In general, we say:

> If a is *positive*, then $\lvert x \rvert > a$ means that x lies to the left of $-a$ *or* to the right of a on the number line.

Try finding the solution for Example 11.

Example 11. Solve $\lvert x \rvert > 2$ and graph the solutions.

Solution: $\lvert x \rvert > 2$ means

$$x < \underline{\quad\quad} \quad or \quad x > \underline{\quad\quad}$$

Check your work on page 134. ▶

Example 12. Solve $\lvert x \rvert \geq \dfrac{1}{2}$ and graph the solutions.

Solution: $\lvert x \rvert \geq \dfrac{1}{2}$ means

$$x \leq -\frac{1}{2} \quad or \quad x \geq \frac{1}{2}$$

If a is positive, then $$\lvert x \rvert > a$$ means $$x < -a \quad or \quad x > a$$	If a is positive, then $$\lvert x \rvert \geq a$$ means $$x \leq -a \quad or \quad x \geq a$$

Example 13. Solve $|x - 2| > 7$ and graph.

Solution: We realize that the quantity within the absolute-value bars must lie to the left of -7 *or* to the right of 7.

$$x - 2 < -7 \qquad \text{or} \qquad x - 2 > 7$$
$$x - 2 + 2 < -7 + 2 \qquad\qquad x - 2 + 2 > 7 + 2$$
$$x < -5 \qquad \text{or} \qquad x > 9$$

Now you try Example 14.

Example 14. Solve $|7x - 4| > 3$ and graph.

Solution: The quantity within the absolute-value bars must lie to the left of _____ *or* to the right of _____ .

$$7x - 4 < \underline{\qquad} \qquad \text{or} \qquad 7x - 4 > \underline{\qquad}$$

or

Check your work on page 134. ▶

Example 15. Solve $\left|2 - \dfrac{x}{3}\right| \geq 4$ and graph.

Solution: $\left|2 - \dfrac{x}{3}\right| \geq 4$ means

$$2 - \frac{x}{3} \leq -4 \qquad\qquad\qquad \text{or} \qquad\qquad\qquad 2 - \frac{x}{3} \geq 4$$

$3 \cdot 2 - \dfrac{3}{1} \cdot \dfrac{x}{3} \leq 3(-4)$ Multiply both sides by LCD: 3. $3 \cdot 2 - \dfrac{3}{1} \cdot \dfrac{x}{3} \geq 3 \cdot 4$

$6 - x \leq -12$ Simplify. $6 - x \geq 12$

$-x \leq -18$ Subtract 6. $-x \geq 6$

$\dfrac{-x}{-1} \geq \dfrac{-18}{-1}$ Divide by -1. $\dfrac{-x}{-1} \leq \dfrac{6}{-1}$

$x \geq 18$ Simplify. $x \leq -6$

Notice that we reversed the directions of the inequalities when we divided both sides by -1.

Now you try Example 16.

Example 16. Solve $|5 - 3x| > 1$ and graph.

Solution: $|5 - 3x| > 1$ means

$$5 - 3x < -1 \quad \text{or} \quad 5 - 3x > 1$$

or

Check your work on page 134. ▶

Example 17. Solve $|x - 2| > -3$.

Solution: Since absolute values are always *positive*, the statement $|x - 2| > -3$ is true for *all* values of the variable. Every real number is a solution.

⫸ Trial Run

Solve.

_____ **1.** $|y| > \dfrac{2}{3}$ _____ **2.** $|m| > -11$

_____ **3.** $|x| \geq 7$ _____ **4.** $|7x| > 14$

_____ **5.** $|2x - 3| \geq 5$ _____ **6.** $|9 - x| \geq 4$

_____ **7.** $\left|\dfrac{5x}{3}\right| > 10$ _____ **8.** $|2 - 3x| > 2$

Answers are on page 134.

A summary of our work with absolute-value equations and inequalities is included here for your reference. We assume that a represents a positive number.

Equality	$|x| = a$ means $x = a$ or $x = -a$	
Less-than inequalities	$|x| < a$ means $-a < x < a$	
	$|x| \leq a$ means $-a \leq x \leq a$	

Greater-than inequalities $\quad |x| > a \quad$ means $\quad x < -a \quad$ or $\quad x > a$

$|x| \geq a \quad$ means $\quad x \leq -a \quad$ or $\quad x \geq a$

The secret to success with absolute-value equations and inequalities is to visualize the number line and immediately rewrite the original statement as an equivalent statement without absolute-value bars. The solutions can then be found using the usual methods for solving first-degree equations and inequalities.

▶ Examples You Completed

Example 1. Solve $|x| = \dfrac{1}{3}$ and graph the solutions.

Solution

$$|x| = \frac{1}{3}$$

$$x = \frac{1}{3} \quad \text{or} \quad x = -\frac{1}{3}$$

Example 3. Solve $|2x - 3| = 7$.

Solution $|2x - 3| = 7$ means

$$2x - 3 = 7 \quad \text{or} \quad 2x - 3 = -7$$
$$2x = 10 \quad \text{or} \quad 2x = -4$$
$$x = 5 \quad \text{or} \quad x = -2$$

Example 4. Solve $|6 - 9x| = 0$.

Solution $|6 - 9x| = 0$ means

$$6 - 9x = 0$$
$$6 - 9x - 6 = 0 - 6$$
$$-9x = -6$$
$$\frac{-9x}{-9} = \frac{-6}{-9}$$
$$x = \frac{2}{3}$$

Example 6. Solve $|x| \leq 5$ and graph.

Solution $|x| \leq 5$ means

$$-5 \leq x \leq 5$$

Example 9. Solve $|2x - 3| \leq 7$ and graph.

Solution: The quantity within the absolute-value bars must lie between (and including) -7 and 7 on the number line.

$$-7 \leq 2x - 3 \leq 7$$
$$-4 \leq \quad 2x \quad \leq 10$$
$$\frac{-4}{2} \leq \quad \frac{2x}{2} \quad \leq \frac{10}{2}$$
$$-2 \leq \quad x \quad \leq 5$$

Example 11. Solve $|x| > 2$ and graph the solutions.

Solution: $|x| > 2$ means

$$x < -2 \quad \text{or} \quad x > 2$$

Example 14. Solve $|7x - 4| \geq 3$ and graph.

Solution: The quantity within the absolute-value bars must lie to the left of -3 *or* to the right of 3.

$$7x - 4 \leq -3 \quad \text{or} \quad 7x - 4 \geq 3$$
$$7x \leq 1 \quad \text{or} \quad 7x \geq 7$$
$$x \leq \frac{1}{7} \quad \text{or} \quad x \geq 1$$

Example 16. Solve $|5 - 3x| > 1$ and graph.

Solution: $|5 - 3x| > 1$ means

$$5 - 3x < -1 \quad \text{or} \quad 5 - 3x > 1$$
$$-3x < -6 \quad \quad -3x > -4$$
$$\frac{-3x}{-3} > \frac{-6}{-3} \quad \quad \frac{-3x}{-3} < \frac{-4}{-3}$$
$$x > 2 \quad \text{or} \quad x < \frac{4}{3}$$

Answers to Trial Runs

page 127 **1.** $x = 5.1$ or $x = -5.1$ **2.** $a = \frac{3}{4}$ or $a = \frac{-3}{4}$ **3.** $x = 6$ or $x = 1$ **4.** $x = \frac{1}{4}$

5. $a = -4$ or $a = -8$ **6.** No solution   **7.** $x = -4$ or $x = 5$ **8.** $x = 0$ or $x = 6$

page 129 **1.** $-3.5 \leq a \leq 3.5$ **2.** $\frac{-4}{5} < y < \frac{4}{5}$ **3.** No solution   **4.** $-3 < x < \frac{3}{8}$

5. $-2 \leq x \leq \frac{3}{2}$ **6.** $-2 < x < 9$ **7.** No solution   **8.** $-7 \leq x \leq 8$

page 132 **1.** $y < \frac{-2}{3}$ or $y > \frac{2}{3}$ **2.** All real numbers   **3.** $x \leq -7$ or $x \geq 7$ **4.** $x < -2$ or $x > 2$

5. $x \leq -1$ or $x \geq 4$ **6.** $x \leq 5$ or $x \geq 13$ **7.** $x < -6$ or $x > 6$ **8.** $x < 0$ or $x > \frac{4}{3}$

EXERCISE SET 2.4

Solve.

_____ **1.** $|x| = 9$

_____ **2.** $|x| = 13$

_____ **3.** $\left|\dfrac{a}{3}\right| = 7$

_____ **4.** $\left|\dfrac{a}{5}\right| = 3$

_____ **5.** $|3x| = 2.1$

_____ **6.** $|5x| = 4.5$

_____ **7.** $|y - 5| = 0$

_____ **8.** $|y + 3| = 0$

_____ **9.** $|2x - 9| = -7$

_____ **10.** $|3x - 5| = -10$

_____ **11.** $|7 - 6x| = 1$

_____ **12.** $|12 - 3x| = 9$

_____ **13.** $|3x - 1| = -2$

_____ **14.** $|3x + 5| = -6$

_____ **15.** $\left|\dfrac{a}{6} + 5\right| = 1$

_____ **16.** $\left|\dfrac{a}{7} + 3\right| = 4$

_____ **17.** $|13 - 3x| = 5$

_____ **18.** $|9 - 3x| = 6$

_____ **19.** $\left|\dfrac{7 - x}{2}\right| = 10$

_____ **20.** $\left|\dfrac{8 - x}{3}\right| = 2$

_____ **21.** $|y| < \dfrac{5}{3}$

_____ **22.** $|y| < \dfrac{7}{2}$

_____ **23.** $\left|\dfrac{a}{7}\right| < 5$

_____ **24.** $\left|\dfrac{a}{3}\right| < 9$

_____ **25.** $|-3x| < 6$

_____ **26.** $|-2x| < 12$

_____ **27.** $|3x - 2| \le 4$

_____ **28.** $|5x - 1| \le 0$

_____ **29.** $\left|5 - \dfrac{x}{3}\right| \le 1$

_____ **30.** $\left|3 - \dfrac{x}{4}\right| \le 2$

_____ **31.** $|2m - 7| \le -8$

_____ **32.** $|4m - 9| \le -10$

_____ **33.** $\left|\dfrac{3x - 1}{4}\right| < 7$

_____ **34.** $\left|\dfrac{4x - 8}{2}\right| < 6$

_____ **35.** $|a| > 2.5$

_____ **36.** $|a| > 3.2$

_____ **37.** $|y - 3| > -7$

_____ **38.** $|y - 2| > -9$

_____ **39.** $|-5x| \ge 10$

_____ **40.** $|-9x| \ge 18$

_____ **41.** $\left|\dfrac{3}{5}x\right| > 6$

_____ **42.** $\left|\dfrac{5}{2}x\right| > 10$

_____ **43.** $|5x - 1| \geq 4$ _____ **44.** $|7x - 1| \geq 13$

_____ **45.** $\left| 1 - \dfrac{m}{3} \right| > 6$ _____ **46.** $\left| 2 - \dfrac{m}{7} \right| > 1$

_____ **47.** $|3x - 7| > -9$ _____ **48.** $|5x - 3| > -12$

_____ **49.** $|7x - 3| \geq 10$ _____ **50.** $|8x - 5| \geq 3$

☆ Stretching the Topics

Solve.

_____ **1.** $|6 - 2[3 - (5 - 4m)]| = 2$

_____ **2.** $4|2x - 2| + 6 < 10$

_____ **3.** $|2x - 8| < 8$ and $|x| + 3 > 4$

Check your answers in the back of your book.

If you can complete **Checkup 2.4**, you are ready to go on to Section 2.5.

CHECKUP 2.4

Solve.

_____ **1.** $|3x| = 15.6$

_____ **2.** $|8 - x| = 9$

_____ **3.** $|5y - 2| = -8$

_____ **4.** $|-4x| < 12$

_____ **5.** $\left|\dfrac{x}{2} - 7\right| \le 5$

_____ **6.** $|3a - 2| \le 7$

_____ **7.** $|3x| > 6$

_____ **8.** $|9 - m| \ge 3$

_____ **9.** $|2x - 1| > 11$

_____ **10.** $\left|\dfrac{3}{2}x - 9\right| > 12$

Check your answers in the back of your book.

If You Missed Problems:	You Should Review Examples:
1–3	1–3
4–6	8–10
7–10	13–16

2.5 Switching from Words to Equations and Inequalities

In Chapter 1 we practiced switching word expressions to expressions containing variables and constants. Now that we have learned to solve first-degree equations and inequalities, we can put those skills to use in solving problems stated in words.

Once again we shall be orderly in our approach and keep the following steps in mind.

> **1.** Read the problem carefully and sort out the given information.
> **2.** Identify the variable.
> **3.** Switch each word expression in the problem to a variable expression.
> **4.** Use an equation or inequality to relate the variable expressions.
> **5.** Solve the equation or inequality.
> **6.** Write a sentence describing the answer to the problem.

Let's try our thinking, then writing approach to solving some word problems.

Example 1. At an hourly wage of $3.50, for how many hours must Beth work so that she will learn $140?

Solution

We Think	*We Write*
The unknown quantity is the hours that Beth works.	Let h = number of hours
Earnings are found by *multiplying* rate per hour times number of hours worked.	$\$3.50h$ = earnings
Beth's earnings must total $140.	$\$3.50h = 140$

$$3.50h = 140$$

$$\frac{3.50h}{3.50} = \frac{140}{3.50}$$

$$h = 40 \text{ hours}$$

Beth must work for 40 hours to earn $140.

Example 2. Ira has scored 80, 62, 75, and 69 on his first four mathematics tests. What is the lowest score that he can earn on his fifth test to have a test average of at least 70?

Solution

We Think	*We Write*
The unknown quantity is Ira's grade on the fifth test.	Let t = Ira's grade on fifth test
To find the average of five tests, we must *add* all the scores and *divide* by 5.	$\dfrac{80 + 62 + 75 + 69 + t}{5}$ = average of five tests
Ira wants an average score of *at least* 70.	$\dfrac{80 + 62 + 75 + 69 + t}{5} \geq 70$

$$\frac{80 + 62 + 75 + 69 + t}{5} \geq 70$$

$$\frac{286 + t}{5} \geq 70$$

$$286 + t \geq 350$$

$$t \geq 64$$

If Ira's fifth score is 64 or higher, he will have an average of at least 70 on the five tests.

Now you try completing Example 3.

Example 3. James wishes to enclose a rectangular chicken yard with 150 feet of fencing. He would like the yard to be twice as long as it is wide. What must be the yard's dimensions?

 Solution

We Think	We Write
The number of feet of fencing is the *perimeter* of the yard.	Perimeter = 2 lengths + 2 widths
We need expressions for length and width. We know less about the width.	Let w = width of yard (feet)
We know that the length is twice the width.	Let _____ = length of yard (feet)
We know that the perimeter is the *sum* of 2 lengths and 2 widths.	2(_____) + 2(_____) = perimeter
The perimeter is 150 feet.	2(_____) + 2(_____) = 150

$$2w + 2(2w) = 150$$

$$2w + 4w = 150$$

$$\underline{} w = 150$$

$$w = \underline{}$$

The width of the yard is _____ feet and the length is _____ feet.

Check your work on page 143. ▶

Example 4. In her purse Anita has 16 coins (nickels, dimes, and quarters), worth $2.05. She has three times as many dimes as nickels. How many of each coin does she have?

Solution

We Think	We Write
First we concentrate on the *number* of coins.	Let n = number of nickels
We know that there are *three times* as many dimes as nickels.	$3n$ = number of dimes
There are 16 coins in all, so the number of quarters is 16 *minus* the number of dimes and nickels.	$16 - (n + 3n)$ = number of quarters $16 - 4n$ = number of quarters
Now we concentrate on the *value* of the coins. Each nickel is worth 5 cents; each dime is worth 10 cents; each quarter is worth 25 cents.	$5 \cdot n$ = value of nickels $10(3n)$ = value of dimes $25(16 - 4n)$ = value of quarters
We know that the total value of the coins is the *sum* of all the coins' values.	$5n + 10(3n) + 25(16 - 4n)$ = total value

We know that the total value is $2.05, or 205 cents.

$$5n + 10(3n) + 25(16 - 4n) = 205$$

$$5n + 10(3n) + 25(16 - 4n) = 205$$

$$5n + 30n + 400 - 100n = 205$$

$$400 - 65n = 205$$

$$-65n = -195$$

$$\frac{-65n}{-65} = \frac{-195}{-65}$$

$$n = 3$$

Number of nickels: $n = 3$

Number of dimes: $3n = 3 \cdot 3 = 9$

Number of quarters: $16 - 4n = 16 - 4 \cdot 3 = 16 - 12 = 4$

Anita's purse contains 3 nickels, 9 dimes, and 4 quarters.

Now let's try working the problem stated at the beginning of this chapter.

Example 5. Mary left Hometown at 6 P.M. driving due east on Route 22 at 55 miles per hour. At the same time, Todd left Anytown driving at 60 miles per hour due west on Route 22. If Hometown is 368 miles due west of Anytown, at what time will Mary and Todd meet?

Solution: Let's use an illustration to help us understand the situation described in this problem.

We Think	*We Write*
The total distance will be the *sum* of Mary's distance and Todd's distance.	Total distance = Mary's distance + Todd's distance
We know that the distance is found by *multiplying* rate per hour times hours driven.	Distance = rate × time
Mary and Todd both drive for the same number of hours.	Let t = hours driven
Mary's rate is 55 mph and Todd's rate is 60 mph.	$55 \cdot t$ = Mary's distance (miles) $60 \cdot t$ = Todd's distance (miles)
The two distances must *sum* to the total distance, 368 miles.	$55t + 60t = 368$

$$55t + 60t = 368$$

$$115t = 368$$

$$\frac{115t}{115} = \frac{368}{115}$$

$$t = 3\frac{23}{115}$$

$$= 3\frac{1}{5} \text{ hours}$$

Mary and Todd must drive for $3\frac{1}{5}$ hours (or 3 hours and 12 minutes) before they meet. If they started at 6 P.M., they will meet at 9:12 P.M..

See if you can complete Example 6.

Example 6. A car rental agency charges a fee of $18 per day, plus 30 cents per mile driven. If Jake rents a car for 1 day, what distance can he drive and keep his total rental charge under $45?

Solution

We Think	We Write
The total charge is the *sum* of the flat fee plus the mileage charge.	Total charge = $18 + mileage charge
The mileage charge depends on the distance driven.	Let x = distance driven (miles)
The mileage charge is found by *multiplying* the rate per mile times the distance driven.	_____ · x = mileage charge (cents)
The flat fee is $18, so we must express the mileage charge in *dollars* also.	_____ x = mileage charge (dollars)
The *total* charge must be *less than* $45.	$18 + $_____ x < $45

$$18 + 0.30x < 45$$
$$0.30x < 27$$
$$\frac{0.30x}{0.30} < \frac{27}{0.30}$$
$$x = \underline{\qquad}$$

Jake must drive fewer than _____ miles.

Check your work on page 143. ▶

Example 7. The Supercal Calculator Company manufactures calculators at a cost of $8.45 each and sells them for $23.95 each. The company's fixed costs are $7300 per month. How many calculators must the company produce and sell to have a profit of at least $2000 this month?

Solution

We Think	We Write
Profit is found by *subtracting* costs from revenue.	Profit = revenue − costs
Revenue is found by *multiplying* selling price times number of units sold.	Revenue = $23.95x$
Costs are found by *adding* variable costs and fixed costs.	Variable costs = $8.45x$ Fixed costs = 7300 Costs = $8.45x + 7300$
Profit = revenue − costs.	Profit = $23.95x - (8.45x + 7300)$ = $23.95x - 8.45x - 7300$ = $15.5x - 7300$
We want profit to be *at least* $2000. Now we solve the inequality.	$15.5x - 7300 \geq 2000$

$$15.5x - 7300 \geq 2000$$

$$15.5x \geq 9300$$

$$\frac{15.5x}{15.5} \geq \frac{9300}{15.5}$$

$$x \geq 600$$

At least 600 calculators must be sold to have a profit of $2000 in one month.

Applying the techniques of algebra to solving word problems can be one of the most rewarding parts of your study in mathematics. You will become successful in that skill if you practice a great deal and learn to be organized in your approach.

⫸ **Trial Run**

Solve.

_____ **1.** Carl rode his bike to the city at a rate of 10 miles per hour and returned immediately on a bus traveling the same route at a rate of 50 miles per hour. If the entire trip took 6 hours, how far is Carl's house from the city?

_____ **2.** The refrigerator repairman charges a service call fee of $29, plus $7.50 per half-hour spent on the job. If James' bill for labor was $66.50, how long did the repairman work on his refrigerator?

Answers are on page 143.

▶ **Examples You Completed**

Example 3

Solution

We Write

Perimeter = 2 lengths + 2 widths

Let w = width of yard

Let $2w$ = length of yard

$2(w) + 2(2w)$ = perimeter

$2(w) + 2(2w) = 150$

$$2w + 2(2w) = 150$$

$$2w + 4w = 150$$

$$6w = 150$$

$$w = 25$$

The width of the yard is 25 feet and the length is 50 feet.

Example 6

Solution

We Write

Total charge = $18 + mileage charge

Let x = distance driven (miles)

$30 \cdot x$ = mileage charge (cents)

$0.30 \cdot x$ = mileage charge (dollars)

$18 + $0.30x < $45

$$18 + 0.30x < 45$$

$$0.30x < 27$$

$$\frac{0.30x}{0.30} < \frac{27}{0.30}$$

$$x < 90$$

Jake must drive fewer than 90 miles.

Answers to Trial Run

page 143   **1.** Carl's house is 50 miles from the city.   **2.** The repairman worked for $2\frac{1}{2}$ hours.

EXERCISE SET 2.5

Solve.

_____ **1.** Steve is saving to buy a car. He has decided he will start looking as soon as he has saved $2000. If he has $1100 in savings now and he can save $150 a month, how many months will it be before Steve can begin looking at cars?

_____ **2.** Mrs. Bealmear, a retiree, will lose some retirement benefits if she earns over $8000 per year working as a baby sitter. If she has already earned $6500, how many more hours may she work at $2.50 per hour without losing any retirement benefits?

_____ **3.** The Band Boosters Club is selling cookies to buy new uniforms. Chocolate chip cookies sell for $3 a box and sugar cookies for $2 a box. They decide that they can sell 650 boxes. How many boxes of each should they sell so that their total sales will be $1700?

_____ **4.** Sara is making a silk flower arrangement of roses and carnations. She needs twice as many roses as carnations. Carnations cost 75 cents each and roses are $1 each. How many of each should she buy if she can spend no more than $18 for the arrangement?

_____ **5.** If Don loses 16 pounds, 9 pounds, and 11 pounds during the first three months of his diet, how much will he have to lose during the fourth month so that he will have an average weight loss of 12 pounds per month?

_____ **6.** If Evelyn earns $37, $58, $61, and $42 during the first 4 days that she sells vegetables from her garden, how much will she have to sell during the fifth day so that her sales average is $50 per day?

_____ **7.** For the toll road, Frank needs 5 more dimes than quarters and the total amount he will have to pay is $3.65. How many of each will he need?

_____ **8.** At the end of the day, Alice's tips consist of three times as many nickels as quarters. If the total of the tips is $8.40, how many of each does she have?

_____ **9.** Mr. Hsu has $5000 invested at 6% interest. How much additional money must he invest at 9% so that his annual income from interest will be the same as if the total amount were invested at 8%?

_____ **10.** Two amounts of money totaling $25,000 are invested in stocks. One amount earns 10% interest and the other earns 13%. Find how much is invested at each amount if the total yearly interest is $2920.

_____ **11.** Mrs. Young is making meat loaf for a banquet. She bought 20 pounds of meat: some sausage and some ground beef. Ground beef is $1.75 a pound and sausage is $1.30 a pound. If the total cost was $31.40, how many pounds of each did she buy?

_____ **12.** Carroll is making ice cream for the math club picnic. He needs twice as much milk as cream. If milk is 90 cents a quart and cream is $2.10 a quart, how many quarts of each did he buy if the total cost was $11.70?

_____ **13.** Two cars are 315 miles apart and are moving toward each other. They will meet in 3 hours. If their speeds differ by 5 miles per hour, what is the speed of each car?

_____ **14.** Richard jogs and walks a total of 10 miles every day. He jogs at 5 miles per hour and walks at 3 miles per hour. If he covers the total distance in 3 hours, how much of the time does he spend walking?

_____ **15.** On a recent vacation, Toni drove to Kentucky Lake. The trip to the lake took her 1 hour less than the trip home. If she averaged 55 miles per hour going and only 45 miles per hour returning, how far does Toni live from the lake?

_____ **16.** Sam rode the bus to see his father but returned by car with a neighbor. On the bus he averaged 10 miles per hour less than when riding with his neighbor. If it took Sam 3 hours to get to his father's house but only 2 hours to return, how far does he live from his father?

_____ **17.** The Book Rack is celebrating National Book Week by having a sale. Charles bought 30 books, some for $2.50 and some for $4. If his total bill was $102, how many of each did he buy?

_____ **18.** Kris is buying bags of candy for Christmas, some for 75 cents and some for $1.25. He bought three times as many bags of 75-cent candy as he did bags costing $1.25. If he spent $17.50 for candy, how many bags of each did he buy?

_____ **19.** Della is buying a living-room suite on the installment plan. She paid $150 as a down payment and is paying $55 a month until it is paid for. If the living-room suite cost $810, for how many months will she be making payments?

_____ **20.** Juan earns a weekly salary of $380 a week and $12 per hour for each hour of overtime. If Juan earned $470 last week, how many hours overtime did he work?

_____ **21.** A cashier at First Federal has a stack of bills—some ten-dollar bills, some five-dollar bills, and some ones. There are 8 fewer five-dollar bills than ten-dollar bills and twice as many one-dollar bills as five-dollar bills. If the total value of the stack is $811, how many bills are there of each denomination?

_____ **22.** Miles has a total of $3.65 in change. He has 2 more nickels than dimes and 3 more quarters than dimes. How many coins of each kind does he have?

_____ **23.** Benson sells vacuum cleaners on a commission. He earns $75 for each vacuum cleaner he sells. How many cleaners can he sell so that he will have a monthly salary of at least $1400?

_____ **24.** If Thor works for an hourly wage of $4.25, how many hours can he work so that his gross pay will be at least $152?

_____ **25.** The perimeter of a triangle can be no more than 35. One side is 7 and the other two sides are the same. Find the possible lengths for the two equal sides.

_____ **26.** The length of one side of a triangle is 9 inches more than another. The sum of the length of these two sides must be greater than the third side, which is 43 inches. Find the possible lengths for the two sides.

_____ **27.** The length of a rectangular playground is to be twice its width. The dimensions must be such that the playground can be enclosed with no more than 360 feet of available fencing. What dimensions can the playground have?

_____ **28.** Janyce has only 134 inches of framing. What will be the dimensions of the largest picture frame she can make if she makes the length 5 inches more than the width?

_____ **29.** The Miller Theater has only 250 seats. If the theater group charges admissions of $5 for adults and $3 for children, find the number of adult tickets they can sell and still cover their production expenses of $1050.

_____ **30.** The Women's Political Caucus is having a barbecue for its 150 members. It will cost them $2.50 to serve a chicken plate and $3.75 to serve a ham plate. If they have only $415, how many ham plates could be served?

_____ **31.** A manufacturer of lighting fixtures produces a certain lamp at a cost of $25.50. The fixed costs for producing lamps each month are $2400. How many lamps can be produced each month if the cost of production is kept below $3420?

_____ **32.** Candy Land sells 5-pound boxes of candy for $13.25. How many boxes must be sold each week if the weekly revenue from the sale of the 5-pound boxes of candy must be greater than $331.25?

_____ **33.** A publishing company finds that the cost of publishing a certain book is $12.50 per copy. The fixed costs for publishing the book are $35,000. How many books must the company publish and sell in order to have a profit of at least $15,000 if each book sells for $25?

☆ Stretching the Topics _____

_____ **1.** Mr. Hickman bought a car, making a down payment of $1000 and financing the remainder. The finance charge is 8.8% per year on the amount financed. If he pays the finance company $15,548 in 48 equal monthly payments, what was the original price of the car?

_____ **2.** The length of a rectangle is 5 feet less than 3 times the width. The sides of an equilateral triangle are the same as the length of the rectangle. If the sum of the two perimeters is 94 feet, find the dimensions of the rectangle.

_____ **3.** The width of a rectangle is 4 centimeters more than one-third of the length. If the perimeter must be no more than 64 centimeters, find the possible lengths for the widths.

Check your answers in the back of your book.

If you can complete **Checkup 2.5**, you are ready to do the Review Exercises.

CHECKUP 2.5

_____ **1.** Irene has 16 feet of antique lace that she wishes to use as a border for a scarf. She would like the scarf to be 2 feet longer than it is wide. What must be the dimensions of the scarf?

_____ **2.** Two cars are 415 miles apart and are on Highway 65 traveling toward each other. One car is traveling at 55 miles per hour and the other is traveling at 65 miles per hour. How long will each car travel before they meet?

_____ **3.** The Little Tot's Babysitting Service charges $5 for the first two hours plus $2.75 an hour for each additional hour. For how many hours may Mrs. Hickman shop and keep her total babysitting charge under $25?

_____ **4.** In the first four basketball games of the season, Lillie scored 23, 19, 26, and 22 points. How many points must she score during the fifth game to have a scoring average of at least 25 points per game?

Check your answers in the back of your book.

If You Missed Problem:	You Should Review Examples:
1	3
2	5
3	6
4	2

Summary

In this chapter we learned to find values of the variable that satisfy first-degree equations or inequalities.

A **first-degree equation** is an equation that can be written in the form

$$ax + b = c$$

When a first-degree equation does not appear in this form we learned to follow these steps to find its solution.

1. Remove fractions by multiplying both sides by the LCD.
2. Remove parentheses and combine like terms.
3. Get variable terms on one side and constant terms on the other by addition or subtraction.
4. Solve for the variable by multiplication and/or division.

First-degree inequalities are inequalities of the form

$$ax + b < c \qquad ax + b > c$$
$$ax + b \leq c \qquad ax + b \geq c$$

To solve **first-degree inequalities**, we use the same steps as those used for solving equations, *except* that when we multiply or divide both sides by a *negative* number, in which case we must *reverse* the direction of the inequality. We learned to graph the solutions of inequalities on the number line.

Finally, we discussed *absolute-value equations and inequalities* and learned to solve them by rewriting them as equivalent statements.

$$|x| = a \qquad \text{means} \qquad x = a \ \text{ or } \ x = -a$$
$$|x| < a \qquad \text{means} \qquad -a < x < a$$
$$|x| > a \qquad \text{means} \qquad x < -a \ \text{ or } \ x > a$$

❑ Speaking the Language of Algebra ─────────────────────

Complete each sentence with the appropriate word or phrase.

1. In a first-degree equation, the variable appears with an exponent of _____ .

2. In solving a first-degree equation, our aim is to _____ _____ _____ .

3. If an equation or inequality contains parentheses, we first _____ _____ and _____ _____ _____ .

4. When graphing the solutions for the inequality $x < a$, there will be a(n) _____ dot at a, and a line extending to the _____ .

5. To say that x is between a and b (but not including a or b) we may write the inequality _____ .

6. When seeking the _____ of the solutions for two inequalities, we look for solutions to one inequality *and* the other inequality. When seeking the _____ of the solutions of two inequalities, we look for the solutions to one inequality *or* the other inequality.

7. If we multiply or divide both sides of an inequality by a *positive* number, the direction of the inequality is _____ . If we multiply or divide both sides of an inequality by a *negative* number, the direction of the inequality is _____ .

8. For the equation $|x| = a$, the solutions are _____ or _____ .
   For the inequality $|x| \leq a$, the solutions are _____ .
   For the inequality $|x| \geq a$, the solutions are _____ or _____ .

REVIEW EXERCISES for Chapter 2: Solving First-Degree Equations and Inequalities

Solve and check.

_____ **1.** $3x - 5 = 16$

_____ **2.** $7 - 4x = 31$

_____ **3.** $\dfrac{x}{6} + 4 = -3$

_____ **4.** $9 + \dfrac{x}{5} = 0$

_____ **5.** $-4x + 2 = 5x + 29$

_____ **6.** $2(x - 4) = -3(2x + 4)$

_____ **7.** $\dfrac{3x - 7}{3} = 6x - 5$

_____ **8.** $4 + \dfrac{5x - 1}{4} = 2x$

_____ **9.** $\dfrac{2x}{7} - 9 = 5x - \dfrac{6}{7}$

_____ **10.** $5(2y - 1) + 3 = 4(2 - y)$

_____ **11.** $4[x - 2(x - 3)] = 3(2x + 8)$

_____ **12.** $F = \dfrac{9}{5}C + 32$ for C

Graph each of the following inequalities on a number line.

13. $m \geq \dfrac{8}{3}$

14. $x < -1$

Find the solution of each of the following by graphing on a number line.

_____ **15.** $x \geq -5$ and $x < 2$

_____ **16.** $x < -3$ and $x \geq -2$

_____ **17.** $x > -1$ or $x > 4$

_____ **18.** $x \geq -5$ or $x < 0$

Solve and graph.

_____ **19.** $\dfrac{-3}{5}x + 2 \geq 8$

_____ **20.** $4(x - 5) < -16$

_____ **21.** $5x - (x + 2) < 5x + 3$

_____ **22.** $3(2y - 1) - 5 \geq 4(y - 3) + 5$

_____ **23.** $5x - 6 < 4 - x$ and $3x - 1 \geq 5x + 3$

_____ **24.** $\dfrac{x - 1}{3} < 5 - x$ and $\dfrac{2x}{3} + 5 > 7$

_____ **25.** $3(2x - 1) \le 7x + 2$ or $2(3x - 1) > 7x + 5$

_____ **26.** $\dfrac{2x - 6}{3} < 3 - x$ or $\dfrac{5x}{3} + 6 < 1$

_____ **27.** $3 \le x - 4 \le 9$

_____ **28.** $-2 \le \dfrac{4 - x}{3} < 0$

Solve.

_____ **29.** $|2x - 3| = 7$

_____ **30.** $|9 - x| = 4$

_____ **31.** $\left|\dfrac{2}{3}x + 4\right| = -9$

_____ **32.** $|3x - 2| \le 13$

_____ **33.** $\left|\dfrac{x}{5} - 3\right| < -1$

_____ **34.** $|2x - 3| > 11$

_____ **35.** $|3x - 7| > -2$

_____ **36.** $\left|9 - \dfrac{3x}{2}\right| \ge 6$

_____ **37.** Mickey sells family-size sailboats. He earns $360 for each one he sells. How many sailboats can he sell each month so that he will have a monthly salary of at least $1600?

_____ **38.** R. J. has two amounts of money invested in stocks. The first amount is $5000 more than the second amount. If the first amount earns 12% interest and the other earns 14% interest, find the amount invested at each rate if the total interest from the two sums last year was $3720.

_____ **39.** Two cars leave from the same point on the interstate traveling in opposite directions. One car is traveling 10 miles per hour faster than the other. If in 4 hours they will be 400 miles apart, at what rate of speed is each traveling?

_____ **40.** Mr. McDonald has 120 feet of fencing to build a loading pen for cattle. The pen must be three times as long as it is wide. Find the largest dimensions the loading pen can have.

Check your answers in the back of your book.

If You Missed Exercises:	You Should Review Examples:	
1, 2	Section 2.1	5
3, 4		6
5		7, 8
6		10
7, 8		13–16
9		11, 12
10, 11		9, 10
12		19
13, 14	Section 2.2	3, 4
15, 16		7–9
17, 18		12–14
19–22	Section 2.3	5–9
23, 24		10, 11
25, 26		12, 13
27, 28		14, 15
29–31	Section 2.4	2–5
32, 33		8–10
34–36		13–17
37–40	Section 2.5	1–6

If you have completed the Review Exercises and corrected your errors, you are ready to take the Practice Test.

PRACTICE TEST for Chapter 2:
Solving First-Degree Equations and Inequalities

Solve.

		SECTION	EXAMPLES
_____	**1.** $10 - 7x = -4$	2.1	5
_____	**2.** $5y - 2 = 8 - 5y$	2.1	7, 8
_____	**3.** $\dfrac{3x}{2} - 1 = 5$	2.1	14
_____	**4.** $2(5a - 1) - 6 = -5(2 - a)$	2.1	9
_____	**5.** $\dfrac{7x - 3}{4} + 1 = 4x$	2.1	16
_____	**6.** $A = \dfrac{h(a + b)}{2}$ for b	2.1	20
_____	**7.** $-4x + 12 \leq 4$	2.3	5
_____	**8.** $6(y - 3) + 5 > 4(1 - y) + 3$	2.3	9
_____	**9.** $0 \leq \dfrac{8x + 1}{5} < 5$	2.3	15
_____	**10.** $|4x - 7| = 2$	2.4	3
_____	**11.** $\left|\dfrac{5 - 2x}{5}\right| = 3$	2.4	5
_____	**12.** $|1 - 2x| < 11$	2.4	10
_____	**13.** $|3x + 2| \geq 5$	2.4	14

Solve by graphing.

		SECTION	EXAMPLES
_____	**14.** $x \leq -2$ and $x > -6$	2.2	8
_____	**15.** $x < 7$ or $x \leq 0$	2.2	12
_____	**16.** $2x - 5 < 3$ and $\dfrac{5 - x}{2} < 4$	2.3	10
_____	**17.** $5(x + 2) < 2x + 7$ or $4 - (x - 3) \leq 3x - 5$	2.3	13
_____	**18.** When Janice rides her bike to work each morning, her rate is 2 miles per hour faster than her rate when she rides home in the evening. If her morning ride takes 1 hour and her evening ride takes $1\frac{1}{4}$ hours, find Janice's rate in the evening.	2.5	5

_____ **19.** The Zeta Zeta fraternity needs $2500 for repairs on its fraternity house. If the members have $1630 in the treasury and an alumnus donates a stereo to be raffled, what is the least number of raffle tickets they must sell at $2 each in order to finance the repairs? 2.5 6

_____ **20.** George has purchased a painting which is 13 inches long and 5 inches wide. What width of framing can he use for the painting if he wishes the framed length to be no more than twice the framed width? 2.5 3

Working with Exponents and Polynomials 3

Marco's rectangular garden is 10 feet longer than it is wide. If he decreases both the length and width of the garden by 3 feet, what variable expression represents the amount of garden area that he will lose?

So far we have dealt with algebraic expressions, equations, and inequalities involving variables raised only to the first power. Now we shall discover how to work with expressions containing variables raised to any integer powers. In this chapter we learn how to:

1. Use the laws of exponents.
2. Identify and simplify polynomials.
3. Operate with polynomials.
4. Switch from word expressions to polynomials.

3.1 **Working with Whole-Number Exponents**

Using the Definition of Exponent

In Chapter 1 we learned that a raised number, called an **exponent,** tells us how many times to use its base as a factor in a product.

Exponent Form		*Factor Form*
3^2	$=$	$3 \cdot 3$
$(-1)^3$	$=$	$(-1)(-1)(-1)$
x^4	$=$	$x \cdot x \cdot x \cdot x$

In general, we agreed that if a is a real number and n is a natural number:

Definition of Exponent

$$a^n = \underbrace{a \cdot a \cdot a \cdot \ldots \cdot a}_{n \text{ factors}}$$

Here a is the **base,** n is the **exponent,** and a^n is called the **nth power of a.** Notice that an exponent applies only to the base to which it is attached unless parentheses indicate otherwise. In other words,

$$2x^3 = 2 \cdot x \cdot x \cdot x \quad \text{but} \quad (2x)^3 = (2x)(2x)(2x) = 8x^3$$

$$-a^2 = -1 \cdot a \cdot a \quad \text{but} \quad (-a)^2 = (-a)(-a) = a^2$$

Try Example 1.

Example 1. Evaluate $x^3 - 5x$ when $x = -2$.

Solution: If $x = -2$, then

$$x^3 - 5x = (\underline{})^3 - 5(\underline{})$$

Check your work on page 169. ▶

Example 2. Evaluate $-x^4 + 2x^2 - x$ when $x = 3$.

Solution: If $x = 3$, then

$$-x^4 + 2x^2 - 2x = -(3)^4 + 2(3)^2$$
$$- 2(3)$$
$$= -81 + 2(9) - 6$$
$$= -81 + 18 - 6$$
$$= -69$$

When a base is written without an exponent, recall that the exponent is "understood" to be 1. You might wonder whether *zero* can ever be used as an exponent. The answer is yes. We shall see shortly how the following definition for zero used as an exponent fits in neatly with the rules for working with exponents.

Definition of Zero Exponent

$$a^0 = 1 \qquad (a \neq 0)$$

Any nonzero base raised to the zero power is equal to 1. Assuming that variable bases are not zero, you try Example 3.

Example 3. Simplify $x^0 + 3^0$.

Solution

$$x^0 + 3^0$$

$$x + 1$$

Example 4. Simplify $(x + 3)^0$.

Solution: Parentheses indicate that $x + 3$ is the base and 0 is the exponent.

$$(x + 3)^0 = 1$$

Check your work on page 169. ▶

Using the Laws of Exponents

In algebra it is often necessary to multiply several powers of the same base. We need a technique for finding a product such as $x^3 \cdot x^5$. Let's write each power as a product of factors. Notice that x appears as a factor eight times on the right.

$$x^3 \cdot x^5 = \overbrace{x \cdot x \cdot x}^{x^3} \cdot \overbrace{x \cdot x \cdot x \cdot x \cdot x}^{x^5}$$

$$= x^8$$

The new exponent 8 is just the *sum* of the exponents in the original product. Will this always work? Let's use two approaches to find another product.

First Approach	*Second Approach*
$x \cdot x^4 \cdot x^2$	$x \cdot x^4 \, x^2$
$= x \cdot x \cdot x \cdot x \cdot x \cdot x \cdot x$	$= x^1 \cdot x^4 \cdot x^2$
$= x^7$	$= x^{1+4+2}$
	$= x^7$

Since our answers are the same using either approach, we are ready to generalize this rule, called the **first law of exponents.**

To *multiply* powers of the same base, we keep the same base and *add* the exponents.

Notice that we apply this law only when the bases are the same.

First Law of Exponents

$$a^m \cdot a^n = a^{m+n}$$

Example 5. Simplify $(xy^2)(xy^3)$.

Solution: We use the commutative and associative properties to rewrite this product with powers of like bases next to each other.

$$(xy^2)(xy^3) = x \cdot x \cdot y^2 \cdot y^3$$
$$= x^1 \cdot x^1 \cdot y^2 \cdot y^3$$
$$= x^{1+1} \cdot y^{2+3}$$
$$= x^2y^5$$

You complete Example 6.

Example 6. Simplify $3x^2(4xy)$.

Solution: First we rearrange the factors.

$$3x^2(4xy) = 3 \cdot 4 \cdot x^2 \cdot x \cdot y$$

Check your work on page 169. ▶

Example 7. Simplify

$$(-5a^2b)(2ab^3)\left(\frac{1}{2}a^4b\right).$$

Solution

$$(-5a^2b)(2ab^3)\left(\frac{1}{2}a^4b\right)$$

$$= -5 \cdot 2 \cdot \frac{1}{2} \cdot a^2 \cdot a \cdot a^4 \cdot b \cdot b^3 \cdot b$$

$$= -5a^7b^5$$

Now you try Example 8.

Example 8. Simplify $a^7 \cdot a^0$.

Solution

$$a^7 \cdot a^0$$

Check your work on page 169. ▶

Example 8 should help you see why the definition $a^0 = 1$ makes sense. Here we multiplied a^7 times a^0 and our answer was a^7. The only real number that leaves every other number unchanged in multiplication is 1, the multiplicative identity.

IIII➡ Trial Run ━━━

_____ **1.** Evaluate $a^4 + 6$ when $a = -3$.

_____ **2.** Evaluate $y^5 - 3y^2 + y$ when $y = -1$.

Simplify.

_____ **3.** $5a^0$

_____ **4.** $(5a)^0$

_____ **5.** $2x^2(5xy)$

_____ **6.** $(4ab)(a^4b^5)$

_____ **7.** $(-2x^2y)(3xy^3)\left(\frac{1}{3}x^4y\right)$

_____ **8.** $7x^3y^2z^0(-2xy)(-3yz)$

Answers are on page 170.

Suppose that we use the first law of exponents to simplify the following expressions.

$(x^2)^3$	$(y^5)^2$
The base is x^2.	The base is y^5.
The exponent is 3.	The exponent is 2.
$(x^2)^3$	$(y^5)^2$
$= x^2 \cdot x^2 \cdot x^2$	$= y^5 \cdot y^5$
$= x^{2+2+2}$	$= y^{5+5}$
$= x^6$	$= y^{10}$

This procedure is called **raising a power to a power.** Notice how the exponent in each answer is related to the exponents in the original problem.

To *raise a power* of a base *to a power*, we keep the same base and *multiply* the exponent.

The **second law of exponents** describes this procedure.

Second Law of Exponents

$$(a^m)^n = a^{m \cdot n}$$

Example 9. Simplify $(x^7)^4$.

Solution

$$(x^7)^4$$
$$= x^{7 \cdot 4}$$
$$= x^{28}$$

You try Example 10.

Example 10. Simplify $(a^6)^5$.

Solution

$$(a^6)^5$$

Check your work on page 170. ▶

Suppose that we wish to find a **power of a product,** as in the following expressions.

$(2y)^4$	$(x^2y)^3$
The base is $2y$.	The base is x^2y.
The exponent is 4.	The exponent is 3.
$(2y)^4$	$(x^2y)^3$
$= (2y)(2y)(2y)(2y)$	$= (x^2y)(x^2y)(x^2y)$
$= 2 \cdot 2 \cdot 2 \cdot 2 \cdot y \cdot y \cdot y \cdot y$	$= x^2 \cdot x^2 \cdot x^2 \cdot y \cdot y \cdot y$
$= 2^4 \cdot y^4$	$= x^6 y^3$
$= 16y^4$	

Can you see a pattern in these examples?

> To raise a *product* to a power, we raise *each factor* to that power.

This is the **third law of exponents** and it can be stated in general as:

Third Law of Exponents
$(a \cdot b)^n = a^n \cdot b^n$

Example 11. Simplify $(-3x)^4$.

Solution

$$(-3x)^4$$
$$= (-3)^4 x^4$$
$$= 81x^4$$

Now you try Example 12.

Example 12. Simplify $(4x^2y^3)^3$.

Solution

$$(4x^2y^3)^3$$
$$= 4^3(x^2)^3(y^3)^3$$

Check your work on page 170. ▶

Student errors in raising products to powers often occur in raising *constant* factors to powers too quickly. You may avoid mistakes if you first write the constant with the appropriate exponent as we have done in our examples.

Example 13. Simplify $(-a^4bc^2)^5$.

Solution

$$(-a^4bc^2)^5$$
$$= (-1 \cdot a^4bc^2)^5$$
$$= (-1)^5(a^4)^5(b^1)^5(c^2)^5$$
$$= -1 \cdot a^{20} \cdot b^5 \cdot c^{10}$$
$$= -a^{20}b^5c^{10}$$

Now complete Example 14.

Example 14. Simplify $(-a^4bc^2)^6$.

Solution

$$(-a^4bc^2)^6$$
$$= (-1 \cdot a^4bc^2)^6$$
$$= (-1)^6(a^4)^6(b^1)^6(c^2)^6$$

$$a^{24}b^6c^{12}$$

Check your work on page 170. ▶

We can use the laws of exponents to simplify more complicated algebraic expressions provided that we review the rules for the order of operations. Remember that we deal with symbols of inclusion (parentheses) first; then we deal with powers, followed by multiplications and/or divisions from left to right, and finally we perform addition and/or subtraction from left to right.

Example 15. Simplify $x^2(xy)^3$.

Solution

$$x^2(xy)^3$$
$$= x^2 \cdot x^3y^3$$
$$= x^5y^3$$

Example 16. Simplify $3a(a^2b)^2 + a^2(a^3b)$.

Solution

$$3a(a^2b)^2 + a^2b(a^3b)$$
$$= 3a(a^2)^2b^2 + a^2b(a^3b)$$
$$= 3a \cdot a^4 \cdot b^2 + a^2 \cdot a^3 \cdot b \cdot b$$
$$= 3a^5b^2 + a^5b^2$$
$$= 4a^5b^2$$

Notice that we combined like terms in the final step.

⫸ **Trial Run**

Simplify.

_____ **1.** $(x^5)^2$ _____ **2.** $(3a)^4$

_____ **3.** $(-5x^3y^4)^2$ _____ **4.** $(-2a^2b^3c)^3$

_____ **5.** $y^2(xy)^5$ _____ **6.** $(2xy)^2 + 3x(xy^2)$

Answers are on page 170.

Since we have learned to raise a product to a power by raising each factor to that power, it seems logical to consider a method for **raising a fraction to a power.**

$\left(\dfrac{x}{y}\right)^2$	$\left(\dfrac{-2}{a}\right)^3$
The base is $\dfrac{x}{y}$.	The base is $\dfrac{-2}{a}$.
The exponent is 2.	The exponent is 3.
$\left(\dfrac{x}{y}\right)^2$	$\left(\dfrac{-2}{a}\right)^3$
$= \dfrac{x}{y} \cdot \dfrac{x}{y}$	$= \dfrac{-2}{a} \cdot \dfrac{-2}{a} \cdot \dfrac{-2}{a}$
$= \dfrac{x \cdot x}{y \cdot y}$	$= \dfrac{(-2)(-2)(-2)}{a \cdot a \cdot a}$
$= \dfrac{x^2}{y^2}$	$= \dfrac{(-2)^3}{a^3}$
	$= \dfrac{-8}{a^3}$

It seems that we could have arrived at these answers more easily just by raising the numerator and denominator to the original power.

To raise a *fraction* to a power, we raise the *numerator* to the power and raise the *denominator* to the power.

Called the **fourth law of exponents,** this rule can be stated in general as follows:

Fourth Law of Exponents

$$\left(\frac{a}{b}\right)^n = \frac{a^n}{b^n} \qquad (b \neq 0)$$

Notice that b cannot be zero because it appears as the denominator and division by zero is undefined.

Example 17. Simplify $\left(\dfrac{2x^2}{5y}\right)^3$.

Solution

$$\left(\frac{2x^2}{5y}\right)^3$$

$$= \frac{(2x^2)^3}{(5y)^3}$$

$$= \frac{2^3(x^2)^3}{5^3y^3}$$

$$= \frac{8x^6}{125y^3}$$

You complete Example 18.

Example 18. Simplify $\left(\dfrac{-xy^3}{2z}\right)^4$.

Solution

$$\left(\frac{-xy^3}{2z}\right)^4$$

$$= \frac{(-1 \cdot xy^3)^4}{(2z)^4}$$

$$= \frac{(-1)^4x^4(y^3)^4}{2^4z^4}$$

Check your work on page 170. ▶

▸ **Trial Run**

Simplify.

_____ 1. $\left(\dfrac{-1}{x}\right)^5$

_____ 2. $\left(\dfrac{3x^2}{2y}\right)^2$

_____ 3. $\left(\dfrac{-ab^2}{3c}\right)^4$

_____ 4. $\left(\dfrac{-3x^0y}{5z^2}\right)^3$

_____ 5. $\left(\dfrac{7a^3b^5}{9c^2}\right)^0$

_____ 6. $\left(\dfrac{x^2y}{-z^3}\right)^7$

Answers are on page 170.

Now we must learn to simplify quotients in which the numerator and denominator contain powers of the same base. Consider the following quotients.

$$\frac{x^5}{x^2}$$

$$= \frac{x \cdot x \cdot x \cdot x \cdot x}{x \cdot x}$$

$$= \frac{x}{x} \cdot \frac{x}{x} \cdot x \cdot x \cdot x$$

$$= 1 \cdot 1 \cdot x \cdot x \cdot x$$

$$= x^3$$

$$\frac{a^3y^4}{ay^3}$$

$$= \frac{a^3}{a} \cdot \frac{y^4}{y^3}$$

$$= \frac{a \cdot a \cdot a}{a} \cdot \frac{y \cdot y \cdot y \cdot y}{y \cdot y \cdot y}$$

$$= \frac{a}{a} \cdot a \cdot a \cdot \frac{y}{y} \cdot \frac{y}{y} \cdot \frac{y}{y} \cdot y$$

$$= 1 \cdot a \cdot a \cdot 1 \cdot 1 \cdot 1 \cdot y$$

$$= a^2y$$

Notice how the exponent on each variable in the answers is related to the exponents in the original expressions.

> To *divide* powers of the same base, we keep the same base and *subtract* the exponent in the denominator from the exponent in the numerator.

This rule for quotients is the **fifth law of exponents.**

> **Fifth Law of Exponents**
>
> $$\frac{a^m}{a^n} = a^{m-n} \qquad (a \neq 0)$$

When variables appear with numerical coefficients, we divide those numbers just as we have always done and then turn our attention to simplifying the variable parts by the fifth law of exponents.

Example 19. Simplify $\frac{6x^2y}{3xy}$.

Solution

$$\frac{6x^2y}{3xy}$$

$$= \frac{6}{3} \cdot \frac{x^2}{x} \cdot \frac{y}{y}$$

$$= 2 \cdot x^{2-1} \cdot y^{1-1}$$

$$= 2 \cdot x^1 \cdot y^0$$

$$= 2 \cdot x \cdot 1$$

$$= 2x$$

Now try Example 20.

Example 20. Simplify $\frac{3a^4b^2c}{2ab}$.

Solution

$$\frac{3a^4b^2c}{2ab}$$

$$= \frac{3}{2} \cdot \frac{a^4}{a} \cdot \frac{b^2}{b} \cdot \frac{c}{1}$$

Check your work on page 170. ▶

⇒ Trial Run

Simplify.

_____ **1.** $\frac{x^7}{x^3}$

_____ **2.** $\frac{-8a^5}{2a^3}$

_____ **3.** $\frac{x^5y^3}{x^3y}$

_____ **4.** $\frac{-9x^3y^5z^2}{9xy^5z^0}$

_____ **5.** $\frac{5a^3b^7}{15ab^2}$

_____ **6.** $\frac{7a^4b^2c}{5a^3b}$

Answers are on page 170.

The ability to work with exponents will be important throughout your study of algebra. We summarize the fundamental definitions and laws of exponents here for your reference. Assume that all variables represent real numbers and that n is a whole number.

Definitions

$$a^n = \underbrace{a \cdot a \cdot a \cdot \ldots \cdot a}_{n \text{ factors}}$$

$$a^0 = 1 \qquad (a \neq 0)$$

Laws of Exponents

(1) $a^m \cdot a^n = a^{m+n}$

(2) $(a^m)^n = a^{m \cdot n}$

(3) $(ab)^n = a^n b^n$

(4) $\left(\dfrac{a}{b}\right)^n = \dfrac{a^n}{b^n} \qquad b \neq 0$

(5) $\dfrac{a^m}{a^n} = a^{m-n} \qquad a \neq 0$

▶ Examples You Completed

Example 1. Evaluate $x^3 - 5x$ when $x = -2$.

Solution: If $x = -2$, then

$$x^3 - 5x = (-2)^3 - 5(-2)$$
$$= -8 + 10$$
$$= 2$$

Example 3. Simplify $x^0 + 3^0$.

Solution

$$x^0 + 3^0$$
$$= 1 + 1$$
$$= 2$$

Example 6. Simplify $3x^2(4xy)$.

Solution: First we rearrange the factors.

$$3x^2(4xy) = 3 \cdot 4 \cdot x^2 \cdot x \cdot y$$
$$= 3 \cdot 4 \cdot x^2 \cdot x^1 \cdot y$$
$$= 12x^{2+1} \cdot y$$
$$= 12x^3 y$$

Example 8. Simplify $a^7 \cdot a^0$.

Solution

$$a^7 \cdot a^0$$
$$= a^{7+0}$$
$$= a^7$$

Example 10. Simplify $(a^6)^5$.

Solution

$$(a^6)^5$$
$$= a^{6 \cdot 5}$$
$$= a^{30}$$

Example 14. Simplify $(-a^4bc^2)^6$.

Solution

$$(-a^4bc^2)^6$$
$$= (-1 \cdot a^4bc^2)^6$$
$$= (-1)^6(a^4)^6(b^1)^6(c^2)^6$$
$$= 1 \cdot a^{24}b^6c^{12}$$
$$= a^{24}b^6c^{12}$$

Example 18. Simplify $\left(\dfrac{-xy^3}{2z}\right)^4$.

Solution

$$\left(\frac{-xy^3}{2z}\right)^4$$
$$= \frac{(-1 \cdot xy^3)^4}{(2z)^4}$$
$$= \frac{(-1)^4x^4(y^3)^4}{2^4z^4}$$
$$= \frac{1 \cdot x^4y^{12}}{16z^4}$$
$$= \frac{x^4y^{12}}{16z^4}$$

Example 12. Simplify $(4x^2y^3)^3$.

Solution

$$(4x^2y^3)^3$$
$$= 4^3(x^2)^3(y^3)^3$$
$$= 4^3x^{2 \cdot 3}y^{3 \cdot 3}$$
$$= 64x^6y^9$$

Example 20. Simplify $\dfrac{3a^4b^2c}{2ab}$.

Solution

$$\frac{3a^4b^2c}{2ab}$$
$$= \frac{3}{2} \cdot \frac{a^4}{a} \cdot \frac{b^2}{b} \cdot c$$
$$= \frac{3}{2} \cdot a^{4-1} \cdot b^{2-1} \cdot c$$
$$= \frac{3}{2} \cdot a^3 \cdot b^1 \cdot c$$
$$= \frac{3a^3bc}{2}$$

Answers to Trial Runs

page 162   **1.** 87   **2.** -5 **3.** 5   **4.** 1   **5.** $10x^3y$ **6.** $4a^5b^6$ **7.** $-2x^7y^5$ **8.** $42x^4y^4z$

page 165   **1.** x^{10} **2.** $81a^4$ **3.** $25x^6y^8$ **4.** $-8a^6b^9c^3$ **5.** x^5y^7 **6.** $7x^2y^2$

page 167   **1.** $\dfrac{-1}{x^5}$ **2.** $\dfrac{9x^4}{4y^2}$ **3.** $\dfrac{a^4b^8}{81c^4}$ **4.** $\dfrac{-27y^3}{125z^6}$ **5.** 1   **6.** $\dfrac{-x^{14}y^7}{z^{21}}$

page 168   **1.** x^4 **2.** $-4a^2$ **3.** x^2y^2 **4.** $-x^2z^2$ **5.** $\dfrac{a^2b^5}{3}$ **6.** $\dfrac{7abc}{5}$

EXERCISE SET 3.1

———— **1.** Evaluate $x^5 - 2x^3 - x$ when $x = 3$.

———— **2.** Evaluate $x^3 - 4x^2 - 3x$ when $x = 5$.

———— **3.** Evaluate $-a^3 + 3a$ when $a = -2$.

———— **4.** Evaluate $-a^5 + 5a$ when $a = -1$.

———— **5.** Simplify $(x + 5)^0$.

———— **6.** Simplify $(x - 7)^0$.

———— **7.** Simplify $a^0 + 3^0$.

———— **8.** Simplify $a^0 - 9^0$.

Simplify using the laws of exponents.

———— **9.** $(-5x^7)(4x)$

———— **10.** $(-7x^5)(9x)$

———— **11.** $(4x^2y^3)(-3x^5y^2)$

———— **12.** $(8x^3y^4)(-2x^2y^5)$

———— **13.** $3x(-x^2y)(-2x^4y^3)$

———— **14.** $-5x(-x^3y)(4x^4y^3)$

———— **15.** $(7xy^2)(-2xy)(-3x^2y)$

———— **16.** $(-2xy^2)(8x^2y^3)(-3xy)$

———— **17.** $(5a^2b^3c)(7abc^4)$

———— **18.** $(9a^2b^3c^5)(3abc^2)$

———— **19.** $(7x^0yz^5)(-6x^2y^3)\left(-\dfrac{1}{3}y^4z^2\right)$

———— **20.** $(8x^2y^0z)(-5x^5y^3)\left(-\dfrac{1}{4}y^5z^2\right)$

———— **21.** $(-8y^2)^2$

———— **22.** $(-5y^3)^2$

———— **23.** $(-2x^2y^3)^5$

———— **24.** $(-3x^4y^3)^3$

———— **25.** $5x(2x^3)^2$

———— **26.** $6x(3x^4)^2$

———— **27.** $9x(7x^5y^2)^0$

———— **28.** $11x(5x^4y^3)^0$

———— **29.** $(5x^4)^2(-2y^5)$

———— **30.** $(2x^2)^3(-y)^4$

———— **31.** $(-6a^2b)^2(2ab^3)^3$

———— **32.** $(-2a^2b)^4(3a^4b^2)^3$

———— **33.** $3a(-2a^4b)^2(a^5b^2)^3$

———— **34.** $5a(-5a^7b)^2(a^2b^3)^4$

———— **35.** $(7x^2yz^5)(-4x^5y^2z)^3$

———— **36.** $(5xy^2z^3)(-3x^4yz^2)^3$

———— **37.** $xy^2(3xy^4) + 2y^3(4x^2y^3)$

———— **38.** $x^2y^3(5x^2y^4) - 3y^5(x^4y^2)$

———— **39.** $-a^2(ab^3)^4 + 3a^2b^5(a^{10}b)$

———— **40.** $-a^5(ab^2)^3 + 7a^4b^4(ab^2)^2$

———— **41.** $\left(-\dfrac{1}{3y}\right)^2$

———— **42.** $\left(-\dfrac{1}{4y}\right)^3$

———— **43.** $\left(\dfrac{5a^2}{3b}\right)^3$

———— **44.** $\left(\dfrac{7a^3}{2b^2}\right)^2$

———— **45.** $\left(\dfrac{-ab^5}{2c^2}\right)^4$

———— **46.** $\left(\dfrac{-a^2b}{3c^3}\right)^4$

_____ 47. $\left(\dfrac{-2x^0y^3}{3z^5}\right)^4$

_____ 48. $\left(\dfrac{-5x^0y^5}{7z^6}\right)^2$

_____ 49. $\left(\dfrac{11x^3y^5}{2z}\right)^0$

_____ 50. $\left(\dfrac{9x^5y^3}{5z^2}\right)^0$

_____ 51. $\dfrac{x^9}{x^5}$

_____ 52. $\dfrac{x^{11}}{x^7}$

_____ 53. $\dfrac{-15a^7}{5a^3}$

_____ 54. $\dfrac{-16a^8}{8a^5}$

_____ 55. $\dfrac{x^9y^4}{x^3y^2}$

_____ 56. $\dfrac{x^{11}y^5}{x^2y^2}$

_____ 57. $\dfrac{-3x^4y^5z^3}{3xy^2z^2}$

_____ 58. $\dfrac{-7x^5y^7z^9}{7x^4y^2z^3}$

_____ 59. $\dfrac{3a^5b^2}{21ab^2}$

_____ 60. $\dfrac{4a^6b^3}{28a^5b^3}$

_____ 61. $\dfrac{9a^6b^5c}{6a^0b^3}$

_____ 62. $\dfrac{12a^7b^4c^2}{8a^3b^0}$

_____ 63. $\left(\dfrac{2a^2b^3c}{ab^2}\right)^3$

_____ 64. $\left(\dfrac{5a^5b^4c}{a^3b^3}\right)^2$

_____ 65. $\left(\dfrac{7ab^3}{b^2}\right)^2\left(\dfrac{a^2b}{a}\right)^3$

_____ 66. $\left(\dfrac{9a^4b^2}{ab}\right)^2\left(\dfrac{a^4b^5}{a^3b^2}\right)^3$

☆ Stretching the Topics _____

_____ 1. Simplify $(-4ab^2c)^3(-ab^3c)^2(-2a^2b^5c)^0$ and evaluate when $a = -1$, $b = 2$, and $c = 1$.

_____ 2. Simplify $(x^{n-4}y)^3(x^6y^{n+4})^2$.

_____ 3. Find the value of a that will make this equation true:
$$(y^{3a+5})(y^a)^4 = y^{8a+12}$$

Check your answers in the back of your book.

If you can complete **Checkup 3.1,** you are ready to go on to Section 3.2.

CHECKUP 3.1

_____ **1.** Evaluate $y^3 - 2y + 1$ when $y = -4$.

_____ **2.** Simplify $12a^0 + 1$.

Simplify using the laws of exponents.

_____ **3.** $3x(-7x^2y)(-2x^5y^2)$

_____ **4.** $(9a^3b^4c^2)(-5a^2b^2c)$

_____ **5.** $(-11y^3)^2$

_____ **6.** $(5x^7)^2(-3x^4)$

_____ **7.** $\left(\dfrac{-9a^2}{3b^3}\right)^3$

_____ **8.** $\left(\dfrac{4a^3b^0}{3c^2}\right)^2$

_____ **9.** $\dfrac{27a^7b^3}{8a^2b^3}$

_____ **10.** $\dfrac{15a^7b^5c}{-5a^3b}$

Check your answers in the back of your book.

If You Missed Problems:	You Should Review Examples:
1	1
2	5
3, 4	5–8
5	11
6	15
7, 8	17, 18
9, 10	19, 20

3.2 **Working with Negative Exponents**

Defining Negative Exponents

Let's use two approaches to simplify the expression

$$\frac{x^2}{x^6}$$

First Approach

$$\frac{x^2}{x^6} = \frac{x \cdot x}{x \cdot x \cdot x \cdot x \cdot x \cdot x}$$

$$= \frac{x}{x} \cdot \frac{x}{x} \cdot \frac{1}{x \cdot x \cdot x \cdot x}$$

$$= 1 \cdot 1 \cdot \frac{1}{x^4}$$

$$= \frac{1}{x^4}$$

Second Approach

Using the fifth law of exponents, we obtain

$$\frac{x^2}{x^6} = x^{2-6}$$

$$= x^{-4}$$

Since we know the first approach is correct and since we want the fifth law of exponents to work for every quotient, we must agree that our two answers mean the same thing.

$$x^{-4} \qquad \text{means} \qquad 1/x^4$$

Similarly, we can say:

Negative Exponent Form	Positive Exponent Form
x^{-3}	$\dfrac{1}{x^3}$
a^{-2}	$\dfrac{1}{a^2}$
2^{-1}	$\dfrac{1}{2}$

These examples should help us agree on the following definition for a negative exponent.

Definition of Negative Exponent

$$a^{-n} = \frac{1}{a^n} \qquad (a \neq 0)$$

Use this definition to complete Example 1.

Example 1. Evaluate 2^{-3}.

Solution

$$2^{-3} = \frac{1}{2^3}$$

$$= $$

Check your work on page 183. ▶

Example 2. Evaluate $(-2)^{-5}$.

Solution

$$(-2)^{-5} = \frac{1}{(-2)^5}$$

$$= \frac{1}{-32}$$

$$= \frac{-1}{32}$$

Now try Example 3.

Example 3. Evaluate $(-2)^{-4}$.

Solution

$$(-2)^{-4} = \frac{1}{(-2)^4}$$

$$= $$

Check your work on page 183. ▶

Example 4. Evaluate -2^{-4}.

Solution: Remember that an exponent applies only to the base to which it is attached, unless parentheses indicate otherwise.

$$-2^{-4} = -1 \cdot 2^{-4}$$

$$= -1 \cdot \frac{1}{2^4}$$

$$= -1 \cdot \frac{1}{16}$$

$$= \frac{-1}{16}$$

Notice from these examples that negative exponent does *not* affect the *sign* of the answer. The negative exponent merely says: "Rewrite the power as 1 divided by the same base raised to the opposite power." The sign of your answer will depend on the sign of the base and whether the final exponent is odd or even.

Example 5. Rewrite $x^{-3}y^{-2}$ without negative exponents.

Solution

$$x^{-3}y^{-2}$$

$$= \frac{1}{x^3} \cdot \frac{1}{y^2}$$

$$= \frac{1}{x^3 y^2}$$

Example 6. Simplify $\dfrac{x^2 y}{x^3 y^5}$ and write the answer with positive exponents.

Solution

$$\frac{x^2 y}{x^3 y^5} = \frac{x^2}{x^3} \cdot \frac{y^1}{y^5}$$

$$= x^{2-3} \cdot y^{1-5}$$

$$= x^{-1} y^{-4}$$

$$= \frac{1}{x} \cdot \frac{1}{y^4}$$

$$= \frac{1}{xy^4}$$

⫸ Trial Run

Evaluate.

_____ **1.** 3^{-2}

_____ **2.** 15^{-1}

_____ **3.** $(-5)^{-3}$

_____ **4.** $(-4)^{-2}$

_____ **5.** -2^{-6}

_____ **6.** $-(-3)^{-3}$

Simplify and write answers with positive exponents.

_____ **7.** $a^{-2}b^{-3}$

_____ **8.** $x^{-4}yz^{-1}$

_____ **9.** $\dfrac{y^4}{y^{10}}$

_____ **10.** $\dfrac{x^3y^2}{x^4y^5}$

Answers are on page 184.

Using the Laws of Exponents

In simplifying expressions containing negative exponents, you will be relieved to know that

> Negative exponents obey all five laws of exponents.

When you simplify expressions with negative exponents, it is wise to *use the laws of exponents first.* If you are asked to express your answer using positive exponents only, *wait* until you have an answer and then make any necessary changes using the definition of a negative exponent.

Let's practice simplifying some expressions, writing answers using positive exponents. You try Example 7.

Example 7. Simplify $x^3 \cdot x^{-7} \cdot x^{-1}$.

Solution: Since the bases are the same, we find the product by _____ the exponents.

$$x^3 \cdot x^{-7} \cdot x^{-1}$$
$$= x^{3+(-7)+(-1)}$$

Check your work on page 183. ▶

Example 8. Simplify $(x^4y^{-2})(x^{-5}y^6)$.

Solution

$$(x^4y^{-2})(x^{-5}y^6)$$
$$= x^4 \cdot x^{-5} \cdot y^{-2} \cdot y^6$$
$$= x^{4+(-5)} \cdot y^{-2+6}$$
$$= x^{-1} \cdot y^4$$
$$= \frac{1}{x} \cdot y^4$$
$$= \frac{y^4}{x}$$

Now try Example 9.

Example 9. Simplify $(x^{-2})^4$.

Solution: To raise a power to a power, we _____ exponents.

$$(x^{-2})^4$$
$$= x^{-2 \cdot 4}$$

Check your work on page 183. ▶

Example 10. Simplify $(5x^{-3}y^{-1})^{-2}$.

Solution: To raise a product to a power, we raise each factor to that power.

$$(5x^{-3}y^{-1})^{-2}$$
$$= 5^{-2}(x^{-3})^{-2}(y^{-1})^{-2}$$
$$= 5^{-2}x^6 y^2$$
$$= \frac{1}{5^2} x^6 y^2$$
$$= \frac{x^6 y^2}{25}$$

Example 11. Simplify $\left(\dfrac{x^2 y^{-4}}{z^{-3}}\right)^{-1}$.

Solution: To raise a quotient to a power, we raise the numerator and denominator to that power.

$$\left(\frac{x^2 y^{-4}}{z^{-3}}\right)^{-1}$$
$$= \frac{(x^2)^{-1} \cdot (y^{-4})^{-1}}{(z^{-3})^{-1}}$$
$$= \frac{x^{-2} y^4}{z^3}$$
$$= \frac{1}{x^2} \cdot \frac{y^4}{z^3}$$
$$= \frac{y^4}{x^2 z^3}$$

Now you try Example 12.

Example 12. Simplify $\dfrac{x^{-8}y^4}{x^{-3}y^{-2}}$.

Solution: To divide powers of the same base, we _____ exponents.

$$\frac{x^{-8}y^4}{x^{-3}y^{-2}} = \frac{x^{-8}}{x^{-3}} \cdot \frac{y^4}{y^{-2}}$$
$$= x^{-8-(-3)} \cdot y^{4-(-2)}$$

Check your work on page 183. ▶

Notice that in each example we simplified the expression *first* by using the laws of exponents. We waited until we had our answer before we rewrote that answer using positive exponents.

||||➡ **Trial Run** ▬▬▬▬▬▬▬▬▬▬▬▬▬▬▬▬▬▬▬▬▬▬▬▬▬▬▬▬▬▬▬▬▬

Simplify and write answers with positive exponents.

_____ 1. $x^4 \cdot x^{-5} \cdot x^{-2}$

_____ 2. $(-3x^5 y^{-3})(2x^{-8} y^5)$

_____ 3. $(x^{-3})^2$

_____ 4. $(x^{-2}y^{-1})^{-4}$

_____ 5. $\left(\dfrac{4x^3 y^{-5}}{z^{-2}}\right)^{-1}$

_____ 6. $\dfrac{x^{-7}y^5}{x^{-2}y^{-3}}$

Answers are on page 184.

Until now we have not encountered an answer with a negative exponent in the denominator of a fraction. How can we express such answers using positive exponents? Let's see what we can do with the fraction $\dfrac{1}{x^{-3}}$.

$$\frac{1}{x^{-3}} = 1 \div x^{-3} \qquad \text{Meaning of fraction bar.}$$

$$= 1 \div \frac{1}{x^3} \qquad \text{Definition of negative exponent.}$$

$$= \frac{1}{1} \cdot \frac{x^3}{1} \qquad \text{Rule for dividing fractions.}$$

$$= \frac{x^3}{1} \qquad \text{Rule for multiplying fractions.}$$

$$= x^3 \qquad \text{Simplifying.}$$

We have discovered that

$$\frac{1}{x^{-3}} = x^3$$

and similarly we can say:

Negative Exponent Form	Positive Exponent Form
$\dfrac{1}{a^{-2}}$	a^2
$\dfrac{1}{2^{-4}}$	2^4
$\dfrac{1}{y^{-1}}$	y

These examples should help you see that, in general:

$$\frac{1}{a^{-n}} = a^n \qquad (a \neq 0)$$

This fact allows us to change answers with negative exponents appearing on factors in the denominator to answers in which all factors have positive exponents. Try rewriting the fraction in Example 13.

Example 13. Rewrite $\dfrac{x^5}{y^{-6}}$ with positive exponents.

Solution

$$\frac{x^5}{y^{-6}} = \frac{x^5}{1} \cdot \frac{1}{y^{-6}}$$

$$=$$

Check your work on page 183. ▶

Example 14. Rewrite $\dfrac{y^4}{x^{-2}z^{-1}}$ with positive exponents.

Solution

$$\frac{y^4}{x^{-2}z^{-1}} = \frac{y^4}{1} \cdot \frac{1}{x^{-2}} \cdot \frac{1}{z^{-1}}$$

$$= y^4 x^2 z^1$$

$$= x^2 y^4 z$$

From our work with negative exponents appearing on factors in the numerators and denominators of fractions, we make the following observations:

Observation	Examples
A factor in the numerator with a negative exponent should be moved to the denominator, and the exponent will be positive.	$\dfrac{3x^{-2}y^{-1}}{5} = \dfrac{3}{5x^2y}$
A factor in the denominator with a negative exponent should be moved to the numerator, and the exponent will be positive.	$\dfrac{10}{x^{-1}} = 10x$ $\dfrac{x^{-2}}{3^{-1}y^{-2}} = \dfrac{3y^2}{x^2}$
A factor with a positive exponent stays where it is.	$\dfrac{2^{-4}a^{-8}b}{c^5} = \dfrac{b}{2^4 a^8 c^5}$ $= \dfrac{b}{16a^8c^5}$

These observations make it much easier to rewrite a simplified fraction using positive exponents. Just be sure that you first use the laws of exponents to simplify the expression.

You try to rewrite the fraction in Example 15.

Example 15. Rewrite $\dfrac{x^{-6}y^3}{a^{-1}z^{-2}}$ using positive exponents.

Solution

$$\dfrac{x^{-6}y^3}{a^{-1}z^{-2}}$$

Check your work on page 183. ▶

Then do Example 16.

Example 17. Simplify $\left(\dfrac{-3x^{-2}}{y^{-3}z}\right)^{-2}$.

Solution

$$\left(\dfrac{-3x^{-2}}{y^{-3}z}\right)^{-2}$$

$$= \dfrac{(-3)^{-2}(x^{-2})^{-2}}{(y^{-3})^{-2}z^{-2}}$$

$$= \dfrac{(-3)^{-2}x^4}{y^6z^{-2}}$$

$$= \dfrac{x^4z^2}{(-3)^2y^6}$$

$$= \dfrac{x^4z^2}{9y^6}$$

Example 16. Simplify $\left(\dfrac{a^7b^3}{c}\right)^{-3}$ and write the answers with positive exponents.

Solution: We first simplify the expression.

$$\left(\dfrac{a^7b^3}{c}\right)^{-3} = \dfrac{(a^7)^{-3}(b^3)^{-3}}{c^{-3}}$$

Check your work on page 184. ▶

Example 18. Simplify $\left(\dfrac{-2x^6y^8}{x^{-2}y^{10}}\right)^{-3}$.

Solution: Since there are like bases within the parentheses, we simplify that quantity first.

$$\left(\dfrac{-2x^6y^8}{x^{-2}y^{10}}\right)^{-3} = [-2x^{6-(-2)}y^{8-10}]^{-3}$$

$$= [-2x^8y^{-2}]^{-3}$$

$$= (-2)^{-3}(x^8)^{-3}(y^{-2})^{-3}$$

$$= (-2)^{-3}x^{-24}y^6$$

$$= \dfrac{y^6}{(-2)^3x^{24}}$$

$$= \dfrac{y^6}{-8x^{24}}$$

$$= \dfrac{-y^6}{8x^{24}}$$

In Examples 17 and 18, notice that the powers of the *numerical factors* were not computed until they had been rewritten with positive exponents. Trying to compute those powers before rewriting them almost always leads to errors.

$$(-3)^{-2} \neq 9 \qquad\qquad (-2)^{-3} \neq 8$$

$$(-3)^{-2} \neq -9 \qquad\qquad (-2)^{-3} \neq -8$$

$$(-3)^{-2} = \dfrac{1}{(-3)^2} = \dfrac{1}{9} \qquad (-2)^{-3} = \dfrac{1}{(-2)^3} = \dfrac{1}{-8} = \dfrac{-1}{8}$$

Example 19. Simplify $\dfrac{(x^{-2}y)^{-3}}{(2x^{-1}y^{-4})^{-1}}$.

Solution

$$\frac{(x^{-2}y)^{-3}}{(2x^{-1}y^{-4})^{-1}}$$
$$= \frac{x^6 y^{-3}}{2^{-1}x^1 y^4}$$
$$= \frac{1}{2^{-1}} \cdot x^{6-1} \cdot y^{-3-4}$$
$$= \frac{x^5 y^{-7}}{2^{-1}}$$
$$= \frac{2x^5}{y^7}$$

You try Example 20.

Example 20. Simplify $\dfrac{(-a^{-3}b^{-1})^{-2}}{(a^{-4}b^2)^{-3}}$.

Solution

$$\frac{(-a^{-3}b^{-1})^{-2}}{(a^{-4}b^2)^{-3}}$$
$$= \frac{(-1 \cdot a^{-3}b^{-1})^{-2}}{(a^{-4}b^2)^{-3}}$$
$$= \frac{(-1)^{-2}a^6 b^2}{a^{12}b^{-6}}$$

Check your work on page 184. ▶

Trial Run

Simplify and write answers with positive exponents.

_____ 1. $\dfrac{x^7}{y^{-9}}$

_____ 2. $\dfrac{x^{-3}z^{-2}}{y^4}$

_____ 3. $\dfrac{a^{-2}b^{-3}}{2^{-1}c^{-1}}$

_____ 4. $\left(\dfrac{a^2 b^5}{c}\right)^{-2}$

_____ 5. $\left(\dfrac{-3x^4 y^{-3}}{x^{-2}y^{-5}}\right)^{-2}$

_____ 6. $\dfrac{(2x^2 z^{-1})^{-5}}{(-x^{-1}z^4)^{-3}}$

Answers are on page 184.

▶ Examples You Completed

Example 1. Evaluate 2^{-3}.

Solution

$$2^{-3} = \frac{1}{2^3}$$
$$= \frac{1}{8}$$

Example 3. Evaluate $(-2)^{-4}$.

Solution

$$(-2)^{-4} = \frac{1}{(-2)^4}$$
$$= \frac{1}{16}$$

Example 7. Simplify $x^3 \cdot x^{-7} \cdot x^{-1}$.

Solution: Since the bases are the same, we find the product by adding the exponents.

$$x^3 \cdot x^{-7} \cdot x^{-1}$$
$$= x^{3+(-7)+(-1)}$$
$$= x^{-5}$$
$$= \frac{1}{x^5}$$

Example 9. Simplify $(x^{-2})^4$.

Solution: to raise a power to a power, we multiply exponents.

$$(x^{-2})^4$$
$$= x^{-2 \cdot 4}$$
$$= x^{-8}$$
$$= \frac{1}{x^8}$$

Example 12. Simplify $\dfrac{x^{-8}y^4}{x^{-3}y^{-2}}$.

Solution: To divide powers of the same base, we subtract exponents.

$$\frac{x^{-8}y^4}{x^{-3}y^{-2}} = \frac{x^{-8}}{x^{-3}} \cdot \frac{y^4}{y^{-2}}$$
$$= x^{-8-(-3)} \cdot y^{4-(-2)}$$
$$= x^{-5} \cdot y^6$$
$$= \frac{1}{x^5} \cdot y^6$$
$$= \frac{y^6}{x^5}$$

Example 13. Rewrite $\dfrac{x^5}{y^{-6}}$ using positive exponents.

Solution

$$\frac{x^5}{y^{-6}} = \frac{x^5}{1} \cdot \frac{1}{y^{-6}}$$
$$= x^5 y^6$$

Example 15. Rewrite $\dfrac{x^{-6}y^3}{a^{-1}z^{-2}}$ using positive exponents.

Solution

$$\frac{x^{-6}y^3}{a^{-1}z^{-2}}$$
$$= \frac{ay^3z^2}{x^6}$$

Example 16. Simplify $\left(\dfrac{a^7b^3}{c}\right)^{-3}$ and write the answer with positive exponents.

Solution: We first simplify the expression.

$$\left(\frac{a^7b^3}{c}\right)^{-3} = \frac{(a^7)^{-3}(b^3)^{-3}}{c^{-3}}$$

$$= \frac{a^{-21}b^{-9}}{c^{-3}}$$

$$= \frac{c^3}{a^{21}b^9}$$

Example 20. Simplify $\dfrac{(-a^{-3}b^{-1})^{-2}}{(a^{-4}b^2)^{-3}}$.

Solution

$$\frac{(-a^{-3}b^{-1})^{-2}}{(a^{-4}b^2)^{-3}}$$

$$= \frac{(-1 \cdot a^{-3}b^{-1})^{-2}}{(a^{-4}b^2)^{-3}}$$

$$= \frac{(-1)^{-2}a^6b^2}{a^{12}b^{-6}}$$

$$= (-1)^{-2}a^{6-12}b^{2-(-6)}$$

$$= (-1)^{-2}a^{-6}b^8$$

$$= \frac{b^8}{(-1)^2a^6}$$

$$= \frac{b^8}{a^6}$$

Answers to Trial Runs

page 177   **1.** $\dfrac{1}{9}$ **2.** $\dfrac{1}{15}$ **3.** $\dfrac{-1}{125}$ **4.** $\dfrac{1}{16}$ **5.** $\dfrac{-1}{64}$ **6.** $\dfrac{1}{27}$ **7.** $\dfrac{1}{a^2b^3}$ **8.** $\dfrac{y}{x^4z}$ **9.** $\dfrac{1}{y^6}$

   10. $\dfrac{1}{xy^3}$

page 178   **1.** $\dfrac{1}{x^3}$ **2.** $\dfrac{-6y^2}{x^3}$ **3.** $\dfrac{1}{x^6}$ **4.** x^8y^4 **5.** $\dfrac{y^5}{4x^3z^2}$ **6.** $\dfrac{y^8}{x^5}$

page 182   **1.** x^7y^9 **2.** $\dfrac{1}{x^3y^4z^2}$ **3.** $\dfrac{2c}{a^2b^3}$ **4.** $\dfrac{c^2}{a^4b^{10}}$ **5.** $\dfrac{1}{9x^{12}y^4}$ **6.** $\dfrac{-z^{17}}{32x^{13}}$

EXERCISE SET 3.2

Evaluate.

_____ **1.** 4^{-2} _____ **2.** 7^{-2}

_____ **3.** $(-3)^{-5}$ _____ **4.** $(-4)^{-3}$

_____ **5.** $(-9)^{-2}$ _____ **6.** $(-3)^{-4}$

_____ **7.** -5^{-2} _____ **8.** -2^{-4}

_____ **9.** $-(-6)^{-1} + 6^{-2}$ _____ **10.** $-(-2)^{-3} + 2^{-4}$

_____ **11.** $\left(\dfrac{3}{4}\right)^{-2}$ _____ **12.** $\left(\dfrac{2}{3}\right)^{-4}$

Simplify and write each answer without negative exponents.

_____ **13.** $x^3 \cdot x^{-5} \cdot x^{-4}$ _____ **14.** $x^{-3} \cdot x^4 \cdot x^{-8}$

_____ **15.** $(a^5 b^{-8})(a^{-3} b^2)$ _____ **16.** $(a^{-7} b^3)(a^{10} b^{-5})$

_____ **17.** $(x^3 y^{-5} z^{-9})(x^{-4} y^5 z^3)$ _____ **18.** $(x^5 y^{-5} z^{-7})(x^{-6} y^2 z^7)$

_____ **19.** $(a^{-4} b^3 c^{-2})(a^5 b^{-7} c^5)$ _____ **20.** $(a^{-9} b^2 c^{-4})(a^{10} b^{-8} c^7)$

_____ **21.** $(x^{-3})^2$ _____ **22.** $(x^{-4})^2$

_____ **23.** $(2x)^{-5}$ _____ **24.** $(3x)^{-3}$

_____ **25.** $9a^{-2}$ _____ **26.** $7a^{-3}$

_____ **27.** $(x^{-5} y^{-1})^{-3}$ _____ **28.** $(x^{-7} y^{-2})^{-2}$

_____ **29.** $(-3x^{-1} y^2)^{-3}$ _____ **30.** $(-2x^{-2} y^3)^{-5}$

_____ **31.** $(7x^{-5} y^2)^2$ _____ **32.** $(8x^{-7} y^3)^2$

_____ **33.** $\dfrac{a^9}{b^{-7}}$ _____ **34.** $\dfrac{a^{10}}{b^{-8}}$

_____ **35.** $\dfrac{y^9}{x^{-1} z^{-3}}$ _____ **36.** $\dfrac{y^7}{x^{-4} z^{-1}}$

_____ **37.** $\left(\dfrac{5}{a^4}\right)^{-3}$ _____ **38.** $\left(\dfrac{6}{a^6}\right)^{-2}$

_____ **39.** $\left(\dfrac{x^6 y^{-3}}{z^{-3}}\right)^{-1}$ _____ **40.** $\left(\dfrac{x^7 y^{-4}}{z^{-5}}\right)^{-1}$

_____ **41.** $\dfrac{x^{-9} y^3}{x^{-4} y^{-3}}$ _____ **42.** $\dfrac{x^{-11} y^5}{x^{-5} y^{-5}}$

185

_____ 43. $\left(\dfrac{a^3b^4}{c^2}\right)^{-3}$

_____ 44. $\left(\dfrac{a^3b^5}{c^4}\right)^{-2}$

_____ 45. $\left(\dfrac{x^9y^5}{x^{-1}y^{-7}}\right)^{-1}$

_____ 46. $\left(\dfrac{x^{10}y^7}{x^{-3}y^{-4}}\right)^{-1}$

_____ 47. $\left(\dfrac{-3x^5y^{-2}}{x^{-3}y^{-7}}\right)^{-2}$

_____ 48. $\left(\dfrac{-5x^7y^{-4}}{x^{-4}y^{-9}}\right)^{-2}$

_____ 49. $\left(\dfrac{-4a^{-5}b^{-4}}{ac^{-3}}\right)^{3}$

_____ 50. $\left(\dfrac{-2a^{-8}b^{-5}}{a^2c^{-4}}\right)^{5}$

_____ 51. $\left(\dfrac{2a^{-2}b}{b^{-3}}\right)^{2}\left(\dfrac{3a^4b^{-1}}{b^3}\right)^{-3}$

_____ 52. $\left(\dfrac{5a^{-1}b^{-2}}{a^{-4}}\right)^{3}\left(\dfrac{9a^{-3}b}{b^{-2}}\right)^{-2}$

_____ 53. $\dfrac{(a^{-1}b^2)^{-2}}{(2a^{-2}b^{-3})^{-1}}$

_____ 54. $\dfrac{(a^2b^{-2})^{-3}}{(3a^{-4}b^3)^{-2}}$

_____ 55. $\dfrac{(-x^{-3}y^{-2})^{-2}}{(x^{-2}y^2)^{-3}}$

_____ 56. $\dfrac{(-2x^{-1}y^{-3})^{-2}}{(x^{-4}y^3)^{-1}}$

☆ Stretching the Topics

_____ 1. Write x^{2a-3b} with positive exponents if $a < 0$ and $b > 0$.

_____ 2. Simplify $\left(\dfrac{-ab^{-2}c^3}{2d^{-3}ef^2}\right)^{-2}\left(\dfrac{2a^{-1}b^3c^{-1}}{-d^2e^{-2}f^2}\right)^{3}$ and write answer with positive exponents.

_____ 3. Simplify $\dfrac{x^{6a+4}y^{3b-2}}{x^{2a-2}y^{b+3}}$ and write answer without a denominator.

Check your answers in the back of your book.

If you can complete **Checkup 3.2,** you are ready to go on to Section 3.3.

CHECKUP 3.2

Evaluate.

_____ **1.** 8^{-2} _____ **2.** $(-4)^{-3}$

Simplify and write each answer with positive exponents.

_____ **3.** $(x^5y^{-7}z^{-3})(x^{-4}y^7z^8)$ _____ **4.** $(x^{-3})^3$

_____ **5.** $(6x)^{-2}$ _____ **6.** $(x^{-3}y^7z^3)^{-2}$

_____ **7.** $\dfrac{x^{-13}y^7}{x^{-8}y^{-7}}$ _____ **8.** $\left(\dfrac{x^5y^{-4}}{z^{-3}}\right)^{-1}$

_____ **9.** $\left(\dfrac{3x^5y^{-2}}{x^{-4}y^3}\right)^{-2}$ _____ **10.** $\dfrac{(3x^{-8}y)^{-1}}{(x^{-6}y^{-2})^{-2}}$

Check your answers in the back of your book.

If You Missed Problems:	You Should Review Examples:
1	1
2	2
3	8
4	9
5, 6	10
7	12
8, 9	18
10	19, 20

3.3 Operating with Polynomials

Now that we have gained skill working with all integers used as exponents, we turn our attention to algebraic expressions that involve *sums* of constant terms and/or variable terms whose variable parts appear with *whole-number exponents*. Such expressions are called **polynomials.**

Definition of a Polynomial. A polynomial in x is a sum of one or more terms of the form

$$ax^n$$

where a is a constant and n is a whole number.

Identifying Polynomials

A polynomial is named according to the number of *terms* it contains.

A **monomial** is a polynomial containing *one* term.
A **binomial** is a polynomial containing *two* terms.
A **trinomial** is a polynomial containing *three* terms.

Polynomials containing more than three terms do not have special names; we label them by stating the number of terms they contain. We also describe a polynomial by naming the variable it contains and by identifying the highest power of that variable that occurs in the polynomial. The highest power appearing in a polynomial of one variable determines the **degree** of the polynomial.

Example	Number of Terms	Degree	Description
x	One	1	First-degree monomial in x
$-3x^2$	One	2	Second-degree monomial in x
$\frac{1}{3}y^5$	One	5	Fifth-degree monomial in y
3	One	0	Constant monomial
$x + 3$	Two	1	First-degree binomial in x
$-3x^2 + 2x$	Two	2	Second-degree binomial in x
$2z - 5z^4$	Two	4	Fourth-degree binomial in z
$x^2 + 5x + 7$	Three	2	Second-degree trinomial in x
$y^3 - 4y^2 + y$	Three	3	Third-degree trinomial in y
$x^3 + 5x^2 - 7x + 2$	Four	3	Third-degree four-term polynomial in x

If a polynomial contains two or more variables, it becomes awkward (and not very useful) to identify its degree. For this reason, we usually label such a polynomial according to its variables and number of terms only.

$$a^2bc \quad \text{is a monomial in } a, b, \text{ and } c$$

$$xy^2 + 3xy \quad \text{is a binomial in } x \text{ and } y$$

$$x^2 - y^2 + z^2 \quad \text{is a trinomial in } x, y, \text{ and } z$$

We have dealt with many algebraic expressions that are *not* polynomials because they cannot be written as *sums* of terms whose variable parts appear with *whole-number* exponents. For example,

$$x^{-3} + 2 \quad \text{is } not \text{ a polynomial}$$

$$\frac{x}{y^2z} \quad \text{is } not \text{ a polynomial}$$

$$\frac{2}{x + 1} \quad \text{is } not \text{ a polynomial}$$

Let's practice naming some polynomials. You try Example 1.

Example 1. Name each polynomial.

15   is a constant _____

$8x^2 - 3x + 4$ is a _____-degree _____ in x

$y^9 - 1$ is a _____-degree _____ in _____

$3xyz$ is a monomial in _____, _____, and _____

Check your work on page 194. ▶

As you may have noticed, we usually write polynomial terms so that the exponents on the variable decrease from left to right. This is called writing a polynomial in **descending powers of the variable.**

Example 2. Rewrite $2x + 5x^2 + 1 - x^3$ in descending powers of the variable.

    Solution

$$2x + 5x^2 + 1 - x^3$$
$$= -x^3 + 5x^2 + 2x + 1$$

Now try Example 3.

Example 3. Rewrite $y - 3 + y^5$ in descending powers of the variable.

    Solution

$$y - 3 + y^5$$
$$=$$

Check your work on page 194. ▶

Adding and Subtracting Polynomials

We learned to add and subtract polynomials in Chapter 1 when we discovered how to **combine like terms.** Recall that like terms are terms that contain exactly the same variable part (same variable raised to same power). We combine like terms by combining the numerical coefficients and keeping the same variable. In adding polynomials, you may use the commutative and associative properties to rearrange the terms if this is helpful.

Example 4. Simplify $(6x^2 + 3x - 1) + (-4x^2 - 5x + 3)$.

Solution

$(6x^2 + 3x - 1) + (-4x^2 - 5x + 3)$

$= 6x^2 + 3x - 1 - 4x^2 - 5x + 3$

$= 6x^2 - 4x^2 + 3x - 5x - 1 + 3$

$= 2x^2 - 2x + 2$

Now you try Example 5.

Example 5. Simplify $(a^3 + 2a) + (5a^2 - 2a + 1)$.

Solution

$(a^3 + 2a) + (5a^2 - 2a + 1)$

$= a^3 + 2a + 5a^2 - 2a + 1$

Check your work on page 194. ▶

Subtraction is performed in a similar manner. From Chapter 1 recall that the subtraction sign between polynomials means that we must find the *opposite* of each term in the second polynomial, and *add*.

Example 6. Simplify $(3x^2 - 4x + 7) - (2x^2 - x + 3)$.

Solution

$(3x^2 - 4x + 7) - (2x^2 - x + 3)$

$= 3x^2 - 4x + 7 - 2x^2 + x - 3$

$= 3x^2 - 2x^2 - 4x + x + 7 - 3$

$= x^2 - 3x + 4$

You complete Example 7.

Example 7. Simplify $(x^3 - 2x^2 + 1) - (2x^2 + 3x + 1)$.

Solution

$(x^3 - 2x^2 + 1) - (2x^2 + 3x + 1)$

$= x^3 - 2x^2 + 1 - 2x^2 - 3x - 1$

Check your work on page 194. ▶

In Chapter 1 we also learned how to multiply a constant times a polynomial using the **distributive property.**

Example 8. Simplify $3(a^2 - 5a) - 5(2a^2 + 4a - 3)$.

Solution

$3(a^2 - 5a) - 5(2a^2 + 4a - 3)$

$= 3a^2 - 15a - 10a^2 - 20a + 15$

$= 3a^2 - 10a^2 - 15a - 20a + 15$

$= -7a^2 - 35a + 15$

You try Example 9.

Example 9. Simplify $-2(5 - x^2) + 4(x^2 - 2x + 3)$.

Solution

$-2(5 - x^2) + 4(x^2 - 2x + 3)$

Check your work on page 194. ▶

Example 10. Simplify $x^2 - 3[x - 2(x^2 - x + 6)]$.

Solution

$\qquad x^2 - 3[x - 2(x^2 - x + 6)]$

$= x^2 - 3[x - 2x^2 + 2x - 12]$ Remove parentheses.

$= x^2 - 3[-2x^2 + 3x - 12]$ Combine like terms within brackets.

$= x^2 + 6x^2 - 9x + 36$ Remove brackets.

$= 7x^2 - 9x + 36$ Combine like terms.

From these examples we see that in simplifying a polynomial expression containing symbols of grouping, we must remember to

1. Deal with symbols of grouping, working from the inside out.
2. Perform multiplications (and/or divisions) from left to right.
3. Perform additions (and/or subtractions) from left to right.

⟩ Trial Run

Simplify.

_____ 1. $(x^3 - 9) + (x^2 - 6x + 9)$

_____ 2. $(5x^2 - 3x + 1) - (3x^2 + x - 2)$

_____ 3. $5(1 - y^2) + 2(y^2 - 3y)$

_____ 4. $2(a^2 - 7a) - 3(-a^2 - a + 1)$

_____ 5. $9x^2 - 2x - 3(2x^2 - 5) + 6$

_____ 6. $7a - 3[2(a - 5b) + 2b]$

Answers are on page 194.

Multiplying Monomials Times Polynomials

In Section 3.1 we learned to find products of monomials using the first law of exponents. Now we must turn our attention to products of monomials and other polynomials. We would like to learn how to find a product such as

$$x(x^2 - 2x + 3)$$

Once again the **distributive property** provides us with the method for performing such multiplications. According to the distributive property for multiplication over addition, we must multiply each term inside the parentheses by the factor outside the parentheses. Sometimes students like to use arrows to organize their work in finding such a product.

$$x(x^2 - 2x + 3) = x(x^2 - 2x + 3)$$
$$= x \cdot x^2 + x(-2x) + x \cdot 3$$
$$= x^3 - 2x^2 + 3x$$

As you can see, once you have used the distributive property, the final steps of simplification merely require that you know how to multiply monomials.

Now you try Example 11.

Example 11. Simplify $-6a(2a^3 - 4a)$.

Solution

$$-6a(2a^3 - 4a)$$
$$= -6a(2a^3 - 4a)$$
$$= -6a(2a^3) - 6a(-4a)$$
$$=$$

Check your work on page 194. ▶

Example 13. Simplify $2a^3b(4a^2b - 7ab - 2ab^2)$.

Solution

$$2a^3b(4a^2b - 7ab - 2ab^2)$$
$$= 2a^3b \cdot 4a^2b + 2a^3b(-7ab)$$
$$+ 2a^3b(-2ab^2)$$
$$=$$

Check your work on page 194. ▶

Example 12. Simplify $(x^2 + xy - y^2)5xy$.

Solution

$$(x^2 + xy - y^2)5xy$$
$$= (x^2 + xy - y^2)5xy$$
$$= 5xy \cdot x^2 + 5xy \cdot xy + 5xy(-y^2)$$
$$= 5x^3y + 5x^2y^2 - 5xy^3$$

Now try Example 13.

Example 14. Simplify $6x - 2x(x - 4)$.

Solution

$$6x - 2x(x - 4)$$
$$= 6x - 2x^2 + 8x$$
$$= -2x^2 + 14x$$

Example 15. Simplify $x^2 - 5[x(x - 1) + 2x(x^2 - x)]$.

Solution

$$x^2 - 5[x(x - 1) + 2x(x^2 - x)]$$
$$= x^2 - 5[x^2 - x + 2x^3 - 2x^2] \qquad \text{Remove parentheses.}$$
$$= x^2 - 5[2x^3 - x^2 - x] \qquad \text{Combine like terms within brackets.}$$
$$= x^2 - 10x^3 + 5x^2 + 5x \qquad \text{Remove brackets.}$$
$$= -10x^3 + 6x^2 + 5x \qquad \text{Combine like terms.}$$

⫸ **Trial Run**

Simplify.

_____ **1.** $a(a^2 - 3a - 10)$ _____ **2.** $-3x(2x^2 - 5x)$

_____ **3.** $(x^2 - 7xy + 12y^2)7x$ _____ **4.** $-5a^2b(7a^2 - 5ab + 3b^2)$

_____ **5.** $7x - 2x[x^2 - x(4x - 2)]$ _____ **6.** $a^2 - 9[a(a - 5) + 3a(a^2 - 1)]$

Answers are on page 194.

▶ **Examples You Completed**

Example 1. Name each polynomial.

$$15 \qquad \text{is a constant monomial}$$

$$8x^2 - 3x + 4 \qquad \text{is a second-degree trinomial in } x$$

$$y^9 - 1 \qquad \text{is a ninth-degree binomial in } y$$

$$3xyz \qquad \text{is a monomial in } x, y, \text{ and } z$$

Example 3. Rewrite $y - 3 + y^5$ in descending powers of the variable.

Solution

$$y - 3 + y^5$$
$$= y^5 + y - 3$$

Example 5. Simplify $(a^3 + 2a) + (5a^2 - 2a + 1)$.

Solution

$$(a^3 + 2a) + (5a^2 - 2a + 1)$$
$$= a^3 + 2a + 5a^2 - 2a + 1$$
$$= a^3 + 5a^2 + 2a - 2a + 1$$
$$= a^3 + 5a^2 + 1$$

Example 7. Simplify $(x^3 - 2x^2 + 1) - (2x^2 + 3x + 1)$.

Solution

$$(x^3 - 2x^2 + 1) - (2x^2 + 3x + 1)$$
$$= x^3 - 2x^2 + 1 - 2x^2 - 3x - 1$$
$$= x^3 - 2x^2 - 2x^2 - 3x + 1 - 1$$
$$= x^3 - 4x^2 - 3x$$

Example 9. Simplify $-2(5 - x^2) + 4(x^2 - 2x + 3)$.

Solution

$$-2(5 - x^2) + 4(x^2 - 2x + 3)$$
$$= -10 + 2x^2 + 4x^2 - 8x + 12$$
$$= 2x^2 + 4x^2 - 8x - 10 + 12$$
$$= 6x^2 - 8x + 2$$

Example 11. Simplify $-6a(2a^3 - 4a)$.

Solution

$$-6a(2a^3 - 4a)$$
$$= -6a(2a^3) - 6a(-4a)$$
$$= -12a^4 + 24a^2$$

Example 13. Simplify $2a^3b(4a^2b - 7ab - 2ab^2)$.

Solution

$$2a^3b(4a^2b - 7ab - 2ab^2)$$
$$= 2a^3b \cdot 4a^2b + 2a^3b(-7ab) + 2a^3b(-2ab^2)$$
$$= 8a^5b^2 - 14a^4b^2 - 4a^4b^3$$

Answers to Trial Runs

page 192   **1.** $x^3 + x^2 - 6x$ **2.** $2x^2 - 4x + 3$ **3.** $-3y^2 - 6y + 5$ **4.** $5a^2 - 11a - 3$
            5. $3x^2 - 2x + 21$ **6.** $a + 24b$

page 193   **1.** $a^3 - 3a^2 - 10a$ **2.** $-6x^3 + 15x^2$ **3.** $7x^3 - 49x^2y + 84xy^2$
            4. $-35a^4b + 25a^3b^2 - 15a^2b^3$ **5.** $6x^3 - 4x^2 + 7x$ **6.** $-27a^3 - 8a^2 + 72a$

EXERCISE SET 3.3

Name each polynomial.

_____ **1.** $5x^5 + 10x^4$

_____ **2.** $3x^7 - 9x^5$

_____ **3.** $x^3 - 1$

_____ **4.** $y^3 - 8$

_____ **5.** -15

_____ **6.** -8

_____ **7.** $4y^2 - 4y + 1$

_____ **8.** $9y^2 - 6y - 1$

_____ **9.** $x^4 - 3x^3 - x^2 + 3x$

_____ **10.** $5x^5 + 10x^4 - x + 2$

Rewrite in descending powers of the variable.

_____ **11.** $3x - 4x^2 + 7 + x^3$

_____ **12.** $5x^2 - 3 + 2x + x^3$

_____ **13.** $8x^2 + 12 + x^4$

_____ **14.** $-9x^3 + 20 + x^6$

_____ **15.** $1 - a + a^2 - a^3$

_____ **16.** $2 - 2a + 3a^2 - 3a^3$

_____ **17.** $5 + 3x$

_____ **18.** $8 - 3y$

Simplify.

_____ **19.** $(x^3 + 3x^2) + (x^2 - 3x + 2)$

_____ **20.** $(x^2 - 3x) + (4x + 2)$

_____ **21.** $(3x^2 + 2x - 1) - (4x^2 - 2x + 5)$

_____ **22.** $(x^2 - 3x - 28) - (2x^2 + 3x - 1)$

_____ **23.** $(4a^2 - 3ab + 2b^2) - (a^2 + 5ab - 14b^2)$

_____ **24.** $(5a^2 - 2ab + b^2) - (a^2 - 7ab - 9b^2)$

_____ **25.** $4(a^2 - 3) - 5(a^2 - 2a - 1)$

_____ **26.** $5(a^2 - 9) - 3(a^2 + 7a - 11)$

_____ **27.** $7(-a^2 + 2a) + 3(2a^2 - 7a + 2)$

_____ **28.** $9(-a^2 + 2a) + 5(3a^2 - 4a + 2)$

_____ **29.** $4(a + b) - 5(2a - b) + 6(a - 2b)$

_____ **30.** $5(a - b) - 4(2a - 3b) + 3(a + 5b)$

_____ **31.** $2(x^2 + 2xy - y^2) - (3x^2 + 4xy - y^2)$

_____ **32.** $3(x^2 - 10xy + 5y^2) - (5x^2 - xy + y^2)$

_____ **33.** $-3(a^2 + a) - 5(3a + 7)$

_____ **34.** $-5(a^2 - 7) - 6(2a + 3)$

_____ **35.** $5(y^2 - 2y - 1) - 2(3y^2 - 5y - 4)$

_____ **36.** $3(x^2 - 2x + 1) - 5(x^2 + x - 1)$

_____ **37.** $3x^2 - 2x + 2(3x^2 - 1) + 9$

_____ **38.** $5x^2 - 7x + 5(4x^2 - 2) + 3$

_____ **39.** $5x^2 - 2(4x^2 - x) - 4(2x + 1)$ _____ **40.** $7x^2 - (3x^2 - 2x) + 3(5x - 1)$

_____ **41.** $4a - 3[2a - 3(a + 2b) - 1] + 5b$

_____ **42.** $7a - 2[3a - 2(a + 5b) + 1] + 3b$

_____ **43.** $9x^2 - 3[2(x^2 - 3x) + 5x] - 4$ _____ **44.** $10x^2 - 5[3(x^2 - 4x) + 2x] - 6$

_____ **45.** $2m(m^2 - 5m + 6)$ _____ **46.** $9m(m^2 - m - 6)$

_____ **47.** $-5xy(3x^2y^2 - xy - 4)$ _____ **48.** $-4xy(5x^2y^2 - xy - 6)$

_____ **49.** $(7a^2 - b^2)2a^2b^2$ _____ **50.** $(a^2 - 2b^2)6a^2b^2$

_____ **51.** $4mn(8m^3 - 2m^2n + mn^2 - 3n^3)$ _____ **52.** $5mn(6m^3 - m^3n + 2mn^2 - 7n^3)$

_____ **53.** $5xy(3x - 4y) - 2xy(4x - 5y)$ _____ **54.** $8xy(2x - y) - xy(x - 2y)$

_____ **55.** $9x - 3x[x^2 - 2x(x - 1)]$ _____ **56.** $8x - 5x[x^2 - 7x(2x - 1)]$

_____ **57.** $a^2 - 7[a(a - 3) + 2a(a^2 - 1)]$

_____ **58.** $2a^2 - 3[a(2a - 1) + a(a^2 - 11)]$

_____ **59.** $-2x^2y^3(3x + 5xy^4) + x^3y(y^2 - 3y^6)$

_____ **60.** $-x^2y^3(8x - 3xy^4) + 2x^3y(4y^2 - 5y^6)$

☆ Stretching the Topics _____

_____ **1.** From $-2x^3 - x^2 + 5x - 4$ subtract the sum of $5x^3 - 4x^2 - 5x + 2$ and $-7x^3 + 3x^2 + x - 7$.

_____ **2.** Simplify $3ab(a^2 - 5ab + 4b^2) - 2ab(6a^2 - 7ab - 8b^2)$ and evaluate when $a = 1$ and $b = -1$.

_____ **3.** Simplify $x^{-2}y^{-1}(3x^2y^3 - x^3y^2 - 2x^4y^{-1}) - xy^{-1}(x^{-1}y^3 - 3y^2 + 4xy)$.

Check your answers in the back of your book.

If you can complete **Checkup 3.3,** you are ready to go on to Section 3.4.

CHECKUP 3.3

Name each polynomial.

_____ **1.** $x^4 - 16$ _____ **2.** $5y^2 - 2y - 7$

Rewrite in descending powers of the variable.

_____ **3.** $x^2 - x^4 + 2 - 2x^3$ _____ **4.** $8a - 9 - 7a^2$

Simplify.

_____ **5.** $4(2a^2 - ab - 3b^2) + 3(-a^2 + 2ab - b^2)$

_____ **6.** $9x^2 - 3(x^2 - 4x) - 5(-x + 2) + 3$

_____ **7.** $11a - 4[2a - 3(a - 7b) - 4b] + 3b$

_____ **8.** $5x(x^2 - 3x + 5)$

_____ **9.** $7mn(3m + 5n) - 2mn(4m - 5n)$

_____ **10.** $5a^2 - 3[2a(a - 1) - 5a(a^2 - 2)]$

Check your answers in the back of your book.

If You Missed Problems:	You Should Review Examples:
1, 2	1
3, 4	2, 3
5, 6	8, 9
7	10
8, 9	11–14
10	15

3.4 Multiplying More Polynomials

Multiplying Binomials

From our work with products of monomials and binomials we know that

$$x(2x + 3) \qquad \text{and} \qquad 5(2x + 3)$$
$$= 2x^2 + 3x \qquad\qquad\qquad = 10x + 15$$

Now suppose that we wish to find the product $(x + 5)(2x + 3)$. According to the distributive property, the second binomial $(2x + 3)$ must be multiplied times each term in the first binomial $(x + 5)$. Then we combine like terms if possible.

$$(x + 5)(2x + 3) = x(2x + 3) + 5(2x + 3)$$
$$= 2x^2 + 3x + 10x + 15$$
$$= 2x^2 + 13x + 15$$

> To multiply binomials, we multiply the first term in the first binomial times each term in the second binomial; then we multiply the second term in the first binomial times each term in the second binomial. Finally, we combine like terms.

You multiply the binomials in Example 1.

Example 1. Multiply $(7x - 1)(x - 2)$.

Solution

$$(7x - 1)(x - 2)$$
$$= 7x(x - 2) - 1(x - 2)$$

Check your work on page 206. ▶

Example 2. Multiply $(5 - 2x)(8 - x)$.

Solution

$$(5 - 2x)(8 - x)$$
$$= 5(8 - x) - 2x(8 - x)$$
$$= 40 - 5x - 16x + 2x^2$$
$$= 2x^2 - 21x + 40$$

Example 3. Find $(3x - 4y)^2$.

Solution

$$(3x - 4y)^2$$
$$= (3x - 4y)(3x - 4y)$$
$$= 3x(3x - 4y) - 4y(3x - 4y)$$
$$= 9x^2 - 12xy - 12xy + 16y^2$$
$$= 9x^2 - 24xy + 16y^2$$

Now you complete Example 4.

Example 4. Find $(a + 6)(a - 6)$.

Solution

$$(a + 6)(a - 6)$$
$$= a(\underline{\hphantom{xxxx}}) + 6(\underline{\hphantom{xxxx}})$$

Check your work on page 206. ▶

Notice that there is no first-degree term (no a-term) in the final answer in Example 4.

Some students prefer to use a slightly shorter method to find products of binomials. As long as we realize that the distributive property still forms the basis for this shorter method, we are safe in using it. This shortcut is sometimes called the **FOIL method** because its steps require that we:

> 1. Multiply the **F**irst terms in the binomials.
> 2. Multiply the **O**uter terms in the binomials.
> 3. Multiply the **I**nner terms in the binomials.
> 4. Multiply the **L**ast terms in the binomials.
> 5. Combine the like terms resulting from those multiplications.

Let's find a product using the FOIL method.

$$\begin{array}{cccc} F & L & F\quad O\quad I\quad L \\ (x\ +\ 5)(2x\ +\ 3) &=& 2x^2\ +\ 3x\ +\ 10x\ +\ 15 \\ I & & \\ O & & \end{array}$$

$$=\ 2x^2\ +\ 13x\ +\ 15$$

Notice that this product is the same product that we found when we did our multiplication the longer way. We are doing exactly the same thing, skipping one step of writing.

You try using the FOIL method in Example 5.

Example 5. Find $(7x\ -\ 3)(2x\ +\ 4)$.

Solution

$$\begin{array}{cc} F & L \\ (7x\ -\ 3)(2x\ +\ 4) \\ I \\ O \end{array}$$

$$\begin{array}{cccc} F & O & I & L \\ =\ \underline{\quad}\ +\ \underline{\quad}\ -\ \underline{\quad}\ -\ \underline{\quad} \end{array}$$

$$=\ \underline{\qquad}\ .$$

Check your work on page 206. ▶

Example 7. Find $(5x\ -\ 2)^2$.

Solution

$$(5x\ -\ 2)^2\ =\ \begin{array}{cc} F & L \\ (5x\ -\ 2)(5x\ -\ 2) \\ I \\ O \end{array}$$

$$\begin{array}{cccc} F & O & I & L \\ =\ \underline{\quad}\ -\ \underline{\quad}\ -\ \underline{\quad}\ +\ \underline{\quad} \end{array}$$

$$=\ \underline{\qquad}$$

Check your work on page 206. ▶

Example 6. Find $(8x\ -\ 3y)(8x\ +\ 3y)$.

Solution

$$\begin{array}{cc} F & L \\ (8x\ -\ 3y)(8x\ +\ 3y) \\ I \\ O \end{array}$$

$$\begin{array}{cccc} F & O & I & L \\ =\ 64x^2\ +\ 24xy\ -\ 24xy\ -\ 9y^2 \end{array}$$

$$=\ 64x^2\ -\ 9y^2$$

Now try Example 7.

Trial Run

Multiply.

_____ **1.** $(2a - 5)(a + 7)$ _____ **2.** $(7x - 2y)^2$

_____ **3.** $(7a - 3)(7a + 3)$ _____ **4.** $(9 - y)(3 - 2y)$

_____ **5.** $(5x + 2y)(5x - 2y)$ _____ **6.** $(3a - 5b)(a + 2b)$

Answers are on page 207.

When we simplify expressions involving several polynomial operations, we must remember to use the rules governing the order of operations. You try to complete Example 8.

Example 8. Simplify $x(2x + 1) + (x + 3)(4x + 2)$.

Solution

$$x(2x + 1) + (x + 3)(4x + 2)$$
$$= 2x^2 + x + 4x^2 + 2x + 12x + 6$$
$$=$$

Check your work on page 206. ▶

Example 9. Simplify $(x - 3)(2x - 1) - (6x + 1)(3x - 2)$.

Solution

$$(x - 3)(2x - 1) - (6x + 1)(3x - 2)$$
$$= 2x^2 - x - 6x + 3$$
$$\quad - (18x^2 - 12x + 3x - 2)$$
$$= 2x^2 - 7x + 3 - (18x^2 - 9x - 2)$$
$$= 2x^2 - 7x + 3 - 18x^2 + 9x + 2$$
$$= -16x^2 + 2x + 5$$

In Example 9, notice that the entire second product was being subtracted from the first product. Therefore, we used parentheses to keep that second product together before taking its opposite and adding.

Example 10. Simplify $(x + 1)^2 - 3(x - 5)^2$.

Solution

$$(x + 1)^2 - 3(x - 5)^2$$
$$= (x + 1)(x + 1) - 3(x - 5)(x - 5)$$
$$= x^2 + x + x + 1 - 3(x^2 - 5x - 5x + 25)$$
$$= x^2 + 2x + 1 - 3(x^2 - 10x + 25)$$
$$= x^2 + 2x + 1 - 3x^2 + 30x - 75$$
$$= -2x^2 + 32x - 74$$

Trying to skip many steps in such problems is not a good idea because it can lead to errors in signs.

Trial Run

Simplify.

_____ **1.** $a(3a + 2) + (a - 1)(a + 2)$

_____ **2.** $(3x - 1)(2x + 5) - (3x - 2)(x + 2)$

_____ **3.** $(x - 3)^2 - 4(x + 5)^2$

_____ **4.** $a(a - 1)(a - 3) - 4(a + 2)(3a - 4)$

Answers are on page 207.

Multiplying Polynomials

Once again the distributive property will be our tool for finding a product such as

$$(x + 3)(x^2 + 8x - 1)$$

We must multiply each term in the first polynomial times each term in the second polynomial and then combine like terms.

Example 11. Find $(x + 3)(x^2 + 8x - 1)$.

Solution

$(x + 3)(x^2 + 8x - 1)$

$= x(x^2 + 8x - 1) + 3(x^2 + 8x - 1)$

$= x^3 + 8x^2 - x + 3x^2 + 24x - 3$

$= x^3 + 11x^2 + 23x - 3$

Now complete Example 12.

Example 12. Find $(x + y)(x^2 - xy + y^2)$.

Solution

$(x + y)(x^2 - xy + y^2)$

$= x(x - xy + y) + y(x^2 - xy + y^2)$

Check your work on page 206. ▶

Example 13. Find $(x - 2)^3$.

Solution

$(x - 2)^3 = (x - 2)(x - 2)(x - 2)$	Meaning of exponent.
$= (x - 2)[(x - 2)(x - 2)]$	Associative property.
$= (x - 2)(x^2 - 2x - 2x + 4)$	Multiplying within brackets.
$= (x - 2)(x^2 - 4x + 4)$	Combining like terms.
$= x(x^2 - 4x + 4) - 2(x^2 - 4x + 4)$	Distributive property.
$= x^3 - 4x^2 + 4x - 2x^2 + 8x - 8$	Distributive property.
$= x^3 - 6x^2 + 12x - 8$	Combining like terms.

Now you complete Example 14.

Example 14. Find $(2x^2 + x - 3)(x^2 - 5x + 1)$.

Solution

$$(2x^2 + x - 3)(x^2 - 5x + 1)$$
$$= 2x^2(x^2 - 5x + 1) + x(x^2 - 5x + 1) - 3(x^2 - 5x + 1)$$

Check your work on page 206. ▶

⫸ Trial Run

Simplify.

_____ **1.** $(x - 1)(x^2 - 3x - 40)$ _____ **2.** $(a - 3)^3$

_____ **3.** $(2x^2 - x + 1)(3x^2 - 2x + 2)$ _____ **4.** $(2x + y)(x^2 - 5xy + 4y^2)$

Answers are on page 207.

Recognizing Special Products

There are certain products of polynomials that crop up over and over again in algebra. Because they occur so frequently, it is worth our while to recognize these special products and learn to quickly compute them.

Let's find a few products and see if we can recognize a pattern.

$(x + 3)(x - 3)$	$(5 - ay)(5 + ay)$	$(6x - 7y)(6x + 7y)$
$= x^2 - 3x + 3x - 9$	$= 25 + 5ay - 5ay + a^2y^2$	$= 36x^2 + 42xy - 42xy - 49y^2$
$= x^2 - 9$	$= 25 - a^2y^2$	$= 36x^2 - 49y^2$

In each of these products, the factors are binomials that are *the sum and the difference of the same two terms*. Each answer is the *difference of the squares* of those same two terms.

In fact, this result will always occur.

> If we multiply the sum of two terms by the difference of the same two terms, the product will be the difference of the squares of those two terms.

We call this special product the **difference of two squares** and we note that for any algebraic expressions A and B:

Product of Sum and Difference of Two Terms

$$(A + B)(A - B) = A^2 - B^2$$

You use this formula to complete Example 15.

Example 15. Find $\left(3x + \dfrac{2}{5}\right)\left(3x - \dfrac{2}{5}\right)$.

Solution

$$\left(3x + \frac{2}{5}\right)\left(3x - \frac{2}{5}\right)$$

$$= (3x)^2 - (\underline{\quad})^2$$

$$= 9x^2 - \underline{\quad}$$

Check your work on page 207. ▶

Then complete Example 16.

Example 16. Find $(8 - xy)(8 + xy)$.

Solution

$$(8 - xy)(8 + xy)$$

$$= (\underline{\quad})^2 - (\underline{\quad})^2$$

$$= \underline{\quad} - x^2y^2$$

Check your work on page 207. ▶

We know how to find the *square* of a binomial by multiplying that binomial times itself. Can you observe a pattern in squaring the following binomials?

$(x + 5)^2$

$= (x + 5)(x + 5)$

$= x^2 + 5x + 5x + 25$

$= x^2 + 10x + 25$

$(7 - x)^2$

$= (7 - x)(7 - x)$

$= 49 - 7x - 7x + x^2$

$= 49 \times 14x + x^2$

$(a + b)^2$

$= (a + b)(a + b)$

$= a^2 + ab + ab + b^2$

$= a^2 + 2ab + b^2$

Look closely at the first example above.

first term	second term	square of first term	twice product of first and second terms	square of second term
↘	↙	↓	↓	↓

$$(x + 5)^2 \quad = \quad x^2 \quad + \quad 2(5 \cdot x) \quad + \quad 5^2$$

$$= \quad x^2 \quad + \quad 10x \quad + \quad 25$$

When we square a binomial, the product will be a trinomial in which:

1. The first term is the square of the first term in the binomial.
2. The third term is the square of the second term in the binomial.
3. The middle term is twice the product of the first and second terms in the binomial.

Using A and B to represent any algebraic expressions, we state the following rule for finding the **square of a binomial.**

Square of Binomial

$$(A + B)^2 = A^2 + 2AB + B^2$$

$$(A - B)^2 = A^2 - 2AB + B^2$$

Such products are called **perfect square trinomials.** To be honest, it is not crucial that students learn the rule for squaring a binomial, since other methods for multiplying binomials are almost as quick as using the rule. However, it is helpful when squaring complicated binomials to have such a rule available.

You practice this rule for squaring a binomial in Example 17.

Example 17. Find $(3x + 8)^2$.

Solution

$$(3x + 8)^2$$
$$= (3x)^2 + 2(3x)(8) + 8^2$$

Check your work on page 207. ▶

Example 18. Find $(6 - 0.5a)$.

Solution

$$(6 - 0.5a)$$
$$= 6^2 - 2(6)(0.5a) + (0.5a)^2$$
$$= 36 - 2(3a) + 0.25a^2$$
$$= 36 - 6a + 0.25a^2$$

Now complete Example 19.

Example 19. Find $(x^2y - z)^2$.

Solution

$$(x^2y - z)^2$$
$$= (x^2y)^2 - 2(x^2y)(z) + z^2$$
$$=$$

Check your work on page 207. ▶

Example 20. Find $[(x + 1) + c]^2$.

Solution: Here $A = x + 1$

$$B = c$$

$$[(x + 1) + c]^2$$
$$= (x + 1)^2 + 2(x + 1)(c) + c^2$$
$$= x^2 + 2x + 1 + 2c(x + 1) + c^2$$
$$= x^2 + 2x + 1 + 2cx + 2c + c^2$$

Example 21. Find $[(x + 3) - 5y][(x + 3) + 5y]$.

Solution: We must multiply the sum and difference of two quantities. Here $A = (x + 3)$ and $B = 5y$. We therefore expect our answer to be the *difference of two squares*, $A^2 - B^2$.

$$[(x + 3) - 5y][(x + 3) + 5y]$$
$$= (x + 3)^2 - (5y)^2$$
$$= x^2 + 6x + 9 - 25y^2$$

Students sometimes forget the rule for squaring binomials and try shortcuts that simply will *not* work.

$$(x + 3)^2 \neq x^2 + 9 \qquad (a - 5)^2 \neq a^2 - 25$$
$$(x + 3)^2 = x^2 + 6x + 9 \qquad (a - 5)^2 = a^2 - 10a + 25$$

Squaring a binomial yields a trinomial; we do *not* square a binomial simply by squaring each term.

⇒ Trial Run

Find the special products.

_____ **1.** $(x - 9)(x + 9)$ _ _____ **2.** $(5x - 7)(5x + 7)$

_____ **3.** $(3a + 7b)(3a - 7b)$ _____ **4.** $(a^2 - b^2)(a^2 + b^2)$

_____ **5.** $(2x - 5)^2$ _____ **6.** $(3a + 2)^2$

_____ **7.** $(9 - 4x)^2$ _____ **8.** $(7x + 5y)^2$

Answers are on page 207.

▶ Examples You Completed

Example 1. Multiply $(7x - 1)(x - 2)$.

Solution

$$(7x - 1)(x - 2)$$
$$= 7x(x - 2) - 1(x - 2)$$
$$= 7x^2 - 14x - x + 2$$
$$= 7x^2 - 15x + 2$$

Example 4. Find $(a + 6)(a - 6)$.

Solution

$$(a + 6)(a - 6)$$
$$= a(a - 6) + 6(a - 6)$$
$$= a^2 - 6a + 6a - 36$$
$$= a^2 - 36$$

Example 5. Find $(7x - 3)(2x + 4)$.

Solution

$$\text{(7x − 3)(2x + 4)}$$

$$= 14x^2 + 28x - 6x - 12$$
$$= 14x^2 + 22x - 12$$

Example 7. Find $(5x - 2)^2$.

Solution

$$(5x - 2)^2 = (5x - 2)(5x - 2)$$

$$= 25x^2 - 10x - 10x + 4$$
$$= 25x^2 - 20x + 4$$

Example 8. Simplify $x(2x + 1) + (x + 3)(4x + 2)$.

Solution

$$x(2x + 1) + (x + 3)(4x + 2)$$
$$= 2x^2 + x + 4x^2 + 2x + 12x + 6$$
$$= 6x^2 + 15x + 6$$

Example 12. Find $(x + y)(x^2 - xy + y^2)$.

Solution

$$(x + y)(x^2 - xy + y^2)$$
$$= x(x^2 - xy + y^2) + y(x^2 - xy + y^2)$$
$$= x^3 - x^2y + xy^2 + x^2y - xy^2 + y^3$$
$$= x^3 + y^3$$

Example 14. Find $(2x^2 + x - 3)(x^2 - 5x + 1)$.

Solution

$$(2x^2 + x - 3)(x^2 - 5x + 1)$$
$$= 2x^2(x^2 - 5x + 1) + x(x^2 - 5x + 1) - 3(x^2 - 5x + 1)$$
$$= 2x^4 - 10x^3 + 2x^2 + x^3 - 5x^2 + x - 3x^2 + 15x - 3$$
$$= 2x^4 - 9x^3 - 6x^2 + 16x - 3$$

Example 15. Find $\left(3x + \dfrac{2}{5}\right)\left(3x - \dfrac{2}{5}\right)$

Solution

$$\left(3x + \frac{2}{5}\right)\left(3x - \frac{2}{5}\right)$$

$$= (3x)^2 - \left(\frac{2}{5}\right)^2$$

$$= 9x^2 - \frac{4}{25}$$

Example 16. Find $(8 - xy)(8 + xy)$.

Solution

$$(8 - xy)(8 + xy)$$

$$= (8)^2 - (xy)^2$$

$$= 64 - x^2y^2$$

Example 17. Find $(3x + 8)^2$.

Solution

$$(3x + 8)^2$$

$$= (3x)^2 + 2(3x)(8) + 8^2$$

$$= 9x^2 + 2(24x) + 64$$

$$= 9x^2 + 48x + 64$$

Example 19. Find $(x^2y - z)^2$.

Solution

$$(x^2y - z)^2$$

$$= (x^2y)^2 - 2(x^2y)(z) + z^2$$

$$= x^4y^2 - 2x^2yz + z^2$$

Answers to Trial Runs

page 201 **1.** $2a^2 + 9a - 35$ **2.** $49x^2 - 28xy + 4y^2$ **3.** $49a^2 - 9$ **4.** $27 - 21y + 2y^2$
5. $25x^2 - 4y^2$ **6.** $3a^2 + ab - 10b^2$

page 202 **1.** $4a^2 + 3a - 2$ **2.** $3x^2 + 9x - 1$ **3.** $-3x^2 - 46x - 91$ **4.** $a^3 - 16a^2 - 5a + 32$

page 203 **1.** $x^3 - 4x^2 - 37x + 40$ **2.** $a^3 - 9a^2 + 27a - 27$ **3.** $6x^4 - 7x^3 + 9x^2 - 4x + 2$
4. $2x^3 - 9x^2y + 3xy^2 + 4y^3$

page 206 **1.** $x^2 - 81$ **2.** $25x^2 - 49$ **3.** $9a^2 - 49b^2$ **4.** $a^4 - b^4$ **5.** $4x^2 - 20x + 25$
6. $9a^2 + 12a + 4$ **7.** $81 - 72x + 16x^2$ **8.** $49x^2 + 70xy + 25y^2$

EXERCISE SET 3.4

Perform the indicated operations and simplify.

_____ **1.** $(4y - 1)(3y - 5)$

_____ **2.** $(5y - 2)(3y - 1)$

_____ **3.** $(6x - 5)(2x + 3)$

_____ **4.** $(5x - 3)(3x + 4)$

_____ **5.** $(9 - 2y)(2 - y)$

_____ **6.** $(8 - 3y)(3 - y)$

_____ **7.** $(4x + 3y)(7x + 2y)$

_____ **8.** $(9x + y)(5x + 7y)$

_____ **9.** $\left(a + \frac{1}{3}b\right)^2$

_____ **10.** $\left(a + \frac{1}{5}b\right)^2$

_____ **11.** $(0.2x - 0.3y)(0.2x + 0.3y)$

_____ **12.** $(0.5x - 0.4y)(0.5x + 0.4y)$

_____ **13.** $(xy - 10)(xy + 5)$

_____ **14.** $(xy - 7)(2xy + 1)$

_____ **15.** $(3xy - 2z)^2$

_____ **16.** $(4xy - 3z)^2$

_____ **17.** $(0.6a - 5b)(0.6a + 5b)$

_____ **18.** $(1.1a - 3b)(1.1a + 3b)$

_____ **19.** $(x^2 - 3y^2)(3x^2 + 8y^2)$

_____ **20.** $(2x^2 - y^2)(4x^2 + 5y^2)$

_____ **21.** $\left(a^2 + \frac{1}{2}b^2\right)^2$

_____ **22.** $\left(\frac{1}{3}a^2 - b^2\right)^2$

_____ **23.** $(x - 3)(x^2 - 2x + 1)$

_____ **24.** $(x - 5)(x^2 - 6x + 9)$

_____ **25.** $-y^2(8 - y)(7 + y)$

_____ **26.** $-y^2(5 - y)(4 + y)$

_____ **27.** $x(2x - 1) + (x - 3)(x + 5)$

_____ **28.** $2x(x - 3) + (x - 2)(x + 4)$

_____ **29.** $(3a + 2)(2a + 5) - (a - 3)(2a + 1)$

_____ **30.** $(7a - 1)(a + 2) - (a - 4)(a + 2)$

_____ **31.** $(x + 1)^2 - 5(x - 3)^2$

_____ **32.** $(x + 2)^2 - 7(x - 1)^2$

_____ **33.** $2(x - 1)(2x - 3) + 3(5x - 2)^2$

_____ **34.** $4(x - 7)(3x + 1) + 2(3x - 4)^2$

_____ **35.** $a(a - 4)(a + 4) - 11a(a - 2)$

_____ **36.** $a(a - 1)(a - 8) - 3a(a - 9)$

_____ **37.** $(a - b)(a + 2b)(a - 5b)$

_____ **38.** $(a - 2b)(a + b)(a - 3b)$

_____ **39.** $(x + 2)(x - 3)^2$

_____ **40.** $(x - 4)(x - 2)^2$

_____ **41.** $(a + 2)^3$

_____ **42.** $(a + 5)^3$

_____ **43.** $(x^2 - 1)(x^3 - 2x^2 + x - 5)$

_____ **44.** $(x^2 - 2)(x^3 - 7x^2 + 3x - 1)$

_____ **45.** $(3x^2 - x - 5)(x^2 + 2x - 1)$

_____ **46.** $(2x^2 - x - 3)(x^2 + 3x - 7)$

_____ **47.** $(a^2 - 5b^2)^2$

_____ **48.** $(a^2 + 7b^2)^2$

_____ **49.** $(x^2 - 2y^2)(x^2 + 2y^2)$

_____ **50.** $(3x^2 - y^2)(3x^2 + y^2)$

_____ **51.** $[(x - y) - 5]^2$

_____ **52.** $[(x - 3y) - 9]^2$

_____ **53.** $[(a + 4) - 3b][(a + 4) + 3b]$

_____ **54.** $[(a + 2) - 5b][(a + 2) + 5b]$

_____ **55.** $(2 - x)^4$

_____ **56.** $(3 - x)^4$

☆ Stretching the Topics ─────────────────────

Perform the operations and simplify.

_____ **1.** $(2a - b - 3c)^2$

_____ **2.** $(4x^{3a} - 7y^b)(2x^{3a} - y^b)$

_____ **3.** $(x^{-1} - y^{-1})(x^{-2} + 2x^{-1}y^1 + 4y^{-2})$

Check your answers in the back of your book.

If you can complete **Checkup 3.4,** you are ready to go on to Section 3.5.

✔ # CHECKUP 3.4

Perform the indicated operations and simplify.

_____ **1.** $(2x - 3)(x + 7)$

_____ **2.** $(a - 10b)^2$

_____ **3.** $(3x - 5y)(3x + 5y)$

_____ **4.** $(x - 3)(x^2 - 5x - 6)$

_____ **5.** $2x(7x - 3y) - (x - 2y)^2$

_____ **6.** $4(a - 1)(a - 9) - (a - 9)(2a - 3)$

_____ **7.** $(x - y)(x + y)(x - 2y)$

_____ **8.** $(a - 5)^3$

_____ **9.** $(x^2 - x + 1)(x^2 - 3x + 2)$

_____ **10.** $[(a + b) - 9]^2$

Check your answers in the back of your book.

If You Missed Problems:	You Should Review Examples:
1	1
2	3, 17, 18
3	4, 15, 16
4	11
5	10
6	8, 9
7, 8	13
9	14
10	20

3.5 Dividing Polynomials

Dividing Polynomials by Monomials

To divide a polynomial by a monomial, we return to the definition of division stated in Chapter 1. Recall that division was defined as multiplication by the reciprocal of the divisor.

Consider the division problem

$$\frac{6x - 12y}{3}$$

and see how the definition of division and the distributive property are used in Approach 1. Then look at Approach 2.

Approach 1	*Approach 2*
$\dfrac{6x - 12y}{3}$	$\dfrac{6x - 12y}{3}$
$= \dfrac{1}{3}(6x - 12y)$	$= \dfrac{6x}{3} - \dfrac{12y}{3}$
$= \dfrac{1}{3}(6x) - \dfrac{1}{3}(12y)$	$= 2x - 4y$
$= 2x - 4y$	

Since our quotient is the same using either approach and Approach 2 is simpler, we agree that

> To divide a polynomial by a monomial, we may divide each term of the polynomial dividend by the monomial divisor.

This technique can be used to divide any polynomial by a monomial and changes such divisions into a sum of monomial fractions that can be simplified using the laws of exponents. Complete Example 1.

Example 1. Find $(6x^3y - 3x^2y^2 + 9xy^2) \div 3xy$.

Solution

$$\frac{6x^3y - 3x^2y^2 + 9xy^2}{3xy}$$

$$= \frac{6x^3y}{3xy} - \frac{3x^2y^2}{3xy} + \frac{9xy^2}{3xy}$$

Check your work on page 220. ▶

Example 2. Find $\dfrac{-2ab^2 + 2a^2b - a^2}{-2a^2b}$.

Solution

$$\frac{-2ab^2 + 2a^2b - a^2}{-2a^2b}$$

$$= \frac{-2ab^2}{-2a^2b} + \frac{2a^2b}{-2a^2b} - \frac{a^2}{-2a^2b}$$

$$= \frac{b}{a} - 1 + \frac{1}{2b}$$

⫸ Trial Run

Divide.

_____ **1.** $(x^4 - 3x^3 + 2x^2 - x) \div x$ _____ **2.** $(8a^3b - 2a^2b^2 + 6ab^3) \div 2ab$

_____ **3.** $\dfrac{x^2 - 9x + 1}{-3x}$ _____ **4.** $\dfrac{x^4y - 3x^3y^2 + 9x^2y^3 - 12xy^4}{3xyz}$

Answers are on page 220.

Dividing Polynomials Using Long Division

When dividing a polynomial by another polynomial that is not a monomial, we must use **long-division** techniques which are similar to long-division techniques used for dividing whole numbers. Let's briefly review long division of whole numbers to remind ourselves exactly what is done in each step.

$$
\begin{array}{r}
26 \\
21\overline{)547} \\
42 \\
\hline
127 \\
126 \\
\hline
1
\end{array}
\qquad\qquad
\begin{array}{r}
27 \\
13\overline{)351} \\
26 \\
\hline
91 \\
91 \\
\hline
0
\end{array}
$$

$$\frac{547}{21} = 26 \text{ remainder } 1 \qquad\qquad \frac{351}{13} = 27$$

or

$$\frac{547}{21} = 26\,\frac{1}{21}$$

To check a long-division problem in arithmetic, recall that we must multiply the divisor times the quotient, then add the remainder to see whether the dividend is obtained. Here

$$21(26) + 1 \overset{?}{=} 547 \qquad\qquad 13(27) + 0 \overset{?}{=} 351$$

$$546 + 1 \overset{?}{=} 547 \qquad\qquad 351 + 0 \overset{?}{=} 351$$

$$547 = 547 \qquad\qquad 351 = 351$$

In general, it will be true that

$$\text{dividend} = (\text{divisor} \times \text{quotient}) + \text{remainder}$$

What is the significance of a *zero* remainder in arithmetic? When no remainder occurs, we know that the divisor divides the dividend exactly. Therefore, the divisor and quotient are *factors* of the dividend. Here, for example,

$$21 \text{ is } \textit{not} \text{ a factor of } 547 \qquad\qquad 13 \text{ } \textit{is} \text{ a factor of } 351$$
$$27 \text{ } \textit{is} \text{ a factor of } 351$$

Let's keep these ideas in mind as we approach long division of polynomials, starting with the quotient $(x^2 + 5x - 14) \div (x - 2)$.

$$x - 2\overline{)x^2 + 5x - 14} \qquad\qquad \text{Set up quotient as long-division problem.}$$

$$\begin{array}{r} x\phantom{{}^2+5x-14} \\ x-2\overline{)x^2+5x-14} \end{array}$$

Divide first term (x) of divisor into first term (x^2) of dividend to get first term (x) of quotient.

$$\begin{array}{r} x\phantom{{}^2+5x-14} \\ x-2\overline{)x^2+5x-14} \\ \underline{x^2-2x} \end{array}$$

Multiply quotient term times *entire* divisor and write product beneath like terms in dividend.

$$\begin{array}{r} x\phantom{{}^2+5x-14} \\ x-2\overline{)x^2+5x-14} \\ \underline{{}^\ominus x^2 \stackrel{\oplus}{-} 2x} \\ 7x-14 \end{array}$$

Subtract carefully (change the sign and add) and bring down the next term of the dividend (-14).

$$\begin{array}{r} x+7 \\ x-2\overline{)x^2+5x-14} \\ \underline{x^2-2x} \\ 7x-14 \\ \underline{{}^\ominus 7x \stackrel{\oplus}{-} 14} \\ 0 \end{array}$$

Divide first term (x) of divisor into first term ($7x$) of new dividend. Then multiply and subtract as before.

We say that

$$\frac{x^2+5x-14}{x-2} = x+7$$

so

$$x^2+5x-14 = (x+7)(x-2)$$

and since the remainder is zero, we conclude that

$$x-2 \text{ is a factor of } x^2+5x-14$$

$$x+7 \text{ is a factor of } x^2+5x-14$$

In performing long division of polynomials, your work will be simplified if you always remember to write the terms of the dividend and divisor in descending powers of the same variable and to leave a space in the dividend for any missing terms. Keep these ideas in mind as you look at Example 3.

Example 3. Divide $5x^2 + x^4 + 10x - 8$ by $x^2 + x - 1$.

Solution: We arrange the terms in descending powers of x and note the missing x^3-term in the dividend.

$$\begin{array}{r} x^2-x+7 \\ x^2+x-1\overline{)x^4+0x^3+5x^2+10x-8} \\ \underline{{}^\ominus x^4 \stackrel{\ominus}{+} x^3 \stackrel{\oplus}{-} x^2} \\ -x^3+6x^2+10x \\ \underline{\stackrel{\oplus}{-}x^3 \stackrel{\oplus}{-} x^2 \stackrel{\oplus}{+}x} \\ 7x^2+9x-8 \\ \underline{{}^\ominus 7x^2 \stackrel{\ominus}{+}7x \stackrel{\oplus}{-} 7} \\ 2x-1 \end{array}$$

Since the degree of the last dividend ($2x - 1$) is less than the degree of the divisor ($x^2 + x - 1$), we stop the long-division process, writing $2x - 1$ as the remainder.

$$\frac{5x^2 + x^4 + 10x - 8}{x^2 + x - 1} = x^2 - x + 7, \quad \text{remainder } 2x - 1$$

$$= x^2 - x + 7 + \frac{2x - 1}{x^2 + x - 1}$$

We have noted four important points to keep in mind when performing long division of polynomials.

1. Write terms of the dividend and divisor in *descending* powers of the same variable.
2. Use a *zero placeholder* for any missing terms in the dividend.
3. Always carefully *subtract* the product of the divisor and the new quotient term after each division.
4. When the difference is an expression with degree less than the degree of the divisor, that difference is the *remainder*.

Example 4. Find $(8a^3 - 10a^2b + 5ab^2 + b^3) \div (2a - b)$.

Solution

$$
\begin{array}{r}
4a^2 - 3ab + b^2 \\
2a - b \overline{)\,8a^3 - 10a^2b + 5ab^2 + b^3} \\
\underline{8a^3 - 4a^2b} \\
-6a^2b + 5ab^2 \\
\underline{-6a^2b + 3ab^2} \\
2ab^2 + b^3 \\
\underline{2ab^2 - b^3} \\
2b^3
\end{array}
$$

$$\frac{8a^3 - 10a^2b + 5ab^2 + b^3}{2a - b} = 4a^2 - 3ab + b^2 + \frac{2b^3}{2a - b}$$

Now you carefully complete Example 5.

Example 5. Find $(4x^4 - 2x^3 + 5x - 1) \div (x + 1)$.

Solution

$$
\begin{array}{r}
4x^3 - 6x^2 \\
x + 1\overline{\smash{)}4x^4 - 2x^3 + 0x^2 + 5x - 1} \\
\underline{\ominus\ominus} \\
4x^4 + 4x^3 \\
-6x^3 + 0x^2 \\
\underline{\oplus\oplus} \\
-6x^3 - 6x^2
\end{array}
$$

$$
\frac{4x^4 - 2x^3 + 5x - 1}{x + 1} = \underline{\hspace{2cm}}
$$

Is $x + 1$ a factor of $4x^4 - 2x^3 + 5x - 1$? $\underline{\hspace{2cm}}$

Check your work on page 220. ▶

⫸ Trial Run

Divide.

$\underline{\hspace{2cm}}$ **1.** $(x^3 - 5x^2 + 2x + 10) \div (x - 2)$

$\underline{\hspace{2cm}}$ **2.** $(2x^3 - 11x^2 + 10x + 3) \div (2x - 3)$

$\underline{\hspace{2cm}}$ **3.** $(3x^4 - 2x^2 + 9) \div (x + 3)$

Answers are on page 220.

Dividing Polynomials Using Synthetic Division

Long division can be a tedious process, but there is a shortcut that can be used under certain circumstances. When a polynomial in x is to be divided by a divisor of the form $x - c$, where c is a constant, we may use the technique of **synthetic division** to find the quotient.

Recall this long-division problem worked earlier.

$$
\begin{array}{r}
x + 7 \leftarrow \text{quotient} \\
\text{divisor} \rightarrow x - 2\overline{\smash{)}x^2 + 5x - 14} \leftarrow \text{dividend} \\
\underline{\ominus\oplus} \\
x^2 - 2x \\
7x - 14 \\
\underline{\ominus\oplus} \\
7x - 14 \\
0 \leftarrow \text{remainder}
\end{array}
$$

This was not a difficult problem, but we had to be very careful with our subtractions. In synthetic division, the subtractions are replaced by additions. Moreover, the polynomial dividend is replaced by its numerical coefficients only, and the binomial divisor is replaced by the constant c (called the **synthetic divisor**).

The steps in finding the quotient $(x^2 + 5x - 14) \div (x - 2)$ by synthetic division are outlined below. Note here that the synthetic divisor is 2.

$$\begin{array}{r|rrr} 2 & 1 & 5 & -14 \\ & & & \\ \hline & & & \end{array}$$
Write the synthetic divisor and the coefficients of the dividend.

$$\begin{array}{r|rrr} 2 & 1 & 5 & -14 \\ & \downarrow & & \\ \hline & 1 & & \end{array}$$
Bring down the first term of the synthetic dividend.

$$\begin{array}{r|rrr} 2 & 1 & 5 & -14 \\ & \downarrow & 2 & \\ \hline & 1 & 7 & \end{array}$$
Multiply the synthetic divisor (2) times the first term in the quotient (1) and write that product (2) under the second term in the dividend (5). Then *add*.

$$\begin{array}{r|rrr} 2 & 1 & 5 & -14 \\ & \downarrow & 2 & 14 \\ \hline & 1 & 7 & 0 \end{array}$$
Multiply the synthetic divisor (2) times the next term in the quotient (7) and write that product (14) under the next term in the dividend (-14). *Add*.

synthetic divisor   coefficients of dividend

coefficients of quotient

The last term in the quotient (0) is the *remainder*. The other terms are the coefficients of the quotient polynomial.

$$x + 7$$

The *degree* of the quotient polynomial is always *one less* than the degree of the dividend.

Using synthetic division, we have found that

$$(x^2 + 5x - 14) \div (x - 2) = x + 7$$

Example 6. Use synthetic division to find the quotient $(x^3 + 3x^2 + 3x + 1) \div (x + 1)$.

Solution: The divisor is $x + 1$, so the synthetic divisor is -1. The quotient will be of *second* degree.

$$\begin{array}{r|rrrr} -1 & 1 & 3 & 3 & 1 \\ & \downarrow & -1 & -2 & -1 \\ \hline & 1 & 2 & 1 & 0 \end{array}$$
$$1x^2 + 2x + 1$$

So

$$(x^3 + 3x^2 + 3x + 1) \div (x + 1) = x^2 + 2x + 1$$

Example 7. Use synthetic division to find the quotient $(x^4 + 5x^2 + 100x - 30) \div (x + 4)$.

Solution: The divisor is $x + 4$, so the synthetic divisor is -4. The quotient will be of *third* degree. Notice the missing x^3-term in the dividend for which we must use a *zero* coefficient.

$$
\begin{array}{r|rrrr}
-4 & 1 & 0 & 5 & 100 & -30 \\
 & \downarrow & -4 & 16 & -84 & -64 \\
\hline
 & 1 & -4 & 21 & 16 & \boxed{-94}
\end{array}
$$

$$x^3 - 4x^2 + 21x + 16 \;\; \text{R-}94$$

$$(x^4 + 5x^2 + 100x - 30) \div (x + 4) = x^3 - 4x^2 + 21x + 16 - \frac{94}{x + 4}$$

For emphasis we repeat that synthetic division can be used only when a polynomial is to be divided by a divisor of the form $x - c$. It can be used only for a divisor that is a first-degree binomial with a leading coefficient of 1.

Now try Example 8.

Example 8. Use synthetic division to find $(x^3 + 2x^2 - 3x - 1) \div (x - 3)$.

Solution: The synthetic divisor is _____ . The quotient will be of _____ degree.

$$
\begin{array}{r|rrrr}
3 & 1 & 2 & -3 & -1 \\
 & \downarrow & 3 & & \\
\hline
 & 1 & \underline{\quad} & \underline{\quad} & \boxed{\underline{\quad}}
\end{array}
$$

$$x^2 + \underline{\quad}x + \underline{\quad}\; \text{R}\underline{\quad}$$

So

$$(x^3 + 2x^2 - 3x - 1) \div (x - 3)$$

$$= x^2 + \underline{\quad}x + \underline{\quad} + \frac{\underline{\quad}}{x - 3}$$

Check your work on page 220. ▶

Example 9. Use synthetic division to find $(y^5 + 1) \div (y + 1)$.

Solution: The synthetic divisor is -1. The quotient will be of fourth degree.

$$
\begin{array}{r|rrrrrr}
-1 & 1 & 0 & 0 & 0 & 0 & 1 \\
 & \downarrow & -1 & 1 & -1 & 1 & -1 \\
\hline
 & 1 & -1 & 1 & -1 & 1 & \boxed{0}
\end{array}
$$

$$y^4 - y^3 + y^2 - y + 1$$

So

$$(y^5 + 1) \div (y + 1)$$

$$= y^4 - y^3 + y^2 - y + 1$$

▥➡ **Trial Run**

Divide using synthetic division.

_____ **1.** $(x^2 - 3x - 4) \div (x - 2)$

_____ **2.** $(2x^4 - x^3 - 20x^2 - 9) \div (x + 3)$

_____ **3.** $(a^4 - 1) \div (a - 1)$

Answers are on page 220.

▶ Examples You Completed

Example 1. Find $(6x^3y - 3x^2y^2 + 9xy^2) \div 3xy$.

Solution

$$\frac{6x^3 - 3x^2y^2 + 9xy^2}{3xy}$$

$$= \frac{6x^3y}{3xy} - \frac{3x^2y^2}{3xy} + \frac{9xy^2}{3xy}$$

$$= 2x^2 - xy + 3y$$

Example 8. Use synthetic division to find $(x^3 + 2x^2 - 3x - 1) \div (x - 3)$.

Solution: The synthetic divisor is 3. The quotient will be of second degree.

$$
\begin{array}{r|rrrr}
3 & 1 & 2 & -3 & -1 \\
  & \downarrow & 3 & 15 & 36 \\
\hline
  & 1 & 5 & 12 & \boxed{35}
\end{array}
$$

$$x^2 + 5x + 12 \quad R35$$

So

$$(x^3 + 2x^2 - 3x - 1) \div (x - 3)$$

$$= x^2 + 5x + 12 + \frac{35}{x - 3}$$

Example 5. Find $(4x^4 - 2x^3 + 5x - 1) \div (x + 1)$.

Solution

$$
\begin{array}{r}
4x^3 - 6x^2 + 6x - 1 \\
x + 1 \overline{)4x^4 - 2x^3 + 0x^2 + 5x - 1} \\
\underline{4x^4 + 4x^3} \\
-6x^3 + 0x^2 \\
\underline{-6x^3 - 6x^2} \\
6x^2 + 5x \\
\underline{6x^2 + 6x} \\
-x - 1 \\
\underline{-x - 1} \\
0
\end{array}
$$

$$(4x^4 - 2x^3 + 5x - 1) \div (x + 1)$$
$$= 4x^3 - 6x^2 + 6x - 1$$

Is $x + 1$ a factor of $4x^4 - 2x^3 + 5x - 1$? yes

Answers to Trial Runs

page 214   **1.** $x^3 - 3x^2 + 2x - 1$ **2.** $4a^2 - ab + 3b^2$ **3.** $-\frac{x}{3} + 3 - \frac{1}{3x}$

4. $\frac{x^3}{3z} - \frac{x^2y}{z} + \frac{3xy^2}{z} - \frac{4y^3}{z}$

page 217   **1.** $x^2 - 3x - 4 + \frac{2}{x - 2}$ **2.** $x^2 - 4x - 1$ **3.** $3x^3 - 9x^2 + 25x - 75 + \frac{234}{x + 3}$

page 219   **1.** $x - 1 - \frac{6}{x - 2}$ **2.** $2x^3 - 7x^2 + x - 3$ **3.** $a^3 + a^2 + a + 1$

EXERCISE SET 3.5

Divide.

_____ **1.** $(x^4 - 7x^3 + 3x^2 + 8x) \div x$

_____ **2.** $(3x^4 - 2x^3 - x^2 + 2x) \div x$

_____ **3.** $(9a^3b - 6a^2b^2 + 15ab^3) \div 3ab$

_____ **4.** $(16a^4b^2 - 8a^3b^3 - 24a^2b^4) \div 4ab$

_____ **5.** $(3x^2 - 6x + 5) \div 3x$

_____ **6.** $(5x^2 - 10x + 1) \div 5x$

_____ **7.** $(2xy - 8xyz + 10x^2y^2) \div (-2xy)$

_____ **8.** $(5xy - 15xyz - 20x^2y^2) \div (-5xy)$

_____ **9.** $(x^6 - 3x^5 + 2x^4 - x^3 + 3x^2) \div x^3$

_____ **10.** $(4x^6 - 8x^5 + 2x^4 - 10x^3 - 5x^2) \div 2x^2$

_____ **11.** $(x^2 - 12x + 32) \div (x - 4)$

_____ **12.** $(x^2 - 9x - 10) \div (x + 1)$

_____ **13.** $(8x^2 + 10x + 3) \div (2x + 1)$

_____ **14.** $(12x^2 - 4x - 1) \div (2x - 1)$

_____ **15.** $(2x^2 + x - 12) \div (2x + 5)$

_____ **16.** $(12x^2 + 17x - 6) \div (4x + 7)$

_____ **17.** $(x^3 - 3x^2 + 5x + 7) \div (x + 1)$

_____ **18.** $(2x^3 + 4x^2 + x - 24) \div (x - 2)$

_____ **19.** $(2a^3 - 7a^2 - 11a + 4) \div (2a - 1)$

_____ **20.** $(3a^3 + a^2 - a - 15) \div (3a - 5)$

_____ **21.** $(x^3 - 3x^2y + 5xy^2 - 6y^3) \div (x - 2y)$

_____ **22.** $(x^3 - x^2y + 2xy^2 + 4y^3) \div (x + y)$

_____ **23.** $(x^4 - 17x^2 + 20x - 6) \div (x^2 + 4x - 3)$

_____ **24.** $(x^4 - 4x^2 + 12x - 9) \div (x^2 - 2x + 3)$

_____ **25.** $(x^3 - 8) \div (x - 2)$

_____ **26.** $(x^3 + 27) \div (x + 3)$

_____ **27.** $(2x^4 - 2x^2 + 1) \div (2x - 4)$

_____ **28.** $(2x^3 - x^2 - 7x + 2) \div (2x + 1)$

_____ **29.** $(8x^4 - 6x^3 + 11x^2 - 9) \div (4x^2 - x - 3)$

_____ **30.** $(10x^4 - 11x^3 + 12x^2 - 4x - 2) \div (2x^2 - x - 1)$

Divide using synthetic division.

_____ **31.** $(x^4 - 2x^3 - x^2 - 1) \div (x + 1)$

_____ **32.** $(x^4 + x^3 - 3x^2 + 7x - 6) \div (x + 2)$

_____ **33.** $(3x^4 - 16x^3 + 6x^2 - 5x) \div (x - 5)$

_____ **34.** $(2x^4 - 6x^3 + 4x - 12) \div (x - 3)$

_____ **35.** $(x^4 + 16) \div (x + 2)$

_____ **36.** $(x^5 + 32) \div (x + 2)$

_____ **37.** $(x^5 + 3x^4 - 10x^3 + x^2 + 3x - 12) \div (x + 5)$

_____ **38.** $(x^5 + 2x^4 + 8x^3 + x^2 + x - 15) \div (x + 4)$

_____ **39.** $(x^6 - 2x^5 + 2x^4 - 2x^3 + 2x^2 - 2) \div (x - 1)$

_____ **40.** $(x^6 - 4x^5 + 5x^4 - 4x^3 + 5x^2 - 4x + 5) \div (x - 2)$

☆ Stretching the Topics ——————————————

Divide.

_____ **1.** $(8x^3y - 6x^2y + 6x^2y^2 - 5xy^2 + xy^3) \div (2x + y)$

_____ **2.** $(x^{3a} + 3x^{2a} - x^a - 3) \div (x^a + 1)$

_____ **3.** Use synthetic division to determine if $n - 2$ is a factor of the polynomial $5n^5 + 10n^4 - 10n^3 - 34n + 33$. Evaluate the polynomial when $n = 2$.

Check your answers in the back of your book.

If you can complete **Checkup 3.5,** you are ready to go on to Section 3.6.

CHECKUP 3.5

Divide.

_____ **1.** $(14x^3y - 7x^2y^2 + 21xy^3) \div 7xy$

_____ **2.** $(x^4 - 2x^3 + 4x^2 - 8x) \div (-2x^2)$

_____ **3.** $(6x^2 - 5x - 21) \div (3x - 7)$

_____ **4.** $(x^3 - x^2 - 17x - 17) \div (x + 3)$

_____ **5.** $(2x^4 - 11x^2 - x^3 - 11x - 2) \div (2x^2 - 5x - 3)$

Divide using synthetic division.

_____ **6.** $(a^5 - 32) \div (a - 2)$

_____ **7.** $(x^3 + x^2 - 10x - 10) \div (x + 3)$

_____ **8.** $(2x^4 + 3x^3 - x^2 - 4) \div (x + 2)$

Check your answers in the back of your book.

If You Missed Problems:	You Should Review Examples:
1	1
2	1, 2
3, 4	5
5	3
6	9
7, 8	6–8

3.6 Switching from Word Expressions to Polynomials

Now that we have learned to operate with polynomials, let's see if we can use our knowledge to translate some word expressions into polynomials.

Example 1. The length of a rectangle is 3 feet longer than twice the width. Write an expression for the area of the rectangle.

Solution

We Think	*We Write*
The area of a rectangle is the *product* of its length and width.	Area = (length)(width)
We know less about the width.	Let w = width (feet)
The length is 3 feet *more* than *twice* the width.	$2w$ = twice the width (feet) $2w + 3$ = length (feet)

We can illustrate this rectangle and label its dimensions.

$2w + 3$

Area is length *times* width.	Area = $(2w + 3)w$ = $2w^2 + 3w$

The expression $2w^2 + 3w$ square feet represents the area of this rectangle.

Now you try Example 2.

Example 2. If the length of each side of a square is increased by 5 meters, write an algebraic expression for the area of the new square.

Solution

We Think	*We Write*
In a square, the length and width are the *same*.	Let s = length of side of old square (meters)
Each side of the new square is 5 meters *more* than each side of the old square.	$s + \underline{\hspace{1cm}}$ = length of side of new square (meters)
The area of a square is found by *squaring* the length of its side.	$(s + \underline{\hspace{1cm}})^2$ = area of new square (square meters)
	$(s + \underline{\hspace{1cm}})^2 = \underline{\hspace{2cm}}$

The expression $\underline{\hspace{2cm}}$ square meters represents the area of the new square.

Check your work on page 230. ▶

Now let's tackle the problem stated at the beginning of this chapter.

Example 3. Marco's rectangular garden is 10 feet longer than it is wide. If he decreases both the length and width of the garden by 3 feet, what variable expression represents the amount of garden area that he will lose?

Solution

We Think	*We Write*

The area lost will be the *difference* between old area and new area.

Area lost = old area − new area

The *area* of a rectangle is length times width.

Area = (length)(width)

For the old garden, we know less about the width. The old length is 10 feet *more* than the old width.

Let x = old width (feet)
$x + 10$ = old length (feet)

Now we know *old area*.

$x(x + 10)$ = old area (square feet)

For the new garden, each dimension is *decreased* by 3 feet.

$x - 3$ = new width (feet)
$(x + 10) - 3$ = new length (feet)
$x + 7$ = new length (feet)

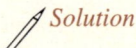

Now we know *new area*.

$(x - 3)(x + 7)$ = new area (square feet)

The area lost is found by *subtracting* new area from old area.

$$x(x + 10) - (x - 3)(x + 7)$$
$$= (x^2 + 10x) - (x^2 + 4x - 21)$$
$$= x^2 + 10x - x^2 - 4x + 21$$
$$= 6x + 21$$

The expression $6x + 21$ square feet represents the garden area that Marco will lose.

Now you complete Example 4.

Example 4. The width of a shoe box is 3 inches more than its height, and its length is 10 inches more than its height. Write an expression for the volume of the box.

Solution

We Think	*We Write*

We recall that the *volume* of a rectangular solid (box) is the *product* of the length, width, and height.

Volume = (length)(width)(height)

We know the least about the height of this box.

Let h = height (inches)

The width is 3 inches *more* than the height.

$h +$ _____ = width (inches)

The length is 10 inches *more* than the height.

$h +$ _____ = length (inches)

The volume is the *product* of the dimensions.

$h(h + \underline{\hspace{1cm}})(h + \underline{\hspace{1cm}})$ cubic inches

$=$

The expression for the volume of the shoe box is $\underline{\hspace{1.5cm}}$ cubic inches.

Check your work on page 230. ▶

A **right triangle** is a triangle in which two of the sides (called the **legs**) meet at a **right angle** (a 90° angle). The two legs are said to be *perpendicular* to each other. The third side, which is opposite the right angle, is called the **hypotenuse** of the right triangle. The parts of a right triangle are often labeled in the following way:

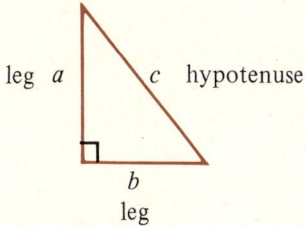

The area of a right triangle is found by multiplying the lengths of the legs and dividing that product by 2.

Area of Right Triangle

$$A = \frac{a \cdot b}{2}$$

See if you can complete Example 5.

Example 5. One leg of a right triangle is 3 inches shorter than the other leg. Write an algebraic expression for the area of the right triangle.

 Solution

We Write	*We Write*
The area of a right triangle depends on the lengths of its legs, *a* and *b*.	$\text{Area} = \dfrac{a \cdot b}{2}$
We know less about the longer leg.	Let a = length of longer leg (inches)
The other leg is 3 inches *shorter*.	$a - \underline{\hspace{1cm}}$ = length of shorter leg (inches)
The area is found by multiplying the lengths of the legs and dividing by 2.	$\text{Area} = \dfrac{a(a -)}{2}$ square inches
	$=$

The expression $\underline{\hspace{1.5cm}}$ square inches represents the area of the right triangle.

Check your work on page 230. ▶

One of the most important mathematical discoveries ever made was that of the Greek mathematician **Pythagoras.** He observed that if we square the length of each leg of a right triangle and add those squares together, their sum will equal the square of the length of the hypotenuse.

Using the usual letters to represent the parts of a right triangle, we state this important result as follows:

Pythagorean Formula

$$a^2 + b^2 = c^2$$

Example 6. If the legs of a right triangle measure 3 inches and 4 inches, find the length of the hypotenuse.

Solution: We must use the Pythagorean formula with $a = 3$ inches and $b = 4$ inches.

$$a^2 + b^2 = c^2$$
$$3^2 + 4^2 = c^2$$
$$9 + 16 = c^2$$
$$25 = c^2$$
$$5 = c \quad \text{or} \quad -5 = c$$

But a dimension cannot be negative, so we conclude that the length of the hypotenuse is 5 inches.

Example 7. One leg of a right triangle is 1 centimeter (cm) more than three times the other leg. Write an algebraic expression for the square of the hypotenuse.

Solution

We Think	*We Write*
We must use the Pythagorean formula.	$a^2 + b^2 = c^2$
We know less about the shorter leg.	Let x = length of shorter leg (cm)
The longer leg is 1 cm *more* than 3 *times* the other.	$3x + 1$ = length of longer leg (cm)
We must *square* the length of each leg and *sum* those squares.	$x^2 + (3x + 1)^2$ $= x^2 + (9x^2 + 6x + 1)$ $= 10x^2 + 6x + 1$

The square of the hypotenuse is the expression $10x^2 + 6x + 1$ square centimeters.

Consecutive integers are integers that immediately follow each other in the list of integers. For example,

$$17, 18, 19 \quad \text{are three consecutive integers}$$
$$-6, -5 \quad \text{are two consecutive integers}$$

You should see that each integer is just *1 more than* the integer immediately preceding it.

Example 8. Write an algebraic expression to represent the sum of any three consecutive integers.

Solution

We Think	*We Write*
We know nothing about the first integer.	Let x = first integer

The next consecutive integer is 1 *more* than the first integer.	$x + 1 =$ second integer
The third consecutive integer is 1 *more* than the second integer.	$(x + 1) + 1 =$ third integer $x + 2 =$ third integer
The *sum* of the three integers is found by adding.	$x + (x + 1) + (x + 2)$ $= 3x + 3$

The expression $3x + 3$ represents the sum of any three consecutive integers.

Now you try to complete Example 9.

Example 9. Write an expression for the product of any three consecutive integers.

／*Solution*

We Think	*We Write*
We know how to represent each of three consecutive integers (from Example 8).	Let $x =$ first integer $x + \underline{\quad} =$ second integer $x + \underline{\quad} =$ third integer
To find the *product* of the three integers, we must _____ .	$x(x + \underline{\quad})(x + \underline{\quad})$ $=$

The expression _____ represents the product of any three consecutive integers.

Check your work on page 230. ▶

▶ **Trial Run**

_____ **1.** If one leg of a right triangle is 8 inches shorter than twice the other leg, write an expression for the square of its hypotenuse. Simplify your answer.

_____ **2.** If two opposite sides of a square are increased by 3 meters and the other two sides are doubled, write an expression for the area of the resulting rectangle. Simplify your answer.

_____ **3.** Write an expression for four times the product of three consecutive integers. Simplify your answer.

Answers are on page 230.

▶ **Examples You Completed**

Example 2

Solution

We Write

Let s = length of side of old square.
$s + 5$ = length of side of new square.

$$(s + 5)^2 = \text{area of new square}$$
$$(s + 5)^2 = s^2 + 10s + 25$$

The expression $s^2 + 10s + 25$ square meters represents the area of the new square.

Example 4

Solution

We Write

Volume = (length)(width)(height)

Let h = height

$h + 3$ = width

$h + 10$ = length

$$h(h + 3)(h + 10)$$
$$= h(h^2 + 13h + 30)$$
$$= h^3 + 13h^2 + 30h$$

The expression for the volume of the shoe box is $h^3 + 13h^2 + 30h$ cubic inches.

Example 5

Solution

We Write

$$\text{Area} = \frac{a \cdot b}{2}$$

Let a = length of longer leg

$a - 3$ = length of shorter leg

$$\text{Area} = \frac{a(a - 3)}{2}$$
$$= \frac{a^2 - 3a}{2}$$

The expression $\dfrac{a^2 - 3a}{2}$ square inches represents the area of the right triangle.

Example 9. Write an expression for the product of any three consecutive integers.

Solution

We Think

We know how to represent each of three consecutive integers (from Example 8).

To find the *product* of the three integers, we must multiply.

We Write

Let x = first integer
$x + 1$ = second integer
$x + 2$ = third integer

$$x(x + 1)(x + 2)$$
$$= x(x^2 + 3x + 2)$$
$$= x^3 + 3x^2 + 2x$$

The expression $x^3 + 3x^2 + 2x$ represents the product of any three consecutive integers.

Answers to Trial Run

page 229   **1.** $5x^2 - 32x + 64$ **2.** $2x^2 + 6x$ **3.** $4x^3 + 12x^2 + 8x$

EXERCISE SET 3.6

_____ **1.** One integer is 12 more than twice another. Write an expression for their product.

_____ **2.** One integer is 4 less than three times another. Write an expression for their product.

_____ **3.** One integer is 7 less than another. Write an expression for the sum of the square of the larger integer and five times the product of the two integers.

_____ **4.** One integer is 2 more than another. Write an expression for the difference of the square of the larger integer and three times the product of the two integers.

_____ **5.** Write an expression for nine times the sum of the squares of two consecutive integers.

_____ **6.** Write an expression for four times the product of two consecutive integers.

_____ **7.** Write an expression for seven times the product of three consecutive integers.

_____ **8.** Write an expression for the sum of the squares of three consecutive integers.

_____ **9.** If the width of a rectangle is 8 feet less than twice the length, write an expression for the area of the rectangle.

_____ **10.** If the length of a rectangle is 6 feet more than three times the width, write an expression for the area.

_____ **11.** The width of a rectangle is 8 inches less than the length. Write an expression for the perimeter of the rectangle.

_____ **12.** The length of a rectangle is 12 inches more than the width. Write an expression for the perimeter of the rectangle.

_____ **13.** If each side of a square is increased by 10 centimeters, write an expression for the difference between the new area and the old area.

_____ **14.** If each side of a square is decreased by 9 centimeters, write an expression for the difference between the old area and the new area.

_____ **15.** One leg of a right triangle is 5 more than twice the other leg. Write an expression for the square of the hypotenuse.

_____ **16.** One leg of a right triangle is one more than four times the other leg. Write an expression for the square of the hypotenuse.

_____ **17.** Write an expression for the area of a right triangle if one leg is 7 less than the other.

_____ **18.** Write an expression for the area of a right triangle if one leg is 8 more than the other.

_____ **19.** If one number is 3 less than twice another, write an expression for the product of the two numbers decreased by three times their sum.

_____ **20.** If one number is 4 more than three times another, write an expression for five times the sum of their squares.

_____ **21.** The Caseys plan to buy a rectangular lot that is 50 feet longer than it is wide. Write an expression for the area of the lot.

_____ **22.** The Brinkleys plan to fence a dog pen that is to be 3 feet longer than twice its width. Write an expression for the area of the pen.

_____ **23.** A flower bed is in the shape of a right triangle with one leg 9 feet shorter than the other leg. Write an expression for the area of the flower bed.

_____ **24.** Donita is building a corner table for her kitchen. The top will be a right triangle with equal legs. Write an expression for the area of the tabletop.

_____ **25.** A homeowner is planning to enlarge the recreation room of her home. The room now measures 4 meters by 5 meters. If she enlarges the room by adding the same length to each side, write an expression for the area of the room after it has been enlarged.

_____ **26.** Mr Abell's garden plot is in the shape of a square. Next year, he plans to enlarge his garden by increasing two opposite sides by 6 feet and the other two sides by 4 feet. Write an expression for the area of the garden plot after it has been enlarged.

_____ **27.** The Hickman's rectangular patio is 8 feet longer than it is wide. If they decide to increase each dimension by 5 feet, write an expression for the amount of patio area they will gain.

_____ **28.** Irene plans to enlarge her kitchen. The room now measures 9 feet by 12 feet. If she enlarges the room by adding the same number of feet to each dimension, write an expression for the amount of area she will gain.

_____ **29.** The floor of a shed is a square. If the height of the shed is 4 feet more than the width of the floor, write an expression for the volume of the shed.

_____ **30.** Mrs. Worth has a square cake pan that is 2 inches deep. She has a larger rectangular cake pan 2 inches deep whose width is 1 inch more than one side of the square pan and whose length is 3 inches more than one side of the square pan. Write an expression for the difference in the volumes of the two pans.

☆ Stretching the Topics _____

_____ **1.** A rectangular cattle pen next to a barn is to be enclosed using 300 feet of fence on three sides and the barn wall on the fourth side. Write an expression for the area that will be enclosed.

_____ **2.** A rectangular piece of cardboard is three times as long as it is wide. A 6-centimeter square is cut from each corner and the sides are turned upward to form a box. Write an expression for the volume of the box.

_____ **3.** Write an expression for the area of a right triangle if the hypotenuse is 25 inches and the perimeter is 60 inches.

Check your answers in the back of your book.

If you can complete **Checkup 3.6,** you are ready to do the Review Exercises.

✓ CHECKUP 3.6

_____ **1.** The length of a rectangle is 3 feet shorter than four times the width. Write an expression for the area of the rectangle.

_____ **2.** Write an expression for the area of a right triangle if one leg is 5 feet more than twice the other leg.

_____ **3.** One leg of a right triangle is 6 inches less than the other. Write an expression for the square of the hypotenuse.

_____ **4.** Write an expression for twice the product of two consecutive integers.

_____ **5.** If the side of a square is increased by 15 centimeters, write an expression for the difference between the old area and the new area.

Check your answers in the back of your book.

If You Missed Problem:	You Should Review Examples:
1	1
2	5
3	7
4	9
5	2, 3

Summary

In this chapter we discussed definitions and rules for working with *integer* **exponents.**

	Symbols	Examples
Definition	$a^n = \underbrace{a \cdot a \cdot \ldots \cdot a}_{n \text{ factors}}$	$(-3)^4 = (-3)(-3)(-3)(-3) = 81$ $x^3 = x \cdot x \cdot x$
Definition	$a^0 = 1 \quad (a \neq 0)$	$x^0 = 1$ $(3x + y)^0 = 1$
Definition	$a^{-n} = \dfrac{1}{a^n} \quad (a \neq 0)$	$(-3)^{-2} = \dfrac{1}{(-3)^2} = \dfrac{1}{9}$ $x^{-1}y^{-3} = \dfrac{1}{xy^3}$
First law of exponents	$a^m \cdot a^n = a^{m+n}$	$x^4 \cdot x^3 = x^7$ $y^5 \cdot y^{-6} = y^{-1} = \dfrac{1}{y}$
Second law of exponents	$(a^m)^n = a^{mn}$	$(x^4)^3 = x^{12}$ $(y^5)^{-6} = y^{-30} = \dfrac{1}{y^{30}}$
Third law of exponents	$(ab)^n = a^n b^n$	$(3x)^2 = 3^2 x^2 = 9x^2$ $(-2x^{-2}y)^{-3} = (-2)^{-3}x^6 y^{-3}$ $= \dfrac{x^6}{(-2)^3 y^3} = \dfrac{-x^6}{8y^3}$
Fourth law of exponents	$\left(\dfrac{a}{b}\right)^n = \dfrac{a^n}{b^n} \quad (b \neq 0)$	$\left(\dfrac{5}{a}\right)^3 = \dfrac{5^3}{a^3} = \dfrac{125}{a^3}$ $\left(\dfrac{xy^{-1}}{2z^{-2}}\right)^2 = \dfrac{x^2 y^{-2}}{2^2 z^{-4}} = \dfrac{x^2 z^4}{4y^2}$
Fifth law of exponents	$\dfrac{a^m}{a^n} = a^{m-n} \quad (a \neq 0)$	$\dfrac{x^7}{x^2} = x^5$ $\dfrac{x^3 y^{-1}}{x^{-2}y} = x^{3-(-2)}y^{-1-1} = x^5 y^{-2} = \dfrac{x^5}{y^2}$

We learned to recognize **polynomials** as sums of constant terms and/or variable terms whose variable parts appear with natural number exponents. To add and subtract polynomials, we used the techniques for combining like terms. To multiply polynomials, we used the distributive property and the laws of exponents. In multiplication we learned to recognize certain special products.

Factors	Special Product	Example
$(A + B)(A - B)$	$= A^2 - B^2$	$(x + 3y)(x - 3y) = x^2 - 9y^2$
$(A + B)^2$	$= A^2 + 2AB + B^2$	$(3x + 5z)^2 = 9x^2 + 30xz + 25z^2$
$(A - B)^2$	$= A^2 - 2AB + B^2$	$(2x - yz)^2 = 4x^2 - 4xyz + y^2z^2$

To divide polynomials, we learned to use long division. When the divisor is of the form $x - c$, we discovered that the method of synthetic division can be used to find a quotient.

Finally, we learned to use polynomials to switch from words to variable expressions, and we noted the very important *Pythagorean formula* for right triangles.

$$a^2 + b^2 = c^2$$

☐ Speaking the Language of Algebra

Complete each sentence with the appropriate word or phrase.

1. In the expression a^n, a is called the _____ and n is called the _____ . The expression a^n tells us to use _____ as a factor in a product _____ times.

2. To multiply powers of the same base, we keep the base and _____ the exponents.

3. To raise a power of a base to another power, we keep the base and _____ the exponents.

4. To raise a product to a power, we raise each _____ to that power.

5. To raise a fraction to a power, we raise the _____ to the power and raise the _____ to the power.

6. To divide powers of the same base, we _____ the exponent in the _____ from the exponent in the _____ .

7. The expression $3x + 2$ is called a _____-degree _____ .

8. The expression $x^2 + 7x - 1$ is called a _____-degree _____ .

9. If the sum of two terms is multiplied times the difference of the same two terms, the product is the _____ _____ _____ _____ of the two terms.

10. According to the Pythagorean formula, if the square of one leg of a right triangle is added to the square of the other leg, the sum equals the _____ _____ _____ _____ .

REVIEW EXERCISES for Chapter 3: Working with Exponents and Polynomials

Simplify using the laws of exponents.

_____ 1. $2x(-5xy^2)(5x^4y^3)$

_____ 2. $(-7y^5)^2$

_____ 3. $(-2x^3)^5(-3x)^2$

_____ 4. $\left(\dfrac{-15x^4}{5y^2}\right)^3$

_____ 5. $\left(\dfrac{2a^3b^0}{3a^2c}\right)^3$

_____ 6. $\dfrac{28a^5b^4}{6a^2b^4}$

_____ 7. Evaluate $(4)^{-2}$.

_____ 8. Evaluate $(-2)^{-3}$.

Simplify and write the answer with positive exponents.

_____ 9. $(x^3y^{-2}z^{-1})(x^{-5}y^2z^4)$

_____ 10. $5x^{-3}$

_____ 11. $(x^{-3}y^5z^{-2})^{-2}$

_____ 12. $\dfrac{x^{-12}y^6}{x^{-9}y^{-6}}$

_____ 13. $\left(\dfrac{2x^3y^{-1}}{x^{-4}y^3}\right)^{-3}$

_____ 14. $\left(\dfrac{3x^{-7}y}{x^{-5}y^{-3}}\right)^2$

Name each polynomial.

_____ 15. $6xy$

_____ 16. $3x^3 - 2x + 1$

Rewrite in descending powers of the variable.

_____ 17. $y^3 - y^2 + 2y^4 + 6$

_____ 18. $3x - 5 - 2x^2$

Perform the indicated operations and simplify.

_____ 19. $7x^2 - 2(x^2 - 3x) + 6(-x + 5) - 7$

_____ 20. $7a - 3[2a - 4(a - 2b) - 6b] + 5b$

_____ 21. $3x(x^2 - 5x + 6)$

_____ 22. $3mn(7m - 2n) - mn(4m - 7n)$

_____ 23. $6a^2 - 2[4a(a - 2) - 3a(a^2 - 1)]$

_____ 24. $-2xy^3(3x - 4x^2y^2) + x^2y(4y^2 - xy^4)$

_____ 25. $(3x - 1)(x + 5)$

_____ 26. $(a - 7b)^2$

_____ **27.** $(4x - 3y)(4x + 3y)$

_____ **28.** $(x - 2)(x^2 - 4x - 3)$

_____ **29.** $4x(2x - y) - (x - 3y)^2$

_____ **30.** $3(a - 1)(a - 7) - (a + 2)(2a - 3)$

_____ **31.** $(a - 2b)(a + b)(2a - b)$

_____ **32.** $(a - 2)^3$

_____ **33.** $(x^2 - x - 1)(x^2 + 2x - 3)$

_____ **34.** $[(a + 2b) + 3]^2$

Divide.

_____ **35.** $(5a^3b - 15a^2b^2 + 10ab^3) \div 5ab$

_____ **36.** $(3x^2 + 17x - 9) \div (x + 6)$

_____ **37.** $(x^3 + 64) \div (x + 4)$

_____ **38.** $(x^3 + 5x^2y - 4xy^2 - 20y^3) \div (x + 2y)$

Divide using synthetic division.

_____ **39.** $(x^3 + 27) \div (x + 3)$

_____ **40.** $(2x^3 - x^2 + x - 2) \div (x - 2)$

_____ **41.** $(3x^4 - x^3 - 2x^2 + 2) \div (x + 1)$

Write the expressions.

_____ **42.** If one leg of a right triangle is five times the other leg, write an algebraic expression to represent the area of the triangle.

_____ **43.** Write an expression for five times the product of two consecutive integers.

_____ **44.** If two opposite sides of a square are increased by 5 centimeters and the other two sides are decreased by 3 centimeters, write an expression for the area of the resulting rectangle.

_____ **45.** If one number is 2 more than three times the other, write an expression for the sum of the two numbers decreased by twice their product.

Check your answers in the back of your book.

If You Missed Exercises:	You Should Review Examples:	
1	Section 3.1	5–7
2, 3		11–16
4, 5		17, 18
6		19, 20
7, 8	Section 3.2	1–4
9		7, 8
10		5
11		9, 10
12		12
13, 14		16–18
15, 16	Section 3.3	1
17, 18		2, 3
19, 20		8–10
21–24		11–15
25–27	Section 3.4	1–7
28		11, 12
29, 30		8, 9
31–33		13, 14
34		21
35	Section 3.5	1, 2
36–38		3–5
39–41		6–9
42	Section 3.6	5
43		8, 9
44		2
45		8, 9

If you have completed the Review Exercises and corrected your errors, you are ready to take the Practice Test.

PRACTICE TEST for Chapter 3:
Working with Exponents and Polynomials

Evaluate.

		SECTION	EXAMPLES

Evaluate.

_____ **1.** $-x^2 + 4x$ when $x = -8$.   3.1   1, 2

_____ **2.** x^{-5} when $x = -2$.   3.2   2

Simplify. Write answers with positive exponents.

_____ **3.** $(-4a^7b)(-2a^5b^2)$ 3.1   5

_____ **4.** $-6x^2y^2(x^4y)\left(\dfrac{1}{3}\,xy\right)$ 3.1   7

_____ **5.** $(-5x^3y)^3$ 3.1   12

_____ **6.** $(-a^4b^2c^3)^3$ 3.1   14

_____ **7.** $\left(\dfrac{-9x^3}{y^2z^3}\right)^2$ 3.1   17

_____ **8.** $\dfrac{-18a^4b^5c^5}{9a^3b^2c^5}$ 3.1   19

_____ **9.** $\left(\dfrac{x^5yz^2}{2x^2yz}\right)^3$ 3.1   17, 19

_____ **10.** $-4x(x^{-3}y^2)(-2x^4y^{-3})$ 3.2   8

_____ **11.** $(a^{-1}b^3c^{-2})^{-3}$ 3.2   10

_____ **12.** $\left(\dfrac{-x^{-3}y^{-1}}{z}\right)^{-3}$ 3.2   17

_____ **13.** $\dfrac{-2x^{-1}y^{-3}}{x^5y^{-7}}$ 3.2   12

_____ **14.** $x^2 - 7(x^2 - 9x) - 3(2 - 4x) - 6x$ 3.3   8

_____ **15.** $5a - 6[2(1 - 3a + a^2) - 4(a^2 - 1)]$ 3.3   10

_____ **16.** $(4x - 7)(3x + 5)$ 3.4   1

_____ **17.** $(3c - 7d)^2$ 3.4   3

_____ **18.** $(9y - 2z)(9y + 2z)$ 3.4   4

_____ **19.** $(x - 5)(3x^2 - 2x - 1)$ 3.4   11

_____ **20.** $(a - 5)^3$ 3.4   13

_____ **21.** $(3x + 1)(x - 2) - (x + 4)(x - 4)$ 3.4   9

_____ **22.** $(-20x^3y^3 - 5x^2y^2 + 10xy) \div (5xy)$ 3.5   1

___(10)__ **23.** $(7x^2 + 3x - 2) \div (x - 3)$ 3.5   6

_____ **24.** $(y^3 + 27) \div (y + 3)$ 3.5   7

_____ **25.** If one leg of a right triangle is 8 inches shorter than the other leg,   3.6   7
write an expression for the square of the hypotenuse. Simplify your
answer.

SHARPENING YOUR SKILLS after Chapters 1–3

		SECTION

_____ **1.** Compare $\dfrac{-11}{2}$ and 4.5 using $<$, $>$, or $=$.   1.1

_____ **2.** Use an inequality to state whether 8.2 is positive or negative.   1.1

_____ **3.** Evaluate $4a^2 - 7a - 20$ when $a = -1$.   1.4

_____ **4.** Evaluate $|8 - 15|$.   1.4

_____ **5.** Name the property illustrated by $(-3) + 8 = 8 + (-3)$.   1.3

Simplify.

_____ **6.** $\dfrac{0}{9 - 4}$ 1.2

_____ **7.** $7 - 11[5 - 3(2)]$ 1.3

_____ **8.** $\dfrac{-1}{7}(7x - 21y) + 3[x - (x - 2y)]$ 1.4

_____ **9.** $-9x^2 + 4x - 3 + 7x^2 - 4x + 1$ 1.4

Solve.

_____ **10.** $-3x + 1 = 7x - 4$ 2.1

_____ **11.** $3(3y - 1) - 9 = 5(4 + y)$ 2.1

_____ **12.** $5(x + 1) > 6(x + 2)$ 2.3

_____ **13.** $0 \leq \dfrac{-y + 5}{2} < 3$ 2.3

_____ **14.** $|9 - x| = 13$ 2.4

_____ **15.** $|5a - 3| \leq 7$ 2.4

Solve by graphing.

_____ **16.** $x < -3$ and $x > \dfrac{3}{2}$ 2.2

_____ **17.** $\dfrac{x - 2}{5} \geq -1$ and $5(2 - x) > 20$ 2.3

_____ **18.** $2(3 - x) \geq 0$ or $6(x - 2) > 3x + 6$ 2.3

_____ **19.** Clara has $500 in a savings account. She finds that her present salary is not adequate to   1.5
meet her expenses, so she has to use $15 per month from her savings. Write an
algebraic expression for the amount left in savings after n months.

_____ **20.** Simon walks and jogs every evening for 2 hours. He jogs at 3 miles per hour and walks   2.5
at 2 miles per hour. If the total distance covered this evening was $5\frac{1}{2}$ miles, how much
time did Simon spend walking?

Factoring 4

A company has found that its *revenue* from the sale of x units of a product is $2x^2 - x + 10$ dollars, while its *costs* for producing x units are $x^2 - 3x + 109$ dollars. Write an expression for the company's *profit* when x units are produced and sold. Then write that expression in factored form and find the profit when the company produces and sells 6 units; 19 units.

Having learned to multiply algebraic expressions, we must now learn how to go backward. We would like to look at an algebraic expression and try to decide what factors might have been multiplied together to yield that expression. We call this process **factoring** an algebraic expression.

Factoring is the process by which we rewrite a sum of terms as a *product*.

Factoring is one of the most useful and important skills to be mastered in the study of algebra. In this chapter we learned how to:

1. Factor out a common factor.

2. Factor special products.

3. Factor trinomials.

4.1 Common Factors

Common Monomial Factors

The first type of polynomial multiplication we studied was multiplication by a monomial, and we learned to use the distributive property to find products such as these:

$$3(x + 2y) \qquad x(5 - x) \qquad -7a(a^2 + 2a - 3)$$

$$= 3x + 6y \qquad = 5x - x^2 \qquad = -7a^3 - 14a^2 + 21a$$

Now we would like to see if a polynomial such as $5a + 5b$ can be rewritten as a *product* of a monomial and a polynomial. We see that both terms contain the numerical factor 5. We say that 5 is a **common factor** for both terms of the polynomial and it is possible for us to write

$$5a + 5b = 5(a + b)$$

The polynomial on the left has now been written as a product of the factors on the right. In other words, we have *factored* our polynomial, and we can check our factors by multiplying to see if their product is actually the original polynomial.

Consider another polynomial to be rewritten in factored form: $6x^2 + 9x$. We must agree to look for *all* factors common to all terms of the polynomial. Here we see the common numerical factor 3 *and* the common variable factor x. In other words, the **greatest common monomial factor** is $3x$, and the factored form for our polynomial becomes

$$6x^2 + 9x = 3x(2x + 3)$$

We repeat here that we should always try to "factor out" the greatest common factor. Keep these ideas in mind:

1. First look for the largest numerical factor common to *all* terms of the polynomial.
2. If *all* terms in the polynomial contain the same variable (or variables), we must factor out the *lowest* power of that variable appearing in the polynomial.

Since factors can always be checked by multiplication, there is no reason for you to factor incorrectly. If the product of your factors is not the original polynomial, you know you have made an error.

You try factoring some polynomials.

Example 1. Factor $2x^9 - 10x^7$.

Solution: The common monomial factor is _____ .

$$2x^9 - 10x^7 = 2x^7(\underline{\quad} - \underline{\quad})$$

Check your work on page 251. ▶

Example 2. Factor $14x^3y + 7xy^2 - 21xy$.

Solution: The common monomial factor is __7xy__ .

$$14x^3y + 7xy^2 - 21xy$$
$$= 7xy(\underline{\quad} + \underline{\quad} - \underline{\quad})$$

Check your work on page 251. ▶

Example 3. Factor $-3x^2 - 6x + 3$.

Solution: Each term contains the numerical factor 3 or -3, but which should be factored out? In general, we must agree that if the first numerical coefficient in the polynomial (sometimes called the **leading coefficient**) is negative, then we'll factor out a negative common factor.

$$-3x^2 - 6x + 3 = -3(x^2 + 2x - 1)$$

Notice that we were very careful about the signs in the second factor.

Now try completing Examples 4 and 5.

Example 4. Factor $4x^2 - 16y^3$.

    Solution: The common monomial factor is _____.

$$= \frac{4x^2 - 16y^3}{}$$

Check your work on page 251. ▶

Example 5. Factor $-x^7y^3 + 4x^5y^4 - 3x^3y^3$.

    Solution: The common monomial factor is _____ .

$$-x^7y^3 + 4x^5y^4 - 3x^3y^3$$
$$= -x^3y^3(\underline{\hspace{2cm}})$$

Check your work on page 251. ▶

Often the terms in a polynomial will contain no common factors. In the expression $3x^2 + 5y$, there is *no* common factor; we cannot rewrite this polynomial as a product.

> When a polynomial cannot be written as a product of factors, we say that the polynomial is **prime** (or **irreducible**).

Although we shall discuss other types of factoring, the common monomial factor is *always* the first thing we look for in factoring polynomials.

⫸ Trial Run

Factor.

_____ **1.** $3x^6 - 4x^8$

_____ **3.** $9x^2z - 45y^2z$

_____ **5.** $-7x^2 + 28x$

_____ **2.** $12a^2 - 18a + 36$

_____ **4.** $10x^3y + 5xy^2 - 15xy$

_____ **6.** $-x^8y^5 + 3x^6y^3 - 5x^4y^2$

Answers are on page 252.

Other Common Factors

Factoring an expression such as

$$a(x + y) + 2(x + y)$$

requires the same skill that you have just mastered. Notice that the factor $(x + y)$ is *common* to both terms. Therefore, we factor the expression as

$$a(x + y) + 2(x + y)$$
$$= (x + y)(a + 2)$$

Now try Example 6.

Example 6. Factor $c(x - 3y) - 5(x - 3y)$.

Solution: The common binomial factor is $(x - 3y)$.

$$c(x - 3y) - 5(x - 3y)$$
$$= (x - 3y)(\underline{\quad} - \underline{\quad})$$

Check your work on page 252. ▶

Example 7. Factor $(b + c)^2 + (b + c)$.

Solution: The common binomial factor is $(b + c)$.

$$(b + c)^2 + (b + c)$$
$$= (b + c)[(b + c) + 1]$$
$$= (b + c)(b + c + 1)$$

Example 8. Factor $4(x - y)^3 - (x - y)^4$.

Solution: The common binomial factor is $(x - y)^3$.

$$4(x - y)^3 - (x - y)^4$$
$$= (x - y)^3[4 - (x - y)]$$
$$= (x - y)^3(4 - x + y)$$

You complete Example 9.

Example 9. Factor $(x - 1)^3 - 4(x - 1)^2 + 3(x - 1)$.

Solution: The common binomial factor is $\underline{\quad}$.

$$(x - 1)^3 - 4(x - 1)^2 + 3(x - 1)$$
$$= (x - 1)[(\underline{x-1})^2 - 4(\underline{x-1}) + \underline{3}]$$

Check your work on page 252. ▶

As you can see, removing common factors requires only that you are familiar with the distributive property and the laws of exponents.

⫸ Trial Run

Factor.

_____ **1.** $x(x + 5) - 3(x + 5)$ _____ **2.** $a(x - y) + b(x - y)$

_____ **3.** $(a + b)^2 + 9(a + b)$ _____ **4.** $(x + y)(m + n) - z(m + n)$

Answers are on page 252.

Factoring by Grouping

Suppose that we use the idea of removing common factors to factor expressions containing *four terms*. In the expression

$$ax + 3x + ay + 3y$$

there is no factor common to all four terms. However, we notice that the first pair of terms contains the common factor x, and the second pair of terms contains the common factor y.

$ax + 3x + ay + 3y = (ax + 3x) + (ay + 3y)$	We can group the expression into two pairs of terms.
$= x(a + 3) + y(a + 3)$	Remove the common factor from each pair of terms. Now the expression is a sum of *two* terms.
$= x(a + 3) + y(a + 3)$	Notice the common binomial factor $(a + 3)$ in both terms.
$= (a + 3)(x + y)$	Remove the common binomial factor $(a + 3)$.

This procedure is called **factoring by grouping** because it requires that we group the terms two by two so that each twosome contains a common factor.

You try completing Example 10.

Example 10. Factor $5x - 5 + xy - y$.

Solution

$$5x - 5 + xy - y$$
$$= (5x - 5) + (xy - y)$$
$$= 5(x - 1) + y(x - 1)$$
$$= (x - 1)(\underline{\hspace{0.7cm}} + \underline{\hspace{0.7cm}})$$

Check your work on page 252. ▶

Example 11. Factor $ax + y + ay + x$.

Solution: We must rearrange the terms.

$$ax + y + ay + x$$
$$= ax + ay + x + y$$
$$= (ax + ay) + (x + y)$$
$$= a(x + y) + 1(x + y)$$
$$= (x + y)(a + 1)$$

Example 12. Factor $ab - 3b - 2a + 6$.

Solution

$$ab - 3b - 2a + 6$$
$$= (ab - 3b) + (-2a + 6)$$
$$= b(a - 3) - 2(a - 3)$$
$$= (a - 3)(b - 2)$$

Notice that we factored -2 from the second pair of terms so that $(a - 3)$ would be a common factor.

Now you try completing Example 13.

Example 13. Factor $x^2 + ax - x - a$.

Solution

$$x^2 + ax - x - a$$
$$= (x^2 + ax) + (-x - a)$$
$$= x(\underline{x + a}) - 1(\underline{x + a})$$
$$= \quad x - 1 \quad x + a$$

Check your work on page 252. ▶

▦➤ Trial Run

Factor.

_____ **1.** $7x + 7y + ax + ay$

_____ **2.** $3ax - 15ay + 2x - 10y$

_____ **3.** $8a^2 + 4a - 2ab - b$

_____ **4.** $12x - 4x^3 - 15 + 5x^2$

Answers are on page 252.

▶ Examples You Completed

Example 1. Factor $2x^9 - 10x^7$.

Solution: The common monomial factor is $2x^7$.

$$2x^9 - 10x^7 = 2x^7(x^2 - 5)$$

Example 4. Factor $4x^2 - 16y^3$.

Solution: The common monomial factor is 4.

$$4x^2 - 16y^3 = 4(x^2 - 4y^3)$$

Example 2. Factor $14x^3y + 7xy^2 - 21xy$.

Solution: The common monomial factor is $7xy$.

$$14x^3y + 7xy^2 - 21xy = 7xy(2x^2 + y - 3)$$

Example 5. Factor $-x^7y^3 + 4x^5y^4 - 3x^3y^3$.

Solution: The common monomial factor is $-x^3y^3$.

$$-x^7y^3 + 4x^5y^4 - 3x^3y^3$$
$$= -x^3y^3(x^4 - 4x^2y + 3)$$

Example 6. Factor $c(x - 3y) - 5(x - 3y)$.

Solution: The common binomial factor is $(x - 3y)$.

$$c(x - 3y) - 5(x - 3y)$$
$$= (x - 3y)(c - 5)$$

Example 9. Factor $(x - 1)^3 - 4(x - 1)^2 + 3(x - 1)$.

Solution: The common binomial factor is $x - 1$.

$$(x - 1)^3 - 4(x - 1)^2 + 3(x - 1)$$
$$= (x - 1)[(x - 1)^2 - 4(x - 1) + 3]$$

Example 10. Factor $5x - 5 + xy - y$.

Solution

$$5x - 5 + xy - y$$
$$= (5x - 5) + (xy - y)$$
$$= 5(x - 1) + y(x - 1)$$
$$= (x - 1)(5 + y)$$

Example 13. Factor $x^2 + ax - x - a$.

Solution

$$x^2 + ax - x - a$$
$$= (x^2 + ax) + (-x - a)$$
$$= x(x + a) - 1(x + a)$$
$$= (x + a)(x - 1)$$

Answers to Trial Runs

page 249 **1.** $x^6(3 - 4x^2)$ **2.** $6(2a^2 - 3a + 6)$ **3.** $9z(x^2 - 5y^2)$ **4.** $5xy(2x^2 + y - 3)$
5. $-7x(x - 4)$ **6.** $-x^4y^2(x^4y^3 - 3x^2y + 5)$

page 250 **1.** $(x + 5)(x - 3)$ **2.** $(x - y)(a + b)$ **3.** $(a + b)(a + b + 9)$ **4.** $(m + n)(x + y - z)$

page 251 **1.** $(x + y)(7 + a)$ **2.** $(x - 5y)(3a + 2)$ **3.** $(2a + 1)(4a - b)$ **4.** $(3 - x^2)(4x - 5)$

EXERCISE SET 4.1

Factor.

_____ **1.** $50x^2 - 10xy + 20y^2$

_____ **2.** $36x^2 - 24xy + 48y^2$

_____ **3.** $2b^3 - 6b^2 - 4b$

_____ **4.** $5b^3 - 10b^2 + 25b$

_____ **5.** $-y^3 + y^2 - 2y$

_____ **6.** $-y^3 - y^2 + 4y$

_____ **7.** $3x^3 - 6x^2y + 12xy^2$

_____ **8.** $5x^3 + 15x^2y + 55xy^2$

_____ **9.** $-a^4b + 8a^2b^2 - 5a^2b^3$

_____ **10.** $-a^5b - 4a^2b^2 + 5a^3b^3$

_____ **11.** $5x^3y^2 - 80x^2y^3$

_____ **12.** $7x^4y^2 - 56x^3y^3$

_____ **13.** $9x^3y - 12x^2y^2 + 24xy^3$

_____ **14.** $27x^3y + 45x^2y^2 - 63xy^3$

_____ **15.** $72a^4 - 40a^3 + 8a^2$

_____ **16.** $72a^5 - 54a^4 - 6a^3$

_____ **17.** $12x^4y^2 + 8x^3y^3 + 4x^2y^4$

_____ **18.** $18x^4y^2 - 12x^3y^3 - 3x^2y^4$

_____ **19.** $9x^4yz + 18x^3y^2z - 27x^2y^3z - 45xy^4z$

_____ **20.** $35xy^4z - 28x^2y^3z - 14x^3y^2z + 7x^4yz$

_____ **21.** $y(y - 1) - 3(y - 1)$

_____ **22.** $y(y + 5) - 4(y + 5)$

_____ **23.** $2x(x + 3) - (x + 3)$

_____ **24.** $3x(x + 8) + (x + 8)$

_____ **25.** $2a(3x - y) - b(3x - y)$

_____ **26.** $5a(2x + y) + 3b(2x + y)$

_____ **27.** $(x - y)^2 - 3(x - y)$

_____ **28.** $(2x + y)^2 + 5(2x + y)$

_____ **29.** $(a + b)(x - y) - 2c(x - y)$

_____ **30.** $(3a - b)(x + y) - 4c(x + y)$

_____ **31.** $(3x - y)(2a + b) - (x - y)(2a + b)$

_____ **32.** $(4x + y)(3a + 2b) - (x + y)(3a + 2b)$

_____ **33.** $5(x - 2)^3 - 4(x - 2)^2 + (x - 2)$

_____ **34.** $9(x - 3)^3 - 2(x - 3)^2 - (x - 3)$

_____ **35.** $4(x + 2y)^3 - 6(x + 2y)^2 - 2(x + 2y)$

_____ **36.** $9(x - y)^3 - 12(x - y)^2 - 3(x - y)$

_____ **37.** $6a + ac + 6b + bc$

_____ **38.** $9x + 9y + xz + yz$

_____ **39.** $x^3 - 7x^2 + 6x - 42$

_____ **40.** $x^3 - 11x^2 + 3x - 33$

_____ **41.** $14ax + 21ay - 10x - 15y$

_____ **42.** $36ax + 45ay - 28x - 35y$

_____ **43.** $70a^2 - 40ab + 21a - 12b$

_____ **44.** $30a^2 - 75ab + 4a - 10b$

_____ **45.** $6a^2x + 3a^2y + 6b^2x + 3b^2y$

_____ **46.** $5a^2x + 10a^2y + 5b^2x + 10b^2y$

_____ **47.** $x^3 + x^2 + x + 1$

_____ **48.** $x^3 - 3x^2 + x - 3$

_____ **49.** $9ab + 7c - 9ac - 7b$ _____ **50.** $8xy + 3z - 8xz - 3y$

_____ **51.** $2ax^2 - 3b - 3bx^2 + 2a$ _____ **52.** $ax + 5by - bx - 5ay$

☆ Stretching the Topics

_____ **1.** Factor $x^{5a} + 4x^{4a}$.

_____ **2.** Factor $x^{3a+2} + 3x^{2a+2} - 5x^2$.

_____ **3.** From the expression $x^{-3} + 5x^{-2} + x^{-1}$, factor out x^{-3}.

Check your answers in the back of your book.

If you can factor the problems in **Checkup 4.1,** you are ready to go on to Section 4.2.

✓ **CHECKUP 4.1**

Factor.

_____ **1.** $35x^2 - 14x + 7$

_____ **2.** $7b^3 - 14b^2 + 28b$

_____ **3.** $5x^2y - 25y^3$

_____ **4.** $-a^6b^2 + 3a^5b^4 + 6a^4b^6$

_____ **5.** $9x^5y^3z - 27x^4y^2z - 18x^3yz$

_____ **6.** $2x(x - 9) - 3(x - 9)$

_____ **7.** $(a + b)^2 - 12(a + b)$

_____ **8.** $6(x - y)^2 - (x - y)^3$

_____ **9.** $2ax + 2bx - 3ay - 3by$

_____ **10.** $x^3 - 7x^2 + x - 7$

Check your answers in the back of your book.

If You Missed Problems:	You Should Review Examples:
1–3	1–4
4	5
5	2
6	6
7	7
8	8
9, 10	10–13

4.2 **Factoring Special Products**

In the section on multiplying polynomials, we learned that when certain kinds of factors are multiplied, they yield certain **special products.** In factoring, it is often possible to spot such a special product and know immediately what its factors are.

Factoring the Difference of Two Squares

From multiplication we know that

$$(x + 3)(x - 3) \qquad (y - x)(y + x) \qquad \left(2a + \frac{5}{7}b\right)\left(2a - \frac{5}{7}b\right)$$

$$= x^2 - 9 \qquad\qquad = y^2 - x^2 \qquad\qquad = 4a^2 - \frac{25}{49}b^2$$

In other words, we recall that the product of the sum and difference of the same two terms is the **difference of the squares** of those terms. Going backwards with this rule, you should agree that a polynomial that is a difference of the squares of two terms can be factored into a product of the sum and difference of those two terms.

Therefore, we see that

$$x^2 - 0.16 \qquad a^2 - 9y^2 \qquad x^2y^2 - z^2$$

$$= (x + 0.4)(x - 0.4) \qquad = (a + 3y)(a - 3y) \qquad = (xy + z)(xy - z)$$

and in general we may say:

Difference of Two Squares

$$A^2 - B^2 = (A + B)(A - B)$$

You try Example 1.

Example 1. Factor $4a^2 - 49y^2$.

Solution

$$4a^2 - 49y^2$$

$$= (\underline{\quad\quad})(\underline{\quad\quad})$$

Check your work on page 261. ▶

Example 2. Factor $5x^2 - 20$.

Solution: First we look for a common monomial factor.

$$5x^2 - 20$$

$$= 5(x^2 - 4) \qquad \text{Factor out 5.}$$

$$= 5(x + 2)(x - 2) \qquad \text{Factor difference of squares.}$$

Now try Example 3.

Example 3. Factor $9 - 81y^2$.

Solution

$$9 - 81y^2$$

$$= 9(\underline{\quad\quad}) \qquad \text{Factor out 9.}$$

$$= 9(\underline{\quad\quad})(\underline{\quad\quad}) \qquad \text{Factor difference of squares.}$$

Check your work on page 261. ▶

Example 4. Factor $-x^3 + x$.

Solution

$$-x^3 + x$$

$$= -x(x^2 - 1) \qquad \text{Factor out } -x.$$

$$= -x(x + 1)(x - 1) \qquad \text{Factor difference of squares.}$$

Let's tackle a problem that looks more complicated.

Example 5. Factor $(x + 1)^2 - 25$.

Solution: Is this polynomial a difference of two squares? Does it fit into the form $A^2 - B^2$?

$$(x + 1)^2 - 25 \qquad \text{We see that}$$
$$\uparrow \qquad \nearrow \qquad A = x + 1$$
$$A^2 - \quad B^2 \qquad\qquad B = 5$$

So our factoring should look like this:

$$A^2 \quad - B^2 = \quad (A \quad + \quad B) \quad (A \quad - \quad B)$$
$$\downarrow \qquad \downarrow \qquad\quad \downarrow \qquad \downarrow \qquad \downarrow \qquad\quad \downarrow$$
$$(x + 1)^2 - 25 = [(x + 1) \quad + \quad 5][(x + 1) \quad - \quad 5]$$
$$= [x + 1 + 5][x + 1 - 5]$$
$$(x + 1)^2 - 25 = (x + 6)(x - 4)$$

You try Example 6, being sure to identify A and B.

Example 6. Factor $4y^2 - (x - 3)^2$.

Solution: Is $4y^2 - (x - 3)^2$ a difference of two squares? _____

Here $A = $ _____

$\qquad B = ($ _____ $)$

So

$$4y^2 - (x - 3)^2$$
$$= [2y + (x - 3)][2y - (x - 3)]$$
$$= (\text{_____})(\text{_____})$$

Check your work on page 262. ▶

Example 7. Factor $x^4 - 16$.

Solution: Is $x^4 - 16$ a difference of two squares? Yes.

Here $A = x^2$

$\qquad B = 4$

So

$$x^4 - 16$$
$$= (x^2 + 4)(x^2 - 4)$$
$$= (x^2 + 4)(x + 2)(x - 2)$$

In Example 7, notice that we were able to factor $x^2 - 4$ as a difference of two squares, but we were *not* able to factor $x^2 + 4$. As long as we are working with real numbers and common factors have been removed,

The *sum* of two squares cannot be factored.
The sum of two squares is a prime polynomial.

Example 8. Factor $x^3 - 3x^2 - x + 3$.

Solution: This polynomial contains four terms, so we must start to factor it by *grouping*.

$$x^3 - 3x^2 - x + 3$$
$$= (x^3 - 3x^2) + (-x + 3) \qquad \text{Group terms two by two.}$$
$$= x^2(x - 3) - 1(x - 3) \qquad \text{Remove factor from each twosome.}$$
$$= (x - 3)(x^2 - 1) \qquad \text{Remove common binomial factor.}$$
$$= (x - 3)(x + 1)(x - 1) \qquad \text{Factor the difference of the two squares.}$$

Trial Run

Factor.

_____ **1.** $x^2 - 36$ _____ **2.** $25a^2 - 9b^2$

_____ **3.** $2y^2 - 98$ _____ **4.** $400 - 4x^2$

_____ **5.** $x^4 - 81y^4$ _____ **6.** $(a + b)^2 - 25c^2$

Answers are on page 262.

Factoring the Difference or Sum of Two Cubes

Just as there is a special rule for factoring the difference of two squares, there is a special rule for factoring the **difference of two cubes.** A cube is a quantity that is the *third* power of some base.

$$x^3 \quad \text{is the cube of } x$$

$$a^3 \quad \text{is the cube of } a$$

$$8 \quad \text{is the cube of 2, because } 8 = 2^3$$

$$1 \quad \text{is the cube of 1, because } 1 = 1^3$$

In this section we seek a method for factoring the difference of any two cubes:

$$A^3 - B^3$$

Recalling that $A - B$ was one of the factors of the difference of two *squares*, $A^2 - B^2$, perhaps we should see whether $A - B$ might be a factor for the difference of two *cubes*, $A^3 - B^3$. If $A - B$ is a factor of $A^3 - B^3$, we should be able to use *long division* to find the other factor. Let's try dividing $A^3 - B^3$ by $A - B$, using long division.

$$
\begin{array}{r}
A^2 + AB + B^2 \\
A - B{\overline{\smash{)}A^3 - 0A^2B + 0AB^2 - B^3}} \\
\underline{A^3 \overset{\oplus}{-} A^2B} \\
A^2B \\
\underline{A^2B \overset{\oplus}{-} AB^2} \\
AB^2 - B^3 \\
\underline{AB^2 \overset{\oplus}{-} B^3} \\
0
\end{array}
$$

Since there is no remainder, we know that $A - B$ divides exactly into $A^3 - B^3$. Therefore, $A - B$ is indeed a factor of $A^3 - B^3$ and the other factor is $A^2 + AB + B^2$. We have discovered the factors for the difference of two cubes.

> **Difference of Two Cubes**
>
> $$A^3 - B^3 = (A - B)(A^2 + AB + B^2)$$

To factor a polynomial that is the difference of two cubes, we first identify A and B, and then substitute into the formula.

Example 9. Factor $y^3 - 1$.

Solution: Here we see that

$$A^3 = y^3 \quad\quad \text{so } A = y$$

$$B^3 = 1 \quad\quad \text{so } B = 1$$

We substitute in the formula.

$$A^3 - B^3 = (A - B)(A^2 + AB + 1)$$

$$y^3 - 1 = (y - 1)(y^2 + y \cdot 1 + 1^2)$$

$$y^3 - 1 = (y - 1)(y^2 + y + 1)$$

Now you try Example 10.

Example 10. Factor $2x^3 - 54y^3$.

Solution: We see a common monomial factor of 2.

$$2x^3 - 54 = 2(x^3 - 27y^3)$$

Now we see a difference of two cubes in which

$$A = \text{\_\_\_\_} \quad \text{and} \quad B = \text{\_\_\_\_}$$

$$2x^3 - 54y^3 = 2(x^3 - 27y^3)$$

$$= 2(x - 3y)$$

$$\cdot [x^2 + x \cdot 3y + (3y)^2]$$

$$2x^3 - 54y^3 = 2(x - 3y)(\text{_____})$$

Check your work on page 262. ▶

The formula for factoring the **sum of two cubes** is similar to the formula for factoring the difference of two cubes. We present it without proof.

Sum of Two Cubes

$$A^3 + B^3 = (A + B)(A^2 - AB + B^2)$$

Example 11. Factor $x^3 + 64$.

Solution: This is a sum of two cubes and we see that

$$A^3 = x^3 \quad\quad \text{so } A = x$$

$$B^3 = 64 \quad\quad \text{so } B = 4$$

We substitute in the formula.

$$A^3 + B^3 = (A + B)(A^2 - AB + B^2)$$

$$x^3 + 64 = (x + 4)(x^2 - x \cdot 4 + 4^2)$$

$$= (x + 4)(x^2 - 4x + 16)$$

Now you try completing Example 12.

Example 12. Factor $27a^3 + 8c^3$.

Solution: This is a sum of two cubes and we see that

$$A^3 = 27a^3 \quad\quad \text{so } A = 3a$$

$$B^3 = 8c^3 \quad\quad \text{so } B = 2c$$

We substitute in the formula.

$$A^3 + B^3 = (A + B)(A^2 - AB + B^2)$$

$$27a^3 + 8c^3 = (3a + 2c)[\text{_____}]$$

$$=$$

Check your work on page 262. ▶

The formulas for factoring the sum or difference of two cubes should be memorized, because there is no quick way to arrive at the correct factors through trial and error. If you write each formula when you use it in the exercises, you will soon find that you have memorized both of them.

Example 13. Factor $125 + (x + 1)^3$.

Solution: Is this a sum of two cubes? Yes, and we see that

$$A^3 = 125 \qquad \text{so } A = 5$$

$$B^3 = (x + 1)^3 \qquad \text{so } B = (x + 1)$$

We substitute in the formula.

$$A^3 + B^3 = (A + B)(A^2 - AB + B^2)$$

$$125 + (x + 1)^3 = [5 + (x + 1)]$$

$$\cdot [5^2 - 5(x + 1) + (x + 1)^2]$$

$$= [5 + x + 1]$$

$$\cdot [25 - 5x - 5 + x^2 + 2x + 1]$$

$$= (x + 6)(x^2 - 3x + 21)$$

We simplified the right-hand side because we could see that it would contain like terms to be combined.

Example 14. Factor $x^6 - a^6$.

Solution: Is this a difference of two *cubes?* Yes, but it is also a difference of two *squares.* Let's treat it as a difference of two squares.

$$x^6 - a^6 = (x^3 + a^3)(x^3 - a^3)$$

Notice on the right-hand side that one factor is a sum of two cubes and the other is a difference of two cubes. We continue.

$$x^6 - a^6 = (x^3 + a^3)(x^3 - a^3)$$

$$= (x + a)(x^2 - ax + a^2)$$

$$\cdot (x - a)(x^2 + ax + a^2)$$

From Example 14 we observe that whenever an expression is a difference of two squares *and* a difference of two cubes, we should begin by factoring it as a difference of two squares.

⫸ Trial Run

Factor.

_____ **1.** $x^3 - 8$

_____ **2.** $125 + y^3$

_____ **3.** $64a^3 - 27b^3$

_____ **4.** $3x^2 - 81y^3$

_____ **5.** $5c^3 + 5$

_____ **6.** $a^6 - 64$

Answers are on page 262.

▶ Examples You Completed

Example 1. Factor $4a^2 - 49y^2$.

Solution

$$4a^2 - 49y^2$$

$$= (2a + 7y)(2a - 7y)$$

Example 3. Factor $9 - 81y^2$.

Solution

$$9 - 81y^2$$

$$= 9(1 - 9y^2)$$

$$= 9(1 + 3y)(1 - 3y)$$

Example 6. Factor $4y^2 - (x - 3)^2$.

Solution: Is $4y^2 - (x - 3)^2$ a difference of two squares? yes.

Here $A = 2y$
$\qquad B = y - 3$

So

$$4y^2 - (x - 3)^2$$
$$= [2y + (x - 3)][2y - (x - 3)]$$
$$= (2y + x - 3)(2y - x + 3)$$

Example 12. Factor $27a^3 + 8c^3$.

Solution: This is a sum of two cubes and we see that

$$A^3 = 27a^3 \qquad \text{so } A = 3a$$
$$B^3 = 8c^3 \qquad \text{so } B = 2c$$

We substitute in the formula.

$$A^3 + B^3 = (A + B)(A^2 - AB + B^2)$$
$$27a^3 + 8c^3 = (3a + 2c)$$
$$\cdot [(3a)^2 - 3a \cdot 2c + (2c)^2]$$
$$= (3a + 2c)(9a^2 - 6ac + 4c^2)$$

Example 10. Factor $2x^3 - 54y^3$.

Solution: We see a common monomial factor of 2.

$$2x^3 - 54y^3 = 2(x^3 - 27y^3)$$

Now we see a difference of two cubes in which

$$A = x \quad \text{and} \quad B = 3y$$
$$2x^3 - 54y^3 = 2(x^3 - 27y^3)$$
$$= 2(x - 3y)[x^2 + x \cdot 3y + (3y)^2]$$
$$= 2(x - 3y)(x^2 + 3xy + 9y^2)$$

Answers to Trial Runs

page 259   **1.** $(x + 6)(x - 6)$ **2.** $(5a + 3b)(5a - 3b)$ **3.** $2(y + 7)(y - 7)$ **4.** $4(10 + x)(10 - x)$
   5. $(x^2 + 9y^2)(x + 3y)(x - 3y)$ **6.** $(a + b + 5c)(a + b - 5c)$

page 261   **1.** $(x - 2)(x^2 + 2x + 4)$ **2.** $(5 + y)(25 - 5y + y^2)$ **3.** $(4a - 3b)(16a^2 + 12ab + 9b^2)$
   4. $3(x - 3y)(x^2 + 3xy + 9y^2)$ **5.** $5(c + 1)(c^2 - c + 1)$
   6. $(a + 2)(a^2 - 2a + 4)(a - 2)(a^2 + 2a + 4)$

EXERCISE SET 4.2

Factor completely.

_____ **1.** $9y^2 - 1$

_____ **2.** $144y^2 - 1$

_____ **3.** $\frac{1}{4}a^2 - \frac{25}{36}$

_____ **4.** $\frac{1}{9}b^2 - \frac{4}{25}$

_____ **5.** $m^2n^2 - 225$

_____ **6.** $m^2n^2 - 16$

_____ **7.** $25x^2y^2 - 16z^2$

_____ **8.** $36x^2y^2 - 49z^2$

_____ **9.** $3x^2 - 0.75$

_____ **10.** $2x^2 - 0.32$

_____ **11.** $6m^3 - 150m$

_____ **12.** $7m^3 - 112m$

_____ **13.** $2x^2y^2 - 72y^4$

_____ **14.** $3x^2y^2 - 48y^4$

_____ **15.** $25x^4 - 121$

_____ **16.** $9x^4 - 64$

_____ **17.** $m^4 - 16$

_____ **18.** $m^4 - 625$

_____ **19.** $3m^4 - 243$

_____ **20.** $5m^4 - 80$

_____ **21.** $(x + y)^2 - 49$

_____ **22.** $(x + y)^2 - 81$

_____ **23.** $(3a + 2b)^2 - 4c^2$

_____ **24.** $(a + 2b)^2 - 9c^2$

_____ **25.** $2x^3y - 8xy^3$

_____ **26.** $3x^3y - 27xy^3$

_____ **27.** $256 - (x + 2y)^2$

_____ **28.** $289 - (x + 3y)^2$

_____ **29.** $(x - 2)^2 - (y + 3)^2$

_____ **30.** $(x + 6)^2 - (y - 1)^2$

_____ **31.** $x^3 - 9x^2 - 9x + 81$

_____ **32.** $x^3 - 4x^2 - 4x + 16$

_____ **33.** $x^3 - 2x^2 - 25x + 50$

_____ **34.** $x^3 - 3x^2 - 4x + 12$

_____ **35.** $x^3 - y^3$

_____ **36.** $m^3 - n^3$

_____ **37.** $a^3 + 8$

_____ **38.** $a^3 + 27$

_____ **39.** $64 - m^3$

_____ **40.** $125 - m^3$

_____ **41.** $x^3y^3 + 1$

_____ **42.** $x^3y^3 + 64$

_____ **43.** $125x^3 - 27y^3$

_____ **44.** $8x^3 - 343y^3$

_____ **45.** $2x^4 + 16x$

_____ **46.** $3x^4 + 81x$

_____ **47.** $a^3b^3 - 64c^3$

_____ **48.** $216a^3b^3 - c^3$

_____ **49.** $4x^5y^2 + 32x^2y^5$

_____ **50.** $5x^4y + 135xy^4$

_____ **51.** $7x^4y^2 - 56xy^5$ _____ **52.** $9x^6y^2 - 72x^3y^5$

_____ **53.** $(x - y)^3 + 125$ _____ **54.** $(x - 3y)^3 + 216$

_____ **55.** $8 - (x - 3)^3$ _____ **56.** $27 - (x + 4)^3$

_____ **57.** $3x^6 - 3y^6$ _____ **58.** $5x^6 - 5y^6$

_____ **59.** $x^4 - x^3 + x - 1$ _____ **60.** $x^4 + 2x^3 - 8x - 16$

☆ Stretching the Topics _____

Factor completely.

_____ **1.** $4x^{2a} - 36y^{2a}$

_____ **2.** $36x^8 - 13x^4 + 1$

_____ **3.** $x^{4a} - x^a$

Check your answers in the back of your book.

If you can do **Checkup 4.2,** then you are ready to go on to Section 4.3.

✓ CHECKUP 4.2

Factor completely.

_____ **1.** $25a^2 - 4$

_____ **2.** $100x^2y^2 - 121z^2$

_____ **3.** $3x^2 - 432$

_____ **4.** $20m^3 - 45mn^2$

_____ **5.** $x^4 - 256$

_____ **6.** $x^3 + 64y^3$

_____ **7.** $3a^3 - 24b^3$

_____ **8.** $(x - 2)^2 - 9y^2$

_____ **9.** $(x - y)^3 + 216$

_____ **10.** $a^6 - b^6$

Check your answers in the back of your book.

If You Missed Problems:	You Should Review Examples:
1	1
2–4	2, 3
5	7
6	11
7	10
8	6
9	13
10	14

4.3 Factoring Trinomials

When we multiplied two first-degree binomials (that were not the sum and difference of the same two terms), recall that our result was usually a second-degree trinomial. Keeping that in mind, we seek a method for factoring trinomials.

Factoring Trinomials (Leading Coefficient of 1)

The FOIL method helped us organize the multiplication of binomials. For example,

$$(x + 3)(x - 5) = x^2 - 5x + 3x - 15$$

$$= x^2 - 2x - 15$$

$$\text{F} \quad \text{O} + \text{I} \quad \text{L}$$

Notice that the first term (the F term) in the trinomial comes from multiplying the first terms of the binomials. The last term (the L term) in the trinomial comes from multiplying the second terms of the binomials. The middle term in the trinomial is the sum of the O and I terms.

With these facts in mind let us see if we can discover what factors were multiplied to yield the trinomial

$$x^2 + 7x + 12$$

We are looking for two binomials. $\qquad (x \qquad)(x \qquad)$

Since the last term of the trinomial is 12, there are several possibilities for the second terms of the binomials.

$$(x + 1)(x + 12)$$
$$(x + 2)(x + 6)$$
$$(x + 3)(x + 4)$$
$$(x - 1)(x - 12)$$
$$(x - 2)(x - 6)$$
$$(x - 3)(x - 4)$$

Since the middle term of the trinomial is positive ($+7x$), we may reject the last three possibilities. We can find out which of the remaining pairs is correct by multiplying.

$$(x + 1)(x + 12) = x^3 + 13x + 12$$
$$(x + 2)(x + 6) = x^2 + 8x + 12$$
$$(x + 3)(x + 4) = x^2 + 7x + 12$$

Only the last pair of factors gives the correct middle term, so

$$x^2 + 7x + 12 = (x + 3)(x + 4)$$

Example 1. Factor $x^2 - 8x + 12$.

Solution

$$x^2 - 8x + 12 = (x \qquad)(x \qquad)$$

Since the last term is again 12, the possible factors are the same as in the preceding example. In this case, however, we reject the *first* three possibilities since their middle terms will be *positive* and we need a *negative* middle term.

$$(x - 1)(x - 12) = x^2 - 13x + 12$$

$$(x - 2)(x - 6) = x^2 - 8x + 12$$

$$(x - 3)(x - 4) = x^2 - 7x + 12$$

We conclude that the correct factorization is

$$x^2 - 8x + 12 = (x - 2)(x - 6)$$

As you can see from these examples, we can best factor trinomials by first concentrating on the first and last terms of the trinomial. Then we can check the possible binomials by multiplying to see when the correct middle term appears. After some practice, you will develop a good sense about likely factors and it will not be necessary to try every possibility.

You try factoring the trinomial in Example 2.

Example 2. Factor $x^2 - 11x + 10$.

Solution: Since the last term (10) is positive, but the middle term $(-11x)$ is negative, we need only consider the possibilities

$$(x - 5)(x - 2)$$
$$(x - 10)(x - 1)$$

After multiplying, we see that

$$x^2 - 11x + 10 = (\underline{\hspace{1cm}})(\underline{\hspace{1cm}}).$$

Check your work on page 276. ▶

Then complete Example 3.

Now consider factoring the trinomial

$$x^2 + 3x - 4$$

We need two binomials.

Since the last term of the trinomial is -4, there are several choices for the second term of the binomials.

We find the correct pair by multiplying.

The correct factorization is

$$x^2 + 3x - 4 = (x + 4)(x - 1)$$

Example 3. Factor $x^2 + 5xy + 6y^2$.

Solution: Since the last term $(6y^2)$ is _____ , and the middle term $(5xy)$ is _____ , we consider the possibilities

$$(x + 6y)(x + \underline{\hspace{0.7cm}})$$
$$(x + 3y)(x + \underline{\hspace{0.7cm}})$$

After multiplying, we see that

$$x^2 + 5xy + 6y^2 = (\underline{\hspace{1cm}})(\underline{\hspace{1cm}})$$

Check your work on page 276. ▶

$(x \quad\quad)(x \quad\quad)$

$(x + 2)(x - 2)$
$(x + 1)(x - 4)$
$(x + 4)(x - 1)$

$(x + 2)(x - 2) = x^2 - 4$
$(x + 1)(x - 4) = x^2 - 3x - 4$
$(x + 4)(x - 1) = x^2 + 3x - 4$

Example 4. Factor $y^2 - 3y - 10$.

Solution: Since the last term of the trinomial is negative (-10), we start out with

$$y^2 - 3y - 10 = (y + \quad)(y - \quad)$$

Considering the factors of 10, the possible binomials are

$$(y + 10)(y - 1)$$
$$(y + 1)(y - 10)$$
$$(y + 5)(y - 2)$$
$$(y + 2)(y - 5)$$

After multiplying, we see that

$$y^2 - 3y - 10 = (y + 2)(y - 5)$$

Now you try Example 5.

Example 5. Factor $a^2 + 3a - 18$.

Solution: Since the last term of the trinomial is negative (-18), we start out with

$$a^2 + 3a - 18 = (a + \quad)(a - \quad)$$

Considering the factors of 18, we find that the possible binomial factors are

$$(a + 18)(a - \underline{\quad})$$
$$(a + 1)(a - \underline{\quad})$$
$$(a + 9)(a - \underline{\quad})$$
$$(a + 2)(a - \underline{\quad})$$
$$(a + 6)(a - \underline{\quad})$$
$$(a + 3)(a - \underline{\quad})$$

After multiplying, we see that

$$a^2 + 3a - 18 = (\underline{\quad})(\underline{\quad})$$

Check your work on page 276. ▶

After a bit of practice you will learn to spot the likely factors without testing each possibility. Look at Example 6.

Example 6. Factor $x^2 - x - 30$.

Solution: Since the last term is negative (-30), we start out with

$$x^2 - x - 30 = (x + \quad)(x - \quad)$$

There are several ways to factor 30, but our middle term of $-1x$ tells us that the factors are just 1 unit apart. The only logical choices are

$$(x + 6)(x - 5)$$
$$(x + 5)(x - 6)$$

After multiplying, we see that

$$x^2 - x - 30 = (x + 5)(x - 6)$$

You try completing Example 7.

Example 7. Factor $20 + 8x - x^2$.

Solution: Since the last term is negative $(-x^2)$, we start out with

$$20 + 8x - x^2 = (\quad + x)(\quad - x)$$

There are several ways to factor 20, but 20 and 1 are too far apart, and 5 and 4 are too close together. The logical choices are

$$(10 + x)(2 - x)$$
$$(2 + x)(10 - x)$$

After multiplying, we see that

$$20 + 8x - x^2 = (\underline{\quad})(\underline{\quad})$$

Check your factorization on page 276. ▶

From our work so far, we notice that in factoring second-degree trinomials:

---

1. If the first and last terms of the trinomial are *positive* and the middle term is *positive*, both terms of both binomial factors will be positive.

$$\text{Example: } x^2 + 5x + 6 = (x + 3)(x + 2)$$

2. If the first and last terms of the trinomial are *positive*, but the middle term is *negative*, the first terms of both binomial factors will be *positive*, but the second terms of both binomial factors will be *negative*.

$$\text{Example: } x^2 - 5x + 6 = (x - 3)(x - 2)$$

3. If the first term of the trinomial is *positive*, but the last term is *negative*, the first terms of both binomials will be *positive*, but one of the second terms will be *positive* and the other second term will be *negative*.

$$\text{Example: } x^2 - 5x - 6 = (x - 6)(x + 1)$$
$$\text{Example: } x^2 + 5x - 6 = (x + 6)(x - 1)$$

If you are thoroughly familiar with the multiplication of binomials, factoring trinomials will be like undoing a puzzle using wise trial and error. Rules are not really necessary since you can (and should) always check your factors by multiplying them.

⫸ Trial Run

Factor.

_____ **1.** $x^2 - 6x + 5$ _____ **2.** $a^2 - 7a - 18$

_____ **3.** $15 + 2x - x^2$ _____ **4.** $x^2 + 14x + 49$

_____ **5.** $x^2 + 11x - 12$ _____ **6.** $24 - 10x + x^2$

Answers are on page 276.

Factoring Trinomials (Leading Coefficient Not 1)

Let's look at more trinomials in which the coefficient of the second-degree term is different from 1. Now there will be more choices to consider, but the basic principles are exactly the same and our choices can still be checked by multiplication.

Example 8. Factor $5x^2 + 7x + 2$.

Solution

There is no common factor, so we know that the first terms in the binomials must multiply to give $5x^2$. The signs are all positive.

$$5x^2 + 7x + 2$$
$$(5x + \quad)(x + \quad)$$

Since the last term is 2, we know there are only two possible pairs of factors.

$$(5x + 2)(x + 1)$$
$$(5x + 1)(x + 2)$$

After checking by multiplication we decide that

$$5x^2 + 7x + 2 = (5x + 2)(x + 1)$$

Example 9. Factor $6x^2 - 29xy - 5y^2$.

Solution

Since the last term is negative, we know what signs the binomial factors must contain.	$6x^2 - 29xy - 5y^2$ $(\_\_\_\_ + \_\_\_\_)(\_\_\_\_ - \_\_\_\_)$
The last term is $-5y^2$, so we know what the second terms of the binomials can be.	$(\_\_\_\_ + 5y)(\_\_\_\_ - y)$ $(\_\_\_\_ - 5y)(\_\_\_\_ + y)$
Since the first term is $6x^2$, there are several possible pairs of factors.	$(3x + 5y)(2x - y)$ $(3x - 5y)(2x + y)$ $(2x + 5y)(3x - y)$ $(2x - 5y)(3x + y)$ $(6x + 5y)(x - y)$ $(6x - 5y)(x + y)$ $(x + 5y)(6x - y)$ $(x - 5y)(6x + y)$

After checking by multiplication we decide that

$$6x^2 - 29xy - 5y^2 = (x - 5y)(6x + y)$$

We found earlier that it will rarely be necessary for you to list *all* possible pairs of factors. In your search, you should never erase a pair that you have tried, even though it is not correct. When there are several possibilities, you will never remember the pairs you have already tested or see a pattern in your work if you erase your factorizations. Usually, you will hit upon the correct pair of factors before exhausting the entire list of possibilities. You should *test each pair as you proceed rather than write them all down and then test*.

You try factoring the trinomial in Example 10.

Example 10. Factor $6x^2 + 29x + 9$.

Solution: We know that all signs must be positive.

$$6x^2 + 29x + 9$$
$$= (\_\_\_\_ + \_\_\_\_)(\_\_\_\_ + \_\_\_\_)$$

Check your work on page 276. ▶

Example 11. Factor $15x^2 - 9x - 42$.

Solution: We spot the common factor 3.

$$15x^2 - 9x - 42$$
$$= 3(5x^2 - 3x - 14)$$
$$= 3(5x + 7)(x - 2)$$

Now try to factor the trinomial in Example 12.

Example 12. Factor $18xy^3 - x^2y - 4xy$.

Solution: We spot the common factor xy.

$$18xy^3 - xy^2 - 4xy$$
$$= xy(_____)$$
$$= xy(_____)(_____)$$

Check your work on page 276. ▶

Example 13. Factor $-8x^3 + 22x^2 - 15x$.

Solution: We spot the common factor $-x$.

$$-8x^3 + 22x^2 - 15x$$
$$= -x(8x^2 - 22x + 15)$$
$$= -x(2x - 3)(4x - 5)$$

➡ Trial Run

Factor completely.

———— **1.** $3y^2 + 36y + 105$ ———— **2.** $30x^2 - x - 42$

———— **3.** $6x^2 + 5x - 56$ ———— **4.** $21x^2 + 20x + 4$

———— **5.** $-2m^3 + 3m^2n - mn^2$ ———— **6.** $12x^3y - 26x^2y^2 - 10xy^3$

Answers are on page 276.

Factoring Perfect Square Trinomials

In Chapter 3 we learned that the square of a binomial is a perfect square trinomial. If we recognize a perfect square trinomial, we may factor it by reversing the formulas learned earlier.

Perfect Square Trinomials

$$A^2 + 2AB + B^2 = (A + B)^2$$
$$A^2 - 2AB + B^2 = (A - B)^2$$

In a perfect square trinomial, the first term must be the square of some quantity A, the last term must be the square of some quantity B, and the middle term must be twice the product of A times B.

Example 14. Factor $x^2 + 10x + 25$.

Solution: Is $x^2 + 10x + 25$ a perfect square trinomial? Yes.

$$A = x \quad \text{and} \quad B = 5$$
$$2AB = 2(x)(5) = 10x$$

Therefore,

$$x^2 + 10x + 25 = (x + 5)^2$$

You try Example 15.

Example 15. Factor $9a^2 - 60ab + 100b^2$.

Solution: Is $9a^2 - 60ab + 100b^2$ a perfect square trinomial? Yes.

$$A = 3a \quad \text{and} \quad B = 10b$$
$$2AB = 2(\underline{3a})(\underline{10b}) = \underline{60ab}$$

Therefore,

$$9a^2 - 60ab + 100b^2 = (\underline{3a} - \underline{10b})^2$$

Check your work on page 276. ▶

➡ Trial Run

Factor completely.

———— **1.** $x^2 + 12x + 36$ ———— **2.** $x^2y^2 - 4xy + 4$

———— **3.** $x^3 + 14x^2 + 49x$ ———— **4.** $-12x^2 - 12x - 3$

———— **5.** $x^2 - 0.6xy + 0.09y^2$ ———— **6.** $a^2 - \dfrac{2}{5}a + \dfrac{1}{25}$

Answers are on page 277.

Factoring More Trinomials

So far we have discovered the general method for factoring a second-degree trinomial into a product of first-degree binomials. Now consider the polynomial

$$(x + y)^2 + 3(x + y) + 2$$

Notice that the square of the variable part $(x + y)$ of the middle term is the variable part of the first term. Letting A represent the quantity $(x + y)$, we can write the original trinomials using the variable A and then factor it.

$$A^2 \ + \ 3A \ + 2 = \ (A \ + 2) \ (A \ + 1)$$
$$(x + y)^2 + 3(x + y) + 2 = [(x + y) + 2][(x + y) + 1]$$
$$= (x + y + 2)(x + y + 1)$$

Polynomials in which the square of the middle term's variable part is the same as the first term's variable part are called **quadratic in form** because they behave like second-degree (quadratic) polynomials.

Example 16. Factor $2(y - h)^2 - 5(y - h) - 12$.

Solution: Notice that the square of the middle term's variable part, $(y - h)$, is the same as the first term's variable part, $(y - h)^2$.

$$2A^2 \ - \ 5A \ - 12 = \ (2A \ + 3) \ (A \ - 4)$$
$$2(y - h)^2 - 5(y - h) - 12 = [2(y - h) + 3][(y - h) - 4]$$
$$= (2y - 2h + 3)(y - h - 4)$$

Try to complete Example 17.

Example 17. Factor $x^{10} + 7x^5 + 6$.

Solution: We notice that the square of the variable part (x^5) of the second term is the variable part (x^{10}) of the first term.

$$x^{10} + 7x^5 + 6$$
$$= (x^5 + \underline{\hspace{1cm}})(x^5 + \underline{\hspace{1cm}})$$

Check your work on page 276. ▶

Example 18. Factor $x^4 - 5x^2 + 4$.

Solution

$$x^4 - 5x^2 + 4$$
$$= (x^2 - 4)(x^2 - 1)$$
$$= (x + 2)(x - 2)(x + 1)(x - 1)$$

Notice that we continued factoring when we saw the difference of two squares.

⫸ **Trial Run**

Factor completely.

_____ **1.** $(x - y)^2 - (x - y) - 56$ _____ **2.** $2(a - 3)^2 + 5(a - 3) - 12$

_____ **3.** $36(x - 2y)^2 + 17(x - 2y) + 2$

_____ **4.** $x^6 + 4x^3 + 4$

_____ **5.** $x^4 - 13x^2 + 36$ _____ **6.** $4x^4 - 21x^2 - 25$

Answers are on page 277.

Using Factored Forms of Polynomials

When it is necessary to *evaluate* a formula or a polynomial expression, often it is easier to make the required substitution in the *factored form* of the polynomial rather than in the original expression.

For example, consider this physics problem.

Example 19. If an object is thrown upward from the ground with a velocity of 1600 feet per second, its distance (s) above the ground after t seconds is given by the formula

$$s = 1600t - 16t^2$$

Find the object's distance above the ground when $t = 50$ seconds.

Solution: We must evaluate the expression on the right-hand side when $t = 50$. Let's use two approaches to find s.

Original Form	**Factored Form**
$s = 1600t - 16t^2$	$s = 1600t - 16t^2$
	$\quad = 16t(100 - t)$
When $t = 50$:	When $t = 50$:
$s = 1600(50) - 16(50)^2$	$s = 16(50)[100 - 50]$
$\quad = 80{,}000 - 16(2500)$	$\quad = 800(50)$
$\quad = 80{,}000 - 40{,}000$	$\quad = 40{,}000$ ft
$\quad = 40{,}000$ ft	

Do you agree that the computation was simpler when the original expression was rewritten in factored form before making the substitution?

You try Example 20.

Example 20. When an amount of money (P) is invested in an account at 8.5% simple interest for t years, the total amount (A) in the account is given by the formula

$$A = P + 0.085Pt$$

Write the expression on the right in factored form and evaluate it when $P = \$6700$ and $t = 10$ years.

Solution

$$A = P + 0.085Pt$$
$$A = P(1 + 0.085t)$$

When $P = 6700$ and $t = 10$,

$$A = 6700[1 + 0.085(10)]$$
$$= 6700[1 + \underline{\quad}]$$
$$= 6700[\underline{\quad}]$$
$$= \$\underline{\quad}$$

Check your work on page 276. ▶

Now let's try the problem stated at the beginning of this chapter.

Example 21. A company has found that its *revenue* from the sale of x units of a product is $2x^2 - x + 10$ dollars, while its *costs* for producing x units are $x^2 - 3x + 109$ dollars. Write an expression for the company's *profit* when x units are produced and sold. Then write that expression in factored form, and find the profit when the company produces and sells 6 units; 19 units.

Solution: Remember from earlier chapters that

$$\text{Profit} = \text{revenue} - \text{costs}$$

Here

$$\text{Profit} = (2x^2 - x + 10) - (x^2 - 3x + 109)$$
$$= 2x^2 - x + 10 - x^2 + 3x - 109$$
$$= x^2 + 2x - 99$$

In factored form,

$$\text{Profit} = (x + 11)(x - 9)$$

When 6 units are sold, $x = 6$ and

$$\text{Profit} = (x + 11)(x - 9)$$
$$= (6 + 11)(6 - 9)$$
$$= 17(-3)$$
$$= -51 \text{ dollars}$$

The company will *lose* $51 when it produces and sells 6 units.

When 19 units are sold, $x = 19$ and

$$\text{Profit} = (x + 11)(x - 9)$$
$$= (19 + 11)(19 - 9)$$
$$= 30(10)$$
$$= 300 \text{ dollars}$$

The company will make $300 profit when it produces and sells 19 units.

⫸ Trial Run

_____ **1.** If a stone is thrown upward from a height of 35 feet with a velocity of 46 feet per second, its height (s) above the ground after t seconds is given by the formula: $s = 35 + 46t - 16t^2$. Rewrite the right-hand expression in factored form and find the stone's height above the ground after 2.5 seconds.

_____ **2.** The perimeter (P) of a rectangle with length l and width w is given by the formula: $P = 2l + 2w$. Rewrite the right-hand expression in factored form and find the perimeter of a rectangle that is 2.37 cm long and 1.23 cm wide.

Answers are on page 277.

▶ **Examples You Completed** _____

Example 2

Solution

$$x^2 - 11x + 10 = (x - 10)(x - 1)$$

Example 3. Factor $x^2 + 5xy + 6y^2$.

Solution: Since the last term ($6y^2$) is positive, and the middle term ($5xy$) is positive, we consider the possibilities

$$(x + 6y)(x + y)$$
$$(x + 3y)(x + 2y)$$

After multiplying, we see that

$$x^2 + 5xy + 6y^2 = (x + 3y)(x + 2y)$$

Example 7

Solution

$$20 + 8x - x^2 = (10 - x)(2 + x)$$

Example 10

Solution

$$6x^2 + 29x + 9$$
$$= (3x + 1)(2x + 9)$$

Example 15. Factor $9a^2 - 6ab + 100b^2$.

Solution: Is $9a^2 - 60ab + 100b^2$ a perfect square trinomial? Yes.

$$A = 3a \quad \text{and} \quad B = 10b$$
$$2AB = 2(3a)(10b) = 60ab$$

Therefore,

$$9a^2 - 60ab + 100b^2 = (3a - 10b)^2$$

Example 17

Solution

$$x^{10} + 7x^5 + 6 = (x^5 + 6)(x^5 + 1)$$

Example 5. Factor $a^2 + 3a - 18$.

Solution: Since the last term of the trinomial is negative (-18), we start out with

$$a^2 + 3a - 18 = (a + \quad)(a - \quad)$$

Considering the factors of 18, we find that the possible binomial factors are

$$(a + 18)(a - 1)$$
$$(a + 1)(a - 18)$$
$$(a + 9)(a - 2)$$
$$(a + 2)(a - 9)$$
$$(a + 6)(a - 3)$$
$$(a + 3)(a - 6)$$

After multiplying, we see that

$$a^2 + 3a - 18 = (a + 6)(a - 3)$$

Example 12

Solution

$$18xy^3 - xy^2 - 4xy$$
$$= xy(18y^2 - y - 4)$$
$$= xy(9y + 4)(2y - 1)$$

Example 20

Solution

$$A = P + 0.085Pt$$
$$= P(1 + 0.085t)$$

When $P = 6700$ and $t = 10$,

$$A = 6700[1 + 0.085(10)]$$
$$= 6700[1 + 0.85]$$
$$= 6700[1.85]$$
$$= \$12,395$$

Answers to Trial Runs _____

page 270   **1.** $(x - 5)(x - 1)$ **2.** $(a - 9)(a + 2)$ **3.** $(5 - x)(3 + x)$ **4.** $(x + 7)(x + 7)$
5. $(x + 12)(x - 1)$ **6.** $(6 - x)(4 - x)$

page 272   **1.** $3(y + 5)(y + 7)$ **2.** $(6x + 7)(5x - 6)$ **3.** $(3x - 8)(2x + 7)$ **4.** $(7x + 2)(3x + 2)$
5. $-m(2m - n)(m - n)$ **6.** $2xy(3x + y)(2x - 5y)$

page 272　**1.** $(x + 6)^2$　　**2.** $(xy - 2)^2$　　**3.** $x(x + 7)^2$　　**4.** $-3(2x + 1)^2$　　**5.** $(x - 0.3y)^2$

6. $\left(a - \dfrac{1}{5}\right)^2$

page 273　**1.** $(x - y - 8)(x - y + 7)$　　**2.** $(2a - 9)(a + 1)$　　**3.** $(9x - 18y + 2)(4x - 8y + 1)$

4. $(x^3 + 2)^2$　　**5.** $(x + 3)(x - 3)(x + 2)(x - 2)$　　**6.** $(2x - 5)(2x + 5)(x^2 + 1)$

page 275　**1.** $s = 50$ ft　　**2.** $P = 7.2$ cm

EXERCISE SET 4.3

Factor the following trinomials. When the trinomial cannot be factored using integers, write "prime."

_____ **1.** $x^2 - 24x + 144$ _____ **2.** $x^2 - 18x + 81$

_____ **3.** $y^2 - y - 30$ _____ **4.** $y^2 - 11y - 26$

_____ **5.** $b^2 + 8b - 33$ _____ **6.** $b^2 + 4b - 21$

_____ **7.** $9a^2 - 24ab + 16b^2$ _____ **8.** $4a^2 + 20ab + 25b^2$

_____ **9.** $x^2 - 10xy + 25y^2$ _____ **10.** $x^2 - 16xy + 64y^2$

_____ **11.** $39 - 16m + m^2$ _____ **12.** $72 - 17m + m^2$

_____ **13.** $x^2y^2 + 2xy - 15$ _____ **14.** $x^2y^2 + 5xy - 24$

_____ **15.** $3x^2 - 3x - 18$ _____ **16.** $5x^2 - 20x + 15$

_____ **17.** $5a^3 - 40a^2 + 80a$ _____ **18.** $4a^3 + 56a^2 + 196a$

_____ **19.** $20y + 6y^2 - 2y^3$ _____ **20.** $27y - 24y^2 - 3y^3$

_____ **21.** $-10ab^4 + 60ab^2 - 90a$ _____ **22.** $-3ab^4 - 12ab^2 - 12a$

_____ **23.** $x^3y - 18x^2y^2 + 80xy^3$ _____ **24.** $x^3y - 15x^2y^2 + 26xy^3$

_____ **25.** $a^3bc - 2a^2b^2c - 63ab^3c$ _____ **26.** $a^3bc - a^2b^2c - 56ab^3c$

_____ **27.** $(a + 1)^2 - 12(a + 1) + 36$ _____ **28.** $(a + 1)^2 - 10(a + 1) + 25$

_____ **29.** $(x + y)^2 - 7(x + y) + 12$ _____ **30.** $(x + y)^2 - 7(x + y) + 10$

_____ **31.** $(x + 5)^2 + 4(x + 5) - 32$ _____ **32.** $(x + 5)^2 - 7(x + 5) - 18$

_____ **33.** $2y^2 + 7y + 3$ _____ **34.** $3y^2 + 16y + 5$

_____ **35.** $10x^2 - 19x + 6$ _____ **36.** $6x^2 - 25x + 25$

_____ **37.** $6a^2 - 5ab - 56b^2$ _____ **38.** $16a^2 + 6ab - 27b^2$

_____ **39.** $30c^2 - 145cd + 45d^2$ _____ **40.** $20c^2 - 68cd + 24d^2$

_____ **41.** $-9x^4 + 33x^3 - 28x^2$ _____ **42.** $-4a^4 + 32a^3 - 63a^2$

_____ **43.** $12m^2 + 13mn - 55n^2$ _____ **44.** $12m^2 - 7mn - 10n^2$

_____ **45.** $24x^3y + 26x^2y - 70xy$ _____ **46.** $105x^3y + 57x^2y - 72xy$

_____ **47.** $5(a - b)^2 + 7(a - b) + 2$ _____ **48.** $7(a - b)^2 + 22(a - b) + 3$

_____ **49.** $25(x + y)^2 - 20(x + y) + 4$ _____ **50.** $16(x - 2y)^2 - 24(x + 2y) + 9$

———— **51.** $8 - 38(x - 2y) + 39(x - 2y)^2$ ———— **52.** $35 - 89(2x + y) + 30(2x + y)^2$

———— **53.** $x^4 - x^2 - 72$ ———— **54.** $x^4 - 6x^2 - 27$

———— **55.** $3x^4 - 27x^2 + 60$ ———— **56.** $2x^4 - 6x^2 - 8$

———— **57.** $3x^6 - 30x^4y^2 + 75x^2y^4$ ———— **58.** $4x^4y^2 - 88x^2y^4 + 484y^6$

———— **59.** $2x^{10} + 16x^5 - 66$ ———— **60.** $3x^{10} - 9x^5 - 84$

———— **61.** The area of a trapezoid is given by $A = \frac{1}{2}\,ah + \frac{1}{2}\,bh$, where h represents the height of the trapezoid and a and b are the lengths of the two bases. Write the expression on the right in factored form and find the area when h is 7 inches, a is 5 inches, and b is 3 inches.

———— **62.** The difference in the areas of two circles is given by the formula $d = \pi R^2 - \pi r^2$. Write the expression on the right in factored form and find the difference in areas when R is 7.5 m and r is 2.5 m. (Leave π in your answer.)

———— **63.** If an object is thrown upward from a height of 30 feet with a velocity of 38 feet per second, its height (s) above the ground after t seconds is given by $s = 30 + 38t - 16t^2$. Write the expression on the right in factored form and find the object's height above the ground after 2.2 seconds.

———— **64.** A company has found that its revenue from the sale of x units of a product is $R = 3x^2 + 13x - 10$ dollars. Write the expression on the right in factored form and find the revenue when 20 units are produced.

———— **65.** The Close Knit Sweater Company has found that its costs for producing x sweaters is $2x^2 + x - 6$ dollars. Write the expression on the right in factored form and find the cost of producing 50 sweaters.

———— **66.** A certain deodorant manufacturer has found that its revenue from the sale of x units of its product is $3x^2 + 3x + 18$ dollars, while its costs for producing x units are $2x^2 - x + 50$ dollars. Write an expression for the company's profit when x units are produced and sold. Then write the expression in factored form, and find the profit when the company produces and sells 92 units.

☆ Stretching the Topics ——————————————————

Factor completely.

———— **1.** $4x^{2a} + 44y^a + 121$

———— **2.** $6(x + y)^{2m} + 19(x + y)^m + 15$

———— **3.** $x^{2a+6} + 12x^{a+3}y^{a-2} + 36y^{2a-4}$

Check your answers in the back of your book.

If you can do the problems in **Checkup 4.3,** you are ready to go on to Section 4.4.

CHECKUP 4.3

Factor.

_____ **1.** $x^2 - 9xy + 18y^2$ 　　　　　 _____ **2.** $2y^2 + 6y - 108$

_____ **3.** $10x^2 - 3xy - 27y^2$ 　　　　 _____ **4.** $49a^2 - 42ab + 9b^2$

_____ **5.** $-6y^3 + 6y^2 + 12y$ 　　　　 _____ **6.** $x^4 - 2x^2 - 63$

_____ **7.** $4x^4 - 37x^2 + 9$ 　　　　　 _____ **8.** $(x + y)^2 + 8(x + y) - 20$

_____ **9.** $2(x + y)^2 - 13(x + y) + 21$

_____ **10.** If an object is thrown upward from the ground with a velocity of 800 feet per second, its distance (s) above the ground after t seconds is given by $s = 800t - 16t^2$. Rewrite the right-hand expression in factored form and find the object's distance above the ground when $t = 25$ seconds.

Check your answers in the back of your book.

If You Missed Problems:	You Should Review Examples:
1	3
2	9–11
4	15
5	12, 13
6, 7	19
8, 9	16
10	19

4.4 Using All Types of Factoring

Factoring is not a random process. To be successful at rewriting expressions in factored form, you should be systematic in your approach and learn to ask yourself the right questions.

1. Do all the terms contain a common factor?
   If so, use the distributive property to factor out that common factor.
2. Does the expression consist of just *two* terms?
   If so, is it a difference of two squares?
   Or is it a difference of two cubes?
   Or is it a sum of two cubes?
3. Does the expression consist of exactly *four* terms?
   If so, can you group the terms two by two?
4. Does the expression consist of exactly *three* terms?
   If so, is it the square of a binomial?
   Or is it the product of two different binomials?

In many cases you will be able to use more than one factoring technique. After each step you should ask yourself these same questions. In other cases you will discover that the expression cannot be factored; it is prime (irreducible).

Example 1. Factor completely $5x^2y + 15x^2 - 20y - 60$.

Solution

$$5x^2 + 15x^2 - 20y - 60$$

$$= 5(x^2y + 3x^2 - 4y - 12) \qquad \text{Remove common factor 5.}$$

$$= 5(x^2y + 3x^2 - 4y - 12) \qquad \text{Mentally try to group the four terms.}$$

$$= 5[x^2(y + 3) - 4(y + 3)] \qquad \text{Factor each pair of terms.}$$

$$= 5(y + 3)(x^2 - 4) \qquad \text{Remove the common binomial factor.}$$

$$= 5(y + 3)(x + 2)(x - 2) \qquad \text{Factor the difference of two squares.}$$

Example 2. Factor completely $-a^3 + 12a^2 - 36a$.

Solution

$$-a^3 + 12a^2 - 36a$$

$$= -a(a^2 - 12a + 36) \qquad \text{Remove common factor } -a.$$

$$= -a(a - 6)^2 \qquad \text{Factor the perfect square trinomial.}$$

EXERCISE SET 4.4

Factor completely.

_____ **1.** $4(x - y)^3 - 3(x - y)^2 - (x - y)$

_____ **2.** $5(a + b)^3 + 13(a + b)^2 - 6(a + b)$

_____ **3.** $2m^3 - 50m$

_____ **4.** $3m^3 - 48m$

_____ **5.** $6a^3 - 30a^2 - 144a$

_____ **6.** $5a^3 - 30a^2 - 135a$

_____ **7.** $a^3bc - 3a^2b^2c - 54ab^3c$

_____ **8.** $x^3yz + 2x^2y^2z - 63xy^3z$

_____ **9.** $3x^6 - 25x^3 + 8$

_____ **10.** $2x^6 - 55x^3 + 27$

_____ **11.** $7(a - b)^2 + 4(a - b) - 3$

_____ **12.** $6(a + b)^2 + (a + b) - 12$

_____ **13.** $x^3 - 5x^2 + x - 5$

_____ **14.** $2x^3 - 3x^2 + 8x - 12$

_____ **15.** $3x^2 - 3x - 270$

_____ **16.** $5x^2 - 5x - 210$

_____ **17.** $-2x^4 + 128x$

_____ **18.** $-3x^4 + 192x$

_____ **19.** $24x^4 - 40x^3y^3 + 56x^2y^4$

_____ **20.** $32x^4 - 80x^3y^3 + 56x^2y^4$

_____ **21.** $81x^2 - 36xy + 4y^2$

_____ **22.** $169x^2 - 130xy + 25y^2$

_____ **23.** $11(x + y)^2 - 35(x + y) + 6$

_____ **24.** $7(x + y)^2 + 26(x + y) - 45$

_____ **25.** $4x^6 - 24x^4y^2 + 20x^2y^4$

_____ **26.** $3x^6 - 24x^4y^2 - 27x^2y^4$

_____ **27.** $x^2 - 30xy + 225y^2$

_____ **28.** $x^2 - 40xy + 144y^2$

_____ **29.** $3x^2 - 6x - 270$

_____ **30.** $4x^2 - 12x - 224$

_____ **31.** $x^3y^3 + 216$

_____ **32.** $x^3y^3 + 125$

_____ **33.** $3b^2 + 5b - 82$

_____ **34.** $4b^2 - 3b + 7$

_____ **35.** $4a^2x - 8a^2y - 4b^2x + 8b^2y$

_____ **36.** $18a^2x - 9a^2y - 18b^2x + 9b^2y$

_____ **37.** $9x^4 - 7x^2 - 16$

_____ **38.** $4x^4 - 21x^2 - 25$

_____ **39.** $24x^3 - 375$

_____ **40.** $54x^3 - 128$

_____ **41.** $36a^3b + 120a^2b^2 + 100ab^3$

_____ **42.** $96a^3b - 144a^2b^2 + 54ab^3$

_____ **43.** $0.25x^2y^2 - 1.21z^2$

_____ **44.** $0.64x^2y^2 - 1.69z^2$

_____ **45.** $4a^6b^4 + 28a^3b^2 + 49$

_____ **46.** $9a^6b^4 - 48a^3b^2 + 64$

_____ **47.** $-18x^2 + 32$ _____ **48.** $-150x^2 + 24$

_____ **49.** $44x^3yz^4 - 100x^3y^2z^3 - 96x^3y^3z^2$ _____ **50.** $35x^4yz^3 - 20x^4y^2z^2 - 55x^4y^3z$

☆ Stretching the Topics _____

Factor completely.

_____ **1.** $2x^{3a-4} + 6x^{2a-2} + 18x^a$

_____ **2.** $a^{3x+1}b - 729ab^{3x+1}$

_____ **3.** $27(x - y)^3 - 64(x + y)^3$

Check your answers in the back of your book.

If you can do the problems in **Checkup 4.4,** you are ready to go on to the Review Exercises.

CHECKUP 4.4

Factor completely.

_____ **1.** $25x^2 - 90xy + 81y^2$

_____ **2.** $3x^2y^2 - 192y^4$

_____ **3.** $4x^5y^2 - 12x^2y^2$

_____ **4.** $15ax - 10ay - 9x + 6y$

_____ **5.** $9x^2 + 48x + 64$

_____ **6.** $x^4 - 35x^2 - 36$

_____ **7.** $7(a - b)^2 + 4(a - b) - 3$

_____ **8.** $x^6 - 64y^6$

_____ **9.** $24x^2 - 46xy + 21y^2$

_____ **10.** $x^5 - x^3 - x^2 + 1$

Check your answers in the back of your book.

If You Missed Problem:	You Should Review Examples:	
1	Section 4.3	15
2	Section 4.2	2
3	Section 4.1	1–3
4	Section 4.1	11
5	Section 4.3	14
6	Section 4.3	18
7	Section 4.3	16
8	Section 4.2	14
9	Section 4.3	9
10	Section 4.2	8, 9

Summary

For reasons that will become very clear in the following chapters, we have spent a great deal of time sharpening our factoring skills in this chapter. We learned that **factoring** is the process by which we rewrite a *sum* (or difference) of terms as a *product*.

In attempting to factor polynomials we learned to check mentally the list of different types of factoring techniques available to us. We agreed to always look first for a common factor and then for a special product before resorting to trial and error.

Type	Polynomial	Factors	Example
Common factor	$AB + AC$	$A(B + C)$	$3x^2 - 6x = 3x(x - 2)$
Difference of two squares	$A^2 - B^2$	$(A + B)(A - B)$	$9x^2 - 25y^2$ $= (3x + 5y)(3x - 5y)$
Difference of two cubes	$A^3 - B^3$	$(A - B)(A^2 + AB + B^2)$	$8a^3 - 1$ $= (2a - 1)(4a^2 + 2a + 1)$
Sum of two cubes	$A^3 + B^3$	$(A + B)(A^2 - AB + B^2)$	$x^3 + 27c^3$ $= (x + 3c)(x^2 - 3cx + 9c^2)$
Four-term polynomial	$AB + AC + DB + DC$	$A(B + C) + D(B + C)$ $= (B + C)(A + D)$	$2x^3 - x^2 + 10x - 5$ $= x^2(2x - 1) + 5(2x - 1)$ $= (2x - 1)(x^2 + 5)$
General trinomial	$Ax^2 + Bx + C$	Trial and error	$6x^2 + 13x - 5$ $= (2x + 5)(3x - 1)$
Perfect square trinomial	$A^2 + 2AB + B^2$ $A^2 - 2AB + B^2$	$(A + B)(A + B)$ $(A - B)(A - B)$	$x^2 + 14x + 49$ $= (x + 7)(x + 7)$

❏ Speaking the Language of Algebra

Complete each sentence with the appropriate word or phrase.

1. Factoring is the process by which we rewrite a sum (or difference) of terms as a _____ .

2. When factoring a polynomial, we always look first for a _____ _____ .

3. To factor a polynomial containing four terms, we begin by _____ the terms.

4. To factor a second-degree trinomial, we use _____ and _____ .

5. If a polynomial cannot be rewritten as a product of factors, we say that it is _____ .

REVIEW EXERCISES for Chapter 4: Factoring

Completely factor each of the following expressions.

_____ 1. $5a^3 - 15a^2 + 10a$

_____ 2. $-x^5y^2 + 5x^4y^3 - 2x^3y^4$

_____ 3. $3x^4y^3z - 6x^3y^2z + 3x^2yz$

_____ 4. $3x(x - 2) + 4(x - 2)$

_____ 5. $(a - 2b)^2 - 3(a - 2b)$

_____ 6. $(2a - 1)(b + 2) - (a + 1)(b + 2)$

_____ 7. $5a + 5b + ac + bc$ _____ 8. $3ax + 3ay - 2bx - 2by$

_____ 9. $x^3 - 9x^2 + 2x - 18$ _____ 10. $6xy + 10z - 6xz - 10y$

_____ 11. $9a^2 - 16$ _____ 12. $4a^2b^2 - 9c^2$

_____ 13. $3x^2 - 48$ _____ 14. $24m^3 - 150mn^2$

_____ 15. $x^4 - 81$ _____ 16. $x^3 + 27y^3$

_____ 17. $27a^3 - 8b^3$ _____ 18. $3x^3y^3z - 24z$

_____ 19. $(x + y)^3 + 27$ _____ 20. $a^6 + 63a^3 - 64$

_____ 21. $x^2 - 12xy + 35y^2$ _____ 22. $3a^2 - 3a - 168$

_____ 23. $-2y^3 - 4y^2 + 30y$ _____ 24. $x^4 - 15x^2 - 16$

_____ 25. $(x + y)^2 + 2(x + y) - 63$ _____ 26. $18x^2 + 23x - 6$

_____ 27. $9x^2 - 30xy + 25y^2$ _____ 28. $36x^4 + 12x^3 - 63x^2$

_____ 29. $3(x - 1)^2 - 22(x - 1) + 24$ _____ 30. $4a^4 - 101a^2 + 25$

Check your answers in the back of your book.

If You Missed Exercises:	You Should Review Examples:	
1–3	Section 4.1	1–5
4–6		6–9
7–10		10–13
11, 12	Section 4.2	1
13, 14		2–4
15		7
16		11, 12
17, 18		9
19		13
20		14
21	Section 4.3	1–5
22, 23		8, 9
24		18
25		16
26–28		9–13
29		16
30		15

If you have completed the Review Exercises and corrected your errors, you are ready to take the Practice Test.

PRACTICE TEST for Chapter 4: Factoring

Completely factor each expression.

		SECTION	EXAMPLE
_____	1. $8w^4 - 4w^2 + 12w$	4.1	2
_____	2. $-5a^2b^3 + 15a^2b - 25ab^2$	4.1	5
_____	3. $7(3x - 4) - 6x(3x - 4)$	4.1	6
_____	4. $5(3y - 2z) + (3y - 2z)^2$	4.1	7
_____	5. $2a + 2b + ac + bc$	4.1	10
_____	6. $5a - ax - 5 + x$	4.1	12, 13
_____	7. $121x^2 - 4y^2$	4.2	1
_____	8. $4x^2 - 324$	4.2	2
_____	9. $a^4 - 1$	4.2	7
_____	10. $(2a - 5b)^2 - 100$	4.2	6
_____	11. $2x^3 + x^2 - 50x - 25$	4.2	8
_____	12. $125x^3 + y^3$	4.2	11
_____	13. $5a^3 - 40$	4.2	10
_____	14. $30x^2 - 13x - 3$	4.3	8
_____	15. $7x^2 - 7xy - 42y^2$	4.3	9, 11
_____	16. $25a^2 - 30ab + 9b^2$	4.3	9
_____	17. $2x^5 - 5x^4 - 12x^3$	4.3	12, 13
_____	18. $40x^2 + 26x - 9$	4.3	10
_____	19. $6(x + y)^2 - 5(x + y) - 4$	4.3	16
_____	20. $x^4 - 37x^2 + 36$	4.3	18

SHARPENING YOUR SKILLS after Chapters 1–4

SECTION

_____ **1.** From the set $\left\{-9,\ -3.2,\ -\sqrt{2},\ \dfrac{1}{3},\ 0,\ \sqrt{3},\ \dfrac{5}{2},\ 4,\ 5.1313\dots\right\}$, list the members that   1.1

belong to the set of rational numbers.

_____ **2.** Rewrite $4x^2 - 3x^4 - 5x + 3x^3$ in descending powers of the variable.   3.3

Evaluate.

_____ **3.** $(x - 7)(3x - 2)$ when $x = 2$.   1.4

_____ **4.** $-5x^2 + 2xy + 9y^2$ when $x = 3$ and $y = -2$.   1.4

_____ **5.** $|2 - \pi|$ 1.5

_____ **6.** $-x^4 - 4$ when $x = -2$ 3.1

_____ **7.** a^{-2} when $a = -4$ 3.2

_____ **8.** $(-3a)^0$ when $a = 2$ 3.1

Simplify.

_____ **9.** $\dfrac{-7(5) - 4[9 - 3(5)]}{-11[15 - 4(4)]}$ 1.3

_____ **10.** $-7(3n - 1) + 6(n + 2)$ 1.4

_____ **11.** $12a - 2[3a - 4(a - 11b) - 9b] + 20b$ 1.4

_____ **12.** $(4x - 3y)(2x + 7y)$ 3.4

_____ **13.** $(x + 2)(x^2 + 4x + 4)$ 3.4

_____ **14.** $(a - 2)^3$ 3.4

_____ **15.** $(x^3 + 125) \div (x + 5)$ 3.5

Simplify and write the answer with positive exponents.

_____ **16.** $4x(-3x^5y^2)\left(\dfrac{-2}{3}x^3y^3\right)$ 3.1

_____ **17.** $\left(\dfrac{-3a^2b^7c^3}{ab^5}\right)^2$ 3.1

_____ **18.** $(x^9y^{-3}z^{-4})(x^{-7}y^8z^4)$ 3.2

_____ **19.** $(2x^{-4}y^6z^2)^{-3}$ 3.2

_____ **20.** $\dfrac{7x^6y^{-3}}{x^{-5}y^2}$ 3.2

Solve.

_____ **21.** $\dfrac{7x - 1}{5} - 4 = 2x$ 2.1

_____ **22.** $7x - (x + 5) \le 3x - 1$ 2.3

_____ **23.** $|13 - m| \ge 2$ 2.4

_____ **24.** Solve and graph $\dfrac{3x - 1}{7} \le 1 + x$ or $4(2x - 1) < 9x + 2.$ 2.3

_____ **25.** If each side of a square is increased by 6 feet, write an expression for the difference   3.6
between the new area and the old area.

Solving Quadratic Equations and Inequalities 5

The three support beams of a picnic shelter form a right triangle. The longest beam measures 25 feet and the shortest is 5 feet less than the other. Find the length of the piece of lumber that would be needed to cut all three beams.

In Chapter 2 we learned to solve first-degree equations. Now that we are thoroughly familiar with the algebra of polynomials and the techniques of factoring, we are ready to solve equations of higher degree. We shall concentrate on equations of the second degree, also called quadratic equations, and in this chapter we learn how to:

1. Use the zero product rule.
2. Solve quadratic equations by factoring.
3. Solve quadratic equations of the form $ax^2 = c$.
4. Solve quadratic equations by completing the square.
5. Use the quadratic formula.
6. Solve quadratic inequalities.
7. Switch from word statements to quadratic equations and inequalities.

5.1 Solving Equations by Factoring

We have solved many first-degree equations that were of the form $ax + b = c$. Now we would like to find techniques for solving equations containing the *square* of a variable, that is, equations of the form $ax^2 + bx + c = 0$.

Using the Zero Product Rule

Suppose you are told that the product of two numbers equals zero. In other words, suppose you know that $A \cdot B = 0$. What conclusion can be drawn from such a statement?

> If the product of two factors is zero, then at least one of the factors must be zero.

This statement is called the **zero product rule** and it can be expressed in symbols as:

> **Zero Product Rule.** If $A \cdot B = 0$, then $A = 0$ or $B = 0$ (or both).

The zero product rule will be our key to solving equations of degree greater than 1. It can be used whenever a product of two factors or three factors or any number of factors is zero.

Example 1. Solve $x(x - 3) = 0$.

Solution: From the zero product rule we know that

$$x = 0 \quad \text{or} \quad x - 3 = 0$$

and we must isolate x in each of these first-degree equations.

$$x = 0 \quad \text{or} \quad x - 3 = 0$$

$$x = 0 \quad \text{or} \quad x = 3$$

Let's check each solution in the original equation.

CHECK	CHECK
$x = 0$	$x = 3$
$x(x - 3) = 0$	$x(x - 3) = 0$
$0(0 - 3) \overset{?}{=} 0$	$3(3 - 3) \overset{?}{=} 0$
$0(-3) \overset{?}{=} 0$	$3(0) \overset{?}{=} 0$
$0 = 0$	$0 = 0$

You complete Example 2.

Example 2. Solve $(2x + 1)(3x - 7) = 0$.

Solution: From the zero product rule we know that

$$2x + 1 = 0 \quad \text{or} \quad 3x - 7 = 0$$

We must isolate x in each of these first-degree equations.

$$2x + 1 = 0 \qquad \text{or} \quad 3x - 7 = 0$$

$$2x = \underline{} \qquad\qquad 3x = \underline{}$$

$$x = \underline{} \qquad\qquad x = \underline{}$$

Check your work on page 307. ▶

Example 3. Solve $8(2 + x)(2 + x) = 0$.

Solution: There are three factors to consider here.

$$8(2 + x)(2 + x) = 0$$

$$8 \neq 0 \quad 2 + x = 0 \quad \text{or} \quad 2 + x = 0$$

$$x = -2 \qquad\qquad x = -2$$

Since 8 can never equal zero, that factor does not contribute a solution. We do not write the same solution twice, so our solution is $x = -2$.

Solve.

———— **1.** $(x + 3)(x - 7) = 0$ ———— **2.** $(5 - a)(5 - a) = 0$

———— **3.** $(5x - 2)(3x + 1) = 0$ ———— **4.** $(8 - 3x)(2 + x) = 0$

———— **5.** $3(x - 4)(x + 4) = 0$ ———— **6.** $10x(x - 5)(3x - 2) = 0$

Answers are on page 308.

Solving Quadratic Equations by Factoring

Each of the equations in the preceding section was presented in factored form. To discover the degree of such equations, we can multiply the factors together. For example,

$$(2x + 1)(3x - 7) = 0 \quad \text{becomes} \quad 6x^2 - 11x - 7 = 0$$

After multiplying the factors we have a **second-degree equation.** All second-degree equations can be written in general form as $ax^2 + bx + c = 0$, where a, b, and c are real numbers, but $a \neq 0$. Such second-degree equations are also called **quadratic equations.**

> **Quadratic Equation**
>
> $ax^2 + bx + c = 0 \quad (a \neq 0)$

Now that we have learned to recognize quadratic equations, we must develop a method for solving such equations. From our work with the zero product rule, you may have realized that to solve a quadratic equation, we should try to *rewrite the quadratic expression in factored form.*

Example 4. Solve $3x^2 + 5x - 2 = 0$.

Solution: We must rewrite the left-hand side in factored form.

$$3x^2 + 5x - 2 = 0$$

$$(3x - 1)(x + 2) = 0$$

Now we use the zero product rule.

$$3x - 1 = 0 \quad \text{or} \quad x + 2 = 0$$

$$3x = 1$$

$$x = \frac{1}{3} \quad \text{or} \quad x = -2$$

Let's check our solutions in the original equation.

<table>
<tr><td>CHECK</td><td>CHECK</td></tr>
</table>

CHECK

$$x = \frac{1}{3}$$

$$3x^2 + 5x - 2 = 0$$

$$3\left(\frac{1}{3}\right)^2 + 5 \cdot \frac{1}{3} - 2 \stackrel{?}{=} 0$$

$$3\left(\frac{1}{9}\right) + \frac{5}{3} - 2 \stackrel{?}{=} 0$$

$$\frac{3}{9} + \frac{15}{9} - \frac{18}{9} \stackrel{?}{=} 0$$

$$\frac{3 + 15 - 18}{9} \stackrel{?}{=} 0$$

$$\frac{0}{9} \stackrel{?}{=} 0$$

$$0 = 0$$

CHECK

$$x = -2$$

$$3x^2 + 5x - 2 = 0$$

$$3(-2)^2 + 5(-2) - 2 \stackrel{?}{=} 0$$

$$3(4) - 10 - 2 \stackrel{?}{=} 0$$

$$12 - 10 - 2 \stackrel{?}{=} 0$$

$$0 = 0$$

You try solving the equation in Example 5.

Example 5. Solve $0 = 5y^2 - 20$.

Solution

$$0 = 5y^2 - 20$$
$$0 = 5(y^2 - 4)$$
$$0 = 5(\underline{\quad\quad})(\underline{\quad\quad})$$

The solutions are $y = \underline{\quad}$ or $y = \underline{\quad}$.

Check your work on page 307. ▶
Now you try Example 6.

Example 6. Solve $7x - 3x^2 = 0$.

Solution

$$7x - 3x^2 = 0$$
$$x(7 - 3x) = 0$$
$$x = \underline{\quad} \quad \text{or} \quad 7 - 3x = 0$$

The solutions are $x = \underline{\quad}$ or $x = \underline{\quad}$.

Check your work on page 307. ▶

In general, we have found that to solve an equation in which a quadratic expression is equal to zero, we:

1. Factor the quadratic expression into a product of first-degree factors.
2. Using the zero product rule, set each of the factors equal to zero, and solve.

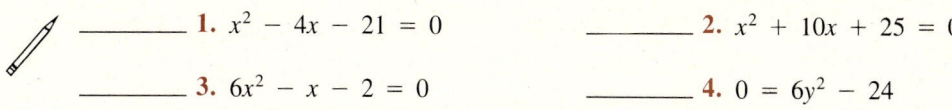

Trial Run

Solve.

_____ **1.** $x^2 - 4x - 21 = 0$ _____ **2.** $x^2 + 10x + 25 = 0$

_____ **3.** $6x^2 - x - 2 = 0$ _____ **4.** $0 = 6y^2 - 24$

Answers are on page 308.

Solving More Quadratic Equations

To use the zero product rule, we must express a quadratic equation as a product of factors equal to zero. What happens when an equation contains a second-degree term on one side but the other side is not zero? We must investigate equations such as $x^2 - 3x = 4$. This is a second-degree equation because it contains a variable raised to the second power. However, it is *not* written in the standard quadratic form $ax^2 + bx + c = 0$. Before we can solve such an equation, we must put it into standard form, with all nonzero terms on one side and zero by itself on the other side. We can accomplish this by adding and/or subtracting the same terms to or from both sides of the original equation.

Example 7. Solve $x^2 - 3x = 4$ and check.

Solution

$$x^2 - 3x = 4$$
$$x^2 - 3x - 4 = 0$$
$$(x - 4)(x + 1) = 0$$
$$x - 4 = 0 \quad \text{or} \quad x + 1 = 0$$
$$x = 4 \quad \text{or} \quad x = -1$$

Let's check both solutions in the original equation.

CHECK

$$x = 4$$
$$x^2 - 3x = 4$$
$$4^2 - 3(4) \stackrel{?}{=} 4$$
$$16 - 12 \stackrel{?}{=} 4$$
$$4 = 4$$

CHECK

$$x = -1$$
$$x^2 - 3x = 4$$
$$(-1)^2 - 3(-1) \stackrel{?}{=} 4$$
$$1 + 3 \stackrel{?}{=} 4$$
$$4 = 4$$

Now you try Example 8. Remember to put the equation into standard form before solving it.

Example 8. Solve $x^2 = 10 - 9x$.

Solution

$$x^2 = 10 - 9x$$
$$x^2 + 9x = 10$$
$$x^2 + 9x - 10 = 0$$

$$x = \underline{\quad\quad} \quad \text{or} \quad x = \underline{\quad\quad}$$

Check your work on page 307. ▶

Example 9. Solve $5y = 3y^2$.

Solution

$$5y = 3y^2$$
$$0 = 3y^2 - 5y$$
$$0 = y(3y - 5)$$
$$y = 0 \quad \text{or} \quad 3y - 5 = 0$$
$$3y = 5$$
$$y = \frac{5}{3}$$

The solutions are $y = 0$ or $y = \frac{5}{3}$.

⫸ Trial Run

Solve.

_____ **1.** $y^2 - 6y = 7$

_____ **2.** $6y^2 = 15y$

_____ **3.** $2x^2 = 19x - 24$

_____ **4.** $-32 + 16x = 2x^2$

Answers are on page 308.

As was the case with first-degree equations, second-degree equations sometimes contain parentheses or other symbols of grouping that must be removed before we can get down to the business of finding solutions.

Suppose, for instance, that we must solve

$$x(x + 3) - 7 = 11$$

It may not be clear to you whether or not this is a quadratic equation at all. At this point, however, that is not important. Whenever we see an equation containing a sum of terms and/or symbols of inclusion, we automatically proceed to remove the parentheses and combine like terms. Then we can decide what kind of equation we are solving and continue with the correct method.

Example 10. Solve $x(x + 3) - 7 = 11$.

Solution

$$x(x + 3) - 7 = 11$$
$$x^2 + 3x - 7 = 11$$
$$x^2 + 3x - 18 = 0$$
$$(x + 6)(x - 3) = 0$$
$$x + 6 = 0 \quad \text{or} \quad x - 3 = 0$$
$$x = -6 \quad \text{or} \quad x = 3$$

CHECK

$$x = -6$$
$$x(x + 3) - 7 = 11$$
$$-6(-6 + 3) - 7 \overset{?}{=} 11$$
$$-6(-3) - 7 \overset{?}{=} 11$$
$$18 - 7 \overset{?}{=} 11$$
$$11 = 11$$

CHECK

$$x = 3$$

$$x(x + 3) - 7 = 11$$

$$3(3 + 3) - 7 \stackrel{?}{=} 11$$

$$3(6) - 7 \stackrel{?}{=} 11$$

$$18 - 7 \stackrel{?}{=} 11$$

$$11 = 11$$

Example 11. Solve $(x - 6)(x + 1) = x(x - 8) + 5$.

   Solution

$(x - 6)(x + 1) = x(x - 8) + 5$	
$x^2 - 5x - 6 = x^2 - 8x + 5$	Remove parentheses.
$-5x - 6 = -8x + 5$	Move x^2-terms to same side.

Notice that the x^2-terms have dropped out. This is a *first-degree equation* that we solve by getting the variable on one side and constant terms on the other.

$3x - 6 = 5$	Add $8x$ to both sides.
$3x = 11$	Add 6 to both sides.
$x = \dfrac{11}{3}$	Divide both sides by 3.

   You try Example 12.

Example 12. Solve $(3x - 1)(x + 7) = -3(x^2 + x + 9)$.

   Solution

$(3x - 1)(x + 7) = -3(x^2 + x + 9)$	
$3x^2 + 20x - 7 = -3x^2 - 3x - 27$	Remove parentheses.
$6x^2 + 20x - 7 = -3x - 27$	Add $3x^2$ to both sides.
$6x^2 + 23x - 7 = -27$	Add $3x$ to both sides.
$6x^2 + 23x + 20 = 0$	Add 27 to both sides.
$(\underline{\hspace{1.5cm}})(\underline{\hspace{1.5cm}}) = 0$	Factor the quadratic expression.

$$x = \underline{\hspace{1cm}} \quad \text{or} \quad x = \underline{\hspace{1cm}}$$

Check your work on page 307. ▶

   In general, we have discovered that any quadratic equation in which the quadratic expression can be factored should be solved using the following steps.

> **Solving Quadratic Equations by Factoring**
>
> **1.** Deal with symbols of grouping and combine like terms.
> **2.** Get all nonzero terms on one side of the equation and zero on the other side.
> **3.** Factor the quadratic expression.
> **4.** Set each factor equal to zero and solve.
> **5.** Check your work by substituting each solution into the original equation.

⟹ Trial Run

Solve.

_____ **1.** $9x^2 - 11x + 9 = 8x^2 - 9x + 12$

_____ **2.** $5x(x - 2) - 6 = 4x^2 - 2(5x + 1)$

_____ **3.** $3(x^2 - 1) - 9 = x(x + 10)$

_____ **4.** $(3x - 2)(x - 4) = -3(x^2 + x - 6)$

Answers are on page 308.

Solving Other Equations by Factoring

Any equation that can be written with a product of factors on one side and zero on the other side can be solved using the zero product rule. You try using the zero product rule in Example 13.

Example 13. Solve

$$(x - 2)(x + 3)(2x - 7) = 0.$$

Solution

$$(x - 2)(x + 3)(2x - 7) = 0$$

$$x - 2 = 0 \quad \text{or} \quad x + 3 = 0$$

$$\text{or}$$

$$2x - 7 = 0$$

$x =$ ____ $x =$ ____ $2x =$ ____

$$x =$$ ____

The three solutions are $x =$ ____ or $x =$
____ or $x =$ ____ .

Check your answers on page 307. ▶

Example 14. Solve $x^3 + 8x^2 = 20x$.

Solution

$$x^3 + 8x^2 = 20x$$

$$x^3 + 8x^2 - 20x = 0$$

$$x(x^2 + 8x - 20) = 0$$

$$x(x + 10)(x - 2) = 0$$

$$x = 0 \text{ or } x + 10 = 0 \quad \text{ or } x - 2 = 0$$

$$x = -10 \qquad x = 2$$

The three solutions are $x = 0$ or $x = -10$ or $x = 2$.

Example 15. Solve

$$x^4 - 29x^2 + 100 = 0.$$

Solution: Since the square of the variable part of the second term (x^2) is identical to the variable part of the first term (x^4), you should recall that we can factor this polynomial just as we factor a quadratic trinomial.

$$x^4 - 29x^2 + 100 = 0$$
$$(x^2 - 25)(x^2 - 4) = 0$$
$$(x + 5)(x - 5)(x + 2)(x - 2) = 0$$
$$x + 5 = 0 \quad \text{or} \quad x - 5 = 0$$
$$x = -5 \qquad x = 5$$

or

$$x + 2 = 0 \quad \text{or} \quad x - 2 = 0$$
$$x = -2 \qquad x = 2$$

Any one of these four solutions satisfies the original equation.

Now you try Example 16.

Example 16. Solve

$$x^3 + 3x^2 - 4x - 12 = 0.$$

Solution

$$x^3 + 3x^2 - 4x - 12 = 0$$

Since this polynomial contains four terms, we can try factoring by grouping.

$$x^2(x + 3) - 4(x + 3) = 0$$
$$(x + 3)(x^2 - 4) = 0$$
$$(x + 3)(\underline{\qquad})(\underline{\qquad}) = 0$$

$$x = \underline{\qquad} \text{ or } x = \underline{\qquad} \text{ or } x = \underline{\qquad}$$

Check your work on page 307. ▶

⫸ Trial Run ▬▬▬▬▬▬▬▬▬▬▬▬▬▬▬▬▬▬▬▬▬

Solve.

_____ **1.** $6x^3 - 21x^2 + 9x = 0$ _____ **2.** $x^4 = 13x^2 - 36$

_____ **3.** $a^3 - 7a^2 - a + 7 = 0$ _____ **4.** $5x^5 - 10x^3 + 5x = 0$

Answers are on page 308.

Finding Equations with Given Solutions

In solving equations by factoring, we found the solutions by setting each first-degree factor equal to zero. Now we would like to reverse our procedure to find an equation for which we know the solutions.

Let's find a quadratic equation with the solutions $x = 2$ and $x = 3$. Throughout this process, think of what the *previous* step would have been in solving the quadratic equation. Work backwards!

$$x = 2 \qquad \text{or} \qquad x = 3$$
$$x - 2 = 0 \qquad \text{or} \qquad x - 3 = 0$$
$$(x - 2)(x - 3) = 0$$
$$x^2 - 5x + 6 = 0$$

We are making use of an important fact.

If r is a *solution* for a polynomial equation, then $x - r$ must be a *factor* of the polynomial.

Example 17. Write a third-degree equation with the solutions $x = 0$, $x = -1$ and $x = 2$.

Solution: We must work backwards.

$$x = 0 \qquad x = -1 \qquad x = 2$$

$$x = 0 \quad x + 1 = 0 \qquad x - 2 = 0$$

$$x(x + 1)(x - 2) = 0$$

$$x(x^2 - x - 2) = 0$$

$$x^3 - x^2 - 2x = 0$$

Example 18. Write a fourth-degree equation with the solutions $x = 1$, $x = -1$, $x = 5$, and $x = -5$.

Solution: We must work backwards.

$$x = 1 \qquad x = -1$$

$$x - 1 = 0 \quad x + 1 = 0$$

$$x = 5 \qquad x = -5$$

$$x - 5 = 0 \quad x + 5 = 0$$

$$(x - 1)(x + 1)(x - 5)(x + 5) = 0$$

$$(x^2 - 1)(x^2 - 25) = 0$$

$$x^4 - 26x^2 + 25 = 0$$

We must point out that in each of these examples, we did not find the *only* equation with the given solutions. Any *constant multiple* of the equations we found will also have the same solutions. For example,

$$x^2 - 5x + 6 = 0 \qquad\qquad 10(x^2 - 5x + 6) = 10(0)$$

$$10x^2 - 50x + 60 = 0$$

has solutions   has solutions

$$x = 2 \quad \text{or} \quad x = 3 \qquad\qquad x = 2 \quad \text{or} \quad x = 3$$

⫸ Trial Run

Find an equation satisfying the given conditions.

_____ **1.** A quadratic equation with solutions $x = 0$ and $x = 5$.

_____ **2.** A third-degree equation with solutions $x = 2$, $x = -2$, and $x = 1$.

Answers are on page 308.

▶ **Examples You Completed** ─────────────────────────────

Example 2. Solve $(2x + 1)(3x - 7) = 0$.

Solution: From the zero product rule we know that

$$2x + 1 = 0 \quad \text{or} \quad 3x - 7 = 0$$

We must isolate x in each of these first-degree equations.

$$
\begin{aligned}
2x + 1 &= 0 & \text{or} \quad 3x - 7 &= 0 \\
2x &= -1 & 3x &= 7 \\
x &= -\frac{1}{2} \quad \text{or} & x &= \frac{7}{3}
\end{aligned}
$$

Example 6. Solve $7x - 3x^2 = 0$.

Solution

$$
\begin{aligned}
7x - 3x^2 &= 0 \\
x(7 - 3x) &= 0 \\
x = 0 \quad \text{or} \quad 7 - 3x &= 0 \\
-3x &= -7 \\
x = 0 \quad \text{or} \quad x &= \frac{7}{3}
\end{aligned}
$$

Example 12

Solution

$$
\begin{aligned}
6x^2 + 23x + 20 &= 0 \\
(3x + 4)(2x + 5) &= 0 \\
3x + 4 = 0 \quad \text{or} \quad 2x + 5 &= 0 \\
3x = -4 \qquad 2x &= -5 \\
x = \frac{-4}{3} \qquad x &= \frac{-5}{2}
\end{aligned}
$$

Example 16. Solve

$$x^3 + 3x^2 - 4x - 12 = 0.$$

Solution

$$
\begin{aligned}
x^3 + 3x^2 - 4x - 12 &= 0 \\
x^2(x + 3) - 4(x + 3) &= 0 \\
(x + 3)(x^2 - 4) &= 0 \\
(x + 3)(x + 2)(x - 2) &= 0 \\
x + 3 = 0 \quad \text{or} \quad x + 2 = 0 \quad \text{or} \quad x - 2 &= 0 \\
x = -3 \quad \text{or} \qquad x = -2 \quad \text{or} \qquad x &= 2
\end{aligned}
$$

Example 5. Solve $0 = 5y^2 - 20$.

Solution

$$
\begin{aligned}
0 &= 5y^2 - 20 \\
0 &= 5(y^2 - 4) \\
0 &= 5(y + 2)(y - 2) \\
5 \neq 0 \quad 0 = y + 2 \quad &\text{or} \quad 0 = y - 2 \\
-2 = y \qquad & \qquad 2 = y
\end{aligned}
$$

The solutions are $y = -2$ or $y = 2$.

Example 8. Solve $x^2 = 10 - 9x$.

Solution

$$
\begin{aligned}
x^2 &= 10 - 9x \\
x^2 + 9x &= 10 \\
x^2 + 9x - 10 &= 0 \\
(x + 10)(x - 1) &= 0 \\
x + 10 = 0 \quad \text{or} \quad x - 1 &= 0 \\
x = -10 \quad \text{or} \qquad x &= 1
\end{aligned}
$$

Example 13. Solve

$$(x - 2)(x + 3)(2x - 7) = 0.$$

Solution

$$
\begin{aligned}
(x - 2)(x + 3)(2x - 7) &= 0 \\
x - 2 = 0 \quad \text{or} \quad x + 3 = 0 \quad \text{or} \quad 2x - 7 &= 0 \\
x = 2 \qquad x = 3 \qquad 2x &= 7 \\
x &= \frac{7}{2}
\end{aligned}
$$

The three solutions are $x = 2$ or $x = -3$ or $x = \frac{7}{2}$.

Answers to Trial Runs

page 299   **1.** $x = -3, x = 7$ **2.** $a = 5$ **3.** $x = \dfrac{2}{5}, x = \dfrac{-1}{3}$ **4.** $x = \dfrac{8}{3}, x = -2$

5. $x = 4, x = -4$ **6.** $x = 0, x = 5, x = \dfrac{2}{3}$

page 301   **1.** $x = 7, x = -3$ **2.** $x = -5$ **3.** $x = \dfrac{-1}{2}, x = \dfrac{2}{3}$ **4.** $y = 2, y = -2$

page 302   **1.** $y = 7, y = -1$ **2.** $y = 0, y = \dfrac{5}{2}$ **3.** $x = 8, x = \dfrac{3}{2}$ **4.** $x = 4$

page 304   **1.** $x = 3, x = -1$ **2.** $x = -2, x = 2$ **3.** $x = 6, x = -1$ **4.** $x = \dfrac{5}{2}, x = \dfrac{-2}{3}$

page 305   **1.** $x = 0, x = \dfrac{1}{2}, x = 3$ **2.** $x = 3, x = -3, x = 2, x = -2$ **3.** $a = 7, a = 1, a = -1$

4. $x = 0, x = -1, x = 1$

page 306   **1.** $x^2 - 5x = 0$ **2.** $x^3 - x^2 - 4x + 4 = 0$

EXERCISE SET 5.1

Solve the following equations.

_____ **1.** $(x - 9)(x + 11) = 0$

_____ **2.** $(x - 7)(x + 12) = 0$

_____ **3.** $(3 - 5x)(3 - 5x) = 0$

_____ **4.** $(4 - 9x)(4 - 9x) = 0$

_____ **5.** $(8x - 1)(3x + 5) = 0$

_____ **6.** $(10x - 3)(3x + 8) = 0$

_____ **7.** $2(9 - 2x)(7 + x) = 0$

_____ **8.** $3(1 - 5x)(6 - x) = 0$

_____ **9.** $(3x - 1)(3x + 1) = 0$

_____ **10.** $(5x - 2)(5x + 2) = 0$

_____ **11.** $8x(x - 2)(3x - 7) = 0$

_____ **12.** $-5x(x - 7)(2x - 9) = 0$

_____ **13.** $x^2 - 11x + 30 = 0$

_____ **14.** $x^2 + 20x + 99 = 0$

_____ **15.** $x^2 + 8x - 20 = 0$

_____ **16.** $x^2 - x - 12 = 0$

_____ **17.** $25 + 10x + x^2 = 0$

_____ **18.** $64 - 16x + x^2 = 0$

_____ **19.** $m^2 + 7m - 44 = 0$

_____ **20.** $m^2 + 3m - 54 = 0$

_____ **21.** $5y^2 + 7y + 2 = 0$

_____ **22.** $7y^2 + 22y + 3 = 0$

_____ **23.** $6a^2 - 34a + 20 = 0$

_____ **24.** $33a^2 - 105a + 18 = 0$

_____ **25.** $a^2 = 16a - 64$

_____ **26.** $a^2 = 15a - 36$

_____ **27.** $5x^2 = 45$

_____ **28.** $3x^2 = 48$

_____ **29.** $2y^2 = 26y$

_____ **30.** $4y^2 = 60y$

_____ **31.** $8m^2 = 15 - 14m$

_____ **32.** $8m^2 = 35 - 18m$

_____ **33.** $2y^2 + 21 = 13y$

_____ **34.** $4y^2 + 15 = 23y$

_____ **35.** $18m + 16 = 9m^2$

_____ **36.** $22 - 25y = 12y^2$

_____ **37.** $7x^2 + 3x - 10 = 5x^2 + 7x + 20$

_____ **38.** $8x^2 - 10x - 16 = 5x^2 - 13x + 20$

_____ **39.** $3x^2 - x = 2(x^2 - 5x)$

_____ **40.** $9x^2 - 2x = 4(2x^2 - 3x)$

_____ **41.** $(x - 5)(x - 6) = x(x + 3) + 7$

_____ **42.** $(x - 11)(x + 9) = x(x - 8) - 3$

_____ **43.** $x(2x - 3) - 4(3x - 5) = 29 - 8x$

_____ **44.** $9x(x - 2) + 6(2x - 1) = 9x + 8$

_____ **45.** $3x(x - 2) + 2(2x^2 - 5) = 2(4x - 5)$

_____ **46.** $13x(x - 8) + 3(x^2 + 8) = 8(3 - 4x)$

_____ **47.** $(3x - 7)(x + 1) = -3(x^2 + 10x + 9)$

_____ **48.** $(4x - 3)(2x - 1) = -8(x^2 - x) + 1$

_____ **49.** $(x - 4)(x + 2)(2x - 3) = 0$

_____ **50.** $(2x - 1)(3x - 2)(x + 5) = 0$

_____ **51.** $25x^3 - 50x^2 + 16x = 0$ _____ **52.** $4x^3 - 20x^2 + 9x = 0$

_____ **53.** $x^4 - 17x^2 + 16 = 0$ _____ **54.** $x^4 - 25x^2 + 144 = 0$

_____ **55.** $3x^4 = 30x^2 - 27$ _____ **56.** $5x^4 = 145x^2 - 500$

_____ **57.** $3x^5 - 24x^3 + 48x = 0$ _____ **58.** $2x^5 - 36x^3 + 162x = 0$

_____ **59.** $2x^5 = 128x^3$ _____ **60.** $3x^5 = 147x^3$

_____ **61.** $x^3 + 3x^2 - 16x - 48 = 0$ _____ **62.** $x^3 + 5x^2 - 4x - 20 = 0$

_____ **63.** $8x^3 - 4x^2 - 18x + 9 = 0$ _____ **64.** $9x^3 - 18x^2 - 4x + 8 = 0$

Find an equation satisfying the given conditions.

_____ **65.** A quadratic equation with solutions $x = -2$ and $x = -9$.

_____ **66.** A quadratic equation with solutions $x = -7$ and $x = 2$.

_____ **67.** A third-degree equation with solutions $x = 0$, $x = 3$, and $x = -3$.

_____ **68.** A third-degree equation with solutions $x = 2$, $x = 1$, and $x = -1$.

_____ **69.** A quadratic equation with solutions $x = \dfrac{2}{3}$ and $x = 5$.

_____ **70.** A quadratic equation with solutions $x = \dfrac{1}{2}$ and $x = \dfrac{2}{5}$.

☆ Stretching the Topics

Solve.

_____ **1.** $(x + 3)(2x + 3) = 77$

_____ **2.** $(x + 3)^2 + (5x + 2)^2 = (7x - 1)^2$

_____ **3.** $3(x - 2)(3x - 1) - 2(2x - 3)(2x + 3) + 5x - 3 = -5(2x + 3) + 27$

Check your answers in the back of your book.

If you can solve the equations in **Checkup 5.1,** you are ready to go on to Section 5.2.

✓ **CHECKUP 5.1**

Solve.

_____ **1.** $(10 - 3x)(1 + 5x) = 0$ _____ **2.** $x^2 - 2x - 63 = 0$

_____ **3.** $6y^2 = 11y + 7$ _____ **4.** $12a^2 - 20 = 22a$

_____ **5.** $(2x - 1)(x + 1) = -x(x - 11) - 4$

_____ **6.** $x(x + 3) + 3(x^2 - 9) = -4(x + 3)$

_____ **7.** $26x^3 - 43x^2 + 6x = 0$ _____ **8.** $x^4 - 40x^2 + 144 = 0$

_____ **9.** $2x^5 - 130x^4 + 128x^3 = 0$

_____ **10.** Write a quadratic equation with solutions $x = -1$ and $x = 4$.

Check your answers in the back of your book.

If You Missed Problems:	You Should Review Exercises:
1	2, 3
2	4
3, 4	7, 8
5, 6	10–12
7	14
8	15
9	16
10	17

5.2 Solving Quadratic Equations of the Form $ax^2 = c$

Although the method of factoring is usually the best way to solve a quadratic equation, there are many quadratic expressions that cannot be factored to yield rational number solutions. In order to develop a method for solving *any* quadratic equation, we take a slight detour here and look at some special kinds of quadratic equations.

Using Square-Root Notation

We learned to solve an equation such as $x^2 - 4 = 0$ by factoring the quadratic expression as a difference of two squares.

$$x^2 - 4 = 0$$

$$(x + 2)(x - 2) = 0$$

$$x + 2 = 0 \qquad \text{or} \qquad x - 2 = 0$$

$$x = -2 \qquad\qquad x = 2$$

However, there is another approach that we might have used on this problem; it is a special kind of quadratic equation in which *the first-degree term is missing*. To solve the equation $x^2 - 4 = 0$, we can rewrite it as

$$x^2 = 4$$

and then ask ourselves: What number, when squared, would equal 4? Our solutions are

$$x = 2 \qquad \text{or} \qquad x = -2$$

because $\qquad (2)^2 = 4 \qquad \text{and} \qquad (-2)^2 = 4$

and they are the same solutions we obtained by factoring.

Let's try using our two approaches to solve $2x^2 = 18$.

First Approach	Second Approach

First Approach

$$2x^2 = 18$$

$$2x^2 - 18 = 0$$

$$2(x^2 - 9) = 0$$

$$2(x + 3)(x - 3) = 0$$

$$x + 3 = 0 \qquad \text{or} \qquad x - 3 = 0$$

$$x = -3 \quad \text{or} \qquad x = 3$$

Second Approach

$$2x^2 = 18$$

$$\frac{2x^2}{2} = \frac{18}{2}$$

$$x^2 = 9$$

$$x = 3 \quad \text{or} \quad x = -3$$

because

$$(3)^2 = 9 \quad \text{and} \quad (-3)^2 = 9$$

Let's practice the second approach on some more equations.

Example 1. Solve $4x^2 = 25$.

Solution

$$4x^2 = 25$$

$$\frac{4x^2}{4} = \frac{25}{4}$$

$$x^2 = \frac{25}{4}$$

$$x = \frac{5}{2} \quad \text{or} \quad x = \frac{-5}{2}$$

because

$$\left(\frac{5}{2}\right)^2 = \frac{25}{4} \quad \text{and} \quad \left(\frac{-5}{2}\right)^2 = \frac{25}{4}$$

You try Example 2.

Example 2. Solve $x^2 - 1 = 0$.

Solution

$$x^2 - 1 = 0$$

$$x^2 = 1$$

$$x = \underline{\quad} \quad \text{or} \quad x = \underline{\quad}$$

because

$$(\underline{\quad})^2 = 1 \quad \text{and} \quad (\underline{\quad})^2 = 1$$

Check your work on page 321. ▶

In each of these examples we solved quadratic equations of the form $ax^2 = c$. After dividing both sides of the equation by the coefficient of the squared term we were left with $x^2 = \frac{c}{a}$.

Then we asked the question: What numbers, when squared, would equal $\frac{c}{a}$? In answering that question, we were actually finding the **square roots** of $\frac{c}{a}$.

Every positive number has two square roots—one positive and one negative. The positive square root of a number is called its **principal square root.**

Principal Square Root. The positive square root of a number n is called its principal square root. It is denoted by $\sqrt{n}$.

Let's practice this notation on some numbers.

Number	Principal Square Root	Square Roots
9	$\sqrt{9} = 3$	$3, -3$
36	$\sqrt{36} = 6$	$6, -6$
$\frac{1}{16}$	$\sqrt{\frac{1}{16}} = \frac{1}{4}$	$\frac{1}{4}, \frac{-1}{4}$
3	$\sqrt{3} = \sqrt{3}$	$\sqrt{3}, -\sqrt{3}$

Notice that the square root of 3 is *not a rational number*. We cannot write $\sqrt{3}$ as a quotient of integers. We can *approximate* $\sqrt{3}$ using a calculator or the table of square roots inside the back cover of this book, but using either of those methods, we can only obtain an estimate ($\sqrt{3} \doteq 1.732051$).* We conclude that $\sqrt{3}$ is an irrational number.

*The symbol $\doteq$ means "is approximately equal to."

Let's practice using our principal-square-root notation to write solutions for some equations of the form $ax^2 = c$.

Example 3. Solve $x^2 = 100$.

Solution

$$x^2 = 100$$
$$x = \sqrt{100} \quad \text{or} \quad x = -\sqrt{100}$$
$$x = 10 \qquad \text{or} \quad x = -10$$

We note that both solutions are rational.

Example 4. Solve $9x^2 - 1 = 0$.

Solution

$$9x^2 - 1 = 0$$
$$9x^2 = 1$$
$$\frac{9x^2}{9} = \frac{1}{9}$$
$$x^2 = \frac{1}{9}$$
$$x = \frac{1}{3} \quad \text{or} \quad x = -\frac{1}{3}$$

Here both solutions are rational.

Now you try Example 5.

Example 5. Solve $x^2 - 49 = 0$.

Solution

$$x^2 - 49 = 0$$
$$x^2 = 49$$
$$x = \sqrt{\rule{1.5em}{0.4pt}} \quad \text{or} \quad x = -\sqrt{\rule{1.5em}{0.4pt}}$$
$$x = \rule{1.5em}{0.4pt} \qquad \text{or} \quad x = \rule{1.5em}{0.4pt}$$

Both solutions are _____ .

Check your work on page 321. ▶

Example 6. Solve $5x^2 = 65$.

Solution

$$5x^2 = 65$$
$$\frac{5x^2}{5} = \frac{65}{5}$$
$$x^2 = 13$$
$$x = \sqrt{13} \quad \text{or} \quad x = -\sqrt{13}$$

Both solutions are irrational.

Suppose that we wish to solve the equation

$$x^2 + 4 = 0$$
$$x^2 = -4$$

Can you think of any numbers whose squares will equal -4? You know that the square of any real number can never be *negative*. Our conclusion is

$$x^2 + 4 = 0 \text{ has } \textit{no real number} \text{ solutions}$$

Although we can write that $x = \sqrt{-4}$ or $x = -\sqrt{-4}$, these solutions are *not real numbers*. We shall discuss numbers that are not real numbers in Chapter 7.

Example 7. Solve $-5x^2 = 35$.

Solution

$$-5x^2 = 35$$

$$\frac{-5x^2}{-5} = \frac{35}{-5}$$

$$x^2 = -7$$

$$x = \sqrt{-7} \quad \text{or} \quad x = -\sqrt{-7}$$

There are *no* real number solutions.

Example 8. Solve $2x^2 + 3 = 5x^2 - 3$.

Solution

$$2x^2 + 3 = 5x^2 - 3$$

$$2x^2 + 3 - 5x^2 = -3$$

$$-3x^2 + 3 = -3$$

$$-3x^2 = -3 - 3$$

$$-3x^2 = -6$$

$$\frac{-3x^2}{-3} = \frac{-6}{-3}$$

$$x^2 = 2$$

$$x = \sqrt{2} \quad \text{or} \quad x = -\sqrt{2}$$

Both solutions are real, but irrational, numbers.

Let's summarize our work with the principal square root as it applies to solving quadratic equations of the form $x^2 = n$.

1. If $x^2 = n$, and n is *positive* and a *perfect square*, the solutions are

$$x = \sqrt{n} \quad \text{or} \quad x = -\sqrt{n}$$

and both solutions are *rational* numbers.

Example: $x^2 = 81$

$$x = \sqrt{81} \quad \text{or} \quad x = -\sqrt{81}$$

$$x = 9 \quad \text{or} \quad x = -9$$

2. If $x^2 = n$, and n is *positive* but *not a perfect square*, the solutions are

$$x = \sqrt{n} \quad \text{or} \quad x = -\sqrt{n}$$

and both solutions are *irrational* numbers.

Example: $x^2 = 5$

$$x = \sqrt{5} \quad \text{or} \quad x = -\sqrt{5}$$

3. If $x^2 = n$, and n is *negative*, there are *no real* solutions.

Example: $x^2 + 3 = 0$

$$x^2 = -3$$

no real solutions

4. If $x^2 = n$, and n is *zero*, the only solution is $x = 0$.

IIII➡ Trial Run

Solve and state the nature of your solutions.

_____ **1.** $x^2 - 2 = 0$ _____ **2.** $16x^2 = 9$

_____ **3.** $7x^2 - 28 = 0$ _____ **4.** $5a^2 = 30$

_____ **5.** $-4x^2 = 36$ _____ **6.** $3x^2 - 7 = 6x^2 + 5$

Answers are on page 322.

Simplifying Square Roots

The set of irrational numbers contains many numbers besides square roots of numbers that are not perfect squares. The algebra of irrational numbers will be discussed thoroughly in Chapter 7. In this chapter we are concerned only with irrational numbers of the form $\sqrt{n}$, and there are a few rules that we might find handy for simplifying square roots of *nonnegative* numbers.
   Let's use two approaches to simplify $\sqrt{36}$.

First Approach	*Second Approach*
$\sqrt{36}$	$\sqrt{36}$
$= 6$	$= \sqrt{4 \cdot 9}$
	$= \sqrt{4} \cdot \sqrt{9}$
	$= 2 \cdot 3$
	$= 6$

Notice that we obtained the same value for $\sqrt{36}$ using both approaches.
We have discovered that for positive factors:

> The square root of a product of factors is equal to the product of the square roots of the factors.

Using A and B to represent positive numbers, we have:

> **Product Rule for Square Roots**
> $$\sqrt{A \cdot B} = \sqrt{A} \cdot \sqrt{B} \qquad \text{where } A \geq 0, B \geq 0$$

   This rule for simplifying square roots is even more useful when the original quantity is not a perfect square. For instance, to simplify $\sqrt{12}$ we must try to write 12 as a product of factors so that at least one factor is a perfect square.

$$\sqrt{12} = \sqrt{4 \cdot 3}$$
$$= \sqrt{4} \cdot \sqrt{3}$$
$$= 2\sqrt{3}$$

Notice that we factored 12 as $4 \cdot 3$ rather than $2 \cdot 6$ because 4 is a perfect square: neither 2 nor 6 is a perfect square. It is customary to write the rational factor before the irrational factor in our final answers. The answer $2\sqrt{3}$ is read "2 times the square root of 3." The number $2\sqrt{3}$ is *irrational* because $\sqrt{3}$ is irrational. Now you try Example 9.

Example 9. Simplify $\sqrt{700}$.

Solution

$$\sqrt{700} = \sqrt{100 \cdot \underline{\qquad}}$$
$$= \sqrt{100} \cdot \sqrt{\underline{\qquad}}$$

Check your work on page 321. ▶

Example 10. Simplify $\sqrt{252}$.

Solution: Sometimes you will not see the largest square factor immediately. Just keep going.

$$\sqrt{252} = \sqrt{4 \cdot 63}$$
$$= \sqrt{4 \cdot 9 \cdot 7}$$
$$= \sqrt{4} \cdot \sqrt{9} \cdot \sqrt{7}$$
$$= 2 \cdot 3\sqrt{7}$$
$$= 6\sqrt{7}$$

Square roots of quotients (fractions) can be simplified in a similar way.

The square root of a quotient of numbers is equal to the quotient of the square roots of those numbers.

Again using A and B to represent positive numbers, we have:

Quotient Rule for Square Roots

$$\sqrt{\frac{A}{B}} = \frac{\sqrt{A}}{\sqrt{B}} \qquad \text{where } A \geq 0, B > 0$$

Example 11. Simplify $\sqrt{\dfrac{9}{16}}$.

Solution

$$\sqrt{\frac{9}{16}} = \frac{\sqrt{9}}{\sqrt{16}}$$
$$= \frac{3}{4}$$

You try Example 12.

Example 12. Simplify $\sqrt{\dfrac{81}{25}}$.

Solution

$$\sqrt{\frac{81}{25}} =$$

Check your work on page 321. ▶

Example 13. Simplify $\sqrt{\dfrac{68}{9}}$.

Solution

$$\sqrt{\frac{68}{9}} = \frac{\sqrt{68}}{\sqrt{9}}$$

$$= \frac{\sqrt{4 \cdot 17}}{3}$$

$$= \frac{\sqrt{4}\sqrt{17}}{3}$$

$$= \frac{2\sqrt{17}}{3}$$

Now you try Example 14.

Example 14. Simplify $\sqrt{\dfrac{48}{49}}$.

Solution

$$\sqrt{\frac{48}{49}} = \frac{\sqrt{48}}{\sqrt{49}}$$

$$=$$

Check your work on page 322. ▶

⫸ **Trial Run**

Simplify.

———— **1.** $\sqrt{48}$

———— **2.** $\sqrt{150}$

———— **3.** $\sqrt{72}$

———— **4.** $\sqrt{76}$

———— **5.** $\sqrt{\dfrac{25}{16}}$

———— **6.** $\sqrt{\dfrac{5}{49}}$

———— **7.** $\sqrt{\dfrac{12}{25}}$

———— **8.** $\sqrt{\dfrac{52}{121}}$

Answers are on page 322.

Our ability to work with square roots now allows us to solve *any* quadratic equation of the form

$$ax^2 = c \qquad (c \geq 0)$$

and to write the solutions in simplified form.

Example 15. Solve $121x^2 = 81$.

Solution

$$121x^2 = 81$$

$$x^2 = \frac{81}{121}$$

$$x = \sqrt{\frac{81}{121}} \quad \text{or} \quad x = -\sqrt{\frac{81}{121}}$$

$$= \frac{\sqrt{81}}{\sqrt{121}} \qquad = \frac{-\sqrt{81}}{\sqrt{121}}$$

$$x = \frac{9}{11} \quad \text{or} \quad x = \frac{-9}{11}$$

Both solutions are rational.

You try Example 16.

Example 16. Solve $2y^2 - 54 = 0$.

Solution

$$2y^2 - 54 = 0$$

$$2y^2 = 54$$

$$y^2 = 27$$

$$y = \sqrt{27} \qquad \text{or} \quad y = -\sqrt{27}$$

$$= \sqrt{\underline{\hspace{0.5cm}} \cdot 3} \qquad = -\sqrt{\underline{\hspace{0.5cm}} \cdot 3}$$

$$y = \underline{\hspace{0.5cm}}\sqrt{3} \quad \text{or} \quad y = \underline{\hspace{0.5cm}}\sqrt{3}$$

Both solutions are $\underline{\hspace{1cm}}$.

Check your work on page 322. ▶

In each of these examples we were forced to simplify the same square root twice. To avoid such duplication, mathematicians sometimes write both solutions at the same time using the symbol $\pm$ before the square root. The $\pm$ symbol is read "plus or minus." It is just a shorthand way of denoting the two different solutions.

Example 17. Solve $18x^2 - 28 = 0$.

Solution

$$18x^2 - 28 = 0$$

$$18x^2 = 28$$

$$x^2 = \frac{28}{18}$$

$$x^2 = \frac{14}{9}$$

$$x = \pm\sqrt{\frac{14}{9}}$$

$$= \pm\frac{\sqrt{14}}{\sqrt{9}}$$

$$= \pm\frac{\sqrt{14}}{3}$$

$$x = \frac{\sqrt{14}}{3} \quad \text{or} \quad x = \frac{-\sqrt{14}}{3}$$

Both solutions are irrational.

You try this notation in completing Example 18.

Example 18. Solve $100y^2 = 243$.

Solution

$$100y^2 = 243$$

$$y^2 = \frac{243}{100}$$

$$y = \pm\sqrt{\frac{243}{100}}$$

$$= \pm\frac{\sqrt{243}}{\sqrt{100}}$$

$$=$$

$$y = \underline{\hspace{1cm}} \quad \text{or} \quad y = \underline{\hspace{1cm}}$$

Both solutions are $\underline{\hspace{1cm}}$.

Check your work on page 322. ▶

Example 19. Solve the formula $V = \dfrac{1}{3}\pi r^2 h$ for r.

Solution

$$V = \frac{1}{3}\pi r^2 h$$

$$3V = \pi r^2 h$$

$$\frac{3V}{\pi h} = r^2$$

$$\pm\sqrt{\frac{3V}{\pi h}} = r$$

Now you try Example 20.

Example 20. Solve the Pythagorean formula, $a^2 + b^2 = c^2$, for a.

Solution

$$a^2 + b^2 = c^2$$

$$a^2 = c^2 - \underline{\hspace{1cm}}$$

$$a = \pm\sqrt{c^2 - \underline{\hspace{1cm}}}$$

Check your work on page 322. ▶

⫸ Trial Run

Solve and state the nature of the solutions.

_____ **1.** $x^2 - 28 = 0$

_____ **2.** $2x^2 = 96$

_____ **3.** $49y^2 = 50$

_____ **4.** $25x^2 - 7 = 0$

_____ **5.** Solve $A = \pi r^2$ for r.

Answers are on page 322.

▶ Examples You Completed

Example 2. Solve $x^2 - 1 = 0$.

Solution

$$x^2 - 1 = 0$$

$$x^2 = 1$$

$$x = 1 \quad \text{or} \quad x = -1$$

$$\text{because}$$

$$(1)^2 = 1 \quad \text{and} \quad (-1)^2 = 1$$

Example 5. Solve $x^2 - 49 = 0$.

Solution

$$x^2 - 49 = 0$$

$$x^2 = 49$$

$$x = \sqrt{49} \quad \text{or} \quad x = -\sqrt{49}$$

$$x = 7 \quad \text{or} \quad x = -7$$

Both solutions are rational.

Example 9. Simplify $\sqrt{700}$.

Solution

$$\sqrt{700} = \sqrt{100 \cdot 7}$$

$$= \sqrt{100} \cdot \sqrt{7}$$

$$= 10\sqrt{7}$$

Example 12. Simplify $\sqrt{\dfrac{81}{25}}$.

Solution

$$\sqrt{\frac{81}{25}} = \frac{\sqrt{81}}{\sqrt{25}}$$

$$= \frac{9}{5}$$

Example 14. Simplify $\sqrt{\dfrac{48}{49}}$.

Solution

$$\sqrt{\frac{48}{49}} = \frac{\sqrt{48}}{\sqrt{49}}$$

$$= \frac{\sqrt{16 \cdot 3}}{7}$$

$$= \frac{\sqrt{16}\sqrt{3}}{7}$$

$$= \frac{4\sqrt{3}}{7}$$

Example 16. Solve $2y^2 - 54 = 0$.

Solution

$$2y^2 - 54 = 0$$

$$2y^2 = 54$$

$$y^2 = 27$$

$$y = \sqrt{27} \quad \text{or} \quad y = -\sqrt{27}$$

$$= \sqrt{9 \cdot 3} \qquad\qquad = -\sqrt{9 \cdot 3}$$

$$y = 3\sqrt{3} \quad \text{or} \quad y = -3\sqrt{3}$$

Both solutions are irrational.

Example 18. Solve $100y^2 = 243$.

Solution

$$100y^2 = 243$$

$$y^2 = \frac{243}{100}$$

$$y = \pm\sqrt{\frac{243}{100}}$$

$$= \pm\frac{\sqrt{243}}{\sqrt{100}}$$

$$= \pm\frac{\sqrt{81 \cdot 3}}{10}$$

$$= \pm\frac{9\sqrt{3}}{10}$$

$$y = \frac{9\sqrt{3}}{10} \quad \text{or} \quad y = \frac{-9\sqrt{3}}{10}$$

Both solutions are irrational.

Example 20. Solve the Pythagorean formula, $a^2 + b^2 = c^2$, for a.

Solution

$$a^2 + b^2 = c^2$$

$$a^2 = c^2 - b^2$$

$$a = \pm\sqrt{c^2 - b^2}$$

Answers to Trial Runs

page 316   **1.** $x = \sqrt{2}, x = -\sqrt{2}$; irrational   **2.** $x = \dfrac{3}{4}, x = \dfrac{-3}{4}$; rational   **3.** $x = 2, x = -2$; rational

           4. $a = \sqrt{6}, a = -\sqrt{6}$; irrational   **5.** No real solutions   **6.** No real solutions

page 319   **1.** $4\sqrt{3}$ **2.** $5\sqrt{6}$ **3.** $6\sqrt{2}$ **4.** $2\sqrt{19}$ **5.** $\dfrac{5}{4}$ **6.** $\dfrac{\sqrt{5}}{7}$ **7.** $\dfrac{2\sqrt{3}}{5}$ **8.** $\dfrac{2\sqrt{13}}{11}$

page 321   **1.** $x = 2\sqrt{7}, x = -2\sqrt{7}$; irrational   **2.** $x = 4\sqrt{3}, x = -4\sqrt{3}$; irrational

           3. $y = \dfrac{5\sqrt{2}}{7}, y = \dfrac{-5\sqrt{2}}{7}$; irrational   **4.** $x = \dfrac{\sqrt{7}}{5}, x = \dfrac{-\sqrt{7}}{5}$; irrational   **5.** $r = \pm\sqrt{\dfrac{A}{\pi}}$

EXERCISE SET 5.2

Simplify the following expressions.

_____ 1. $\sqrt{54}$

_____ 2. $\sqrt{75}$

_____ 3. $\sqrt{242}$

_____ 4. $\sqrt{288}$

_____ 5. $\sqrt{180}$

_____ 6. $\sqrt{300}$

_____ 7. $\sqrt{507}$

_____ 8. $\sqrt{578}$

_____ 9. $\sqrt{225}$

_____ 10. $\sqrt{256}$

_____ 11. $\sqrt{\dfrac{9}{16}}$

_____ 12. $\sqrt{\dfrac{25}{49}}$

_____ 13. $\sqrt{\dfrac{121}{4}}$

_____ 14. $\sqrt{\dfrac{169}{81}}$

_____ 15. $\sqrt{\dfrac{7}{25}}$

_____ 16. $\sqrt{\dfrac{3}{121}}$

_____ 17. $\sqrt{\dfrac{8}{9}}$

_____ 18. $\sqrt{\dfrac{27}{16}}$

_____ 19. $\sqrt{\dfrac{98}{289}}$

_____ 20. $\sqrt{\dfrac{72}{225}}$

Solve the following equations and state the nature of the solutions.

_____ 21. $x^2 - 8 = 0$

_____ 22. $x^2 - 12 = 0$

_____ 23. $x^2 + 9 = 0$

_____ 24. $x^2 + 25 = 0$

_____ 25. $2x^2 = 98$

_____ 26. $3x^2 = 75$

_____ 27. $27y^2 - 3 = 0$

_____ 28. $16y^2 - 4 = 0$

_____ 29. $225x^2 = 49$

_____ 30. $81x^2 = 16$

_____ 31. $32a^2 - 2 = 0$

_____ 32. $18x^2 - 2 = 0$

_____ 33. $6 - 5x^2 = 11$

_____ 34. $9 - 7x^2 = 16$

_____ 35. $3x^2 + 22 = 5x^2 + 2$

_____ 36. $8x^2 + 21 = 10x^2 + 9$

_____ 37. $2(x^2 - 5) = 11(x^2 - 10)$

_____ 38. $3(x^2 - 2) = 7(x^2 - 8)$

_____ 39. $4x(x - 6) = 12(5 - 2x)$

_____ 40. $3x(x - 5) = 15(4 - x)$

_____ 41. Solve $x = 3a^2 - y$ for a.

_____ 42. Solve $x = 5a^2 + 2y$ for a.

_____ **43.** Solve $A = 6s^2$ for s.

_____ **44.** Solve $A = \dfrac{1}{2} h^2$ for h.

_____ **45.** Solve $S = \dfrac{1}{2} gt^2$ for t.

_____ **46.** Solve $S = 4\pi r^2$ for r.

_____ **47.** Solve $V = \pi r^2 h$ for r.

_____ **48.** Solve $E = mc^2$ for c.

☆ Stretching the Topics

Solve.

_____ **1.** $4x^4 - 17x^2 + 15 = 0$

_____ **2.** $x^4 - 16 = 0$

_____ **3.** Solve for x in terms of a: $x^4 - 10a^2x^2 + 9a^4 = 0$

Check your answers in the back of your book.

If you can complete **Checkup 5.2,** you are ready to go on to Section 5.3.

CHECKUP 5.2

Simplify.

_____ **1.** $\sqrt{12}$ 　　　_____ **2.** $\sqrt{175}$ 　　　_____ **3.** $\sqrt{\dfrac{25}{81}}$

_____ **4.** $\sqrt{\dfrac{125}{144}}$ 　　　　　　_____ **5.** $\sqrt{\dfrac{32}{25}}$

Solve the following equations and state the nature of the solutions.

_____ **6.** $x^2 - 24 = 0$ 　　　　　_____ **7.** $x^2 + 36 = 0$

_____ **8.** $25x^2 = 81$ 　　　　　_____ **9.** $18x^2 - 40 = 0$

_____ **10.** Solve $a^2 + b^2 = c^2$ for b.

Check your answers in the back of your book.

If You Missed Problems:	You Should Review Examples:
1, 2	9, 10
3, 4	11, 12
5	13, 14
6	16
7	7
8, 9	15–18
10	19, 20

5.3 Solving Quadratic Equations by Other Methods

Solving Quadratic Equations by Completing the Square

In Section 5.2 we became thoroughly familiar with the method for solving quadratic equations of the form

$$ax^2 = c$$

A similar method can be used to solve an equation such as

$$(x + 1)^2 = 4$$

By treating $(x + 1)$ as the variable quantity, we solve the equation.

$$(x + 1)^2 = 4$$
$$x + 1 = \pm\sqrt{4}$$
$$x + 1 = \pm 2$$

$$x + 1 = 2 \quad \text{or} \quad x + 1 = -2$$
$$x = 1 \qquad\qquad x = -3$$

Notice that we solved for x in the last step. You should check to see that both solutions satisfy the original equation.

Example 1. Solve $(x + 6)^2 - 7 = 0$.

Solution

$$(x + 6)^2 - 7 = 0$$
$$(x + 6)^2 = 7$$
$$x + 6 = \pm\sqrt{7}$$
$$x = -6 \pm\sqrt{7}$$
$$x = -6 + \sqrt{7} \quad \text{or} \quad x = -6 - \sqrt{7}$$

Here both solutions are irrational.

You try Example 2.

Example 2. Solve $(x - 2)^2 = 6$.

Solution

$$(x - 2)^2 = 6$$

$$x = \underline{\qquad} \quad \text{or} \quad x = \underline{\qquad}$$

Both solutions are _____ .

Check your work on page 337. ▶

Notice what these equations have in common. Each of them was expressed as "the square of a first-degree binomial equals a constant." If we can rewrite a quadratic equation in this form, we can use this method to solve it.

Remember that the *square of a binomial is a perfect square trinomial* in which

The first term is the *square* of the first binomial term.
The last term is the *square* of the second binomial term.
The middle term is *twice the product* of the first and second binomial terms.

Example 3. Solve $x^2 + 2x + 1 = 5$.

Solution: You should see that the quantity on the left is a perfect square trinomial.

$$x^2 + 2x + 1 = 5$$
$$(x + 1)^2 = 5$$
$$x + 1 = \pm\sqrt{5}$$
$$x = -1 \pm \sqrt{5}$$
$$x = -1 + \sqrt{5} \quad \text{or} \quad x = -1 - \sqrt{5}$$

You try Example 4.

Example 4. Solve $x^2 - 10x + 25 = 3$.

Solution

$$x^2 - 10x + 25 = 3$$
$$(x - 5)^2 = 3$$

$$x = \underline{\hspace{2cm}} \quad \text{or} \quad x = \underline{\hspace{2cm}}$$

Both solutions are _____ .

Check your work on page 337. ▶

Let's try this method on the equation

$$x^2 + 6x = 1$$

Wait a minute. The left-hand side here is not a trinomial at all, much less a *perfect square trinomial*. Let's take a closer look at the left-hand side of this equation and see if we can make it a perfect square trinomial.

$x^2 + 6x$	We certainly need a constant term.
$x^2 + 6x + \underline{\hspace{1cm}}$	We want this trinomial to be the square of a binomial.
$(x + \underline{\hspace{1cm}})^2$	We know the middle term of the trinomial ($6x$) must be *twice* the product of the first and second binomial terms.
$(x + 3)^2$	Since the first term is x, the second term must be 3, because twice $3x$ is $6x$.
$x^2 + 6x + 9$	We square this binomial to decide on the constant term.

Now we return to the original equation

$$x^2 + 6x = 1$$

We know we must add 9 to the left-hand side to make it a perfect square trinomial, but if we add 9 on the left, we must also add 9 on the right.

$$x^2 + 6x + 9 = 1 + 9$$
$$x^2 + 6x + 9 = 10$$
$$(x + 3)^2 = 10$$
$$x + 3 = \pm\sqrt{10}$$
$$x = -3 \pm \sqrt{10}$$
$$x = -3 + \sqrt{10} \quad \text{or} \quad x = -3 - \sqrt{10}$$

This technique is called **completing the square** and it can be diagrammed for the last problem in the following way:

$$x^2 + 6x \qquad = 1$$
$$x^2 + 6x + 9 = 1 + 9$$
$$\text{half} \diagdown \diagup \text{square}$$
$$(x + 3)^2 = 10$$

Solving a Quadratic Equation by Completing the Square

1. Keep the variable terms on one side and combine the constant terms on the other side.
2. Be sure that the coefficient of the squared term is 1.
3. Find half of the coefficient of x, square it, and add that number to both sides of the equation.
4. Rewrite the perfect square trinomial as the square of a binomial.
5. Solve the resulting equation $(x + r)^2 = c$.

Example 5. Solve $x^2 - 14x + 3 = 8$ by completing the square.

Solution

$$x^2 - 14x + 3 = 8$$
$$x^2 - 14x \qquad = 5$$
$$x^2 - 14x + 49 = 5 + 49$$
$$\text{half} \diagdown \diagup \text{square}$$
$$(x - 7)^2 = 54$$
$$x - 7 = \pm\sqrt{54}$$
$$x - 7 = \pm\sqrt{9 \cdot 6}$$
$$x - 7 = \pm 3\sqrt{6}$$
$$x = 7 \pm 3\sqrt{6}$$
$$x = 7 + 3\sqrt{6} \quad \text{or} \quad x = 7 - 3\sqrt{6}$$

Now you carefully complete Example 6.

Example 6. Solve $x^2 + 4x = -2$ by completing the square.

Solution

$$x^2 + 4x = -2$$
$$x^2 + 4x + \underline{\quad} = -2 + \underline{\quad}$$
$$\text{half} \diagdown \diagup \text{square}$$
$$(x + \underline{\quad})^2 = 2$$

$$x = \underline{\quad\quad} \quad \text{or} \quad x = \underline{\quad\quad}$$

Check your work on page 337. ▶

Example 7. Solve $4y^2 - 20y = 2$ by completing the square.

Solution: Notice that the coefficient of the squared term is not 1. We must divide both sides of the equation by 4 to make the leading coefficient 1.

$$4y^2 - 20y = 2$$

$$y^2 - 5y = \frac{2}{4} \qquad \text{Divide both sides by 4.}$$

$$y^2 - 5y + \frac{25}{4} = \frac{2}{4} + \frac{25}{4} \qquad \begin{array}{l}\text{Add the necessary constant term} \\ \text{to complete the square.}\end{array}$$

half \qquad square

$$\left(y - \frac{5}{2}\right)^2 = \frac{27}{4}$$

$$y - \frac{5}{2} = \pm\sqrt{\frac{27}{4}} \qquad \text{Take the square root of both sides.}$$

$$y - \frac{5}{2} = \pm\frac{3\sqrt{3}}{2} \qquad \text{Simplify the radical expression.}$$

$$y = \frac{5}{2} \pm \frac{3\sqrt{3}}{2} \qquad \text{Isolate the variable.}$$

$$y = \frac{5 \pm 3\sqrt{3}}{2} \qquad \text{Combine fractions on right.}$$

$$y = \frac{5 + 3\sqrt{3}}{2} \quad \text{or} \quad y = \frac{5 - 3\sqrt{3}}{2} \qquad \text{Write solutions.}$$

⁞⊪➤ Trial Run

Solve by completing the square.

———— **1.** $x^2 - 8x = 1$ ⠀⠀⠀⠀ ———— **2.** $x^2 + 20x + 50 = 0$

———— **3.** $z^2 - 24z - 1 = 5$ ⠀⠀⠀⠀ ———— **4.** $2x^2 + 6x + 5 = 6$

Answers are on page 338.

Using the Quadratic Formula

The method of completing the square can be used to develop a method that allows us to solve any quadratic equation.

Consider the standard quadratic equation

$$ax^2 + bx + c = 0$$

and see if we can use the method of completing the square to solve for x in terms of a, b, and c.

$$ax^2 + bx = -c \qquad \text{Move constants to the right side.}$$

$$x^2 + \frac{bx}{a} = \frac{-c}{a} \qquad \begin{array}{l}\text{Divide both sides by } a \text{ so that} \\ \text{the coefficient of } x^2 \text{ is 1.}\end{array}$$

$$x^2 + \frac{bx}{a} + \frac{b^2}{4a^2} = \frac{-c}{a} + \frac{b^2}{4a^2}$$

Take half the coefficient of x, square it, and add the square to both sides.

half   square

$$\left(x + \frac{1}{2} \cdot \frac{b}{a}\right)^2 = \frac{-c}{a} + \frac{b^2}{4a^2}$$

$$\left(x + \frac{b}{2a}\right)^2 = \frac{-4ac}{4a^2} + \frac{b^2}{4a^2}$$

Add the fractions on the right using $4a^2$ as the common denominator.

$$\left(x + \frac{b}{2a}\right)^2 = \frac{b^2 - 4ac}{4a^2}$$

$$x + \frac{b}{2a} = \pm\sqrt{\frac{b^2 - 4ac}{4a^2}}$$

Take the square root of both sides of the equation.

$$x + \frac{b}{2a} = \pm\frac{\sqrt{b^2 - 4ac}}{\sqrt{4a^2}}$$

Simplify the square-root expression on the right.

$$x + \frac{b}{2a} = \pm\frac{\sqrt{b^2 - 4ac}}{2a}$$

$$x = \frac{-b}{2a} \pm \frac{\sqrt{b^2 - 4ac}}{2a}$$

Subtract $\frac{b}{2a}$ from both sides to isolate x on the left.

$$x = \frac{-b \pm \sqrt{b^2 - 4ac}}{2a}$$

Combine the fractions on the right.

Our solutions for $ax^2 + bx + c = 0$ are

$$x = \frac{-b + \sqrt{b^2 - 4ac}}{2a} \quad \text{or} \quad x = \frac{-b - \sqrt{b^2 - 4ac}}{2a}$$

We have derived a formula that allows us to find the solutions of a quadratic equation using only the coefficients of the terms in the original equation. This important and useful formula is called the **quadratic formula.**

Quadratic Formula. If $ax^2 + bx + c = 0$, then

$$x = \frac{-b \pm \sqrt{b^2 - 4ac}}{2a}$$

are the solutions.

You must memorize the quadratic formula unless you intend to complete the square whenever you solve a quadratic equation that does not factor easily. As with every new formula, you will learn it best if you write it down every time you use it.

Let's practice solving a few quadratic equations using the quadratic formula. Be sure to identify a, b, and c after writing the equation in the form $ax^2 + bx + c = 0$.

Example 8. Solve $x^2 + 5x - 6 = 0$ using the quadratic formula.

Solution

$$x^2 + 5x - 6 = 0$$

Here $a = 1$, $b = 5$, and $c = -6$.

$$x = \frac{-b \pm \sqrt{b^2 - 4ac}}{2a}$$

$$x = \frac{-5 \pm \sqrt{5^2 - 4(1)(-6)}}{2(1)}$$

$$= \frac{-5 \pm \sqrt{25 + 24}}{2}$$

$$= \frac{-5 \pm \sqrt{49}}{2}$$

$$x = \frac{-5 \pm 7}{2}$$

$$x = \frac{-5 + 7}{2} \quad \text{or} \quad x = \frac{-5 - 7}{2}$$

$$= \frac{2}{2} \qquad\qquad = \frac{-12}{2}$$

$$x = 1 \qquad \text{or} \quad x = -6$$

Example 9. Solve $x^2 - 2x = 1$ using the quadratic formula.

Solution

$$x^2 - 2x = 1$$

We put the equation into standard form.

$$x^2 - 2x - 1 = 0$$

Here $a = 1$, $b = -2$, and $c = -1$.

$$x = \frac{-b \pm \sqrt{b^2 - 4ac}}{2a}$$

$$x = \frac{-(-2) \pm \sqrt{(-2)^2 - 4(1)(-1)}}{2(1)}$$

$$= \frac{2 \pm \sqrt{4 + 4}}{2}$$

$$= \frac{2 \pm \sqrt{8}}{2}$$

$$= \frac{2 \pm \sqrt{4 \cdot 2}}{2}$$

$$= \frac{2 \pm 2\sqrt{2}}{2}$$

$$= \frac{2(1 \pm \sqrt{2})}{2}$$

$$x = 1 \pm \sqrt{2}$$

$$x = 1 + \sqrt{2} \quad \text{or} \quad x = 1 - \sqrt{2}$$

In Example 8 you should notice that the original trinomial could have been factored. The solutions found by factoring are the same as those found using the quadratic formula, but the method of factoring would be much quicker. *Whenever a second-degree equation can be factored, you should solve it by factoring.* You should use the quadratic formula only when you cannot see any way in which to factor the quadratic expression.

Now try Example 10.

Example 10. Solve $3x^2 + 5x + 1 = 0$ using the quadratic formula.

Solution

$$3x^2 + 5x + 1 = 0$$

The polynomial does not factor. Here $a = 3$, $b = 5$, and $c = 1$.

$$x = \frac{-b \pm \sqrt{b^2 - 4ac}}{2a}$$

$$x = \frac{-5 \pm \sqrt{5^2 - 4(3)(1)}}{2(3)}$$

$$= \frac{-5 \pm \sqrt{\underline{\quad} - \underline{\quad}}}{6}$$

$$= \frac{-5 \pm \sqrt{\underline{\quad}}}{6}$$

$$x = \frac{-5 + \sqrt{\underline{\quad}}}{6} \quad \text{or}$$

$$x = \frac{-5 - \sqrt{\underline{\quad}}}{6}$$

Both solutions are _____ .

Check your work on page 337. ▶

Example 11. Solve $2y^2 - y + 3 = 0$ using the quadratic formula.

Solution

$$2y^2 - y + 3 = 0$$

The polynomial does not factor. Here $a = 2$, $b = -1$, and $c = 3$.

$$y = \frac{-b \pm \sqrt{b^2 - 4ac}}{2a}$$

$$y = \frac{-(-1) \pm \sqrt{(-1)^2 - 4(2)(3)}}{2(2)}$$

$$= \frac{1 \pm \sqrt{1 - 24}}{4}$$

$$= \frac{1 \pm \sqrt{-23}}{4}$$

Since $\sqrt{-23}$ is not a real number, this quadratic equation has *no* real solutions.

Example 12. Solve the formula $d = vt - 16t^2$ for t, using the quadratic formula.

Solution: We put the equation in standard form and identify a, b, and c.

$$d = vt - 16t^2$$

$$16t^2 - vt + d = 0$$

Add $16t^2$, subtract vt to put equation in standard form.

$$t = \frac{-b \pm \sqrt{b^2 - 4ac}}{2a}$$

Here $a = 16$, $b = -v$, and $c = d$.

$$t = \frac{-(-v) \pm \sqrt{(-v)^2 - 4(16)(d)}}{2(16)}$$

Substitute into the quadratic formula.

$$t = \frac{v \pm \sqrt{v^2 - 64d}}{32}$$

Simplify.

$$t = \frac{v + \sqrt{v^2 - 64d}}{32} \quad \text{or} \quad t = \frac{v - \sqrt{v^2 - 64d}}{32}$$

〰️➡ **Trial Run**

Solve using the quadratic formula.

_____ **1.** $x^2 - 4x + 4 = 0$ _____ **2.** $x^2 - 4x = 21$

_____ **3.** $2y^2 = 5y - 1$ _____ **4.** $x^2 - 3x = 1$

_____ **5.** $5a^2 + 2a = 4$ _____ **6.** $2x^2 + x + 1 = 0$

Answers are on page 338.

Working with the Discriminant

In using the quadratic formula, we discovered that the solutions of a quadratic equation might be rational numbers or irrational numbers or not real numbers. The kind of solutions obtained depends entirely on the quantity under the radical in the quadratic formula.

$$x = \frac{-b \pm \sqrt{b^2 - 4ac}}{2a}$$

That quantity $b^2 - 4ac$ is called the **discriminant** for a quadratic equation and it can be used to decide what kind of solution the equation will have.

If $b^2 - 4ac$ is *zero*, then the square root is zero and the only solution to the equation is $x = \dfrac{-b}{2a}$.

If $b^2 - 4ac$ is *positive and a perfect square*, then the square root is a rational number, and the solutions to the equation are both rational numbers.

If $b^2 - 4ac$ is *positive but not a perfect square*, then the square root is an irrational number, and the solutions to the equation are both irrational numbers.

If $b^2 - 4ac$ is *negative*, then the square root is not a real number, and the equation has no real solutions.

We can summarize these facts about the discriminant in the following table.

Discriminant $b^2 - 4ac$	Kind of Solution for $ax^2 + bx + c = 0$	Example
$b^2 - 4ac = 0$	One rational solution	$x^2 + 6x + 9 = 0$ $b^2 - 4ac = 6^2 - 4(1)(9)$ $= 36 - 36$ $= 0$ *Solution:* $x = -3$
$b^2 - 4ac > 0$ and perfect square	Two different rational solutions	$x^2 + 3x - 4 = 0$ $b^2 - 4ac = 3^2 - 4(1)(-4)$ $= 9 + 16$ $= 25$ *Solutions:* $x = -4, x = 1$
$b^2 - 4ac > 0$ and *not* perfect square	Two different irrational solutions	$x^2 + 4x - 3 = 0$ $b^2 - 4ac = 4^2 - 4(1)(-3)$ $= 16 + 12$ $= 28$ *Solutions:* $x = -2 + \sqrt{7}, x = -2 - \sqrt{7}$
$b^2 - 4ac < 0$	No real solutions	$x^2 - 2x + 4 = 0$ $b^2 - 4ac = (-2)^2 - 4(1)(4)$ $= 4 - 16$ $= -12$ *Solutions:* $x = 1 + \sqrt{-3}, x = 1 - \sqrt{-3}$ not real numbers

You practice using the discriminant to describe the solutions for the equations in Examples 13 and 14.

Example 13. Describe the solutions for $2x^2 - 5x - 3 = 0$.

Solution

$$2x^2 - 5x - 3 = 0$$
$$b^2 - 4ac = (-5)^2 - 4(2)(-3)$$

There are _____ solutions.

Check your work on page 337. ▶

Example 14. Describe the solutions for $x^2 + x + 1 = 0$.

Solution

$$x^2 + x + 1 = 0$$
$$b^2 - 4ac = 1^2 - 4(1)(1)$$

There are _____ solutions.

Check your work on page 337. ▶

⫸ Trial Run

Describe the solutions for each equation.

_____ **1.** $x^2 - 3x - 10 = 0$ _____ **2.** $x^2 + 5x + 2 = 0$

_____ **3.** $x^2 - 10x + 25 = 0$ _____ **4.** $x^2 + 2x + 5 = 0$

Answers are on page 338.

Choosing the Best Method

Having learned four ways to solve quadratic equations, you may wonder how to choose the most direct method to use. The most efficient method depends on the particular quadratic equation you wish to solve. In any case, first set the quadratic expression equal to zero and look at the expression $ax^2 + bx + c$.

If the Quadratic Expression:	Then the Equation Can Be Solved by:	Example
Factors readily using integers	Factoring and setting each factor equal to zero	$3x^2 + 2x - 1 = 0$ $(3x - 1)(x + 1) = 0$ $3x - 1 = 0$ or $x + 1 = 0$ $x = \dfrac{1}{3}$ or $x = -1$
Contains no first-degree term ($b = 0$)	Isolating the second-degree term, then taking square roots	$2x^2 - 6 = 0$ $2x^2 = 6$ $x^2 = 3$ $x = \pm\sqrt{3}$ $x = \sqrt{3}$ or $x = -\sqrt{3}$
Has a first-degree term with an even coefficient	Completing the square	$x^2 - 4x + 2 = 0$ $x^2 - 4x + 4 = -2 + 4$ $(x - 2)^2 = 2$ $x - 2 = \pm\sqrt{2}$ $x = 2 \pm \sqrt{2}$ $x = 2 + \sqrt{2}$ or $x = 2 - \sqrt{2}$
Does not meet any special conditions	Using the quadratic formula $$x = \frac{-b \pm \sqrt{b^2 - 4ac}}{2a}$$	$x^2 + 3x - 5 = 0$ $x = \dfrac{-2 \pm \sqrt{3^2 - 4(1)(-5)}}{2(1)}$ $= \dfrac{-2 \pm \sqrt{9 + 20}}{2}$ $= \dfrac{-2 \pm \sqrt{29}}{2}$ $x = \dfrac{-2 + \sqrt{29}}{2}$ or $x = \dfrac{-2 - \sqrt{29}}{2}$

In reality, few students choose to solve a quadratic equation by completing the square unless instructed to do so. If a quadratic expression does not factor or the first-degree term is not missing, most students turn to the quadratic formula.

▶ Examples You Completed

Example 2. Solve $(x - 2)^2 = 6$.

Solution

$$(x - 2)^2 = 6$$
$$x - 2 = \pm\sqrt{6}$$
$$x = 2 \pm \sqrt{6}$$
$$x = 2 + \sqrt{6} \quad \text{or} \quad x = 2 - \sqrt{6}$$

Both solutions are irrational.

Example 4. Solve $x^2 - 10x + 25 = 3$.

Solution

$$x^2 - 10x + 25 = 3$$
$$(x - 5)^2 = 3$$
$$x - 5 = \pm\sqrt{3}$$
$$x = 5 \pm \sqrt{3}$$
$$x = 5 + \sqrt{3} \quad \text{or} \quad x = 5 - \sqrt{3}$$

Both solutions are irrational.

Example 6. Solve $x^2 + 4x = -2$ by completing the square.

Solution

$$x^2 + 4x \qquad = -2$$
$$x^2 + 4x + 4 = -2 + 4$$

half ╲╱ square

$$(x + 2)^2 = 2$$
$$x + 2 = \pm\sqrt{2}$$
$$x = -2 \pm \sqrt{2}$$
$$x = -2 + \sqrt{2} \quad \text{or} \quad x = -2 - \sqrt{2}$$

Example 10. Solve $3x^2 + 5x + 1 = 0$ using the quadratic formula.

Solution

$$3x^2 + 5x + 1 = 0$$

Here $a = 3$, $b = 5$, and $c = 1$.

$$x = \frac{-b \pm \sqrt{b^2 - 4ac}}{2a}$$
$$x = \frac{-5 \pm \sqrt{5^2 - 4(3)(1)}}{2(3)}$$
$$= \frac{-5 \pm \sqrt{25 - 12}}{6}$$
$$= \frac{-5 \pm \sqrt{13}}{6}$$
$$x = \frac{-5 + \sqrt{13}}{6} \quad \text{or} \quad x = \frac{-5 - \sqrt{13}}{6}$$

Both solutions are irrational.

Example 13. Describe the solutions for $2x^2 - 5x - 3 = 0$.

Solution

$$2x^2 - 5x - 3 = 0$$
$$b^2 - 4ac = (-5)^2 - 4(2)(-3)$$
$$= 25 + 24$$
$$= 49$$

There are two different rational solutions.

Example 14. Describe the solutions for $x^2 + x + 1 = 0$.

Solution

$$x^2 + x + 1 = 0$$
$$b^2 - 4ac = 1^2 - 4(1)(1)$$
$$= 1 - 4$$
$$= -3$$

There are no real solutions.

Answers to Trial Runs

page 330   **1.** $x = 4 + \sqrt{17}, x = 4 - \sqrt{17}$ **2.** $x = -10 + 5\sqrt{2}, x = -10 - 5\sqrt{2}$

3. $z = 12 + 5\sqrt{6}, z = 12 - 5\sqrt{6}$ **4.** $x = \dfrac{-3 + \sqrt{11}}{2}, x = \dfrac{-3 - \sqrt{11}}{2}$

page 333   **1.** $x = 2$ **2.** $x = 7, x = -3$ **3.** $y = \dfrac{5 + \sqrt{17}}{4}, y = \dfrac{5 - \sqrt{17}}{4}$

4. $x = \dfrac{3 + \sqrt{13}}{2}, x = \dfrac{3 - \sqrt{13}}{2}$ **5.** $a = \dfrac{-1 + \sqrt{21}}{5}, a = \dfrac{-1 - \sqrt{21}}{5}$

6. $x = \dfrac{1 \pm \sqrt{-7}}{4}$; no real solutions

page 336   **1.** Two different rational solutions   **2.** Two different irrational solutions
3. One rational solution   **4.** No real solutions

EXERCISE SET 5.3

Solve by completing the square.

_____ **1.** $x^2 + 2x - 15 = 0$ _____ **2.** $x^2 - 2x - 8 = 0$

_____ **3.** $x^2 + 4x = 0$ _____ **4.** $x^2 - 10x = 0$

_____ **5.** $3x^2 + x - 3 = 0$ _____ **6.** $3x^2 + 7x + 3 = 0$

_____ **7.** $4x^2 + 9x + 4 = 0$ _____ **8.** $5x^2 - 11x + 5 = 0$

_____ **9.** $x^2 + 4x + 7 = 0$ _____ **10.** $x^2 - 2x + 5 = 0$

Solve by using the quadratic formula.

_____ **11.** $x^2 - 4x + 3 = 0$ _____ **12.** $x^2 - 5x + 6 = 0$

_____ **13.** $x^2 + 8x = 0$ _____ **14.** $x^2 + 7x = 0$

_____ **15.** $x^2 + x = 11$ _____ **16.** $x^2 + x = 3$

_____ **17.** $2x^2 + x - 2 = 0$ _____ **18.** $2x^2 - 3x - 1 = 0$

_____ **19.** $25x^2 = 10x + 2$ _____ **20.** $9x^2 = 6x + 1$

Solve by using the best method.

_____ **21.** $x^2 - 9x + 20 = 0$ _____ **22.** $x^2 - 7x + 12 = 0$

_____ **23.** $2x^2 - 5x - 3 = 0$ _____ **24.** $2x^2 - x - 6 = 0$

_____ **25.** $2x^2 + x + 4 = 0$ _____ **26.** $2x^2 - x + 1 = 0$

_____ **27.** $x^2 + 3x - 17 = 0$ _____ **28.** $x^2 + 2x - 7 = 0$

_____ **29.** $6x^2 + 8x + 1 = 0$ _____ **30.** $9x^2 + 12x - 1 = 0$

_____ **31.** $x^2 + 6x + 9 = 0$ _____ **32.** $x^2 + 10x + 25 = 0$

_____ **33.** $x^2 - 2 = 0$ _____ **34.** $x^2 - 5 = 0$

_____ **35.** $x^2 - 4 = 2x$ _____ **36.** $x^2 - 7 = 2x$

_____ **37.** $3x^2 + 7x + 1 = 0$ _____ **38.** $3x^2 + 5x + 1 = 0$

_____ **39.** $x^2 - 2x + 5 = 0$ _____ **40.** $x^2 - 4x + 5 = 0$

_____ **41.** $x^2 + 2 = 4x$ _____ **42.** $x^2 + 6 = 6x$

_____ **43.** $x(4x - 3) = 5 - 2x$ _____ **44.** $3x(x - 3) = 4 - 7x$

_____ **45.** $-4(x^2 + 4) = -12x - 9$ _____ **46.** $-2(x^2 + 9) = -12x - 10$

_____ **47.** Solve $M = ax^2 + cx$ for x.

_____ **48.** Solve $P = kx^2 - nx$ for x.

_____ **49.** Solve $A = \pi r^2 + 2\pi rh$ for r.

_____ **50.** Solve $A = 2s^2 + 4sh$ for s.

Describe the solutions for each equation after inspecting the discriminant.

_____ **51.** $x^2 + 64 = 16x$

_____ **52.** $x^2 + 25 = 10x$

_____ **53.** $2x^2 - 9x - 5 = 0$

_____ **54.** $3x^2 + 13x + 4 = 0$

_____ **55.** $x^2 - 4x - 6 = 0$

_____ **56.** $x^2 - 7x - 1 = 0$

_____ **57.** $2x^2 - 3x - 5 = 0$

_____ **58.** $4x^2 - x + 2 = 0$

_____ **59.** $6x^2 - 7x = 3$

_____ **60.** $8x^2 + 10x = 3$

☆ Stretching the Topics _____

_____ **1.** Solve by completing the square. $x^2 + (2\sqrt{3})x - 24 = 0$

_____ **2.** Solve by using the quadratic formula. $x^2 + 6\sqrt{5}x + 45 = 0$

_____ **3.** Use the discriminant to describe the solutions for the equation $\sqrt{6}x^2 - 4x - 2\sqrt{6} = 0$.

Check your answers in the back of your book.

If you can solve the equations in **Checkup 5.3,** you are ready to go on to Section 5.4.

CHECKUP 5.3

Solve by completing the square.

_____ **1.** $y^2 - 2y - 9 = 2$

Solve using the quadratic formula.

_____ **2.** $2a^2 - 2a = -1$

Solve using the best method.

_____ **3.** $x^2 - 2x - 4 = 0$

_____ **4.** $x^2 + 3x = 1$

_____ **5.** $5a^2 - 8a + 2 = 0$

_____ **6.** $2x^2 + 2x - 3 = 0$

_____ **7.** Solve $A = 2\pi r^2 + 2\pi rh$ for r.

_____ **8.** Describe the solutions for $5x^2 + 14x = 3$.

Check your answers in the back of your book.

If You Missed Problems:	You Should Review Examples:
1	5
2	7
3–5	8–10
6	11
7	12
8	13, 14

5.4 **Solving Quadratic Inequalities**

Now that we have a thorough understanding of the methods for solving quadratic equations, we turn our attention to **quadratic inequalities.** We learned how to solve first-degree inequalities in Chapter 2, but second-degree inequalities require a slightly different approach.

Using the Case Method

We would like to solve inequalities such as

$$x^2 + 3x - 10 > 0$$

You recall that quadratic (second-degree) equations required the use of factoring whenever possible. Perhaps we should consider factoring as a good starting place in dealing with quadratic inequalities also. Let's rewrite the quadratic expression in our inequality in factored form.

$$x^2 + 3x - 10 > 0$$
$$(x + 5)(x - 2) > 0$$

This inequality now says that we want the product of the quantities $(x + 5)$ and $(x - 2)$ to be greater than zero. In other words, we want that product of factors to be *positive*. When will the product of two factors be positive? If both factors are positive *or* if both factors are negative.

We agree, then, that there are two different cases to be considered here.

> *Case 1:* Suppose that both factors are positive.
> *Case 2:* Suppose that both factors are negative.

Each of these cases must be treated separately, using our knowledge of solutions of pairs of inequalities.

Case 1: Suppose that both factors are *positive.*

$$x + 5 > 0 \quad \text{and} \quad x - 2 > 0$$
$$x > -5 \quad \text{and} \quad x > 2$$

Both conditions must be met at the same time. Our solutions from Case 1 are

$$x > 2$$

Case 2: Suppose that both factors are *negative.*

$$x + 5 < 0 \quad \text{and} \quad x - 2 < 0$$
$$x < -5 \quad \text{and} \quad x < 2$$

Both conditions must be met at the same time. Our solutions from Case 2 are

$$x < -5$$

Since the inequality $x^2 + 3x - 10 > 0$ is satisfied by either Case 1 solutions *or* Case 2 solutions, our final solutions are $x > 2$ or $x < -5$.

Example 1. Solve $3y^2 - 8y \leq 35$.

Solution: We first get the quadratic expression on one side and write it in factored form.

$$3y^2 - 8y - 35 \leq 0$$

$$(3y + 7)(y - 5) \leq 0$$

Here we have a product that is *less than* or equal to zero. In other words, the product of $(3y + 7)$ and $(y - 5)$ is a *negative* number or zero. When will the product of two factors be negative? When one factor is positive and the other factor is negative. Since we do not know which is positive and which is negative, we must consider two cases.

Case 1: Suppose that the first factor is positive (or zero) and the second factor is negative (or zero).

$$3y + 7 \geq 0 \quad \text{and} \quad y - 5 \leq 0$$

$$3y \geq -7$$

$$y \geq \frac{-7}{3} \quad \text{and} \quad y \leq 5$$

Both conditions must be met at the same time, so our solutions from Case 1 are

$$\frac{-7}{3} \leq y \leq 5$$

Case 2: Suppose that the first factor is negative (or zero) and the second factor is positive (or zero).

$$3y + 7 \leq 0 \quad \text{and} \quad y - 5 \geq 0$$

$$3y \leq -7$$

$$y \leq \frac{-7}{3} \quad \text{and} \quad y \geq 5$$

Both conditions must be met at the same time, so Case 2 gives *no solutions*.

Our original inequality $3y^2 - 8y \leq 35$ will be satisfied by Case 1 solutions *or* Case 2 solutions, so our final solutions are

$$\frac{-7}{3} \leq y \leq 5$$

In other words, if we choose any y value from the interval between $\frac{-7}{3}$ and 5, our original inequality will be satisfied.

Example 2. Solve $1 - x^2 > 0$.

Solution

$$1 - x^2 > 0$$

$$(1 + x)(1 - x) > 0$$

We want our product of factors to be *positive*.

Case 1: Suppose that both factors are positive.

$$1 + x > 0 \quad \text{and} \quad 1 - x > 0$$

$$-x > -1$$

$$x > -1 \quad \text{and} \quad x < 1$$

Case 2: Suppose that both factors are negative.

$$1 + x < 0 \quad \text{and} \quad 1 - x < 0$$

$$-x < -1$$

$$x < -1 \quad \text{and} \quad x > 1$$

Solutions from Case 1 are   There are *no solutions* from Case 2.

$$-1 < x < 1$$

Our final solutions for the inequality $1 - x^2 > 0$ are

$$-1 < x < 1$$

▥▶ Trial Run

Solve each inequality by considering the cases.

_____ **1.** $x^2 + x - 12 > 0$ _____ **2.** $x^2 - 2x \leq 24$

_____ **3.** $2x^2 - x \geq 0$ _____ **4.** $4 - x^2 > 0$

Answers are on page 348.

Using Critical Points on the Number Line

As you may have noticed, in each of the last examples there were certain key points (or **critical points**) on the number line where we concentrated our attention.

Inequality	Factored Form	Critical Points
$x^2 + 3x - 10 > 0$	$(x + 5)(x - 2) > 0$	$x = -5, x = 2$
$3y^2 - 8y \leq 35$	$(3y + 7)(y - 5) \leq 0$	$y = \dfrac{-7}{3}, y = 5$
$1 - x^2 > 0$	$(1 + x)(1 - x) > 0$	$x = -1, x = 1$

The critical points for an inequality separate the number line into distinct intervals. Each of our inequalities was satisfied by all the numbers in one or more of these intervals. Perhaps we could use a number-line approach to solve quadratic inequalities without using cases at all.

For the inequality

$$x^2 + 3x - 10 > 0$$

$$(x + 5)(x - 2) > 0$$

the critical points on the number line are

$$x = -5 \quad \text{and} \quad x = 2$$

We separate the number line into three distinct intervals determined by the critical points, using *open* dots at each critical point because this is a $>$ inequality (rather than a $\geq$ inequality).

Remember, to satisfy our original inequality, the *product* of the factors must be *positive*. We must check the *signs* of each of the factors $(x + 5)$ and $(x - 2)$ in each of the intervals to determine in which intervals their product is positive.

To decide on the sign of a factor in a particular interval, we simply choose some number in the interval, substitute it into each factor, and draw a conclusion about that factor's *sign* in that interval. We organize our work in the following diagram.

Notice in interval I that both factors are negative, so their product is *positive*. In interval II, the factor $(x + 5)$ is positive, but the factor $(x - 2)$ is negative, so their product is *negative*. In interval III both factors are positive, so their product is *positive*.

What is the solution? Since we wanted the product to be *positive,* the original inequality is satisfied in interval I or in interval III. We can describe those intervals as follows:

$$\text{Interval I:} \quad x < -5$$
$$\text{Interval III:} \quad x > 2$$

Our solutions are

$$x < -5 \quad or \quad x > 2$$

which are the same solutions obtained earlier using the case method.

Example 3. Solve $x^2 + 10x + 21 \leq 0$ using the critical-point method.

Solution

$$x^2 + 10x + 21 \leq 0$$
$$(x + 7)(x + 3) \leq 0$$

The critical points are

$$x = -7 \quad \text{and} \quad x = -3$$

and we use *solid* dots because this is a $\leq$ inequality (rather than a $<$ inequality).

Since we want the product to be *negative* or zero (≤ 0), the solutions lie in interval II.

$$-7 \leq x \leq -3$$

Example 4. Solve $2x^2 + 3x \geq -1$.

Solution

$$2x^2 + 3x \geq -1$$

$$2x^2 + 3x + 1 \geq 0$$

$$(2x + 1)(x + 1) \geq 0$$

Critical points are

$$x = \frac{-1}{2} \quad \text{and} \quad x = -1$$

The dots should be solid, and we want the product to be positive or zero (≥ 0).

The product is positive or zero in interval I or interval III, so the solutions are

$$x \leq -1 \quad \text{or} \quad x \geq -\frac{1}{2}$$

You try completing Example 5.

Example 5. Solve $x^2 < 2x - 1$.

Solution

$$x^2 < 2x - 1$$

$$x^2 - 2x + 1 < 0$$

$$(x - 1)(x - 1) < 0$$

The only critical point is

$$x = \underline{\hphantom{xxx}}$$

The dot should be _____ and we want the product to be _____ (< 0).

$(x - 1)(x - 1)$

The product is *never* negative, so there is _____ for this inequality.

Check your work on page 348. ▶

It should not surprise you that there was no solution for the inequality in Example 5.

$$x^2 - 2x + 1 < 0$$

$$(x - 1)(x - 1) < 0$$

$$(x - 1)^2 < 0$$

We know that the square of a real quantity can *never* be *negative*.
On the other hand, if we had been asked to solve

$$x^2 - 2x + 1 \geq 0$$

$$(x - 1)^2 \geq 0$$

the solutions would have been *all* real numbers, since $(x - 1)^2$ is *always* positive or zero.

⟫ **Trial Run**

Solve each inequality by the critical-point method.

_____ **1.** $x^2 - 5x < 0$

_____ **2.** $3x^2 + 11x - 4 \geq 0$

_____ **3.** $6 - x - x^2 < 0$ _____ **4.** $x^2 \le 25$

Answers are on page 348.

▶ Example You Completed

Example 5. Solve $x^2 < 2x - 1$.

Solution

$$x^2 - 2x + 1 < 0$$
$$(x - 1)(x - 1) < 0$$

The only critical point is

$$x = 1$$

The dot should be open and we want the product to be negative (< 0).

	I	II
$x - 1$	$-$	$+$
$x - 1$	$-$	$+$
	-1 0 1 2	
$(x-1)(x-1)$	$+$	$+$

The product is *never* negative, so there is no solution for this inequality.

Answers to Trial Runs

page 345   **1.** $x < -4$ or $x > 3$ **2.** $-4 \le x \le 6$ **3.** $x \le 0$ or $x \ge \frac{1}{2}$ **4.** $-2 < x < 2$

page 347   **1.** $0 < x < 5$ **2.** $x \le -4$ or $x \ge \frac{1}{3}$ **3.** $x < -3$ or $x > 2$ **4.** $-5 \le x \le 5$

EXERCISE SET 5.4

Solve the following inequalities.

_____ 1. $x^2 - 49 < 0$ _____ 2. $x^2 - 16 < 0$

_____ 3. $4x^2 > 25$ _____ 4. $9x^2 > 4$

_____ 5. $x^2 + 3x \geq 0$ _____ 6. $x^2 + 9x \geq 0$

_____ 7. $2x^2 - x \leq 0$ _____ 8. $3x^2 - 2x \leq 0$

_____ 9. $x^2 + 3x + 2 < 0$ _____ 10. $x^2 + 7x + 10 < 0$

_____ 11. $x^2 + 3x \leq 4$ _____ 12. $x^2 - 2x \leq 8$

_____ 13. $x^2 + 4x > 5$ _____ 14. $x^2 - x > 42$

_____ 15. $x^2 + 21 \leq 10x$ _____ 16. $x^2 + 20 \leq 9x$

_____ 17. $25 - x^2 > 0$ _____ 18. $36 - x^2 > 0$

_____ 19. $3x^2 + 17x > 6$ _____ 20. $6x^2 - x > 12$

_____ 21. $6x^2 - 13x \leq 8$ _____ 22. $3x^2 + 17x \leq 6$

_____ 23. $4x^2 - 20x + 25 \geq 0$ _____ 24. $9x^2 - 12x + 4 \geq 0$

_____ 25. $3x(x + 5) \geq 5(x + 1) - 4x$ _____ 26. $2x(x + 5) \geq 3(x + 6) - 2x$

_____ 27. $(3x - 1)(2x + 5) < -x(6 + x) + 1$

_____ 28. $(4x - 1)(2x - 3) < 3(4 - x) + 1$

☆ Stretching the Topics _____

Solve the inequalities.

_____ 1. $x^2 - 2 \geq 0$

_____ 2. $x^3 < 4x$

_____ 3. $y^3 - y^2 > 9y - 9$

Check your answers in the back of your book.

If you can solve the inequalities in **Checkup 5.4,** you are ready to go on to Section 5.5.

CHECKUP 5.4

Solve the following inequalities.

_____ **1.** $x^2 + 8x + 15 \geq 0$ _____ **2.** $x^2 - 4x < 0$

_____ **3.** $x^2 + 9x > 36$ _____ **4.** $2x^2 + 9 < 19x$

_____ **5.** $4x^2 \geq 25x + 21$ _____ **6.** $81 < 9x - x^2$

Check your answers in the back of your book.

If You Missed Problems:	You Should Review Example:
1, 2	3
3, 4	4
5, 6	5

5.5 Switching from Word Statements to Quadratic Equations

Now that we have learned to solve quadratic equations and inequalities, it is time to put that skill to use in solving word problems.

Example 1. Mr. Green's rectangular garden is 10 feet longer than it is wide. If the area of the garden is 264 square feet, what are the dimensions of the garden?

Solution

We Think	*We Write*
The area of a rectangle is found by *multiplying* length times width.	Area = (length)(width)
We know less about the width.	Let w = width
The length is 10 feet *more than* the width.	Let $w + 10$ = length
We can illustrate the garden with dimensions labeled.	

$$w + 10$$

w

We Think	*We Write*
Area is the product of length times width.	Area = $(w + 10)w$
Here the area is 264 square feet.	$(w + 10)w = 264$
	$w^2 + 10w = 264$

Now we have a quadratic equation to solve.

$$w^2 + 10w = 264$$

$$w^2 + 10w - 264 = 0$$

$$(w + 22)(w - 12) = 0$$

$$w + 22 = 0 \qquad \text{or} \qquad w - 12 = 0$$

$$w = -22 \qquad\qquad w = 12$$

Since width cannot be *negative*, we conclude that

$$\text{Width: } w \qquad = 12 \text{ feet}$$

$$\text{Length: } w + 10 = 12 + 10$$

$$= 22 \text{ feet}$$

The garden is 12 feet wide and 22 feet long.

Example 2. What must be the width of a frame around a square painting measuring 11 inches on each side if the framed painting is to cover exactly 289 square inches on a wall?

Solution

We Think	*We Write*
The area covered by the framed painting will be the *product* of its length and width.	Area = (framed length)(framed width)
We do not know the width of the frame.	Let x = width of frame

We can use an illustration to find an expression for the length and width of the framed painting.

Now we have expressions for framed length and framed width.

Framed length = $2x + 11$
Framed width = $2x + 11$

Area is the *product* of framed length and width.

Area = $(2x + 11)(2x + 11)$

We know that the area covered by the framed painting is 289 square inches.

$(2x + 11)(2x + 11) = 289$
$4x^2 + 44x + 121 = 289$

Now we have a quadratic equation to solve.

$$4x^2 + 44x + 121 = 289$$

$$4x^2 + 44x - 168 = 0$$

$$4(x^2 + 11x - 42) = 0$$

$$4(x + 14)(x - 3) = 0$$

$$x + 14 = 0 \qquad \text{or} \qquad x - 3 = 0$$
$$x = -14 \qquad\qquad\qquad x = 3$$

Only a positive solution makes sense here. The frame must be 3 inches wide.

Now you help solve Example 3, which is the problem stated at the beginning of this chapter.

Example 3. The three support beams of a picnic shelter form a right triangle. The longest beam measures 25 feet and the shortest is 5 feet less than the other. Find the length of the piece of lumber that would be needed to cut all three beams.

Solution

We Think	We Write

A right triangle has two legs and a hypotenuse. The hypotenuse is always the longest side of a right triangle.

The length of the piece of lumber must equal the *sum* of the lengths of the two legs and the hypotenuse.

Length of lumber = first leg
          + second leg
          + hypotenuse

The shorter leg is 5 feet *less* than the other leg.

Let x = longer leg length
$x - \underline{\hspace{1cm}}$ = shorter leg length

According to the Pythagorean theorem, the square of the hypotenuse of a right triangle is equal to the *sum* of the squares of its legs.

$$(25)^2 = x^2 + (x - \underline{\quad 5 \quad})^2$$
$$625 = x^2 + x^2 - \underline{\quad\quad}x + \underline{\quad\quad}$$
$$625 = \underline{\quad\quad}x^2 - \underline{\quad\quad}x + \underline{\quad\quad}$$

Now we have a quadratic equation to solve.

$$2x^2 - 10x + 25 = 625$$
$$2x^2 - 10x - 600 = 0$$
$$2(x^2 - 5x - 300) = 0$$
$$2(x - \underline{\quad\quad})(x + \underline{\quad\quad}) = 0$$
$$x = \underline{\quad\quad} \quad \text{or} \quad x = -\underline{\quad\quad}$$

Only a positive length makes sense here.

$$\text{Length of longer leg:} \quad x = \underline{\quad\quad} \text{ ft}$$
$$\text{Length of shorter leg:} \; x - 5 = \underline{\quad\quad} - 5$$
$$= \underline{\quad\quad} \text{ ft}$$

Length of hypotenuse: 25 ft

We have not yet answered the question. How long must the piece of lumber be?

$$\text{Length of lumber} = \text{first leg} + \text{second leg} + \text{hypotenuse}$$
$$= \underline{\quad\quad} + \underline{\quad\quad} + 25$$
$$= \underline{\quad\quad} \text{ ft}$$

The piece of lumber for all three beams must be \underline{\quad\quad} feet long.

Check your work on page 356. ▶

Example 4. A company has found that when x units of a product are manufactured and sold, its revenue is $x^2 + 100x$ dollars and its costs are $240x + 500$ dollars. How many units must be produced and sold to make a profit of $1000?

    Solution

We Think	*We Write*
We know that profit is found by *subtracting* costs from revenue.	Profit = revenue − costs
We have expressions for revenue and costs.	Profit = $(x^2 + 100x) - (240x + 500)$
We want profit to *equal* $1000.	$(x^2 + 100x) - (240x + 500) = 1000$

We have a quadratic equation to solve.

$$x^2 + 100x - (240x + 500) = 1000$$
$$x^2 + 100x - 240x - 500 = 1000$$
$$x^2 - 140x - 500 = 1000$$
$$x^2 - 140x - 1500 = 0$$
$$(x - 150)(x + 10) = 0$$
$$x - 150 = 0 \quad \text{or} \quad x + 10 = 0$$
$$x = 150 \quad\quad\quad\quad x = -10$$

Only a positive number of units makes sense here. The company must manufacture and sell 150 units of its product to make a profit of $1000 dollars.

⫸ Trial Run

Solve.

_____ **1.** A gardener has decided to decrease the area of his square garden to 2700 square yards by decreasing the length by 6 yards and the width by 2 yards. What were the dimensions of the square garden?

_____ **2.** The total revenue from producing and selling q units of a product is given by the expression $q^2 + 50q$. How many units must be produced and sold to have $5000 in revenue?

Answers are on page 356.

▶ Example You Completed

Example 3

Solution

We Write

Let $x =$ longer leg
$x - 5 =$ shorter leg

$(25)^2 = x^2 + (x - 5)^2$
$625 = x^2 + x^2 - 10x + 25$
$625 = 2x^2 - 10x + 25$

Now we have a quadratic equation to solve.

$$2x^2 - 10x + 25 = 625$$

$$2x^2 - 10x - 600 = 0$$

$$2(x^2 - 5x - 300) = 0$$

$$2(x - 20)(x + 15) = 0$$

$$x - 20 = 0 \quad \text{or} \quad x + 15 = 0$$

$$x = 20 \qquad\qquad x = -15$$

Only a positive length makes sense here.

Length of longer leg: $\quad x = 20$ ft

Length of shorter leg: $x - 5 = 20 - 5$

$= 15$ ft

Length of hypotenuse: 25 ft

We have not yet answered the question. How long must the piece of lumber be?

Length of lumber = first leg + second leg + hypotenuse

$= 20 + 15 + 25$

$= 60$ ft

The piece of lumber for all three beams must be 60 feet long.

Answers to Trial Run

page 356   **1.** 56 yards by 56 yards   **2.** 50

EXERCISE SET 5.5

_____ **1.** The length of a rectangle is 5 feet less than twice its width. If the area is 88 square feet, find the dimensions of the rectangle.

_____ **2.** The width of a rectangle is 18 feet less than three times the length and the area is 48 square feet. Find the dimensions of the rectangle.

_____ **3.** Mr. Mrogenski plans to buy a rectangular lot that is 50 feet longer than it is wide. If the area of the lot is 30,000 square feet, find its dimensions.

_____ **4.** Mr. Collins is building a cattle loading pen that is to be 15 feet longer than it is wide. He has enough fence to enclose an area of 1750 square feet. Find the dimensions of the pen.

_____ **5.** Ella has a vegetable garden that is a square. She plans to increase the area of the garden by increasing the length by 10 feet and decreasing the width by 2 feet. If the area of the enlarged garden is 1900 square feet, find its dimensions. By how many square feet did Ella increase the size of her garden?

_____ **6.** Eudena has a square family room that she plans to enlarge by increasing the length by 6 feet and increasing the width by 2 feet. The area of the enlarged family room will be 252 square feet. Find the dimensions of the new room. By how many square feet did Eudena increase the size of her room?

_____ **7.** One square has a side 5 centimeters shorter than the side of a second square. The area of the larger is four times the area of the smaller. Find the dimensions of each square.

_____ **8.** One square has a side 8 centimeters longer than the side of a second square. The area of the larger is nine times the area of the smaller. Find the dimensions of each square.

_____ **9.** Find the dimensions of a room that is 6 feet longer than it is wide and can be covered by 216 square feet of carpeting.

_____ **10.** The length of a yard is 6 feet less than twice the width. If it takes 1008 square feet of sodding to cover the yard, find the dimensions.

_____ **11.** Jim wishes to build an open tin storage box. The box is to have a square bottom, sides 4 inches high, and a volume of 576 cubic inches. Find the dimensions of the box. What size piece of tin should Jim buy if he wishes to build the box from one piece by cutting squares with 4-inch sides from each corner?

_____ **12.** Olivia has a square piece of cardboard that she wishes to make into an open box by cutting a 3-inch square out of each corner and folding up the sides. If the box will have a volume of 675 cubic inches, find the dimensions of the piece of cardboard.

_____ **13.** Mr. Olsen is plowing a field that is 60 rods on each side. He wishes to plow the field in two different time periods. How wide a strip must he plow around the field the first time so that the second time he will have only 1600 square rods to plow?

_____ **14.** What must be the width of a sidewalk around a swimming pool that is 54 feet by 34 feet if the total area allowed for the pool and walk is 2400 square feet?

_____ **15.** The product of two consecutive integers is 24 more than nine times the larger. Find the two numbers.

_____ **16.** The sum of the squares of two consecutive integers is 73 more than the product of the two. Find the two numbers.

_____ **17.** One leg of a right triangle is 7 centimeters less than the other leg and the hypotenuse is 17 centimeters. Find the lengths of the two legs.

_____ **18.** One leg of a right triangle is 3 centimeters more than the other leg and the hypotenuse is 15 centimeters. Find the lengths of the two legs.

_____ **19.** The altitude of a triangle is 4 feet less than twice the base and the area is 63 square feet. Find the altitude of the triangle.

_____ **20.** The base of a triangle is 6 centimeters less than four times the altitude and the area is 104 square centimeters. Find the base of the triangle.

_____ **21.** A television antenna was bent over by a storm in such a way that the part bending over was 13 feet. The part left standing was 7 feet more than the distance from the foot of the antenna to the point where the broken part touched the ground. What was the height of the antenna before the storm?

_____ **22.** There are two pictures on Margaret's wall; one is a square and one is a rectangle. The length of the rectangular picture is 3 inches more than one side of the square, and the width of the rectangular picture is 1 inch less than one side of the square. If the two pictures occupy 109 square inches of area, find the dimensions of the rectangular picture.

_____ **23.** In a craft shop the total cost in dollars for producing Christmas wreaths is given by the expression $x^2 - 8x + 25$, where x represents the number of wreaths. If the shop can spend $45 on this project, how many wreaths can be produced?

_____ **24.** The Basketball Boosters Club is planning a trip for fans to the tournament in San Diego. The profit to be made by the club in dollars is given by the expression $x^2 + 5x - 150$ when x represents the number of people. How many people must make the trip if the club wants to make a profit of $600?

_____ **25.** The Bronco Boot Company has determined that the revenue received from selling x pairs of boots is given in dollars by the expression $x^2 - 29x + 390$. If the company wishes the revenue for one week to be $600, how many pairs of boots must be sold?

_____ **26.** A manufacturer of toy chests has found that when x number of chests are manufactured and sold, its revenue is $x^2 + 220x$ dollars and its costs are $300x + 2500$ dollars. How many units must be produced and sold in order to make a profit of $4000?

_____ **27.** The distance an object falls from rest is given by $s = 16t^2$, where s is the distance and t is the time in seconds. How long does it take an object dropped from a height of 144 feet to reach the ground?

_____ **28.** The relationship of the distance s to the velocity v and the time t in seconds when an object is thrown downward is given by $s = vt + 16t^2$. How long does it take an object thrown downward from a height of 156 feet with a velocity of 4 feet per second to hit the ground?

☆ Stretching the Topics _____

Solve.

_____ **1.** The length of a rectangular room is 12 feet less than twice its width. The width of a carpet on the floor is 4 feet less than the width of the room. The length of the carpet is 6 feet more than its width. If 152 square feet of the floor are not covered by carpet, find the dimensions of the room.

_____ **2.** Find three consecutive integers such that the square of the second is 51 less than the sum of the squares of the other two.

Check your answers in the back of your book.

If you can complete **Checkup 5.5,** you are ready to do the Review Exercises.

CHECKUP 5.5

_____ **1.** The length of Mr. Schroder's rectangular garden 8 feet less than twice the width. If the area of the garden is 192 square feet, what are the dimensions of the garden?

_____ **2.** What must be the width of a border around a rectangular quilt that is 7 feet wide and 8 feet long if the total area the quilt must cover is 72 square feet?

_____ **3.** The hypotenuse of a right triangle is 2 centimeters more than one of the legs. If the other leg is 8 centimeters, find the length of the hypotenuse.

Check your answers in the back of your book.

If You Missed Problem:	You Should Review Example:
1	1
2	2
3	3

Summary

In this chapter we discussed solutions for **quadratic** (second-degree) **equations** and **inequalities.** We learned several methods for solving quadratic equations of the form $ax^2 + bx + c = 0$ and agreed to use the most appropriate method for a particular equation.

Method	When a Good Choice	Example
Factoring	When quadratic expression can be factored using integers	$2x^2 - 5x = 0$ $(2x + 1)(x - 3) = 0$ $2x + 1 = 0$ or $x - 3 = 0$ $x = \dfrac{-1}{2}$ $\quad x = 3$
Taking square roots	When first-degree term is missing ($b = 0$)	$9x^2 - 2 = 0$ $9x^2 = 2$ $x^2 = \dfrac{2}{9}$ $x = \sqrt{\dfrac{2}{9}}$ or $x = -\sqrt{\dfrac{2}{9}}$ $x = \dfrac{\sqrt{2}}{3}$ $\quad x = \dfrac{-\sqrt{2}}{3}$
Completing the square	When coefficient of middle term is even, but quadratic expression does not factor	$x^2 - 2x - 5 = 0$ $x^2 - 2x = 5$ $x^2 - 2x + 1 = 5 + 1$ $(x - 1)^2 = 6$ $x - 1 = \pm\sqrt{6}$ $x = 1 \pm \sqrt{6}$ $x = 1 + \sqrt{6}$ or $x = 1 - \sqrt{6}$
Quadratic formula	When other methods cannot be used	$x^2 + 3x + 1 = 0$ $x = \dfrac{-3 \pm \sqrt{3^2 - 4(1)(1)}}{2(1)}$ $= \dfrac{-3 \pm \sqrt{5}}{2}$ $x = \dfrac{-3 + \sqrt{5}}{2}$ or $\dfrac{-3 - \sqrt{5}}{2}$

After solving equations by factoring, we learned to write equations with given solutions using the fact that if r is a solution for a polynomial equation, then $x - r$ is a factor of the polynomials.

Our work with the quadratic formula

$$x = \frac{-b \pm \sqrt{b^2 - 4ac}}{2a}$$

led us to a discussion of square roots and we learned two rules for simplifying square roots.

<table>
<tr><td>

Product Rule for Square Roots

$$\sqrt{A \cdot B} = \sqrt{A} \cdot \sqrt{B} \quad \begin{array}{l} A \geq 0 \\ B \geq 0 \end{array}$$

Example: $\sqrt{150} = \sqrt{25 \cdot 6}$
$$= 5\sqrt{6}$$

</td><td>

Quotient Rule for Square Roots

$$\sqrt{\frac{A}{B}} = \frac{\sqrt{A}}{\sqrt{B}} \quad \begin{array}{l} A \geq 0 \\ B > 0 \end{array}$$

Example: $\sqrt{\dfrac{3}{16}} = \dfrac{\sqrt{3}}{\sqrt{16}}$
$$= \frac{\sqrt{3}}{4}$$

</td></tr>
</table>

The quadratic formula also led us to a discussion of the **discriminant,** $b^2 - 4ac$, from which we can determine what kind of solutions a quadratic equation has.

Discriminant	Kind of Solution
$b^2 - 4ac = 0$	One rational solution
$b^2 - 4ac > 0$ and perfect square	Two different rational solutions
$b^2 - 4ac > 0$, *not* perfect square	Two different irrational solutions
$b^2 - 4ac < 0$	No real number solutions

In our work with **quadratic inequalities** we learned to find solutions by cases and by critical points that separate the real number line into distinct intervals. Determining the sign of each factor of the quadratic expression in each interval allowed us to determine the intervals on which the inequality was satisfied.

Finally, we learned to switch from word statements to quadratic equations.

☐ Speaking the Language of Algebra

Complete each sentence with the appropriate word or phrase.

1. An equation of the form $ax^2 + bx + c = 0$ is called a second-degree or _____ equation.

2. According to the zero product rule, if a product of factors is zero, then _____ .

3. If $x = r$ is a solution for a polynomial equation, we know that _____ is a _____ of the polynomial.

4. If $ax^2 + bx + c = 0$, the solutions for the equation are $x =$ _____ .

5. The expression $b^2 - 4ac$ is called the _____ . If $b^2 - 4ac < 0$, the equation $ax^2 + bx + c = 0$ has _____ _____ solutions.

6. If $(x - r)(x - s) > 0$, we know that $x - r$ and $x - s$ are both _____ or both _____ .

REVIEW EXERCISES for Chapter 5: Solving Quadratic Equations and Inequalities

Solve by factoring.

_____ 1. $(3x - 5)(x + 7) = 0$

_____ 2. $x^2 - 5x - 24 = 0$

_____ 3. $2x^2 + 9x - 35 = 0$

_____ 4. $x^2 + 12x + 36 = 0$

_____ 5. $0 = 7x^2 - 63$

_____ 6. $32x - 8x^2 = 0$

_____ 7. $3a^2 = 14a - 8$

_____ 8. $10a^2 + 6 = 32a$

_____ 9. $3x^2 = 48$

_____ 10. $7y = 5y^2$

_____ 11. $(2x - 1)(x - 5) = -x(x - 14) - 23$

_____ 12. $x(x - 5) + 4(x^2 - 4) = 2(13x - 11)$

_____ 13. $12x^3 + 31x^2 - 15x = 0$

_____ 14. $x^4 - 29x^2 + 100 = 0$

_____ 15. $3x^5 + 3x^4 - 168x^3 = 0$

_____ 16. Write a quadratic equation with solutions $x = 8$ and $x = -3$.

Simplify the following radical expressions.

_____ 17. $\sqrt{75}$

_____ 18. $\sqrt{162}$

_____ 19. $\sqrt{\dfrac{5}{9}}$

_____ 20. $\sqrt{\dfrac{18}{49}}$

Solve the following equations and state the nature of the roots.

_____ 21. $x^2 - 32 = 0$

_____ 22. $x^2 + 49 = 0$

_____ 23. $9x^2 = 25$

_____ 24. $16x^2 - 1 = 0$

Solve by completing the square.

_____ 25. $x^2 - 2x - 48 = 0$

_____ 26. $x^2 - 3x - 6 = 0$

_____ 27. $4x^2 - 20x + 25 = 0$

_____ 28. $x^2 + 4x - 1 = 0$

Solve by using the quadratic formula.

_____ 29. $9x^2 - x = 1$

_____ 30. $3y^2 - 2y - 5 = 0$

_____ 31. $x^2 + 2x = 6$

_____ 32. $4x^2 - 3x = 2$

_____ 33. Solve $2d = n^2 - 3n$ for n.

_____ 34. Solve $S = gt + 16t^2$ for t.

Use the discriminant to describe the solutions for the following equations.

_____ **35.** $2x^2 + x - 2 = 0$ _____ **36.** $3x^2 + 2x - 1 = 0$

_____ **37.** $x^2 + 81 = 0$ _____ **38.** $x^2 = 18x - 81$

Solve the following inequalities.

_____ **39.** $x^2 < 144$ _____ **40.** $x^2 - 3x \geq 0$

_____ **41.** $2x^2 - 9x \leq 5$ _____ **42.** $x^3 - 4x > 0$

Solve.

_____ **43.** Mr. Crawford is building a rectangular table that is 24 inches longer than it is wide. If the area of the table is 5040 square inches, find its dimensions.

_____ **44.** One square has a side 6 centimeters longer than the side of a second square. The area of the larger is four times the area of the smaller. Find the dimensions of each square.

_____ **45.** Find the width of a frame around a picture that is 12 inches by 14 inches if the total area allowed for the framed picture is 360 square inches.

_____ **46.** The altitude of a triangle is 5 feet less than the base and the area is 42 square feet. Find the altitude of the triangle.

Check your answers in the back of your book.

If You Missed Exercise:	You Should Review Examples:	
1	SECTION 5.1	1–3
2–6		4–6
7–10		7–9
11, 12		10–12
13		14
14		15
15		16
16		17
17, 18	SECTION 5.2	9, 10
19, 20		11–14
21–24		15–18
25–28	SECTION 5.3	1–7
29–32		8–11
33, 34		12
35–38		13, 14
39–42	SECTION 5.4	3–5
43–46	SECTION 5.5	1–4

If you have completed the Review Exercises and corrected your errors, you are ready to take the Practice Test.

PRACTICE TEST for Chapter 5: Solving Quadratic Equations and Inequalities

Solve.

		SECTION	EXAMPLE
_____	**1.** $x^2 + 5x - 36 = 0$	5.1	4
_____	**2.** $9x^2 = 30x - 25$	5.1	8
_____	**3.** $7x^2 = 21x$	5.1	9
_____	**4.** $5x^3 - 7x^2 - 6x = 0$	5.1	14
_____	**5.** $(x + 6)(x - 1) = 8$	5.1	11, 12
_____	**6.** $x^4 - 20x^2 + 64 = 0$	5.1	15
_____	**7.** $2x^3 - x^2 - 2x + 1 = 0$	5.1	16
_____	**8.** Write a quadratic equation with the solutions $x = 4$ and $x = -4$.	5.1	17

Solve.

		SECTION	EXAMPLE
_____	**9.** $81a^2 = 4$	5.2	15
_____	**10.** $9x^2 - 20 = 0$	5.2	16
_____	**11.** $x^2 + 7x + 3 = 0$	5.3	9, 10
_____	**12.** $3x^2 - 2x - 2 = 0$	5.3	9, 10
_____	**13.** $y^2 + 9y + 25 = 0$	5.3	11
_____	**14.** Solve $K = 3ay^2 - y$ for y.	5.3	12
_____	**15.** Use the discriminant to describe the solutions for $x^2 + 16 = 8x$.	5.3	13, 14

Solve.

		SECTION	EXAMPLE
_____	**16.** $2x^2 - 7x \leq 0$	5.4	4
_____	**17.** $5x^2 > 6x + 8$	5.4	5
_____	**18.** The length of one leg of a right triangle is 5 inches. The hypotenuse is 1 inch longer than the other leg. Find the lengths of the missing parts of the triangle.	5.5	3

SHARPENING YOUR SKILLS after Chapters 1–5

Name _____ **Date** _____

_____ **1.** Use an inequality to state that $x - 2$ is a negative number.

SECTION 1.1

_____ **2.** Compare $\dfrac{-16}{3}$ and $-\sqrt{36}$ using $<$, $>$, or $=$.

1.1

Simplify.

_____ **3.** $\dfrac{-12\left(\dfrac{1}{4}\right) + 8[5 - (-7)]}{-7[36 - 9(4)]}$

1.3

_____ **4.** $|8 - 12| - |7 - 5|$

1.4

_____ **5.** -4^{-2}

3.2

_____ **6.** $(x + 3)^0$

3.1

_____ **7.** $\dfrac{-1}{4}(8x - 16y) + 3[x - (x + 2y)]$

1.4

_____ **8.** $(9x^2yz^5)(-7x^4y^3z)$

3.1

_____ **9.** $\left(\dfrac{-3x^5y^0}{5z^3}\right)^2$

3.1

_____ **10.** $\left(\dfrac{x^{12}y^{-2}}{x^{-5}y^7}\right)^{-1}$

3.2

_____ **11.** $-5x^5y^2(3xy^4 + 2y^3) + 3x^3y^4(x^2y - 2x^3y^2)$

3.3

_____ **12.** $(a - 2b)(a + 5b)(a - 7b)$

3.4

_____ **13.** $[(x + y) - 6]^2$

3.4

_____ **14.** $(10a^3b - 20a^2b^2 + 15ab^3) \div 5ab$

3.5

_____ **15.** $(x^3 - 9x + 10) \div (x - 2)$

3.5

Factor completely.

_____ **16.** $a^2 + 12ab + 27b^2$

4.3

_____ **17.** $5x^4 - 80$

4.3

_____ **18.** $27x^3 - 64$

4.2

_____ **19.** $x^3 - 5x^2 + 9x - 45$

4.2

_____ **20.** $-18x^4 + 9x^3 + 27x^2$

4.3

369

_____ **21.** $x^2 - 18x + 81$ 4.3

_____ **22.** Evaluate $A = P + Prt$ when $P = \$5000$, $r = 9.5\%$, and $t = 4$ years.   4.3

Solve.

_____ **23.** $4[3x + (x - 7)] = 3(x - 8) + 9$ 2.1

_____ **24.** $x < 3$ or $x \le 0$ 2.2

_____ **25.** $3x - 4 > 2$ and $2(x - 3) < -10$ 2.3

_____ **26.** $-1 \le 2(-x + 4) - 3 < 7$ 2.3

_____ **27.** $|3x - 1| < 11$ 2.4

_____ **28.** Solve $ax + by = ab$ for a.   2.1

_____ **29.** Salesman Henri earns a base salary of \$700 per month plus a \$30 commission for each water bed he sells. How many water beds must he sell to earn at least \$1000 this month?   2.5

_____ **30.** If two opposite sides of a square are increased by 3 inches and the other two sides are decreased by 2 inches, write an expression for the area of the resulting rectangle. Simplify your answer.   3.6

Working with Rational Expressions 6

Mr. Romano uses two hoses to fill his swimming pool. By itself, the larger hose can fill the pool in 7 hours, while the smaller hose requires 10 hours by itself. While using both hoses to fill his pool last weekend, Mr. Romano discovered after 2 hours that the pool drain had been left open. If the pool normally drains in 12 hours, how long after he closed the drain was the pool filled?

From Chapter 1 we recall that a **rational number** is any real number that can be expressed as a fraction with an integer numerator and a *nonzero* integer denominator. In this chapter we discuss the algebra of **rational expressions,** which are fractions with variables and/or constants in the numerator and *nonzero* variable and/or constant expressions in the denominator. For the most part, we shall concentrate on **rational algebraic expressions,** which are fractions with a polynomial in the numerator and a *nonzero* polynomial in the denominator. In this chapter we learn how to:

1. State restrictions on the domain of the variable in a rational expression.
2. Reduce rational expressions to lowest terms.
3. Multiply and divide rational expressions.
4. Add and subtract rational expressions.
5. Simplify complex fractions.
6. Solve fractional equations and inequalities.
7. Switch from words to fractional equations.

6.1 Simplifying Rational Algebraic Expressions

Before we learn to simplify rational algebraic expressions, it is important that we discuss thoroughly the fact that the denominators of such expressions can *never* be zero.

Restricting the Domain of a Variable

When dealing with an algebraic fraction such as $\dfrac{5}{x}$, we know that its numerical value depends on the value that we choose to substitute for the variable. We may choose *almost* any value for x in evaluating this fraction, but there is one number that we may *not* substitute for x. That number is zero, because

$$\text{if } x = 0 \quad \text{then} \quad \frac{5}{0} \text{ is undefined}$$

The set of all possible replacements for a variable in an expression is called the **domain** of the variable. Not allowing the replacement of a particular value for the variable is called **restricting the domain** of the variable.

> For a rational algebraic expression, we must restrict the domain of the variable to exclude any values that will make the denominator equal zero.

To determine the necessary restrictions on the domain of the variable in a rational expression, we must decide what values would make the denominator zero. Then we exclude those values from the domain.

Rational Expression	Denominator	Restrictions
$\dfrac{11}{3x}$	$3x$	$x \neq 0$
$\dfrac{2a}{a-3}$	$a-3$	$a \neq 3$
$\dfrac{x+2}{x(x+6)}$	$x(x+6)$	$x \neq 0, x \neq -6$
$\dfrac{3x+5}{x^2+3x+2}$	$x^2 + 3x + 2$ $= (x+2)(x+1)$	$x \neq -2, x \neq -1$

Notice that we do not concern ourselves with the nature of the numerator of a fraction. The numerator of a fraction can certainly be zero, as long as the denominator is *not* zero.

> If the numerator of a fraction is zero but the denominator is not zero, then the value of the fraction is zero.
>
> If the denominator of a fraction is zero, then the fraction is undefined.

Example 1. Restrict the domain for $\dfrac{7x + 1}{2x^3 - 7x^2 - 15x}$.

Solution: We factor the denominator.

$$\frac{7x + 1}{2x^3 - 7x^2 - 15x} = \frac{7x + 1}{x(2x^2 - 7x - 15)}$$

$$= \frac{7x + 1}{x(2x + 3)(x - 5)}$$

The denominator will be zero whenever any one of its factors is zero.

$$x = 0 \qquad 2x + 3 = 0 \qquad x - 5 = 0$$

$$2x = -3$$

$$x = 0 \qquad x = \frac{-3}{2} \qquad x = 5$$

Restrictions: $x \neq 0$, $x \neq \dfrac{-3}{2}$, $x \neq 5$.

You complete Example 2.

Example 2: Restrict the domain for $\dfrac{y + 3}{y^2 - 4}$.

Solution: We factor the denominator.

$$\frac{y + 3}{y^2 - 4} =$$

Restrictions:

Check your work on page 378. ▶

Reducing Rational Algebraic Expressions

We say that two fractions are *equal* if they represent the same number. We can decide if two fractions are equal using the following definition.

Equality of Fractions

$$\frac{A}{B} = \frac{C}{D} \qquad \text{provided that} \qquad A \cdot D = B \cdot C$$

where $B \neq 0$, $D \neq 0$.

For example,

$$\frac{2}{3} = \frac{8}{12} \qquad \text{because} \qquad 2 \cdot 12 = 3 \cdot 8$$

$$\frac{15}{20} = \frac{3}{4} \qquad \text{because} \qquad 15 \cdot 4 = 20 \cdot 3$$

We know from arithmetic that numerical fractions can be **reduced** by dividing numerator and denominator by any factors common to both.

Fraction	Factored Form	Reduced Form
$\dfrac{3}{12}$	$\dfrac{3 \cdot 1}{3 \cdot 2 \cdot 2} = \dfrac{\overset{1}{\cancel{3}} \cdot 1}{\underset{1}{\cancel{3}} \cdot 2 \cdot 2}$	$\dfrac{1}{4}$
$\dfrac{10}{15}$	$\dfrac{2 \cdot 5}{3 \cdot 5} = \dfrac{2 \cdot \overset{1}{\cancel{5}}}{3 \cdot \underset{1}{\cancel{5}}}$	$\dfrac{2}{3}$
$\dfrac{42}{28}$	$\dfrac{2 \cdot 3 \cdot 7}{2 \cdot 2 \cdot 7} = \dfrac{\overset{1}{\cancel{2}} \cdot 3 \cdot \overset{1}{\cancel{7}}}{\underset{1}{\cancel{2}} \cdot 2 \cdot \underset{1}{\cancel{7}}}$	$\dfrac{3}{2}$

To reduce a fraction, divide the numerator and denominator by any nonzero *factors* common to both. A fraction has been *reduced to lowest terms* when the numerator and denominator contain no common factor (except the number 1).

Letting A, B, and C represent algebraic expressions, we may say

Reducing Fractions

$$\frac{A \cdot C}{B \cdot C} = \frac{A}{B} \qquad B \neq 0, C \neq 0$$

To reduce a fraction with a monomial numerator and denominator, we may use either the laws of exponents or the method of common factors to reduce the fraction to lowest terms. Let's use both approaches to reduce

$$\frac{-3x^4 y^3 z}{5x^2 y^5 z}$$

after noting that the restrictions are $x \neq 0$, $y \neq 0$, $z \neq 0$.

First Approach

Using the laws of exponents, we have

$$\frac{-3x^4 y^3 z}{5x^2 y^5 z} = \frac{-3x^{4-2} y^{3-5} z^{1-1}}{5}$$

$$= \frac{-3x^2 y^{-2} z^0}{5}$$

$$= \frac{-3x^2}{5y^2}$$

Second Approach

Using common factors, we have

$$\frac{-3x^4 y^3 z}{5x^2 y^5 z} = \frac{-3 \cdot x^2 \cdot x^2 \cdot y^3 \cdot z}{5 \cdot x^2 \cdot y^3 \cdot y^2 \cdot z}$$

$$= \frac{-3 \cdot \overset{1}{\cancel{x^2}} \cdot x^2 \cdot \overset{1}{\cancel{y^3}} \cdot \overset{1}{\cancel{z}}}{5 \cdot \underset{1}{\cancel{x^2}} \cdot \underset{1}{\cancel{y^3}} \cdot y^2 \cdot \underset{1}{\cancel{z}}}$$

$$= \frac{-3x^2}{5y^2}$$

Our answers are the same using either approach.

You complete Example 3.

Example 3. Use the laws of exponents to reduce $\dfrac{2x^3y^7z}{x^2yz^2}$.

Solution: We note that the restrictions are $x \neq 0$, $y \neq 0$, $z \neq 0$.

$$\frac{2x^3y^7z}{x^2yz^2} = 2x^{3-2}y^{7-1}z^{1-2}$$

Check your work on page 378. ▶

Then complete Example 4.

Example 4. Use common factors to reduce $\dfrac{2x^3y^7z}{x^2yz^2}$.

Solution: We note that the restrictions are $x \neq 0$, $y \neq 0$, $z \neq 0$.

$$\frac{2x^3y^7z}{x^2yz^2} = \frac{2 \cdot x^2 \cdot x \cdot y \cdot y^6 \cdot z}{x^2 \cdot y \cdot z \cdot z}$$

$$= \underline{\hspace{2cm}}$$

Check your work on page 379. ▶

In dealing with a fraction whose numerator and/or denominator are *not* monomials, we must rewrite the numerator and denominator in completely factored form before looking for common factors. Moreover, we must *state restrictions* on the domain of the variable *before* dividing numerator and denominator by any common factors.

For example, suppose that we reduce this fraction.

$$\frac{3x + 15}{x^2 + 7x + 10}$$

$$= \frac{3(x + 5)}{(x + 5)(x + 2)} \qquad x \neq -5, x \neq -2$$

$$= \frac{3\cancel{(x + 5)}^{\,1}}{\cancel{(x + 5)}_{\,1}(x + 2)}$$

$$= \frac{3}{x + 2} \qquad x \neq -5, x \neq -2$$

The reduced version of our fraction is equivalent to the original fraction, *provided* that we avoid those values of the variable that have been excluded from the domain. Either version of the fraction will yield the same numerical value when any number (except -5 or -2) is substituted for the variable x.

Steps for Reducing Fractions

1. Factor numerator and denominator completely.
2. State necessary restrictions on the domain of the variable.
3. Divide numerator and denominator by any factors common to both.

Stating restrictions should become an automatic procedure whenever we are working with rational algebraic expressions.

Example 5. Reduce $\dfrac{x^2 - 3x}{x^2 - 9}$.

Solution

$$\dfrac{x^2 - 3x}{x^2 - 9}$$

$$= \dfrac{x(x - 3)}{(x + 3)(x - 3)} \quad x \neq -3, x \neq 3$$

$$= \dfrac{x\cancel{(x - 3)}^{\;1}}{(x + 3)\cancel{(x - 3)}_{\;1}}$$

$$= \dfrac{x}{x + 3} \quad x \neq -3, x \neq 3$$

Now you complete Example 6.

Example 6. Reduce $\dfrac{x + 4}{x^2 + 8x + 16}$.

Solution

$$\dfrac{x + 4}{x^2 + 8x + 16}$$

$$= \qquad\qquad x \neq \underline{\quad\quad}$$

$$= \qquad\qquad x \neq \underline{\quad\quad}$$

Check your work on page 379. ▶

In Example 5 you should not be tempted to "cross out" the x's in the numerator and denominator of $\dfrac{x}{x + 3}$. In that denominator, x is a *term*, not a factor. *We can reduce a fraction only by dividing out common factors.*

Example 7. Reduce $\dfrac{x^3 - 8}{3x - 6}$.

Solution

$$\dfrac{x^3 - 8}{3x - 6}$$

The numerator is a difference of two cubes.

$$= \dfrac{(x - 3)(x^2 + 2x + 4)}{3(x - 2)} \quad x \neq 2$$

$$= \dfrac{\cancel{(x - 2)}^{\;1}(x^2 + 2x + 4)}{3\cancel{(x - 2)}_{\;1}}$$

$$= \dfrac{x^2 + 2x + 4}{3} \quad x \neq 2$$

You complete Example 8.

Example 8. Reduce $\dfrac{x^2 + 11x + 30}{2x^3 + 11x^2 - 6x}$.

Solution

$$\dfrac{x^2 + 11x + 30}{2x^3 + 11x^2 - 6x}$$

$$= \dfrac{(x + 5)(x + 6)}{x(2x^2 + 11x - 6)}$$

$$= \dfrac{(x + 5)(x + 6)}{x(\underline{\quad\quad})(\underline{\quad\quad})} \quad x \neq 0$$

$$x \neq \underline{\quad\quad}$$

$$x \neq \underline{\quad\quad}$$

Check your work on page 379. ▶

From Example 8, note that we agree to *leave a reduced fraction with numerator and denominator in factored form.*

Example 9. Reduce $\dfrac{x^2 + 2xy - x - 2y}{x^2 + 5xy + 6y^2}$.

Solution

$$\frac{x^2 + 2xy - x - 2y}{x^2 + 5xy + 6y^2}$$

$$= \frac{x(x + 2y) - 1(x + 2y)}{(x + 2y)(x + 3y)} \qquad \begin{array}{l} x \neq -2y \\ x \neq -3y \end{array}$$

$$= \frac{(x + 2y)(x - 1)}{(x + 2y)(x + 3y)}$$

$$= \frac{x - 1}{x + 3y} \qquad \begin{array}{l} x \neq -2y, \\ x \neq -3y \end{array}$$

Example 10. Reduce

$$\frac{5(x - 3)^2 - 10x(x - 3)}{(x - 3)^4}.$$

Solution

$$\frac{5(x - 3)^2 - 10x(x - 3)}{(x - 3)^4}$$

$$= \frac{5(x - 3)[(x - 3) - 2x]}{(x - 3)^4} \qquad x \neq 3$$

$$= \frac{5(x - 3)(x - 3 - 2x)}{(x - 3)^4}$$

$$= \frac{5(x - 3)(-x - 3)}{(x - 3)^4}$$

$$= \frac{5(-x - 3)}{(x - 3)^3}$$

$$= \frac{-5(x + 3)}{(x - 3)^3} \qquad x \neq 3$$

At first glance, fractions such as

$$\frac{x - 2}{2 - x} \qquad \text{and} \qquad \frac{4 - y}{y - 4}$$

may appear to be in lowest terms. Suppose that we take a closer look at these fractions.

Write numerators and denominators with the variable first.	$\dfrac{x - 2}{-x + 2}$	$\dfrac{-y + 4}{y - 4}$
Factor -1 from the binomial in which the leading coefficient is negative.	$\dfrac{x - 2}{-1(x - 2)}$	$\dfrac{-1(y - 4)}{y - 4}$
State restrictions.	$x \neq 2$	$y \neq 4$
Divide numerator and denominator by the common factor.	$\dfrac{x - 2}{-1(x - 2)}$	$\dfrac{-1(y - 4)}{y - 4}$
	$= \dfrac{1}{-1}$	$= \dfrac{-1}{1}$
Simplify your answer.	-1	-1

Indeed, it is always the case that

$$\frac{A - B}{B - A} = -1 \qquad A \neq B$$

and this is often a handy fact to use in reducing fractions.

Example 11. Reduce $\dfrac{25 - x^2}{x^2 - 2x - 15}$.

Solution

$$\dfrac{25 - x^2}{x^2 - 2x - 15}$$

$$= \dfrac{(5 + x)(5 - x)}{(x - 5)(x + 3)} \quad x \neq 5,\ x \neq -3$$

$$= \dfrac{(5 + x)\cancel{(5 - x)}^{-1}}{\cancel{(x - 5)}_{1}(x + 3)}$$

$$= \dfrac{-(5 + x)}{x + 3} \quad x \neq 5,\ x \neq -3$$

Now you try Example 12.

Example 12. Reduce $\dfrac{5 - 10x}{2x^2 + 7x - 4}$.

Solution

$$\dfrac{5 - 10x}{2x^2 + 7x - 4}$$

$$= \qquad\qquad\qquad x \neq \underline{\quad\quad}$$

$$\qquad\qquad\qquad\qquad x \neq \underline{\quad\quad}$$

Check your work on page 379. ▶

IIII➡ Trial Run

State restrictions and reduce the fractions.

_____ 1. $\dfrac{14a^2bc^4}{-7a^3bc}$

_____ 2. $\dfrac{18xy - 14x}{2x}$

_____ 3. $\dfrac{81x^4}{9x^2 + 9x}$

_____ 4. $\dfrac{x^2 - 49}{4x + 28}$

_____ 5. $\dfrac{y - 5}{y^2 - 10y + 25}$

_____ 6. $\dfrac{x^2 - 5x - 36}{2x^2 - 19x + 9}$

_____ 7. $\dfrac{x^2 - 4xy - 3x + 12y}{x^2 - 3xy - 4y^2}$

_____ 8. $\dfrac{16 - a^2}{3a^2 - 7a - 20}$

Answers are on page 379.

▶ Examples You Completed

Example 2. Restrict the domain for $\dfrac{y + 3}{y^2 - 4}$.

Solution: We factor the denominator.

$$\dfrac{y + 3}{y^2 - 4} = \dfrac{y + 3}{(y + 2)(y - 2)}$$

The denominator will be zero whenever any one of its factors is zero.

$$y + 2 = 0 \qquad y - 2 = 0$$
$$y = -2 \qquad\quad y = 2$$

Restrictions: $y \neq -2,\ y \neq 2$.

Example 3. Use the laws of exponents to reduce $\dfrac{2x^3y^7z}{x^2yz^2}$.

Solution: We note that the restrictions are $x \neq 0,\ y \neq 0,\ z \neq 0$.

$$\dfrac{2x^3y^7z}{x^2yz^2} = 2x^{3-2}y^{7-1}z^{1-2}$$

$$= 2x^1y^6z^{-1}$$

$$= \dfrac{2xy^6}{z}$$

Example 4. Use common factors to reduce $\dfrac{2x^3y^7z}{x^2yz^2}$.

Solution: We note that the restrictions are $x \neq 0$, $y \neq 0$, $z \neq 0$.

$$\frac{2x^3y^7z}{x^2yz^2} = \frac{\overset{1}{\cancel{2x^2}} \cdot x \cdot \overset{1}{\cancel{y}} \cdot y^6 \cdot \overset{1}{\cancel{z}}}{\underset{1}{\cancel{x^2}}\underset{1}{\cancel{y}} \cdot \underset{1}{\cancel{z}} \cdot z}$$

$$= \frac{2xy^6}{z}$$

Example 6. Reduce $\dfrac{x + 4}{x^2 + 8x + 16}$.

Solution

$$\frac{x + 4}{x^2 + 8x + 16}$$

$$= \frac{x + 4}{(x + 4)(x + 4)} \qquad x \neq -4$$

$$= \frac{1}{x + 4} \qquad x \neq -4$$

Example 8. Reduce $\dfrac{x^2 + 11x + 30}{2x^3 + 11x^2 - 6x}$.

Solution

$$\frac{x^2 + 11x + 30}{2x^3 + 11x^2 - 6x}$$

$$= \frac{(x + 5)(x + 6)}{x(2x^2 + 11x - 6)}$$

$$= \frac{(x + 5)\overset{1}{\cancel{(x + 6)}}}{x(2x - 1)\underset{1}{\cancel{(x + 6)}}} \qquad x \neq 0$$

$$\qquad\qquad x \neq \frac{1}{2}$$

$$\qquad\qquad x \neq -6$$

$$= \frac{x + 5}{x(2x - 1)} \qquad x \neq 0$$

$$\qquad\qquad x \neq \frac{1}{2}$$

$$\qquad\qquad x \neq -6$$

Example 12. Reduce $\dfrac{5 - 10x}{2x^2 + 7x - 4}$.

Solution

$$\frac{5 - 10x}{2x^2 + 7x - 4}$$

$$= \frac{5\overset{-1}{\cancel{(1 - 2x)}}}{\underset{1}{\cancel{(2x - 1)}}(x + 4)} \qquad x \neq \frac{1}{2}$$

$$\qquad\qquad x \neq -4$$

$$= \frac{-5}{x + 4} \qquad x \neq \frac{1}{2}$$

$$\qquad\qquad x \neq -4$$

Answers to Trial Run

page 378 **1.** $\dfrac{-2c^3}{a}$; $a \neq 0$, $b \neq 0$, $c \neq 0$ **2.** $9y - 7$; $x \neq 0$ **3.** $\dfrac{9x^3}{x + 1}$; $x \neq 0$, $x \neq -1$

4. $\dfrac{x - 7}{4}$; $x \neq -7$ **5.** $\dfrac{1}{y - 5}$; $y \neq 5$ **6.** $\dfrac{x + 4}{2x - 1}$; $x \neq 9$, $x \neq \dfrac{1}{2}$

7. $\dfrac{x - 3}{x + y}$; $x \neq 4y$, $x \neq -y$ **8.** $\dfrac{-(4 + a)}{3a + 5}$; $a \neq 4$, $a \neq \dfrac{-5}{3}$

EXERCISE SET 6.1

State restrictions on the domain of the variables and reduce the fractions.

_____ **1.** $\dfrac{-10a^3b^2}{5ab^5}$

_____ **2.** $\dfrac{-9a^5b^3}{3a^2b^7}$

_____ **3.** $\dfrac{8a^2b^3c^2}{24a^4b^5c}$

_____ **4.** $\dfrac{6a^3b^4c^3}{18a^8b^3c^4}$

_____ **5.** $\dfrac{x^3y^2z^7}{x^5y^4z^7}$

_____ **6.** $\dfrac{x^5y^3z^2}{x^7y^3z^4}$

_____ **7.** $\dfrac{12xy - 6x}{3x}$

_____ **8.** $\dfrac{15xy - 10x}{5x}$

_____ **9.** $\dfrac{5y}{25y^2 - 20y}$

_____ **10.** $\dfrac{3y}{9y^2 - 15y}$

_____ **11.** $\dfrac{-12a^2b - 15abc}{18ab}$

_____ **12.** $\dfrac{-10a^2b - 18abc}{22ab}$

_____ **13.** $\dfrac{9x + 9y}{18x + 18y}$

_____ **14.** $\dfrac{5x + 5y}{15x + 15y}$

_____ **15.** $\dfrac{3x - 3y}{6x + 6y}$

_____ **16.** $\dfrac{4x - 4y}{12x + 12y}$

_____ **17.** $\dfrac{x^2 - 81}{3x + 27}$

_____ **18.** $\dfrac{x^2 - 64}{4x + 32}$

_____ **19.** $\dfrac{x^2 + 9x + 20}{x^2 - 16}$

_____ **20.** $\dfrac{x^2 + 7x + 12}{x^2 - 9}$

_____ **21.** $\dfrac{x^2 + 6x + 8}{x^2 + 7x + 10}$

_____ **22.** $\dfrac{x^2 + 8x + 15}{x^2 + 5x + 6}$

_____ **23.** $\dfrac{25y^2 - 225}{5y^2 - 50y + 105}$

_____ **24.** $\dfrac{16y^2 - 144}{4y^2 - 20y + 24}$

_____ **25.** $\dfrac{2b - 1}{4b^2 - 4b + 1}$

_____ **26.** $\dfrac{3b + 1}{9b^2 + 6b + 1}$

_____ **27.** $\dfrac{3x^2 + 2x}{6x^2 - 17x - 14}$

_____ **28.** $\dfrac{5x^2 - 4x}{5x^2 - 14x + 8}$

_____ **29.** $\dfrac{a^2 + 12a + 35}{2a^3 + 9a^2 - 5a}$

_____ **30.** $\dfrac{a^2 - 6a - 27}{2a^3 + 3a^2 - 9a}$

_____ **31.** $\dfrac{x^3 + 27}{4x^2 + 7x - 15}$

_____ **32.** $\dfrac{x^3 + 8}{8x^2 + 13x - 6}$

Reduce each fraction. Do not state restrictions.

_____ **33.** $\dfrac{3a^2 + 15a + 75}{a^3 - 125}$

_____ **34.** $\dfrac{4a^2 + 16a + 64}{a^3 - 64}$

_____ **35.** $\dfrac{16a^4 - 1}{20a^2 + 5}$

_____ **36.** $\dfrac{81a^4 - 16}{27a^2 + 12}$

_____ **37.** $\dfrac{x^3 - x^2 + x - 1}{x^4 - 1}$

_____ **38.** $\dfrac{x^3 - 2x^2 + 4x - 8}{x^4 - 16}$

_____ **39.** $\dfrac{3x^2 + 14xy - 5y^2}{3x^2 + 2xy - y^2}$

_____ **40.** $\dfrac{6x^2 + xy - y^2}{2x^2 - xy - y^2}$

_____ **41.** $\dfrac{2a - b}{b^2 - 4ab + 4a^2}$

_____ **42.** $\dfrac{3a - b}{b^2 - 6ab + 9a^2}$

_____ **43.** $\dfrac{x^2 - 25}{20 - 19x + 3x^2}$

_____ **44.** $\dfrac{x^2 - 49}{14 - 23x + 3x^2}$

_____ **45.** $\dfrac{3x^2 + 2xy - y^2}{4y^2 - 17xy + 15x^2}$

_____ **46.** $\dfrac{2x^2 - 5xy - 7y^2}{14y^2 - 39xy + 10x^2}$

_____ **47.** $\dfrac{a^2 - 5a - ab + 5b}{3a^2 - 75}$

_____ **48.** $\dfrac{a^2 - 6a - ab + 6b}{2a^2 - 72}$

_____ **49.** $\dfrac{4x^3 - 9x + 16x^2 - 36}{2x^2 + 11x + 12}$

_____ **50.** $\dfrac{25x^3 - 16x + 50x^2 - 32}{5x^2 + 6x - 8}$

_____ **51.** $\dfrac{4(x - 2)^2 + 8x(x - 2)}{(x - 2)^4}$

_____ **52.** $\dfrac{3(x - 5)^2 + 6x(x - 5)}{(x - 5)^4}$

_____ **53.** $\dfrac{a^4 - 29a^2 + 100}{a^2 - 3a - 10}$

_____ **54.** $\dfrac{a^4 - 20a^2 + 64}{a^2 - 2a - 8}$

☆ Stretching the Topics _____

Reduce the fractions.

_____ **1.** $\dfrac{6x^3 - 24x^2 + 24x}{12x^3 - 4x^4 - x^5}$

_____ **2.** $\dfrac{x^{4n} - x^{2n} - 12}{x^{3n} - 2x^{2n} + 3x^n - 6}$

_____ **3.** State the restriction on the variable for $\dfrac{3x + 4y}{x^3 - 4x^2y - 5xy^2}$.

Check your answers in the back of your book.

If you can reduce the fractions in **Checkup 6.1,** you are ready to go on to Section 6.2.

CHECKUP 6.1

State restrictions and reduce the fractions.

_____ 1. $\dfrac{-34x^5y^{10}z^3}{17x^4y^{10}z^5}$

_____ 2. $\dfrac{9y}{18y^2 - 27y}$

_____ 3. $\dfrac{9x + 9y}{18x - 18y}$

_____ 4. $\dfrac{-10a^2b - 15ab}{20ab}$

_____ 5. $\dfrac{x^2 - 3x - 54}{x^2 - 81}$

_____ 6. $\dfrac{3x^2 - 18x}{x^2 - 36}$

Reduce the fractions. Do not state restrictions.

_____ 7. $\dfrac{36x^2 - 23xy - 3y^2}{4x^2 + xy - 3y^2}$

_____ 8. $\dfrac{3a - 2b}{4b^2 - 12ab + 9a^2}$

_____ 9. $\dfrac{a^4 - 625}{a^3 - 5a^2 + 25a - 125}$

_____ 10. $\dfrac{8(x - y)^2 + 24x(x - y)}{4x^2 - 5xy + y^2}$

Check your answers in the back of your book.

If You Missed Problems:	You Should Review Examples:
1	3, 4
2–6	5–8
7	9
8	11, 12
9	9
10	10

6.2 Multiplying and Dividing Rational Expressions

Multiplying Rational Expressions

A look at some products of numerical fractions might give us a hint about how to multiply rational expressions.

$$\frac{1}{3} \cdot \frac{2}{5} = \frac{1 \cdot 2}{3 \cdot 5} \qquad\qquad \frac{5}{6} \cdot \frac{3}{10} = \frac{5}{2 \cdot 3} \cdot \frac{3}{2 \cdot 5}$$

$$= \frac{2}{15} \qquad\qquad\qquad = \frac{\overset{1}{\cancel{5}}}{2 \cdot \underset{1}{\cancel{3}}} \cdot \frac{\overset{1}{\cancel{3}}}{2 \cdot \underset{1}{\cancel{5}}}$$

$$= \frac{1 \cdot 1}{2 \cdot 2}$$

$$= \frac{1}{4}$$

Notice that we reduced in our second example even before we found the product of the numerators and denominators. In general, we agree that

> To find the product of fractions, we multiply the numerators and multiply the denominators, reducing if possible.

Or, using *A, B, C,* and *D* to represent algebraic expressions, we say:

***

Product of Rational Expressions

$$\frac{A}{B} \cdot \frac{C}{D} = \frac{A \cdot C}{B \cdot D} \qquad B \neq 0, D \neq 0$$

***

Whenever we multiply fractions, we should first factor the numerators and denominators completely, so that we can easily spot common factors.

Example 1. Find $\dfrac{3x + 6}{x^2 - 9} \cdot \dfrac{5x - 15}{x^2 + 5x + 6}$.

Solution

$$\dfrac{3x + 6}{x^2 - 9} \cdot \dfrac{5x - 15}{x^2 + 5x + 6}$$

$$= \dfrac{3\cancel{(x + 2)}}{\cancel{(x - 3)}(x + 3)} \cdot \dfrac{5\cancel{(x - 3)}}{(x + 3)\cancel{(x + 2)}}$$

$$x \neq 3, x \neq -3, x \neq -2$$

$$= \dfrac{3 \cdot 5}{(x + 3)(x + 3)}$$

$$= \dfrac{15}{(x + 3)^2} \quad x \neq 3, x \neq -3, x \neq -2$$

Example 2. Find

$$\dfrac{x^2 + 4x + 3}{x^2 + 2x + 1} \dfrac{x^2 - 7x - 8}{x^2 - 5x - 24}.$$

Solution

$$\dfrac{x^2 + 4x + 3}{x^2 + 2x + 1} \cdot \dfrac{x^2 - 7x - 8}{x^2 - 5x - 24}$$

$$= \dfrac{\cancel{(x + 1)}\cancel{(x + 3)}}{\cancel{(x + 1)}\cancel{(x + 1)}} \cdot \dfrac{\cancel{(x - 8)}\cancel{(x + 1)}}{\cancel{(x + 3)}\cancel{(x - 8)}}$$

$$x \neq -1, x \neq -3, x \neq 8$$

$$= 1 \qquad x \neq -1, x \neq -3, x \neq 8$$

Notice that we do *not* multiply the final variable factors in the numerator and denominator. They are best left in factored form.

Now you complete Example 3.

Example 3. Find $\dfrac{x^2 - 9x}{6x^4} \cdot \dfrac{x^2}{x^2 - 9} \cdot \dfrac{4x + 12}{9 + 17x - 2x^2}$.

Solution

$$\dfrac{x^2 - 9x}{6x^4} \cdot \dfrac{x^2}{x^2 - 9} \cdot \dfrac{4x + 12}{9 + 17x - 2x^2}$$

$$= \dfrac{x(x - 9)}{6x^4} \cdot \dfrac{x^2}{(x + 3)(x - 3)} \cdot \dfrac{4(x + 3)}{(9 - x)(1 + 2x)}$$

$$x \neq \underline{\quad}, x \neq \underline{\quad},$$
$$x \neq \underline{\quad}, x \neq \underline{\quad},$$
$$x \neq \underline{\quad}$$

$$= \underline{\quad\quad}$$

Check your work on page 390. ▶

Dividing Rational Expressions

Once again, we recall from arithmetic the technique for dividing numerical fractions.

$$\dfrac{5}{6} \div \dfrac{2}{7} = \dfrac{5}{6} \cdot \dfrac{7}{2}$$

$$= \dfrac{5 \cdot 7}{6 \cdot 2}$$

$$= \dfrac{35}{12}$$

Before stating the rule for dividing rational expressions, we must recall the names of the parts of a division problem.

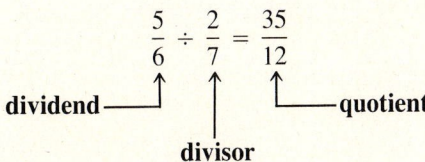

$$\frac{5}{6} \div \frac{2}{7} = \frac{35}{12}$$

dividend ——— divisor ——— quotient

Now the rule for division can be stated as follows:

> To divide rational expressions, we invert the divisor and multiply.

Letting A, B, C, and D represent algebraic expressions, we say:

> **Division of Rational Expressions**
>
> $$\frac{A}{B} \div \frac{C}{D} = \frac{A}{B} \cdot \frac{D}{C} \qquad B \neq 0, D \neq 0, C \neq 0$$

Notice that we place restrictions on D because it is the denominator of the divisor *before* it is inverted. Restrictions are also necessary on C because it is the denominator *after* the divisor is inverted. Rather than get bogged down stating restrictions in these more complicated expressions, let us agree that values of the variable that make the denominators equal zero will be understood to be excluded. Until further notice, *we will not state restrictions unless specifically asked to do so.*

Example 4. Find $\dfrac{x}{x + 1} \div \dfrac{x^2}{2x + 2}$.

Solution

$$\frac{x}{x + 1} \div \frac{x^2}{2x + 2}$$

$$= \frac{x}{x + 1} \cdot \frac{2x + 2}{x^2}$$

$$= \frac{\overset{1}{\cancel{x}}}{\cancel{x + 1}} \cdot \frac{2\cancel{(x + 1)}}{\cancel{x^2}}$$

$$= \frac{2}{x}$$

Example 5. Find $\dfrac{x^2 y}{x^3 + 5x^2} \div 5xy^2$.

Solution

$$\frac{x^2 y}{x^3 + 5x^2} \div \frac{5xy^2}{1}$$

$$= \frac{\overset{1}{\cancel{x^2 y}}}{\underset{1}{\cancel{x^2}(x + 5)}} \cdot \frac{1}{\underset{y}{5x\cancel{y^2}}}$$

$$= \frac{1}{5xy(x + 5)}$$

Now you try Example 6.

Example 6. Find

$$\frac{x^2 + 6x + 8}{x^2 - 4} \div \frac{5x + 15}{x^2 - 2x}.$$

Solution

$$\frac{x^2 + 6x + 8}{x^2 - 4} \div \frac{5x + 15}{x^2 - 2x}$$

$$= \frac{x^2 + 6x + 8}{x^2 - 4} \cdot \frac{x^2 - 2x}{5x + 15}$$

$$= \underline{\hspace{2cm}}$$

Check your work on page 390. ▶

Example 7. Find

$$\frac{2x^2 - 11x - 21}{x^2 + 10x + 25} \div \frac{98 - 2x^2}{20x + 4x^2}.$$

Solution

$$\frac{2x^2 - 11x - 21}{x^2 + 10x + 25} \div \frac{98 - 2x^2}{20x + 4x^2}$$

$$= \frac{2x^2 - 11x - 21}{x^2 + 10x + 25} \cdot \frac{20x + 4x^2}{98 - 2x^2}$$

$$= \frac{(2x + 3)(x - 7)}{(x + 5)(x + 5)} \cdot \frac{4x(5 + x)}{2(49 - x^2)}$$

$$= \frac{(2x + 3)\overset{-1}{\cancel{(x - 7)}}}{(x + 5)\underset{1}{\cancel{(x + 5)}}} \cdot \frac{4x\overset{1}{\cancel{(5 + x)}}}{2(7 + x)\underset{1}{\cancel{(7 - x)}}}$$

$$= \frac{-2x(2x + 3)}{(x + 5)(x + 7)}$$

◾▶ **Trial Run** ▬▬▬▬▬▬▬▬▬▬▬▬▬▬▬

Perform the indicated operation.

_____ 1. $\dfrac{2x + 2}{x + 3} \cdot \dfrac{1}{x + 1}$

_____ 2. $\dfrac{x^3 - 8}{6x - 12} \cdot \dfrac{36}{3x^3 + 6x^2 + 12x}$

_____ 3. $\dfrac{x^2 + 6x + 5}{x^2 + 10x + 25} \cdot \dfrac{x^2 + 2x - 15}{x^2 - 6x - 7}$

_____ 4. $\dfrac{16 - x^2}{2x^2 - 11x + 12} \cdot \dfrac{6x^2 - 9x}{9x^2 + 37x + 4}$

_____ 5. $\dfrac{x^3 - 49x}{xy} \div (x + 7)$

_____ 6. $\dfrac{x^2 + x - 6}{x^2 - 9} \div \dfrac{5x - 10}{x^2 + 2x}$

_____ 7. $\dfrac{x + 2}{4x - 8} \div \dfrac{x^2 - 7x - 18}{2x - 4}$

Answers are on page 391.

As always, symbols of grouping tell us which operation to perform first.

Example 8. Find $\left[\dfrac{x^2 - 1}{6x + 12} \cdot \dfrac{x^2 + 2x}{x^3 + 2x^2 + x} \right] \div \dfrac{x^2 + 3x - 4}{x^2 + 5x + 4}$.

Solution

$$\left[\frac{x^2 - 1}{6x + 12} \cdot \frac{x^2 + 2x}{x^3 + 2x^2 + x} \right] \div \frac{x^2 + 3x - 4}{x^2 + 5x + 4}$$

$$= \left[\frac{(x + 1)(x - 1)}{6(x + 2)} \cdot \frac{x(x + 2)}{x(x^2 + 2x + 1)} \right] \cdot \frac{x^2 + 5x + 4}{x^2 + 3x - 4}$$

$$= \left[\frac{\cancel{(x + 1)}(x - 1)}{6\cancel{(x + 2)}} \cdot \frac{\cancel{x}\cancel{(x + 2)}}{\cancel{x}\cancel{(x + 1)}(x + 1)} \right] \cdot \frac{(x + 4)(x + 1)}{(x + 4)(x - 1)}$$

$$= \frac{\cancel{x - 1}}{6\cancel{(x + 1)}} \cdot \frac{\cancel{(x + 4)}\cancel{(x + 1)}}{\cancel{(x + 4)}\cancel{(x - 1)}}$$

$$= \frac{1}{6}$$

Now you complete Example 9.

Example 9. Find $\dfrac{3x^2}{(2x + 1)^2} \div \left[\dfrac{6x^2 - 3x}{5x} \div \dfrac{4x^2 - 1}{20x^3} \right]$

Solution

$$\frac{3x^2}{(2x + 1)^2} \div \left[\frac{6x^2 - 3x}{5x} \div \frac{4x^2 - 1}{20x^3} \right]$$

$$= \frac{3x^2}{(2x + 1)^2} \div \left[\frac{6x^2 - 3x}{5x} \cdot \frac{20x^3}{4x^2 - 1} \right]$$

$$= \frac{3x^2}{(2x + 1)^2} \div \left[\frac{3x(\underline{\hspace{1.5cm}})}{5x} \cdot \frac{20x^3}{(\underline{\hspace{1cm}})(\underline{\hspace{1cm}})} \right]$$

Check your work on page 391. ▶

⟾ Trial Run

Perform the indicated operations.

_____ **1.** $\left[\dfrac{xy^2}{4x^2 + 4x} \div \dfrac{x^2y}{3x^2 - 3}\right] \cdot \dfrac{4x}{3y}$

_____ **2.** $\left[\dfrac{x^2 - 25}{10x} \cdot \dfrac{50}{3x + 15}\right] \div \dfrac{2x - 10}{3x}$

_____ **3.** $\left[\dfrac{x^3 - 8}{5x - 15} \div \dfrac{x^2 + x - 6}{15x^2 + 40x - 15}\right] \div \dfrac{2x^3 + 4x^2 + 8x}{4x^2 - 13x + 3}$

Answers are on page 391.

▶ Examples You Completed

Example 3. Find $\dfrac{x^2 - 9x}{6x^4} \cdot \dfrac{x^2}{x^2 - 9} \cdot \dfrac{4x + 12}{9 + 17x - 2x^2}$.

Solution

$$\frac{x^2 - 9x}{6x^4} \cdot \frac{x^2}{x^2 - 9} \cdot \frac{4x + 12}{9 + 17x - 2x^2}$$

$$= \frac{\overset{1}{\cancel{x}}(\overset{-1}{\cancel{x - 9}})}{\underset{3x}{\cancel{6x^4}}} \cdot \frac{\overset{1}{\cancel{x^2}}}{(x + 3)(x - 3)} \cdot \frac{\overset{2}{\cancel{4}}(\overset{1}{\cancel{x + 3}})}{\underset{1}{(9 - x)(1 + 2x)}} \qquad \begin{array}{l} x \neq 0,\ x \neq -3, \\ x \neq 3,\ x \neq 9, \\ x \neq -\dfrac{1}{2} \end{array}$$

$$= \frac{-2}{3x(x - 3)(1 + 2x)}$$

Example 6. Find $\dfrac{x^2 + 6x + 8}{x^2 - 4} \div \dfrac{5x + 15}{x^2 - 2x}$.

Solution

$$\frac{x^2 + 6x + 8}{x^2 - 4} \div \frac{5x + 15}{x^2 - 2x}$$

$$= \frac{x^2 + 6x + 8}{x^2 - 4} \cdot \frac{x^2 - 2x}{5x + 15}$$

$$= \frac{(x + 4)\overset{1}{\cancel{(x + 2)}}}{\underset{1}{\cancel{(x + 2)}}\underset{1}{\cancel{(x - 2)}}} \cdot \frac{x\overset{1}{\cancel{(x - 2)}}}{5(x + 3)}$$

$$= \frac{x(x + 4)}{5(x + 3)}$$

Example 9. Find $\dfrac{3x^2}{(2x+1)^2} \div \left[\dfrac{6x^2-3x}{5x} \div \dfrac{4x^2-1}{20x^3} \right]$.

Solution

$$\dfrac{3x^2}{(2x+1)^2} \div \left[\dfrac{6x^2-3x}{5x} \div \dfrac{4x^2-1}{20x^3} \right]$$

$$= \dfrac{3x^2}{(2x+1)^2} \div \left[\dfrac{6x^2-3x}{5x} \cdot \dfrac{20x^3}{4x^2-1} \right]$$

$$= \dfrac{3x^2}{(2x+1)^2} \div \left[\dfrac{\overset{1}{3x}(2x-\overset{1}{\cancel{1}})}{\underset{11}{\cancel{5x}}} \cdot \dfrac{\overset{4}{20x^3}}{(2x+1)(2x-\underset{1}{\cancel{1}})} \right]$$

$$= \dfrac{3x^2}{(2x+1)^2} \div \dfrac{12x^3}{2x+1}$$

$$= \dfrac{\overset{1}{\cancel{3x^2}}}{(2x+1)(2x+\underset{1}{\cancel{1}})} \cdot \dfrac{\overset{1}{2x+1}}{\underset{4x}{\cancel{12x^3}}}$$

$$= \dfrac{1}{4x(2x+1)}$$

Answers to Trial Runs

page 388   **1.** $\dfrac{2}{x+3}$ **2.** $\dfrac{2}{x}$ **3.** $\dfrac{x-3}{x-7}$ **4.** $\dfrac{-3}{9x+1}$ **5.** $\dfrac{x-7}{y}$ **6.** $\dfrac{x(x+2)}{5(x-3)}$ **7.** $\dfrac{1}{2(x-9)}$

page 390   **1.** $\dfrac{x-1}{x}$ **2.** $\dfrac{5}{2}$ **3.** $\dfrac{(3x-1)(4x-1)}{2x}$

Name _____ Date _____

EXERCISE SET 6.2

Perform the indicated operations.

_____ 1. $\dfrac{2x + 18}{3x^2} \cdot \dfrac{9x}{x + 9}$

_____ 2. $\dfrac{5x + 20}{4x^2} \cdot \dfrac{12x}{x + 4}$

_____ 3. $\dfrac{a^2 + b^2}{6} \cdot \dfrac{42}{7a + 7b}$

_____ 4. $\dfrac{a^2 + b^2}{19} \cdot \dfrac{38}{2a + 2b}$

_____ 5. $\dfrac{x^2 - 25}{3x - 21} \cdot \dfrac{2x - 14}{x^2 + 10x + 25}$

_____ 6. $\dfrac{x^2 - 36}{2x - 10} \cdot \dfrac{3x - 15}{x^2 - 12x + 36}$

_____ 7. $\dfrac{x^2 - 81y^2}{x^2 + 9y^2} \cdot \dfrac{3x^2 + 27y^2}{x - 9y}$

_____ 8. $\dfrac{x^2 - 64y^2}{x^2 + 8y^2} \cdot \dfrac{5x^2 + 40y^2}{x - 8y}$

_____ 9. $\dfrac{(2x + 5)^2}{64yz^3} \cdot \dfrac{40y^2z}{4x^2 - 25}$

_____ 10. $\dfrac{(3x - 2)^2}{81yz^3} \cdot \dfrac{72y^5z}{9x^2 - 4}$

_____ 11. $\dfrac{2x^2 + 3x + 1}{2x^2 + 5x + 3} \cdot \dfrac{2x^2 + 15x + 18}{2x^2 + 13x + 6}$

_____ 12. $\dfrac{20x^2 - 7x - 3}{4x^2 - 7x - 2} \cdot \dfrac{3x^2 - 5x - 2}{15x^2 - 4x - 3}$

_____ 13. $\dfrac{6x^2 + xy - y^2}{2x^2 - 9xy - 5y^2} \cdot \dfrac{x^2 - 25y^2}{15x - 5y}$

_____ 14. $\dfrac{8x^2 + 18xy - 5y^2}{16x^2 - 8xy + y^2} \cdot \dfrac{4x^2 + 3xy - y^2}{12x + 30y}$

_____ 15. $\dfrac{x^3 - 27}{9x - 18} \cdot \dfrac{36}{2x^2 + 6x + 18}$

_____ 16. $\dfrac{x^3 - 64}{7x - 28} \cdot \dfrac{42}{3x^2 + 12x + 48}$

_____ 17. $\dfrac{25 - x^2}{3x^2 + 7x - 40} \cdot \dfrac{3x^2 + 12x}{2x^3 - 2x^2 - 40x}$

_____ 18. $\dfrac{49 - x^2}{5x^2 + 32x - 21} \cdot \dfrac{5x^2 - 30x}{2x^3 - 26x^2 + 84x}$

_____ 19. $\dfrac{6 - 23x - 4x^2}{3x^2 + 10x - 48} \cdot \dfrac{3x^2 + x - 24}{8x^2 - 6x + 1}$

_____ 20. $\dfrac{9 - 3x - 2x^2}{2x^2 + 11x - 21} \cdot \dfrac{2x^2 + 9x - 35}{2x^2 + 11x + 15}$

_____ 21. $\dfrac{3x^3 - 75x}{9x^4} \cdot \dfrac{3x - 24}{x^2 - 3x - 40} \cdot \dfrac{x^2}{5 - 6x + x^2}$

_____ 22. $\dfrac{2x^3 - 98x}{10x^4} \cdot \dfrac{x^2}{x^2 - 2x - 63} \cdot \dfrac{5x - 45}{14 - 9x + x^2}$

_____ 23. $\dfrac{x^4 - 13x^2 + 36}{x^4 - 14x^2 - 32} \cdot \dfrac{x^3 - 4x^2 + 2x - 8}{x^3 + 3x^2 - 4x - 12}$

_____ 24. $\dfrac{x^4 - 26x^2 + 25}{x^4 - 6x^2 - 27} \cdot \dfrac{x^3 - 3x^2 + 3x - 9}{x^3 + 5x^2 - x - 5}$

_____ 25. $\dfrac{x^5}{x + 5} \div \dfrac{x^2}{3x + 15}$

_____ 26. $\dfrac{x^6}{x - 6} \div \dfrac{x^3}{2x - 12}$

_____ 27. $\dfrac{x^2 y^5}{x + 9} \div 9x^2 y^3$

_____ 28. $\dfrac{x^3 y^4}{x + 5} \div 5x^3 y$

_____ 29. $\dfrac{x^2 + 7x + 10}{63x} \div \dfrac{7x + 14}{9x}$

_____ 30. $\dfrac{x^2 - 12x + 32}{60x} \div \dfrac{5x - 20}{12x}$

393

_____ 31. $\dfrac{2x^2 - 19x + 24}{x^2 - 64} \div \dfrac{3x - 24}{x^2 + 8x}$

_____ 32. $\dfrac{3x^2 - 22x + 7}{x^2 - 49} \div \dfrac{5x - 35}{x^2 + 7x}$

_____ 33. $\dfrac{x + 3}{5x - 20} \div \dfrac{x^2 - 9x - 36}{3x - 12}$

_____ 34. $\dfrac{x + 7}{8x - 56} \div \dfrac{x^2 + 4x - 21}{4x - 28}$

_____ 35. $\dfrac{x^2 - 8x + 15}{x^2 - 25} \div \dfrac{x^2 - 9}{x + 5}$

_____ 36. $\dfrac{x^2 - 7x - 18}{x^2 - 81} \div \dfrac{x^2 + 4x + 4}{x + 9}$

_____ 37. $\dfrac{6x^2 - 11x + 5}{x^3 - 1} \div \dfrac{6x^2 - 23x + 15}{3x^3 + 3x^2 + 3x}$

_____ 38. $\dfrac{3x^2 - 19x + 28}{x^3 - 64} \div \dfrac{3x^2 - 22x + 35}{5x^3 + 20x^2 + 80x}$

_____ 39. $\dfrac{xy - y^2}{x^2 - 2xy + y^2} \div \dfrac{5xy - 3y^2}{5x^2 - 8xy + 3y^2}$

_____ 40. $\dfrac{x^2 - 7xy}{x^2 - 14xy + 49y^2} \div \dfrac{3x^2 + 2xy}{3x^2 - 19xy - 14y^2}$

_____ 41. $\dfrac{25 - x^2}{x^2 - 11x + 30} \div \dfrac{2x^2 + 7x - 15}{2x^2 - 15x + 18}$

_____ 42. $\dfrac{9 - x^2}{x^2 - 11x + 24} \div \dfrac{4x^2 + 13x + 3}{4x^2 - 31x - 8}$

_____ 43. $\dfrac{4x^2 + 19x - 5}{6x^2 - 11x - 2} \div \dfrac{6 - 19x - 20x^2}{5x^2 - 4x - 12}$

_____ 44. $\dfrac{3x^2 + 26x - 9}{3x^2 - 14x + 8} \div \dfrac{1 + 5x - 24x^2}{8x^2 - 31x - 4}$

_____ 45. $\left[\dfrac{x^2 - 25}{10x} \cdot \dfrac{50}{3x + 15}\right] \div \dfrac{2x - 10}{3x}$

_____ 46. $\left[\dfrac{x^2 - 9}{3x} \cdot \dfrac{45}{4x - 12}\right] \div \dfrac{5x + 15}{7x}$

_____ 47. $\dfrac{4x^2 - 36}{x^2 + 2x - 15} \cdot \left[\dfrac{2x^2 + 11x + 5}{16x^2 + 16x - 96} \div \dfrac{6x^2 + 7x + 2}{3x^2 - 4x - 4}\right]$

_____ 48. $\dfrac{9x^2 - 1}{3x^2 - 26x - 9} \cdot \left[\dfrac{x^2 - 13x + 36}{2x^2 - 21x + 55} \div \dfrac{3x^2 - 13x + 4}{x^2 - 10x + 25}\right]$

_____ 49. $\left[\dfrac{8x^3 - 27}{(x + 2)^2} \cdot \dfrac{4x^2 + 8x}{2x^2 - 7x + 6}\right] \div \dfrac{8x^4 + 12x^3 + 18x^2}{x^2 - 4}$

_____ 50. $\left[\dfrac{x^3 + 125}{(x + 3)^2} \cdot \dfrac{8x^2 + 24x}{x^2 + 2x - 15}\right] \div \dfrac{6x^4 - 30x^3 + 150x^2}{x^2 - 9}$

☆ Stretching the Topics

Perform the indicated operation.

_____ **1.** $\dfrac{10x^3 - 4x^2 + 15x - 6}{10x^2 - 24x + 8} \cdot \dfrac{2x^3 - 4x^2 - 3x + 6}{4x^4 - 9}$

_____ **2.** $\dfrac{2x^{2a} - 2x^a - 12}{6x^{2a} + 18x^a} \div \dfrac{x^{3a} - 3x^{2a} - 4x^a + 12}{3x^{2a} - 5x^a - 2}$

_____ **3.** $\dfrac{2x^{3a} + 6x^{2a}}{x^{2b+c}} \cdot \dfrac{x^{a+2b} - 3x^{2b}}{x^{2a+c} - 9x^c}$

Check your answers in the back of your book.

If you can complete **Checkup 6.2,** you are ready to go on to Section 6.3.

CHECKUP 6.2

Perform the indicated operation. Leave the answers in factored form.

_____ 1. $\dfrac{4x + 32}{5x^2} \cdot \dfrac{15x}{x + 8}$

_____ 2. $\dfrac{x^2 - 49}{8x - 24} \cdot \dfrac{x^2 - 3x}{x^2 + 3x - 28}$

_____ 3. $\dfrac{(2x - 7)^2}{36y^4z^2} \cdot \dfrac{30y^3z^2}{4x^2 - 49}$

_____ 4. $\dfrac{7x^2 + 5xy - 2y^2}{x^2 - 8xy - 9y^2} \cdot \dfrac{x^2 - 81y^2}{14xy - 4y^2}$

_____ 5. $\dfrac{x^2 - 225}{x^7y^3} \div \dfrac{3x + 45}{x^2y^6}$

_____ 6. $\dfrac{x^7y^5}{x^3 + 4x^2} \div 4x^5y^2$

_____ 7. $\dfrac{5x^2 - 49x + 36}{x^2 - 81} \div \dfrac{25x^2 - 16}{x + 9}$

_____ 8. $\dfrac{5x^2 - 47x - 30}{x^3 + 8} \div \dfrac{5x^2 - 32x - 21}{7x^3 - 14x^2 + 28x}$

_____ 9. $\dfrac{2x^2 - 6xy}{x^2 - xy - 6y^2} \div \dfrac{5x^2y - 10x^3}{2x^2 + 3xy - 2y^2}$

_____ 10. $\left[\dfrac{x^2 - 169}{8x} \cdot \dfrac{16}{3x - 39} \right] \div \dfrac{6x + 78}{9x}$

Check your answers in the back of your book.

If You Missed Problems:	You Should Review Examples:
1	1
2, 3	2, 3
4	2
5	4
6	5
7	6
8, 9	7
10	8

6.3 Adding and Subtracting Rational Expressions

Combining Fractions with Like Denominators

From earlier work with addition and subtraction of algebraic expressions, you remember that only *like terms* could be combined into a single term. The same conditions hold for addition and subtraction of rational algebraic expressions. Only *like fractions* can be combined into a single fraction. But what do we mean by like fractions?

> Fractions are called *like fractions* if they have the *same denominator*.

Now we recall the rule for adding or subtracting like fractions.

> To add (or subtract) fractions having the same denominator, we add (or subtract) the numerators and keep the same denominator.
>
> $$\frac{A}{C} + \frac{B}{C} = \frac{A + B}{C} \qquad C \neq 0$$

You try Examples 1 and 2, remembering to try reducing *after* you have found the sum or difference.

Example 1. Find $\dfrac{7}{2x} - \dfrac{1}{2x}$.

Solution

$$\frac{7}{2x} - \frac{1}{2x}$$

$$= \frac{7 - 1}{2x}$$

Check your work on page 408. ▶

Example 2. Find $\dfrac{x}{x + 3} + \dfrac{2x}{x + 3}$.

Solution

$$\frac{x}{x + 3} + \frac{2x}{x + 3}$$

$$= \frac{x + 2x}{x + 3}$$

Check your work on page 408. ▶

Example 3. Find $\dfrac{8x}{2x+1} + \dfrac{4}{2x+1}$.

Solution

$$\dfrac{8x}{2x+1} + \dfrac{4}{2x+1}$$

$$= \dfrac{8x+4}{2x+1}$$

$$= \dfrac{4\cancel{(2x+1)}^{\;1}}{\cancel{2x+1}_{\;1}}$$

$$= 4$$

Now complete Example 4.

Example 4. Find $\dfrac{2x+3}{x-1} + \dfrac{4-x}{x-1}$.

Solution

$$\dfrac{2x+3}{x-1} + \dfrac{4-x}{x-1}$$

$$= \dfrac{2x+3+4-x}{x-1}$$

Check your work on page 408. ▶

You will avoid unnecessary errors in signs when subtracting fractions if you will remember to use *parentheses* around the numerator being subtracted.

Try Example 5.

Example 5. Find $\dfrac{4x-7}{x^2-4} - \dfrac{3x-9}{x^2-4}$.

Solution

$$\dfrac{4x-7}{x^2-4} - \dfrac{3x-9}{x^2-4}$$

$$= \dfrac{4x-7-(3x-9)}{x^2-4}$$

$$= \dfrac{4x-7-3x+9}{x^2-4}$$

Check your work on page 408. ▶

Example 6. Find $\dfrac{5(x-3)}{2x+11} - \dfrac{2(3x+1)}{2x+11}$.

Solution

$$\dfrac{5(x-3)}{2x+11} - \dfrac{2(3x+1)}{2x+11}$$

$$= \dfrac{5(x-3) - 2(3x+1)}{2x+11}$$

$$= \dfrac{5x-15-6x-2}{2x+11}$$

$$= \dfrac{-x-17}{2x+11}$$

Example 7. Find $\dfrac{(x+1)(x-6)}{2x+1} - \dfrac{x(4-x)}{2x+1} + \dfrac{1}{2x+1}$.

Solution

$$\dfrac{(x+1)(x-6)}{2x+1} - \dfrac{x(4-x)}{2x+1} + \dfrac{1}{2x+1}$$

$$= \dfrac{(x+1)(x-6) - x(4-x) + 1}{2x+1}$$

$$= \dfrac{x^2 - 5x - 6 - 4x + x^2 + 1}{2x+1}$$

$$= \dfrac{2x^2 - 9x - 5}{2x+1}$$

$$= \dfrac{\overset{1}{\cancel{(2x+1)}}(x-5)}{\underset{1}{\cancel{2x+1}}}$$

$$= x - 5$$

⫸ Trial Run

Perform the indicated operation and reduce if possible.

————— **1.** $\dfrac{5}{3x} + \dfrac{7}{3x}$

————— **2.** $\dfrac{2x}{x+7} + \dfrac{5x}{x+7}$

————— **3.** $\dfrac{6x}{3x-2} - \dfrac{4}{3x-2}$

————— **4.** $\dfrac{x}{x^2-16} + \dfrac{4}{x^2-16}$

————— **5.** $\dfrac{6x+3}{x^2-25} - \dfrac{5x+8}{x^2-25}$

————— **6.** $\dfrac{(2x-3)(x-1)}{x-2} + \dfrac{x(x+3)}{x-2} - \dfrac{11}{x-2}$

Answers are on page 409.

Building Fractions

We have learned to reduce fractions by *dividing* the numerator and denominator by any common nonzero factor. It is also possible to build a new fraction from an old fraction by *multiplying* the numerator and denominator by any nonzero fraction.

For example,

$$\dfrac{1}{2} = \dfrac{1 \cdot 5}{2 \cdot 5} = \dfrac{5}{10}$$

Here we multiplied the numerator and denominator by 5 to change $\dfrac{1}{2}$ to the equivalent fraction

$\dfrac{5}{10}$. Notice that in performing that multiplication we were actually multiplying the entire fraction by $\dfrac{5}{5}$ or 1.

The process of **building fractions** is really the opposite of reducing fractions and uses the same basic property. Letting A, B, and C represent algebraic expressions, we note that

$$\boxed{\begin{array}{l} \textbf{Building Fractions} \\[6pt] \dfrac{A}{B} = \dfrac{A \cdot C}{B \cdot C} \qquad B \neq 0,\ C \neq 0 \end{array}}$$

In the next section you will need some skill at building fractions to new fractions having specified denominators. Let's practice a bit here.

Suppose that we wish to change $\dfrac{3}{4}$ to a fraction with a denominator of 24.

$$\dfrac{3}{4} = \dfrac{?}{24} \qquad$$ We must find a numerator for the fraction on the right.

$$\dfrac{3}{4} = \dfrac{?}{4(6)} \qquad$$ The denominator on the right is 6 times the denominator on the left.

$$\dfrac{3}{4} = \dfrac{3(6)}{4(6)} \qquad$$ The numerator on the right must *also* be 6 times the numerator on the left.

$$\dfrac{3}{4} = \dfrac{18}{24} \qquad$$ We have found our new fraction.

Let's state this procedure in words.

To change an old fraction to an equivalent fraction with a new denominator, we must decide by what factor the old denominator was multiplied to arrive at the new denominator. Then we must multiply the old numerator by that same factor to arrive at the new numerator.

$$\dfrac{4}{3x} = \dfrac{?}{3x(x+1)} \qquad$$ By what factor was the old denominator multiplied to arrive at the new denominator? The factor must be $(x+1)$.

$$\dfrac{4}{3x} = \dfrac{4(x+1)}{3x(x+1)} \qquad$$ We must multiply the old numerator by $(x+1)$ also.

Example 8. Find the new fraction.

$$\frac{5}{x^2} = \frac{?}{2x^5y}$$

Solution

$$\frac{5}{x^2} = \frac{?}{2x^5y}$$

$$\frac{5}{x^2} = \frac{?}{x^2(2x^3y)}$$

$$\frac{5}{x^2} = \frac{5(2x^3y)}{x^2(2x^3y)}$$

$$\frac{5}{x^2} = \frac{10x^3y}{2x^5y}$$

Example 9. Find the new fraction.

$$\frac{x}{x + 3} = \frac{?}{x(x + 1)(x + 3)}$$

Solution

$$\frac{x}{x + 3} = \frac{?}{x(x + 1)(x + 3)}$$

$$= \frac{x \cdot x(x + 1)}{x(x + 1)(x + 3)}$$

$$= \frac{x^2(x + 1)}{x(x + 1)(x + 3)}$$

Example 10. Find the new fraction.

$$\frac{x + 1}{3x} = \frac{?}{6x^2(2x + 5)}$$

Solution

$$\frac{x + 1}{3x} = \frac{?}{6x^2(2x + 5)}$$

$$= \frac{?}{3x \cdot 2x(2x + 5)}$$

$$\frac{x + 1}{3x} = \frac{(x + 1)(2x)(2x + 5)}{3x \cdot 2x(2x + 5)}$$

$$\frac{x + 1}{3x} = \frac{2x(x + 1)(2x + 5)}{6x^2(2x + 5)}$$

You complete Example 11.

Example 11. Find the new fraction.

$$\frac{4}{x + 1} = \frac{?}{3(x + 1)^2}$$

Solution

$$\frac{4}{x + 1} = \frac{?}{3(x + 1)^2}$$

Check your work on page 408. ▶

⫸ **Trial Run**

Find the new fraction.

_____ 1. $\dfrac{-2}{7} = \dfrac{?}{91}$

_____ 2. $\dfrac{5}{xy} = \dfrac{?}{3x^2y^3}$

_____ 3. $\dfrac{x}{2x + 1} = \dfrac{?}{(2x + 1)(x - 3)}$

_____ 4. $\dfrac{7x}{x - 3} = \dfrac{?}{2x^2(x + 3)(x - 3)}$

_____ 5. $\dfrac{x + 2}{x + 3} = \dfrac{?}{5(x + 3)^2}$

_____ 6. $\dfrac{x + 3}{2x + 3} = \dfrac{?}{2x^2 - 7x - 15}$

Answers are on page 409.

Combining Fractions with Different Denominators

We have agreed that we can combine fractions by addition and subtraction only when their denominators are exactly the same. To find sums of fractions that do not share the same denominator, we must somehow *make* the denominators match by finding a **common denominator.** A common denominator is a number into which each of the original denominators will divide exactly (with zero remainder).

The **lowest common denominator (LCD)** for a group of fractions is the *smallest* number into which each of the original denominators will divide exactly (with no remainder).

To work the addition problem $\dfrac{1}{2} + \dfrac{4}{3}$, we must convert each of the original fractions to a new fraction having the LCD as its denominator. To do this, we must use the method of building fractions that we have just learned. For $\dfrac{1}{2}$ and $\dfrac{4}{3}$, the LCD is $2 \cdot 3$.

$$\frac{1}{2} + \frac{4}{3} = \frac{1(\;\;)}{2 \cdot 3} + \frac{4(\;\;)}{2 \cdot 3}$$

$$= \frac{1 \cdot 3}{2 \cdot 3} + \frac{4 \cdot 2}{2 \cdot 3}$$

$$= \frac{3}{6} + \frac{8}{6}$$

$$= \frac{3 + 8}{6}$$

$$= \frac{11}{6}$$

It should be clear that in finding the LCD for a group of fractions, we must consider the *factors* of the original denominators. **The LCD must contain any factor that appears in any of the original denominators.** If there is overlap among factors, we must use any factor the *most* number of times that it appears in any *single* denominator.

Consider the sum

$$\frac{1}{25x} + \frac{4}{5x^2} = \frac{1}{5 \cdot 5 \cdot x} + \frac{4}{5 \cdot x \cdot x}$$

The only denominator factors are 5 and x. But 5 appears *twice* in the first denominator and x appears *twice* in the second denominator. Each of those factors must therefore appear *twice* in the LCD.

LCD: $5 \cdot 5 \cdot x \cdot x$

$$\frac{1}{25x} + \frac{4}{5x^2} = \frac{1}{5 \cdot 5 \cdot x} + \frac{4}{5 \cdot x \cdot x}$$

$$= \frac{1(x)}{5 \cdot 5 \cdot x \cdot x} + \frac{4(5)}{5 \cdot 5 \cdot x \cdot x}$$

$$= \frac{x}{25x^2} + \frac{20}{25x^2}$$

$$= \frac{x + 20}{25x^2}$$

> The lowest common denominator for a group of fractions is the product of all the different factors appearing in the original denominators, with any single factor appearing the *most* number of times that it appears in any *single* denominator.

The first task in adding and subtracting fractions is finding the lowest common denominator. Obviously, factoring plays an important role, and the first step is to factor the original denominators completely.

You try Example 12.

Example 12. Find $\dfrac{6}{xy} + \dfrac{5}{2x^2}$.

Solution

$\dfrac{6}{xy} + \dfrac{5}{2x^2}$ LCD: $2x^2y$.

$= \dfrac{6(\underline{\quad})}{2x^2y} + \dfrac{5(\underline{\quad})}{2x^2y}$ Multiply each fraction by its building factor.

$= \dfrac{\underline{\quad}}{2x^2y} + \dfrac{\underline{\quad}}{2x^2y}$ Simplify each numerator.

$= \dfrac{\underline{\quad} + \underline{\quad}}{2x^2y}$ Combine numerators.

Check your work on page 408. ▶

Then try Example 13.

Example 13. Find $\dfrac{5}{x+3} + \dfrac{x}{3}$.

Solution: The denominators are in factored form.

$\dfrac{5}{x+3} + \dfrac{x}{3}$ LCD: $3(x+3)$.

$= \dfrac{5(\underline{\quad})}{3(x+3)} + \dfrac{x(\underline{\quad\quad})}{3(x+3)}$

Check your work on page 409. ▶

Example 14. Find $\dfrac{2}{x(x+1)^2} + \dfrac{7}{x^3(x+1)}$.

Solution: Note how many times each factor appears in a single denominator. The LCD is $x^3(x+1)^2$.

$\dfrac{2}{x(x+1)^2} + \dfrac{7}{x^3(x+1)}$

$= \dfrac{2x^2}{x^3(x+1)^2} + \dfrac{7(x+1)}{x^3(x+1)^2}$ Multiply each fraction by its building factor.

$= \dfrac{2x^2}{x^3(x+1)^2} + \dfrac{7x+7}{x^3(x+1)^2}$ Simplify each numerator.

$= \dfrac{2x^2+7x+7}{x^3(x+1)^2}$ Combine numerators.

Example 15. Find $x + \dfrac{2}{x}$.

Solution

$x + \dfrac{2}{x}$

$= \dfrac{x}{1} + \dfrac{2}{x}$ LCD: x.

$= \dfrac{x \cdot x}{x} + \dfrac{2}{x}$

$= \dfrac{x^2}{x} + \dfrac{2}{x}$

$= \dfrac{x^2+2}{x}$

Before doing a few more examples, let's summarize the steps to be used in adding or subtracting fractions with different denominators.

> **Steps for Adding and Subtracting Fractions**
>
> **1.** Write all the denominators in factored form.
> **2.** Choose the LCD as the product of all the different factors of the original denominators, with each factor appearing the most number of times that it appears in any single denominator.
> **3.** Change each of the original fractions to a new fraction having the LCD as its denominator.
> **4.** Simplify each numerator by multiplying its factors.
> **5.** Combine the numerators, keeping the common denominator (in factored form).
> **6.** Reduce if possible.

Example 16. Find $\qquad - \qquad$.

Solution: First we must factor the denominators.

$$\frac{x + 1}{x^2 + 5x + 6} - \frac{2x}{x^2 - 9}$$

$$= \frac{x + 1}{(x + 3)(x + 2)} - \frac{2x}{(x + 3)(x - 3)} \qquad \text{LCD: } (x + 3)(x + 2)(x - 3).$$

$$= \frac{(x + 1)(x - 3)}{(x + 3)(x + 2)(x - 3)} - \frac{2x(x + 2)}{(x + 3)(x + 2)(x - 3)} \qquad \begin{array}{l}\text{Multiply each fraction by}\\\text{its building factors.}\end{array}$$

$$= \frac{x^2 - 2x - 3}{(x + 3)(x + 2)(x - 3)} - \frac{2x^2 + 4x}{(x + 3)(x + 2)(x - 3)} \qquad \text{Simplify each numerator.}$$

$$= \frac{x^2 - 2x - 3 - (2x^2 + 4x)}{(x + 3)(x + 2)(x - 3)} \qquad \text{Combine numerators.}$$

$$= \frac{x^2 - 2x - 3 - 2x^2 - 4x}{(x + 3)(x + 2)(x - 3)} \qquad \text{Remove parentheses.}$$

$$= \frac{-x^2 - 6x - 3}{(x + 3)(x + 2)(x - 3)} \qquad \begin{array}{l}\text{Simplify by combining}\\\text{like terms.}\end{array}$$

Notice that we *never* multiply the factors in the LCD; it is always left in factored form so that we can decide whether the final fraction can be reduced.

You try Example 17.

Example 17. Find $\dfrac{3}{x^2 - 2x - 3} - \dfrac{4}{x^2 + x}$.

Solution: First we factor the denominators.

$$\frac{3}{x^2 - 2x - 3} - \frac{4}{x^2 + x}$$

$$= \frac{3}{(x - 3)(x + 1)} - \frac{4}{x(x + 1)} \qquad \text{LCD: } x(x - 3)(x + 1).$$

$$= \frac{3(\underline{\quad\quad})}{x(x - 3)(x + 1)} - \frac{4(\underline{\quad\quad})}{x(x - 3)(x + 1)} \qquad \begin{array}{l}\text{Multiply each fraction by}\\\text{its building factor.}\end{array}$$

$$= \frac{3x}{x(x - 3)(x + 1)} - \frac{4x - 12}{x(x - 3)(x + 1)} \qquad \text{Simplify each numerator.}$$

$$= \frac{3x - (4x - 12)}{x(x - 3)(x + 1)}$$ Combine numerators.

$$= \frac{\rule{3cm}{0.4pt}}{x(x - 3)(x + 1)}$$ Remove parentheses.

$$= \frac{\rule{3cm}{0.4pt}}{x(x - 3)(x + 1)}$$ Simplify by combining like terms.

Check your work on page 409. ▶

Example 18. Find $\dfrac{3}{2x - 2} + \dfrac{5}{1 - x}$.

Solution: Factor the denominators first.

$$\frac{3}{2x - 2} + \frac{5}{1 - x}$$

$$= \frac{3}{2(x - 1)} + \frac{5}{1 - x}$$

Recalling the relationship between $x - 1$ and $1 - x$, we may rewrite the second denominator

$$\frac{3}{2(x - 1)} + \frac{5}{-1(x - 1)}$$ Rewrite $1 - x$ as $-1(x - 1)$.

$$= \frac{3}{2(x - 1)} + \frac{-5}{x - 1}$$ Let the numerator carry the sign. LCD: $2(x - 1)$.

$$= \frac{3}{2(x - 1)} + \frac{-5(2)}{2(x - 1)}$$ Multiply each fraction by its building factor.

$$= \frac{3}{2(x - 1)} + \frac{-10}{2(x - 1)}$$ Simplify each numerator.

$$= \frac{3 - 10}{2(x - 1)}$$ Combine the numerators.

$$= \frac{-7}{2(x - 1)}$$ Simplify by combining like terms.

⫸ Trial Run

Combine the fractions and reduce if possible.

_____ 1. $\dfrac{5}{x} + \dfrac{3}{4x^3}$ _____ 2. $\dfrac{4}{x + 5} - \dfrac{x}{5}$

_____ 3. $\dfrac{3}{x(x + 2)^2} - \dfrac{1}{x^3(x + 2)}$ _____ 4. $\dfrac{x + 1}{(x - 2)(x - 7)} + \dfrac{4}{x^2 - 49}$

_____ 5. $\dfrac{x}{x^2 + 8x + 16} + \dfrac{5}{x^2 - 16} - \dfrac{2}{3x - 12}$

_____ 6. $\dfrac{4}{3x - 6} + \dfrac{3}{2 - x}$

Answers are on page 409.

▶ Examples You Completed

Example 1. Find $\dfrac{7}{2x} - \dfrac{1}{2x}$.

Solution

$$\frac{7}{2x} - \frac{1}{2x}$$

$$= \frac{7 - 1}{2x}$$

$$= \frac{6}{2x}$$

$$= \frac{2 \cdot 3}{2x}$$

$$= \frac{3}{x}$$

Example 2. Find $\dfrac{x}{x + 3} + \dfrac{2x}{x + 3}$.

Solution

$$\frac{x}{x + 3} + \frac{2x}{x + 3}$$

$$= \frac{x + 2x}{x + 3}$$

$$= \frac{3x}{x + 3}$$

Example 4. Find $\dfrac{2x + 3}{x - 1} + \dfrac{4 - x}{x - 1}$.

Solution

$$\frac{2x + 3}{x - 1} + \frac{4 - x}{x - 1}$$

$$= \frac{2x + 3 + 4 - x}{x - 1}$$

$$= \frac{x + 7}{x - 1}$$

Example 5. Find $\dfrac{4x - 7}{x^2 - 4} - \dfrac{3x - 9}{x^2 - 4}$.

Solution

$$\frac{4x - 7}{x^2 - 4} - \frac{3x - 9}{x^2 - 4}$$

$$= \frac{4x - 7 - (3x - 9)}{x^2 - 4}$$

$$= \frac{4x - 7 - 3x + 9}{x^2 - 4}$$

$$= \frac{x + 2}{x^2 - 4}$$

$$= \frac{\overset{1}{\cancel{x + 2}}}{\cancel{(x + 2)}(x - 2)}$$

$$= \frac{1}{x - 2}$$

Example 11. Find the new fraction.

$$\frac{4}{x + 1} = \frac{?}{3(x + 1)^2}$$

Solution

$$\frac{4}{x + 1} = \frac{?}{3(x + 1)^2}$$

$$= \frac{?}{3(x + 1)(x + 1)}$$

$$= \frac{4 \cdot 3(x + 1)}{3(x + 1)(x + 1)}$$

$$\frac{4}{x + 1} = \frac{12(x + 1)}{3(x + 1)(x + 1)}$$

Example 12. Find $\dfrac{6}{xy} + \dfrac{5}{2x^2}$.

Solutions

$$\frac{6}{xy} + \frac{5}{2x^2}$$

$$= \frac{6}{x \cdot y} + \frac{5}{2 \cdot x \cdot x} \qquad \text{LCD: } 2x^2y.$$

$$= \frac{6(2x)}{2x^2y} + \frac{5(y)}{2x^2y}$$

$$= \frac{12x}{2x^2y} + \frac{5y}{2x^2y}$$

$$= \frac{12x + 5y}{2x^2y}$$

Example 13. Find $\dfrac{5}{x+3} + \dfrac{x}{3}$.

Solution

$$\frac{5}{x+3} + \frac{x}{3} \qquad \text{LCD: } 3(x+3)$$

$$= \frac{5(3)}{3(x+3)} + \frac{x(x+3)}{3(x+3)}$$

$$= \frac{15}{3(x+3)} + \frac{x^2+3x}{3(x+3)}$$

$$= \frac{15+x^2+3x}{3(x+3)}$$

$$= \frac{x^2+3x+15}{3(x+3)}$$

Example 17. Find $\dfrac{3}{x^2-2x-3} - \dfrac{4}{x^2+x}$.

Solution

$$\frac{3}{x^2-2x-3} - \frac{4}{x^2+x}$$

$$= \frac{3}{(x-3)(x+1)} - \frac{4}{x(x+1)} \qquad \text{LCD: } x(x-3)(x+1).$$

$$= \frac{3(x)}{x(x-3)(x+1)} - \frac{4(x-3)}{x(x-3)(x+1)} \qquad \text{Multiply each fraction by its building factor.}$$

$$= \frac{3x}{x(x-3)(x+1)} - \frac{4x-12}{x(x-3)(x+1)} \qquad \text{Simplify each numerator.}$$

$$= \frac{3x-(4x-12)}{x(x-3)(x+1)} \qquad \text{Combine numerators.}$$

$$= \frac{3x-4x+12}{x(x-3)(x+1)} \qquad \text{Remove parentheses.}$$

$$= \frac{-x+12}{x(x-3)(x+1)} \qquad \text{Simplify by combining like terms.}$$

Answers to Trial Runs

page 401   **1.** $\dfrac{4}{x}$ **2.** $\dfrac{7x}{x+7}$ **3.** 2 **4.** $\dfrac{1}{x-4}$ **5.** $\dfrac{1}{x+5}$ **6.** $3x+4$

page 403   **1.** $\dfrac{-26}{91}$ **2.** $\dfrac{15xy^2}{3x^2y^3}$ **3.** $\dfrac{x(x-3)}{(2x+1)(x-3)}$ **4.** $\dfrac{14x^3(x+3)}{2x^2(x+3)(x-3)}$ **5.** $\dfrac{5(x+2)(x+3)}{5(x+3)^2}$ **6.** $\dfrac{(x+3)(x-5)}{(2x+3)(x-5)}$

page 407   **1.** $\dfrac{20x^2+3}{4x^3}$ **2.** $\dfrac{-x^2-5x+20}{5(x+5)}$ **3.** $\dfrac{3x^2-x-2}{x^3(x+2)^2}$ **4.** $\dfrac{x^2+12x-1}{(x-2)(x-7)(x+7)}$ **5.** $\dfrac{x^2-13x+28}{3(x+4)^2(x-4)}$ **6.** $\dfrac{-5}{3(x-2)}$

EXERCISE SET 6.3

Combine the fractions and, if possible, reduce.

_____ 1. $\dfrac{5}{7} + \dfrac{11}{7} - \dfrac{2}{7}$

_____ 2. $\dfrac{8}{3} + \dfrac{11}{3} - \dfrac{4}{3}$

_____ 3. $\dfrac{5}{3x} + \dfrac{1}{3x}$

_____ 4. $\dfrac{5}{4x} + \dfrac{11}{4x}$

_____ 5. $\dfrac{y}{x} + \dfrac{3}{x}$

_____ 6. $\dfrac{a}{b} + \dfrac{5}{b}$

_____ 7. $\dfrac{3x}{2x - 1} - \dfrac{5x}{2x - 1}$

_____ 8. $\dfrac{4x}{3x + 2} - \dfrac{7x}{3x + 2}$

_____ 9. $\dfrac{12x}{4x + 3} + \dfrac{9}{4x + 3}$

_____ 10. $\dfrac{15x}{5x - 2} - \dfrac{6}{5x - 2}$

_____ 11. $\dfrac{x}{x^2 - 81} + \dfrac{9}{x^2 - 81}$

_____ 12. $\dfrac{x}{x^2 - 100} - \dfrac{10}{x^2 - 100}$

_____ 13. $\dfrac{5x - 1}{2x - 3} - \dfrac{x + 5}{2x - 3}$

_____ 14. $\dfrac{10x - 5}{3x - 4} - \dfrac{x + 7}{3x - 4}$

_____ 15. $\dfrac{5x - 9}{x^2 - 16} - \dfrac{4x - 5}{x^2 - 16}$

_____ 16. $\dfrac{9x - 7}{x^2 - 9} - \dfrac{8x - 10}{x^2 - 9}$

_____ 17. $\dfrac{2a^2}{a^2 + 7a} + \dfrac{11a}{a^2 + 7a} - \dfrac{21}{a^2 + 7a}$

_____ 18. $\dfrac{2a^2}{a^2 + 3a} + \dfrac{a}{a^2 + 3a} - \dfrac{15}{a^2 + 3a}$

_____ 19. $\dfrac{2x^2 - 3x}{2x^2 - 9x + 4} + \dfrac{4x - 1}{2x^2 - 9x + 4}$

_____ 20. $\dfrac{x^2 - 11x}{x^2 - 3x - 4} + \dfrac{x + 24}{x^2 - 3x - 4}$

_____ 21. $\dfrac{(3x - 5)(x - 1)}{2x - 3} - \dfrac{x(x + 3)}{2x - 3} + \dfrac{7}{2x - 3}$

_____ 22. $\dfrac{(2x + 3)(x + 7)}{x + 5} + \dfrac{x(3x + 2)}{x + 5} - \dfrac{51}{x + 5}$

_____ 23. $\dfrac{3}{x} - \dfrac{5}{x^2} - \dfrac{2}{5x^2}$

_____ 24. $\dfrac{5}{x} + \dfrac{6}{x^2} - \dfrac{7}{3x^2}$

_____ 25. $\dfrac{2}{x} - \dfrac{3}{y} + \dfrac{2}{5x} - \dfrac{1}{3y}$

_____ 26. $\dfrac{3}{x} - \dfrac{2}{y} + \dfrac{1}{4x} - \dfrac{4}{3y}$

_____ 27. $\dfrac{3}{x} - \dfrac{2}{x + 1}$

_____ 28. $\dfrac{2}{x} + \dfrac{3}{x - 2}$

_____ 29. $\dfrac{x - 5}{x^2 - 3x} + \dfrac{5}{2x^2}$

_____ 30. $\dfrac{x + 1}{x^2 + 5x} - \dfrac{4}{3x^2}$

_____ **31.** $\dfrac{7}{x + 2} - \dfrac{5}{(x + 2)^2}$

_____ **32.** $\dfrac{11}{x - 3} - \dfrac{4}{(x - 3)^2}$

_____ **33.** $\dfrac{3}{x + 4} - \dfrac{5}{x - 4}$

_____ **34.** $\dfrac{4}{x + 5} - \dfrac{3}{x - 5}$

_____ **35.** $\dfrac{4}{3x + 3} - \dfrac{5}{x + 1}$

_____ **36.** $\dfrac{7}{4x - 4} - \dfrac{4}{x - 1}$

_____ **37.** $\dfrac{x}{x^2 - 25} - \dfrac{1}{x^2 - 5x}$

_____ **38.** $\dfrac{x}{x^2 - 49} - \dfrac{3}{x^2 - 7x}$

_____ **39.** $\dfrac{1}{x^3 - 4x^2} - \dfrac{16}{x^5 - 4x^4}$

_____ **40.** $\dfrac{1}{x^3 - 5x^2} - \dfrac{25}{x^5 - 5x^4}$

_____ **41.** $\dfrac{x + 6}{x^2 + 2x - 8} + \dfrac{3}{x + 4}$

_____ **42.** $\dfrac{x - 8}{x^2 - 11x + 18} - \dfrac{4}{x - 9}$

_____ **43.** $\dfrac{2x - 6}{2x^2 - 7x + 3} - \dfrac{5}{2x^2 - x}$

_____ **44.** $\dfrac{3x - 21}{3x^2 - 22x + 7} - \dfrac{4}{3x^2 - x}$

_____ **45.** $\dfrac{5}{x + 3} + \dfrac{1}{x - 2} - \dfrac{3x}{x^2 + x - 6}$

_____ **46.** $\dfrac{4}{3x + 1} - \dfrac{3}{x - 1} + \dfrac{2x}{3x^2 - 2x - 1}$

_____ **47.** $\dfrac{4}{x - 2} - \dfrac{3}{x + 5} - \dfrac{28}{x^2 + 3x - 10}$

_____ **48.** $\dfrac{5}{x - 1} - \dfrac{2}{x + 5} - \dfrac{12}{x^2 + 4x - 5}$

_____ **49.** $\dfrac{2x + 1}{x^2 + x - 20} + \dfrac{3x + 8}{x^2 + 3x - 10} - \dfrac{3x - 10}{x^2 - 6x + 8}$

_____ **50.** $\dfrac{x - 4}{2x^2 - 11x + 5} - \dfrac{2x - 1}{x^2 - 8x + 15} + \dfrac{x + 3}{2x^2 - 7x + 3}$

_____ **51.** $\dfrac{1}{x - 2} - \dfrac{6x}{x^3 - 8}$

_____ **52.** $\dfrac{1}{x - 3} - \dfrac{9x}{x^3 - 27}$

Stretching the Topics _____

Perform the indicated operations.

_____ **1.** $\dfrac{3x^a}{5y^{2b}} - \dfrac{4}{10x^a y^b} + \dfrac{y^b}{2x^{2a}}$

_____ **2.** $\dfrac{x^a}{x^a + 8} - \dfrac{8 - 7x^a}{x^{2a} + 8x^a}$

_____ **3.** $\left(x - 1 + \dfrac{1}{x + 1} \right)\left(\dfrac{1}{x} - \dfrac{1}{x^2} \right)$

Check your answers in the back of your book.

If you can complete **Checkup 6.3,** you are ready to go on to Section 6.4.

✔ CHECKUP 6.3

Combine the fractions and, if possible, reduce.

_____ 1. $\dfrac{7}{3x^2} - \dfrac{4}{3x^2}$

_____ 2. $\dfrac{4x}{5x-3} + \dfrac{6x}{5x-3}$

_____ 3. $\dfrac{3x+5}{x^2-4} - \dfrac{4x+3}{x^2-4}$

_____ 4. $\dfrac{3x(x+3)}{12x^2-17x+6} + \dfrac{4x-10}{12x^2-17x+6}$

_____ 5. $\dfrac{3}{2xy^2} + \dfrac{5}{3x^2y}$

_____ 6. $\dfrac{4}{x-3} + \dfrac{7}{x}$

_____ 7. $\dfrac{3}{x-2} + \dfrac{4}{x+2}$

_____ 8. $\dfrac{3}{2x^2(x-1)} - \dfrac{1}{x(x-1)^2}$

_____ 9. $\dfrac{x+8}{x^2-2x-63} - \dfrac{1}{x+7}$

_____ 10. $\dfrac{4}{x-6} - \dfrac{5}{x+2} + \dfrac{2(x-14)}{x^2-4x-12}$

Check your answers in the back of your book.

If You Missed Problem:	You Should Review Examples:
1	1
2	2
3	3
4	4–6
5	12
6, 7	13
8	14
9, 10	16, 17

6.4 Simplifying Complex Fractions

A **complex fraction** is a fraction that contains fractions in the numerator and/or denominator. Success with complex fractions depends on skills already learned in adding, subtracting, multiplying, and dividing fractions. As before, we assume that the variables have been restricted to avoid division by zero.

To simplify a complex fraction such as

$$\frac{\dfrac{x^2}{y^3}}{\dfrac{3x}{y^8}}$$

we must realize that this is, after all, a *division* problem.

Example 1. Find $\dfrac{\dfrac{x^2}{y^3}}{\dfrac{3x}{y^8}}$.

Solution

$$\frac{\dfrac{x^2}{y^3}}{\dfrac{3x}{y^8}} = \frac{x^2}{y^3} \div \frac{3x}{y^8}$$

$$= \frac{x^2}{y^3} \cdot \frac{y^8}{3x}$$

$$= \frac{\overset{x}{\cancel{x^2}}\ \overset{y^5}{\cancel{y^8}}}{\underset{1}{\cancel{y^3}}\ \underset{1}{\cancel{3x}}}$$

$$= \frac{xy^5}{3}$$

Example 2. Find $\dfrac{\dfrac{3x^2 - 12}{x^3}}{\dfrac{5x + 10}{x^2 + x}}$.

Solution

$$\frac{\dfrac{3x^2 - 12}{x^3}}{\dfrac{5x + 10}{x^2 + x}} = \frac{3x^2 - 12}{x^3} \div \frac{5x + 10}{x^2 + x}$$

$$= \frac{3(x^2 - 4)}{x^3} \cdot \frac{x^2 + x}{5x + 10}$$

$$= \frac{3\cancel{(x + 2)}(x - 2)}{\underset{x^2}{\cancel{x^3}}} \cdot \frac{\overset{1}{\cancel{x}}(x + 1)}{5\underset{1}{\cancel{(x + 2)}}}$$

$$= \frac{3(x - 2)(x + 1)}{5x^2}$$

To simplify a complex fraction, we rewrite it as a division problem, and divide.

IIII➡ Trial Run

Simplify the complex fractions.

_____ **1.** $\dfrac{\dfrac{8a^3}{3b^2}}{\dfrac{16a^5}{15b^8}}$

_____ **2.** $\dfrac{\dfrac{1}{x^2 - 9}}{\dfrac{1}{2x + 6}}$

_____ **3.** $\dfrac{\dfrac{x^3 - 8}{x^2 - 2x}}{3}$

_____ **4.** $\dfrac{\dfrac{2x^2 - 32}{x^5}}{\dfrac{3x - 12}{x^3 + 4x^2}}$

Answers are on page 421.

Sometimes a complex fraction contains a numerator and/or denominator that is a sum or difference of fractional expressions. We must develop a method for simplifying such complex fractions.

Example 3. Simplify $\dfrac{x - \dfrac{1}{x}}{x - 1}$.

Solution

$$\frac{x - \dfrac{1}{x}}{x - 1} = \left[x - \frac{1}{x} \right] \div [x - 1]$$

In order to use the usual method for division, each of the bracketed quantities must be a *single fraction*.

$$\left[x - \frac{1}{x} \right] \div [x - 1] = \left[\frac{x^2}{x} - \frac{1}{x} \right] \div \left[\frac{x - 1}{1} \right]$$

$$= \frac{x^2 - 1}{x} \div \frac{x - 1}{1}$$

$$= \frac{x^2 - 1}{x} \cdot \frac{1}{x - 1}$$

$$= \frac{(x + 1)\cancel{(x - 1)}}{x} \cdot \frac{1}{\cancel{x - 1}}$$

$$= \frac{x + 1}{x}$$

Simplifying Complex Fractions (Method 1)

1. Rewrite the complex fraction as a division problem.
2. Express each part of the division problem as a single fraction.
3. Proceed with the method for dividing fractions.

It is important to stress the fact that each part of the division problem *must* be expressed as a single fraction before we proceed with the actual division. In other words, before we can invert the divisor and multiply, that divisor *must* be a single fraction.

Now you complete Example 4.

Example 4. Simplify $\dfrac{2x + \dfrac{5x - 3}{x}}{\dfrac{x}{3} - \dfrac{3}{x}}$.

Solution

$$\frac{2x + \dfrac{5x - 3}{x}}{\dfrac{x}{3} - \dfrac{3}{x}} = \left[\frac{2x}{1} + \frac{5x - 3}{x}\right] \div \left[\frac{x}{3} - \frac{3}{x}\right]$$

$$= \left[\frac{2x \cdot x}{x} + \frac{5x - 3}{x}\right] \div \left[\frac{x \cdot x}{3x} - \frac{3 \cdot 3}{3x}\right]$$

$$= \left[\frac{2x^2}{x} + \frac{5x - 3}{x}\right] \div \left[\frac{x^2}{3x} - \frac{9}{3x}\right]$$

$$= \frac{2x^2 + 5x - 3}{x} \div \frac{x^2 - 9}{3x}$$

$$= \frac{2x^2 + 5x - 3}{x} \cdot \frac{3x}{x^2 - 9}$$

Check your work on page 421. ▶

There is a second approach that some prefer for simplifying complex fractions. It is especially useful if the same denominators appear in the numerator and denominator of the complex fraction. Example 4 is a good candidate for the second approach. In this method, the numerator and denominator are *both* multiplied by the quantity that is the LCD for *all* the fractions in the numerator and denominator. Let's rework Example 4 by the second method.

$$\frac{2x + \dfrac{5x - 3}{x}}{\dfrac{x}{3} - \dfrac{3}{x}}$$

Choose the LCD for *both* numerator and denominator. LCD: $3x$.

$$= \frac{3x\left[2x + \dfrac{5x - 3}{x}\right]}{3x\left[\dfrac{x}{3} - \dfrac{3}{x}\right]}$$

Multiply numerator and denominator by $3x$.

$$= \frac{3x(2x) + \dfrac{3\overset{1}{\cancel{x}}}{1}\left(\dfrac{5x - 3}{\cancel{x}}\right)}{\dfrac{\overset{1}{\cancel{3}}x}{1}\left(\dfrac{x}{\underset{1}{\cancel{3}}}\right) - \dfrac{3\overset{1}{\cancel{x}}}{1}\left(\dfrac{3}{\underset{1}{\cancel{x}}}\right)}$$

Use the distributive property.

$$= \frac{6x^2 + 15x - 9}{x^2 - 9}$$

Remove parentheses.

$$= \frac{3(2x^2 + 5x - 3)}{(x + 3)(x - 3)}$$

Factor numerator and denominator.

$$= \frac{3(2x - 1)\overset{1}{\cancel{(x + 3)}}}{\underset{1}{\cancel{(x + 3)}}(x - 3)}$$

Factor numerator completely.

$$= \frac{3(2x - 1)}{x - 3)}$$

Reduce the fraction.

Notice that this result is the same as that obtained earlier.

Simplifying Complex Fractions (Method 2)

1. Choose the LCD for *both* the numerator and the denominator.
2. Multiply both numerator and denominator by that LCD.
3. Simplify the numerator and denominator, and reduce if possible.

You may use either method to simplify a complex fraction. Some students always choose the second method, but for fractions containing several different denominators, the first method is actually less complicated. Consider Example 5.

Example 5. Simplify $\dfrac{\dfrac{3}{x} + \dfrac{2}{x - 1}}{x + \dfrac{4x^2 - 5x}{x + 2}}$.

Solution

$$\frac{\dfrac{3}{x} + \dfrac{2}{x - 1}}{x + \dfrac{4x^2 - 5x}{x + 2}} = \left[\frac{3}{x} + \frac{2}{x - 1} \right] \div \left[\frac{x}{1} + \frac{4x^2 - 5x}{x + 2} \right]$$

$$= \left[\frac{3(x - 1)}{x(x - 1)} + \frac{2 \cdot x}{x(x - 1)} \right] \div \left[\frac{x(x + 2)}{x + 2} + \frac{4x^2 - 5x}{x + 2} \right]$$

$$= \left[\frac{3x - 3 + 2x}{x(x + 1)} \right] \div \left[\frac{x^2 + 2x + 4x^2 - 5x}{x + 2} \right]$$

$$= \left[\frac{5x - 3}{x(x - 1)} \right] \div \left[\frac{5x^2 - 3x}{x + 2} \right]$$

$$= \frac{5x - 3}{x(x - 1)} \cdot \frac{x + 2}{5x^2 - 3x}$$

$$= \frac{\overset{1}{\cancel{5x - 3}}}{x(x - 1)} \cdot \frac{x + 2}{x\cancel{(5x - 3)}}_{1}$$

$$= \frac{x + 2}{x^2(x - 1)}$$

⮚ Trial Run

Simplify the complex fractions.

_____ **1.** $\dfrac{x - \dfrac{4}{x}}{x + 2}$

_____ **2.** $\dfrac{3 - \dfrac{1}{x}}{9 - \dfrac{1}{x^2}}$

_____ **3.** $\dfrac{\dfrac{1}{x^2} - \dfrac{1}{16}}{\dfrac{1}{x} + \dfrac{1}{4}}$

_____ **4.** $\dfrac{\dfrac{-2}{x + 5} + \dfrac{1}{2}}{\dfrac{1}{x} + \dfrac{3}{x + 4}}$

Answers are on page 422.

A complex fraction can appear in disguise when it contains negative exponents. Recall the definition of a negative exponent:

$$a^{-n} = \frac{1}{a^n}$$

Example 6. Simplify $\dfrac{x + 2x^{-1}}{x - 4x^{-3}}$.

Solution

$$\dfrac{x + 2x^{-1}}{x - 4x^{-3}}$$

$$= \dfrac{x + 2 \cdot \dfrac{1}{x}}{x - 4 \cdot \dfrac{1}{x^3}}$$

$$= \dfrac{x + \dfrac{2}{x}}{x - \dfrac{4}{x^3}}$$

Let's multiply numerator and denominator by the LCD for both. LCD: x^3.

$$\dfrac{x^3 \left[x + \dfrac{2}{x} \right]}{x^3 \left[x - \dfrac{4}{x^3} \right]}$$

$$= \dfrac{x^3 \cdot x + \dfrac{x^3}{1} \cdot \dfrac{2}{x}}{x^3 \cdot x - \dfrac{x^3}{1} \cdot \dfrac{4}{x^3}}$$

$$= \dfrac{x^4 + 2x^2}{x^4 - 4}$$

$$= \dfrac{x^2 \cancel{(x^2 + 2)}^{\,1}}{(x^2 - 2)\cancel{(x^2 + 2)}_{\,1}}$$

$$= \dfrac{x^2}{x^2 - 2}$$

Now complete Example 7.

Example 7. Simplify $\dfrac{xy^{-1}}{y^{-1} + x}$.

Solution

$$\dfrac{xy^{-1}}{y^{-1} + x}$$

$$= \dfrac{x \cdot \dfrac{1}{y}}{\dfrac{1}{y} + x}$$

$$= \dfrac{\dfrac{x}{y}}{\dfrac{1}{y} + x}$$

Multiply numerator and denominator by the LCD for both. LCD: y.

$$\dfrac{y \left[\dfrac{x}{y} \right]}{y \left[\dfrac{1}{y} + x \right]}$$

$$=$$

Check your work on page 421. ▶

⫸ Trial Run ━━━━━━━━━━━━━━━━━━━━━━━━━━━━━━━━━━━━

Simplify.

_____ 1. $\dfrac{1 + 3x^{-1}}{1 - 9x^{-2}}$

_____ 2. $\dfrac{1 + 2x^{-1} - 3x^{-2}}{1 + 5x^{-1} + 6x^{-2}}$

_____ 3. $\dfrac{x^{-1} - y^{-1}}{x^{-1} + y^{-1}}$

_____ 4. $\dfrac{x + 3x^{-1}}{x - 9x^{-3}}$

Answers are on page 422.

▶ Examples You Completed

Example 4. Simplify $\dfrac{2x + \dfrac{5x - 3}{x}}{\dfrac{x}{3} - \dfrac{3}{x}}$.

Solution

$$\frac{2x + \dfrac{5x - 3}{x}}{\dfrac{x}{3} - \dfrac{3}{x}} = \left[\frac{2x}{1} + \frac{5x - 3}{x}\right] \div \left[\frac{x}{3} - \frac{3}{x}\right]$$

$$= \left[\frac{2x \cdot x}{x} + \frac{5x - 3}{x}\right] \div \left[\frac{x \cdot x}{3x} - \frac{3 \cdot 3}{3x}\right]$$

$$= \left[\frac{2x^2}{x} + \frac{5x - 3}{x}\right] \div \left[\frac{x^2}{3x} - \frac{9}{3x}\right]$$

$$= \frac{2x^2 + 5x - 3}{x} \div \frac{x^2 - 9}{3x}$$

$$= \frac{2x^2 + 5x - 3}{x} \cdot \frac{3x}{x^2 - 9}$$

$$= \frac{(2x - 1)\cancel{(x + 3)}}{\cancel{x}} \cdot \frac{\overset{1}{\cancel{3x}}}{\cancel{(x + 3)}(x - 3)}$$

$$= \frac{3(2x - 1)}{x - 3}$$

Example 7. Simplify $\dfrac{xy^{-1}}{y^{-1} + x}$.

Solution

$$\frac{xy^{-1}}{y^{-1} + x} = \frac{x \cdot \dfrac{1}{y}}{\dfrac{1}{y} + x}$$

$$= \frac{\dfrac{x}{y}}{\dfrac{1}{y} + x}$$

$$= \frac{y\left[\dfrac{x}{y}\right]}{y\left[\dfrac{1}{y} + x\right]}$$

$$= \frac{x}{1 + xy}$$

Answers to Trial Runs

page 415   **1.** $\dfrac{5b^6}{2a^2}$ **2.** $\dfrac{2}{x - 3}$ **3.** $\dfrac{3(x + 2x + 4)}{x}$ **4.** $\dfrac{2(x + 4)^2}{3x^3}$

page 419 1. $\dfrac{x-2}{x}$ 2. $\dfrac{x}{3x+1}$ 3. $\dfrac{4-x}{4x}$ 4. $\dfrac{x(x+4)}{8(x+5)}$

page 420 1. $\dfrac{x}{x-3}$ 2. $\dfrac{x-1}{x+2}$ 3. $\dfrac{y-x}{y+x}$ 4. $\dfrac{x^2}{x^2-3}$

EXERCISE SET 6.4

Simplify the fractions.

_____ 1. $\dfrac{\dfrac{3}{8}}{\dfrac{15}{4}}$

_____ 2. $\dfrac{\dfrac{5}{9}}{\dfrac{10}{3}}$

_____ 3. $\dfrac{\dfrac{x^7}{y^2}}{\dfrac{3x^5}{y}}$

_____ 4. $\dfrac{\dfrac{x^5}{y^3}}{\dfrac{4x^2}{y^2}}$

_____ 5. $\dfrac{\dfrac{x+y}{3x^2}}{\dfrac{x^2-y^2}{15x^4}}$

_____ 6. $\dfrac{\dfrac{2x-y}{5y^2}}{\dfrac{4x^2-y^2}{20y^5}}$

_____ 7. $\dfrac{\dfrac{1}{x^2-16}}{\dfrac{5}{3x-12}}$

_____ 8. $\dfrac{\dfrac{1}{x^2-49}}{\dfrac{2}{4x+28}}$

_____ 9. $\dfrac{\dfrac{x^2-2x}{x^2-4}}{x^\cdot}$

_____ 10. $\dfrac{\dfrac{x^2-5x}{x^2-25}}{x^3}$

_____ 11. $\dfrac{\dfrac{x^2-3x}{2}}{x^3-27}$

_____ 12. $\dfrac{\dfrac{x^2+5x}{4}}{x^3+125}$

_____ 13. $\dfrac{\dfrac{9x^2-16}{7x^2}}{\dfrac{3x-4}{14x^5}}$

_____ 14. $\dfrac{\dfrac{4x^2-9}{3x^2}}{\dfrac{2x+3}{18x^3}}$

_____ 15. $\dfrac{3+\dfrac{1}{5}}{1-\dfrac{3}{7}}$

_____ 16. $\dfrac{6+\dfrac{2}{3}}{1-\dfrac{5}{9}}$

_____ 17. $\dfrac{x-\dfrac{9}{x}}{x+3}$

_____ 18. $\dfrac{x-\dfrac{25}{x}}{x-5}$

_____ 19. $\dfrac{\dfrac{1}{x}}{1+\dfrac{1}{x}}$

_____ 20. $\dfrac{1-\dfrac{1}{x}}{\dfrac{1}{x}}$

_____ 21. $\dfrac{4 - \dfrac{1}{x}}{16 - \dfrac{1}{x^2}}$

_____ 22. $\dfrac{7 + \dfrac{1}{x}}{49 - \dfrac{1}{x^2}}$

_____ 23. $\dfrac{8 - \dfrac{1}{a^3}}{2 - \dfrac{1}{a}}$

_____ 24. $\dfrac{64 + \dfrac{1}{a^3}}{4 + \dfrac{1}{a}}$

_____ 25. $\dfrac{\dfrac{1}{3} - \dfrac{1}{4}}{\dfrac{1}{2} + \dfrac{1}{8}}$

_____ 26. $\dfrac{\dfrac{4}{5} - \dfrac{1}{5}}{\dfrac{7}{10} - \dfrac{1}{2}}$

_____ 27. $\dfrac{\dfrac{1}{x^2} - \dfrac{1}{81}}{\dfrac{1}{x} + \dfrac{1}{9}}$

_____ 28. $\dfrac{\dfrac{1}{x^2} - \dfrac{1}{100}}{\dfrac{1}{x} - \dfrac{1}{10}}$

_____ 29. $\dfrac{10 - \dfrac{x + 3}{x^2}}{\dfrac{x}{3} - \dfrac{1}{5}}$

_____ 30. $\dfrac{4 - \dfrac{17x + 15}{x^2}}{\dfrac{x}{3} + \dfrac{1}{4}}$

_____ 31. $\dfrac{x + \dfrac{x - 30}{x}}{\dfrac{x}{5} - \dfrac{5}{x}}$

_____ 32. $\dfrac{x + \dfrac{x - 72}{x}}{\dfrac{x}{8} - \dfrac{8}{x}}$

_____ 33. $\dfrac{x + \dfrac{21}{x + 10}}{x + 6 - \dfrac{12}{x + 7}}$

_____ 34. $\dfrac{x + \dfrac{12}{x + 7}}{x + 8 + \dfrac{4}{x + 3}}$

_____ 35. $\dfrac{\dfrac{2x}{x^2 - 4} + \dfrac{3}{x - 2}}{\dfrac{5}{x + 2} + \dfrac{4}{x - 2}}$

_____ 36. $\dfrac{\dfrac{3x}{x^2 - 25} + \dfrac{1}{x - 5}}{\dfrac{4}{x - 5} + \dfrac{3}{x + 5}}$

_____ 37. $\dfrac{\dfrac{a}{a + 4} - \dfrac{3}{a + 2}}{\dfrac{2a + 7}{a^2 + 6a + 8} - \dfrac{1}{a + 2}}$

_____ 38. $\dfrac{\dfrac{a}{a + 3} - \dfrac{2}{a + 1}}{\dfrac{3a + 5}{a^2 + 4a + 3} - \dfrac{1}{a + 3}}$

_____ 39. $\dfrac{\dfrac{a + b}{a - b} - \dfrac{a - b}{a + b}}{\dfrac{1}{a - b} + \dfrac{1}{a + b}}$

_____ 40. $\dfrac{\dfrac{a - b}{a + b} + \dfrac{a + b}{a - b}}{\dfrac{a}{a + b} + \dfrac{b}{a - b}}$

———— 41. $\dfrac{\dfrac{3}{x-2} - \dfrac{2}{x-1}}{\dfrac{5}{x-1} - \dfrac{4}{x+2}}$

———— 42. $\dfrac{\dfrac{4}{x-3} - \dfrac{1}{x+4}}{\dfrac{2}{x+4} - \dfrac{3}{x+3}}$

———— 43. $\dfrac{1 + 4x^{-1}}{1 - 16x^{-2}}$

———— 44. $\dfrac{1 + 5x^{-1}}{1 - 25x^{-2}}$

———— 45. $\dfrac{x^{-3} + 8}{x^{-1} + 2}$

———— 46. $\dfrac{x^{-3} + 125}{x^{-1} + 5}$

———— 47. $\dfrac{2 + 9x^{-1} - 5x^{-2}}{1 + 3x^{-1} - 10x^{-2}}$

———— 48. $\dfrac{3 - 2x^{-1} - x^{-2}}{1 + 2x^{-1} - 3x^{-2}}$

———— 49. $\dfrac{x^{-1} - 2y^{-1}}{x^{-1} + 2y^{-1}}$

———— 50. $\dfrac{3x^{-1} + y^{-1}}{3x^{-1} - y^{-1}}$

———— 51. $\dfrac{4x^2y^{-3}}{y^{-3} + 2x^2}$

———— 52. $\dfrac{5xy^{-1}}{y^{-1} + 5x}$

———— 53. $\dfrac{x + 4x^{-1}}{x - 16x^{-3}}$

———— 54. $\dfrac{x + 5x^{-1}}{x - 25x^{-3}}$

☆ Stretching the Topics

———— 1. $\dfrac{\dfrac{x+1}{x-4} + \dfrac{x-5}{x+5}}{\dfrac{x-1}{x+5} - \dfrac{x-2}{x+3}}$

———— 2. $\dfrac{\dfrac{x+d-1}{x+d} - \dfrac{x-1}{x}}{d}$

———— 3. $\dfrac{\dfrac{12x^{2n}}{x^{2n}-25} - \dfrac{3}{x^n-5}}{\dfrac{4}{x^n+5} + \dfrac{15}{x^{2n}-25}}$

Check your answers in the back of your book.

If you can simplify the fractions in **Checkup 6.4,** you are ready to do Section 6.5.

✔ **CHECKUP 6.4**

Simplify the fractions.

_____ 1. $\dfrac{\dfrac{x^{10}}{y^4}}{\dfrac{5x^3}{y^7}}$

_____ 2. $\dfrac{\dfrac{2x-3}{3x^2}}{\dfrac{4x^2-9}{15x^5}}$

_____ 3. $\dfrac{\dfrac{x^2+11x}{x^2-121}}{x^2}$

_____ 4. $\dfrac{x-\dfrac{81}{x}}{x+9}$

_____ 5. $\dfrac{8-\dfrac{27}{a^3}}{2-\dfrac{3}{a}}$

_____ 6. $\dfrac{\dfrac{1}{x^2}-\dfrac{1}{49}}{\dfrac{1}{x}-\dfrac{1}{7}}$

_____ 7. $\dfrac{x-\dfrac{35}{x+2}}{\dfrac{x}{7}-\dfrac{7}{x}}$

_____ 8. $\dfrac{\dfrac{6}{a+3}-\dfrac{4}{a-4}}{\dfrac{2}{a-4}+\dfrac{5}{a+3}}$

_____ 9. $\dfrac{12+11x^{-1}+2x^{-2}}{3x^{-1}-13x^{-2}-10x^{-3}}$

_____ 10. $\dfrac{x^{-1}+3y^{-1}}{x^{-2}-9y^{-2}}$

Check your answers in the back of your book.

If You Missed Problems:	You Should Review Examples:
1	1
2, 3	2
4	3
5, 6	4
7, 8	5
9, 10	6, 7

6.5 Solving Fractional Equations and Inequalities

To solve fractional equations and inequalities, we recall previously learned skills.

Solving Fractional Equations

To solve a fractional equation, we try to write it as a first degree or second-degree equation. To find solutions to equations such as

$$\frac{x}{3} + \frac{2x+1}{5} = \frac{1}{2} \quad \text{or} \quad \frac{3}{x+1} - \frac{5}{x} = \frac{19}{x^2+x}$$

we must first find a way to eliminate the denominators.

> To eliminate the denominators in a fractional equation, we may multiply both sides of the equation by the LCD for *all* the fractions in the equation.

Of course, we may multiply both sides of an equation only by a nonzero quantity, so we must *state restrictions* after selecting the LCD.

Example 1. Solve $\dfrac{2}{x} - \dfrac{5}{4} = \dfrac{9}{2x}$.

Solution

$$\frac{2}{x} - \frac{5}{4} = \frac{9}{2x}$$

$$\frac{2}{x} - \frac{5}{2 \cdot 2} = \frac{9}{2 \cdot x} \qquad \text{Factor all denominators.}$$

The LCD is $2 \cdot 2 \cdot x = 4x$, and $x \neq 0$ Choose LCD. State restrictions.

$$\frac{4x}{1}\left[\frac{2}{x}\right] - \frac{4x}{1}\left[\frac{5}{4}\right] = \frac{4x}{1}\left[\frac{9}{2x}\right] \qquad \text{Multiply both sides by LCD: } 4x.$$

$$\frac{\overset{1}{4\cancel{x}}}{1}\left[\frac{2}{\cancel{x}}\right] - \frac{\overset{1}{\cancel{4}x}}{1}\left[\frac{5}{\cancel{4}}\right] = \frac{\overset{2}{\cancel{4}x}}{1}\left[\frac{9}{2\cancel{x}}\right] \qquad \text{Reduce before finding products.}$$

$$8 - 5x = 18 \qquad \text{Find products.}$$

$$-5x = 10 \qquad \text{Isolate } x\text{-term.}$$

$$x = -2 \qquad \text{Isolate } x.$$

Since the restriction was $x \neq 0$, the solution $x = -2$ is acceptable.

Example 2. Solve $\dfrac{y}{2} = \dfrac{x}{x-3}$ for x.

Solution

$$\frac{y}{2} = \frac{x}{x-3} \qquad \begin{array}{l} \text{LCD: } 2(x-3) \\ x \neq 3 \end{array} \qquad \text{Choose LCD. State restrictions.}$$

$$\frac{\overset{1}{\cancel{2}}(x-3)}{1}\left[\frac{y}{\cancel{2}}\right] = \frac{2(\overset{1}{\cancel{x-3}})}{1}\left[\frac{x}{\cancel{x-3}}\right]$$ Multiply both sides by LCD: $2(x-3)$.

$$(x-3)y = 2x$$ Find products.

$$xy - 3y = 2x$$ Remove parentheses.

$$xy - 2x = 3y$$ Isolate x-terms.

$$x(y-2) = 3y$$ Factor x from terms on left.

$$x = \frac{3y}{y-2} \qquad y \neq 2, x \neq 3$$ Divide by $y-2$ to isolate x.

You complete Example 3.

Example 3. Solve $\dfrac{3}{x+1} - \dfrac{5}{x} = \dfrac{19}{x^2+x}$.

Solution

$$\frac{3}{x+1} - \frac{5}{x} = \frac{19}{x^2+x}$$

$$\frac{3}{x+1} - \frac{5}{x} = \frac{19}{x(x+1)}$$ LCD: $x(x+1)$ and $x \neq$ _____ , $x \neq$ _____ .

$$\frac{x(\overset{1}{\cancel{x+1}})}{1}\left[\frac{3}{\cancel{x+1}}\right] - \frac{\overset{1}{\cancel{x}}(x+1)}{1}\left[\frac{5}{\cancel{x}}\right] = \frac{\overset{1}{\cancel{x}}\overset{1}{(\cancel{x+1})}}{1}\left[\frac{19}{\underset{1}{\cancel{x}}\underset{1}{(\cancel{x+1})}}\right]$$

$$3x - 5(x+1) = 19$$

Since the restrictions were $x \neq$ _____ , $x \neq$ _____ , the solution $x =$ _____ is acceptable.

Check your work on page 435. ▶

Example 4. Solve $\dfrac{x+1}{x-5} = \dfrac{20}{x^2-25} + \dfrac{2}{x+5}$.

Solution

$$\frac{x+1}{x-5} = \frac{20}{x^2-25} + \frac{2}{x+5}$$

$$\frac{x+1}{x-5} = \frac{20}{(x+5)(x-5)} + \frac{2}{x+5}$$ LCD: $(x+5)(x-5)$ and $x \neq 5$, $x \neq -5$.

$$\frac{(x+5)(\overset{1}{\cancel{x-5}})}{1}\left[\frac{x+1}{\cancel{x-5}}\right] = \frac{\overset{1}{\cancel{x+5}}\overset{1}{(\cancel{x-5})}}{1}\left[\frac{20}{\underset{1}{\cancel{x+5}}\underset{1}{(\cancel{x-5})}}\right]$$

$$+ \frac{\overset{1}{(\cancel{x+5})}(x-5)}{1}\left[\frac{2}{\cancel{x+5}}\right]$$

$$(x + 5)(x + 1) = 20 + 2(x - 5)$$
$$x^2 + 6x + 5 = 20 + 2x - 10$$
$$x^2 + 6x + 5 = 10 + 2x$$

We recognize that this is a second-degree (quadratic) equation that must be written with zero on one side of the equation.

$$x^2 + 4x - 5 = 0$$
$$(x + 5)(x - 1) = 0$$
$$x + 5 = 0 \qquad \text{or} \qquad x - 1 = 0$$
$$x = -5 \qquad \text{or} \qquad x = 1$$

Since the restrictions were $x \neq 5$, $x \neq -5$, one of our solutions is unacceptable. The solution $x = -5$ must be rejected because it will *not* satisfy the original equation. The only solution is

$$x = 1$$

Example 4 should help you realize the importance of stating restrictions on the domain of the variable as soon as you have chosen your LCD. Having done so, you can easily see whether your solutions are acceptable. Solutions that are obtained correctly but do not satisfy the original equation are called **extraneous solutions.**

⟫ Trial Run

Solve.

_____ 1. $\dfrac{x}{3} + \dfrac{5}{6} = \dfrac{9x}{2}$

_____ 2. $\dfrac{3x - 2}{x^2} - \dfrac{4}{3x} = 0$

_____ 3. $\dfrac{5}{2x} = \dfrac{2}{x + 7}$

_____ 4. $\dfrac{x}{x - 3} + \dfrac{x - 6}{x - 3} = 1$

_____ 5. $\dfrac{x - 5}{x - 3} = \dfrac{18}{x^2 - 9} + \dfrac{5}{x + 3}$

_____ 6. $\dfrac{x + y}{x} = \dfrac{2y}{5}$ for x.

Answers are on page 436.

Solving Fractional Inequalities

Just as the signs of the factors in a *product* determine whether it is positive or negative, you should recall that the signs of the numerator and denominator of a *fraction* determine whether the fraction is positive or negative.

> If its numerator is positive and its denominator is positive, a fraction will be *positive*.
>
> If its numerator is negative and its denominator is negative, a fraction will be *positive*.
>
> If its numerator is positive and its denominator is negative, a fraction will be *negative*.
>
> If its numerator is negative and its denominator is positive, a fraction will be *negative*.

Let's keep these facts in mind as we take a look at some fractional inequalities.

Example 5. Solve $\dfrac{x}{3} > 0$.

Solution

$$\frac{x}{3} > 0$$

We wish this fraction to be *positive*. Since the denominator is positive 3, the numerator must also be *positive*. The solution is

$$x > 0$$

Now you complete Example 6.

Example 6. Solve $\dfrac{5}{3 - x} < 0$.

Solution

$$\frac{5}{3 - x} < 0$$

We wish this fraction to be _____ . Since the numerator is *positive* 5, the denominator must be _____ . The solution is found by

$$3 - x < 0$$

$$-x < \text{\_\_\_\_}$$

$$\frac{-1x}{-1} > \frac{\text{====}}{-1}$$

$$x > \text{\_\_\_\_}$$

Check your work on page 436. ▶

Example 7. Solve $\dfrac{-6}{2x + 1} \le 0$.

Solution

$$\frac{-6}{2x + 1} \le 0$$

We wish this fraction to be *negative* (or zero). Since the numerator is a *negative* constant, the denominator must be *positive*. Shall we say

$$2x + 1 \ge 0?$$

Can we allow the denominator to be greater than or *equal* to zero? The denominator of a fraction can *never equal zero*. We find our only solutions from

$$2x + 1 > 0$$

$$2x > -1$$

$$x > \frac{-1}{2}$$

We conclude that if $x > \dfrac{-1}{2}$, then $\dfrac{-6}{2x + 1} < 0$. This fraction can never *equal* zero.

From Example 7 and our earlier work with fractions we must remember:

> If the numerator of a fraction is zero and the denominator is *not* zero, the fraction will equal zero.
>
> If the denominator of a fraction is zero, the fraction will be undefined.

▌▶ **Trial Run**

Solve.

_____ **1.** $\dfrac{x}{2} < 0$

_____ **2.** $\dfrac{3}{x + 2} > 0$

_____ **3.** $\dfrac{x - 1}{-5} \geq 0$

_____ **4.** $\dfrac{8}{3 - 4x} \geq 0$

Answers are on page 436.

What happens when a variable appears in more than one factor of the numerator and/or denominator of a fractional inequality? As you may have guessed, fractional inequalities containing several variable factors can be treated using the **critical-point method** discussed in Chapter 5. We approach them in the same way that we approached inequalities that contained *products* of variable factors.

Example 8. Solve $\dfrac{7}{x(x - 3)} < 0$.

Solution

$$\frac{7}{x(x - 3)} < 0$$

We locate critical points on the number line by deciding what values of the variable will make each factor zero. Here critical points are $x = 0$ and $x = 3$. Our dots should be *open*, and we want our fraction to be *negative*.

Since the fraction is negative in interval II, we write our solutions as

$$0 < x < 3$$

Now you try Example 9.

Example 9. Solve $\dfrac{x}{x + 2} > 0$.

Solution

$$\frac{x}{x + 2} > 0$$

Our critical points are $x =$ _____ and $x =$ _____ . The dots should be _____ , and we want the fraction to be _____ .

Since the fraction is positive in interval _____ or interval _____ , we write our solutions as

$$x < \text{\_\_\_\_} \qquad \text{or} \qquad x > \text{\_\_\_\_} .$$

Check your work on page 436. ▶

   The key to solving fractional inequalities lies in knowing that the sign of a fraction is determined by the signs of the factors in its numerator and denominator. To solve an inequality such as

$$\frac{x}{x-2} \geq \frac{2}{3}$$

we must rewrite it in a form that will allow us to make judgments about *signs*. In other words, we must *rewrite the inequality with a single fraction on one side and zero on the other side*.

> We may *not* multiply both sides of any inequality by a *variable* quantity.

Why? Because we do not know the *sign* of such a quantity, so we cannot judge whether or not to reverse the direction of the inequality.

Example 10. Solve $\dfrac{x}{x-2} \geq \dfrac{2}{3}$.

Solution

$$\frac{x}{x-2} \geq \frac{2}{3}$$

$$\frac{x}{x-2} - \frac{2}{3} \geq 0 \qquad \begin{array}{l}\text{Subtract } \frac{2}{3} \text{ from both sides.}\\[4pt]\text{LCD: } 3(x-2) \text{ and } x \neq 2.\end{array}$$

$$\frac{3 \cdot x}{3(x-2)} - \frac{2(x-2)}{3(x-2)} \geq 0 \qquad \begin{array}{l}\text{Build each fraction to a fraction}\\\text{containing the LCD.}\end{array}$$

$$\frac{3x}{3(x-2)} - \frac{2x-4}{3(x-2)} \geq 0 \qquad \text{Simplify each numerator.}$$

$$\frac{3x - 2x + 4}{3(x - 2)} \geq 0 \qquad \text{Combine the fractions.}$$

$$\frac{x + 4}{3(x - 2)} \geq 0 \qquad \text{Simplify the numerator by combining like terms.}$$

Critical points are

$$x = -4 \qquad \text{(solid dot)} \qquad \text{Solid dot because if } x = -4 \text{ the inequality is satisfied.}$$

$$x = 2 \qquad \text{(open dot)} \qquad \text{Open dot because if } x = 2 \text{ the fraction is undefined.}$$

We want the fraction to be positive (or zero).

Since the fraction is positive in intervals I and III, our solutions are

$$x \leq -4 \qquad \text{or} \qquad x > 2$$

⟫ Trial Run

Solve.

_____ 1. $\dfrac{x}{x - 3} > 0$

_____ 2. $\dfrac{5}{x(x - 2)} < 0$

_____ 3. $\dfrac{x - 2}{x} \geq \dfrac{1}{2}$

_____ 4. $\dfrac{x - 2}{x + 1} < 2$

Answers are on page 436.

▶ Examples You Completed

Example 3. Solve $\dfrac{3}{x + 1} - \dfrac{5}{x} = \dfrac{19}{x^2 + x}$.

Solution

$$\frac{3}{x + 1} - \frac{5}{x} = \frac{19}{x^2 + x}$$

$$\frac{3}{x + 1} - \frac{5}{x} = \frac{19}{x(x + 1)} \qquad \text{LCD: } x(x + 1) \text{ and } x \neq 0, x \neq -1.$$

$$\frac{x(x+1)}{1}\left[\frac{3}{x+1}\right] - \frac{x(x+1)}{1}\left[\frac{5}{x}\right] \cdot = \frac{x(x+1)}{1}\left[\frac{19}{x(x+1)}\right]$$

$$3x - 5(x+1) = 19$$

$$3x - 5x - 5 = 19$$

$$-2x - 5 = 19$$

$$-2x = 24$$

$$x = -12$$

Since the restrictions were $x \neq 0$, $x \neq -1$, the solution $x = -12$ is acceptable.

Example 6. Solve $\dfrac{5}{3-x} < 0$.

Solution

$$\frac{5}{3-x} < 0$$

We wish this fraction to be negative. Since the numerator is *positive* 5, the denominator must be negative. The solution is found by

$$3 - x < 0$$

$$-x < -3$$

$$\frac{-1x}{-1} > \frac{-3}{-1}$$

$$x > 3$$

Example 9. Solve $\dfrac{x}{x+2} > 0$.

Solution

$$\frac{x}{x+2} > 0$$

Our critical points are $x = 0$ and $x = -2$. The dots should be open, and we want the fraction to be positive.

Since the fraction is positive in interval I or interval III, we write our solutions as

$$x < -2 \qquad \text{or} \qquad x > 0$$

Answers to Trial Runs

page 431 **1.** $x = \dfrac{1}{5}$ **2.** $x = \dfrac{6}{5}$ **3.** $x = -35$ **4.** No solution **5.** $x = 9, x = -2$

6. $x = \dfrac{-5y}{5 - 2y}$

page 433 **1.** $x < 0$ **2.** $x > -2$ **3.** $x \leq 1$ **4.** $x < \dfrac{3}{4}$

page 435 **1.** $x < 0$ or $x > 3$ **2.** $0 < x < 2$ **3.** $x < 0$ or $x \geq 4$ **4.** $x < -4$ or $x > -1$

EXERCISE SET 6.5

Solve the following equations.

_____ 1. $\dfrac{2x}{7} = \dfrac{x-2}{6}$

_____ 2. $\dfrac{3x}{5} = \dfrac{x-1}{4}$

_____ 3. $\dfrac{7}{x} - \dfrac{3}{2} = \dfrac{5}{2x}$

_____ 4. $\dfrac{11}{x} - \dfrac{4}{3} = \dfrac{1}{3x}$

_____ 5. $\dfrac{2x-1}{x^2} - \dfrac{5}{3x} = 0$

_____ 6. $\dfrac{3x+1}{x^2} - \dfrac{7}{5x} = 0$

_____ 7. $\dfrac{11}{4x} = \dfrac{5}{x+4}$

_____ 8. $\dfrac{3}{7x} = \dfrac{2}{x+7}$

_____ 9. $\dfrac{4}{x+7} = \dfrac{5}{x^2-49}$

_____ 10. $\dfrac{3}{x-5} = \dfrac{2}{x^2-25}$

_____ 11. $\dfrac{4}{2x-1} + \dfrac{1}{2x+1} = \dfrac{2}{4x^2-1}$

_____ 12. $\dfrac{5}{3x+1} - \dfrac{2}{3x-1} = \dfrac{1}{9x^2-1}$

_____ 13. $\dfrac{3}{x^2-x-12} - \dfrac{2}{x-4} = \dfrac{1}{2x+6}$

_____ 14. $\dfrac{1}{x^2-6x-7} - \dfrac{1}{x-7} = \dfrac{2}{3x+3}$

_____ 15. $\dfrac{5}{x+1} = \dfrac{3x-2}{x^2-15x-16}$

_____ 16. $\dfrac{3}{x-2} = \dfrac{7x-4}{x^2-8x+12}$

_____ 17. $\dfrac{5x+3}{x-7} = 2$

_____ 18. $\dfrac{6x+5}{x-9} = 3$

_____ 19. $\dfrac{a-2}{a+5} = \dfrac{a-1}{a+1}$

_____ 20. $\dfrac{a-7}{a-2} = \dfrac{a+3}{a+2}$

_____ 21. $\dfrac{2x}{2x-3} - \dfrac{6}{2x-3} = 2$

_____ 22. $\dfrac{3x}{3x-1} - \dfrac{4}{3x-1} = 4$

_____ 23. $\dfrac{x}{x-3} - \dfrac{1}{2x+3} = \dfrac{2x^2+9}{2x^2-3x-9}$

_____ 24. $\dfrac{x}{x+5} - \dfrac{2}{3x-1} = \dfrac{3x^2+5}{3x^2+1+x-5}$

_____ 25. $\dfrac{x+2}{x-3} = \dfrac{30}{x^2-9} + \dfrac{1}{x+3}$

_____ 26. $\dfrac{x+4}{x-2} + \dfrac{20}{x^2-4} = -\dfrac{5}{x-2}$

_____ 27. $\dfrac{x}{x+1} + \dfrac{2x}{x-1} = \dfrac{10}{x^2-1}$

_____ 28. $\dfrac{3x}{x-2} + \dfrac{2x}{x+2} = \dfrac{-}{x^2-4}$

_____ 29. $\dfrac{y+1}{2y-5} = \dfrac{2y+5}{y-1}$

_____ 30. $\dfrac{2y-1}{3y+4} = \dfrac{3y-4}{2y+1}$

_____ 31. $\dfrac{x}{2x-6} - \dfrac{3}{x^2-6x+9} = \dfrac{x-2}{3x-9}$

_____ 32. $\dfrac{x}{3x-6} - \dfrac{4}{x^2-4} = \dfrac{3}{2x-4}$

_____ 33. $\dfrac{3x - y}{4x} = \dfrac{2 - xy}{4}$ for y.

_____ 34. $\dfrac{2m - n}{3m} = \dfrac{1 - mn}{3}$ for n.

_____ 35. $\dfrac{1}{2x} + \dfrac{2}{y} = \dfrac{1}{z}$ for y.

_____ 36. $\dfrac{2}{a} - \dfrac{3}{b} = \dfrac{1}{c}$ for a.

_____ 37. $A = P(1 + rt)$ for t.

_____ 38. $S = \dfrac{1}{2} g(2t - 1)$ for t.

Solve the following inequalities.

_____ 39. $\dfrac{x}{5} < 0$

_____ 40. $\dfrac{x}{7} \geq 0$

_____ 41. $\dfrac{5}{x - 3} > 0$

_____ 42. $\dfrac{9}{x - 7} < 0$

_____ 43. $\dfrac{x + 3}{-7} \geq 0$

_____ 44. $\dfrac{x + 6}{-5} \geq 0$

_____ 45. $\dfrac{-3}{2x - 5} \leq 0$

_____ 46. $\dfrac{-1}{3x - 7} \leq 0$

_____ 47. $\dfrac{6 - 2x}{5} \geq 0$

_____ 48. $\dfrac{9 - 3x}{4} \geq 0$

_____ 49. $\dfrac{9}{5 - x} \leq 0$

_____ 50. $\dfrac{10}{7 - x} \leq 0$

_____ 51. $\dfrac{x}{x - 4} > 0$

_____ 52. $\dfrac{x}{x - 6} > 0$

_____ 53. $\dfrac{3}{x(x - 1)} < 0$

_____ 54. $\dfrac{5}{x(x - 5)} < 0$

_____ 55. $\dfrac{x - 3}{x} \geq \dfrac{1}{3}$

_____ 56. $\dfrac{x - 5}{2x} \geq \dfrac{1}{4}$

_____ 57. $\dfrac{x - 3}{x - 1} < 2$

_____ 58. $\dfrac{x - 5}{x + 3} < 3$

☆ Stretching the Topics

Solve for x.

_____ 1. $\dfrac{2x}{x^2 - 4} - \dfrac{2x}{x^2 + x - 2} + \dfrac{x}{x^2 - 3x + 2} = 0$

————— **2.** $\dfrac{x - 1}{x^2 + 3x + 2} + \dfrac{x + 1}{x^2 + x - 2} = \dfrac{x + 3}{x^2 - 1}$

————— **3.** $\dfrac{10k^2}{x^2 - kx - 6k^2} + \dfrac{4x - 5k}{x - 3k} = \dfrac{4x + 3k}{x + 2k}$

Check your answers in the back of your book.

If you can solve the equations and inequalities in **Checkup 6.5,** you are ready to go on to Section 6.6.

✓ **CHECKUP 6.5**

Solve.

_____ 1. $\dfrac{x + 5}{2} = \dfrac{x - 4}{3}$

_____ 2. $\dfrac{6}{a} - \dfrac{4}{9} = \dfrac{8}{9a}$

_____ 3. $\dfrac{5}{2b + 1} = \dfrac{-4}{b - 1}$

_____ 4. $\dfrac{3}{a - 5} + \dfrac{4a}{a^2 - a - 20} = 0$

_____ 5. $\dfrac{x - 2}{x - 4} - \dfrac{16}{x^2 - 16} = \dfrac{1}{x + 4}$

_____ 6. $\dfrac{x}{5} = \dfrac{y}{y - 1}$ for y

_____ 7. $\dfrac{6}{x - 5} > 0$

_____ 8. $\dfrac{x}{x - 10} \le 0$

_____ 9. $\dfrac{x - 6}{x} \ge 3$

_____ 10. $\dfrac{x - 1}{x + 1} > \dfrac{1}{2}$

Check your answers in the back of your book.

If You Missed Problems:	You Should Review Examples:
1, 2	1
3–5	3, 4
6	2
7	9
8	7
9, 10	10

6.6 Switching from Words to Fractional Equations

There are many instances in which the information given in a word problem requires the use of a fractional equation.

Working with Ratio and Proportion

A **ratio** is a fraction that is used to compare two numbers. Ratios are handy tools to use in solving many everyday mathematical problems.

Suppose that your take-home pay is $975 per month and you save $150 from that amount. Then the ratio of your saving to your take-home pay is given by

$$\frac{\text{Saving}}{\text{Take-home pay}} = \frac{\$150}{\$975}$$

$$= \frac{2}{13}$$

Suppose that your take-home pay is increased to $1170 but you still wish to maintain the *same* ratio of saving to take-home pay. To decide how much you should save (x), you must solve an equation.

$$\frac{\text{Saving}}{\text{Take-home pay}} \qquad \frac{2}{13} = \frac{x}{1170}$$

Such a fractional equation, which states that two ratios are equal, is called a **proportion.** It is most easily solved using the definition of equality of fractions that was stated earlier in this chapter.

$$\frac{A}{B} = \frac{C}{D} \quad \text{if and only if} \quad A \cdot D = B \cdot C \qquad (B \neq 0, D \neq 0)$$

Example 1. Solve the proportion $\dfrac{2}{13} = \dfrac{x}{\$1170}$.

Solution

$$\frac{2}{13} = \frac{x}{1170}$$

$$13 \cdot x = 2(1170)$$

$$13x = 2340$$

$$x = \frac{2340}{13}$$

$$x = 180$$

You should save $180 from your take-home pay of $1170.

Example 2. The admissions policy at Hometown University requires that 7 in-state students be admitted for every 2 out-of-state students. If 1854 freshmen have been admitted in the required ratio, how many are in-state students and how many are from other states?

Solution

We Think	We Write

The ratio of in-state to out-of-state students is 7 to 2.

$$\frac{\text{In-state}}{\text{Out-of-state}} = \frac{7}{2}$$

We do not know how many of either type student there are.

Let x = number of in-state students

We know that the total is 1854.

Let $1854 - x$ = number of out-of-state students

The ratio is 7 to 2.

$$\frac{x}{1854 - x} = \frac{7}{2}$$

$$\frac{x}{1854 - x} = \frac{7}{2}$$

$$7(1854 - x) = 2x$$

$$12{,}978 - 7x = 2x$$

$$12{,}978 = 9x$$

$$\frac{12{,}978}{9} = x$$

$$1442 = x$$

In-state students: $x = 1442$

Out-of-state students: $1854 - x = 1854 - 1442$

$$= 412$$

There are 1442 in-state students and 412 out-of-state students in the freshman class.

Working with Distance Problems

In earlier chapters we solved distance problems using the fact that

$$\boxed{\begin{array}{c} \text{Distance} = \text{rate} \times \text{time} \\ d = r \cdot t \end{array}}$$

Now let us consider distance problems that lead to fractional equations.

Example 3. On long trips, Bob drives his car at a constant rate that is 5 miles per hour faster than his brother George's constant rate. If Bob can drive 312 miles in the same length of time that George drives 282 miles, how fast does each one drive?

Solution

We Think	We Write

We know that distance equals rate times time.

Distance = (rate)(time)

We know the brothers' distances.

Bob's distance = 312 miles
George's distance = 282 miles

Solution

We Think	*We Write*
We realize that the total work done (1 lawn) will equal the *sum* of the amounts of work (parts of lawn) done by each sister.	Linda's work + Nancy's work = 1 lawn
Each sister's work equals her rate multiplied by her time.	Linda's work = (rate)(time) Nancy's work = (rate)(time)
The time is the same, but we do not know what it is.	Let t = time worked (hours)
If Linda can rake 1 lawn in 4 hours, she can rake $\frac{1}{4}$ of the lawn in 1 hour.	Linda's rate = $\frac{1}{4}$ lawn per hour
If Nancy can rake 1 lawn in 2 hours, she can rake $\frac{1}{2}$ of the lawn in 1 hour.	Nancy's rate = $\frac{1}{2}$ lawn per hour
Each sister's work is the *product* of her rate and her time.	Linda's work = $\frac{1}{4} \cdot t$ Nancy's work = $\frac{1}{2} \cdot t$
The sum of the sisters' work must equal 1 whole job (the whole lawn).	$\frac{1}{4} \cdot t + \frac{1}{2} \cdot t = 1$

$$\frac{1}{4} \cdot \frac{t}{1} + \frac{1}{2} \cdot \frac{t}{1} = 1$$

$$\frac{t}{4} + \frac{t}{2} = 1 \qquad \text{LCD: 4.}$$

$$\frac{4}{1} \cdot \frac{t}{4} + \frac{4}{1} \cdot \frac{t}{2} = 4 \cdot 1$$

$$t + 2t = 4$$

$$3t = 4$$

$$t = \frac{4}{3}$$

If the sisters rake together, they can complete the lawn in $1\frac{1}{3}$ hours (or 1 hour 20 minutes). Notice that the time needed is less than either sister's time alone. This should *always* be true.

Now let's consider the problem stated at the beginning of this chapter.

Example 6. Mr. Romano uses two hoses to fill his swimming pool. By itself, the larger hose can fill the pool in 7 hours, while the smaller hose requires 10 hours by itself. While using both hoses to fill his pool last weekend, Mr. Romano discovered after 2 hours that the pool drain had been left open. If the pool normally drains in 12 hours, how long after he closed the drain would the pool be filled?

Solution

We Think	*We Write*
We know each hose's rate.	Large hose rate $= \dfrac{1}{7}$ pool per hour
	Small hose rate $= \dfrac{1}{10}$ pool per hour
We know the drain's rate, but it acts *against* the filling of the pool.	Drain rate $= \dfrac{-1}{12}$ pool per hour
We know the draining time.	Time drain was open $= 2$ hours
We seek the unknown amount of time *after* the drain was closed.	Let $t =$ time after closing drain.
The hoses had been filling for two hours *before* the drain was closed.	$t + 2 =$ large-hose time $t + 2 =$ small-hose time
Each hose's work and the drain's work are found by multiplying rate times time.	$\dfrac{1}{7}(t + 2) =$ large-hose work
	$\dfrac{1}{10}(t + 2) =$ small-hose work
	$\dfrac{-1}{12}(2) =$ drain work
The total work is the *sum* of all work done and must equal 1 filled pool.	$\dfrac{1}{7}(t + 2) + \dfrac{1}{10}(t + 2) - \dfrac{1}{12}(2) = 1$

$$\frac{t + 2}{7} + \frac{t + 2}{10} - \frac{2}{12} = 1$$

$$\frac{t + 2}{7} + \frac{t + 2}{10} - \frac{1}{6} = 1 \qquad \text{LCD: 210.}$$

$$\frac{\overset{30}{\cancel{210}}}{1}\left[\frac{t + 2}{\underset{1}{\cancel{7}}}\right] + \frac{\overset{21}{\cancel{210}}}{1}\left[\frac{t + 2}{\underset{1}{\cancel{10}}}\right] - \frac{\overset{35}{\cancel{210}}}{1}\left[\frac{1}{\underset{1}{\cancel{6}}}\right] = 210(1)$$

$$30(t + 2) + 21(t + 2) - 35 = 210$$

$$30t + 60 + 21t + 42 - 35 = 210$$

$$51t + 67 = 210$$

$$51t = 143$$

$$t = \frac{143}{51}$$

$$t = 2.8$$

The pool will be filled about 2.8 hours after Mr. Romano closes the drain or about 4.8 hours after he began filling it.

▌▌▶ Trial Run

Solve.

_____ **1.** Gary has decided that he would like the ratio of his saving to his spending to be 3 to 10. If his take-home pay is $1300 per month, how much should he save and how much should he spend?

_____ **2.** It takes train A 1 hour longer to travel 400 miles than it takes train B. If train B travels 20 miles per hour faster than train A, find each train's rate.

_____ **3.** Mr. Kretchmar can wax the gym floor in 6 hours and Mr. Freeman can do the same job in 4 hours. If Mr. Kretchmar starts working at 7 A.M. and is joined by Mr. Freeman at 8 A.M., when will the floor be finished?

Answers are on page 449.

Answers to Trial Run

page 449   **1.** Save $300; spend $1000   **2.** Train A: 80 mph; train B: 100 mph   **3.** 10 A.M.

EXERCISE SET 6.6

_____ 1. On a road map, 1 inch represents 15 miles. If the distance between Nashville and Louisville is 185 miles, how many inches are needed to represent this on the map?

_____ 2. If an $8000 investment earns $480 in 6 months, how much will $15,000 earn invested at the same rate for the same period of time?

_____ 3. If Kyle drives 660 miles in $1\frac{1}{2}$ days, how many days would it take him to drive 1760 miles at the same rate?

_____ 4. The sales tax on a $395 television set is $19.75. If the sales tax on a refrigerator is $39.50, what is the selling price of the refrigerator?

_____ 5. Ann and Merideth each owns an interest in the Specialty Shop. The ratio of Ann's income from the investment to Merideth's income is 4 to 5. Together one month they received $1530. How much did each receive that month?

_____ 6. In a certain math class the ratio of success to failure was 20 to 3. If 100 people passed the class, what was the total enrollment?

_____ 7. At the baseball cap factory, Burt and Johnny together can sew bills on 55 caps in 1 hour. The ratio of the number Burt can sew to the number Johnny can do is 2 to 3. Find how many caps each can complete in 1 hour.

_____ 8. Maria jogs 2 miles per hour faster than her sister Lola. If Maria can jog 9 miles in the same length of time that Lola can jog 6 miles, how fast does each one jog?

_____ 9. It takes Gretchen 6 minutes longer to ride her bike 6 miles than it does for George to ride the same distance. Find the rate of speed at which each rides if George rides 2 miles per hour faster than Gretchen.

_____ 10. Lucy drives her car 15 miles per hour faster than Mitch rides his motorcycle. If Lucy covers 275 miles in the same amount of time Mitch travels 200 miles, what is the rate of speed of each vehicle?

_____ 11. When Michelle and Carolyn leave for work, they each drive at the same rate. If Michelle travels 20 miles in 6 minutes less time than Carol drives 25 miles, how long does it take each of them to get to work?

_____ 12. Lyle took his boat 144 miles up the Ohio River and then back again. The total time for the trip was 15 hours. Find the speed each way if the speed down the river was 8 miles per hour faster than the speed up the river.

_____ 13. Victoria took the train part of the way to college and then the bus the remainder of the trip. The train traveled 30 miles per hour faster than the bus. The train and the bus each traveled 400 miles, but the time spent on the train was 3 hours less than the time on the bus. Find the speed of each.

_____ 14. Denise can clean the dog pens at the Humane Shelter in 8 hours and Barry can do it in 6 hours. How long will it take them to clean the pens if they work together?

———— **15.** Gus can wallpaper Bernice's house in 5 days. If Bernice helps him, they can do the work in 3 days. How long would it take Bernice to paper the house by herself?

———— **16.** Mel can clean all the ovens at the Dipsy Donut Shop in 5 hours and Gus can do the same job in 3 hours. If Mel starts cleaning and Gus joins him in an hour, how long will it be before the ovens are clean?

———— **17.** Suzie agreed to make draperies for her mother's apartment. She estimated it would take her 15 hours, but with her mother helping her they were finished in 8 hours. How long would it have taken her mother to make the draperies?

———— **18.** It takes 3 hours to fill a cattle watering tank using a pump and a hose. When a pump is used alone, the tank can be filled in 5 hours. If the pump were not working, how long would it take to fill the tank using only the hose?

———— **19.** A bathtub can be filled by the faucet in 5 minutes, but it takes 7 minutes for the drain to empty the tub. If the tub is empty and both are left open, how long would it take before the tub is overflowing?

———— **20.** Mr. Donosky uses two pipes to fill his reservoir. Pipe A takes twice as long as pipe B. A third pipe empties the tank in 2 hours. It takes 4 hours to fill the tank to half its capacity when all three are open. How long would it take pipe A to fill the tank if the other two were closed?

☆ Stretching the Topics _____

———— **1.** Two pipes are pouring water into a reservoir and one is taking water out. The first pipe can fill it in 20 days, the second pipe can fill it in n days and the third pipe can empty it in 12 days. What must be the value of n in order that the reservoir be filled in 90 days?

———— **2.** One plane few 382 miles with the help of a tailwind in $\frac{3}{4}$ of the time that another plane flew 480 miles in the opposite direction. If each plane can fly 500 miles per hour in still air, find the rate of the tailwind to the nearest tenth of a mile per hour.

Check your answers in the back of your book.

If you can complete **Checkup 6.6,** you are ready to do the Review Exercises.

CHECKUP 6.6

_____ **1.** In the Hillsboro City School System, 2 out of 5 teachers belong to a union. How many teachers belong to the union if the school system has 700 teachers?

_____ **2.** Working together, Erma and Bob can install the plumbing in their new house in 16 hours. Working alone, it would take Erma 40 hours to do it. How long would it take Bob to do the installation working alone?

_____ **3.** It takes Richard the same amount of time to drive 420 miles as it takes Sam to drive 385 miles. If Richard's speed is 5 miles per hour faster than Sam's speed, how fast does Sam drive?

Check your answers in the back of your book.

If You Missed Problem:	You Should Review Example:
1	1
2	5
3	2

Summary

In this chapter we learned to reduce, add, subtract, multiply, and divide rational algebraic expressions.

If We Wish to:	We Must:	Example
Reduce a fraction	Divide numerator and denominator by any common factors	$\dfrac{x^2 - 4}{3x + 6} = \dfrac{(x + 2)(x - 2)}{3(x + 2)} = \dfrac{x - 2}{3}$
Multiply fractions	Multiply numerators and multiply denominators, reducing if possible. $$\frac{A}{B} \cdot \frac{C}{D} = \frac{AC}{BD}$$	$\dfrac{x + 2}{3x + 3} \cdot \dfrac{x^2 - 1}{x^2 + 5x + 6}$ $= \dfrac{\overset{1}{\cancel{x + 2}}}{3\underset{1}{\cancel{(x + 1)}}} \cdot \dfrac{\overset{1}{\cancel{(x + 1)}}(x - 1)}{\underset{1}{\cancel{(x + 2)}}(x + 3)} = \dfrac{x - 1}{3(x + 3)}$
Divide fractions	Invert the divisor and multiply. $$\frac{A}{B} \div \frac{C}{D} = \frac{A}{B} \cdot \frac{D}{C} = \frac{AD}{BC}$$	$\dfrac{-x^2 y}{3} \div \dfrac{xy^2}{15} = \dfrac{-x^2 y}{3} \cdot \dfrac{15}{xy^2} = \dfrac{-5x}{y}$
Add or subtract fractions	Choose the LCD for the fractions. Change all fractions to fractions having the LCD, then add or subtract numerators, keeping the LCD; reduce if possible	$\dfrac{3}{x^2} + \dfrac{2}{5x + 10} - \dfrac{1}{x^2 + 2x}$ $= \dfrac{3}{x^2} + \dfrac{2}{5(x + 2)} - \dfrac{1}{x(x + 2)}$ LCD: $5x^2(x + 2)$ $= \dfrac{15(x + 2)}{5x^2(x + 2)} + \dfrac{2x^2}{5x^2(x + 2)} - \dfrac{5x}{5x^2(x + 2)}$ $= \dfrac{15x + 30 + 2x^2 - 5x}{5x^2(x + 2)} = \dfrac{2(x^2 + 5x + 15)}{5x^2(x + 2)}$
Simplify complex fractions	*Method 1:* Rewrite the fraction as a division problem	$\dfrac{\dfrac{1}{x} + \dfrac{1}{2}}{\dfrac{2}{x + 2}}$ $= \left[\dfrac{1}{x} + \dfrac{2}{2} \right] \div \left[\dfrac{2}{x + 2} \right]$ $= \left[\dfrac{2}{2x} + \dfrac{x}{2x} \right] \div \left[\dfrac{2}{x + 2} \right]$ $= \dfrac{2 + x}{2x} \cdot \dfrac{x + 2}{2}$ $= \dfrac{(x + 2)^2}{4x}$

Method 2: Multiply both numerator and denominator by the LCD for both

$$\frac{\dfrac{1}{x} + \dfrac{3}{5}}{\dfrac{2}{x^2}} \qquad \text{LCD: } 5x^2$$

$$= \frac{5x^2\left(\dfrac{1}{x} + \dfrac{3}{5}\right)}{5x^2\left(\dfrac{2}{x^2}\right)}$$

$$= \frac{5x + 3x^2}{10}$$

Throughout our work with algebraic fractions, we were careful to state **restrictions** on the domain of the variable in order to avoid division by zero. In solving fractional equations we agreed to multiply both sides of the equation by the LCD for all the fractions *after* stating restrictions. Using this procedure, we were able to reject **extraneous** solutions.

To solve **fractional inequalities** we use the same critical point method used in solving quadratic inequalities. Once again we were careful to avoid solutions that made denominators equal zero.

Finally, we used our knowledge of fractional equations to solve problems stated in words. In particular we learned to solve proportions, distance problems, and work problems.

❑ Speaking the Language of Algebra _____

Complete each sentence with the appropriate word or phrase.

1. A fraction whose numerator and denominator are polynomials is called a _____ _____ _____ .

2. We restrict the domain of the variable in a rational algebraic expression in order to avoid _____ _____ _____ .

3. To reduce fractions we may _____ the numerator and denominator by any _____ _____ .

4. To look for the LCD for a group of fractions, we first _____ each of the original denominators.

5. To solve a fractional equation, we multiply both sides of the equation by the _____ .

6. If a fraction is positive and we know its numerator is negative, we conclude that its denominator is _____ .

7. An equation that states that two ratios are equal is called a _____ .

8. The amount of work done by a person or a machine can be found by multiplying the _____ times the _____ .

REVIEW EXERCISES for Chapter 6: Working with Rational Expressions

State the restrictions and reduce the fractions

_____ **1.** $\dfrac{-9x^5y^3z}{6x^2y^3z^4}$

_____ **2.** $\dfrac{15x}{15x^2 - 30x}$

_____ **3.** $\dfrac{3x + 3y}{12x - 12y}$

_____ **4.** $\dfrac{x^2 + 2x - 35}{x^2 - 49}$

_____ **5.** $\dfrac{21x^2 - 17xy + 2y^2}{7x^2 + 13xy - 2y^2}$

_____ **6.** $\dfrac{a^3 - 8b^3}{3a^2 - 11ab + 10b^2}$

_____ **7.** $\dfrac{4(x - y)^2 + 3y(x - y)}{4x^2 - 5xy + y^2}$

_____ **8.** $\dfrac{64 - x^2}{2x^2 - 19x + 24}$

Simplify.

_____ **9.** $\dfrac{3x + 15}{21x^2} \cdot \dfrac{14x}{x + 5}$

_____ **10.** $\dfrac{x^2 - 36}{5x + 20} \cdot \dfrac{x^2 + 4x}{x^2 - 4x - 12}$

_____ **11.** $\dfrac{(3x - 5y)^2}{2x^2 - 7xy - 4y^2} \cdot \dfrac{3x^2 - 7xy - 20y^2}{9x^2 - 25y^2}$

_____ **12.** $\dfrac{x^2 - 121y^2}{x^9y^2} \div \dfrac{5x - 55y}{x^5y^3}$

_____ **13.** $\dfrac{3x^2 - 6xy}{3x^2 + 2xy - y^2} \div \dfrac{6x^2y - 3x^3}{9x^2 + 8xy - y^2}$

_____ **14.** $\left(\dfrac{x^2 - 144}{7x} \cdot \dfrac{14}{2x - 24}\right) \div \dfrac{3x + 36}{8x}$

_____ **15.** $\dfrac{5x}{2x - 3} + \dfrac{3x}{2x - 3}$

_____ **16.** $\dfrac{4x + 1}{x^2 - 9} - \dfrac{3x - 2}{x^2 - 9}$

_____ **17.** $\dfrac{5}{x + 10} + \dfrac{3}{x}$

_____ **18.** $\dfrac{2x}{x - 5} + \dfrac{x}{x + 5}$

_____ **19.** $\dfrac{3x + 6}{x^2 - x - 56} - \dfrac{2}{x - 8}$

_____ **20.** $\dfrac{1}{x - 4} - \dfrac{3}{x + 2} + \dfrac{2(2x - 11)}{x^2 - 2x - 8}$

_____ **21.** $\dfrac{\dfrac{x^{12}}{y^2}}{\dfrac{3x^9}{y}}$

_____ **22.** $\dfrac{\dfrac{3x - 1}{4x^2}}{\dfrac{9x^2 - 1}{12x^3}}$

_____ **23.** $\dfrac{x - \dfrac{64}{x}}{x - 8}$

_____ **24.** $\dfrac{\dfrac{1}{x^2} - \dfrac{1}{25}}{\dfrac{1}{x} - \dfrac{1}{5}}$

_____ 25. $\dfrac{1 - 8a^{-3}}{1 - 2a^{-1}}$

_____ 26. $\dfrac{6 - 7x^{-1} - 5x^{-2}}{3 - 5x^{-1}}$

Solve.

_____ 27. $\dfrac{x + 3}{4} = \dfrac{x - 5}{3}$

_____ 28. $\dfrac{2}{a - 6} - \dfrac{4a}{a^2 + a - 42} = 0$

_____ 29. $\dfrac{x + 5}{x - 3} + \dfrac{2x}{x^2 - 9} = \dfrac{1}{x + 3}$

_____ 30. $\dfrac{3}{x - 9} > 0$

_____ 31. $\dfrac{2x - 5}{x} \geq -3$

_____ 32. $\dfrac{x - 2}{x + 1} < \dfrac{1}{2}$

_____ 33. On the architect's drawing of Lillian's house, $\frac{1}{4}$ inch represents 2 feet. If the length of her living room on the drawing is $2\frac{1}{2}$ inches, find the actual length of the living room.

_____ 34. Alex can clean the ovens at the Pizza Parlor in 6 hours. If Ron helps him, they can do the work in $2\frac{1}{2}$ hours. How long would it take Ron to clean the ovens by himself?

_____ 35. Becky drives her car 10 miles per hour faster than Mac drives his truck. If Becky drives 385 miles in the same amount of time Mac drives 315 miles, find the rate of speed of each vehicle.

Check your answers in the back of your book.

If You Missed Exercises:	You Should Review Examples:	
1	SECTION 6.1	3, 4
2–6		5–8
7		10
8		11, 12
9–11	SECTION 6.3	1–3
12, 13		4–7
14		8, 9
15, 16	SECTION 6.3	3–5
17, 18		13–15
19, 20		16–18
21, 22	SECTION 6.4	1, 2
23, 24		3–5
25, 26		6, 7
27	SECTION 6.5	1
28, 29		3, 4
30		5–7
31, 32		8–10
33	SECTION 6.6	2
34		5
35		3, 4

If you have completed the Review Exercises and corrected your errors, you are ready to take the Practice Test.

PRACTICE TEST for Chapter 6: Working with Rational Expressions

State the restrictions on the variables and reduce the fractions.

		SECTION	EXAMPLES

_____ 1. $\dfrac{27x^4y^5c}{18x^3y^5c^3}$ *← first 2 problems*   6.1   3, 4

_____ 2. $\dfrac{x^2 + 2x - 15}{x^2 + 3x - 10}$ 6.1   8

_____ 3. $\dfrac{y^3 - 1}{y^2 - 1}$ 6.1   7

_____ 4. $\dfrac{m^2 + 2mn - m - 2n}{m^2 - mn - 6n^2}$ 6.1   9

Simplify.

_____ 5. $\dfrac{x^2 + 6x - 16}{3x^2 - 9x} \cdot \dfrac{6x - 18}{x^2 - x - 2}$ #3 #4 has Division   6.2   1

_____ 6. $\dfrac{x^2 - 49}{5x} \cdot \dfrac{50}{2x - 14} \cdot \dfrac{4x^3}{5x + 35}$ 6.2   3

_____ 7. $\dfrac{3a^2 - 75b^2}{a^4b^5} \div \dfrac{4a - 20b}{ab^6}$ 6.2   4, 5

_____ 8. $\dfrac{x^2 - 4xy + 4y^2}{x^3 - 2x^2y} \div \dfrac{x^2 + xy - 6y^2}{7x + 21y}$ 6.2   6, 7

_____ 9. $\dfrac{6x - 5}{x^2 - 49} - \dfrac{5x - 12}{x^2 - 49}$ 6.3   5

_____ 10. $\dfrac{3}{5x + 2} - \dfrac{1}{5x}$ #5   6.3   13

_____ 11. $\dfrac{y}{y + 6} - \dfrac{6}{6 - y}$ 6.3   18

_____ 12. $\dfrac{x + 1}{x^2 + 7x + 12} + \dfrac{2}{x + 3}$ #6   6.3   16

_____ 13. $\dfrac{10}{3x - 4} - \dfrac{x}{x + 6} + \dfrac{x + 2}{3x^2 + 14x - 24}$ 6.3   16, 17

_____ 14. $\dfrac{\dfrac{1 - 2x}{7x^2}}{\dfrac{1 - 4x^2}{2x}}$ 6.4   2

_____ 15. $\dfrac{2x + 7 - \dfrac{15}{x}}{\dfrac{x}{5} - \dfrac{5}{x}}$ #7   6.4   4

_____ 16. $\dfrac{2 - x^{-1} - 10x^{-2}}{1 + 2x^{-1}}$ 6.4   6, 7

Solve.

_____ 17. $\dfrac{8x + 1}{15} = \dfrac{2 - x}{6}$ 6.5   1

_____ 18. $\dfrac{2x - 6}{x^2 - 6x - 7} - \dfrac{1}{x - 7} = 0$ #8   6.5   3, 4

_____ 19. $\dfrac{5x - 1}{x + 4} \le 0$ 6.5   8, 9

_____ 20. A large and small hose together can fill a pool in 7 hours. If the large hose alone can fill the pool in 10 hours, how long will it take the small hose alone to fill the pool?   6.6   6

SHARPENING YOUR SKILLS after Chapters 1–6

	SECTION
1. Use an inequality to state whether $-\sqrt{13}$ is positive or negative.	1.1
2. Evaluate $(2x - 1)(4x - 3)$ when $x = \dfrac{1}{2}$.	1.4
3. Evaluate $-a^3$ when $a = -3$.	3.1
4. Evaluate -4^{-2}.	3.1
5. Simplify $\sqrt{768}$.	5.2
6. Use the discriminant to describe the solutions for $3x^2 + 3x = 1$.	5.3
7. Write a quadratic equation with the solutions $x = 3$ and $x = -2$.	5.1

Simplify the expressions.

8. $9[5x - (3x - y)] - [4(x + y) - 4y]$	1.4
9. $a^3b(a^4b^2)^3 + 3ab^7(a^3b^4)^2$	3.1
10. $\left(\dfrac{a^{-3}b}{4b^{-2}}\right)^2 \left(\dfrac{a^2b^{-1}}{3b^5}\right)^{-3}$	3.2
11. $a(a^2 - 7) - 2(a - 3)$	3.3
12. $-y^2(7 - y)(9 - y)$	3.4
13. $(x^5 - 9x^4 + 5x^3 - 11x^2) \div x^3$	3.5
14. $(x^3 - 4x^2y - 11xy^2 - 6y^3) \div (x + y)$	3.5

Factor completely.

15. $-3x^6y^3 + 21x^4y^2 - 27x^2y$	4.1
16. $(a - b)^2 - 5(a - b)$	4.1
17. $9x^2y^2 - 121$	4.2
18. $10y^2 - 7y - 12$	4.3
19. $120 - 9y - 3y^2$	4.3

Solve.

20. $\dfrac{9y - 2}{5} + 8 = -2y$	2.1

_____ **21.** $|2x - 11| = -7$ 2.4

_____ **22.** $\left| 2 - \dfrac{m}{5} \right| > 3$ 2.4

_____ **23.** $-3x(x + 8)(9x - 5) = 0$ 5.1

_____ **24.** $30m^2 = 7m + 2$ 5.3

_____ **25.** $121x^2 = 64$ 5.2

_____ **26.** $x(x - 5) = 4 - 3x$ 5.3

_____ **27.** $2(y - 1) - 9 \le 8(y + 2) + 13$ 2.3

_____ **28.** $-2 \le -a + 4 < 7$ 2.3

_____ **29.** $5x^2 + 18x \le 8$ 5.4

_____ **30.** Solve $M = 2ax^2 + x$ for x.   5.3

Working with Rational Exponents and Radicals

7

One leg of a right triangle is 1 foot longer than the other. If the shorter leg is increased by 4 feet and the length of the longer leg is multiplied by 6, the length of the hypotenuse of the enlarged triangle is five times the length of the original hypotenuse. Find the dimensions of both triangles.

In Chapter 3 we learned to work with integer exponents, and in Chapter 5 we discussed square roots of numbers. Now it is time to discuss both exponents and roots in more depth and see how the two concepts are related. We must discuss the use of rational numbers as exponents and learn to deal with roots other than square roots. In this chapter we learn to:

1. Work with rational exponents.
2. Relate rational exponents to radicals.
3. Simplify radical expressions.
4. Operate with radical expressions.
5. Rationalize denominators.
6. Solve radical equations.
7. Work with complex numbers.

7.1 Introducing Rational Exponents and Radicals

Working with Rational Exponents

From earlier work with exponents and square roots, you should recall the following definitions.

Exponent Definitions	Principal Square Root
$a^n = \underbrace{a \cdot a \cdot \ldots \cdot a}_{n \text{ factors}}$	
$a^0 = 1 \qquad (a \neq 0)$	$\sqrt{a} = b$ if $b^2 = a$
$a^{-n} = \dfrac{1}{a^n} \qquad (a \neq 0)$	where $b \geq 0$

You should also recall the five basic laws of exponents that allowed us to perform operations with integer powers of bases.

Laws of Exponents

$$(1) \quad a^m \cdot a^n = a^{m+n}$$

$$(2) \quad (a^m)^n = a^{mn}$$

$$(3) \quad (ab)^n = a^n b^n$$

$$(4) \quad \left(\frac{a}{b}\right)^n = \frac{a^n}{b^n} \qquad (b \neq 0)$$

$$(5) \quad \frac{a^m}{a^n} = a^{m-n} \qquad (a \neq 0)$$

You will be relieved to know that these same laws of exponents "work" for any *rational numbers* used as exponents. (Remember that rational numbers are real numbers that are expressible as fractions of integers.)

> The five laws of exponents can be used for rational exponents as well as for integer exponents.

For example,

$$(1) \quad x^{2/3} \cdot x^{5/3} = x^{(2/3)+(5/3)} = x^{7/3}$$

$$(2) \quad (x^{3/4})^8 = x^{(3/4)\cdot(8/1)} = x^6$$

$$(3) \quad (xy)^{1/2} = x^{1/2} \cdot y^{1/2}$$

$$(4) \quad \left(\frac{a}{b}\right)^{2/3} = \frac{a^{2/3}}{b^{2/3}}$$

$$(5) \quad \frac{x^{3/4}}{x^{1/4}} = x^{(3/4)-(1/4)} = x^{2/4} = x^{1/2}$$

It is also true that negative rational exponents can be interpreted in the same way that we interpreted integer exponents.

$$a^{-p/q} = \frac{1}{a^{p/q}} \qquad (a \neq 0)$$

For example,

$$x^{-2/3} = \frac{1}{x^{2/3}} \qquad \text{and} \qquad \frac{x^{5/6}}{x^{7/6}} = x^{(5/6)-(7/6)} = x^{-2/6} = x^{-1/3} = \frac{1}{x^{1/3}}$$

You may be concerned by the fact that we have not yet given any meaning to a rational exponent. We shall arrive at an interpretation for a rational exponent in the third part of this section. For now, however, we need only know that rational exponents obey the five laws of exponents, so that expressions involving rational exponents should be simplified using exactly the same techniques we used for integer exponents.

You try Example 1.

Example 1. Simplify $x^{3/5} \cdot x^{4/5} \cdot x^{-2/5}$.

Solution: To multiply, we must _____ exponents.

$$x^{3/5} \cdot x^{4/5} \cdot x^{-2/5}$$
$$= x^{(3/5)+(4/5)-(2/5)}$$

Check your work on page 472. ▶

Example 2. Simplify $x^{1/3} \cdot x^{-1/6} \cdot x^{3/4}$.

Solution: We must add exponents, but here we must use an LCD for the fractional exponents.

$$x^{1/3} \cdot x^{-1/6} \cdot x^{3/4}$$
$$= x^{(1/3)-(1/6)+(3/4)}$$
$$= x^{(4/12)-(2/12)+(9/12)}$$
$$= x^{11/12}$$

Now try Example 3.

Example 3. Simplify $(x^{3/4}y^{-1/8})^4$.

Solution: To raise a product to a power, we must raise each factor to that power.

$$(x^{3/4}y^{-1/8})^4$$
$$= (x^{3/4})^4(y^{-1/8})^4$$
$$= x^{(3/4)\cdot(4/1)}y^{-(1/8)\cdot(4/1)}$$

Check your work on page 472.

Example 4. Simplify $\left(\dfrac{x^{1/6}z^{-2}}{y^{3/4}}\right)^{1/3}$.

Solution: To raise a fraction to a power, we raise the numerator and denominator to that power.

$$\left(\frac{x^{1/6}z^{-2}}{y^{3/4}}\right)^{1/3}$$
$$= \frac{(x^{1/6}z^{-2})^{1/3}}{(y^{3/4})^{1/3}}$$
$$= \frac{x^{(1/6)\cdot(1/3)}z^{-(2/1)\cdot(1/3)}}{y^{(3/4)\cdot(1/3)}}$$
$$= \frac{x^{1/18}z^{-2/3}}{y^{1/4}}$$
$$= \frac{x^{1/18}}{y^{1/4}z^{2/3}}$$

Example 5. Simplify $\dfrac{x^{7/8}y^{-1/12}}{x^{-3/4}y^{9/2}}$.

Solution: To divide, we must subtract exponents.

$$\frac{x^{7/8}y^{-1/12}}{x^{-3/4}y^{9/2}} = x^{(7/8)-(-3/4)}y^{-(1/12)-(9/2)}$$

$$= x^{(7/8)+(6/8)}y^{-(1/12)-(54/12)}$$

$$= x^{13/8}y^{-55/12}$$

$$= \frac{x^{13/8}}{y^{55/12}}$$

Remember that we agreed in our earlier work with positive and negative exponents that it is less confusing if we use the laws of exponents *first*. Once we arrive at our simplified expression, we then use the definition of a negative exponent to rewrite the expression using positive exponents.

⏩ Trial Run

Simplify.

_____ **1.** $x^{3/4} \cdot x^{9/4}$

_____ **2.** $a^{1/2} \cdot a^{3/4} \cdot a^{-1/4}$

_____ **3.** $x^{2/5} \cdot x^{3/10} \cdot x^{-1/2}$

_____ **4.** $(x^{2/3}y^{-1/2})^6$

_____ **5.** $\left(\dfrac{x^{5/2}y^{-15}}{a^{10}}\right)^{1/5}$

_____ **6.** $\dfrac{x^{5/8}y^{5/6}}{x^{3/4}y^{1/12}}$

Answers are on page 472.

Introducing Radical Expressions

Earlier we were reminded of the square-root notation used in Chapter 5. We recalled that if a is a positive number or zero, then

$$\sqrt{a} = b \qquad \text{means} \qquad b^2 = a$$

We also agreed that $\sqrt{a}$ would always represent the positive square root of a, called the *principal square root* of a. The important restriction that a must be a *positive* number or zero was necessary because

The square root of a negative number is *not* a real number.

When we are asked to find the *square root* of a, we are being asked to find a number whose *square* is the number a. Suppose that we are asked to find a number whose *cube* (or third power) is the number a, or a number whose *fourth* power is a and so on. Here we are looking for a **cube root,** or a **fourth root,** and so on.

Root	Symbol	Examples	Reasons
Square root	$\sqrt{a}$	$\sqrt{4} = 2$	$2^2 = 4$
Cube root	$\sqrt[3]{a}$	$\sqrt[3]{8} = 2$	$2^3 = 8$
		$\sqrt[3]{-8} = -2$	$(-2)^3 = -8$
Fourth root	$\sqrt[4]{a}$	$\sqrt[4]{16} = 2$	$2^4 = 16$
		$\sqrt[4]{-16}$ is not real	
Fifth root	$\sqrt[5]{a}$	$\sqrt[5]{32} = 2$	$2^5 = 32$
		$\sqrt[5]{-32} = -2$	$(-2)^5 = -32$

Do you see the pattern? The *odd* root of any real number will be real, but the *even* root of a negative number is not a real number.

Roots of Real Numbers

$$\sqrt[n]{a} = b \qquad \text{if} \qquad b^n = a$$

where *n* is a natural number and *a* is real.

1. If *a* is positive, then

$$\sqrt[n]{a} \text{ is real and positive}$$

2. If *a* is negative, and

if *n* is odd, then $\sqrt[n]{a}$ is real and negative

if *n* is even, then $\sqrt[n]{a}$ is *not* a real number

Let's pause a moment and identify the parts of a **radical expression.**

The natural number *n* is the **index;** it indicates what root we are trying to find. If no index number appears, it is understood to be a 2. The $\sqrt{}$ symbol is called the **radical** and the quantity within the radical is called the **radicand.**

Let's practice using radical notation by finding the values of some radical expressions.

Example 6. Find $\sqrt[3]{1000}$.

Solution

$\sqrt[3]{1000} = 10$ because $10^3 = 1000$.

You try Example 7.

Example 7. Find $\sqrt[4]{16}$.

Solution

$\sqrt[4]{16} = \underline{\hspace{1cm}}$ because $(\underline{\hspace{1cm}})^4 = 16$.

Check your work on page 472. ▶

Example 8. Find $\sqrt[3]{\dfrac{-1}{27}}$.

Solution

$$\sqrt[3]{\dfrac{-1}{27}} = \dfrac{-1}{3} \text{ because } \left(\dfrac{-1}{3}\right)^3 = \dfrac{-1}{27}.$$

Now try Example 9.

Example 9. Find $\sqrt[5]{\dfrac{1}{32}}$.

Solution

$$\sqrt[5]{\dfrac{1}{32}} = \underline{\quad\quad} \text{ because } (\underline{\quad\quad})^5 = \dfrac{1}{32}.$$

Check your work on page 472. ▶

Example 10. Find $\sqrt[6]{-1}$.

Solution: $\sqrt[6]{-1}$ is not real. An even root of a negative number is not a real number.

You try Example 11.

Example 11. Find $\sqrt[7]{-1}$.

Solution

$$\sqrt[7]{-1} = \underline{\quad\quad} \text{ because } (\underline{\quad\quad})^7 = -1.$$

Check your work on page 472. ▶

⫸ Trial Run

Find the value of each radical expression.

_____ **1.** $\sqrt[5]{243}$

_____ **2.** $\sqrt[3]{-125}$

_____ **3.** $\sqrt{-9}$

_____ **4.** $\sqrt[5]{-32}$

_____ **5.** $\sqrt[4]{\dfrac{1}{16}}$

_____ **6.** $\sqrt[3]{\dfrac{-1}{8}}$

Answers are on page 472.

Relating Rational Exponents and Radicals

Suppose that we return to rational exponents and see if we can come up with an interpretation for exponents that are fractions. From the second law of exponents we know that

$$(a^{1/2})^2 = a \qquad\qquad (a^{1/3})^3 = a$$

So $a^{1/2}$ is the number whose square is a. But we have already agreed that $\sqrt{a}$ is the number whose square is a. Therefore, we conclude that

$$a^{1/2} = \sqrt{a}$$

So $a^{1/3}$ is the number whose cube is a. But we have already agreed that $\sqrt[3]{a}$ is the number whose cube is a. Therefore, we conclude that

$$a^{1/3} = \sqrt[3]{a}$$

In fact, for any natural number n, we define

$$\boxed{\,a^{1/n} = \sqrt[n]{a} \qquad \text{provided that the root is real}\,}$$

To avoid having to make restrictions upon a when n is even, we often precede a list of exercises with the statement: *Assume that all radical expressions represent real numbers.* Let's make this assumption as we practice switching back and forth between radical and exponential expressions.

Example 12. Write $\sqrt[3]{4x}$ in exponential form.

   Solution

$$\sqrt[3]{4x}$$
$$= (4x)^{1/3}$$

You try Example 13.

Example 13. Write $3\sqrt[5]{y}$ in exponential form.

   Solution

$$3\sqrt[5]{y}$$
$$=$$

Check your work on page 472. ▶

Example 14. Write $(x + y)^{1/3}$ in radical form.

   Solution

$$(x + y)^{1/3}$$
$$= \sqrt[3]{x + y}$$

Now try Example 15.

Example 15. Write $x^{1/3} + y^{1/3}$ in radical form.

   Solution

$$x^{1/3} + y^{1/3}$$
$$=$$

Check your work on page 472. ▶

Example 16. Write $9^{1/2} + 16^{1/2}$ in radical form and simplify if possible.

   Solution

$$9^{1/2} + 16^{1/2}$$
$$= \sqrt{9} + \sqrt{16}$$
$$= 3 + 4$$
$$= 7$$

You complete Example 17.

Example 17. Write $(9 + 16)^{1/2}$ in radical form and simplify if possible.

   Solution

$$(9 + 16)^{1/2}$$
$$= \sqrt{9 + 16}$$

Check your work on page 472. ▶

From Examples 16 and 17, we note that

$$\sqrt{9} + \sqrt{16} \neq \sqrt{9 + 16}$$
$$3 + 4 \neq \sqrt{25}$$
$$7 \neq 5$$

and we state in general that *the nth root of a sum is **not** equal to the sum of the nth roots.*

$$\sqrt[n]{a + b} \neq \sqrt[n]{a} + \sqrt[n]{b}$$

Now that we know that

$$a^{1/n} = \sqrt[n]{a}$$

perhaps we can use this definition and the laws of exponents to find meaning for expressions such as $x^{2/3}$ and $a^{3/4}$.

For $x^{2/3}$,

$$x^{2/3} = (x^{1/3})^2 \qquad \text{or} \qquad x^{2/3} = (x^2)^{1/3}$$
$$= (\sqrt[3]{x})^2 \qquad\qquad\qquad = \sqrt[3]{x^2}$$

For $a^{3/4}$,

$$a^{3/4} = (a^{1/4})^3 \qquad \text{or} \qquad a^{3/4} = (a^3)^{1/4}$$
$$= (\sqrt[4]{a})^3 \qquad\qquad\qquad = \sqrt[4]{a^3}$$

For each of these examples we arrived at two possible interpretations. Indeed, if we assume that our radicals represent real numbers, then either interpretation is acceptable.

Definition of Rational Exponent

$$a^{m/n} = \sqrt[n]{a^m} = (\sqrt[n]{a})^m$$

where m and n are natural numbers, and $a \geq 0$ whenever n is *even*.

Notice that in switching between fractional exponent form and radical form:

$$a^{m/n} = \sqrt[n]{a^m}$$

The denominator of the fractional exponent tells us the index of the radical.

The numerator of the fractional exponent tells us the exponent on the base.

Example 18. Write $8^{2/3}$ in radical form and simplify.

Solution

$$8^{2/3} = \sqrt[3]{8^2} \qquad \text{or} \qquad 8^{2/3} = (\sqrt[3]{8})^2$$
$$= \sqrt[3]{64} \qquad\qquad\qquad = 2^2$$
$$= 4 \qquad\qquad\qquad\quad = 4$$

Example 19. Write $32^{7/5}$ in radical form and simplify.

Solution

$$32^{7/5} = (\sqrt[5]{32})^7 \qquad \text{or} \qquad 32^{7/5} = \sqrt[5]{32^7}$$
$$= 2^7 \qquad\qquad\qquad\qquad = \sqrt[5]{34,359,738,368}$$
$$= 128 \qquad\qquad\qquad\qquad = 128$$

You should agree that in Example 19 it was much easier to find the root first, then the power. Quite often this is the case.

Example 20. Write $(x^2 - 1)^{4/7}$ radical form.

Solution

$$(x^2 - 1)^{4/7} = \sqrt[7]{(x^2 - 1)^4}$$

or

$$(x^2 - 1)^{4/7} = (\sqrt[7]{x^2 - 1})^4$$

You try Example 21.

Example 21. Write $3x^{5/11}$ in radical form.

Solution

$$3x^{5/11} = 3\underline{\qquad}$$

or

$$3x^{5/11} = 3(\underline{\qquad})$$

Check your work on page 472. ▶

Example 22. Write $(x - 5)^{-2/3}$ in radical form.

Solution

$$(x - 5)^{-2/3} = \frac{1}{(x - 5)^{2/3}}$$

$$= \frac{1}{\sqrt[3]{(x - 5)^2}}$$

or

$$(x - 5)^{-2/3} = \frac{1}{(\sqrt[3]{x - 5})^2}$$

Example 23. Write $\frac{1}{\sqrt[3]{x^2y}}$ in exponential form.

Solution

$$\frac{1}{\sqrt[3]{x^2y}} = \frac{1}{(x^2y)^{1/3}}$$

$$= \frac{1}{x^{2/3}y^{1/3}}$$

Now try Example 24.

Example 24. Write $-\sqrt{x}$ in exponential form. Assume that $x \geq 0$.

Solution

$$-\sqrt{x} = -1 \cdot \sqrt{x}$$

$$= -x^{\overline{}}$$

Check your work on page 472. ▶

Example 25. Write $\sqrt{-x}$ in exponential form. Assume that $x \leq 0$.

Solution

$$\sqrt{-x} = (-x)^{1/2}$$

⫸ **Trial Run**

Assume that all radical expressions represent real numbers.

_____ **1.** Simplify $27^{2/3}$.

_____ **2.** Write $x^{3/8}$ in radical form.

_____ **3.** Write $\frac{1}{\sqrt[5]{x^2y^3}}$ in exponential form.

_____ **4.** Write $-\sqrt[3]{a}$ in exponential form.

_____ **5.** Write $(x - 3)^{3/4}$ in radical form.

_____ **6.** Write $(x - y)^{-2/5}$ in radical form.

Answers are on page 472.

▶ **Examples You Completed**

Example 1. Simplify $x^{3/5} \cdot x^{4/5} \cdot x^{-2/5}$.

Solution: To multiply, we must add exponents.

$$x^{3/5} \cdot x^{4/5} \cdot x^{-2/5}$$
$$= x^{(3/5)+(4/5)-(2/5)}$$
$$= x^{5/5}$$
$$= x$$

Example 3

Solution

$$(x^{3/4})^4(y^{-1/8})^4$$
$$= x^{(3/4)\cdot(4/1)}y^{-(1/8)\cdot(4/1)}$$
$$= x^3 y^{-1/2}$$
$$= \frac{x^3}{y^{1/2}}$$

Example 7. Find $\sqrt[4]{16}$.

Solution

$$\sqrt[4]{16} = 2 \text{ because } (2)^4 = 16.$$

Example 9. Find $\sqrt[5]{\dfrac{1}{32}}$.

Solution

$$\sqrt[5]{\frac{1}{32}} = \frac{1}{2} \text{ because } \left(\frac{1}{2}\right)^5 = \frac{1}{32}.$$

Example 11. Find $\sqrt[7]{-1}$.

Solution

$$\sqrt[7]{-1} = -1 \text{ because } (-1)^7 = -1.$$

Example 13. Write $3\sqrt[5]{y}$ in exponential form.

Solution

$$3\sqrt[5]{y}$$
$$= 3y^{1/5}$$

Example 15. Write $x^{1/3} + y^{1/3}$ in radical form.

Solution

$$x^{1/3} + y^{1/3}$$
$$= \sqrt[3]{x} + \sqrt[3]{y}$$

Example 17. Write $(9 + 16)^{1/2}$ in radical form.

Solution

$$(9 + 16)^{1/2}$$
$$= \sqrt{9 + 16}$$
$$= \sqrt{25}$$
$$= 5$$

Example 21. Write $3x^{5/11}$ in radical form.

Solution

$$3x^{5/11} = 3\sqrt[11]{x^5}$$
$$\text{or}$$
$$3x^{5/11} = 3(\sqrt[11]{x})^5$$

Example 24. Write $-\sqrt{x}$ in exponential form. Assume that $x \geq 0$.

Solution

$$-\sqrt{x} = -1 \cdot \sqrt{x}$$
$$= -x^{1/2}$$

Answers to Trial Runs

page 466   **1.** x^3 **2.** a **3.** $x^{1/5}$ **4.** $\dfrac{x^4}{y^3}$ **5.** $\dfrac{x^{1/2}}{y^3 a^2}$ **6.** $\dfrac{y^{3/4}}{x^{1/8}}$

page 468   **1.** 3   **2.** -5 **3.** Not a real number   **4.** -2 **5.** $\dfrac{1}{2}$ **6.** $-\dfrac{1}{2}$

page 471   **1.** 9   **2.** $\sqrt[8]{x^3}$ or $(\sqrt[8]{x})^3$ **3.** $\dfrac{1}{x^{2/5}y^{2/5}}$ **4.** $-a^{1/3}$ **5.** $\sqrt[4]{(x-3)^3}$ or $(\sqrt[4]{x-3})^3$

   6. $\dfrac{1}{\sqrt[5]{(x-y)^2}}$ or $\dfrac{1}{(\sqrt[5]{x-y})^2}$

EXERCISE SET 7.1

Simplify.

_____ 1. $x^{2/5} \cdot x^{8/5}$ _____ 2. $x^{3/4} \cdot x^{5/4}$

_____ 3. $a^{2/3} \cdot a^{7/3} \cdot a^{-4/3}$ _____ 4. $a^{2/7} \cdot a^{-1/7} \cdot a^{4/7}$

_____ 5. $(x^{2/3})^6$ _____ 6. $(x^{3/4})^8$

_____ 7. $(x^{3/4} \cdot y^{-3})^{4/3}$ _____ 8. $(x^{3/5} \cdot y^{-6})^{5/3}$

_____ 9. $(a^{3/4} \cdot b^{-1/3})^{-2}$ _____ 10. $(a^{5/6} \cdot b^{-2/5})^{-3}$

_____ 11. $x^{2/3} \cdot x^{2/5}$ _____ 12. $x^{1/3} \cdot x^{3/4}$

_____ 13. $\left(\dfrac{x^{3/5}y^{-3/10}}{z^{2/5}}\right)^{5/6}$ _____ 14. $\left(\dfrac{x^{3/4}y^{-1/8}}{z^{5/6}}\right)^{4/5}$

_____ 15. $\dfrac{x^{3/4}y^{2/3}}{x^{1/4}y^{1/3}}$ _____ 16. $\dfrac{x^{5/6}y^{3/5}}{x^{1/6}y^{2/5}}$

_____ 17. $\dfrac{a^{1/4}b^3}{a^{5/8}b^{1/2}}$ _____ 18. $\dfrac{a^{2/3}b^4}{a^{5/6}b^{2/5}}$

_____ 19. $(x^{4/3} \cdot y^{-3/5} \cdot z^{6/7})^{1/12}$ _____ 20. $(x^{3/4} \cdot y^{-5/2} \cdot z^{15/2})^{1/15}$

Find the value of each of the following.

_____ 21. $\sqrt[3]{-27}$ _____ 22. $\sqrt[3]{-8}$

_____ 23. $\sqrt[4]{81}$ _____ 24. $\sqrt[4]{256}$

_____ 25. $\sqrt[4]{-16}$ _____ 26. $\sqrt[4]{-1}$

_____ 27. $\sqrt[3]{\dfrac{-64}{125}}$ _____ 28. $\sqrt[3]{\dfrac{-8}{27}}$

_____ 29. $\sqrt[5]{32}$ _____ 30. $\sqrt[5]{243}$

_____ 31. $\sqrt{\dfrac{25}{81}}$ _____ 32. $\sqrt{\dfrac{36}{49}}$

Write each of the following in exponential form.

_____ 33. $\sqrt[4]{2x}$ _____ 34. $\sqrt{3x}$

_____ 35. $5\sqrt{a}$ _____ 36. $3\sqrt{y}$

_____ 37. $\sqrt{x^2 + 16}$ _____ 38. $\sqrt{x^2 + 25}$

_____ 39. $-\sqrt[5]{y}$ _____ 40. $-\sqrt[3]{x}$

_____ 41. $\sqrt[3]{-a}$ _____ 42. $\sqrt[5]{-b}$

_____ 43. $\sqrt[3]{x} + \sqrt[3]{y}$ _____ 44. $\sqrt{a} + \sqrt{b}$

_____ 45. $\sqrt[5]{x^2 y}$ _____ 46. $\sqrt[4]{x^3 y}$

_____ 47. $\sqrt[5]{(x - y)^2}$ _____ 48. $\sqrt[3]{(x + y)^2}$

_____ 49. $\dfrac{1}{\sqrt[3]{x^4 y^2}}$ _____ 50. $\dfrac{1}{\sqrt[5]{x^4 y^3}}$

Write each of the following in radical form and simplify as much as possible. Assume that all radical expressions represent real numbers.

_____ 51. $(3y^3)^{1/5}$ _____ 52. $(2x^3)^{1/4}$

_____ 53. $4x^{1/2}$ _____ 54. $16x^{1/4}$

_____ 55. $(a + b)^{1/3}$ _____ 56. $(a - b)^{1/2}$

_____ 57. $x^{1/5} + y^{1/5}$ _____ 58. $x^{1/6} - y^{1/6}$

_____ 59. $(9 + 16)^{1/2}$ _____ 60. $(25 + 144)^{1/2}$

_____ 61. $(4 + 9)^{1/2}$ _____ 62. $(25 + 4)^{1/2}$

_____ 63. $y^{2/3}$ _____ 64. $x^{5/8}$

_____ 65. $(x - 3)^{2/3}$ _____ 66. $(x + 1)^{3/4}$

_____ 67. $(a - b)^{-3/5}$ _____ 68. $(a + b)^{-2/3}$

_____ 69. $8^{2/3}$ _____ 70. $32^{2/5}$

_____ 71. $81^{-3/4}$ _____ 72. $27^{-2/3}$

_____ 73. $8^{1/3} + 27^{1/3}$ _____ 74. $25^{1/2} + 100^{1/2}$

_____ 75. $\left(\dfrac{25}{36}\right)^{-1/2}$ _____ 76. $\left(\dfrac{4}{49}\right)^{-1/2}$

☆ Stretching the Topics

_____ 1. Find the value of $25^{-3/2} \cdot 16^{3/4} \cdot 8^{-4/3}$.

_____ 2. Simplify $\left(\dfrac{64^{-3}}{27n^{-6}}\right)^{2/3}$.

_____ 3. Simplify $\dfrac{27^{3/4} x^{5/6} y^{1/6}}{27^{1/12} x^{-1/6} y^{-1/4}}$.

Check your answers in the back of your book.

If you can complete **Checkup 7.1,** you are ready to go on to Section 7.2.

CHECKUP 7.1

Simplify.

_____ 1. $a^{3/4} \cdot a^{1/2} \cdot a^{-1/4}$

_____ 2. $(x^{1/10} \cdot y^{-2/5})^{-5}$

_____ 3. $\dfrac{x^{3/4} \cdot y^{2/5}}{x^{1/2} \cdot y^{7/2}}$

_____ 4. $(x^{1/9} \cdot y^{-15/4} \cdot z^{5/6})^{3/5}$

Find the value.

_____ 5. $\sqrt{-81}$

_____ 6. $\sqrt[5]{\dfrac{-1}{32}}$

Write in exponential form.

_____ 7. $\sqrt[7]{x^2 y^3}$

_____ 8. $\sqrt{(x + y)^3}$

Write in radical form and simplify if possible.

_____ 9. $(x - y)^{2/3}$

_____ 10. $64^{5/6}$

Check your answers in the back of your book.

If You Missed Problems:	You Should Review Examples:
1, 2	1–3
3, 4	3–5
5	10
6	8, 9
7	12, 13
8	14
9, 10	18–20

7.2 Working with Radical Expressions

Simplifying Radical Expressions

In Chapter 5 we learned to simplify radicals that were square roots of perfect squares. Let's look again at $\sqrt{a^2}$.

Suppose that $a > 0$	Suppose that $a < 0$
$\sqrt{4^2} = \sqrt{16} = 4$	$\sqrt{(-4)^2} = \sqrt{16} = 4$
$\sqrt{\left(\frac{1}{3}\right)^2} = \sqrt{\frac{1}{9}} = \frac{1}{3}$	$\sqrt{\left(\frac{-1}{3}\right)^2} = \sqrt{\frac{1}{9}} = \frac{1}{3}$
$\sqrt{11^2} = \sqrt{121} = 11$	$\sqrt{(-11)^2} = \sqrt{121} = 11$

Because we know that the principal square root of a number is always nonnegative, we have found that $\sqrt{a^2}$ is a only when a is *not* negative (see the first column in the table). If a is negative, then $\sqrt{a^2}$ is the *opposite* of a (see the second column). In other words,

$$\sqrt{a^2} = \begin{cases} a & \text{if } a \text{ is positive or zero} \\ -a & \text{if } a \text{ is negative} \end{cases}$$

Doesn't this definition seem familiar? From Chapter 1 you should recognize it as the definition for the **absolute value of** ***a.***

$$\sqrt{a^2} = |a|$$

As a matter of fact, this definition should be used whenever we work with an *even n*th root of an *n*th power. Of course, if *n* is *odd,* it is not correct to use absolute value notation.

$$\sqrt[5]{2^5} = \sqrt[5]{32} = 2$$
$$\sqrt[3]{(-4)^3} = \sqrt[3]{-64} = -4$$

We summarize our discussion by noting that

$$\sqrt[n]{a^n} = \begin{cases} a & \text{if } n \text{ is odd} \\ |a| & \text{if } n \text{ is even} \end{cases}$$

Example 1. Simplify $\sqrt[7]{x^7}$.

   Solution: Here n is odd.

$$\sqrt[7]{x^7} = x$$

You try Example 2.

Example 2. Simplify $\sqrt[4]{y^4}$.

   Solution: Here n is _____ .

$$\sqrt[4]{y^4} = \underline{}$$

Check your work on page 483. ▶

Example 3. Simplify $\sqrt{(x - 3)^2}$.

   Solution: Here n is even.

$$\sqrt{(x - 3)^2} = |x - 3|$$

Now complete Example 4.

Example 4. Simplify $\sqrt[5]{(a - 1)^5}$.

   Solution: Here n is _____ .

$$\sqrt[5]{(a - 1)^5} = \underline{}$$

Check your work on page 483. ▶

In Chapter 5 we also learned to simplify square roots whose radicands were products or quotients. These same properties can be used to simplify radicals with any index, n.

$$\sqrt[n]{a \cdot b} = \sqrt[n]{a} \cdot \sqrt[n]{b}$$

$$\sqrt[n]{\frac{a}{b}} = \frac{\sqrt[n]{a}}{\sqrt[n]{b}} \qquad (b \neq 0)$$

where it is understood that each radical expression represents a real number.

We say that a radical expression is in **simplest form** when no factor appears under the radical with an exponent that is greater than or equal to the index on the radical. To simplify a radical expression with index n, we follow these steps.

Simplifying Radical Expressions

1. Write the radicand as a product of factors, with the most possible factors written as nth powers.
2. Find the nth roots of all the factors that are nth powers.
3. Leave the remaining factors in the radicand.

For simplicity's sake, we shall assume throughout the following examples that *if the index is even, then all variable factors in the radicand represent nonnegative numbers*. Such an assumption allows us to dispense with absolute value notation.

Example 5. Simplify $\sqrt[4]{32x^5y}$.

Solution

$\sqrt[4]{32x^5y}$

$= \sqrt[4]{16 \cdot 2 \cdot x^4 \cdot x \cdot y}$ Look for perfect fourth powers.

$= \sqrt[4]{16} \cdot \sqrt[4]{x^4} \cdot \sqrt[4]{2xy}$

$= 2x\sqrt[4]{2xy}$ Remove fourth roots.

Example 6. Simplify $\sqrt[3]{\dfrac{8x^3}{y^6}}$.

Solution

$\sqrt[3]{\dfrac{8x^3}{y^6}}$

$= \dfrac{\sqrt[3]{8x^3}}{\sqrt[3]{y^6}}$ Separate the radicals.

$= \dfrac{\sqrt[3]{8} \cdot \sqrt[3]{x^3}}{\sqrt[3]{y^3} \cdot \sqrt[3]{y^3}}$ Look for perfect cubes.

$= \dfrac{2x}{y \cdot y}$ Remove cube roots.

$= \dfrac{2x}{y^2}$ Simplify.

A quotient of two radical expressions should be written as a *single* radical whenever the numerator and denominator contain common factors. Once the numerator and denominator have no more common factors, they should be considered as the quotient of two *separate* radical expressions.

You try Example 7.

Example 7. Simplify $\sqrt[3]{\dfrac{81x^4}{8y^3}}$.

Solution

$\sqrt[3]{\dfrac{81x^4}{8y^3}} = \dfrac{\sqrt[3]{81x^4}}{\sqrt[3]{8y^3}}$ Separate the radicals.

$= \dfrac{\sqrt[3]{27 \cdot 3 \cdot x^3 \cdot x}}{\sqrt[3]{8 \cdot y^3}}$ Look for perfect cubes.

$= \dfrac{\sqrt[3]{27} \; \sqrt[3]{x^3} \; \sqrt[3]{3x}}{\sqrt[3]{8} \; \sqrt[3]{y^3}}$

$=$ Remove cube roots.

Check your work on page 483. ▶

Example 8. Simplify $\dfrac{\sqrt[5]{-64x^8}}{\sqrt[5]{2x}}$.

Solution

$\dfrac{\sqrt[5]{-64x^6}}{\sqrt[5]{2x}} = \sqrt[5]{\dfrac{-64x^8}{2x}}$ Write as one radical.

$= \sqrt[5]{-32x^7}$ Reduce.

$= \sqrt[5]{-32 \cdot x^5 \, x^2}$

$= -2x\sqrt[5]{x^2}$ Remove fifth roots.

Now try Example 9.

Example 9. Simplify $\sqrt[7]{(x + 3)^8}$.

Solution

$\sqrt[7]{(x + 3)^8}$

$= \sqrt[7]{(x + 3)^7 \cdot (x + 3)}$

$=$

Check your work on page 483. ▶

Example 10. Simplify $\sqrt{x^2 - 10x + 25}$.

Solution

$\sqrt{x^2 - 10x + 25}$

$= \sqrt{(x - 5)^2}$

$= x - 5$

IIII➡ **Trial Run**

Simplify. Assume that any variable expression under a radical with an even index represents a nonnegative number.

_____ **1.** $\sqrt{48x^3y^2}$

_____ **2.** $\sqrt[3]{\dfrac{54a^2}{125b^3}}$

_____ **3.** $\dfrac{\sqrt[5]{-96x^9}}{\sqrt[5]{3x^3}}$

_____ **4.** $\sqrt{\dfrac{49x^5}{64y^6}}$

_____ **5.** $\sqrt[6]{(x-2)^7}$

_____ **6.** $\sqrt{x^2 - 4x + 4}$

Answers are on page 483.

Adding and Subtracting Radical Expressions

Very early in our study of algebra we learned that in addition and subtraction, we may combine only *like terms*. The same rule applies to radical expressions; we may combine only **like radicals.**

> Like radicals have the same index *and* the same radicand.

For example,

Like Radical Terms	Not Like Radical Terms
$\sqrt{3}$ and $6\sqrt{3}$	$\sqrt{5}$ and $\sqrt{3}$
$4\sqrt[3]{2}$ and $-3\sqrt[3]{2}$	$\sqrt[3]{2}$ and $\sqrt{2}$
$\sqrt{2x}$ and $5\sqrt{2x}$	$3\sqrt{2x}$ and $2\sqrt{3x}$
$x\sqrt{y}$ and $-2x\sqrt{y}$	$x\sqrt{y}$ and $y\sqrt{x}$

We learned to combine like variable terms by combining their numerical coefficients. Similarly,

> To combine like radicals, we combine their numerical coefficients and keep the same radical.

You try combining the radicals in Example 11.

Example 11. Simplify

$$\sqrt{2} + 5\sqrt{2} - 8\sqrt{2}.$$

Solution

$$\sqrt{2} + 5\sqrt{2} - 8\sqrt{2}$$
$$= 1\sqrt{2} + 5\sqrt{2} - 8\sqrt{2}$$
$$= \underline{\quad}\sqrt{2}$$

Check your work on page 483. ▶

Example 13. Simplify

$$6 + x\sqrt[3]{4} + 7 - 5x\sqrt[3]{4}.$$

Solution

$$6 + x\sqrt[3]{4} + 7 - 5x\sqrt[3]{4}$$
$$= 6 + 7 + x\sqrt[3]{4} - 5x\sqrt[3]{4}$$
$$= \underline{\quad} - \underline{\quad}\sqrt[3]{4}$$

Check your work on page 483. ▶

Example 12. Simplify

$$\sqrt{2} - 3\sqrt{5} - 9\sqrt{2} + 3\sqrt{5}.$$

Solution

$$\sqrt{2} - 3\sqrt{5} - 9\sqrt{2} + 3\sqrt{5}$$
$$= 1\sqrt{2} - 9\sqrt{2} - 3\sqrt{5} + 3\sqrt{5}$$
$$= -8\sqrt{2}$$

Now try Example 13.

Example 14. Simplify

$$2(\sqrt{5} + 1) - 3(8 - 2\sqrt{5}).$$

Solution

$$2(\sqrt{5} + 1) - 3(8 - 2\sqrt{5})$$
$$= 2\sqrt{5} + 2 - 24 + 6\sqrt{5}$$
$$= 2\sqrt{5} + 6\sqrt{5} + 2 - 24$$
$$= 8\sqrt{5} - 22$$

Sometimes we shall be asked to combine radicals that do not appear to be alike. In such cases we must try to simplify each radical expression before deciding whether or not they are like radicals.

Try Example 15.

Example 15. Simplify $\sqrt{50} + \sqrt{8}$.

Solution

$$\sqrt{50} + \sqrt{8}$$
$$= \sqrt{25 \cdot 2} + \sqrt{4 \cdot 2}$$
$$= \underline{\quad}\sqrt{2} + \underline{\quad}\sqrt{2}$$
$$= \underline{\quad}\sqrt{2}$$

Check your work on page 483. ▶

Example 16. Simplify

$$-6\sqrt[3]{24x} + 10\sqrt[3]{81x}.$$

Solution

$$-6\sqrt[3]{24x} + 10\sqrt[3]{81x}$$
$$= -6\sqrt[3]{8 \cdot 3x} + 10\sqrt[3]{27 \cdot 3x}$$
$$= -6 \cdot 2\sqrt[3]{3x} + 10 \cdot 3\sqrt[3]{3x}$$
$$= -12\sqrt[3]{3x} + 30\sqrt[3]{3x}$$
$$= 18\sqrt[3]{3x}$$

Example 17. Simplify $\sqrt{\dfrac{8}{9}} + \dfrac{5\sqrt{2}}{2} - \sqrt{32}$.

Solution

$$\sqrt{\frac{8}{9}} + \frac{5\sqrt{2}}{2} - \sqrt{32}$$

$$= \sqrt{\frac{8}{9}} + \frac{5\sqrt{2}}{2} - \sqrt{16 \cdot 2}$$

$$= \frac{\sqrt{4 \cdot 2}}{3} + \frac{5\sqrt{2}}{2} - 4\sqrt{2}$$

$$= \frac{2\sqrt{2}}{3} + \frac{5\sqrt{2}}{2} - \frac{4\sqrt{2}}{1} \qquad \text{LCD: } 3 \cdot 2 = 6.$$

$$= \frac{2 \cdot 2\sqrt{2}}{6} + \frac{3 \cdot 5\sqrt{2}}{6} - \frac{6 \cdot 4\sqrt{2}}{6}$$

$$= \frac{4\sqrt{2} + 15\sqrt{2} - 24\sqrt{2}}{6}$$

$$= \frac{-5\sqrt{2}}{6}$$

Example 18. Simplify $4\sqrt[5]{-32x^6} + 8x\sqrt[5]{-x}$.

Solution

$$4\sqrt[5]{-32x^6} + 8x\sqrt[5]{-x}$$

$$= 4\sqrt[5]{-32 \cdot x^5 \cdot x} + 8x\sqrt[5]{-1 \cdot x}$$

$$= 4(-2)x\sqrt[5]{x} + 8x(-1)\sqrt[5]{x}$$

$$= -8x\sqrt[5]{x} - 8x\sqrt[5]{x}$$

$$= -16x\sqrt[5]{x}$$

▐▶ **Trial Run**

Simplify.

_____ **1.** $\sqrt{3} + 7\sqrt{3} - 5\sqrt{3}$ _____ **2.** $\sqrt{2} - 3\sqrt{5} + 4\sqrt{2} + 2\sqrt{5}$

_____ **3.** $5x\sqrt{6} - 3x\sqrt{6} - 7x\sqrt{6}$ _____ **4.** $3(\sqrt{2} - 4\sqrt[3]{2}) - (\sqrt{2} - 3\sqrt[3]{2})$

_____ **5.** $\sqrt{18} + \sqrt{2}$ _____ **6.** $5x\sqrt{8} + 3\sqrt{32x^2}$

_____ **7.** $4\sqrt[3]{24} - 5\sqrt[3]{81}$ _____ **8.** $\sqrt{\dfrac{27}{16}} + \sqrt{\dfrac{48}{25}}$

Answers are on page 483.

▶ **Examples You Completed** _____

Example 2. Simplify $\sqrt[4]{y^4}$.

Solution: Here n is even.
$$\sqrt[4]{y^4} = |y|$$

Example 7. Simplify $\sqrt[3]{\dfrac{81x^4}{8y^3}}$.

Solution

$$\sqrt[3]{\frac{81x^4}{8y^3}} = \frac{\sqrt[3]{81x^4}}{\sqrt[3]{8y^3}}$$

$$= \frac{\sqrt[3]{27 \cdot 3 \cdot x^3 \cdot x}}{\sqrt[3]{8 \cdot y^3}}$$

$$= \frac{\sqrt[3]{27}\ \sqrt[3]{x^3}\ \sqrt[3]{3x}}{\sqrt[3]{8}\ \sqrt[3]{y^3}}$$

$$= \frac{3x\sqrt[3]{3x}}{2y}$$

Example 13. Simplify
$$6 + x\sqrt[3]{4} + 7 - 5x\sqrt[3]{4}.$$

Solution

$$6 + x\sqrt[3]{4} + 7 - 5x\sqrt[3]{4}$$
$$= 6 + 7 + x\sqrt[3]{4} - 5x\sqrt[3]{4}$$
$$= 13 - 4x\sqrt[3]{4}$$

Example 4. Simplify $\sqrt[5]{(a-1)^5}$.

Solution: Here n is odd.
$$\sqrt[5]{(a-1)^5} = a - 1$$

Example 9. Simplify $\sqrt[7]{(x+3)^8}$.

Solution

$$\sqrt[7]{(x+3)^8}$$
$$= \sqrt[7]{(x+3)^7 \cdot (x+3)}$$
$$= (x+3)\sqrt[7]{x+3}$$

Example 11. Simplify $\sqrt{2} + 5\sqrt{2} - 8\sqrt{2}$.

Solution

$$\sqrt{2} + 5\sqrt{2} - 8\sqrt{2}$$
$$= 1\sqrt{2} + 5\sqrt{2} - 8\sqrt{2}$$
$$= -2\sqrt{2}$$

Example 15. Simplify $\sqrt{50} + \sqrt{8}$.

Solution

$$\sqrt{50} + \sqrt{8}$$
$$= \sqrt{25 \cdot 2} + \sqrt{4 \cdot 2}$$
$$= 5\sqrt{2} + 2\sqrt{2}$$
$$= 7\sqrt{2}$$

Answers to Trial Runs _____

page 480   **1.** $4xy\sqrt{3x}$ **2.** $\dfrac{3a\sqrt[3]{2a}}{5b}$ **3.** $-2x\sqrt[5]{x}$ **4.** $\dfrac{7x^2\sqrt{x}}{8y^3}$ **5.** $(x-2)\sqrt[6]{(x-2)}$ **6.** $x - 2$

page 482   **1.** $3\sqrt{3}$ **2.** $5\sqrt{2} - \sqrt{5}$ **3.** $-5x\sqrt{6}$ **4.** $2\sqrt{2} - 9\sqrt[3]{2}$ **5.** $4\sqrt{2}$ **6.** $22x\sqrt{2}$
7. $-7\sqrt[3]{3}$ **8.** $\dfrac{31\sqrt{3}}{20}$

EXERCISE SET 7.2

Simplify. Assume that any variable expression under a radical with an even index represents a nonnegative number.

_____ **1.** $\sqrt{12b^2}$ _____ **2.** $\sqrt{50a^2}$

_____ **3.** $\sqrt{24x^5y^7}$ _____ **4.** $\sqrt{45x^3y^5}$

_____ **5.** $9a\sqrt{16a^{12}}$ _____ **6.** $4b\sqrt{25b^8}$

_____ **7.** $\sqrt{24c^{10}d^3}$ _____ **8.** $\sqrt{45c^6d^5}$

_____ **9.** $\sqrt[3]{-40a^5b^4}$ _____ **10.** $\sqrt[3]{-128a^7b^8}$

_____ **11.** $\sqrt{\dfrac{48a^5}{9b^2}}$ _____ **12.** $\sqrt{\dfrac{63a^9}{16b^4}}$

_____ **13.** $\sqrt[3]{\dfrac{16x^4}{27y^6}}$ _____ **14.** $\sqrt[3]{\dfrac{54x^7}{125y^3}}$

_____ **15.** $\sqrt{\dfrac{144x^7}{25y^4}}$ _____ **16.** $\sqrt{\dfrac{121x^5}{36y^2}}$

_____ **17.** $\dfrac{\sqrt[5]{64x^{10}}}{\sqrt[5]{2x^3}}$ _____ **18.** $\dfrac{\sqrt[5]{486x^9}}{\sqrt[5]{2x^2}}$

_____ **19.** $\dfrac{\sqrt{147x^5}}{\sqrt{3x}}$ _____ **20.** $\dfrac{\sqrt{162x^7}}{\sqrt{2x}}$

_____ **21.** $\dfrac{\sqrt[3]{32x^5}}{\sqrt[3]{2x}}$ _____ **22.** $\dfrac{\sqrt[3]{162x^7}}{\sqrt[3]{3x^2}}$

_____ **23.** $\sqrt[4]{32x^9y^{10}}$ _____ **24.** $\sqrt[3]{72x^4y^3}$

_____ **25.** $\sqrt[5]{(x-3)^6}$ _____ **26.** $\sqrt[4]{(x-5)^5}$

_____ **27.** $\sqrt{x^2-6x+9}$ _____ **28.** $\sqrt{x^2+14x+49}$

_____ **29.** $\sqrt{4x^2+4x+1}$ _____ **30.** $\sqrt{4x^2+12x+9}$

_____ **31.** $\sqrt{5}+9\sqrt{5}-3\sqrt{5}$ _____ **32.** $\sqrt{3}-6\sqrt{3}+8\sqrt{3}$

_____ **33.** $\sqrt{3}-2\sqrt{7}+5\sqrt{3}+8\sqrt{7}$ _____ **34.** $2\sqrt{2}-5\sqrt{6}-8\sqrt{2}+9\sqrt{6}$

_____ **35.** $3x\sqrt{2}-4x\sqrt{2}+x\sqrt{2}$ _____ **36.** $5x\sqrt{10}-3x\sqrt{10}-2x\sqrt{10}$

_____ **37.** $9\sqrt[3]{5}-13\sqrt[3]{5}$ _____ **38.** $8\sqrt[3]{7}-19\sqrt[3]{7}$

_____ **39.** $8-2\sqrt{x}+7-5\sqrt{x}$ _____ **40.** $9-3\sqrt{x}-12+5\sqrt{x}$

_____ **41.** $3(\sqrt{5}-y)-2(3\sqrt{5}+4y)$ _____ **42.** $6(\sqrt{5}-2y)+3(4\sqrt{5}-3y)$

_____ **43.** $4(\sqrt{a} - 2\sqrt[3]{b}) - 3(2\sqrt{a} + 5\sqrt[3]{b})$

_____ **44.** $9(\sqrt{a} - \sqrt[3]{b}) - 4(\sqrt{a} - 2\sqrt[3]{b})$

_____ **45.** $8(\sqrt[5]{2} - 3\sqrt[3]{2}) - 7(2\sqrt[5]{2} - 3\sqrt[3]{2})$

_____ **46.** $7(\sqrt[4]{3} - 2\sqrt{3}) - 5(2\sqrt[4]{3} - 6\sqrt{3})$

_____ **47.** $\sqrt{12} + \sqrt{3}$ _____ **48.** $\sqrt{50} + \sqrt{2}$

_____ **49.** $\sqrt{20} - \sqrt{80} + \sqrt{45}$ _____ **50.** $\sqrt{24} - \sqrt{54} + \sqrt{600}$

_____ **51.** $7x\sqrt{12} - 2\sqrt{75x^2}$ _____ **52.** $9x\sqrt{20} - 3\sqrt{80x^2}$

_____ **53.** $5\sqrt[3]{16} - 4\sqrt[3]{-54}$ _____ **54.** $3\sqrt[3]{40} - 7\sqrt[3]{-5}$

_____ **55.** $2\sqrt[5]{x^6y^7} - x\sqrt[5]{32xy^7}$ _____ **56.** $3\sqrt[3]{x^5y^6} - x\sqrt[3]{27x^2y^6}$

_____ **57.** $\sqrt{\dfrac{50}{9}} - \sqrt{\dfrac{98}{25}}$ _____ **58.** $\sqrt{\dfrac{75}{16}} - \sqrt{\dfrac{12}{49}}$

_____ **59.** $\sqrt{72} - \sqrt{75} + \sqrt{98}$ _____ **60.** $\sqrt{500} - \sqrt{27} - \sqrt{48}$

_____ **61.** $b\sqrt[3]{24a^5b} + 3a\sqrt[3]{81a^2b^4}$ _____ **62.** $a\sqrt[3]{128a^2b^4} - 2b\sqrt[3]{54a^5b}$

☆ Stretching the Topics _____

Simplify.

_____ **1.** $\sqrt{x^{4m+1}}$

_____ **2.** $\sqrt[3]{x^{6m}y^{9n}}$

_____ **3.** $3x\sqrt{2x^{-2}y^5z^{-4}} - \dfrac{2y\sqrt{50y^2z^{-2}}}{z}$

Check your answers in the back of your book.

If you can simplify the radical expressions in **Checkup 7.2,** you are ready to go on to Section 7.3.

✔ **CHECKUP 7.2**

Simplify. Assume that any variable expression under an even radical represents a positive number.

_____ **1.** $\sqrt{44x^5y^2}$

_____ **2.** $\sqrt[3]{\dfrac{-27a^4}{8b^3}}$

_____ **3.** $\sqrt[4]{(x+1)^5}$

_____ **4.** $3x\sqrt{2} - 7x\sqrt{2} - 5x\sqrt{2}$

_____ **5.** $3(\sqrt{7} - 5x) - 2(3x - \sqrt{7})$

_____ **6.** $\sqrt{63x^2} + \sqrt{28x^2} - \sqrt{175x^2}$

_____ **7.** $\sqrt[3]{80} - \sqrt[3]{270}$

_____ **8.** $\sqrt[3]{-3} - \sqrt{75} + \sqrt[3]{375} + \sqrt{243}$

_____ **9.** $\sqrt{\dfrac{50}{49}} - \sqrt{\dfrac{8}{9}}$

_____ **10.** $6\sqrt[3]{x^7y^4} - x^2\sqrt[3]{27xy^4}$

Check your answers in the back of your book.

If You Missed Problems:	You Should Review Examples:
1, 2	5, 6
3	9, 10
4	11, 12
5	14
6–8	15, 16
9	17
10	18

7.3 Multiplying and Dividing Radical Expressions

Multiplying and Dividing Radicals with Same Index

It follows from our work with simplifying radicals that if $\sqrt[n]{a}$ and $\sqrt[n]{b}$ represent *real numbers,* then

$$\sqrt[n]{a} \cdot \sqrt[n]{b} = \sqrt[n]{ab}$$

$$\frac{\sqrt[n]{a}}{\sqrt[n]{b}} = \sqrt[n]{\frac{a}{b}} \qquad (b \neq 0)$$

To multiply radicals with the same index, we may multiply the radicands and keep the same index.

To divide radicals with the same index, we may divide the radicands and keep the same index.

You try Example 1.

Example 1. Simplify $\sqrt{2} \cdot \sqrt{6}$.

Solution

$$\sqrt{2} \cdot \sqrt{6}$$
$$= \sqrt{2 \cdot 6}$$
$$= \sqrt{12}$$
$$= \sqrt{\underline{\qquad} \cdot 3}$$
$$= \underline{\qquad} \sqrt{\underline{\qquad}}$$

Check your work on page 499. ▶

Example 2. Simplify $\sqrt[3]{-2x} \cdot \sqrt[3]{32x^2}$.

Solution

$$\sqrt[3]{-2x} \cdot \sqrt[3]{32x^2}$$
$$= \sqrt[3]{-2x \cdot 32x^2}$$
$$= \sqrt[3]{-64x^3}$$
$$= -4x$$

Example 3. Simplify $\sqrt[5]{\frac{10x}{3}} \cdot \sqrt[5]{\frac{-3x}{5}}$.

Solution

$$\sqrt[5]{\frac{10x}{3}} \cdot \sqrt[5]{\frac{-3x}{5}}$$

$$= \sqrt[5]{\frac{10x}{3} \cdot \frac{-3x}{5}} \qquad \text{Multiply radicands.}$$

$$= \sqrt[5]{\frac{-30x^2}{15}} \qquad \text{Simplify products.}$$

$$= \sqrt[5]{-2x^2} \qquad \text{Simplify quotient.}$$

$$= \sqrt[5]{-1 \cdot 2x^2}$$

$$= -\sqrt[5]{2x^2}$$

Now try Example 4.

Example 4. Simplify $\sqrt{3x} \cdot \sqrt{3x}$.

Solution

$$\sqrt{3x} \cdot \sqrt{3x}$$
$$= \sqrt{3x \cdot 3x}$$

Check your work on page 499. ▶

From Example 4, we note that if $\sqrt{a}$ is a *real* number, then a is greater than or equal to zero and

$$\sqrt{a}\,\sqrt{a} = (\sqrt{a})^2 = a$$

Example 5. Simplify $\dfrac{\sqrt[3]{2000x^4y}}{\sqrt[3]{2x}}$.

Solution

$$\frac{\sqrt[3]{2000x^4y}}{\sqrt[3]{2x}}$$

$$= \sqrt[3]{\frac{2000x^4y}{2x}}$$

$$= \sqrt[3]{1000x^3y}$$

$$= 10x\sqrt[3]{y}$$

Now complete Example 6. ▶

Example 6. Simplify $\dfrac{\sqrt{5(x+1)^3}}{\sqrt{x+1}}$.

Solution

$$\frac{\sqrt{5(x+1)^3}}{\sqrt{x+1}}$$

$$= \sqrt{\frac{5(x+1)^3}{x+1}}$$

$$= \sqrt{5(x+1)^2}$$

$$=$$

Check your work on page 499. ▶

▦➡ Trial Run

Simplify. Assume that any variable expression under a radical with an even index represents a nonnegative number.

_____ **1.** $\sqrt{3} \cdot \sqrt{15}$

_____ **2.** $\sqrt[3]{-10a} \cdot \sqrt[3]{4a^2}$

_____ **3.** $\sqrt[4]{8x} \cdot \sqrt[4]{2x^3}$

_____ **4.** $\sqrt[5]{\dfrac{12x}{5}} \cdot \sqrt[5]{\dfrac{-5x^2}{4}}$

_____ **5.** $\dfrac{\sqrt{500x^5y}}{\sqrt{5x^2}}$

_____ **6.** $\dfrac{\sqrt[3]{3(x-1)^5}}{\sqrt[3]{(x-1)^2}}$

Answers are on page 500.

To find products of expressions containing sums and/or differences of radicals, we must use the **distributive property** for multiplication over addition and carry out the multiplication of radicals in the usual way.

You try finding the product in Example 7.

Example 7. Simplify $\sqrt{2}(1 + \sqrt{3})$.

Solution

$$\sqrt{2}(1 + \sqrt{3})$$
$$= \sqrt{2} \cdot 1 + \sqrt{2} \cdot \sqrt{3}$$
$$= \underline{\quad\quad} + \underline{\quad\quad}$$

Check your work on page 499. ▶

Example 8. Simplify $3\sqrt{11}(\sqrt{11} - 2)$.

Solution

$$3\sqrt{11}(\sqrt{11} - 2)$$
$$= 3\sqrt{11} \cdot \sqrt{11} - 2 \cdot 3\sqrt{11}$$
$$= 3 \cdot 11 - 6\sqrt{11}$$
$$= 33 - 6\sqrt{11}$$

You try Example 9.

Example 9. Simplify $\sqrt[3]{x}(8 + \sqrt[3]{x^2})$.

Solution

$$\sqrt[3]{x}(8 + \sqrt[3]{x^2})$$
$$= 8\sqrt[3]{x} + \sqrt[3]{x} \cdot \sqrt[3]{x^2}$$
$$= 8\sqrt[3]{x} + \sqrt[3]{\underline{\quad\quad}}$$
$$= 8\sqrt[3]{x} + \underline{\quad\quad}$$

Check your work on page 499. ▶

Example 10. Simplify

$$(2 - \sqrt{x})(3 + \sqrt{x}).$$

Solution

$$(2 - \sqrt{x})(3 + \sqrt{x})$$
$$= 2(3 + \sqrt{x}) - \sqrt{x}(3 + \sqrt{x})$$
$$= 6 + 2\sqrt{x} - 3\sqrt{x} - \sqrt{x} \cdot \sqrt{x}$$
$$= 6 + 2\sqrt{x} - 3\sqrt{x} - x$$
$$= 6 - \sqrt{x} - x$$

It may have occurred to you that Example 10 could have been worked using the FOIL method from Chapter 3. Indeed, this is true; let's try the FOIL method on some products.

Example 11. Find $(\sqrt{x} + \sqrt{y})^2$.

Solution

$$(\sqrt{x} + \sqrt{y})^2$$

$$= (\sqrt{x} + \sqrt{y})(\sqrt{x} + \sqrt{y})$$

$$= \sqrt{x}\sqrt{x} + \sqrt{x}\sqrt{y} + \sqrt{y}\sqrt{x} + \sqrt{y}\sqrt{y}$$
$$= x + \sqrt{xy} + \sqrt{xy} + y$$
$$= x + 2\sqrt{xy} + y$$

Example 12. Find

$$(\sqrt{x} + 5\sqrt{3})(2\sqrt{x} - \sqrt{3}).$$

Solution

$$(\sqrt{x} + 5\sqrt{3})(2\sqrt{x} - \sqrt{3})$$
$$= \sqrt{x} \cdot 2\sqrt{x} - \sqrt{x} \cdot \sqrt{3}$$
$$\quad + 5\sqrt{3} \cdot 2\sqrt{x} - 5\sqrt{3} \cdot \sqrt{3}$$
$$= 2x - \sqrt{3x} + 10\sqrt{3x} - 5 \cdot 3$$
$$= 2x + 9\sqrt{3x} - 15$$

Note that the product in Example 11 could have also been found using the rule for squaring a binomial

$$(\sqrt{x} + \sqrt{y})^2 = (\sqrt{x})^2 + 2\sqrt{x}\sqrt{y} + (\sqrt{y})^2$$
$$= x + 2\sqrt{xy} + y$$

Now complete Example 13.

Example 13. Find

$$(\sqrt{7} + \sqrt{2})(\sqrt{7} - \sqrt{2}).$$

Solution: We notice that we are multiplying the sum and difference of the same two terms.

$$(\sqrt{7} + \sqrt{2})(\sqrt{7} - \sqrt{2})$$
$$= (\sqrt{7})^2 - (\sqrt{2})^2$$
$$=$$
$$=$$

Check your work on page 499. ▶

Example 14. Find $(3 - \sqrt{x + 1})^2$.

Solution

$$(3 - \sqrt{x + 1})^2$$
$$= 3^2 - 2 \cdot 3\sqrt{x + 1} + (\sqrt{x + 1})^2$$
$$= 9 - 6\sqrt{x + 1} + x + 1$$
$$= x - 6\sqrt{x + 1} + 10$$

This time we used the rule for squaring a binomial.

In Example 13 we multiplied the irrational number $\sqrt{7} + \sqrt{2}$ times the irrational number $\sqrt{7} - \sqrt{2}$ and the product was the *rational* number 5. Such irrational numbers as $\sqrt{7} + \sqrt{2}$ and $\sqrt{7} - \sqrt{2}$ are sometimes called **real conjugates** of each other. The product of an irrational number and its real conjugate is always a *rational* number.

⫸ Trial Run

_____ **1.** $\sqrt{3}(2 + \sqrt{5})$

_____ **2.** $3\sqrt{7}(\sqrt{7} - 2)$

_____ **3.** $\sqrt[3]{x}(2 - \sqrt[3]{x^2})$

_____ **4.** $(1 - \sqrt{x})(3 + 2\sqrt{x})$

_____ **5.** $(3\sqrt{2} - \sqrt{y})(5\sqrt{2} + 2\sqrt{y})$

_____ **6.** $(\sqrt{x} - 7)(\sqrt{x} + 7)$

Answers are on page 500.

Rationalizing Denominators Containing One Term

You should recall from Chapter 1 that a number is called an **irrational number** when it cannot be expressed as a fraction of integers. Although we can *approximate* irrational numbers, we cannot find their *exact* value. For instance, from a table or calculator we could find the following approximations:

$$\sqrt{3} \doteq 1.732$$
$$\sqrt[3]{2} \doteq 1.259$$
$$1 + \sqrt{5} \doteq 1 + 2.236 \doteq 3.236$$
$$\sqrt{7} - \sqrt{2} \doteq 2.646 - 1.414 \doteq 1.232$$

The symbol $\doteq$ means "is approximately equal to."

Surely you would agree that these are not convenient numbers to work with, especially if they appear in denominator of a fraction. For this reason (and for other reasons that will be clear to you in later mathematics courses), mathematicians usually agree that

No radical should appear in the denominator of the final form of a radical expression.

Somehow we must remove the radical from the denominator of an expression such as $\dfrac{1}{\sqrt{3}}$ and change the irrational number in the denominator to a *rational* number. This process is therefore called **rationalizing the denominator.** But how can we accomplish the desired change?

> We rationalize the denominator of a fraction by multiplying the numerator and denominator by an appropriate rationalizing factor.

To get rid of the $\sqrt{3}$ in the denominator of the fraction $\dfrac{1}{\sqrt{3}}$, we choose the rationalizing factor to be $\sqrt{3}$ (since $\sqrt{3} \cdot \sqrt{3} = 3$).

$$\frac{1}{\sqrt{3}} = \frac{1}{\sqrt{3}} \cdot \frac{\sqrt{3}}{\sqrt{3}}$$

$$= \frac{1\sqrt{3}}{\sqrt{3} \cdot \sqrt{3}}$$

$$= \frac{\sqrt{3}}{3}$$

We have indeed rid the denominator of its irrational number; we have rationalized the denominator.

You try Example 15.

Example 15. Rationalize the denominator of $\dfrac{5}{\sqrt{2x}}$.

Solution

$$\frac{5}{\sqrt{2x}} \cdot \frac{\sqrt{\rule{1.5em}{0pt}}}{\sqrt{\rule{1.5em}{0pt}}}$$

$$= \frac{5\sqrt{\rule{1.5em}{0pt}}}{\sqrt{2x} \cdot \sqrt{\rule{1.5em}{0pt}}}$$

$$=$$

Check your work on page 499. ▶

Example 16. Rationalize the denominator of $\dfrac{1 + \sqrt{a}}{\sqrt{a}}$.

Solution

$$\frac{1 + \sqrt{a}}{\sqrt{a}} \cdot \frac{\sqrt{a}}{\sqrt{a}}$$

$$= \frac{(1 + \sqrt{a})\sqrt{a}}{\sqrt{a} \cdot \sqrt{a}}$$

$$= \frac{\sqrt{a} + \sqrt{a} \cdot \sqrt{a}}{a}$$

$$= \frac{\sqrt{a} + a}{a}$$

Now try Example 17.

Example 17. Rationalize the denominator of $\dfrac{3}{\sqrt{a+b}}$.

Solution

$$\frac{3}{\sqrt{a+b}} \cdot \frac{\sqrt{a+b}}{\sqrt{a+b}}$$

$$= \frac{3\sqrt{a+b}}{\sqrt{a+b} \cdot \sqrt{a+b}}$$

$$=$$

Check your work on page 499. ▶

Example 18. Rationalize the denominator of $\dfrac{x+1}{\sqrt{x^2-1}}$.

Solution

$$\frac{x+1}{\sqrt{x^2-1}} \cdot \frac{\sqrt{x^2-1}}{\sqrt{x^2-1}}$$

$$= \frac{(x+1)\sqrt{x^2-1}}{\sqrt{x^2-1} \cdot \sqrt{x^2-1}}$$

$$= \frac{(x+1)\sqrt{x^2-1}}{x^2-1}$$

$$= \frac{\overset{1}{\cancel{(x+1)}}\sqrt{x^2-1}}{\underset{1}{\cancel{(x+1)}}(x-1)}$$

$$= \frac{\sqrt{x^2-1}}{x-1}$$

To rationalize the denominator of the fraction

$$\frac{5}{\sqrt[3]{2}}$$

we know that the rationalizing factor must be a cube root and that we are aiming for the *cube root of a perfect cube* as our new denominator.

$$\frac{5}{\sqrt[3]{2}} = \frac{5}{\sqrt[3]{2^1}} \cdot \frac{\sqrt[3]{(\quad)}}{\sqrt[3]{(\quad)}} = \frac{5\sqrt[3]{(\quad)}}{\sqrt[3]{2^3}}$$

Since the original radicand is 2^1, it is logical for us to aim for a radicand of 2^3. The missing radicand must be 2^2.

$$\frac{5}{\sqrt[3]{2}} = \frac{5}{\sqrt[3]{2^1}} \cdot \frac{\sqrt[3]{2^2}}{\sqrt[3]{2^2}}$$

$$= \frac{5\sqrt[3]{4}}{\sqrt[3]{2^3}}$$

$$= \frac{5\sqrt[3]{4}}{2}$$

Choosing the rationalizing factor is not difficult if you decide *first* what you are aiming for.

Example 19. Rationalize the denominator of $\dfrac{1}{\sqrt[3]{3x^2}}$.

Solution: We want the cube root of a perfect cube in the new denominator. The rationalizing factor is $\sqrt[3]{3^2x}$.

$$\frac{1}{\sqrt[3]{3x^2}} \cdot \frac{\sqrt[3]{3^2x}}{\sqrt[3]{3^2x}}$$

$$= \frac{\sqrt[3]{9x}}{\sqrt[3]{3^3x^3}}$$

$$= \frac{\sqrt[3]{9x}}{3x}$$

Example 20. Rationalize the denominator of $\dfrac{a}{\sqrt[5]{4a^3}}$.

Solution: We want the fifth root of a perfect fifth power in the new denominator. The rationalizing factor is $\sqrt[5]{2^3a^2}$.

$$\frac{a}{\sqrt[5]{2^2a^3}} \cdot \frac{\sqrt[5]{2^3a^2}}{\sqrt[5]{2^3a^2}}$$

$$= \frac{a\sqrt[5]{8a^2}}{\sqrt[5]{2^5a^5}}$$

$$= \frac{a\sqrt[5]{8a^2}}{2a}$$

$$= \frac{\sqrt[5]{8a^2}}{2}$$

⫸ Trial Run

Rationalize the denominators. Assume that variable expressions in radicals with an even index represent nonnegative numbers.

_____ **1.** $\dfrac{3}{\sqrt{5}}$

_____ **2.** $\dfrac{x}{\sqrt{2x}}$

_____ **3.** $\dfrac{2}{\sqrt{a+3}}$

_____ **4.** $\dfrac{x+2}{\sqrt{x^2-4}}$

_____ **5.** $\dfrac{6}{\sqrt[3]{2x}}$

_____ **6.** $\dfrac{1}{\sqrt[4]{x^2y^3}}$

Answers are on page 500.

To rationalize the denominator of a radical expression whose radicand is a fraction, we should use the fact that

$$\sqrt[n]{\frac{a}{b}} = \frac{\sqrt[n]{a}}{\sqrt[n]{b}}$$

before we attempt the simplifying and rationalizing processes.

Example 21. Simplify $\sqrt{\dfrac{12x^5}{y}}$.

Solution

$$\sqrt{\frac{12x^5}{y}} = \frac{\sqrt{12x^5}}{\sqrt{y}}$$

$$= \frac{\sqrt{4 \cdot 3 \cdot x^2 \cdot x^2 \cdot x}}{\sqrt{y}}$$

$$= \frac{2x^2\sqrt{3x}}{\sqrt{y}} \cdot \frac{\sqrt{y}}{\sqrt{y}}$$

$$= \frac{2x^2\sqrt{3xy}}{y}$$

Example 22. Simplify $\sqrt[3]{\dfrac{16a^4b^3}{9c}}$.

Solution

$$\sqrt[3]{\frac{16a^4b^3}{9c}} = \frac{\sqrt[3]{16a^4b^3}}{\sqrt[3]{9c}}$$

$$= \frac{\sqrt[3]{8 \cdot 2 \cdot a^3 \cdot a \cdot b^3}}{\sqrt[3]{3^2c}}$$

$$= \frac{2ab\sqrt[3]{2a}}{\sqrt[3]{3^2c}} \cdot \frac{\sqrt[3]{3c^2}}{\sqrt[3]{3c^2}}$$

$$= \frac{2ab\sqrt[3]{6ac^2}}{3c}$$

Of course, if the numerator and denominator of a fractional radicand contain common factors, the fraction should be *reduced* by the usual methods *before* we attempt the process of simplifying and rationalizing.

You try to complete Example 23.

Example 23. Simplify $\dfrac{\sqrt{5x^5y^2}}{\sqrt{20x^2y^3}}$.

Solution

$$\frac{\sqrt{5x^5y^2}}{\sqrt{20x^2y^3}} = \sqrt{\frac{5x^5y^2}{20x^2y^3}}$$

$$= \sqrt{\frac{x^3}{4y}}$$

$$= \frac{\sqrt{x^3}}{\sqrt{4y}}$$

$$=$$

Check your work on page 500. ▶

Example 24. Simplify $\dfrac{\sqrt[3]{24a^4b^2}}{\sqrt[3]{15a^5b^4}}$.

Solution

$$\frac{\sqrt[3]{24a^4b^2}}{\sqrt[3]{15a^5b^4}} = \sqrt[3]{\frac{24a^4b^2}{15a^5b^4}}$$

$$= \sqrt[3]{\frac{8}{5ab^2}}$$

$$= \frac{\sqrt[3]{8}}{\sqrt[3]{5ab^2}}$$

$$= \frac{2}{\sqrt[3]{5ab^2}} \cdot \frac{\sqrt[3]{5^2a^2b}}{\sqrt[3]{5^2a^2b}}$$

$$= \frac{2\sqrt[3]{25a^2b}}{5ab}$$

Simplifying Fractions Containing Radicals

1. Reduce the fraction if possible by treating it as a single radical.
2. Rewrite the result as a fraction of separate radicals.
3. Simplify the remaining radicals and rationalize the denominator if necessary.

⫸ Trial Run

Simplify and rationalize denominators. Assume that variables are positive.

_____ **1.** $\sqrt{\dfrac{20x^3}{3y}}$ _____ **2.** $\sqrt{\dfrac{26a^2x}{13ax^2}}$

_____ **3.** $\dfrac{\sqrt[3]{10x}}{\sqrt[3]{9x}}$ _____ **4.** $\dfrac{\sqrt{20xy^4}}{\sqrt{50x^2y}}$

Answers are on page 500.

Rationalizing Denominators Containing Two Terms

Rationalizing the denominator of a fraction such as

$$\frac{-3}{1 + \sqrt{5}}$$

requires a slightly different approach. If we multiply numerator and denominator by $1 - \sqrt{5}$, the real conjugate of the denominator, look at the result.

$$\frac{-3}{1 + \sqrt{5}} \cdot \frac{1 - \sqrt{5}}{1 - \sqrt{5}} = \frac{-3(1 - \sqrt{5})}{(1 + \sqrt{5})(1 - \sqrt{5})} \qquad \text{Multiply numerator and denominator by } 1 - \sqrt{5}.$$

$$= \frac{-3(1 - \sqrt{5})}{1^2 - (\sqrt{5})^2} \qquad \text{Multiply denominators.}$$

$$= \frac{-3(1 - \sqrt{5})}{1 - 5} \qquad \text{Simplify squares.}$$

$$= \frac{-3(1 - \sqrt{5})}{-4} \qquad \text{Simplify the denominator.}$$

$$= \frac{3(1 - \sqrt{5})}{4} \qquad \text{Divide numerator and denominator by } -1.$$

We have indeed rationalized the denominator.

> To rationalize a denominator that is the *sum* or *difference* of two terms (one or both terms being a square root), we multiply the numerator and denominator by the *real conjugate* of the denominator.

This technique relies on the fact that when we multiply the sum and difference of the same two terms, the product is the *difference of the squares* of the two terms.
    You try Example 25.

Example 25. Rationalize the denominator of $\dfrac{2}{\sqrt{7} - \sqrt{3}}$.

Solution: The rationalizing factor is ____ + ____ .

$$\dfrac{2}{\sqrt{7} - \sqrt{3}} \cdot \dfrac{\sqrt{7} + \sqrt{3}}{\sqrt{7} + \sqrt{3}}$$

Multiply by real conjugate of denominator.

$$= \dfrac{2(\sqrt{7} + \sqrt{3})}{(\sqrt{7})^2 - (\sqrt{3})^2}$$

$$= \dfrac{2(\sqrt{7} + \sqrt{3})}{\underline{} - \underline{}} \quad \text{Simplify squares.}$$

$$= \dfrac{2(\sqrt{7} + \sqrt{3})}{\underline{}} \quad \text{Simplify the denominator.}$$

$$= \dfrac{\sqrt{7} + \sqrt{3}}{\underline{}} \quad \text{Reduce the fraction.}$$

Check your work on page 500. ▶

Example 26. Rationalize the denominator of $\dfrac{\sqrt{y}}{\sqrt{x} + \sqrt{y}}$.

Solution: The rationalizing factor is $\sqrt{x} - \sqrt{y}$.

$$\dfrac{\sqrt{y}}{\sqrt{x} + \sqrt{y}} \cdot \dfrac{\sqrt{x} - \sqrt{y}}{\sqrt{x} - \sqrt{y}}$$

Multiply by real conjugate of denominator.

$$= \dfrac{\sqrt{y}(\sqrt{x} - \sqrt{y})}{(\sqrt{x} + \sqrt{y})(\sqrt{x} - \sqrt{y})}$$

$$= \dfrac{\sqrt{xy} - \sqrt{y^2}}{(\sqrt{x})^2 - (\sqrt{y})^2} \quad \begin{array}{l}\text{Use distribu-}\\\text{tive property}\\\text{in numerator.}\end{array}$$

$$= \dfrac{\sqrt{xy} - y}{x - y} \quad \text{Simplify.}$$

Contrast Example 26 with Example 27.

Example 27. Rationalize the denominator of $\dfrac{\sqrt{y}}{\sqrt{x + y}}$.

Solution: Here the denominator is a *single* radical.

$$\dfrac{\sqrt{y}}{\sqrt{x + y}} \cdot \dfrac{\sqrt{x + y}}{\sqrt{x + y}} \quad \text{Multiply by the single radical } \sqrt{x + y}.$$

$$= \dfrac{\sqrt{y}(\sqrt{x + y})}{x + y} \quad \text{Simplify the numerator.}$$

$$= \dfrac{\sqrt{y(x + y)}}{x + y} \quad \text{Simplify the denominator.}$$

Students often confuse these last two examples. We must be careful to decide whether our denominator is a single radical term *or* a sum or difference of radical terms before we choose the correct rationalizing factor.

⫸ **Trial Run**

Rationalize the denominator. Assume that any variable expression under a radical with an even index represents a nonnegative number.

____ **1.** $\dfrac{5}{1 - \sqrt{2}}$

____ **2.** $\dfrac{1}{\sqrt{5} + \sqrt{2}}$

____ **3.** $\dfrac{\sqrt{a}}{\sqrt{a} + 1}$

____ **4.** $\dfrac{\sqrt{x}}{\sqrt{x} - \sqrt{2}}$

_____ 5. $\dfrac{\sqrt{x}}{\sqrt{x} - 2}$ _____ 6. $\dfrac{\sqrt{6} - 3\sqrt{2}}{2\sqrt{6} - \sqrt{2}}$

Answers are on page 500.

▶ Examples You Completed

Example 1. Simplify $\sqrt{2} \cdot \sqrt{6}$.

Solution

$$\sqrt{2} \cdot \sqrt{6}$$
$$= \sqrt{2 \cdot 6}$$
$$= \sqrt{12}$$
$$= \sqrt{4 \cdot 3}$$
$$= 2\sqrt{3}$$

Example 4. Simplify $\sqrt{3x} \cdot \sqrt{3x}$.

Solution

$$\sqrt{3x} \cdot \sqrt{3x}$$
$$= \sqrt{3x \cdot 3x}$$
$$= \sqrt{9x^2}$$
$$= 3x$$

Example 5. Simplify $\sqrt{2}(1 + \sqrt{3})$.

Solution

$$\sqrt{2}(1 + \sqrt{3})$$
$$= \sqrt{2} \cdot 1 + \sqrt{2} \cdot \sqrt{3}$$
$$= \sqrt{2} + \sqrt{6}$$

Example 6. Simplify $\dfrac{\sqrt{5(x + 1)^3}}{\sqrt{x + 1}}$.

Solution

$$\frac{\sqrt{5(x + 1)^3}}{\sqrt{x + 1}}$$
$$= \sqrt{\frac{5(x + 1)^3}{x + 1}}$$
$$= \sqrt{5(x + 1)^2}$$
$$= (x + 1)\sqrt{5}$$

Example 9. Simplify $\sqrt[3]{x}(8 + \sqrt[3]{x^2})$.

Solution

$$\sqrt[3]{x}(8 + \sqrt[3]{x^2})$$
$$= 8\sqrt[3]{x} + \sqrt[3]{x} \cdot \sqrt[3]{x^2}$$
$$= 8\sqrt[3]{x} + \sqrt[3]{x^3}$$
$$= 8\sqrt[3]{x} + x$$

Example 13. Find $(\sqrt{7} + \sqrt{2})(\sqrt{7} - \sqrt{2})$.

Solution: We notice that we are multiplying the sum and difference of the same two terms.

$$(\sqrt{7} + \sqrt{2})(\sqrt{7} - \sqrt{2})$$
$$= (\sqrt{7})^2 - (\sqrt{2})^2$$
$$= 7 - 2$$
$$= 5$$

Example 15. Rationalize the denominator of $\dfrac{5}{\sqrt{2x}}$.

Solution

$$\frac{5}{\sqrt{2x}} \cdot \frac{\sqrt{2x}}{\sqrt{2x}}$$
$$= \frac{5\sqrt{2x}}{\sqrt{2x} \cdot \sqrt{2x}}$$
$$= \frac{5\sqrt{2x}}{2x}$$

Example 17. Rationalize the denominator of $\dfrac{3}{\sqrt{a + b}}$.

Solution

$$\frac{3}{\sqrt{a + b}} \cdot \frac{\sqrt{a + b}}{\sqrt{a + b}}$$
$$= \frac{3\sqrt{a + b}}{\sqrt{a + b} \cdot \sqrt{a + b}}$$
$$= \frac{3\sqrt{a + b}}{a + b}$$

Example 23. Simplify $\dfrac{\sqrt{5x^5y^2}}{\sqrt{20x^2y^3}}$.

Solution

$$\frac{\sqrt{5x^5y^2}}{\sqrt{20x^2y^3}} = \sqrt{\frac{5x^5y^2}{20x^2y^3}}$$

$$= \sqrt{\frac{x^3}{4y}}$$

$$= \frac{\sqrt{x^3}}{\sqrt{4y}}$$

$$= \frac{x\sqrt{x}}{2\sqrt{y}} \cdot \frac{\sqrt{y}}{\sqrt{y}}$$

$$= \frac{x\sqrt{xy}}{2y}$$

Example 25. Rationalize the denominator of $\dfrac{2}{\sqrt{7} - \sqrt{3}}$.

Solution: The rationalizing factor is $\sqrt{7} + \sqrt{3}$.

$$\frac{2}{\sqrt{7} - \sqrt{3}} \cdot \frac{\sqrt{7} + \sqrt{3}}{\sqrt{7} + \sqrt{3}}$$

$$= \frac{2(\sqrt{7} + \sqrt{3})}{(\sqrt{7})^2 - (\sqrt{3})^2}$$

$$= \frac{2(\sqrt{7} + \sqrt{3})}{7 - 3}$$

$$= \frac{2(\sqrt{7} + \sqrt{3})}{4}$$

$$= \frac{\sqrt{7} + \sqrt{3}}{2}$$

Answers to Trial Runs

page 490   **1.** $3\sqrt{5}$ **2.** $-2a\sqrt[3]{5}$ **3.** $2x$ **4.** $-\sqrt[5]{3x^3}$ **5.** $10x\sqrt{xy}$ **6.** $(x-1)\sqrt[3]{3}$

page 492   **1.** $2\sqrt{3} + \sqrt{15}$ **2.** $21 - 6\sqrt{7}$ **3.** $2\sqrt[3]{x} - x$ **4.** $3 - \sqrt{x} - 2x$ **5.** $30 + \sqrt{2y} - 2y$
6. $x - 49$

page 495   **1.** $\dfrac{3\sqrt{5}}{5}$ **2.** $\dfrac{\sqrt{2x}}{2}$ **3.** $\dfrac{2\sqrt{a+3}}{a+3}$ **4.** $\dfrac{\sqrt{x^2-4}}{x-2}$ **5.** $\dfrac{3\sqrt[3]{4x^2}}{x}$ **6.** $\dfrac{\sqrt[4]{x^2y}}{xy}$

page 497   **1.** $\dfrac{2x\sqrt{15xy}}{3y}$ **2.** $\dfrac{\sqrt{2ax}}{x}$ **3.** $\dfrac{\sqrt[3]{30}}{3}$ **4.** $\dfrac{y\sqrt{10xy}}{5x}$

page 498   **1.** $-5 - 5\sqrt{2}$ **2.** $\dfrac{\sqrt{5} - \sqrt{2}}{3}$ **3.** $\dfrac{a - \sqrt{a}}{a-1}$ **4.** $\dfrac{x + \sqrt{2x}}{x-2}$ **5.** $\dfrac{\sqrt{x(x-2)}}{x-2}$
6. $\dfrac{3 - 5\sqrt{3}}{11}$

EXERCISE SET 7.3

Simplify. Assume that any variable expression under a radical with an even index represents a nonnegative number.

_____ **1.** $\sqrt{7} \cdot \sqrt{14}$ _____ **2.** $\sqrt{5} \cdot \sqrt{10}$

_____ **3.** $\sqrt{3a} \cdot \sqrt{6a^3}$ _____ **4.** $\sqrt{24a} \cdot \sqrt{2a^3}$

_____ **5.** $\sqrt[3]{-6x} \cdot \sqrt[3]{9x^2}$ _____ **6.** $\sqrt[3]{-10x^2} \cdot \sqrt[3]{4x}$

_____ **7.** $\sqrt[4]{27x^2} \cdot \sqrt[4]{3x^6}$ _____ **8.** $\sqrt[4]{4x^6} \cdot \sqrt[4]{4x^2}$

_____ **9.** $\sqrt[3]{\dfrac{9x^2}{2}} \cdot \sqrt[3]{\dfrac{-16x}{3}}$ _____ **10.** $\sqrt[3]{\dfrac{4x}{3}} \cdot \sqrt[3]{\dfrac{-375}{2}}$

_____ **11.** $\sqrt[5]{\dfrac{24x}{5}} \cdot \sqrt[5]{\dfrac{20x^2}{3}}$ _____ **12.** $\sqrt[5]{\dfrac{21x^3}{2}} \cdot \sqrt[5]{\dfrac{6x^2}{7}}$

_____ **13.** $\dfrac{\sqrt{128x^5y}}{\sqrt{2x^2}}$ _____ **14.** $\dfrac{\sqrt{180x^3y}}{\sqrt{5x}}$

_____ **15.** $\sqrt{5x} \cdot \sqrt{5x}$ _____ **16.** $\sqrt{7y} \cdot \sqrt{7y}$

_____ **17.** $\sqrt{6x} \cdot \sqrt{15xy}$ _____ **18.** $\sqrt{10x} \cdot \sqrt{20xy}$

_____ **19.** $\sqrt{18x^3yz} \cdot \sqrt{8xy^4z}$ _____ **20.** $\sqrt{21xy^3z^5} \cdot \sqrt{14xyz^2}$

_____ **21.** $\sqrt{2}(\sqrt{6} + \sqrt{2})$ _____ **22.** $\sqrt{3}(\sqrt{15} + \sqrt{3})$

_____ **23.** $\sqrt{3x}(\sqrt{x} - \sqrt{3})$ _____ **24.** $\sqrt{5y}(\sqrt{y} - \sqrt{5})$

_____ **25.** $\sqrt{6xy}(\sqrt{3x} + \sqrt{2y})$ _____ **26.** $\sqrt{10xy}(\sqrt{5x} + \sqrt{2y})$

_____ **27.** $\sqrt[3]{a}(3 - \sqrt[3]{a^2})$ _____ **28.** $\sqrt[3]{2}(a - \sqrt[3]{4})$

_____ **29.** $(1 - \sqrt{x})(2 - \sqrt{x})$ _____ **30.** $(3 - \sqrt{x})(2 + \sqrt{x})$

_____ **31.** $(\sqrt{2} - 3\sqrt{y})(5\sqrt{2} + \sqrt{y})$ _____ **32.** $(3\sqrt{5} - 2\sqrt{y})(\sqrt{5} + \sqrt{y})$

_____ **33.** $(\sqrt{x} - 3)^2$ _____ **34.** $(\sqrt{x} - 7)^2$

_____ **35.** $(\sqrt{2x} - 3)(\sqrt{2x} + 3)$ _____ **36.** $(\sqrt{3x} - 4)(\sqrt{3x} + 4)$

_____ **37.** $(4\sqrt{x} + 2\sqrt{5})(4\sqrt{x} - 2\sqrt{5})$ _____ **38.** $(3\sqrt{7} - 8\sqrt{x})(3\sqrt{7} + 8\sqrt{x})$

_____ **39.** $(x - \sqrt{2y})(x + \sqrt{2y})$ _____ **40.** $(\sqrt{3x} - y)(\sqrt{3x} + y)$

_____ **41.** $(3\sqrt{x} - 5\sqrt{y})^2$ _____ **42.** $(4\sqrt{x} - 7\sqrt{y})^2$

_____ **43.** $\dfrac{4}{\sqrt{6}}$ _____ **44.** $\dfrac{3}{\sqrt{15}}$

_____ 45. $\dfrac{\sqrt{3}}{\sqrt{6}}$

_____ 46. $\dfrac{\sqrt{2}}{\sqrt{10}}$

_____ 47. $\dfrac{\sqrt{2x}}{\sqrt{3x}}$

_____ 48. $\dfrac{\sqrt{5x}}{\sqrt{2x}}$

_____ 49. $\sqrt{\dfrac{18x^3}{5y}}$

_____ 50. $\sqrt{\dfrac{28x^5}{3y}}$

_____ 51. $\sqrt{\dfrac{54x^4y^2z^2}{10xz^3}}$

_____ 52. $\sqrt{\dfrac{32x^3y^4}{12xyz}}$

_____ 53. $\dfrac{4}{\sqrt{a}+5}$

_____ 54. $\dfrac{3}{\sqrt{a}+2}$

_____ 55. $\dfrac{x-3}{\sqrt{x^2-9}}$

_____ 56. $\dfrac{x+5}{\sqrt{x^2-25}}$

_____ 57. $\dfrac{3}{\sqrt[3]{5}}$

_____ 58. $\dfrac{4}{\sqrt[3]{3}}$

_____ 59. $\dfrac{64}{\sqrt[4]{3}}$

_____ 60. $\dfrac{10x}{\sqrt[4]{2}}$

_____ 61. $\dfrac{6}{\sqrt[3]{3x}}$

_____ 62. $\dfrac{10}{\sqrt[3]{5x^2}}$

_____ 63. $\sqrt[4]{\dfrac{3}{4y}}$

_____ 64. $\sqrt[4]{\dfrac{2}{9x^3}}$

_____ 65. $\sqrt[3]{\dfrac{5}{4y^2}}$

_____ 66. $\sqrt[3]{\dfrac{7}{3x}}$

_____ 67. $\dfrac{3}{2-\sqrt{3}}$

_____ 68. $\dfrac{2}{4-\sqrt{5}}$

_____ 69. $\dfrac{1}{\sqrt{7}+\sqrt{2}}$

_____ 70. $\dfrac{1}{\sqrt{5}+\sqrt{3}}$

_____ 71. $\dfrac{\sqrt{x}}{\sqrt{x}-2}$

_____ 72. $\dfrac{\sqrt{3}}{\sqrt{3}-x}$

_____ 73. $\dfrac{\sqrt{a}}{\sqrt{a}-3}$

_____ 74. $\dfrac{\sqrt{a}}{\sqrt{a}+5}$

_____ 75. $\dfrac{\sqrt{6}}{\sqrt{3}+\sqrt{2}}$

_____ 76. $\dfrac{\sqrt{15}}{\sqrt{5}-\sqrt{3}}$

_____ 77. $\dfrac{\sqrt{10}-2\sqrt{2}}{4\sqrt{10}-\sqrt{2}}$

_____ 78. $\dfrac{\sqrt{3}-2\sqrt{6}}{2\sqrt{3}-\sqrt{6}}$

☆ Stretching the Topics

_____ **1.** Simplify $(\sqrt[3]{x} - \sqrt[3]{2})(\sqrt[3]{x^2} + \sqrt[3]{2x} + \sqrt[3]{4})$.

_____ **2.** Factor $9x - 4y$ as the difference of two squares.

_____ **3.** Simplify $\dfrac{\sqrt{x + 2}}{\sqrt{x + 2} - 5}$.

Check your answers in the back of your book.

If you can complete **Checkup 7.3,** you are ready to do Section 7.4.

CHECKUP 7.3

Simplify. Assume that any variable expression under a radical with an even index represents a nonnegative number.

——————— **1.** $\sqrt{6x} \cdot \sqrt{10x^2}$ ——————— **2.** $\sqrt{\dfrac{48x^3y}{3x}}$

——————— **3.** $\sqrt{5}(\sqrt{10} + \sqrt{5})$ ——————— **4.** $(2\sqrt{5} - \sqrt{10})(2\sqrt{5} + \sqrt{10})$

——————— **5.** $(\sqrt{x} - 6)^2$ ——————— **6.** $\dfrac{6}{\sqrt{15}}$

——————— **7.** $\dfrac{\sqrt{6x}}{\sqrt{10x}}$ ——————— **8.** $\dfrac{15x}{\sqrt[3]{-3x}}$

——————— **9.** $\sqrt{\dfrac{45x^3}{2y}}$ ——————— **10.** $\dfrac{\sqrt{8}}{\sqrt{2} + \sqrt{3}}$

Check your answers in the back of your book.

If You Missed Problems:	You Should Review Examples:
1	1, 2
2	3
3–5	7–12
6, 7	15, 16
8	19
9	21
10	25

7.4 Solving Radical Equations

We have now acquired enough skill in working with radicals to enable us to solve equations containing variables within a radical expression. Such equations are called **radical equations** and some are very easily solved. For example, to solve $\sqrt{x} = 2$ we may square both sides of the equation to eliminate the radical.

$$(\sqrt{x})^2 = 2^2$$

$$x = 4$$

Is squaring both sides of an equation an acceptable manipulation? To answer that question, we must decide whether the solutions to the equation obtained after squaring are *exactly* the same as the solutions to the original equation.

Let's square both sides of another radical equation and check to see if our solution satisfies the original equation.

CHECK

$$\sqrt{x} = -3 \qquad\qquad x = 9$$

$$(\sqrt{x})^2 = (-3)^2 \qquad\qquad \sqrt{9} \overset{?}{=} -3$$

$$x = 9 \qquad\qquad 3 \neq -3$$

Here the solution to the squared equation does *not* satisfy the original equation. In fact, the original equation has *no* solution because the principal square root of a number is never negative.

Sometimes squaring both sides of an equation "works" and sometimes it does not. If we square both sides of an equation, we must always *check* our solutions in the original equation. If the original equation has a solution, it will always appear as a solution to the squared equation.

> We may square both sides of an equation, provided that we always check any solutions in the original equation.

A solution to a squared equation that does *not* satisfy the original equation is called an **extraneous solution.**

You try Example 1.

Example 1. Solve $\sqrt{2x + 1} = 5$.

Solution

$$\sqrt{2x + 1} = 5$$

$$(\sqrt{2x + 1})^2 = 5^2$$

$$2x + 1 = \underline{\qquad}$$

$$x = \underline{\qquad}$$

Check your work on page 514. ▶

CHECK

$$x = \underline{\qquad}$$

$$2(\underline{\qquad}) + 1 \overset{?}{=} 5$$

$$\underline{\qquad} + 1 \overset{?}{=} 5$$

$$\underline{\qquad} \overset{?}{=} 5$$

$$5 = 5$$

The solution is $x = \underline{\qquad}$.

Example 2. Solve $5 + \sqrt[3]{x} = 2$.

Solution: We isolate the radical and raise both sides to the *third* power.

$$5 + \sqrt[3]{x} = 2$$
$$\sqrt[3]{x} = -3$$
$$(\sqrt[3]{x})^3 = (-3)^3$$
$$x = -27$$

CHECK

$$x = -27$$
$$5 + \sqrt[3]{-27} \stackrel{?}{=} 2$$
$$5 + (-3) \stackrel{?}{=} 2$$
$$2 = 2$$

The solution is $x = -27$.

Most of our radical equations will contain square roots of variable expressions. If a radical equation contains just one such radical expression, we follow these steps.

Solving Equations Containing One Radical

1. Isolate the radical expression on one side of the equation.
2. Square both sides of the equation *carefully*.
3. Solve the resulting equation.
4. Check the solutions in the original equation.

You try Example 3.

Example 3. Solve $\sqrt{x + 2} - x = 0$.

Solution

$$\sqrt{x + 2} - x = 0$$

$$\sqrt{x + 2} = x \qquad \text{Isolate the radical.}$$

$$(\sqrt{x + 2})^2 = x^2 \qquad \text{Square both sides.}$$

$$x + 2 = x^2$$

$$0 = x^2 - x - 2 \qquad \text{Rewrite quadratic equation in standard form.}$$

$$0 = (\underline{\hspace{1cm}})(\underline{\hspace{1cm}}) \qquad \text{Factor.}$$

$$x + \underline{\hspace{0.5cm}} = 0 \qquad \text{or} \qquad x - \underline{\hspace{0.5cm}} = 0$$

$$x = \underline{\hspace{0.5cm}} \qquad \text{or} \qquad x = \underline{\hspace{0.5cm}}$$

CHECK

$$x = -1$$

$$\sqrt{-1 + 2} - (-1) \overset{?}{=} 0$$

Substitute -1 for x in the original equation.

$$\sqrt{1} + 1 \overset{?}{=} 0$$

$$1 + 1 \overset{?}{=} 0$$

$$2 \neq 0$$

-1 is *not* a solution.

CHECK

$$x = 2$$

$$\sqrt{2 + 2} - 2 \overset{?}{=} 0$$

Substitute 2 for x in the original equation.

$$\sqrt{4} - 2 \overset{?}{=} 0$$

$$\underline{\hspace{1cm}} - 2 \overset{?}{=} 0$$

$$\underline{\hspace{1cm}} = 0$$

2 is a solution.

The only solution is $x = \underline{\hspace{1cm}}$.

Check your work on page 514. ▶

Example 4. Solve $x + \sqrt{3 - x} = 2x + 3$.

Solution

$$x + \sqrt{3 - x} = 2x + 3$$

$$\sqrt{3 - x} = x + 3$$

Isolate the radical.

$$(\sqrt{3 - x})^2 = (x + 3)^2$$

Square both sides.

$$3 - x = x^2 + 6x + 9$$

$$0 = x^2 + 7x + 6$$

Rewrite quadratic equation in standard form.

$$0 = (x + 6)(x + 1)$$

Factor.

$$x + 6 = 0 \qquad x + 1 = 0$$

$$x = -6 \qquad x = -1$$

CHECK

$$x = -6$$

$$-6 + \sqrt{3 - (-6)} \overset{?}{=} 2(-6) + 3$$

Substitute -6 for x in the original equation.

$$-6 + \sqrt{9} \overset{?}{=} -12 + 3$$

$$-6 + 3 \overset{?}{=} -9$$

$$-3 \neq -9$$

-6 is *not* a solution.

CHECK

$$x = -1$$

$$-1 + \sqrt{3 - (-1)} \stackrel{?}{=} 2(-1) + 3$$

Substitute -1 for x in the original equation.

$$-1 + \sqrt{4} \stackrel{?}{=} -2 + 3$$

$$-1 + 2 \stackrel{?}{=} 1$$

$$1 = 1$$

-1 is a solution.

The only solution is $x = -1$.

⇒ Trial Run

Solve.

_____ **1.** $\sqrt{x} = 3$

_____ **2.** $\sqrt{3x + 4} = 5$

_____ **3.** $9 + \sqrt{2x - 1} = 6$

_____ **4.** $\sqrt[3]{x - 1} = -3$

_____ **5.** $\sqrt{x + 12} - x = 0$

_____ **6.** $\sqrt{7x + 14} = x + 2$

Answers are on page 514.

If a radical equation contains *two* square roots of variable expressions, it is most easily solved if we:

> **Solving Radical Equations**
>
> **1.** Isolate one radical expression on one side of the equation.
> **2.** Square both sides of the equation carefully.
> **3.** If the equation still contains a radical expression, isolate it and square both sides again.
> **4.** Solve the resulting equation.
> **5.** Check the solutions in the original equation.

You try Example 5.

Example 5. Solve $\sqrt{x + 1} - \sqrt{5x - 3} = 0$.

Solution

$$\sqrt{x + 1} - \sqrt{5x - 3} = 0$$

$$\sqrt{x + 1} = \sqrt{5x - 3}$$

Isolate one radical.

$$(\sqrt{x + 1})^2 = (\sqrt{5x - 3})^2$$

Square both sides.

$$x + 1 = 5x - 3$$

Solve the first-degree equation.

CHECK

$$x = 1$$

$$\sqrt{1 + 1} - \sqrt{5(1) - 3} \stackrel{?}{=} 0$$

$$\sqrt{2} - \sqrt{5 - 3} \stackrel{?}{=} 0$$

$$\sqrt{2} - \sqrt{2} \stackrel{?}{=} 0$$

$$0 = 0$$

The solution is $x = $ _____ .

Check your work on page 514. ▶

Example 6. Solve $\sqrt{2x - 5} - \sqrt{x - 2} = 2$.

Solution

$$\sqrt{2x - 5} - \sqrt{x - 2} = 2$$

$$\sqrt{2x - 5} = 2 + \sqrt{x - 2} \qquad \text{Isolate one radical.}$$

$$(\sqrt{2x - 5})^2 = (2 + \sqrt{x - 2})^2 \qquad \text{Square both sides.}$$

$$2x - 5 = (2 + \sqrt{x - 2})(2 + \sqrt{x - 2})$$

$$2x - 5 = 4 + 4\sqrt{x - 2} + x - 2$$

$$2x - 5 = 2 + x + 4\sqrt{x - 2} \qquad \text{Combine like terms.}$$

$$x - 7 = 4\sqrt{x - 2} \qquad \text{Isolate the radical.}$$

We must square both sides *again*.

$$(x - 7)^2 = (4\sqrt{x - 2})^2 \qquad \text{Square both sides.}$$

$$x^2 - 14x - 49 = 16(x - 2)$$

$$x^2 - 14x + 49 = 16x - 32 \qquad \text{Solve the quadratic equation.}$$

$$x^2 - 30x + 81 = 0$$

$$(x - 27)(x - 3) = 0$$

$$x - 27 = 0 \quad \text{or} \quad x - 3 = 0$$

$$x = 27 \qquad\qquad x = 3$$

CHECK

$$x = 27$$

$$\sqrt{2(27) - 5} - \sqrt{27 - 2} \stackrel{?}{=} 2$$

$$\sqrt{54 - 5} - \sqrt{25} \stackrel{?}{=} 2$$

$$\sqrt{49} - 5 \stackrel{?}{=} 2$$

$$7 - 5 \stackrel{?}{=} 2$$

$$2 = 2$$

CHECK

$$x = 3$$

$$\sqrt{2(3) - 5} - \sqrt{3 - 2} \stackrel{?}{=} 2$$

$$\sqrt{6 - 5} - \sqrt{1} \stackrel{?}{=} 2$$

$$\sqrt{1} - 1 \stackrel{?}{=} 2$$

$$1 - 1 \stackrel{?}{=} 2$$

$$0 \neq 2$$

The only solution is $x = 27$.

As you can see, solving radical equations can be a lengthy process, but the skills required are not new ones. Skipping steps or squaring carelessly are students' usual sources of error. You must be systematic and careful as you proceed.

Let's solve the problem stated at the beginning of this chapter.

Example 7. One leg of a right triangle is 1 foot longer than the other. If the shorter leg is increased by 4 feet and the length of the longer leg is multiplied by 6, the length of the hypotenuse of the enlarged right triangle is five times the length of the original hypotenuse. Find the dimensions of both triangles.

Solution: We can let x represent the length of the shorter leg of the original triangle and illustrate both triangles.

We Think	*We Write*
In every right triangle with leg lengths a and b and hypotenuse of length c we can use the Pythagorean theorem.	$c^2 = a^2 + b^2$ or $$c = \sqrt{a^2 + b^2}$$
We know expressions for the lengths of the sides for both triangles, so we can write expressions for the hypotenuses.	Original hypotenuse: $$\sqrt{x^2 + (x + 1)^2}$$ Enlarged hypotenuse: $$\sqrt{[6(x + 1)]^2 + (x + 4)^2}$$
We know that the length of the enlarged hypotenuse is five times the length of the original hypotenuse.	$$\sqrt{[6(x + 1)]^2 + (x + 4)^2}$$ $$= 5\sqrt{x^2 + (x + 1)^2}$$

We can solve this radical equation.

$$\sqrt{[6(x+1)]^2 + (x+4)^2} = 5\sqrt{x^2 + (x+1)^2}$$

$$\sqrt{36(x^2 + 2x + 1) + x^2 + 8x + 16} = 5\sqrt{x^2 + x^2 + 2x + 1}$$

$$\sqrt{36x^2 + 72x + 36 + x^2 + 8x + 16} = 5\sqrt{2x^2 + 2x + 1}$$

$$\sqrt{37x^2 + 80x + 52} = 5\sqrt{2x^2 + 2x + 1}$$

$$(\sqrt{37x^2 + 80x + 52})^2 = (5\sqrt{2x^2 + 2x + 1})^2$$

$$37x^2 + 80x + 52 = 25(2x^2 + 2x + 1)$$

$$37x^2 + 80x + 52 = 50x^2 + 50x + 25$$

$$0 = 13x^2 - 30x - 27$$

$$0 = (13x + 9)(x - 3)$$

$$13x + 9 = 0 \qquad \text{or} \qquad x - 3 = 0$$

$$13x = -9 \qquad\qquad\qquad x = 3$$

$$x = \frac{-9}{13}$$

Since length cannot be negative, we reject $x = \dfrac{-9}{13}$ and find that $x = 3$ satisfies the equation.

Original Triangle	*Enlarged Triangle*
$x = 3$ ft	$x + 4 = 7$ ft
$x + 1 = 4$ ft	$6(x + 1) = 24$ ft
$\sqrt{x^2 + (x+1)^2} = 5$ ft	$\sqrt{[6(x+1)]^2 + (x+4)^2} = 25$ ft

The dimensions of the original triangle are 3 feet, 4 feet, and 5 feet; the dimensions of the enlarged triangle are 7 feet, 24 feet, and 25 feet.

⫸ Trial Run

Solve.

_____ **1.** $\sqrt{x} = \sqrt{3x - 2}$ _____ **2.** $\sqrt{2x - 3} = \sqrt{4x - 11}$

_____ **3.** $\sqrt{5x - 2} + \sqrt{3x} = 0$ _____ **4.** $\sqrt{x - 3} = 3 - \sqrt{x}$

Answers are on page 514.

▶ Examples You Completed

Example 1. Solve $\sqrt{2x + 1} = 5$.

Solution

$$\sqrt{2x + 1} = 5$$
$$(\sqrt{2x + 1})^2 = 5^2$$
$$2x + 1 = 25$$
$$2x = 24$$
$$x = 12$$

CHECK

$$x = 12$$
$$\sqrt{2(12) + 1} \overset{?}{=} 5$$
$$\sqrt{24 + 1} \overset{?}{=} 5$$
$$\sqrt{25} \overset{?}{=} 5$$
$$5 = 5$$

Example 3. Solve $\sqrt{x + 2} - x = 0$.

Solution

$$\sqrt{x + 2} - x = 0$$
$$\sqrt{x + 2} = x$$
$$(\sqrt{x + 2})^2 = x^2$$
$$x + 2 = x^2$$
$$0 = x^2 - x - 2$$
$$0 = (x + 1)(x - 2)$$
$$x + 1 = 0 \quad \text{or} \quad x - 2 = 0$$
$$x = -1 \qquad\qquad x = 2$$

The only solution is $x = 2$.

CHECK

$$x = -1$$
$$\sqrt{-1 + 2} - (-1) \overset{?}{=} 0$$
$$\sqrt{1} + 1 \overset{?}{=} 0$$
$$1 + 1 \overset{?}{=} 0$$
$$2 \neq 0$$

CHECK

$$x = 2$$
$$\sqrt{2 + 2} - 2 \overset{?}{=} 0$$
$$\sqrt{4} - 2 \overset{?}{=} 0$$
$$2 - 2 \overset{?}{=} 0$$
$$0 = 0$$

Example 5. Solve $\sqrt{x + 1} - \sqrt{5x - 3} = 0$.

Solution

$$\sqrt{x + 1} - \sqrt{5x - 3} = 0$$
$$\sqrt{x + 1} = \sqrt{5x - 3}$$
$$(\sqrt{x + 1})^2 = (\sqrt{5x - 3})^2$$
$$x + 1 = 5x - 3$$
$$1 = 4x - 3$$
$$4 = 4x$$
$$1 = x$$

CHECK

$$x = 1$$
$$\sqrt{1 + 1} - \sqrt{5(1) - 3} \overset{?}{=} 0$$
$$\sqrt{2} - \sqrt{5 - 3} \overset{?}{=} 0$$
$$\sqrt{2} - \sqrt{2} \overset{?}{=} 0$$
$$0 = 0$$

The solution is $x = 1$.

Answers to Trial Runs

page 510   **1.** $x = 9$ **2.** $x = 7$ **3.** No solution   **4.** $x = -26$ **5.** $x = 4$ **6.** $x = 5$ or $x = -2$

page 513   **1.** $x = 1$ **2.** $x = 4$ **3.** No solution   **4.** $x = 4$

EXERCISE SET 7.4

Solve.

_____ 1. $\sqrt{x} = 5$

_____ 2. $\sqrt{x} = 7$

_____ 3. $\sqrt{2x} - 3 = 7$

_____ 4. $\sqrt{3x} + 2 = 8$

_____ 5. $\sqrt{2x - 3} = 3$

_____ 6. $\sqrt{3x - 5} = 7$

_____ 7. $\sqrt{5x + 1} + 3 = 7$

_____ 8. $\sqrt{7x + 1} + 2 = 10$

_____ 9. $\sqrt{4x - 7} + 3 = 0$

_____ 10. $\sqrt{3x - 5} + 2 = 0$

_____ 11. $8 - \sqrt{x - 3} = 7$

_____ 12. $12 - \sqrt{x - 5} = 11$

_____ 13. $13 - \sqrt{2x - 3} = 10$

_____ 14. $15 - \sqrt{3x + 1} = 11$

_____ 15. $\sqrt[3]{x - 1} = -2$

_____ 16. $\sqrt[3]{x - 1} = -4$

_____ 17. $\sqrt[3]{5x - 7} = 2$

_____ 18. $\sqrt[3]{2x - 5} = 1$

_____ 19. $\sqrt{2x + 8} - x = 0$

_____ 20. $\sqrt{4x + 21} - x = 0$

_____ 21. $\sqrt{9x - 17} = x - 1$

_____ 22. $\sqrt{1 - 2x} = x + 7$

_____ 23. $\sqrt{x} = \sqrt{5x - 8}$

_____ 24. $\sqrt{3x} = \sqrt{2x + 5}$

_____ 25. $\sqrt{7x - 5} + \sqrt{2x} = 0$

_____ 26. $\sqrt{6x - 10} + \sqrt{4x} = 0$

_____ 27. $\sqrt{7x - 1} = \sqrt{2x + 4}$

_____ 28. $\sqrt{3x - 7} = \sqrt{2x + 8}$

_____ 29. $\sqrt{x - 5} = 5 - \sqrt{x}$

_____ 30. $\sqrt{x - 7} = 7 - \sqrt{x}$

_____ 31. $\sqrt{3x - 2} = 5 + \sqrt{3x - 7}$

_____ 32. $\sqrt{x - 5} = 4 + \sqrt{x + 3}$

_____ 33. $\sqrt[3]{x^2 + x} = \sqrt[3]{20 - 7x}$

_____ 34. $\sqrt[3]{x^2 + 4x} = \sqrt[3]{x + 10}$

_____ 35. $\sqrt{3x - 5} + \sqrt{x - 1} = 2$

_____ 36. $\sqrt{3x + 9} - \sqrt{2x + 7} = 1$

_____ 37. $\sqrt{2y + 3} - \sqrt{y + 1} = 1$

_____ 38. $\sqrt{4x - 3} - \sqrt{2x - 5} = 2$

_____ 39. One leg of a right triangle is 1 centimeter longer than the other. If the shorter leg is increased by 3 centimeters and the length of the longer leg is doubled, the length of the hypotenuse of the enlarged right triangle is twice the length of the original hypotenuse. Find the dimensions of the original triangle.

_____ 40. One leg of a right triangle is 7 inches longer than the other. If the shorter leg is increased by 4 inches and the longer leg remains the same, the length of the hypotenuse of the enlarged right triangle is 2 more than the length of the original hypotenuse. Find the dimensions of the original triangle.

_____ **41.** The radius of a sphere is given by the formula $r = \sqrt[3]{\dfrac{3V}{4\pi}}$. Find the volume

(V) if the radius of the sphere is 9 inches.

_____ **42.** The radius of a cone is given by the formula $r = \sqrt{\dfrac{3V}{\pi h}}$. Find the volume (V)

of the cone if the radius is 6 feet and the height is 10 feet.

_____ **43.** Solve the formula $d = \sqrt{w^2 + l^2}$ for w.

_____ **44.** Solve the formula $r = \sqrt{\dfrac{A}{\pi}}$ for A.

☆ Stretching the Topics _____

_____ **1.** Solve $\sqrt{x + 2} = \sqrt[4]{9x - 2}$.

_____ **2.** Solve $\dfrac{\sqrt{x - 2}}{x - 2} = \dfrac{x - 5}{\sqrt{x - 2}}$.

_____ **3.** The time T for one period of a simple pendulum is given by the formula

$T = 2\pi\sqrt{\dfrac{l}{g}}$. Solve for l, the length of the pendulum's string.

Check your answers in the back of your book.

If you can complete **Checkup 7.4,** you are ready to do Section 7.5.

CHECKUP 7.4

Solve.

_____ 1. $\sqrt{3x - 5} = 4$

_____ 2. $\sqrt[3]{6x} + 2 = 4$

_____ 3. $\sqrt{5x + 1} + 2 = 0$

_____ 4. $9 - \sqrt{x - 5} = 7$

_____ 5. $7 - \sqrt{4x - 15} = 8$

_____ 6. $\sqrt[3]{x + 1} = -2$

_____ 7. $\sqrt{2x + 63} - x = 0$

_____ 8. $\sqrt{15x - 9} = x + 3$

_____ 9. $\sqrt{9x - 1} = \sqrt{3x + 11}$

_____ 10. $\sqrt{x + 2} + \sqrt{x - 1} = 3$

Check your answers in the back of your book.

If You Missed Problems:	You Should Review Examples:
1	1
2	2
3–8	3, 4
9, 10	5, 6

7.5 **Working with Complex Numbers**

In our work with quadratic equations and square roots, we have agreed that *the square root of a negative number is not a real number*. It is time now to look at a new set of numbers in which the square root of a negative number has meaning.

This new set of numbers is called the set of **complex numbers.**

Recognizing Complex Numbers

We would like to define the set of complex numbers so that *every* quadratic equation will have solutions within that set. Let's look at a few typical quadratic equations and see what kind of solutions occur.

Example 1. Solve $x^2 + 5x - 6 = 0$.

Solution

$$x^2 + 5x - 6 = 0$$
$$(x + 6)(x - 1) = 0$$
$$x + 6 = 0 \quad \text{or} \quad x - 1 = 0$$
$$x = -6 \qquad\qquad x = 1$$

Both solutions are *real* (and rational).

Example 2. Solve $x^2 - 7 = 0$.

Solution

$$x^2 - 7 = 0$$
$$x^2 = 7$$
$$x = \pm\sqrt{7}$$
$$x = \sqrt{7} \quad \text{or} \quad x = -\sqrt{7}$$

Both solutions are *real* (but irrational).

Example 3. Solve $x^2 + 5 = 0$.

Solution

$$x^2 + 5 = 0$$
$$x^2 = -5$$
$$x = \pm\sqrt{-5}$$
$$x = \sqrt{-5} \quad \text{or} \quad x = -\sqrt{-5}$$

Since $\sqrt{-5}$ is not a real number, there are *no real* solutions.

Example 4. Solve $x^2 + x + 4 = 0$.

Solution: We must use the quadratic formula with $a = 1$, $b = 1$, and $c = 4$.

$$x = \frac{-1 \pm \sqrt{1^2 - 4(1)(4)}}{2(1)}$$
$$= \frac{-1 \pm \sqrt{1 - 16}}{2}$$
$$x = \frac{-1 \pm \sqrt{-15}}{2}$$

There are *no real* solutions.

We wish to define the set of complex numbers to include solutions of the types obtained in all four of our examples. First we shall use the letter i to denote the square root of -1.

Definition of i

$$i = \sqrt{-1}$$

Using this notation, we may rewrite the solutions for Examples 3 and 4.

$$x = \sqrt{-5} \quad \text{or} \quad x = -\sqrt{-5}$$
$$= \sqrt{5(-1)} \qquad x = -\sqrt{5(-1)}$$
$$= \sqrt{5} \cdot \sqrt{-1} \qquad = -\sqrt{5} \cdot \sqrt{-1}$$
$$x = \sqrt{5}i \quad \text{or} \quad x = -\sqrt{5}i$$

$$x = \frac{-1 \pm \sqrt{-15}}{2}$$
$$= \frac{-1 \pm \sqrt{15(-1)}}{2}$$
$$= \frac{-1 \pm \sqrt{15} \cdot \sqrt{-1}}{2}$$
$$x = \frac{-1 \pm \sqrt{15}i}{2}$$

We call the number i the **imaginary unit** to emphasize the fact that i is *not a real number*.

Example 5. Write $\sqrt{-9}$ using the imaginary unit.

Solution
$$\sqrt{-9} = \sqrt{9(-1)}$$
$$= \sqrt{9} \cdot \sqrt{-1}$$
$$= 3i$$

Now you try Example 6.

Example 6. Write $1 + \sqrt{-3}$ using the imaginary unit.

Solution
$$1 + \sqrt{-3} = 1 + \sqrt{3(\underline{\quad})}$$
$$= 1 + \sqrt{3} \cdot \sqrt{\underline{\quad}}$$
$$= 1 + \sqrt{3} \underline{\quad}$$

Check your work on page 529. ▶

Each of the numbers we have written is called a **complex number.**

Definition of Complex Number. A complex number is a number that can be written in the standard form

$$a + bi$$

where a and b are real numbers and $i = \sqrt{-1}$. In any complex number $a + bi$,

a is called the **real part.**
b is called the **imaginary part.**
i is called the **imaginary unit.**

Now try Example 7.

Example 7. Identify the parts of $3 + i$.

Solution
$$3 + i$$
$$= 3 + 1i$$

The real part is _____ .
The imaginary part is _____ .

Check your work on page 529. ▶

Example 8. Identify the parts of -6.

Solution
$$-6$$
$$= -6 + 0i$$

The real part is -6.
The imaginary part is 0.

From Example 8 we see that

> Every real number belongs to the set of complex numbers. The imaginary part of a real number is always 0.

Example 9. Write $5 - \sqrt{-32}$ in standard complex form and identify its parts.

Solution

$$5 - \sqrt{-32}$$
$$= 5 - \sqrt{32}\sqrt{-1}$$
$$= 5 - \sqrt{16 \cdot 2}\sqrt{-1}$$
$$= 5 - 4\sqrt{2}\,i$$

The real part is 5.
The imaginary part is $-4\sqrt{2}$.

Now you try Example 10.

Example 10. Write $\sqrt{300}$ in standard complex form and identify its parts.

Solution

$$\sqrt{300}$$
$$= \sqrt{\underline{\quad} \cdot 3}$$
$$= \underline{\quad}\sqrt{3}$$
$$= \underline{\quad}\sqrt{3} + 0i$$

The real part is $\underline{\quad}\sqrt{3}$.
The imaginary part is $\underline{\quad}$.

Check your work on page 529. ▶

For two complex numbers to be equal, their real parts must be equal and their imaginary parts must be equal.

> **Equality of Complex Numbers.** If $a + bi = c + di$, then
> $$a = c \quad \text{and} \quad b = d$$

Example 11. Solve $2x + i = 3 - 4yi$ for real values of x and y.

Solution

$$2x + i = 3 - 4yi$$
$$2x + 1i = 3 + (-4y)i$$

Real Parts **Imaginary Parts**

$$2x = 3 \qquad\qquad 1 = -4y$$

$$x = \frac{3}{2} \qquad\qquad \frac{-1}{4} = y$$

Now you try Example 12.

Example 12. Solve $x + yi = \sqrt{3} - \sqrt{7}i$ for real values of x and y.

Solution

$$x + yi = \sqrt{3} - \sqrt{7}\,i$$
$$x + yi = \sqrt{3} + (\underline{\quad})i$$

Real Parts **Imaginary Parts**

$$x = \underline{\quad} \qquad\qquad y = \underline{\quad}$$

Check your work on page 529. ▶

To add (or subtract) complex numbers, we do what is completely natural. We must add (or subtract) the real parts, and add (or subtract) the imaginary parts and attach the imaginary unit.

Addition of Complex Numbers

$(a + bi) + (c + di) = (a + c) + (b + d)i$

You complete Example 13.

Example 13. Find $(3 + 2i) + (7 + 4i)$.

Solution

$(3 + 2i) + (7 + 4i)$

$= (3 + \underline{}) + (2 + \underline{})i$ Add real parts; add imaginary parts.

$= \underline{} + \underline{}i$

Check your work on page 530. ▶

Example 14. Find

$(4 - \sqrt{3}i) - (1 + \sqrt{3}i)$.

Solution

$(4 - \sqrt{3}i) - (1 + \sqrt{3}i)$

$= (4 - 1) + (-\sqrt{3} - \sqrt{3})i$ Add real parts; add imaginary parts.

$= 3 + (-2\sqrt{3})i$

$= 3 - 2\sqrt{3}i$

As you have probably noticed, the method used for simplifying sums or differences of complex numbers is similar to the methods for combining like terms. You may think of the *i*-terms as being like variable terms.

⟫ Trial Run

Write each number in standard complex form.

_____ **1.** $\sqrt{-36}$

_____ **2.** $\sqrt{25}$

_____ **3.** $3 - \sqrt{-8}$

_____ **4.** $\sqrt{12} + \sqrt{-16}$

Simplify.

_____ **5.** $(7 - \sqrt{2}i) + (4 - 3\sqrt{2}i)$

_____ **6.** $[(3 + 2i) - (6 + i)] + (4 + i)$

Solve for real values of x and y.

_____ **7.** $2x - 3i = 6 + 3yi$

Answers are on page 530.

Multiplying Complex Numbers

A complex number whose real part is zero is called a **pure imaginary number.** Some pure imaginary numbers are

$$i \qquad 5i \qquad -\sqrt{3}i \qquad \frac{1}{8}i$$

To multiply pure imaginary numbers, we may use the laws of exponents. For example,

$$i \cdot i \qquad\qquad 5i \cdot i^2 \qquad\qquad (2i)^4$$
$$= i^2 \qquad\qquad = 5i^3 \qquad\qquad = 2^4 i^4$$
$$= 16i^4$$

But let's look more closely at the powers of i. Remembering that $i = \sqrt{-1}$, we know that i is the number whose square is -1.

Since $i = \sqrt{-1}$, we know $i^2 = -1$

Now we can investigate some powers of i.

Power of i	Simplest Form
i^1	i
i^2	-1
i^3	$i^2 \cdot i = (-1)i = -i$
i^4	$i^2 \cdot i^2 = (-1)(-1) = 1$

The powers of i occur in cycles of four. To evaluate any power of i, we can break it down into a product of some of these four basic powers, using i^4 the most possible times (since $i^4 = 1$ is easy to work with).

Example 15. Simplify i^5.

Solution

$$i^5$$
$$= i^4 \cdot i$$
$$= 1 \cdot i$$
$$= i$$

Now you try Example 16.

Example 16. Simplify i^7.

Solution

$$i^7$$
$$= i^4 \cdot (\underline{})$$
$$= 1 \cdot (\underline{})$$
$$= \underline{}$$

Check your work on page 530. ▶

Example 17. Simplify $5i^{101}$.

Solution

$$5i^{101}$$
$$= 5 \cdot (i^4)^{25} \cdot i$$
$$= 5 \cdot (1)^{25} \cdot i$$
$$= 5 \cdot 1 \cdot i$$
$$= 5i$$

Now try Example 18.

Example 18. Simplify $\sqrt{2}i^{18}$.

Solution

$$\sqrt{2}i^{18}$$
$$= \sqrt{2}\,(i^4)^4 \cdot (\underline{})$$
$$= \sqrt{2}\,(1)^4(\underline{})$$
$$= \sqrt{2}\,(1)(\underline{})$$
$$= \underline{}$$

Check your work on page 530. ▶

The product of complex numbers is found using a process similar to the *distributive property* for multiplication over addition, together with the facts just learned about the *powers of i*.
You try Example 19.

Example 19. Simplify $-3(4 - 8i)$.

Solution

$-3(4 - 8i)$

$= -3 \cdot 4 - 3(-8i)$ Use distributive property.

$= \underline{} + \underline{} i$ Simplify products.

Check your work on page 530. ▶

Example 20. Simplify $-6i(5 - i)$.

Solution

$-6i(5 - i)$

$= -6i \cdot 5 - 6i(-i)$ Use distributive property.

$= -30i + 6i^2$ Simplify products.

$= -30i + 6(-1)$ Substitute -1 for i^2.

$= -30i - 6$ Simplify.

$= -6 - 30i$ Put in standard form.

Sometimes a complex number does not appear in standard complex form $a + bi$. However,

> We must write complex numbers in standard form using the imaginary unit before we try to operate with them.

Example 21. Simplify $(3 + \sqrt{-9}) + (-2 + \sqrt{-25})$.

Solution

$$(3 + \sqrt{-9}) + (-2 + \sqrt{-25})$$

$= (3 + 3i) + (-2 + 5i)$ Put each number in standard form.

$= 1 + 8i$ Add.

Example 22. Simplify $\sqrt{-4}(5 + \sqrt{-9})$.

Solution

$$\sqrt{-4}(5 + \sqrt{-9})$$

$= 2i(5 + 3i)$ Put each number in standard form.

$= 10i + 6i^2$ Use distributive property.

$= 10i + 6(-1)$

$= -6 + 10i$ Put in standard form.

Make careful note of the fact that a complex number *must* be switched to standard form before it is used in performing any operation.

Right	Wrong
$\sqrt{-9} \cdot \sqrt{-4}$	$\sqrt{-9} \cdot \sqrt{-4}$
$= 3i \cdot 2i$	$\neq \sqrt{(-9)(-4)}$
$= 6i^2$	$\neq \sqrt{36}$
$= 6(-1)$	$\neq 6$
$= -6$	

Let's practice multiplying complex numbers in standard form, using methods similar to those used in the multiplication of binomials.

You complete Example 23.

Example 23. Simplify $(3 + 4i)(1 + 2i)$.

Solution: We can use the FOIL technique here.

$(3 + 4i)(1 + 2i)$

$= 3 + 6i + 4i + 8i^2$

$= 3 + \underline{\quad}i + 8(\underline{\quad})$

$= \underline{\quad} + \underline{\quad}i$

Check your work on page 530. ▶

Example 24. Simplify

$(5 - \sqrt{-3})(2 + \sqrt{-12})$.

Solution: We first put the numbers in standard complex form. Then use FOIL.

$(5 - \sqrt{-3})(2 + \sqrt{-12})$

$= (5 - \sqrt{3}i)(2 + 2\sqrt{3}i)$

$= 10 + 10\sqrt{3}i - 2\sqrt{3}i - 2\sqrt{3}\sqrt{3}i^2$

$= 10 + 8\sqrt{3}i - 2(3)i^2$

$= 10 + 8\sqrt{3}i - 6(-1)$

$= 10 + 8\sqrt{3}\,i + 6$

$= 16 + 8\sqrt{3}i$

Now try Example 25.

Example 25. Simplify $(3 - \sqrt{2}i)^2$.

Solution

$(3 - \sqrt{2}i)^2$

$= 9 - 2 \cdot 3\sqrt{2}i + (\sqrt{2}i)^2$

Check your work on page 530. ▶

Example 26. Simplify

$(\sqrt{7} + 3i)(\sqrt{7} - 3i)$.

Solution

$(\sqrt{7} + 3i)(\sqrt{7} - 3i)$

$= \sqrt{7} \cdot \sqrt{7} - 3\sqrt{7}i + 3\sqrt{7}i - 9i^2$

$= 7 - 9i^2$

$= 7 - 9(-1)$

$= 7 + 9$

$= 16$

$= 16 + 0i$

Notice that the product of two complex numbers is always a complex number. Sometimes the product is a complex number that is *real,* as in Example 26. In Example 26, we multiplied $(\sqrt{7} + 3i)(\sqrt{7} - 3i)$ and the product was the real number 16. Indeed, whenever we multiply $(a + bi)(a - bi)$, the product will be a *real number*.

$$(a + bi)(a - bi) = a^2 - b^2i^2$$
$$= a^2 - b^2(-1)$$
$$= a^2 + b^2$$

As you can see, the imaginary unit does not appear in the product. The product is *real*.

For the complex number $a + bi$, the complex number $a - bi$ is called its (complex) **conjugate.** The product of a complex number and its conjugate is always a *real number* (namely, $a^2 + b^2$).

Complex Number	Conjugate	Product
$3 + 2i$	$3 - 2i$	$(3 + 2i)(3 - 2i)$
		$= 3^2 - 4i^2$
		$= 9 - 4(-1)$
		$= 13$
$6 - \sqrt{3}i$	$6 + \sqrt{3}i$	$(6 - \sqrt{3}i)(6 + \sqrt{3}i)$
		$= 6^2 - 3i^2$
		$= 36 - 3(-1)$
		$= 39$
$0 - i$	$0 + i$	$(0 - i)(0 + i)$
		$= 0 - i^2$
		$= -(-1)$
		$= 1$

⫸ Trial Run

Simplify.

_____ **1.** i^{16} _____ **2.** $7i^{65}$

_____ **3.** $-5(2 - 3i)$ _____ **4.** $4i(6 - i)$

_____ **5.** $\sqrt{-9}(3 + \sqrt{-25})$ _____ **6.** $(2 - 3i)(5 + i)$

_____ **7.** $(5 - \sqrt{3}i)^2$ _____ **8.** $(\sqrt{12} - \sqrt{-8})(\sqrt{3} + \sqrt{-2})$

_____ **9.** $(\sqrt{2} - 11i)(\sqrt{2} + 11i)$ _____ **10.** Write the conjugate of $1 + \sqrt{5}i$.

Answers are on page 530.

Dividing Complex Numbers

To divide any complex number by a *real* number, we use the same methods as we used to divide a binomial by a constant. We shall always write our answers in standard complex form.

Example 27. Simplify $\dfrac{3 + 2i}{6}$.

Solution

$$\dfrac{3 + 2i}{6} = \dfrac{1}{6}(3 + 2i)$$

$$= \dfrac{3}{6} + \dfrac{2}{6}i$$

$$= \dfrac{1}{2} + \dfrac{1}{3}i$$

You try Example 28.

Example 28. Simplify $\dfrac{4 - \sqrt{3}i}{2}$.

Solution

$$\dfrac{4 - \sqrt{3}i}{2} = \dfrac{1}{2}(4 - \sqrt{3}i)$$

$$=$$

Check your work on page 530. ▶

To divide a complex number by a *pure imaginary* number, we first eliminate the imaginary unit from the denominator. Remembering that $i = \sqrt{-1}$, we shall use a technique similar to that used for rationalizing denominators. Recalling the fact that $i \cdot i = i^2 = -1$, we simplify such quotients by multiplying numerator and denominator by i.

Example 29. Simplify $\dfrac{5}{2i}$.

Solution

$$\dfrac{5}{2i} = \dfrac{5}{2i} \cdot \dfrac{i}{i}$$

$$= \dfrac{5i}{2i^2}$$

$$= \dfrac{5i}{2(-1)}$$

$$= \dfrac{-5}{2}i$$

$$= 0 - \dfrac{5}{2}i \quad \text{Standard complex form.}$$

Example 30. Simplify $\dfrac{1 - 5i}{i}$.

Solution

$$\dfrac{1 - 5i}{i} = \dfrac{1 - 5i}{i} \cdot \dfrac{i}{i}$$

$$= \dfrac{(1 - 5i)i}{i^2}$$

$$= \dfrac{i - 5i^2}{-1}$$

$$= \dfrac{i - 5(-1)}{-1}$$

$$= \dfrac{i + 5}{-1}$$

$$= \dfrac{i}{-1} + \dfrac{5}{-1}$$

$$= -5 - i \quad \text{Standard complex form.}$$

If a fraction contained a denominator that was a *sum* (or difference) of radicals, recall that we rationalized that denominator using the *difference* (or sum) of those radicals as the rationalizing factor. Using the same reasoning,

> We can simplify a quotient whose denominator is $a + bi$ by multiplying the numerator and denominator by the *conjugate*, $a - bi$.

Let's use the conjugate of the denominator to simplify some quotients, writing our answers in standard complex form.

Example 31. Simplify $\dfrac{3}{2 + i}$.

Solution

$$\frac{3}{2 + i} = \frac{3}{2 + i} \cdot \frac{2 - i}{2 - i}$$

$$= \frac{3(2 - i)}{(2 + i)(2 - i)}$$

$$= \frac{3(2 - i)}{4 - i^2}$$

$$= \frac{3(2 - i)}{4 - (-1)}$$

$$= \frac{3(2 - i)}{5}$$

$$= \frac{6 - 3i}{5}$$

$$= \frac{6}{5} - \frac{3}{5} i$$

Example 32. Simplify $\dfrac{26i}{3 - 2i}$.

Solution

$$\frac{26i}{3 - 2i} = \frac{26i}{3 - 2i} \cdot \frac{3 + 2i}{3 + 2i}$$

$$= \frac{26i(3 + 2i)}{(3 - 2i)(3 + 2i)}$$

$$= \frac{26i(3 + 2i)}{9 - 4i^2}$$

$$= \frac{26i(3 + 2i)}{9 - 4(-1)}$$

$$= \frac{\overset{2}{\cancel{26}}i(3 + 2i)}{\underset{1}{\cancel{13}}}$$

$$= 2i(3 + 2i)$$

$$= 6i + 4i^2$$

$$= -4 + 6i$$

Usually, when simplifying quotients of complex numbers, it is wise to delay removing parentheses in the numerator just in case the new denominator contains a common factor. This did not happen in Example 31 but it did happen in Example 32.

Example 33. Simplify $\dfrac{3 + \sqrt{-3}}{4 + \sqrt{-3}}$.

Solution: We must write complex numbers in standard form before we begin to operate.

$$\frac{3 + \sqrt{-3}}{4 + \sqrt{-3}} = \frac{3 + \sqrt{3}i}{4 + \sqrt{3}i} \cdot \frac{4 - \sqrt{3}i}{4 - \sqrt{3}i}$$
Write complex numbers in standard form. Multiply by the conjugate of the denominator.

$$= \frac{12 - 3\sqrt{3}i + 4\sqrt{3}i - 3i^2}{16 - 3i^2}$$
Multiply numerators; multiply denominators.

$$= \frac{12 + \sqrt{3}i - 3(-1)}{16 - 3(-1)}$$
Combine like terms. Substitute -1 for i^2.

$$= \frac{12 + \sqrt{3}i + 3}{16 + 3}$$
Simplify.

$$= \frac{15 + \sqrt{3}i}{19} \qquad \text{Combine like terms.}$$

$$= \frac{15}{19} + \frac{\sqrt{3}}{19}\, i \qquad \text{Write the answer in standard form.}$$

▥▶ Trial Run

Simplify, writing answers in standard complex form.

_____ **1.** $\dfrac{9 - 12i}{3}$ _____ **2.** $\dfrac{10 - \sqrt{2}i}{5}$

_____ **3.** $\dfrac{-12}{\sqrt{3}i}$ _____ **4.** $\dfrac{\sqrt{2} - 3i}{i}$

_____ **5.** $\dfrac{1}{3 + i}$ _____ **6.** $\dfrac{17i}{3 - 5i}$

_____ **7.** $\dfrac{11}{3 - \sqrt{-2}}$ _____ **8.** $\dfrac{4 + \sqrt{5}i}{2 - \sqrt{5}i}$

Answers are on page 530.

▶ Examples You Completed

Example 6. Write $1 + \sqrt{-3}$ using the imaginary unit.

Solution

$$1 + \sqrt{3} = 1 + \sqrt{3(-1)}$$
$$= 1 + \sqrt{3} \cdot \sqrt{-1}$$
$$= 1 + \sqrt{3}i$$

Example 7. Identify the parts of $3 + i$.

Solution

$$3 + i$$
$$= 3 + 1i$$

The real part is 3.
The imaginary part is 1.

Example 10. Write $\sqrt{300}$ in standard complex form and identify its parts.

Solution

$$\sqrt{300}$$
$$= \sqrt{100 \cdot 3}$$
$$= 10\sqrt{3}$$
$$= 10\sqrt{3} + 0i$$

The real part is $10\sqrt{3}$.
The imaginary part is 0.

Example 12. Solve $x + yi = \sqrt{3} - \sqrt{7}i$ for real values of x and y.

Solution

$$x + yi = \sqrt{3} - \sqrt{7}i$$
$$x + yi = \sqrt{3} + (-\sqrt{7})i$$

Real Parts   **Imaginary Parts**

$$x = \sqrt{3} \qquad\qquad y = -\sqrt{7}$$

Example 13. Find $(3 + 2i) + (7 + 4i)$.

Solution

$$(3 + 2i) + (7 + 4i)$$
$$= (3 + 7) + (2 + 4)i$$
$$= 10 + 6i$$

Example 16. Simplify i^7.

Solution

$$i^7$$
$$= i^4 \cdot i^3$$
$$= 1 \cdot (-i)$$
$$= -i$$

Example 18. Simplify $\sqrt{2}i^{18}$.

Solution

$$\sqrt{2}i^{18}$$
$$= \sqrt{2}(i^4)^4 \cdot (i^2)$$
$$= \sqrt{2}(1)^4(-1)$$
$$= \sqrt{2}(1)(-1)$$
$$= -\sqrt{2}$$

Example 19. Simplify $-3(4 - 8i)$.

Solution

$$-3(4 - 8i)$$
$$= -3 \cdot 4 - 3 \cdot (-8i)$$
$$= -12 + 24i$$

Example 23. Simplify $(3 + 4i)(1 + 2i)$.

Solution: We can use the FOIL technique here.

$$(3 + 4i)(1 + 2i)$$
$$= 3 + 6i + 4i + 8i^2$$
$$= 3 + 10i + 8(-1)$$
$$= -5 + 10i$$

Example 25. Simplify $(3 - \sqrt{2}i)^2$.

Solution

$$(3 - \sqrt{2}i)^2$$
$$= 9 - 2 \cdot 3\sqrt{2}i + (\sqrt{2}i)^2$$
$$= 9 - 6\sqrt{2}i + 2i^2$$
$$= 9 - 6\sqrt{2}i + 2(-1)$$
$$= 7 - 6\sqrt{2}i$$

Example 28. Simplify $\dfrac{4 - \sqrt{3}i}{2}$.

Solution

$$\frac{4 - \sqrt{3}i}{2} = \frac{1}{2}(4 - \sqrt{3}i)$$
$$= \frac{4}{2} - \frac{\sqrt{3}}{2}i$$
$$= 2 - \frac{\sqrt{3}}{2}i$$

Answers to Trial Runs

page 522 **1.** $0 + 6i$ **2.** $5 + 0i$ **3.** $3 - 2\sqrt{2}i$ **4.** $2\sqrt{3} + 4i$ **5.** $11 - 4\sqrt{2}i$ **6.** $1 + 2i$
7. $x = 3, y = -1$

page 526 **1.** 1 **2.** $7i$ **3.** $-10 + 15i$ **4.** $4 + 24i$ **5.** $-15 + 9i$ **6.** $13 - 13i$
7. $22 - 10\sqrt{3}i$ **8.** 10 **9.** 123 **10.** $1 - \sqrt{5}i$

page 529 **1.** $3 - 4i$ **2.** $2 - \dfrac{\sqrt{2}}{5}i$ **3.** $0 + 4\sqrt{3}i$ **4.** $-3 - \sqrt{2}i$ **5.** $\dfrac{3}{10} - \dfrac{1}{10}i$

6. $\dfrac{-5}{2} + \dfrac{3}{2}i$ **7.** $3 + \sqrt{2}i$ **8.** $\dfrac{1}{3} + \dfrac{2\sqrt{5}}{3}i$

EXERCISE SET 7.5

Write each of the following numbers in standard complex form.

_____ **1.** $\sqrt{-16}$

_____ **2.** $\sqrt{-64}$

_____ **3.** $\sqrt{-21}$

_____ **4.** $\sqrt{-15}$

_____ **5.** $\sqrt{-50}$

_____ **6.** $\sqrt{-45}$

_____ **7.** $-\sqrt{-75}$

_____ **8.** $-\sqrt{-18}$

_____ **9.** $4 - \sqrt{-54}$

_____ **10.** $3 - \sqrt{-60}$

_____ **11.** $18 + \sqrt{-49}$

_____ **12.** $32 + \sqrt{-81}$

Find real values for x and y so that each of the following equations is true.

_____ **13.** $4x - 2yi = 4 + 8i$

_____ **14.** $5x - 3yi = 10 + 15i$

_____ **15.** $2x - 4i = -8 - 12yi$

_____ **16.** $3x - 9i = -9 - 18yi$

_____ **17.** $4x - 5i = -2 + 3yi$

_____ **18.** $2x - 7i = -7 + 5yi$

_____ **19.** $14y + 3i = 21 + xi$

_____ **20.** $15y + 4i = 45 + xi$

_____ **21.** $(3x - 1) - i = 5 - (y + 1)i$

_____ **22.** $(2x - 3) - 3i = 3 - (y + 2)$

_____ **23.** $(2x - 1) - (3y + 2)i = 7 + 2i$

_____ **24.** $(x - 9) - (2y - 1)i = 8 + 9i$

Simplify each of the following.

_____ **25.** i^5

_____ **26.** i^9

_____ **27.** i^{12}

_____ **28.** i^{20}

_____ **29.** $-i^7$

_____ **30.** $-i^{11}$

_____ **31.** i^{73}

_____ **32.** i^{37}

Simplify each of the following. Write answers in standard complex form.

_____ **33.** $(2 + 3i) + (3 + 6i)$

_____ **34.** $(5 + 4i) + (7 + 2i)$

_____ **35.** $(9 - 5i) - (9 + 4i)$

_____ **36.** $(9 - 6i) - (8 + 2i)$

_____ **37.** $(12 - 3\sqrt{-8}) - (3 - \sqrt{-2})$

_____ **38.** $(14 - \sqrt{-27}) - (11 + 5\sqrt{-3})$

_____ **39.** $(-3 + \sqrt{-49}) + (8 - 7i)$

_____ **40.** $(-7 + \sqrt{-36}) + (10 - 6i)$

_____ **41.** $[(7 - i) - (3 + 2i)] - (6 - 4i)$

_____ **42.** $[(5 - i) - (6 + 2i)] - (4 + 9i)$

_____ **43.** $(7 - 4i) - [(-2 + i) - (3 + 7i)]$

_____ **44.** $(8 - 3i) - [(-1 + i) - (2 + 6i)]$

531

_____ **45.** $-6(3 + 2i)$ _____ **46.** $-8(5 + 7i)$

_____ **47.** $3i(7 + i)$ _____ **48.** $9i(2 - i)$

_____ **49.** $\sqrt{-25}(2 + \sqrt{-4})$ _____ **50.** $\sqrt{-64}(3 + \sqrt{-9})$

_____ **51.** $(7 - 2i)(3 + 5i)$ _____ **52.** $(8 - 3i)(6 + 2i)$

_____ **53.** $(3 - \sqrt{5}i)^2$ _____ **54.** $(4 + \sqrt{3}i)^2$

_____ **55.** $(\sqrt{50} - \sqrt{-3})(\sqrt{2} - \sqrt{-48})$ _____ **56.** $(\sqrt{32} + \sqrt{-5})(\sqrt{2} + \sqrt{-45})$

_____ **57.** $(\sqrt{3} - 4i)(\sqrt{3} + 4i)$ _____ **58.** $(\sqrt{5} - 3i)(\sqrt{5} + 3i)$

_____ **59.** $(-3 - 4i)(2 - 5i)$ _____ **60.** $(-6 - 2i)(3 - 4i)$

_____ **61.** $\dfrac{6 - 15i}{3}$ _____ **62.** $\dfrac{10 - 25i}{5}$

_____ **63.** $\dfrac{4 - 10i}{5}$ _____ **64.** $\dfrac{3 + 14i}{7}$

_____ **65.** $\dfrac{12 - \sqrt{3}i}{4}$ _____ **66.** $\dfrac{21 - \sqrt{7}i}{3}$

_____ **67.** $\dfrac{5}{i}$ _____ **68.** $\dfrac{6}{i}$

_____ **69.** $\dfrac{-20}{\sqrt{5}i}$ _____ **70.** $\dfrac{-16}{\sqrt{2}i}$

_____ **71.** $\dfrac{-\sqrt{3} - i}{2i}$ _____ **72.** $\dfrac{-\sqrt{2} - 4i}{3i}$

_____ **73.** $\dfrac{39}{2 - 3i}$ _____ **74.** $\dfrac{10}{4 - 2i}$

_____ **75.** $\dfrac{2}{7 - i}$ _____ **76.** $\dfrac{5}{8 - i}$

_____ **77.** $\dfrac{13i}{7 - \sqrt{3}i}$ _____ **78.** $\dfrac{22i}{3 - \sqrt{2}i}$

_____ **79.** $\dfrac{-1}{2 - \sqrt{-25}}$ _____ **80.** $\dfrac{-2}{2 - \sqrt{-9}}$

_____ **81.** $\dfrac{5 + 4i}{3 + 6i}$ _____ **82.** $\dfrac{2 + i}{5 - 6i}$

_____ **83.** $\dfrac{2 - \sqrt{-5}}{2 + \sqrt{-5}}$ _____ **84.** $\dfrac{4 - \sqrt{-2}}{4 + \sqrt{-2}}$

☆ Stretching the Topics

_____ 1. Simplify $(1 - i\sqrt{3})^3$.

_____ 2. Factor $x^2 + 9$ as the difference of two squares.

_____ 3. Solve $(4 - 5i) + (x + yi) = 0$ for x and y. What is the additive inverse of $4 - 5i$?

Check your answers in the back of your book.

If you can complete **Checkup 7.5,** you are ready to go on to Section 7.6.

✔ CHECKUP 7.5

Write in standard form.

_____ **1.** $\sqrt{-121}$

_____ **2.** $9 - \sqrt{-28}$

_____ **3.** Solve $3x - 8i = 12 + 2yi$ for real values of x and y.

_____ **4.** Evaluate i^{34}.

Simplify.

_____ **5.** $(8 - 3i) - (2 + 2i)$

_____ **6.** $(4 - 3\sqrt{-12}) - (1 + \sqrt{-27})$

_____ **7.** $(2 - 3i)(4 + 3i)$

_____ **8.** $(5 - 2i)^2$

_____ **9.** $\dfrac{3 - 2i}{i}$

_____ **10.** $\dfrac{1 - 3i}{1 + 2i}$

Check your answers in the back of your book.

If You Missed Problems:	You Should Review Examples:
1	5
2	9
3	11, 12
4	15–18
5, 6	21
7, 8	25, 26
9	30
10	33

7.6 Solving Equations with Complex Solutions

Now that we have learned to work within the set of complex numbers, we can solve many equations that have *no real solutions*.

Example 1. Solve $x^2 + 4 = 0$.

Solution

$$x^2 + 4 = 0$$
$$x^2 = -4$$
$$x = \pm\sqrt{-4}$$
$$x = \pm 2i$$
$$x = 2i \quad \text{or} \quad x = -2i$$

Both solutions are complex numbers.

Example 2. Solve $x^2 + x + 2 = 0$.

Solution: We must use the quadratic formula. Here $a = 1$, $b = 1$, and $c = 2$.

$$x = \frac{-1 \pm \sqrt{1^2 - 4(1)(2)}}{2(1)} \quad \text{Substitute into the quadratic formula.}$$

$$= \frac{-1 \pm \sqrt{1 - 8}}{2}$$

$$= \frac{-1 \pm \sqrt{-7}}{2} \quad \text{Simplify.}$$

$$= \frac{-1 \pm \sqrt{7}i}{2} \quad \text{Write numerator in standard form.}$$

$$= \frac{-1}{2} + \frac{\sqrt{7}}{2}i \quad \text{Write answers in standard form.}$$

$$x = \frac{-1}{2} + \frac{\sqrt{7}}{2}i \quad \text{or} \quad x = \frac{-1}{2} - \frac{\sqrt{7}}{2}i$$

Both solutions are complex numbers.

Example 3. Solve $x^3 = 8$.

Solution

$$x^3 = 8$$
$$x^3 - 8 = 0$$

We factor the difference of two cubes.

$$(x - 2)(x^2 + 2x + 4) = 0$$

$$x - 2 = 0 \quad \text{or} \quad x^2 + 2x + 4 = 0$$

$$x = 2 \quad \text{or} \quad x = \frac{-2 \pm \sqrt{2^2 - 4(1)(4)}}{2}$$

$$= \frac{-2 \pm \sqrt{4 - 16}}{2}$$

$$= \frac{-2 \pm \sqrt{-12}}{2}$$

$$= \frac{-2 \pm 2\sqrt{3}i}{2} \quad \text{Simplify the radical.}$$

$$= \frac{\overset{1}{\cancel{2}}(-1 \pm \sqrt{3}i)}{\underset{1}{\cancel{2}}} \quad \text{Factor numerator and reduce.}$$

$$x = 2 \quad \text{or} \quad x = -1 \pm \sqrt{3}i$$

The solutions are

$$x = 2 \quad \text{or} \quad x = -1 + \sqrt{3}i \quad \text{or} \quad x = -1 - \sqrt{3}i$$

The equation has three complex solutions; one of the complex solutions is real.

You complete Example 4.

Example 4. Solve $x^4 + x^2 - 6 = 0$.

Solution

$$x^4 - x^2 - 6 = 0$$
$$(x^2 - 3)(x^2 + 2) = 0$$

$$x^2 - 3 = 0 \qquad \text{or} \quad x^2 + 2 = 0$$
$$x^2 = \underline{\quad} \qquad\qquad\qquad x^2 = \underline{\quad}$$
$$x = \pm\sqrt{\underline{\quad}} \qquad\qquad x = \pm\sqrt{\underline{\quad}}$$
$$\qquad\qquad\qquad\qquad\qquad x = \pm\sqrt{\underline{\quad}}i$$

The solutions are

$$x = \sqrt{\underline{\quad}} \quad \text{or} \quad x = -\sqrt{\underline{\quad}} \quad \text{or} \quad x = \sqrt{\underline{\quad}}i \quad \text{or} \quad x = -\sqrt{\underline{\quad}}i$$

The equation has _____ complex solutions; _____ of the complex solutions are real.

Check your work on page 538. ▶

⫸ Trial Run

Find all complex solutions.

_____ **1.** $x^2 + 5 = 0$ \qquad\qquad _____ **2.** $x^2 - x + 1 = 0$

_____ **3.** $x^2 + 5x = x - 7$ \qquad\qquad _____ **4.** $x^4 - 4 = 0$

Answers are on page 538.

▶ Example You Completed

Example 4. Solve $x^4 - x^2 - 6 = 0$.

Solution

$$x^4 - x^2 - 6 = 0$$
$$(x^2 - 3)(x^2 + 2) = 0$$

$$x^2 - 3 = 0 \qquad \text{or} \quad x^2 + 2 = 0$$
$$x^2 = 3 \qquad\qquad\qquad x^2 = -2$$
$$x = \pm\sqrt{3} \qquad\qquad x = \pm\sqrt{-2}$$
$$\qquad\qquad\qquad\qquad x = \pm\sqrt{2}i$$

$$x = \sqrt{3} \quad \text{or} \quad x = -\sqrt{3} \quad \text{or} \quad x = \sqrt{2}i \quad x = -\sqrt{2}i$$

The equation has four complex solutions; two of the complex solutions are real.

Answers to Trial Run

page 538 **1.** $x = \pm\sqrt{5}i$ \quad **2.** $x = \dfrac{1}{2} \pm \dfrac{\sqrt{3}}{2}i$ \quad **3.** $x = -2 \pm \sqrt{3}i$ \quad **4.** $x = \pm\sqrt{2}$ or $x = \pm\sqrt{2}i$

EXERCISE SET 7.6

Solve.

_____ 1. $x^2 + 9 = 0$

_____ 2. $x^2 + 49 = 0$

_____ 3. $4x^2 + 25 = 0$

_____ 4. $9x^2 + 16 = 0$

_____ 5. $x^2 + 3 = 0$

_____ 6. $x^2 + 5 = 0$

_____ 7. $x^2 + x + 1 = 0$

_____ 8. $x^2 - 2x + 2 = 0$

_____ 9. $3x^2 - 8x + 7 = 0$

_____ 10. $5x^2 - 10x + 7 = 0$

_____ 11. $2x^2 + 4 = 3x$

_____ 12. $6x^2 + 2 = 5x$

_____ 13. $x(x + 5) = 2x - 3$

_____ 14. $x(2x + 1) = 3x - 1$

_____ 15. $5x^2 + 6x + 4 = 0$

_____ 16. $3x^2 + 4x + 3 = 0$

_____ 17. $x^3 + 8 = 0$

_____ 18. $x^3 + 27 = 0$

_____ 19. $x^3 = 125$

_____ 20. $x^3 = 64$

_____ 21. $8x^3 - 1 = 0$

_____ 22. $27x^3 - 8 = 0$

_____ 23. $x^4 - 81 = 0$

_____ 24. $x^4 - 16 = 0$

_____ 25. $x^4 = 121$

_____ 26. $x^4 = 100$

_____ 27. $x^4 + 13x^2 + 36 = 0$

_____ 28. $x^4 + 10x^2 + 9 = 0$

_____ 29. $x^4 + 3x^2 = 10$

_____ 30. $x^4 + 4x^2 = 21$

_____ 31. $3 + \dfrac{4}{x} = \dfrac{-2}{x^2}$

_____ 32. $\dfrac{9}{x^2} + 1 = \dfrac{3}{x}$

_____ 33. $3x + \dfrac{1}{x} = 2$

_____ 34. $x + \dfrac{3}{x} = 3$

_____ 35. $\dfrac{1}{x} = \dfrac{3x}{x - 2}$

_____ 36. $\dfrac{x}{x - 3} = \dfrac{5}{x}$

☆ Stretching the Topics _____

Solve.

_____ 1. $\sqrt{x - 1} = 3i$

_____ 2. $2ix^2 - 5x - 2i = 0$

_____ 3. Write a quadratic equation that has $2i$ and $-3i$ as its solutions.

Check your answers in the back of your book.
If you can solve the equations in **Checkup 7.6,** you are ready to do the Review Exercises.

✔ **CHECKUP 7.6**

Solve.

_____ **1.** $x^2 + 25 = 0$

_____ **2.** $x^2 + 24 = 0$

_____ **3.** $x^2 - x + 25 = 0$

_____ **4.** $5x^2 + 9 = 12x$

_____ **5.** $x^3 + 125 = 0$

_____ **6.** $64x^3 = 27$

_____ **7.** $x^4 = 100$

_____ **8.** $x^4 + 25x^2 + 144 = 0$

_____ **9.** $x + \dfrac{7}{x} = 2$

_____ **10.** $\dfrac{2x}{x - 1} = \dfrac{3}{x}$

Check your answers in the back of your book.

If You Missed Problems:	You Should Review Example:
1, 2	1
3, 4	2
5, 6	3
7, 8	4
9, 10	2

Summary

In this chapter we learned to work with fractional exponents and radicals.

Fractional Exponent form	Radical Form	Examples
$a^{1/n}$	$\sqrt[n]{a}$	$(-8)^{1/3} = \sqrt[3]{-8} = -2$
		$(x + 3)^{1/2} = \sqrt{x + 3}$
$a^{-1/n}$	$\dfrac{1}{\sqrt[n]{a}}$	$(25)^{-1/2} = \dfrac{1}{\sqrt{25}} = \dfrac{1}{5}$
		$(-27)^{-1/3} = \dfrac{1}{\sqrt[3]{-27}} = \dfrac{-1}{3}$
$a^{m/n}$	$\sqrt[n]{a^m}$ or $(\sqrt[n]{a})^m$	$x^{2/3} = \sqrt[3]{x^2}$
		$32^{7/5} = (\sqrt[5]{32})^7 = 2^7 = 128$

If a is a real number and n is a natural number, we say the $\sqrt[n]{a} = b$ if $b^n = a$. Moreover, we noted that

1. If a is positive, then $\sqrt[n]{a}$ is real and positive.
2. If a is negative and n is *odd*, then $\sqrt[n]{a}$ is real and negative.
3. If a is negative and n is *even*, then $\sqrt[n]{a}$ is *not* a real number; instead, $\sqrt[n]{a}$ is an imaginary number.

We also discovered that

$$\sqrt[n]{a^n} = \begin{cases} a & \text{if } n \text{ is odd} \\ |a| & \text{if } n \text{ is even} \end{cases}$$

We learned to simplify radical expressions and operate with radical expressions using several techniques.

If We Want to:	We Must:	Examples
Simplify radical expressions	Use $\sqrt[n]{a \cdot b} = \sqrt[n]{a} \sqrt[n]{b}$ or Use $\sqrt[n]{\dfrac{a}{b}} = \dfrac{\sqrt[n]{a}}{\sqrt[n]{b}}$	$\sqrt{72x^3} = \sqrt{36 \cdot 2 \, x^2 \cdot x}$ $= 6x\sqrt{2x}$ $\sqrt{\dfrac{12a^3}{25c^4}} = \dfrac{\sqrt{12a^3}}{\sqrt{25c^4}} = \dfrac{2a\sqrt{3a}}{5c^2}$
Add or subtract radical expressions	Combine like radicals	$\sqrt{12x} + \sqrt{3x} - \sqrt{75x}$ $= 2\sqrt{3x} + \sqrt{3x} - 5\sqrt{3x}$ $= -2\sqrt{3x}$
Multiply or divide radical expressions	Use $\sqrt[n]{a} \cdot \sqrt[n]{b} = \sqrt[n]{ab}$	$\sqrt{7a} \cdot \sqrt{14a} = \sqrt{98a^2} = \sqrt{49 \cdot 2 \cdot a^2}$ $= 7a\sqrt{2}$ $\dfrac{\sqrt[3]{24x^4}}{\sqrt[3]{3x}} = \sqrt[3]{\dfrac{24x^4}{3x}} = \sqrt[3]{8x^3} = 2x$

Multiply expressions containing sums or differences of radicals	Use the distributive property	$\sqrt{x}(3 + \sqrt{x}) = 3\sqrt{x} + x$ $(\sqrt{x} - \sqrt{3})(\sqrt{x} + 5\sqrt{3})$ $\quad = x + 5\sqrt{3x} - \sqrt{3x} - 5 \cdot 3$ $\quad = x + 4\sqrt{3x} - 15$
Rationalize the denominator of a fraction	Multiply numerator and denominator by an appropriate rationalizing factor	$\dfrac{7}{\sqrt{3x}} = \dfrac{7}{\sqrt{3x}} \cdot \dfrac{\sqrt{3x}}{\sqrt{3x}} = \dfrac{7\sqrt{3x}}{3x}$ $\dfrac{7}{\sqrt{3} + x} = \dfrac{7}{\sqrt{3} + x} \cdot \dfrac{\sqrt{3} - x}{\sqrt{3} - x}$ $\quad = \dfrac{7(\sqrt{3} - x)}{3 - x^2}$

Radical equations are equations in which a variable expression appears in a radicand. To solve such equations, we isolate the radical and raise both sides to the appropriate power. Whenever we square both sides of a radical equation, we must remember to check our solutions in the original equation and reject **extraneous solutions.**

In this chapter we expanded our work with numbers to include square roots of negative numbers and discovered the set of **complex numbers.** Every complex number can be written in the form

$$a + bi$$

where a is real, b is real, and $i = \sqrt{-1}$. In any complex number, a is called the *real part,* b is called the *imaginary part,* and i is called the *imaginary unit.* Every *real number* is a complex number of the form $a + 0i$. A complex number of the form $0 + bi$ is called a **pure imaginary number.** The **conjugate** of a complex number $a + bi$ is the complex number $a - bi$.

	Method	Examples
Equality of complex numbers	If $a + bi = c + di$ then $a = c$ and $b = d$	If $x + yi = 3 + 2i$ then $x = 3$ and $y = 2$
Addition of complex numbers	$(a + bi) + (c + di)$ $= (a + c) + (b + d)i$	$(3 + 2i) + (7 - i) = 10 + i$
Powers of i	$i^1 = i$ $i^2 = -1$ $i^3 = -i$ $i^4 = 1$	$i^{10} = (i^4)^2 \cdot i^2$ $\quad = (1)^2 \cdot i^2$ $\quad = -1$
Multiplication of complex numbers	Use the distributive property	$3i(1 - i) = 3i - 3i^2 = 3 + 3i$ $(2 - 5i)(1 + 4i) = 2 + 8i - 5i - 20i^2$ $\quad = 2 + 3i + 20$ $\quad = 22 + 3i$

Division of complex numbers	Multiply numerator and denominator by the conjugate of the denominator	$\dfrac{3 + 2i}{4 - 3i} = \dfrac{(3 + 2i)(4 + 3i)}{(4 - 3i)(4 + 3i)}$
		$= \dfrac{12 + 9i + 8i + 6i^2}{16 - 9i^2}$
		$= \dfrac{12 + 17i - 6}{16 + 9}$
		$= \dfrac{6 + 17i}{25}$
		$= \dfrac{6}{25} + \dfrac{17}{25}i$

After becoming familiar with complex numbers, we used familiar techniques to solve equations with solutions that were complex numbers.

❑ Speaking the Language of Algebra

Complete each sentence with the appropriate word or phrase.

1. The square root of a negative number is *not* a _____ number.

2. In the radical expression $\sqrt[n]{a}$, we call n the _____ and a the _____ .

3. In adding and subtracting radical expressions, we may combine _____ _____ .

4. The process by which we rid the denominator of radical expressions is called _____ _____ _____ .

5. In solving radical equations, we must check all solutions and reject any that are _____ .

6. Every number of the form $a + bi$ is called a _____ _____ , where a is _____ , b is _____ , and $i =$ _____ .

7. For the number $a + bi$, a is called the _____ part, b is called the _____ part, and i is called the _____ _____ .

8. If two complex numbers are equal, we know their _____ _____ are equal and their _____ _____ are equal.

9. For the complex number $a + bi$, the complex number $a - bi$ is called its _____ .

10. To divide complex numbers, we multiply numerator and denominator by the _____ of the _____ .

REVIEW EXERCISES for Chapter 7: Working with Rational Exponents and Radicals

Simplify.

_____ **1.** $x^{1/2} \cdot x^{3/4} \cdot x^{-1/4}$

_____ **2.** $(x^{1/6} \cdot y^{-2/3})^{12}$

_____ **3.** $(x^{1/2} \cdot y^{7/6} \cdot z^{-2/3})^{3/2}$

_____ **4.** $\dfrac{x^{5/6} \cdot y^{2/5}}{x^{2/3} \cdot y^{3/10}}$

Find the value.

_____ **5.** $\sqrt[3]{-125}$

$\frac{1}{3}$
_____ **6.** $\sqrt[4]{\dfrac{1}{81}}$

Write in exponential form.

_____ **7.** $\sqrt[5]{a^2 b^3}$ $a^{\frac{2}{5}} \ b^{\frac{3}{5}}$

_____ **8.** $\dfrac{1}{\sqrt[5]{(x-y)^2}}$ $\dfrac{1}{\sqrt[5]{x} - \sqrt[5]{y}}$ $\times$

Write in radical form and simplify if possible.

$\sqrt[4]{(a+b)^3}$

_____ **9.** $27^{2/3}$

_____ **10.** $(a+b)^{3/4}$

Simplify. Assume that any variable expression under an even radical represents a positive number.

$3xy\sqrt{2x}$

_____ **11.** $\sqrt{18x^3y^2}$

_____ **12.** $\sqrt[3]{\dfrac{-24x^5}{125y^3}}$

$4(x)$
_____ **13.** $4(\sqrt{7} - 2x) - (2\sqrt{7} + 3x)$

_____ **14.** $\sqrt{20x^2} + \sqrt{45x^2} - \sqrt{80x^2}$

_____ **15.** $\sqrt[3]{16} - \sqrt{75} + \sqrt[3]{54} + \sqrt{48}$

_____ **16.** $\sqrt{\dfrac{50}{81}} - \sqrt{\dfrac{98}{9}}$

_____ **17.** $\sqrt{10x} \cdot \sqrt{15x^3}$

_____ **18.** $\sqrt{\dfrac{27a^3b}{3a}}$

_____ **19.** $\sqrt{3}(\sqrt{6} + \sqrt{3})$

_____ **20.** $(4\sqrt{5} - \sqrt{2})(4\sqrt{5} + \sqrt{2})$

_____ **21.** $(\sqrt{2x} - 3)^2$

_____ **22.** $\dfrac{4}{\sqrt{6a}}$

_____ **23.** $\dfrac{3}{\sqrt[3]{2x^2}}$

_____ **24.** $\sqrt{\dfrac{50x^3}{3y}}$

_____ **25.** $\dfrac{\sqrt{3x^3y}}{\sqrt{12x^2y^4}}$

_____ **26.** $\dfrac{\sqrt{6}}{\sqrt{2} - \sqrt{3}}$

Solve.

_____ **27.** $\sqrt{5x} - 7 = 3$

_____ **28.** Solve the formula $r = \sqrt{\dfrac{v}{\pi h}}$ for v.

_____ **29.** $5 - \sqrt{x - 2} = 4$

_____ **30.** $\sqrt[3]{2x - 1} = -3$

_____ **31.** $\sqrt{x - 1} = x - 3$

_____ **32.** $\sqrt{3x - 5} = \sqrt{3 - x}$

Write in standard complex number form.

_____ **33.** $\sqrt{-144}$

_____ **34.** $2 + \sqrt{-25}$

_____ **35.** $6 - \sqrt{-12}$

_____ **36.** $\sqrt{-4} + \sqrt{-49}$

_____ **37.** Solve $4x - 8i = 12 + 2yi$ for real values of x and y.

_____ **38.** Simplify i^{63}.

Simplify.

_____ **39.** $(3 - 2\sqrt{-18}) - (2 + 3\sqrt{-8})$

_____ **40.** $(5 - 2i)(3 + 4i)$

_____ **41.** $\dfrac{1 - 2i}{3i}$

_____ **42.** $\dfrac{2 - 3i}{1 + i}$

Solve.

_____ **43.** $x^2 + 27 = 0$

_____ **44.** $3x^2 - 2x + 1 = 0$

_____ **45.** $x^2 - 2x = -4$

_____ **46.** $x^3 - 64 = 0$

_____ **47.** $x + \dfrac{1}{x} = -1$

_____ **48.** $\dfrac{5x}{x + 1} = \dfrac{-2}{x}$

Check your answers in the back of your book.

If You Missed Exercises:	You Should Review	Examples:
1	SECTION 7.1	1, 2
2, 3		3, 4
4		5
5, 6		6–9
7		12, 13
8		22
9, 10		18–20
11, 12	SECTION 7.2	5–8
13–15		11–16
16		17
17	SECTION 7.3	1–3
18		5
19		7, 8
20		10
21		11
22		15
23		19, 20
24, 25		23, 24
26		25, 26
27–30	SECTION 7.4	1, 2
31		3, 4
32		5
33–36	SECTION 7.5	5–10
37		11, 12
38		15–18
39		21
40		23
41		30
42		31–33
43	SECTION 7.6	1
44–48		2–4

If you have completed the Review Exercises and corrected your errors, you are ready to take the Practice Test.

PRACTICE TEST for Chapter 7: Working with Rational Exponents and Radicals

		SECTION	EXAMPLE
_____	**1.** Write $\sqrt[4]{x^3 y}$ in exponential form.	7.1	12, 13
_____	**2.** Write $3(x - y)^{2/5}$ in radical form.	7.1	20, 21
_____	**3.** Write $(a^2 - b^2)^{-1/3}$ in radical form.	7.1	22
_____	**4.** Write $\sqrt{-49}$ in standard complex number form.	7.5	10
_____	**5.** Write $5 - \sqrt{-75}$ in standard complex number form.	7.5	9

Simplify. Assume that any variable expression under a radical with an even index represents a positive number.

		SECTION	EXAMPLE
_____	**6.** $\sqrt[3]{216}$	7.1	6, 8
_____	**7.** $\left(\dfrac{8}{27}\right)^{2/3}$	7.1	18, 19
_____	**8.** $x^{3/5} \cdot x^{1/2} \cdot x^{3/10}$	7.1	1, 2
_____	**9.** $(a^{3/5} \cdot b^{-2/3})^{-15}$	7.1	3
_____	**10.** $\dfrac{x^{3/4} y^{7/10}}{x^{3/8} y^{1/2}}$	7.1	5
_____	**11.** $(x^{-3/5} \cdot y^{6/5} \cdot z^{-3})^{5/12}$	7.1	4
_____	**12.** $\sqrt{\dfrac{63x^3}{49y^2}}$	7.2	6
_____	**13.** $5\sqrt{8x} - \sqrt{242x} + 2\sqrt{98x}$	7.2	15, 16
_____	**14.** $\sqrt[3]{81} - \sqrt{12} - \sqrt[3]{192} + \sqrt{108}$	7.2	13, 14
_____	**15.** $\sqrt{\dfrac{45x^2}{4}} - \sqrt{\dfrac{5x^2}{36}}$	7.2	17
_____	**16.** $\sqrt{12x} \cdot \sqrt{18x^4}$	7.3	2
_____	**17.** $(4 - \sqrt{7})(3 + 2\sqrt{7})$	7.3	8, 10
_____	**18.** $(\sqrt{3x} - 5)^2$	7.3	11
_____	**19.** $\dfrac{\sqrt{63x^5}}{\sqrt{7x}}$	7.3	5
_____	**20.** $\sqrt{\dfrac{200x^3}{7y}}$	7.3	21, 22

_____ 21. $\dfrac{\sqrt[3]{3}}{\sqrt[3]{25x^2}}$ 7.3 23, 24

_____ 22. $\dfrac{10}{\sqrt{7} + \sqrt{3}}$ 7.3 25, 26

_____ 23. $(3 - \sqrt{-9})(1 - \sqrt{-49})$ 7.5 21

_____ 24. $\dfrac{4 - i}{4i}$ 7.5 30

_____ 25. $\dfrac{5 + 2i}{3 - i}$ 7.5 32, 33

_____ 26. Solve $4x - 9i = 8 + 21yi$ for real values of x and y. 7.5 11

Solve.

_____ 27. $3 + \sqrt{x - 9} = 5$ 7.4 1, 2

_____ 28. $\sqrt{8 - 7x} = x - 2$ 7.4 3, 4

_____ 29. $3a^2 - 2a + 1 = 0$ 7.6 2

_____ 30. Solve the formula $r = \sqrt{\dfrac{S}{4\pi}}$ for S. 7.4 1

SHARPENING YOUR SKILLS after Chapters 1–7

Evaluate each expression.   SECTION

_____ **1.** $(x - 9)(-x + 7)$ when $x = 3$ 1.4

_____ **2.** $|5 - \sqrt{10}|$ 1.4

_____ **3.** $-x^4y^3$ when $x = -1$ and $y = -2$ 3.1

_____ **4.** $(-8)^{-2}$ 3.2

_____ **5.** $\sqrt{\dfrac{18}{25}}$ 5.2

Simplify.

_____ **6.** $12 - (-5)\left[\dfrac{1 + 4(11 - 6)}{-12 + 3(-3)}\right]$ 1.3

_____ **7.** $-8[2(3x - 1) + 9x] + \dfrac{1}{7}(14x)$ 1.4

_____ **8.** $(10x^0y^3z)(-3x^5y^7)\left(\dfrac{-1}{6}yz^4\right)$ 3.1

_____ **9.** $\left(\dfrac{-4x^2y^{-3}}{x^{-1}y^{-2}}\right)^{-1} \cdot \left(\dfrac{6x^4y^{-2}}{x^3y}\right)^3$ 3.2

_____ **10.** $a^2 - 9[a(a + 6) + 3a(a^2 - 2)]$ 3.3

_____ **11.** $(x + 5)(x - 1)^2$ 3.4

_____ **12.** $(21x^2 - 20x + 5) \div (3x - 2)$ 3.5

_____ **13.** $\dfrac{a^2 - 6a - 55}{5a^2 + 23a - 10}$ 6.1

_____ **14.** $\dfrac{x^2 - 169}{x^3y^8} \div \dfrac{2x + 26}{x^5y^3}$ 6.2

_____ **15.** $\dfrac{x}{x^2 - 36} - \dfrac{3}{x^2 - 6x}$ 6.3

_____ **16.** $\dfrac{6 - \dfrac{5x + 25}{x^2}}{\dfrac{3}{5} + \dfrac{1}{x}}$ 6.4

Factor completely.

——————— **17.** $6x^4yz - 12x^3y^2z - 30x^2y^3z - 4xy^4z$ 4.1

——————— **18.** $3x^5y^2 + 81x^2y^5$ 4.2

——————— **19.** $3a^2 - 17a + 10$ 4.3

——————— **20.** $2x^4 - 14x^2 - 36$ 4.3

——————— **21.** $x^4 - 2x^3 + 8x - 16$ 4.4

Solve.

——————— **22.** $3(x + 2) + 4(x - 6) = 3x - 22$ 2.1

——————— **23.** $|7x - 2| \geq 5$ 2.4

——————— **24.** $x(13x - 4) + 3 = 3x(x + 1)$ 5.1

——————— **25.** $3x(3x - 5) = 1 - 9x$ 5.3

——————— **26.** $x^2 - 18x + 81 \geq 0$ 5.4

——————— **27.** $\dfrac{x}{x - 4} - \dfrac{12}{x - 4} = 3$ 6.5

——————— **28.** $\dfrac{x}{x - 9} > 0$ 6.5

——————— **29.** Two amounts of money totaling $5000 are invested in savings certificates. One amount   2.5
earns 12% and the other earns 9%. Find how much is invested at each amount if the
total yearly interest is $517.50.

——————— **30.** Lee is building a 30-foot by 40-foot swimming pool bordered by a concrete walkway. If   5.5
the total area allowed for the pool and walk is 1824 square feet, find the width of the
walkway.

Graphing First-Degree Equations and Inequalities

8

The distance required to stop a moving car varies directly as the square of the speed of the car. If the distance required to stop a car traveling 20 miles per hour is 30 feet, what distance would be required to stop a car traveling 40 miles per hour?

Until now we have solved equations and inequalities containing only one variable. In everyday algebra problems, however, it is often necessary to deal with more than one variable. In this chapter we concentrate on first-degree equations and inequalities containing two variables and learn to:

1. Graph first-degree equations using the arbitrary-point method.
2. Graph first-degree equations using the intercept method.
3. Work with distance and slope.
4. Graph first-degree equations using the slope-intercept method.
5. Find equations of lines.
6. Graph first-degree inequalities.
7. Work with function notation.
8. Work with variation.
9. Switch from word statements to functions.

8.1 Graphing First-Degree Equations

Suppose that a theater owner sells adults' tickets at $3 each and children's tickets at $2 each. To compute the dollars received in one day, the owner must multiply the number sold of each kind of ticket times the price of the ticket. Since we cannot assume that the number of adults' tickets sold is the same as the number of children's tickets sold, we must use *two* variables here. We can let

$$x = \text{number of adults' tickets sold}$$

$$y = \text{number of children's tickets sold}$$

Then

$$3x = \text{dollars received from adults' tickets}$$

$$2y = \text{dollars received from childrens' tickets}$$

and

$$3x + 2y = \text{total dollars received}$$

If the theater owner knows that a total of $360 must be received each day in order for the theater to break even, then we may state this fact using the equation

$$3x + 2y = 360$$

This is an example of a **first-degree equation in two variables.**

Standard Form for First-Degree Equation in x and y

$$Ax + By = C$$

where A, B, and C are constants

We turn our attention now to finding real solutions for first-degree equations in two variables.

Finding Ordered Pairs

In our example we were looking for possible values of x and y that would satisfy the equation

$$3x + 2y = 360$$

Suppose that the theater owner expects to sell 100 adults' tickets. How many children's tickets must be sold? Here we are saying that x is 100 and we must find the corresponding value for y by substituting 100 for x in our equation.

$$3x + 2y = 360$$
$$3(100) + 2y = 360$$
$$300 + 2y = 360$$
$$2y = 60$$
$$y = 30$$

The theater owner must sell 30 children's tickets. We now know that $x = 100$ and $y = 30$ satisfies our equation. We write this solution as an **ordered pair:**

$$(x, y) = (100, 30)$$

Let's find some more ordered pairs that satisfy our equation. Suppose that:

90 children's tickets are sold.   No adults' tickets are sold.

$$y = 90$$ $$x = 0$$

$$3x + 2(90) = 360$$ $$3(0) + 2y = 360$$

$$3x + 180 = 360$$ $$2y = 360$$

$$3x = 180$$ $$y = 180$$

$$x = 60$$ When $x = 0$, $y = 180$

When $y = 90$, $x = 60$ $(x, y) = (0, 180)$

$(x, y) = (60, 90)$

You should agree that there are many ordered pairs that satisfy our theater ticket equation. Let's find some ordered pairs of numbers that satisfy the equation

$$4x + 5y = 5$$

To find such ordered pairs, we could substitute several x-values and find the corresponding y-values. Our best move, however, might be to first *isolate* y in the equation.

$$4x + 5y = 5$$

$$5y = 5 - 4x$$

$$y = \frac{5 - 4x}{5}$$

Now we may choose any x-value and find the corresponding y-value.

x	$y = \dfrac{5 - 4x}{5}$	(x, y)
$x = 0$	$y = \dfrac{5 - 4(0)}{5} = \dfrac{5}{5} = 1$	$(0, 1)$
$x = 2$	$y = \dfrac{5 - 4(2)}{5} = \dfrac{5 - 8}{5} = \dfrac{-3}{5}$	$\left(2, \dfrac{-3}{5}\right)$
$x = -1$	$y = \dfrac{5 - 4(-1)}{5} = \dfrac{5 + 4}{5} = \dfrac{9}{5}$	$\left(-1, \dfrac{9}{5}\right)$

Example 1. Find three ordered pairs of numbers that satisfy $x + 2y = 3$.

Solution: First we isolate y.

$$x + 2y = 3$$
$$2y = 3 - x$$
$$y = \frac{3 - x}{2}$$

x	$y = \dfrac{3 - x}{2}$	(x, y)
-3	$y = \dfrac{3 - (-3)}{2}$	
	$= \dfrac{3 + 3}{2}$	
	$= \dfrac{6}{2} = 3$	$(-3, 3)$
0	$y = \dfrac{3 - 0}{2}$	
	$= \dfrac{3}{2}$	$\left(0, \dfrac{3}{2}\right)$
3	$y = \dfrac{3 - 3}{2}$	
	$= \dfrac{0}{2} = 0$	$(3, 0)$

You try Example 2.

Example 2. Find three ordered pairs of numbers that satisfy $5x - y = 0$.

Solution: First we isolate _____ .

$$5x - y = 0$$
$$-y = -5x$$
$$y = 5x$$

x	$y = 5x$	(x, y)
-1	$y = 5(\underline{})$	
	$= \underline{}$	$(-1, \underline{})$
0	$y = 5(\underline{})$	
	$= \underline{}$	$(0, \underline{})$
2	$y = 5(\underline{})$	
	$= \underline{}$	$(2, \underline{})$

Check your work on page 572. ▶

To find ordered pairs of real numbers that satisfy a first-degree equation in two variables, we have used the following procedure.

> **Finding Ordered Pairs**
>
> **1.** Isolate y in the original equation.
> **2.** Choose any x-values and find the corresponding y-values.
> **3.** Write each solution as an ordered pair (x, y).

In this method of treating first-degree equations containing x and y, we call x the **independent variable** and y the **dependent variable,** because the value of y *depends* on the value we arbitrarily choose for x. Although we usually find just a few such ordered pairs, there is always an infinite number of ordered pairs that will satisfy each equation.

Graphing First-Degree Equations Using Arbitrary Points

When we solved first-degree equations and inequalities containing one variable in Chapter 2, we often illustrated our solutions using the real number line. To illustrate ordered pairs of real numbers satisfying equations containing *two* variables, however, the real number line alone will not work. Instead, we must use *two* real number lines − a horizontal number line for the independent variable x and a vertical number line for the dependent variable y.

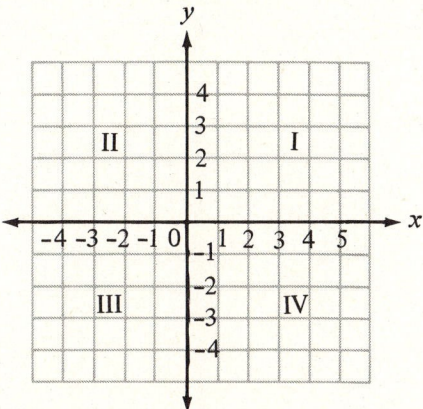

The horizontal number line is called the **x-axis** and the vertical number line is called the **y-axis.** The point where the axes intersect is called the **origin;** it is the zero point for each axis. All points to the *right* of the origin on the x-axis correspond to *positive* x-values, while points to the *left* of the origin correspond to *negative* x-values. Similarly, all points *above* the origin on the y-axis correspond to *positive* y-values, while points *below* the origin correspond to *negative* y-values.

We call the plane determined by this pair of axes the **Cartesian coordinate plane** (or the **real plane**). The axes divide this plane into four distinct regions called **quadrants,** labeled with Roman numerals in our illustration.

We can use the Cartesian coordinate plane to graph ordered pairs of real numbers. To locate a point such as

$$(3, 2)$$

we must realize that (3, 2) names the point where x is 3 *and* y is 2.

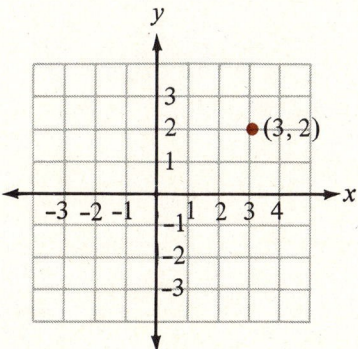

Every point in the plane has two parts.

> *x-coordinate* (or *abscissa*):   the x-value in an ordered pair
> *y-coordinate* (or *ordinate*):   the y-value in an ordered pair

Try graphing the points in Example 3.

Example 3. Graph the points $(-1, 2)$, $(2, -3)$, and $(-5, -5)$.

Solution

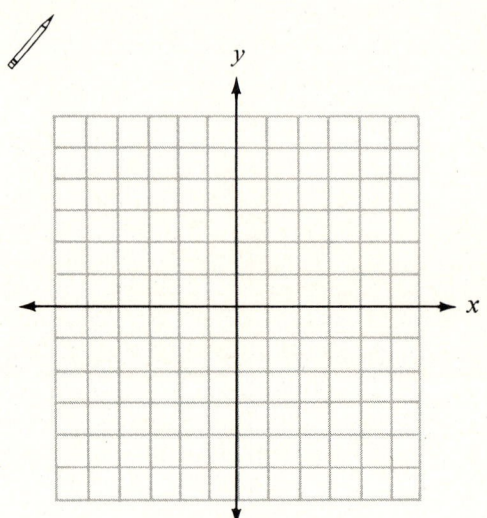

Example 4. Graph the points $(0, -3)$, $\left(0, \dfrac{7}{2}\right)$, $(0, 0)$, $(2, 0)$, and $(-3, 0)$.

Solution

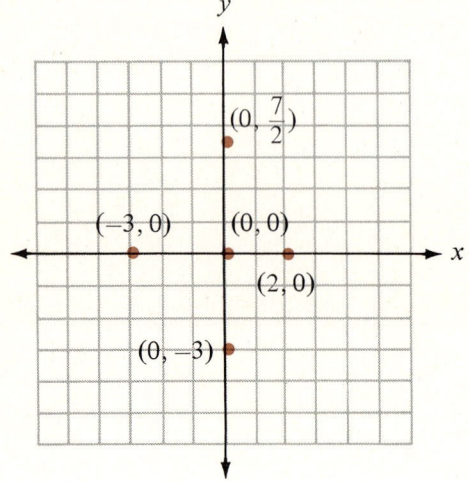

Check your work on page 572. ▶

From Example 4 we note some useful facts.

> If a point has an *x*-coordinate of zero, that point lies on the *y*-axis.
> If a point has a *y*-coordinate of zero, that point lies on the *x*-axis.
> The point $(0, 0)$ is the origin.

We also note that every ordered pair of real numbers corresponds to one and only one point in the Cartesian plane. Moreover, each point in the plane corresponds to one and only one ordered pair of real numbers. To plot points with irrational coordinates, we should estimate their location as accurately as possible.

Example 5. Graph the points $(\sqrt{3}, 0)$, $(-1, \sqrt{2})$, and $(\sqrt{2}, -\sqrt{5})$.

Solution

⫸ **Trial Run**

Find three ordered pairs of real numbers that satisfy each equation. There are lots of correct answers.

_____ **1.** $y = 3x - 1$ _____ **2.** $2x + y = 3$ _____ **3.** $3x - 5y = 1$

4. Graph and label the points:

$A(0, -2)$

$B\left(\dfrac{-1}{2}, -3\right)$

$C\left(\dfrac{3}{2}, -2\right)$

$D(4, 0)$

$E(3, \sqrt{2})$

$F(-\sqrt{5}, 3)$

Answers are on page 574.

Earlier in this chapter we learned to find ordered pairs that would satisfy an equation containing two variables. Now that we have learned to plot ordered pairs in the Cartesian coordinate plane, we can graph the solutions for an equation in x and y.

For example, for the equation

$$y = 5 - x$$

we can find several ordered pairs.

x	$y = 5 - x$		(x, y)
-2	$y = 5 - (-2)$	$= 7$	$(-2, 7)$
0	$y = 5 - 0$	$= 5$	$(0, 5)$
1	$y = 5 - 1$	$= 4$	$(1, 4)$
$1\dfrac{1}{2}$	$y = 5 - 1\dfrac{1}{2}$	$= 3\dfrac{1}{2}$	$\left(1\dfrac{1}{2}, 3\dfrac{1}{2}\right)$
$3 + \sqrt{2}$	$y = 5 - (3 + \sqrt{2}) = 2 - \sqrt{2}$		$(3 + \sqrt{2}, 2 - \sqrt{2})$
5	$y = 5 - 5$	$= 0$	$(5, 0)$
8	$y = 5 - 8$	$= -3$	$(8, -3)$

Now we can graph these ordered pairs in the Cartesian plane.

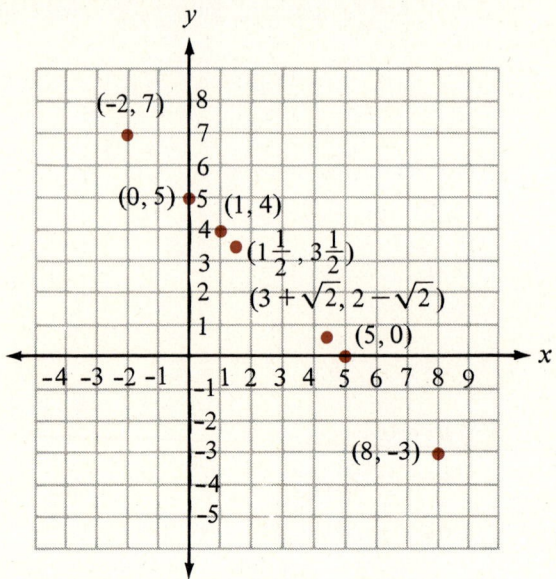

It is more than a coincidence that all these points seem to lie on a straight line. As a matter of fact, any ordered pair that satisfies the equation $y = 5 - x$ will lie on this line, and the coordinates of any point that lies on this line will satisfy that equation. In fact:

> The graph of ordered pairs satisfying a first-degree equation of the form
> $$Ax + By = C$$
> (where A and B are not *both* zero) is always a straight line.

For this reason, we sometimes call an equation of the form $Ax + By = C$ a **linear equation.**

There are several methods for graphing a linear equation. We have already discovered the method of "choose an x, find the corresponding y." This is called the **arbitrary point method.** Of course, we know that any *two* points will determine a straight line, but to double check our accuracy, we shall use *three* points. To obtain a good spread of points, let's agree to choose one positive x-value, one negative x-value, and zero.

Example 6. Use three arbitrary points to graph $2x - y = 6$.

Solution: First we isolate y.

$$2x - y = 6$$
$$-y = -2x + 6$$
$$y = 2x - 6$$

x	$y = 2x - 6$	(x, y)
-1	$y = 2(-1) - 6$	
	$= -2 - 6 = -8$	$(-1, -8)$
0	$y = 2(0) - 6$	
	$= -6$	$(0, -6)$
3	$y = 2(3) - 6$	
	$= 6 - 6 = 0$	$(3, 0)$

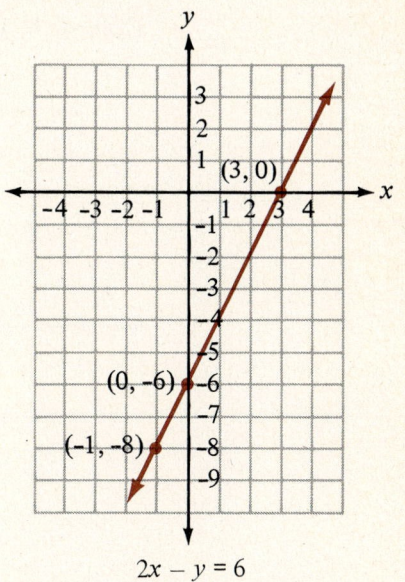

$2x - y = 6$

Notice that our points lie on a straight line that continues indefinitely in either direction.

Now try Example 7.

Example 7. Use three arbitrary points to graph $y = 3x$.

Solution: Here y is already isolated.

x	$y = 3x$	(x, y)
-2	$y = 3(\underline{}) = \underline{}$	$(-2, \underline{})$
0	$y = 3(\underline{}) = \underline{}$	$(0, \underline{})$
1	$y = 3(\underline{}) = \underline{}$	$(1, \underline{})$

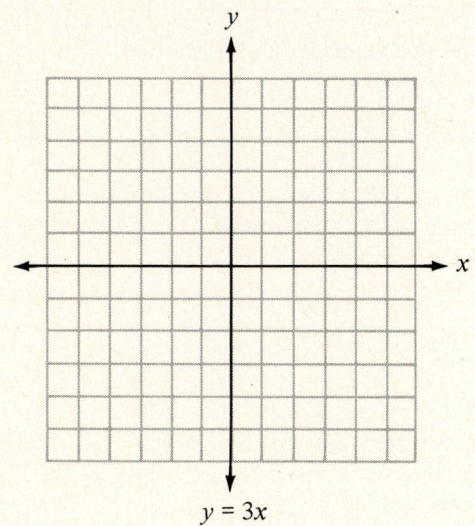

$y = 3x$

Check your work on page 573. ▶

Example 8. Use three arbitrary points to graph $3x + 2y = 6$.

Solution: First we isolate y.

$$3x + 2y = 6$$

$$2y = -3x + 6$$

$$y = \frac{-3x + 6}{2}$$

x	$y = \dfrac{-3x + 6}{2}$	(x, y)
-2	$y = \dfrac{-3(-2) + 6}{2}$	
	$= \dfrac{6 + 6}{2} = 6$	$(-2, 6)$
0	$y = \dfrac{-3(0) + 6}{2}$	
	$= \dfrac{6}{2} = 3$	$(0, 3)$
3	$y = \dfrac{-3(3) + 6}{2}$	
	$= \dfrac{-9 + 6}{2} = \dfrac{-3}{2}$	$\left(3, \dfrac{-3}{2}\right)$

$3x + 2y = 6$

We have now developed a procedure for graphing linear equations of the form $Ax + By = C$ using the method of arbitrary points.

Graphing Lines by the Arbitrary-Point Method

1. Isolate y in the original equation.
2. Choose at least three x-values and find the corresponding y-values by substitution.
3. Write each solution as an ordered pair (x, y).
4. Plot the point in the Cartesian plane corresponding to each ordered pair.
5. Connect the points with a straight line. (If your points do not lie on a straight line, check each of your substitutions.)

⫸ **Trial Run**

Graph each equation using three arbitrary points.

1. $x + y = 5$

2. $y = 3x - 2$

3. $2x - y = 3$

4. $y = -9x$

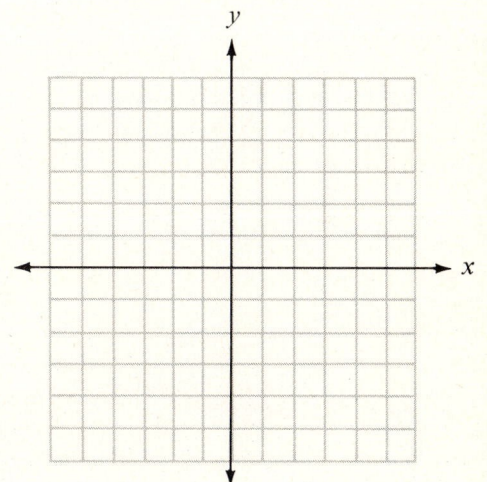

Answers are on page 574.

Graphing First-Degree Equations Using Intercepts

If you look back at the linear equations we have graphed, you will notice that each straight line crosses the *x*-axis at some point and also crosses the *y*-axis at some point. These points are called the **intercepts** for a graph.

> *x-intercept:* the *x*-coordinate of the point at which a graph crosses the *x*-axis.
> *y-intercept:* the *y*-coordinate of the point at which a graph crosses the *y*-axis.

Earlier we observed that every point on the x-axis has a y-coordinate of zero and every point on the y-axis has an x-coordinate of zero. Putting these observations together with the idea of intercepts, we conclude that

The x-intercept for a graph is the x-coordinate of the point that has a y-coordinate of 0.
The y-intercept for a graph is the y-coordinate of the point that has an x-coordinate of 0.

In other words, to find the x-intercept for the graph of an equation, we must let $y = 0$ in the equation. To find the y-intercept, we must let $x = 0$ in the equation. When finding intercepts, it is not necessary to begin by isolating either variable. Let's find the intercepts for the graph of

$$2x - 5y = 10$$

x-INTERCEPT: Let $y = 0$ y-INTERCEPT: Let $x = 0$

$2x - 5y = 10$ $2x - 5y = 10$

$2x - 5(0) = 10$ $2(0) - 5y = 10$

$2x = 10$ $-5y = 10$

$x = 5$ $y = -2$

The graph crosses the x-axis   The graph crosses the y-axis
at the point $(5, 0)$.   at the point $(0, -2)$.

Because there is less chance of error when we substitute 0 for variable, we trust these points to be accurate and graph the line.

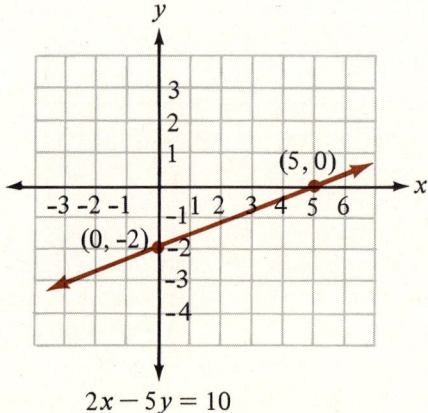

$$2x - 5y = 10$$

Now try Example 9.

Example 9. Use intercepts to graph $2x + 3y = 4$.

Solution

$$2x + 3y = 4$$

x-INTERCEPT: Let $y = 0$ y-INTERCEPT: Let $x = 0$

$2x + 3(\underline{}) = 4$ $2(\underline{}) + 3y = 4$

$2x = 4$ $3y = 4$

$x = \underline{}$ $y = \underline{}$

The graph crosses the *x*-axis   The graph crosses the *y*-axis
at (_____ , _____).   at (_____ , _____)

$$2x + 3y = 4$$

Check your work on page 573. ▶

Don't you agree that the intercept method is a handy method to use? Will it always be the best method? Look at Example 10.

Example 10. Use intercepts to graph $x - y = 0$.

Solution

$$x - y = 0$$

x-INTERCEPT: Let $y = 0$. y-INTERCEPT: Let $x = 0$.

$$x - 0 = 0 \qquad\qquad 0 - y = 0$$

$$x = 0 \qquad\qquad y = 0$$

$$(0, 0) \qquad\qquad (0, 0)$$

We seem to be in trouble here because the x-intercept and y-intercept both occur at the origin. When this happens, we must locate another point on the graph by choosing an arbitrary x-value and finding the corresponding y-value.

Let $x = 2$.

$$x - y = 0$$

$$2 - y = 0$$

$$-y = -2$$

$$y = 2$$

$$(2, 2)$$

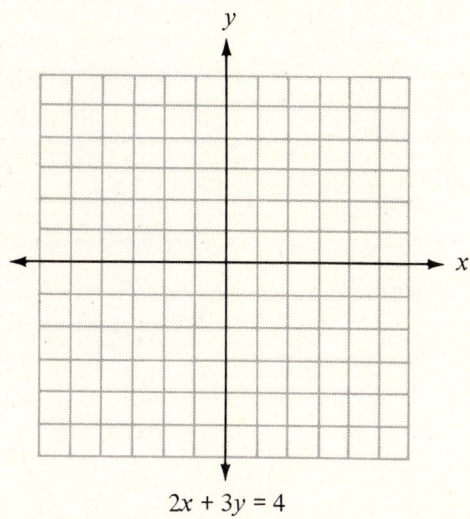

$$x - y = 0$$

To graph a linear equation of the form $Ax + By = C$ using the intercept method, we have used the following procedure.

> **Graphing Lines by the Intercept Method**
>
> **1.** Find the *x*-intercept by letting $y = 0$ in the original equation.
> **2.** Find the *y*-intercept by letting $x = 0$ in the original equation.
> **3.** Write the ordered pair corresponding to each intercept.
> **4.** Plot the intercept points and join with a straight line.

⫸ Trial Run

Graph each equation using intercepts.

1. $5x - 3y = 15$

2. $-2x - 3y = 5$

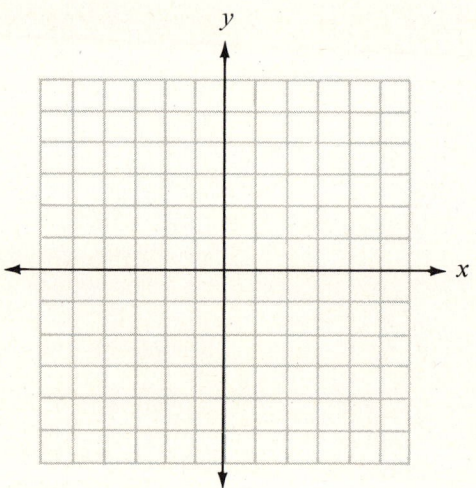

3. $-3x + 8y = 12$

4. $4x + y = 6$

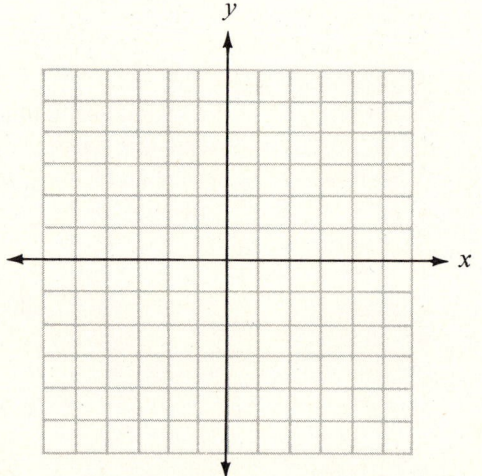

Answers are on page 575.

Graphing Constant Equations

So far we have discussed graphs of linear equations of the form $Ax + By = C$, where A and B are not both zero, but we have not considered equations of this form in which $A = 0$ or $B = 0$. In other words, we have not discussed linear equations in which the x-term or y-term is missing.

Suppose that we wish to graph an equation such as

$$y = 2$$

which states that $y = 2$, *no matter what x is*. Let's graph some ordered pairs satisfying this equation.

x	y	(x, y)
-3	2	$(-3, 2)$
0	2	$(0, 2)$
2	2	$(2, 2)$

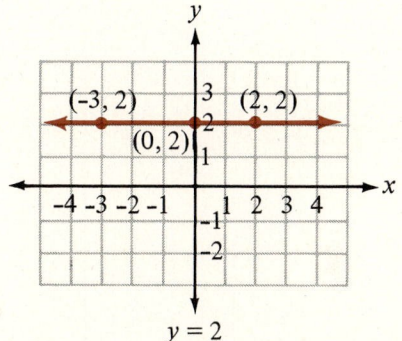

$y = 2$

The graph of $y = 2$ is a straight line, but in particular, it is a **horizontal** line (parallel to the x-axis) crossing the y-axis at $(0, 2)$.

> The graph of $y = c$ (where c is a constant) is a horizontal line (parallel to the x-axis) crossing the y-axis at $(0, c)$.

Having made this observation, we agree to graph equations of the form $y = c$ *by inspection*, without plotting several points.

Try Example 11.

Example 11. Graph $3y = -9$ by inspection.

Solution: We isolate _____ .

$$3y = -9$$
$$y = -3$$

The graph is a _____ line, crossing the _____-axis at (0, _____).

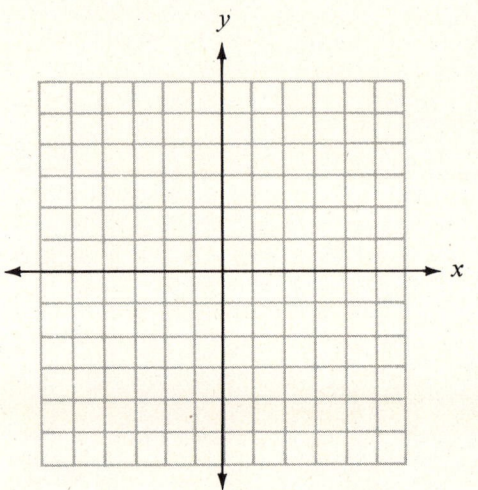

Check your work on page 573. ▶

Example 12. Graph $2y + 3 = 3$ by inspection.

Solution: We isolate y.

$$2y + 3 = 3$$
$$2y = 0$$
$$y = 0$$

The graph is a horizontal line, crossing the y-axis at (0, 0).

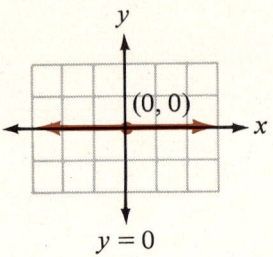

Notice from Example 12 that the graph of $y = 0$ is the x-axis.

To graph an equation such as

$$x = 1$$

we realize that $x = 1$, *no matter what y is*. Some possible ordered pairs for this equation are

(1, −2)

(1, 0)

(1, 3)

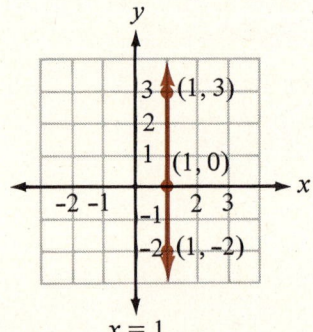

Our graph is a **vertical** line (parallel to the y-axis) crossing the x-axis at (1, 0).

The graph of $x = k$ (where k is a constant) is a vertical line (parallel to the y-axis) crossing the x-axis at $(k, 0)$.

We may graph linear equations of the form $x = k$ by inspection without plotting several points.
Try Example 13.

Example 13. Graph $2x + 5 = 0$ by inspection.

   Solution: We isolate _____ .

$$2x + 5 = 0$$
$$2x = -5$$
$$x = \frac{-5}{2}$$

The graph is a _____ line, crossing the _____ -axis at (_____ , 0).

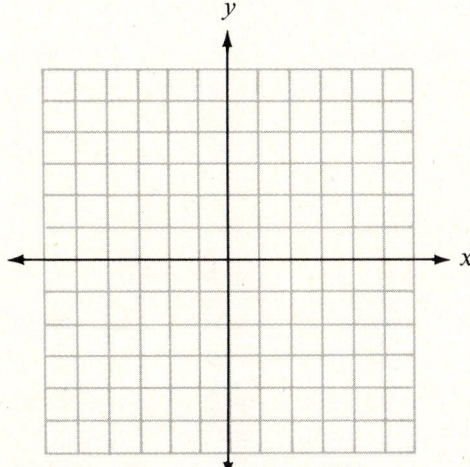

Check your work on page 573. ▶

Example 14. Graph $3(x - 1) = 5x - 3$ by inspection.

   Solution: We isolate x.

$$3(x - 1) = 5x - 3$$
$$3x - 3 = 5x - 3$$
$$-2x = 0$$
$$x = 0$$

The graph is a vertical line, crossing the x-axis at $(0, 0)$.

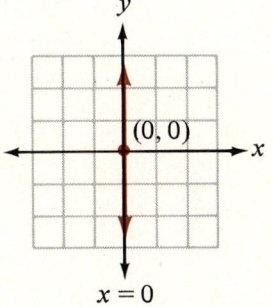

$x = 0$

Notice from Example 14 that the graph of $x = 0$ is the y-axis.

▐▐▐▶ **Trial Run**

Graph each equation by inspection.

1. $y = 3$

2. $2x - 3 = 0$

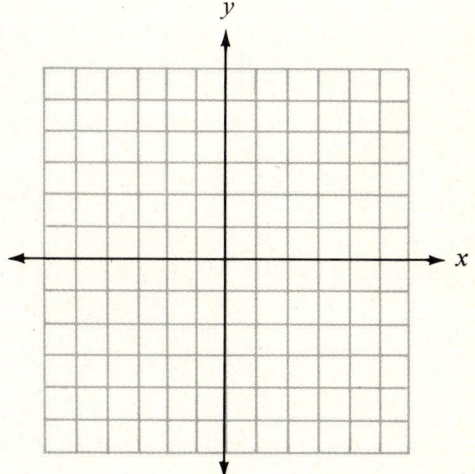

3. $3x + 13 = 10$ **4.** $9 + 4y = 1$

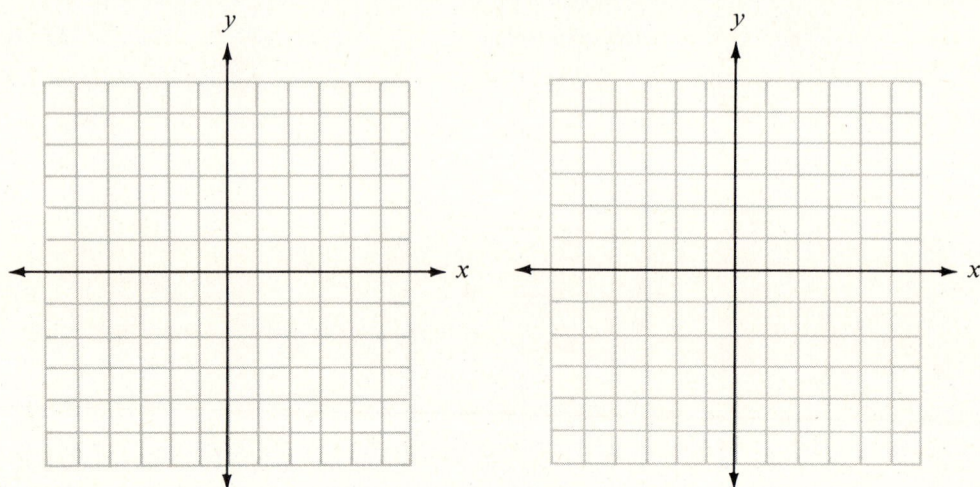

Answers are on page 575.

Let's summarize the methods we have discussed for graphing linear equations of the form $Ax + By = C$.

1. *Arbitrary-point method.* If y can be easily isolated in the given equation, we find three points by the method of "choose an x, find the corresponding y."
2. *Intercept method.* If we do not wish to isolate y in the given equation, we can find the x-intercept (by letting $y = 0$) and find the y-intercept (by letting $x = 0$).
3. *Method for constant equations.* If the equation is of the form $y = c$ (where c is a constant), we know the graph is a horizontal line crossing the y-axis at $(0, c)$. If the equation is of the form $x = k$ (where k is a constant), we know that the graph is a vertical line crossing the x-axis at $(k, 0)$.

▶ Examples You Completed

Example 2. Find three ordered pairs of numbers that satisfy $5x - y = 0$.

Solution: First we isolate y.

$$5x - y = 0$$
$$-y = -5x$$
$$y = 5x$$

x	$y = 5x$	(x, y)
-1	$y = 5(-1)$	
	$= -5$	$(-1, -5)$
0	$y = 5(0)$	
	$= 0$	$(0, 0)$
2	$y = 5(2)$	
	$= 10$	$(2, 10)$

Example 3. Graph the points $(-1, 2)$, $(2, -3)$, and $(-5, -5)$.

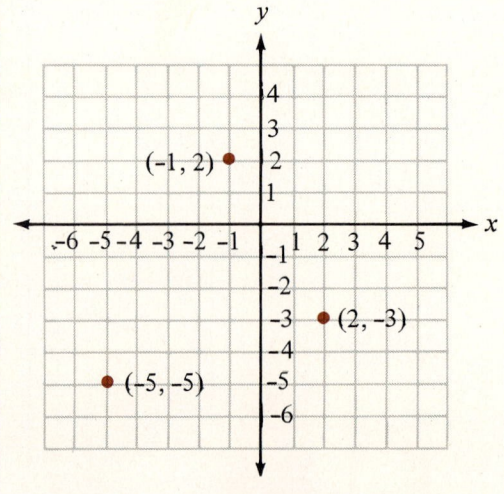

Example 7. Use three arbitrary points to graph $y = 3x$.

Solution: Here y is already isolated.

x	$y = 3x$	(x, y)
-2	$y = 3(-2) = -6$	$(-2, -6)$
0	$y = 3(0) = 0$	$(0, 0)$
1	$y = 3(1) = 3$	$(1, 3)$

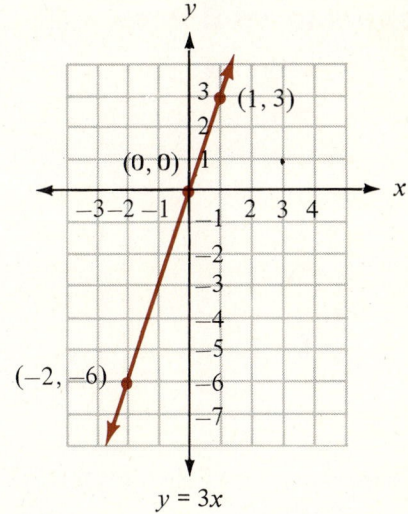

$y = 3x$

Example 9. Use intercepts to graph $2x + 3y = 4$.

Solution

$$2x + 3y = 4$$

x-INTERCEPT: Let $y = 0$.

$$2x + 3(0) = 4$$
$$2x = 4$$
$$x = 2$$

The graph crosses the x-axis at $(2, 0)$.

y-INTERCEPT: Let $x = 0$.

$$2(0) + 3y = 4$$
$$3y = 4$$
$$y = \frac{4}{3}$$

The graph crosses the y-axis at $\left(0, \frac{4}{3}\right)$.

$2x + 3y = 4$

Example 11. Graph $3y = -9$ by inspection.

Solution: We isolate y.

$$3y = -9$$
$$y = -3$$

The graph is a horizontal line, crossing the y-axis at $(0, -3)$.

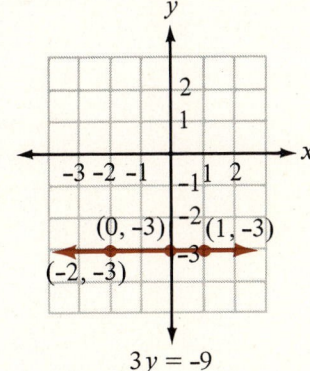

$3y = -9$

Example 13. Graph $2x + 5 = 0$ by inspection.

Solution: We isolate x.

$$2x + 5 = 0$$
$$2x = -5$$
$$x = \frac{-5}{2}$$

The graph is a vertical line, crossing the x-axis at $\left(\frac{-5}{2}, 0\right)$.

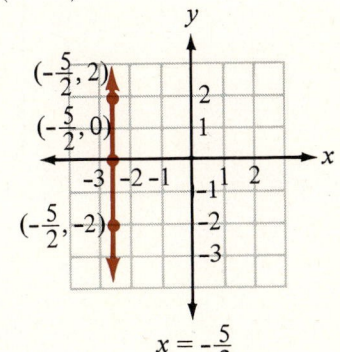

$x = -\frac{5}{2}$

Answers to Trial Runs

page 561 **4.**

page 565 **1.**

2.

3.

4.

page 568   **1.**

2.

3.

4.

page 571   **1.**

2.

3.

4.

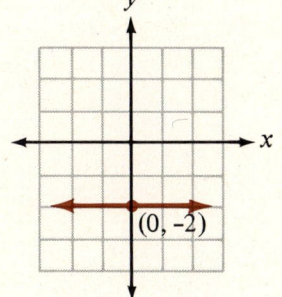

slope

$y = mx + b$ ← y-intercept

$y = 3x + 2$

M-slope rise/run

EXERCISE SET 8.1

Graph each equation using the arbitrary-point method.

1. $x + y = 6$ **2.** $x + y = 2$

3. $2y = 16x$ **4.** $3y = 15x$

5. $y = 5x - 2$ **6.** $y = 3x - 1$

7. $2x + y = 1$ **8.** $3x + y = 2$

9. $x + 4y = 10$ **10.** $x + 6y = 7$

11. $2x - y = -5$ **12.** $3x - y = -8$

Graph each equation using the intercept method.

13. $x + y = -3$ **14.** $x + y = -2$

15. $x - y = 5$ **16.** $x - y = 4$

17. $x + 2y = 5$ **18.** $x + 4y = 6$

19. $3x - y = 7$ **20.** $2x - y = -3$

21. $-5x + 2y = 15$ **22.** $-7x + 3y = 14$

23. $-x - y = 3$ **24.** $-x - y = 1$

Graph each of the following using any method you choose.

25. $3x + y = 6$ **26.** $5x + y = 5$

27. $6y - 16 = 0$ **28.** $8y - 28 = 0$

29. $y = 10x$ **30.** $y = 7x$

31. $2x - 7 = 0$ **32.** $3x - 5 = 0$

33. $5x + 3y = -15$ **34.** $3x + 4y = 12$

35. $9(x - 4) = 2x - 1$ **36.** $8(x - 3) = 5x - 7$

37. $3x - 4y = -14$ **38.** $2x - 3y = -10$

39. $3x - 5y = 0$ **40.** $2x - 9y = 0$

☆ Stretching the Topics _____

_____ **1.** Graph and write the equation for the horizontal line through the point $(-2, 3)$.

_____ **2.** Graph $\dfrac{1}{5}x + \dfrac{3}{10}y = 2$.

_____ **3.** Graph $0.2x - 0.5y = 1.5$.

Check your answers in the back of your book.

If you can complete **Checkup 8.1,** you are ready to go on to Section 8.2.

✓ **CHECKUP 8.1**

Graph each equation using the arbitrary-point method.

1. $3x - y = 4$

2. $4x + y = 0$

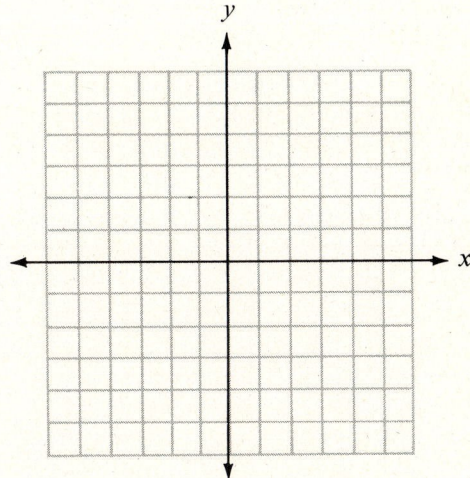

Graph each equation using the intercept method.

3. $2x - y = 4$

4. $3x - y = 5$

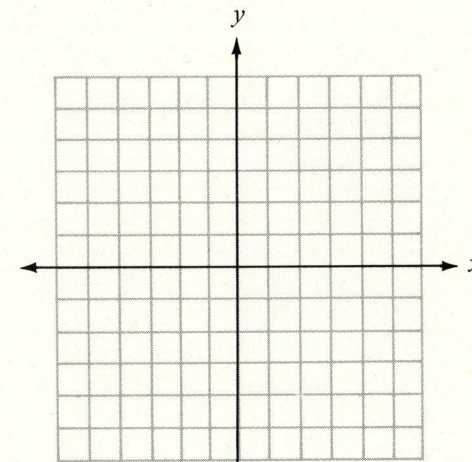

Graph each equation using any method you choose.

5. $y = \dfrac{2}{3}x$

6. $3y + 7 = 0$

7. $2x = 8$

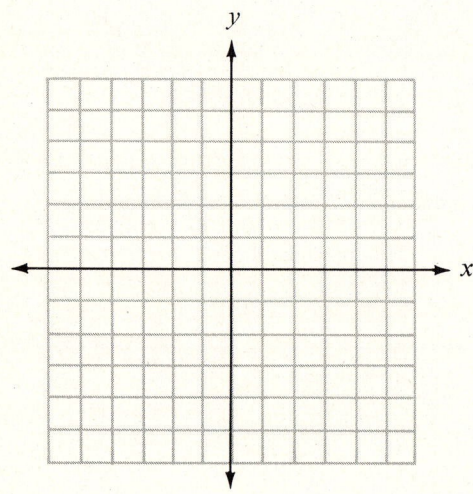

8. $-2x + 3y = -12$

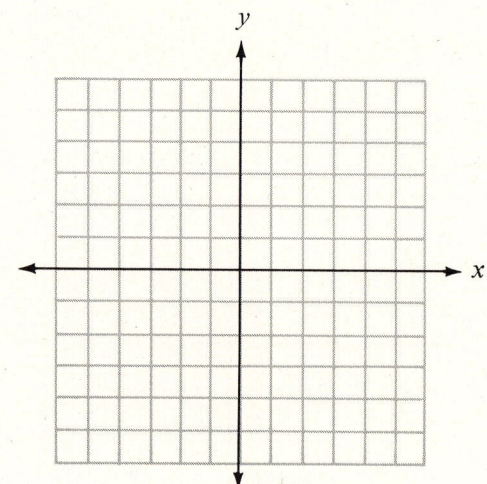

Check your answers in the back of your book.

If You Missed Problems:	You Should Review Examples:
1, 2	6–8
3, 4	9, 10
5	7
6	11
7	13
8	9

8.2 Working with Distance and Slope

Finding the Distance Between Two Points

Sometimes we wish to find the **distance between two points** on a line. For instance, suppose that we wish to find the distance between the points (1, 3) and (4, 7). First let's plot these points in the plane and treat the line segment between them as the hypotenuse of a right triangle.

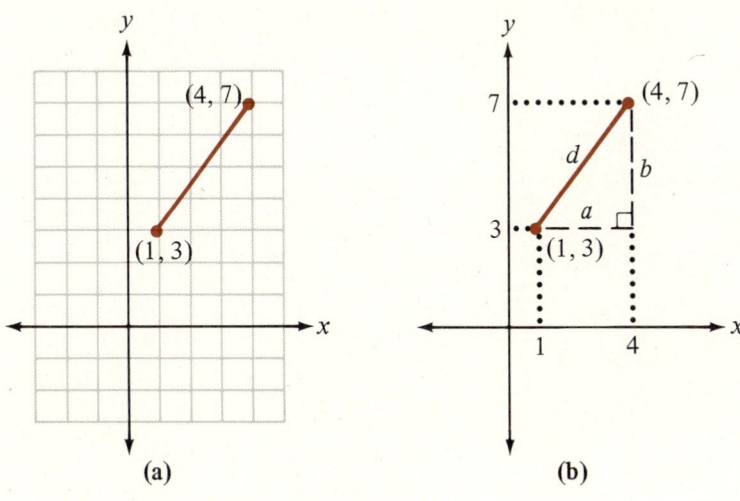

(a)   (b)

Earlier we discussed the Pythagorean theorem for a right triangle, which states that

$$a^2 + b^2 = d^2$$

Here a is the difference between x-values of 4 and 1, so $a = 3$. Similarly, b is the difference between y-values of 7 and 3, so $b = 4$. Therefore, $a^2 + b^2 = d^2$ becomes

$$3^2 + 4^2 = d^2$$

$$9 + 16 = d^2$$

$$25 = d^2$$

$$d = \pm\sqrt{25}$$

$$d = \pm 5$$

But since d represents distance, it cannot be negative.

$$d = 5$$

The distance between the points (1, 3) and (4, 7) is 5 units.

Having seen how to use the Pythagorean theorem to find the distance between two particular points, let's see if we can arrive at a formula for finding the distance between *any* two points. Suppose that we represent any two points by

$$(x_1, y_1) \quad \text{and} \quad (x_2, y_2)$$

where the subscripts of 1 and 2 help us identify the coordinates of the first and second points. Let's plot these points in the plane and construct our right triangle.

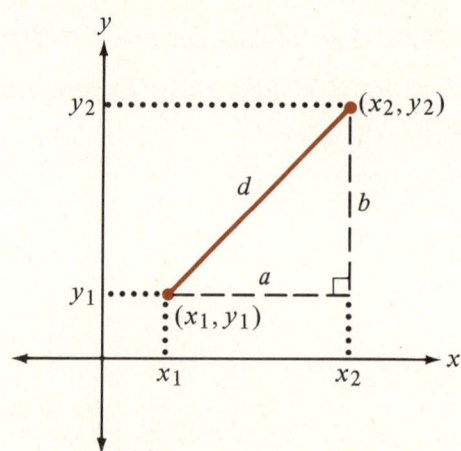

Since $x_2 > x_1$ and $y_2 > y_1$, we can see that $a = x_2 - x_1$ and $b = y_2 - y_1$. So $d^2 = a^2 + b^2$ becomes

$$d^2 = (x_2 - x_1)^2 + (y_2 - y_1)^2$$

$$d = \sqrt{(x_2 - x_1)^2 + (y_2 - y_1)^2}$$

This is the general formula for the distance between two points.

Distance Formula. If (x_1, y_1) and (x_2, y_2) are two points, then the distance between them is found by

$$d = \sqrt{(x_2 - x_1)^2 + (y_2 - y_1)^2}$$

Example 1. Find the distance between $(-2, 1)$ and $(3, -4)$.

Solution: Let $(x_1, y_1) = (-2, 1)$
$(x_2, y_2) = (3, -4)$

$d = \sqrt{(x_2 - x_1)^2 + (y_2 - y_1)^2}$

$\quad = \sqrt{[3 - (-2)]^2 + (-4 - 1)^2}$

$\quad = \sqrt{5^2 + (-5)^2}$

$\quad = \sqrt{25 + 25}$

$\quad = \sqrt{50}$

$\quad = \sqrt{25 \cdot 2}$

$\quad = 5\sqrt{2}$

Now you try Example 2.

Example 2. Find the distance between $(0, 3)$ and $(7, 5)$.

Solution: Let $(x_1, y_1) = (0, 3)$
$(x_2, y_2) = (7, 5)$

$d = \sqrt{(x_2 - x_1)^2 + (y_2 - y_1)^2}$

$\quad = \sqrt{(7 - 0)^2 + (5 - 3)^2}$

Check your work on page 595. ▶

You might wonder whether it matters which of the given points is labeled (x_1, y_1) and which is labeled (x_2, y_2). You can verify for yourself that the choice of labels does *not* matter; the distance between the same two points will be the same.

Finding the Midpoint of a Line Segment

The **midpoint** of a line segment joining two points (x_1, y_1) and (x_2, y_2) is the point located halfway between those two points. To find the coordinates of the midpoint, we again construct our right triangle and draw perpendicular lines from the midpoint (x_M, y_M) to both legs of the right triangle.

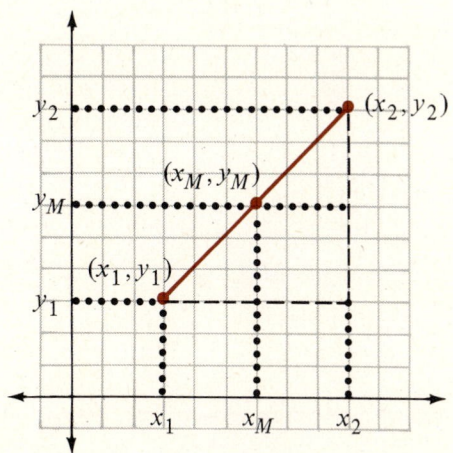

In geometry it can be proved that the perpendicular lines from the midpoint divide each leg in half, so

$$x_M - x_1 = x_2 - x_M \quad \text{and} \quad y_M - y_1 = y_2 = -y_M$$

$$2x_M = x_1 + x_2 \qquad\qquad 2y_M = y_1 + y_2$$

$$x_M = \frac{x_1 + x_2}{2} \qquad\qquad y_M = \frac{y_1 + y_2}{2}$$

We have now discovered how to find the coordinates of the midpoint of the line segment joining two points.

Finding the Midpoint of a Line Segment. The midpoint (x_M, y_M) of the line segment joining the points (x_1, y_1) and (x_2, y_2) is found by

$$(x_M, y_M) = \left(\frac{x_1 + x_2}{2}, \ \frac{y_1 + y_2}{2} \right)$$

As you can see, the *x*-coordinate of the midpoint is just the *average* of the *x*-coordinates of the endpoints, and the *y*-coordinate of the midpoint is the *average* of the *y*-coordinates of the endpoints.

Example 3. Find the midpoint of the line segment joining $(2, -3)$ and $(6, 1)$.

Solution: Let $(x_1, y_1) = (2, -3)$
$(x_2, y_2) = (6, 1)$

$$x_M = \frac{x_1 + x_2}{2} \quad \text{and} \quad y_M = \frac{y_1 + y_2}{2}$$

$$= \frac{2 + 6}{2} \qquad\qquad = \frac{-3 + 1}{2}$$

$$= \frac{8}{2} \qquad\qquad\quad = \frac{-2}{2}$$

$$x_M = 4 \qquad\qquad\quad y_M = -1$$

$$(x_M, y_M) = (4, -1)$$

You try Example 4.

Example 4. Find the midpoint of the line segment joining $(-2, 0)$ and $(3, 0)$.

Solution: Let $(x_1, y_1) = (-2, 0)$
$(x_2, y_2) = (3, 0)$

$$x_M = \frac{x_1 + x_2}{2} \quad \text{and} \quad y_M = \frac{y_1 + y_2}{2}$$

$$(x_M, y_M) = (\underline{\quad}, \underline{\quad})$$

Check your work on page 595. ▶

⟹ Trial Run

Find the distance between the points.

_____ **1.** $(0, 2)$ and $(4, 8)$ _____ **2.** $(-4, -1)$ and $(1, 11)$

Find the midpoint of the line segment joining the points.

_____ **3.** $(3, 3)$ and $(-6, 7)$ _____ **4.** $(0, 1)$ and $(0, -5)$

Answers are on page 596.

Defining the Slope of the Line

Let's take a look at the graphs of several linear equations.

$y = 2x$

$y = 1 - 2x$

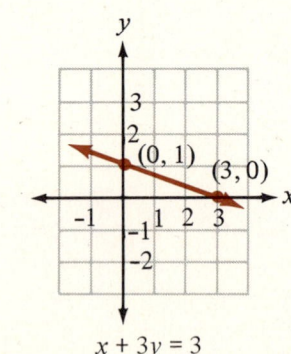

$x + 3y = 3$

If we wish to describe each of these lines in terms of their steepness, we might say that the first two lines appear to be equally steep but in opposite directions. The third line seems less steep than the other lines.

We can measure the steepness or **slope** of a line by locating any two points on that line and then comparing the units of change in the y-values (sometimes called the **rise**) to the units of change in the x-values (sometimes called the **run**).

Measuring Slope

$$\frac{\text{rise}}{\text{run}} \quad \text{or} \quad \frac{\text{vertical change}}{\text{horizontal change}} \quad \text{or} \quad \frac{\text{change in } y}{\text{change in } x}$$

To find the numerical value of the slope of a line joining two points (x_1, y_1) and (x_2, y_2), we construct our familiar triangle. The change in y-values is $y_2 - y_1$ and the change in x-values is $x_2 - x_1$.

It is customary to use the letter m to represent slope.

Definition of Slope. If (x_1, y_1) and (x_2, y_2) are two points on a line, then the slope of that line is defined by

$$m = \frac{y_2 - y_1}{x_2 - x_1}$$

provided that $x_2 \neq x_1$.

Let's use the slope definition to compute the slopes of some lines. It will be helpful to sketch each line before calculating its slope. You try Example 5.

Example 5. Find the slope of the line through (1, 3) and (4, 7).

Solution

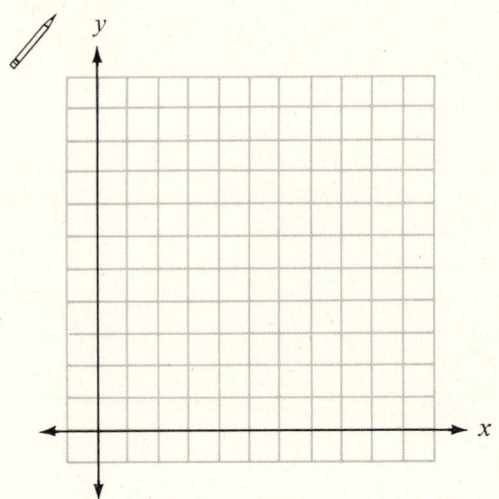

Let $(x_1, y_1) = (1, 3)$

$(x_2, y_2) = (4, 7)$

Then $m = \dfrac{y_2 - y_1}{x_2 - x_1}$

$m = \underline{\hspace{1cm}}$

Check your work on page 595. ▶

Notice that the line in Example 5 slopes *upward* from left to right, and to move from the first point to the second point, we must go to the *right* 3 units and *up* 4 units.

Example 6. Find the slope of the line joining $(-5, 1)$ and $(-3, -2)$.

Solution

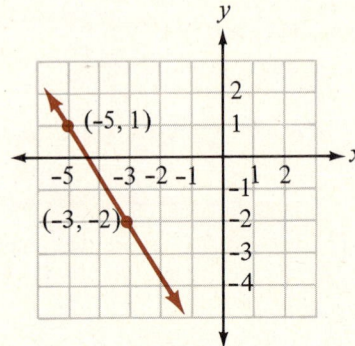

Let $(x_1, y_1) = (-5, 1)$

$(x_2, y_2) = (-3, -2)$

Then $m = \dfrac{y_2 - y_1}{x_2 - x_1}$

$= \dfrac{-2 - 1}{-3 - (-5)}$

$m = \dfrac{-3}{2}$

Notice that the line in Example 6 slopes *downward* from left to right, and to move from the first point to the second point, we must go to the *right* 2 units and *down* 3 units.

From Examples 5 and 6 we observe that:

> If the slope of a line is *positive*, then that line slopes *upward* from left to right.
> If the slope of a line is *negative*, then that line slopes *downward* from left to right.

Now try Example 7.

Example 7. Find the slope of the line joining $(1, 4)$ and $(-2, 4)$.

Solution

Let $(x_1, y_1) = (1, 4)$

$(x_2, y_2) = (-2, 4)$

Then $m =$

$m =$ _____

Check your work on page 595. ▶

Example 8. Find the slope of the line joining $(2, 3)$ and $(2, -1)$.

Solution

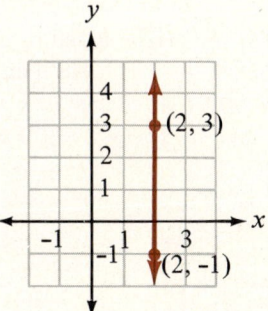

Let $(x_1, y_1) = (2, 3)$

$(x_2, y_2) = (2, -1)$

Then $m = \dfrac{-1 - 3}{2 - 2}$

$= \dfrac{-4}{0}$

m is undefined.

From Examples 7 and 8 we observe that

> The slope of a horizontal line is zero.
> The slope of a vertical line is undefined.

⫸ Trial Run

Find the slope of the line joining the given points.

_____ **1.** $(2, 1)$ and $(3, 0)$ _____ **2.** $(-3, 5)$ and $(4, 5)$

_____ **3.** $(9, -3)$ and $(3, 1)$ _____ **4.** $(7, -1)$ and $(7, 5)$

Answers are on page 596.

Graphing a Line Using the Slope

If we know one point on a line and the slope of that line, we can find another point on that line using the fact that

$$\text{Slope} = \frac{\text{vertical change}}{\text{horizontal change}}$$

For instance, if we know that the point (2, 1) is on a line with slope $\frac{3}{4}$, we know that we can start at the point (2, 1) and move to the *right* 4 units and *up* 3 units to arrive at another point on the same line.

$$(x_1, y_1) = (2, 1)$$

$$m = \frac{3}{4} \begin{array}{l} \uparrow \\ \rightarrow \end{array}$$

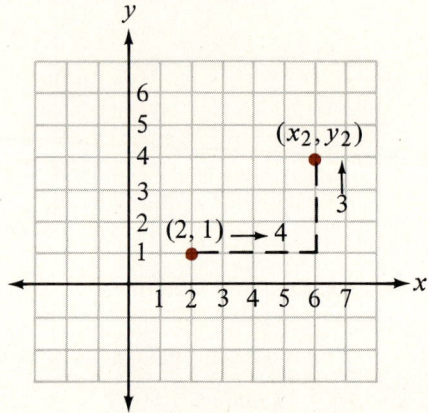

The coordinates of the point (x_2, y_2) can be found by arithmetic. If we begin at a point with x-coordinate of 2 and move 4 units to the right, we'll arrive at a point with x-coordinate of 6. Similarly, if we begin at a point with y-coordinate of 1 and move *up* 3 units, we'll arrive at a point with y-coordinate of 4. Therefore, $(x_2, y_2) = (6, 4)$, and we can graph the line.

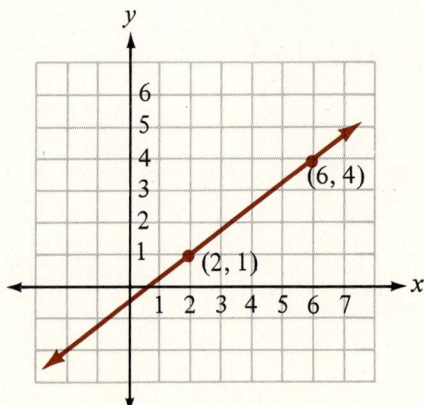

Example 9. Graph the line through $(3, -2)$ with slope of $\dfrac{-1}{4}$.

Solution

$$(x_1, y_1) = (3, -2)$$

$$m = \frac{-1}{4} \begin{array}{c} \downarrow \\ \rightarrow \end{array}$$

We begin at $(3, -2)$ and move to the *right* 4 units and *down* 1 unit.

You try Example 10.

Example 10. Graph the line through the origin with slope of 3.

Solution

$$(x_1, y_1) = (0, 0)$$

$$m = 3 = \frac{3}{1} \begin{array}{c} \uparrow \\ \rightarrow \end{array}$$

We begin at (_____ , _____) and move to the _____ 1 unit and _____ 3 units.

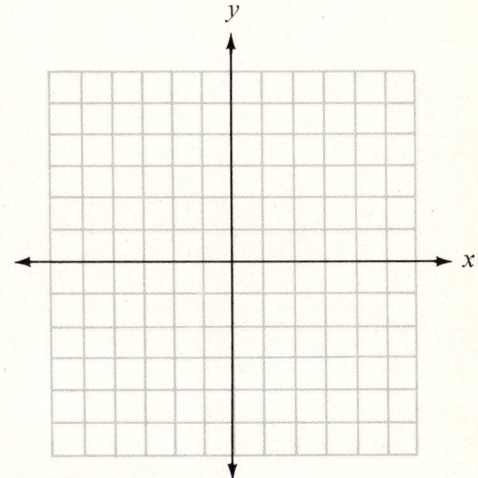

Check your work on page 596. ▶

Then try Example 11.

Example 11. Graph the line through $(5, -2)$ with slope of zero.

Solution

$$(x_1, y_1) = (5, -2)$$

$$m = 0$$

We recall that if the slope of a line is zero, that line is _____ .

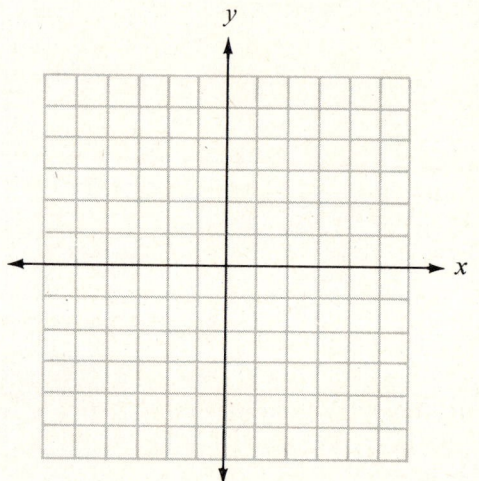

What is the *equation* for this line? Because it is a *horizontal* line, its equation must be $y =$ _____ .

Check your work on page 596. ▶

Example 12. Graph the line through $(5, -2)$ with undefined slope.

Solution

$$(x_1, y_1) = (5, -2)$$

$$m \text{ is undefined}$$

We recall that if the slope of a line is undefined, that line is vertical.

What is the *equation* for this line? Because it is a *vertical* line, its equation must be $x = 5$.

⫸ Trial Run

Graph the line through the given point with the given slope.

1. $(0, 0);\ m = \dfrac{5}{2}$

2. $(4, 2);\ m = 0$

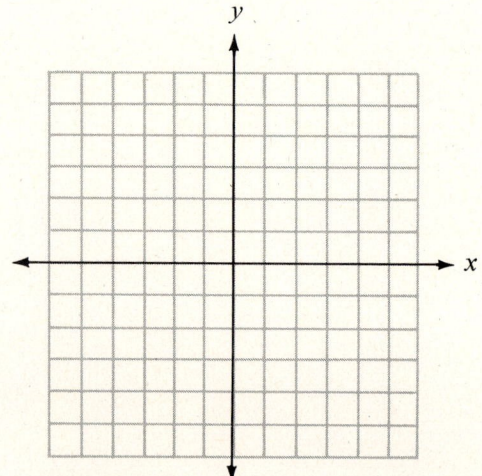

3. $(-2, -4)$; $m = -1$ **4.** $(-3, 2)$; m is undefined

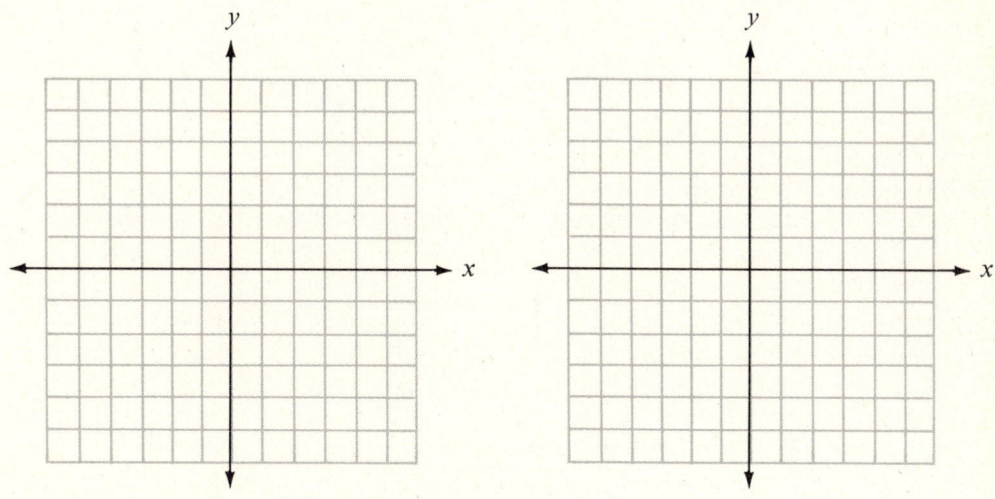

Answers are on page 597.

Graphing Lines by the Slope-Intercept Method

Take another look at the graphs of linear equations presented at the beginning of this section. For each graph, the y-intercept has been found by the usual method. Let's find the slope of each line using the points labeled on each graph.

$(x_1, y_1) = (0, 0)$

$(x_2, y_2) = (1, 2)$

$$m = \frac{2 - 0}{1 - 0}$$

$$= \frac{2}{1}$$

$$m = 2$$

and

y-intercept is 0

$(x_1, y_1) = (0, 1)$

$(x_2, y_2) = (-1, 3)$

$$m = \frac{3 - 1}{-1 - 0}$$

$$= \frac{2}{-1}$$

$$m = -2$$

and

y-intercept is 1

$(x_1, y_1) = (0, 1)$

$(x_2, y_2) = (3, 0)$

$$m = \frac{0 - 1}{3 - 0}$$

$$= \frac{-1}{3}$$

$$m = \frac{-1}{3}$$

and

y-intercept is 1

Now let's rewrite each line's equation in the same form, with y isolated on the left-hand side and the x-term and constant term on the right-hand side.

$y = 2x$	$y = 1 - 2x$	$x + 3y = 3$
becomes	becomes	becomes
$y = 2x + 0$	$y = -2x + 1$	$y = \dfrac{-1}{3}x + 1$

Now we summarize the information about these equations and their slopes and y-intercepts.

Equation	Slope	y-Intercept
$y = 2x + 0$	2	0
$y = -2x + 1$	-2	1
$y = \dfrac{-1}{3}x + 1$	$\dfrac{-1}{3}$	1

It is not just a coincidence that when y is isolated in a linear equation, the slope of its graph matches the coefficient of the x-term and the y-intercept matches the constant term. In fact (unless the coefficient of y is zero), every linear equation can be written in this form.

When a linear equation is written with y isolated on one side, it is said to be in **slope-intercept form** because we can read the slope and the y-intercept directly from the equation.

Slope-Intercept Form. If a linear equation is written in the form

$$y = mx + b$$

then the slope of the line is m and the y-intercept is b.

Slope-intercept form is useful because it allows us to graph a line just by looking at the equation without substituting any values for either variable. We just begin at the y-intercept point and use the slope to move to a second point.

Example 13. Use slope-intercept form to graph $2x - 5y = 5$.

Solution: First we isolate y.

$$2x - 5y = 5$$
$$-5y = -2x + 5$$
$$y = \frac{-2x}{-5} + \frac{5}{-5}$$
$$y = \frac{2}{5}x - 1$$

Slope: $m = \dfrac{2}{5} \begin{smallmatrix} \uparrow \\ \rightarrow \end{smallmatrix}$

y-intercept: $b = -1$

$$(0, -1)$$

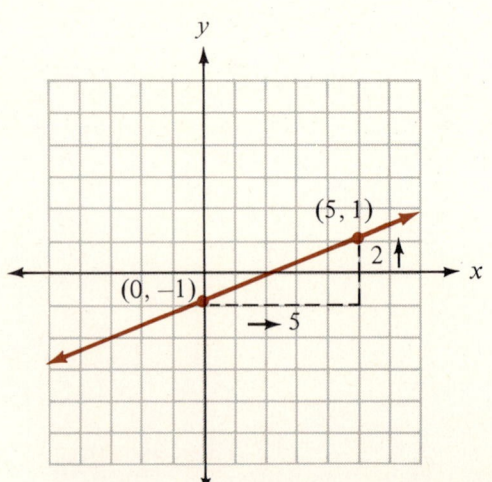

You try Example 14.

Example 14. Use slope-intercept form to graph $x + y = 0$.

Solution: First we isolate y.

$$x + y = 0$$

$$y = -x$$

$$y = -1x + 0$$

Slope: $m = -1 = \dfrac{-1}{1} \begin{array}{c} \downarrow \\ \rightarrow \end{array}$

y-intercept: $b = 0$

$$(0, 0)$$

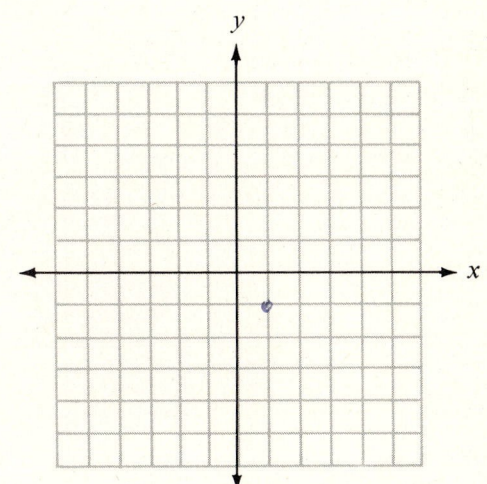

Check your graph on page 597. ▶

Example 15. Use slope-intercept form to graph $2y + 5 = 0$.

Solution: First we isolate y.

$$2y + 5 = 0$$

$$2y = -5$$

$$y = \dfrac{-5}{2}$$

$$y = 0x - \dfrac{5}{2}$$

Slope: $m = 0 = \dfrac{0}{1}$

y-intercept: $b = \dfrac{-5}{2}$

$$\left(0, \dfrac{-5}{2}\right)$$

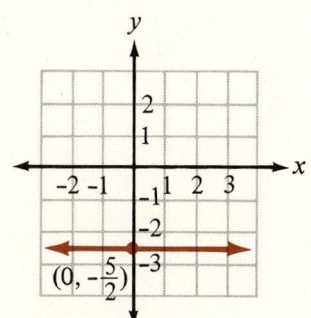

In Example 15 we knew our graph would be a horizontal line as soon as we arrived at the constant equation $y = \dfrac{-5}{2}$. The slope-intercept information confirmed that fact.

⫸ Trial Run

Use the slope-intercept method to graph each equation.

1. $y = -3x + 6$

2. $x - y = 4$

3. $x + 4y = -8$

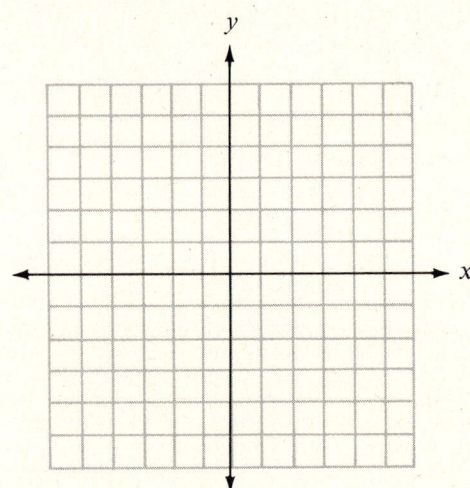

4. $3x - 4y = 12$

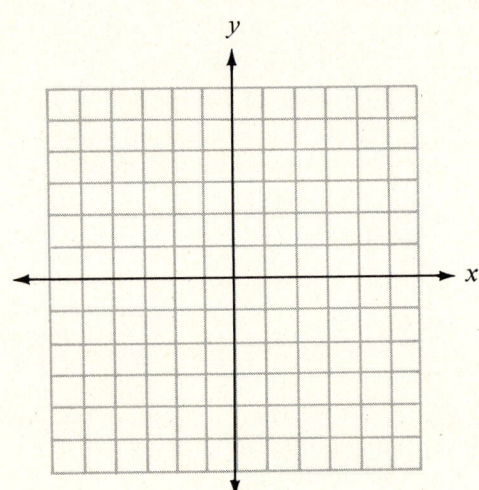

Answers are on page 597.

▶ **Examples You Completed**

Example 2. Find the distance between $(0, 3)$ and $(7, 5)$.

Solution: Let $(x_1, y_1) = (0, 3)$
$\qquad\qquad (x_2, y_2) = (7, 5)$

$$d = \sqrt{(x_2 - x_1)^2 + (y_2 - y_1)^2}$$
$$= \sqrt{(7 - 0)^2 + (5 - 3)^2}$$
$$= \sqrt{(7)^2 + (2)^2}$$
$$= \sqrt{49 + 4}$$
$$= \sqrt{53}$$

Example 4. Find the midpoint of the line segment joining $(-2, 0)$ and $(3, 0)$.

Solution: Let $(x_1, y_1) = (-2, 0)$
$\qquad\qquad (x_2, y_2) = (3, 0)$

$$x_M = \frac{x_1 + x_2}{2} \quad \text{and} \quad y_M = \frac{y_1 + y_2}{2}$$
$$= \frac{-2 + 3}{2} \qquad\qquad = \frac{0 + 0}{2}$$
$$= \frac{1}{2} \qquad\qquad\qquad = 0$$

$$(x_M, y_M) = \left(\frac{1}{2}, 0\right)$$

Example 5. Find the slope of the line through $(1, 3)$ and $(4, 7)$.

Solution

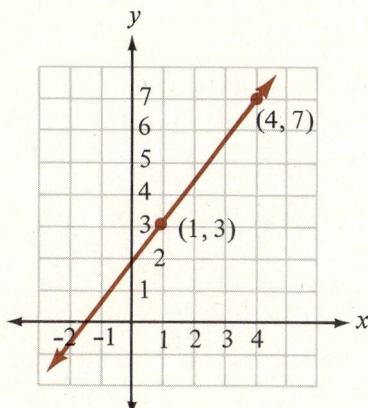

Let $(x_1, y_1) = (1, 3)$
$\qquad (x_2, y_2) = (4, 7)$

Then $\quad m = \dfrac{y_2 - y_1}{x_2 - x_1}$

$$= \frac{7 - 3}{4 - 1}$$

$$m = \frac{4}{3}$$

Example 7. Find the slope of the line joining $(1, 4)$ and $(-2, 4)$.

Solution

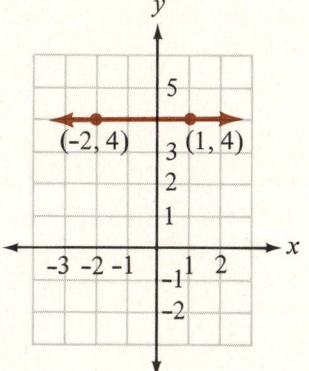

Let $(x_1, y_1) = (1, 4)$
$\qquad (x_2, y_2) = (-2, 4)$

Then $\quad m = \dfrac{4 - 4}{-2 - 1}$

$$= \frac{0}{-3}$$

$$m = 0$$

Example 10. Graph the line through the origin with slope of 3.

Solution

$$(x_1, y_1) = (0, 0)$$

$$m = 3 = \frac{3}{1} \begin{array}{c}\uparrow \\ \rightarrow\end{array}$$

We begin at $(0, 0)$ and move to the right 1 unit and up 3 units.

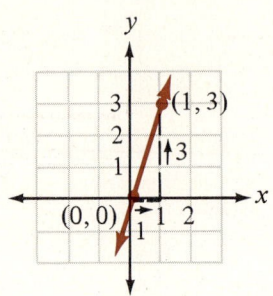

Example 11. Graph the line through $(5, -2)$ with slope of zero.

Solution

$$(x_1, y_1) = (5, -2)$$

$$m = 0$$

We recall that if the slope of a line is zero, that line is horizontal.

What is the *equation* for this line? Because it is a *horizontal* line, its equation must be $y = -2$.

Example 14. Use slope-intercept form to graph $x + y = 0$.

Solution: First we isolate y.

$$x + y = 0$$

$$y = -x$$

$$y = -1x + 0$$

$$\text{Slope: } m = -1 = \frac{-1}{1} \begin{array}{c}\downarrow \\ \rightarrow\end{array}$$

$$y\text{-intercept: } b = 0$$

$$(0, 0)$$

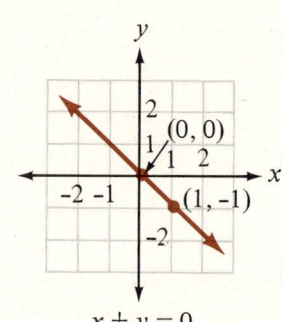

$$x + y = 0$$

Answers to Trial Runs

page 584   **1.** $2\sqrt{13}$ **2.** 13   **3.** $\left(\dfrac{-3}{2}, 5\right)$ **4.** $(0, -2)$

page 587   **1.** -1 **2.** 0   **3.** $\dfrac{-2}{3}$ **4.** Undefined

page 590   **1.**

2.

3.

4.

page 594   **1.**

2.

3.

4.

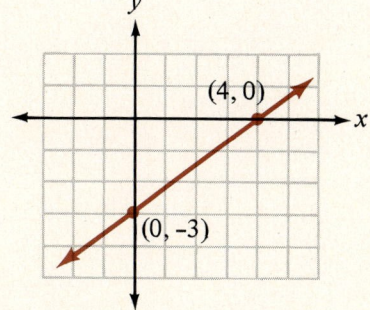

EXERCISE SET 8.2

Find the distance between the given points.

_____ **1.** $(-3, 2)$ and $(0, 6)$

_____ **2.** $(-7, -2)$ and $(-1, 6)$

_____ **3.** $(2, -5)$ and $(5, 1)$

_____ **4.** $(4, -7)$ and $(6, -3)$

_____ **5.** $(-1, 2)$ and $(0, 4)$

_____ **6.** $(1, -2)$ and $(2, 1)$

_____ **7.** $(3, 6)$ and $(-2, -6)$

_____ **8.** $(4, -10)$ and $(-4, 5)$

_____ **9.** $\left(1\frac{1}{2}, 3\frac{1}{2}\right)$ and $\left(1, 2\frac{1}{2}\right)$

_____ **10.** $\left(\frac{1}{3}, 1\frac{1}{3}\right)$ and $(-1, 2)$

_____ **11.** $(-5, 2)$ and $(-3, 7)$

_____ **12.** $(-4, 3)$ and $(-1, 8)$

Find the midpoint of the line segment joining the given points.

_____ **13.** $(3, 7)$ and $(9, 15)$

_____ **14.** $(1, 6)$ and $(7, 10)$

_____ **15.** $(-5, 3)$ and $(7, -7)$

_____ **16.** $(-8, 4)$ and $(12, -10)$

_____ **17.** $(0, -3)$ and $(-5, 9)$

_____ **18.** $(0, -4)$ and $(-3, 10)$

_____ **19.** $(-6, 8)$ and $(-6, 3)$

_____ **20.** $(-5, 11)$ and $(-5, 6)$

_____ **21.** $(12, -3)$ and $(-4, 3)$

_____ **22.** $(9, -4)$ and $(-7, 4)$

_____ **23.** $\left(-1\frac{1}{2}, 0\right)$ and $\left(1, 3\frac{1}{2}\right)$

_____ **24.** $\left(-2\frac{1}{2}, 5\frac{1}{2}\right)$ and $(0, -2)$

Find the slope of the line through the given points.

_____ **25.** $(0, 5)$ and $(-3, 0)$

_____ **26.** $(0, -7)$ and $(4, 0)$

_____ **27.** $(5, -1)$ and $(5, 4)$

_____ **28.** $(2, -7)$ and $(2, 3)$

_____ **29.** $(6, -3)$ and $(8, -5)$

_____ **30.** $(9, -7)$ and $(5, -3)$

_____ **31.** $(3, 5)$ and $(-2, 5)$

_____ **32.** $(-4, -3)$ and $(5, -3)$

_____ **33.** $\left(\frac{1}{2}, 3\right)$ and $\left(\frac{2}{3}, \frac{-1}{4}\right)$

_____ **34.** $\left(\frac{3}{4}, \frac{5}{6}\right)$ and $\left(\frac{-1}{2}, \frac{1}{3}\right)$

_____ **35.** $(2, -3)$ and $(5, 6)$

_____ **36.** $(1, -2)$ and $(-3, 6)$

Graph the line through the given point with the given slope.

37. $(1, 3)$; $m = 2$

38. $(2, 5)$; $m = 3$

39. $(5, -1)$; $m = -3$

40. $(4, -3)$; $m = -2$

41. $(-2, 3)$; $m = \dfrac{-1}{2}$

42. $(-1, -1)$; $m = \dfrac{-1}{4}$

43. $(3, 5)$; $m = 0$

44. $(1, -4)$; $m = 0$

45. $(-8, 2)$; $m = \dfrac{3}{2}$

46. $(-6, 3)$; $m = \dfrac{4}{3}$

47. $(5, -4)$; m is undefined

48. $(3, -6)$; m is undefined

Use the slope-intercept method to graph each of the following.

49. $y = -2x + 1$

50. $y = -3x + 4$

51. $x - y = 9$

52. $x - y = 6$

53. $x + 3y = 15$

54. $x + 6y = 12$

55. $3x - 2y = 6$

56. $5x - 2y = 10$

57. $-3x + 4y = -2$

58. $-2x + 9y = -6$

59. $3x + 5y = 0$

60. $4x + 3y = 0$

☆ Stretching the Topics _____

_____ **1.** Find the perimeter of a triangle with vertices $A(3, -1)$, $B(-5, 5)$, and $C(3, 5)$.

_____ **2.** If one endpoint of a line segment is $\left(\dfrac{1}{3}, \dfrac{2}{3}\right)$ and the midpoint is $\left(\dfrac{-2}{3}, \dfrac{-5}{4}\right)$, find the coordinates of the other endpoint.

_____ **3.** Find the slope and y-intercept of the graph of $4x - 3[2(5x - 1) - 4(7y - 3)] = 81$.

Check your answers in the back of your book.

If you can complete **Checkup 8.2**, you are ready to go on to Section 8.3.

CHECKUP 8.2

_____ **1.** Find the distance between $(-3, 1)$ and $(0, 5)$.

_____ **2.** Find the midpoint of the line segment joining $(9, -9)$ and $(5, -7)$.

Find the slope of the line through the given points.

_____ **3.** $(16, -4)$ and $(9, 3)$ _____ **4.** $(1, 5)$ and $(-3, 8)$

Graph the line through the given point with the given slope.

5. $(2, -7)$; $m = 2$

6. $(-5, 8)$; $m = \dfrac{-3}{2}$

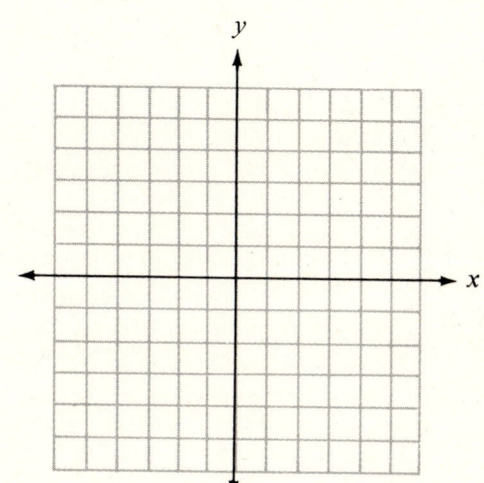

7. $(5, -1)$; $m = 0$

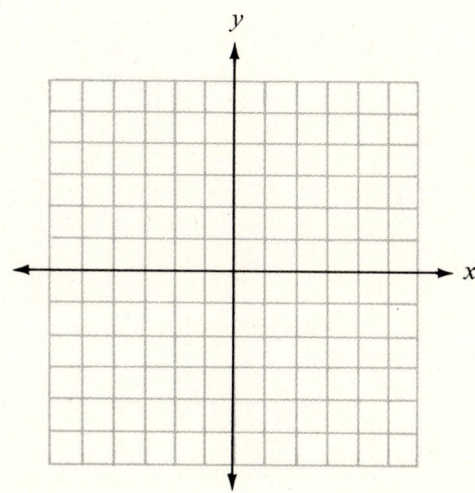

Use the slope-intercept method to graph each of the following.

8. $y = 2x - 3$

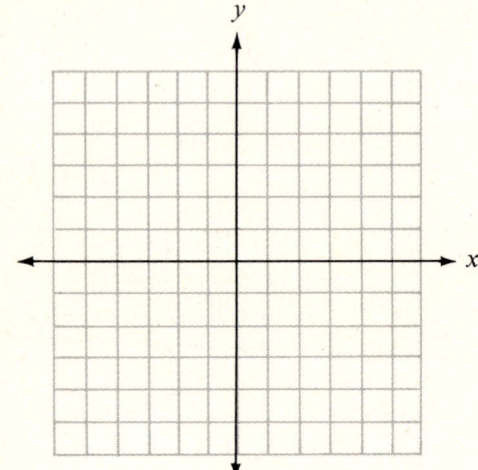

9. $3x + 5y = 10$

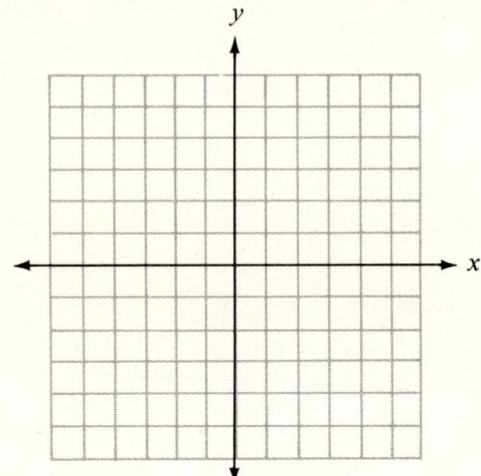

10. $3x - 4y = 0$

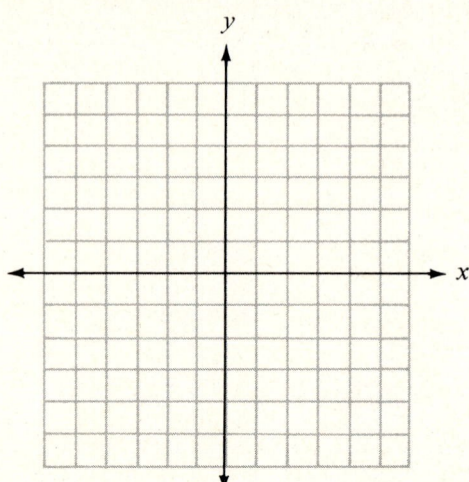

Check your answers in the back of your book.

If You Missed Problems:	You Should Review Examples:
1	1, 2
2	3, 4
3, 4	5–8
5, 6	9, 10
7	11
8–10	13–15

8.3 Finding Equations of Lines

We have discussed several methods for graphing the solutions of first-degree equations in two variables. Now we would like to discover ways to find the equation that corresponds to a given straight line graph.

Finding Constant Equations

Suppose that you were asked to find an equation for each of these graphs.

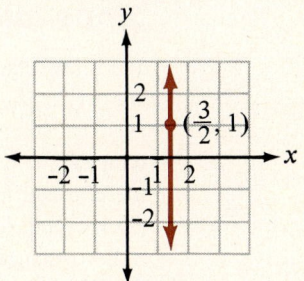

We know the equation for a *horizontal* line is always

$$y = c$$

where c is a constant. Since this line contains the point $(0, -1)$, the value of c must be the y-coordinate at that point. Thus c must be -1, and the equation of this line is

$$y = -1$$

We know that the equation for a *vertical* line is always

$$x = k$$

where k is a constant. Since this line contains the point $\left(\dfrac{3}{2}, 1\right)$, the value of k must be the x-coordinate at that point. Thus k must be $\dfrac{3}{2}$, and the equation of this line is

$$x = \dfrac{3}{2}$$

You try Examples 1 and 2.

Example 1. Find the equation of the horizontal line through $(-3, 5)$.

   Solution: A horizontal line has an equation ____ = ____ . For the given point, the y-coordinate is ____ , so the equation must be

$$\text{\_\_\_\_} = \text{\_\_\_\_}$$

Check your work on page 613. ▶

Example 2. Find the equation of the vertical line through $(-3, 5)$.

   Solution: A vertical line has an equation ____ = ____ , so the equation must be

$$\text{\_\_\_\_} = \text{\_\_\_\_}$$

Check your work on page 613. ▶

⫸ **Trial Run** ━━━

Find the equation of each line.

_____ **1.** Vertical line through $(4, -2)$

_____ **2.** Horizontal line through $(0, -2)$

_____ **3.** The x-axis

_____ **4.** The y-axis

Answers are on page 614.

Finding Equations of Lines Using the Slope-Intercept Method

We know that if a linear equation is written in the form

$$y = mx + b$$

then m, the coefficient of the x-term, represents the *slope* of the graph of the equation, and the constant term b represents the *y-intercept*. This slope-intercept form for a linear equation allows us to find the equation for a line if we know its slope (m) and its y-intercept (b).

For instance, if we know that a line has a y-intercept of -2 and a slope of 3, then we know that $m = 3$ and $b = -2$. The equation of the line must be

$$y = 3x - 2$$

Example 3. Find the equation of the line through the origin with slope of $\dfrac{-2}{5}$.

Solution: The graph crosses the y-axis at $(0, 0)$, so

$$b = 0$$

and

$$m = \frac{-2}{5}$$

The equation of the line is

$$y = \frac{-2}{5}x + 0$$

$$y = \frac{-2}{5}x \qquad \text{Slope-intercept form.}$$

To write this equation in standard form, $Ax + By = C$, we first must eliminate the fractions.

$$y = \frac{-2}{5}x$$

$$5y = -2x \qquad \text{Multiply by LCD: 5.}$$

$$2x + 5y = 0 \qquad \text{Standard form.}$$

You try Example 4.

Example 4. Find the equation of the line through $(0, -1)$ with slope of $\dfrac{1}{2}$.

Solution: The graph crosses the y-axis at $(\text{\_\_\_\_} , \text{\_\_\_\_})$, so

$$b = \text{\_\_\_\_}$$

and

$$m = \text{\_\_\_\_}$$

The equation of the line is

$$y = \text{\_\_\_\_}x - \text{\_\_\_\_} \qquad \text{Slope-intercept form.}$$

To write this equation in standard form, we first must eliminate the fractions.

$$y = \frac{1}{2}x - 1$$

$$2y = x - 2 \qquad \text{Multiply by LCD: 2.}$$

$$-x + 2y = -2$$

$$x - 2y = 2 \qquad \text{Standard form.}$$

Check your work on page 613. ▶

Either of the two forms we have shown is an acceptable way of writing a first-degree equation in two variables. Slope-intercept form is preferable for its graphing information, but we shall find standard form useful when solving systems of equations in Chapter 10.

> **Slope-intercept form:** $y = mx + b$
> **Standard linear form:** $Ax + By = C$

⫸ Trial Run

Find the equation of the line with the given slope and y-intercept. Write each equation in slope-intercept form and in standard linear form.

_____ **1.** $m = -4, b = 2$

_____ **2.** $m = 0, b = 4$

_____ **3.** $m = \dfrac{2}{3}, b = \dfrac{-5}{6}$

Answers are on page 614.

Finding Equations of Lines Using the Point-Slope Method

We now have a method for finding the equation of a line when we know its slope m and y-intercept b. Unfortunately, both those pieces of information may not be available. Let's see if we can develop methods for finding the equation for a line when we are given just the *slope and one point* on the line or when we are given *any two points* on the line.

Recall that the slope of a line is defined by

$$m = \frac{y_2 - y_1}{x_2 - x_1}$$

Suppose that we let (x, y) represent *any* point on a line that we know passes through the point $(2, 5)$ and has a slope of 4. Here we are saying that

$$(x_1, y_1) = (2, 5)$$

$$(x_2, y_2) = (x, y)$$

$$m = 4$$

But

$$m = \frac{y_2 - y_1}{x_2 - x_1} \qquad \text{Definition of slope.}$$

so

$$4 = \frac{y - 5}{x - 2} \qquad \text{Substitution.}$$

$$4(x - 2) = y - 5 \qquad \text{Multiply both sides by } x - 2.$$

$$4x - 8 = y - 5 \qquad \text{Simplify.}$$

$$4x - 3 = y \qquad \text{Isolate } y.$$

$$y = 4x - 3 \qquad \text{Slope-intercept form.}$$

We have found the equation of the line through $(2, 5)$ with slope of 4.

In general, if m is the slope of a line passing through a specific point (x_1, y_1), then

$$m = \frac{y - y_1}{x - x_1}$$

$$m(x - x_1) = y - y_1 \qquad \text{Multiply both sides by } x - x_1.$$

$$y - y_1 = m(x - x_1)$$

This equation is called the **point-slope form** for a linear equation.

Point-Slope Form. If (x_1, y_1) is a point on a line with slope m, then the equation of that line can be found by

$$y - y_1 = m(x - x_1)$$

Example 5. Find the equation of the line through $(2, -4)$ with slope of $\frac{1}{3}$.

Solution

$$(x_1, y_1) = (2, -4)$$

$$m = \frac{1}{3}$$

$$y - y_1 = m(x - x_1)$$

$$y - (-4) = \frac{1}{3}(x - 2)$$

$$y + 4 = \frac{1}{3}x - \frac{2}{3}$$

$$y = \frac{1}{3}x - \frac{2}{3} - 4$$

$$y = \frac{1}{3}x - \frac{2}{3} - \frac{12}{3}$$

$$y = \frac{1}{3}x - \frac{14}{3} \qquad \text{Slope-intercept form.}$$

or

$$3y = x - 14$$

$$-x + 3y = -14$$

$$x - 3y = 14 \qquad \text{Standard form.}$$

You complete Example 6.

Example 6. Find the equation of the line with x-intercept of -1 and slope of -2.

Solution

$$(x_1, y_1) = (-1, 0)$$

$$m = -2$$

$$y - y_1 = m(x - x_1)$$

$$y - 0 = -2[x - (-1)]$$

$$y = \underline{\hspace{1.5cm}} \qquad \text{Slope-intercept form.}$$

or

$$\underline{\hspace{1.5cm}} \qquad \text{Standard form.}$$

Check your work on page 613. ▶

Then try Example 7.

Example 7. Find the equation of the line through $(7, -1)$ with zero slope.

Solution: A line with zero slope is a _____ line. Therefore, the equation must be $y = c$, where c is a constant. Since the given point has a y-coordinate of _____ , the equation must be

$$y = \underline{\hspace{1cm}}$$

Check your work on page 614. ▶

Example 8. Find the equation of the line through $(7, -1)$ with undefined slope.

Solution: A line with undefined slope is a *vertical* line. Therefore, the equation must be $x = k$, where k is a constant. Since the given point has an x-coordinate of 7, the equation must be

$$x = 7$$

⫸ Trial Run

Find the equation of the line through the given point with the given slope.

_____ **1.** $(5, 1)$; $m = 3$

_____ **2.** $(-3, 2)$; $m = 0$

_____ **3.** $(-2, 3)$; $m = \dfrac{-1}{2}$

_____ **4.** $(-3, 2)$; m is undefined

Answers are on page 614.

Finding Equations of Lines from Two Points

It is very useful to be able to find the equation for a line when given just two points on the line. Actually, we already have the tools necessary for finding an equation from two points.

For instance, suppose that we wish to find the equation for the line through $(-1, 5)$ and $(2, 3)$. Both methods for finding equations of lines require the *slope* of the line, so let's find the slope here.

$$(x_1, y_1) = (-1, 5)$$

$$(x_2, y_2) = (2, 3)$$

$$m = \frac{3 - 5}{2 - (-1)}$$

$$m = \frac{-2}{3}$$

Now that we know the slope, we may use it with either of the given points to find the equation using the *point-slope method*.

Let's use $(x_1, y_1) = (-1, 5)$ Or use $(x_2, y_2) = (2, 3)$

$$m = \frac{-2}{3} \qquad\qquad\qquad m = \frac{-2}{3}$$

$$y - y_1 = m(x - x_1) \qquad\qquad y - y_2 = m(x - x_2)$$

$$y - 5 = \frac{-2}{3}[x - (-1)] \qquad\qquad y - 3 = \frac{-2}{3}(x - 2)$$

$$y - 5 = \frac{-2}{3}[x + 1] \qquad\qquad y - 3 = \frac{-2}{3}x + \frac{4}{3}$$

$$y - 5 = \frac{-2}{3}x - \frac{2}{3} \qquad\qquad y = \frac{-2}{3}x + \frac{4}{3} + 3$$

$$y = \frac{-2}{3}x - \frac{2}{3} + 5 \qquad\qquad y = \frac{-2}{3}x + \frac{4}{3} + \frac{9}{3}$$

$$y = \frac{-2}{3}x - \frac{2}{3} + \frac{15}{3} \qquad\qquad y = \frac{-2}{3}x + \frac{13}{3}$$

$$y = \frac{-2}{3}x + \frac{13}{3}$$

We arrive at the same equation using either of the given points. *It does not matter which point we use; the equation for the line will be the same.*

Now you try Example 9.

Example 9. Find the equation of the line through $(-4, -7)$ and $(-6, -2)$.

Solution

$$(x_1, y_1) = (-4, -7)$$

$$(x_2, y_2) = (-6, -2)$$

$$m = \frac{-2 - (-7)}{-6 - (-4)}$$

$$m = \frac{-5}{2}$$

$$y - y_1 = m(x - x_1)$$

$$y - (\underline{\hspace{1cm}}) = \frac{-5}{2} [x - (\underline{\hspace{1cm}})]$$

Or, in standard form,

Check your work on page 614. ▶

Example 10. Find the equation of the line with x-intercept of -2 and y-intercept of 6.

Solution: Remember that intercepts tell us points.

x-intercept: $(x_1, y_1) = (-2, 0)$

y-intercept: $(x_2, y_2) = (0, 6)$

$$m = \frac{6 - 0}{0 - (-2)}$$

$$= \frac{6}{2}$$

$$m = 3$$

Since the y-intercept is given, we can use slope-intercept form immediately.

$$y = mx + b$$

$$y = 3x + 6$$

Example 11. Find the equation of the line through $(5, -3)$ and $(-2, -3)$.

Solution

$$(x_1, y_1) = (5, -3)$$

$$(x_2, y_2) = (-2, -3)$$

$$m = \frac{-3 - (-3)}{-2 - 5}$$

$$= \frac{0}{-7}$$

$$= 0$$

Since the slope is zero, this is a horizontal line with equation

$$y = -3$$

You complete Example 12.

Example 12. Find the equation of the line through $(11, -1)$ and $(11, 13)$.

Solution

$$(x_1, y_1) = (11, -1)$$

$$(x_2, y_2) = (11, 13)$$

$$m = \frac{13 - (-1)}{11 - 11}$$

$$=$$

m is _____ . Since the slope is _____ this is a _____ line with equation:

Check your work on page 614. ▶

Trial Run

Find the equation of the line through the given points.

———— **1.** (2, 3) and (1, 5)   ———— **2.** (4, 5) and (-2, 5)

———— **3.** (-5, -8) and (0, 2)   ———— **4.** (10, 3) and (10, -6)

Answers are on page 614.

Finding Equations of Parallel and Perpendicular Lines

Let's graph the equations $y = 2x + 3$ and $y = 2x - 1$ in the same Cartesian plane, using the slope-intercept method.

(1) $y = 2x + 3$ (2) $y = 2x - 1$

$m = \dfrac{2}{1}$ $m = \dfrac{2}{1}$

$b = 3$ $b = -1$

(0, 3)   (0, -1)

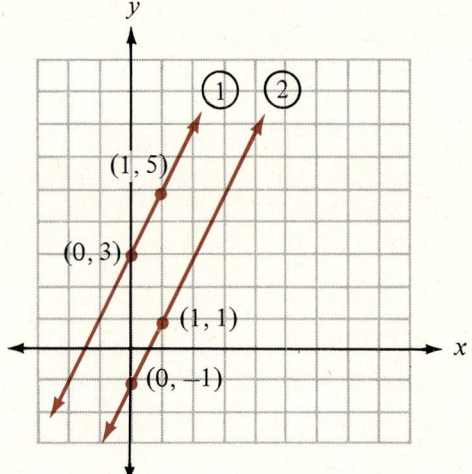

From the equations we see that these lines have the *same slope* but *different y-intercepts*. From the graph we see that the lines are **parallel.**

> If two lines have the same slope but different *y*-intercepts, the lines are *parallel*.

If we want to find the equation of a line that is parallel to another line, we may "borrow" the slope and use it in the appropriate formula for finding the equation of the new line.

Example 13. Find the equation of the line through the point $(1, 2)$ and parallel to the line $y = 3x + 4$.

Solution: We must "borrow" the slope from the line $y = 3x + 4$.

$$m = 3$$

$$(x_1, y_1) = (1, 2)$$

$$y - y_1 = m(x - x_1)$$

$$y - 2 = 3(x - 1)$$

$$y - 2 = 3x - 3$$

$$y = 3x - 1$$

Example 14. Find the equation of the line through the point $(0, 5)$ and parallel to the line $2x + 3y = 1$.

Solution: We must "borrow" the slope from the line $2x + 3y = 1$. To read the slope, we must isolate y.

$$2x + 3y = 1$$

$$3y = -2x + 1$$

$$y = \frac{-2}{3}x + \frac{1}{3}$$

$$m = \frac{-2}{3}$$

The given point $(0, 5)$ is the y-intercept point, so

$$b = 5$$

The equation we need is

$$y = \frac{-2}{3}x + 5$$

Although the proof is beyond the scope of this course, a relationship also exists between the slopes of **perpendicular** lines.

If two lines are *perpendicular*, their slopes are *negative reciprocals* of each other. If the slope of a given line is m, then the slope of any line perpendicular to the given line is $\frac{-1}{m}$.

Example 15. Find the equation of the line perpendicular to the line $x + 2y = 3$ and passing through the point $(5, 1)$.

Solution: We need the negative reciprocal of the slope of the given line $x + 2y = 3$.

$$2y = -x + 3 \qquad \text{Isolate } y \text{ term.}$$

$$y = \frac{-1}{2}x + \frac{3}{2} \qquad \text{Divide by 2.}$$

$$m = \frac{-1}{2} \qquad \text{Read the slope of the given line.}$$

$$\frac{-1}{m} = 2 \qquad \text{Find the negative reciprocal of the slope.}$$

$$y - 1 = 2(x - 5) \qquad \text{Use point-slope method for equation of line.}$$

$$y - 1 = 2x - 10 \qquad \text{Simplify.}$$

$$y = 2x - 9 \qquad \text{Equation of perpendicular line.}$$

Perhaps a graph of the two equations in Example 15 will convince you that they represent perpendicular lines.

Equation (1)

$$x + 2y = 3$$

$$y = \frac{-1}{2}x + \frac{3}{2}$$

$$m = \frac{-1}{2}$$

$$b = \frac{3}{2}$$

Equation (2)

$$y = 2x - 9$$

$$m = 2$$

$$b = -9$$

⫸ Trial Run

Find the equation of the line satisfying the given conditions.

_____ **1.** With y-intercept of -7 and parallel to $y = 4x + 1$

_____ **2.** Through $(2, 4)$ and parallel to $x + y = 2$

_____ **3.** Through $(3, -2)$ and perpendicular to $y = 3x - 1$

Answers are on page 614.

Let's summarize the methods that we have developed for finding equations of straight lines. The method to be used depends, of course, on the information we are given.

Given	Equation to Use	Name
Slope: m y-intercept: b	$y = mx + b$	Slope-intercept form
Slope: m Point: (x_1, y_1)	$y - y_1 = m(x - x_1)$	Point-slope form
Two points: (x_1, y_1) and (x_2, y_2)	$m = \dfrac{y_2 - y_1}{x_2 - x_1}$	Slope definition
	$y - y_1 = m(x - x_1)$	Point-slope form
Line is horizontal	$y = c$, a constant	Constant equation
Line is vertical	$x = k$, a constant	Constant equation

If we are told that a line is *parallel* to a given line, $y = mx + b$, we know that the slope of the parallel line is also m. If we are told that a line is *perpendicular* to a given line, $y = mx + b$, we know that the slope of the perpendicular line is $\dfrac{-1}{m}$.

▶ Examples You Completed

Example 1. Find the equation of the horizontal line through $(-3, 5)$.

Solution: A horizontal line has an equation $y = c$. For the given point, the y-coordinate is 5, so the equation must be

$$y = 5$$

Example 4. Find the equation of the line through $(0, -1)$ with slope of $\dfrac{1}{2}$.

Solution: The graph crosses the y-axis at $(0, -1)$, so

$$b = -1$$

and

$$m = \frac{1}{2}$$

The equation of the line is

$$y = \frac{1}{2}x - 1 \quad \text{Slope-intercept form.}$$

Example 2. Find the equation of the vertical line through $(-3, 5)$.

Solution: A vertical line has an equation $x = k$. For the given point, the x-coordinate is -3, so the equation must be

$$x = -3$$

Example 6. Find the equation of the line with x-intercept of -1 and slope of -2.

Solution

$$(x_1, y_1) = (-1, 0)$$

$$m = -2$$

$$y - y_1 = m(x - x_1)$$

$$y - 0 = -2[x - (-1)]$$

$$y = -2[x + 1]$$

$$y = -2x - 2 \quad \text{Slope-intercept form.}$$

or

$$2x + y = -2 \quad \text{Standard form.}$$

Example 7. Find the equation of the line through $(7, -1)$ with zero slope.

Solution: A line with zero slope is a horizontal line. Therefore, the equation must be $y = c$, where c is a constant. Since the given point has a y-coordinate of -1, the equation must be

$$y = -1$$

Example 12. Find the equation of the line through $(11, -1)$ and $(11, 13)$.

Solution

$$(x_1, y_1) = (11, -1)$$

$$(x_2, y_2) = (11, 13)$$

$$m = \frac{13 - (-1)}{11 - 11}$$

$$= \frac{14}{0}$$

m is undefined. Since the slope is undefined this is a vertical line with equation

$$x = 11$$

Example 9. Find the equation of the line through $(-4, -7)$ and $(-6, -2)$.

Solution

$$(x_1, y_1) = (-4, -7)$$

$$(x_2, y_2) = (-6, -2)$$

$$m = \frac{-2 - (-7)}{-6 - (-4)}$$

$$m = \frac{-5}{2}$$

$$y - y_1 = m(x - x_1)$$

$$y - (-7) = \frac{-5}{2} [x - (-4)]$$

$$y + 7 = \frac{-5}{2} [x + 4]$$

$$y + 7 = \frac{-5}{2} x - 10$$

$$y = \frac{-5}{2} x - 17$$

Or, in standard form

$$2y = -5x - 34$$

$$5x + 2y = -34$$

Answers to Trial Runs

page 603 **1.** $x = 4$ **2.** $y = -2$ **3.** $y = 0$ **4.** $x = 0$

page 605 **1.** $y = -4x + 2$ or $4x + y = 2$ **2.** $y = 4$ **3.** $y = \frac{2}{3}x - \frac{5}{6}$ or $4x - 6y = 5$

page 607 **1.** $y = 3x - 14$ or $3x - y = 14$ **2.** $y = 2$ **3.** $y = \frac{-1}{2}x + 2$ or $x + 2y = 4$
4. $x = -3$

page 610 **1.** $y = -2x + 7$ or $2x + y = 7$ **2.** $y = 5$ **3.** $y = 2x + 2$ or $2x - y = -2$
4. $x = 10$

page 612 **1.** $y = 4x - 7$ or $4x - y = 7$ **2.** $y = -x + 6$ or $x + y = 6$
3. $y = \frac{-1}{3}x - 1$ or $x + 3y = -3$

EXERCISE SET 8.3

Find the equation of the horizontal line through the given point.

_____ 1. $(4, -5)$

_____ 2. $(8, -3)$

_____ 3. $\left(-1\frac{1}{2}, \frac{1}{2}\right)$

_____ 4. $\left(2\frac{1}{2}, \frac{3}{4}\right)$

Find the equation of the vertical line through the given point.

_____ 5. $(2, -6)$

_____ 6. $(3, -4)$

_____ 7. $\left(-\frac{2}{3}, 4\right)$

_____ 8. $\left(-1\frac{1}{4}, 2\right)$

Find the equation of the line with the given slope and y-intercept.

_____ 9. $m = 4, b = 2$

_____ 10. $m = 3, b = 5$

_____ 11. $m = -2, b = 1$

_____ 12. $m = -3, b = 4$

_____ 13. $m = \frac{1}{3}, b = \frac{-2}{3}$

_____ 14. $m = \frac{1}{4}, b = \frac{-3}{4}$

_____ 15. $m = 0, b = 10$

_____ 16. $m = 0, b = 8$

_____ 17. $m = \frac{1}{2}, b = \frac{-5}{4}$

_____ 18. $m = \frac{3}{2}, b = \frac{-1}{10}$

_____ 19. $m = 7, b = -2$

_____ 20. $m = 5, b = -1$

Find the equation of the line through the given point with the given slope.

_____ 21. $(-1, 5), m = 8$

_____ 22. $(-2, 3), m = 6$

_____ 23. $(2, 7), m = -1$

_____ 24. $(3, 5), m = -2$

_____ 25. $(-6, -3), m = 0$

_____ 26. $(5, 2), m = 0$

_____ 27. $(-1, 4), m = \frac{3}{4}$

_____ 28. $(6, -3), m = \frac{2}{5}$

_____ 29. $(-2, 1), m$ is undefined

_____ 30. $(-6, 4), m$ is undefined

_____ 31. $\left(\frac{-1}{4}, \frac{3}{4}\right), m = \frac{-2}{3}$

_____ 32. $\left(\frac{1}{7}, \frac{-5}{7}\right), m = \frac{-1}{2}$

Find the equation of the line through the given points.

_____ 33. $(0, 2), (1, 0)$

_____ 34. $(0, 4), (-8, 0)$

_____ 35. $(-1, 7), (-4, 12)$ _____ 36. $(4, 6), (7, -3)$

_____ 37. $(9, -8), (-1, -8)$ _____ 38. $(6, 8), (-3, 8)$

_____ 39. $(-3, 7), (-1, 2)$ _____ 40. $(1, 11), (-2, 4)$

_____ 41. $(6, -5), (6, -10)$ _____ 42. $(-3, 4), (-3, 9)$

_____ 43. $\left(\frac{5}{2}, \frac{7}{5}\right), \left(\frac{1}{2}, \frac{-3}{5}\right)$ _____ 44. $\left(\frac{7}{3}, \frac{-11}{8}\right), \left(\frac{1}{3}, \frac{-7}{8}\right)$

Find the equation of the line through the given point and parallel to the given line.

_____ 45. $(-3, 2); y = -2x + 5$ _____ 46. $(-2, 1); y = 3x + 4$

_____ 47. $(-1, 1); 2x - y = 4$ _____ 48. $(-3, 3); 3x + y = -2$

_____ 49. $(0, 4); x + 3y = -4$ _____ 50. $(2, 0); x + 5y = 1$

Find the equation of the line through the given point and perpendicular to the given line.

_____ 51. $(4, 1); y = 3x$ _____ 52. $(3, 2); y = 5x$

_____ 53. $(-2, 0); x + 2y = 1$ _____ 54. $(-4, 0); x - 3y = 2$

_____ 55. $(3, -2); 4x - 2y = 5$ _____ 56. $(5, -1); 3x - 5y = 1$

_____ 57. Find the equation of the line with x-intercept of 3 and y-intercept of -5.

_____ 58. Find the equation of the line with x-intercept of -2 and y-intercept of -7.

_____ 59. Find the equation of the line with x-intercept of 2 and y-intercept of 4.

_____ 60. Find the equation of the line with x-intercept of 5 and y-intercept of 1.

☆ Stretching the Topics _____

_____ 1. Find the equation of the line through $(-4, 10)$ and parallel to the line through the points $(0, 5)$ and $(-8, -8)$.

_____ 2. Find the equation of the line through the midpoint of and perpendicular to the line segment with endpoints $(5, -1)$ and $(-3, 1)$.

_____ 3. Find the equations of the sides of a triangle whose vertices are $(1, -8)$, $(4, 2)$, and $(-3, 5)$.

Check your answers in the back of your book.

If you can complete **Checkup 8.3,** you are ready to go on to Section 8.4.

✓ CHECKUP 8.3

_____ **1.** Find the equation of the horizontal line through $(-3, 7)$.

_____ **2.** Find the equation of the vertical line through $\left(\dfrac{1}{2}, \dfrac{-2}{3}\right)$.

Find the equation of the line with the given slope and y-intercept.

_____ **3.** $m = -4, b = 2$ _____ **4.** $m = \dfrac{-2}{3}, b = 0$

Find the equation of the line through the given point with the given slope.

_____ **5.** $(-1, 1), m = -4$ _____ **6.** $(-3, 5), m$ is undefined

Find the equation of the line through the given points.

_____ **7.** $(1, -1)$ and $(3, 9)$ _____ **8.** $\left(\dfrac{4}{5}, \dfrac{-1}{2}\right)$ and $\left(\dfrac{-6}{5}, \dfrac{3}{2}\right)$

_____ **9.** Find the equation of the line through the point $(-1, 2)$ and parallel to $3x - y = 5$.

_____ **10.** Find the equation of the line through the point $(0, -5)$ and perpendicular to $x + 4y = 0$.

Check your answers in the back of your book.

If You Missed Problems:	You Should Review Examples:
1, 2	1, 2
3, 4	3, 4
5, 6	5–8
7, 8	9–12
9	13, 14
10	15

8.4 Graphing First-Degree Inequalities

Graphing a first-degree equation in two variables allows us to ''picture'' the set of all ordered pairs satisfying that equation. It is also handy to be able to picture the set of all ordered pairs that satisfy a first-degree **inequality** in two variables.

In each case we will graph the *equation* corresponding to the given *inequality* before we graph the inequality itself.

For example to find all points that satisfy the inequality $y \leq 2$, we must first graph the equation $y = 2$. We know the graph of $y = 2$ is a horizontal line crossing the y-axis at $(0, 2)$. To satisfy the inequality $y \leq 2$, we want all points in the plane whose y-coordinates are *less than 2 or equal 2*. All such points lie *on* the line $y = 2$ or *below* that line. We shade the appropriate region.

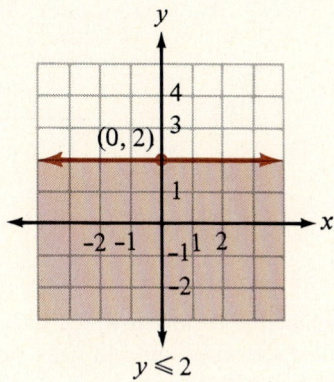

$$y \leqslant 2$$

Example 1. Graph $x > -1$.

Solution: We want all points in the plane whose x-coordinates are greater than -1. First we graph the vertical line $x = -1$. All x-values that are greater than -1 lie to the *right* of this line, but *not* on the line itself. Just as we used an *open* dot on a number line to show that a point was acting as a boundary but was not included in the graph itself, we can use a *dashed* line here to indicate that the vertical line $x = -1$ is acting as a boundary but is not included. Then we shade the region to the right of that line.

$$x > -1$$

In graphing linear inequalities, it is wise to decide right away whether the graph of the corresponding equation should be solid or dashed.

> If the given inequality is of the $\leq$ or $\geq$ type, the graph of the corresponding boundary equation is a *solid* line. If the given inequality is of the $<$ or $>$ type, the graph of the corresponding boundary equation is a *dashed* line.

To graph the inequality

$$y < 4x - 1$$

we must first graph the equation

$$y = 4x - 1$$

using a *dashed* line. This equation is in slope-intercept form, so we know that the line has a slope of 4 and a y-intercept of -1.

To determine the region in which $y < 4x - 1$ we can look at some points on either side of the line and see where ordered pairs satisfying the inequality must lie.

Above the line?		Below the line?	
$(0, 2)$:	$2 \overset{?}{<} 4(0) - 1$	$(3, 0)$:	$0 \overset{?}{<} 4(3) - 1$
	$2 \overset{?}{<} -1$ No		$0 \overset{?}{<} 11$ Yes
$(-2, 1)$:	$1 \overset{?}{<} 4(-2) - 1$	$(0, -2)$:	$-2 \overset{?}{<} 4(0) - 1$
	$1 \overset{?}{<} 9$ No		$-2 \overset{?}{<} -1$ Yes

Points *below* the line $y = 4x - 1$ satisfy the less-than inequality $y < 4x - 1$, but points above the line do not. Our graph is the shaded region *below* the line.

$$y < 4x - 1$$

Example 2. Graph $x + 3y > 0$.

Solution: We isolate y.

$$x + 3y > 0$$

$$3y > -x$$

$$y > \frac{-1}{3}x$$

Now we graph the corresponding equation $y = \frac{-1}{3}x$, which is a dashed line with slope of $\frac{-1}{3}$, crossing the y-axis at $(0, 0)$. Testing points above and below the line, we discover that this greater than inequality is satisfied by points *above* the line.

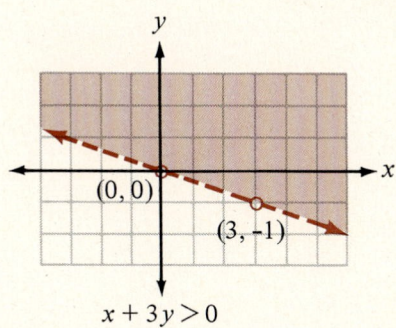

$x + 3y > 0$

In general, we may use the following steps to graph linear inequalities.

Graphing Linear Inequalities

1. Isolate y in the given inequality.
2. Decide whether the boundary line is solid or dashed and shade the appropriate region.

   If $y < mx + b$, the line is *dashed* and we shade *below* the line.

   If $y \leq mx + b$, the line is *solid* and we shade *below* the line.

   If $y > mx + b$, the line is *dashed* and we shade *above* the line.

   If $y \geq mx + b$, the line is *solid* and we shade *above* the line.

Example 3. Graph the inequality $3x - 5y < 10$.

Solution: We isolate y.

$$3x - 5y < 10$$

$$-5y < -3x + 10$$

$$y > \frac{3}{5}x - 2 \quad \text{Remember to reverse the direction of inequality. Why?}$$

The corresponding equation is

$$y = \frac{3}{5}x - 2$$

so we draw a *dashed* line with slope of $\frac{3}{5}$ and y-intercept of -2. Then we shade *above* the line, since the inequality is $y > \frac{3}{5}x - 2$.

You try Example 4.

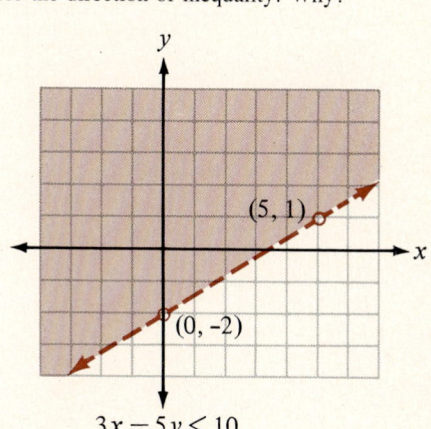

$3x - 5y < 10$

Example 4. Graph the inequality $y \leq 2 - 3x$.

Solution

$$y \leq 2 - 3x$$

$$y \leq -3x + 2$$

The corresponding equation is $y = -3x + 2$ so we draw a _____ line with slope of _____ and y-intercept of _____ . Then we shade _____ the line.

Check your work on page 623. ▶

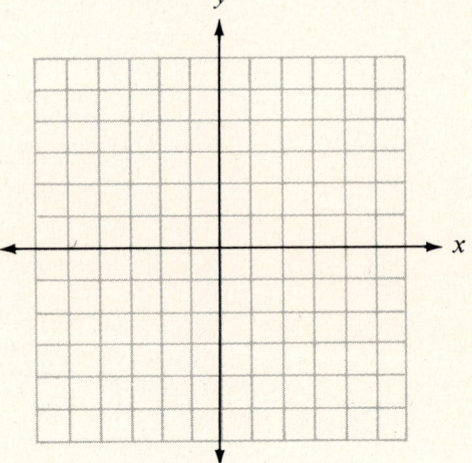

⫸ Trial Run

Graph the inequalities.

1. $y < -2$

2. $x \geq \dfrac{-1}{2}$

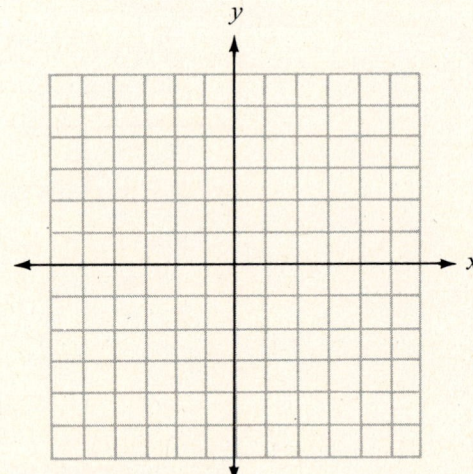

3. $y \geq 2x - 1$

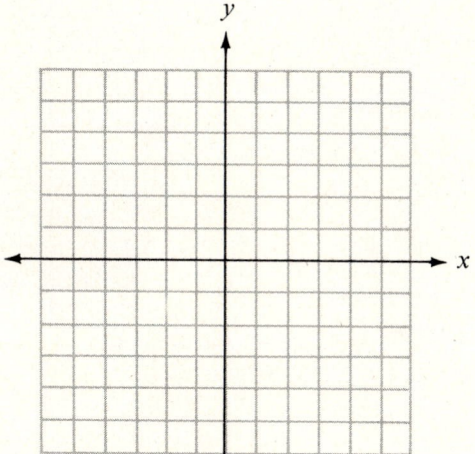

4. $2x - 3y > 6$

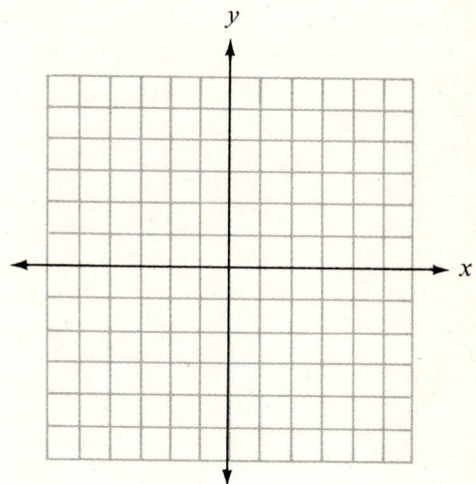

Answers are on page 623.

▶ Example You Completed

Example 4. Graph the inequality $y \leq 2 - 3x$.

Solution

$$y \leq 2 - 3x$$

$$y \leq -3x + 2$$

The corresponding equation is $y = -3x + 2$, so we draw a solid line with slope of -3 and y-intercept of 2. Then we shade below the line.

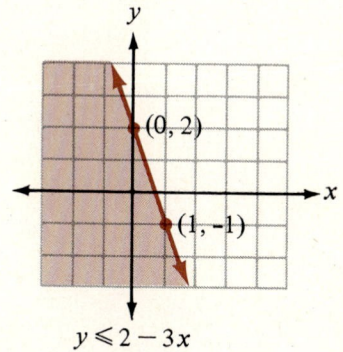

Answers to Trial Run

page 622   **1.**

2.

3.

4.

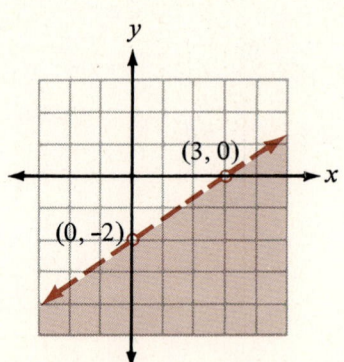

EXERCISE SET 8.4

Graph the inequalities.

1. $y > 5$

2. $y < -1$

3. $x \leq 2$

4. $x \geq -4$

5. $y \geq 3x - 2$

6. $y \geq -4x + 1$

7. $x + y < -2$
$y < -x - 2$

8. $x + y < 4$

9. $y \leq \dfrac{2}{3}x$

10. $y \leq \dfrac{3}{4}x$

11. $2x - y > 1$

12. $3x - y > -2$

13. $x > 3$

14. $x \geq -2$

15. $x + 2y > -4$

16. $x + 3y > -9$

17. $3x - 2y > 8$

18. $2x - 3y > 6$

19. $3x + 5y \geq 7$

20. $2x + 3y \geq 8$

☆ Stretching the Topics _____

Graph.

1. $7 - [5 - (4 - y)] > 2 - x$

2. $-5 \leq x < 8$

3. $|y| > 5$

Check your answers in the back of your book.

If you can complete **Checkup 8.4,** you are ready to go on to Section 8.5.

CHECKUP 8.4

Graph the inequalities.

1. $x \le 4$

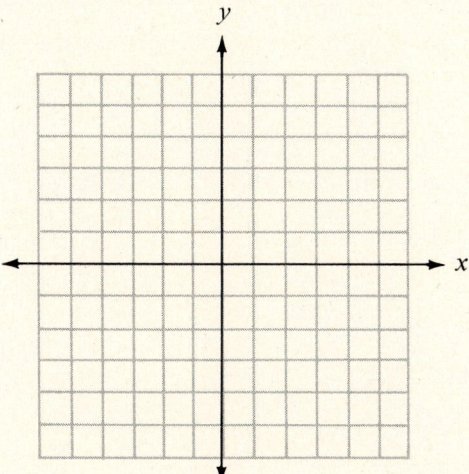

2. $x + 2y > -2$

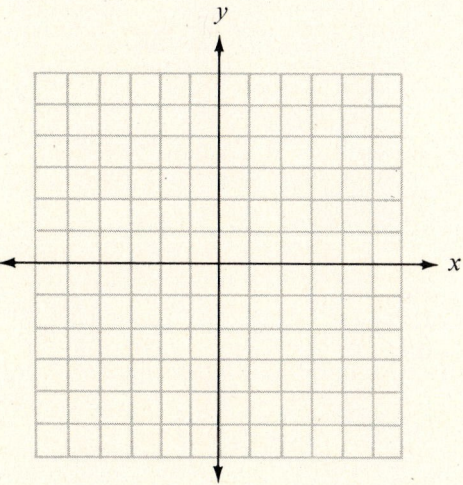

3. $2x - 3y > -3$

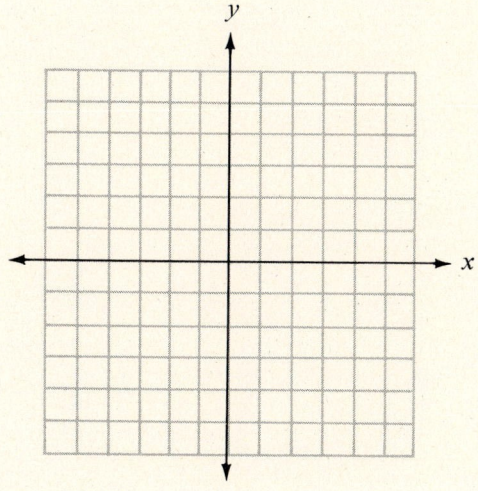

4. $y \le 4 - 2x$

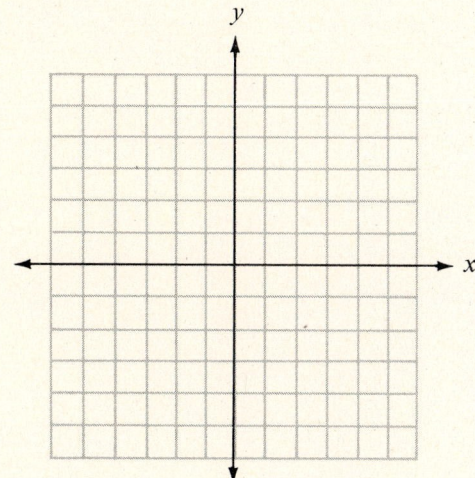

Check your answers in the back of your book.

If You Missed Problem:	You Should Review Example:
1	1
2	2
3	3
4	4

8.5 Working with Relations and Functions

Recognizing Functions

In this chapter we have been working with sets of ordered pairs and their graphs. A set of ordered pairs is sometimes called a **relation.** We have already noted that in each ordered pair (x, y), x is called the **independent variable** and y is called the **dependent variable.** The set of all permissible values for the independent variable is called the **domain** for the relation, and the set of all permissible values for the dependent variable is called the **range** for the relation.

Let's identify each of these parts for the relation described by

$$y = 2x + 1$$

Some ordered pairs in our relation might be

$$\left\{ (-2, -3), (0, 1), \left(\frac{1}{2}, 2\right), (3, 7), (4.3, 9.6), (\sqrt{3}, 2\sqrt{3} + 1), \ldots \right\}$$

The independent variable is x; the dependent variable is y. The domain and range are the set of all real numbers. Or, using notation discussed in Chapter 1, we may write

$$\text{Domain:} \quad x \in R$$

$$\text{Range:} \quad y \in R$$

We can use a graph to represent the ordered pairs in this relation in yet another way. Every ordered pair in the relation appears as a point on the line graphed here.

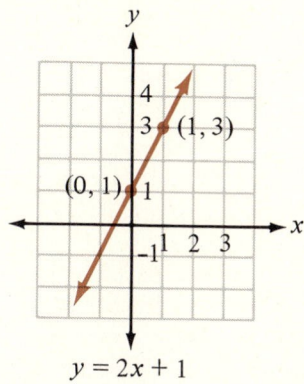

$$y = 2x + 1$$

In this relation, the equation $y = 2x + 1$ provides a rule for arriving at the ordered pairs. To write the relation correctly, we must write $\{(x, y) : y = 2x + 1\}$. However, we shall usually refer to such a set of ordered pairs by stating only its equation.

Every set of ordered pairs graphed in this chapter has been a relation. In some of the relations, each x in the domain has been paired with *one and only one* y in the range. When that occurs, we say that the relation is a **function.**

Relation: any set of ordered pairs.
> *Domain:* the set of all replacement values for the independent variable in a relation.
> *Range:* the set of all replacement values for the dependent variable in a relation.

Function: a set of ordered pairs in which each element in the domain is paired with one and only one element in the range.

Is the set of ordered pairs described by $y = 2x + 1$ a function? Certainly, because if we substitute any x-value into this equation, we arrive at one and only one y-value.

Example 1. Discuss the relation $y = 3$.

Solution: Some ordered pairs in this relation are

$$\left\{ (-2, 3), (0, 3), \left(\frac{1}{2}, 3\right), (\sqrt{2}, 3), \ldots \right\}$$

No matter what real number we substitute for x, the y-value is always 3.

Domain: $x \in R$

Range: $y = 3$

Is this a function? Is any x-value paired with more than one y-value? No. Therefore,

$y = 3$ is a function

and we can easily graph it.

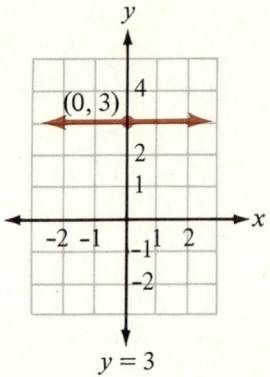

$y = 3$

Example 2. Discuss the relation $x = 2$.

Solution: Some ordered pairs in this relation are

$$\left\{ (2, -5), \left(2, \frac{-1}{3}\right), (2, 0), (2, \sqrt{6}), \ldots \right\}$$

No matter what real number we substitute for y, the x-value is always 2.

Domain: $x = 2$

Range: $y \in R$

Is this a function? The only x-value in the domain is 2, and it is paired with infinitely many different y-values. Therefore,

$x = 2$ is *not* a function

but we can graph it.

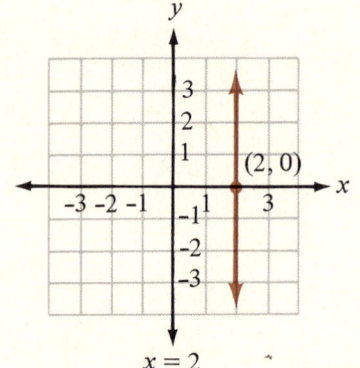

$x = 2$

Notice how the *graph* of a relation can help us decide whether that relation is a function. We may locate on the x-axis any x-value in the domain. Then we may see how many points on our graph correspond to that x-value. If any x-value corresponds to more than one point on the graph, that x is being paired with more than one y, and our relation is *not* a function. This technique is called the **vertical-line test.**

Vertical-Line Test. If a vertical line drawn through any x-value in the domain of a relation crosses the graph of the relation in more than one point, the relation is *not* a function.

Example 3. Use the vertical-line test to decide whether each graph represents a function. Give the domain and range for each.

Function?   Yes
Domain: $x \in R$
Range: $y \in R$

Function?   Yes
Domain: $x \in R$
Range: $y \in R, y \geq 0$

Function?   No
Domain: $x \in R, x \geq 0$
Range: $y \in R$

Function?   No
Domain: $x = -2$
Range: $y \in R$

Function?   Yes
Domain: $x \in R$
Range: $y \in R, y \geq 1$

Function?   No
Domain: $x \in R$
Range: $y \in R$

⫸ **Trial Run**

For each relation, give the domain and range and tell whether it is a function.

_____ **1.** $\{(-1, 3), (1, 5), (0, 0), (1, 4), (-1, 2)\}$
         Domain: _____   Range: _____   Function? _____

_____ **2.** $\{(-2, 4), (-1, 0), (0, 4)\}$
         Domain: _____   Range: _____   Function? _____

_____ **3.** $\{(x, y) : y = x - 1\}$
         Domain: _____   Range: _____   Function? _____

Use the vertical line test to tell which graphs represent functions. Also give the domain and range.

_____ 4.

Do: _____
Ra: _____

_____ 5.

Do: _____
Ra: _____

_____ 6.

Do: _____
Ra: _____

Answers are on page 635.

Using Function Notation

When a set of ordered pairs represents a function, we sometimes say that *the dependent variable is a function of the independent variable*. To state the fact that *y* is a function of *x*, mathematicians invented **function notation.**

> **Function Notation**
>
> $y = f(x)$ is read
>
> "y is a function of x"
>
> or
>
> "y equals f of x"

As a matter of fact, any letter can be used to name a function. When dealing with several functions of x in the same discussion, we must use different letters to name the different functions.

Function notation can be used whenever we wish to describe a function. For example,

$$y = 2x + 1 \qquad \text{can be written} \qquad f(x) = 2x + 1$$

$$y = 3 \qquad \text{can be written} \qquad g(x) = 3$$

$$y = x^2 - 2 \qquad \text{can be written} \qquad F(x) = x^2 - 2$$

Function notation provides us with a quick way of finding ordered pairs. For example, if $f(x) = 7 - 3x$, then to find $f(2)$ we must evaluate $7 - 3x$ when x is 2.

$f(x) = 7 - 3x$	(x, y)
$f(2) = 7 - 3(2) = 1$	$(2, 1)$
$f(0) = 7 - 3(0) = 7$	$(0, 7)$
$f(\sqrt{2}) = 7 - 3\sqrt{2}$	$(\sqrt{2}, 7 - 3\sqrt{2})$

Function notation is also useful in finding function values when the independent variable x is replaced by other *variable* expressions. For example, if

$$F(x) = 5x - 2$$

$F(3)$	$F(a)$	$F(2x)$	$F(x + 3)$	$F(x + h)$
$= 5(3) - 2$	$= 5a - 2$	$= 5(2x) - 2$	$= 5(x + 3) - 2$	$= 5(x + h) - 2$
$= 13$		$= 10x - 2$	$= 5x + 13$	$= 5x + 5h - 2$

In each case, the expression replacing x on the left was substituted for x on the right.

You complete Example 4.

Example 4. If $f(x) = x^2 - 3x + 1$, find $f(0)$, $f(-2)$, $f(3a)$, and $f(x + h)$.

Solution

$$f(x) = x^2 - 3x + 1$$

$f(0)$	$f(-2)$	$f(3a)$	$f(x + h)$
$= (0)^2 - 3(0) + 1$	$= (-2)^2 - 3(-2) + 1$	$= (3a)^2 - 3(3a) + 1$	$= (\underline{\quad})^2 - 3(\underline{\quad}) + 1$
$= 0 - 0 + 1$	$= \underline{\quad} + \underline{\quad} + 1$	$= \underline{\quad} - \underline{\quad} + 1$	$= \underline{\qquad} + 1$
$= \underline{\quad}$	$= \underline{\quad}$	$= \underline{\quad}$	

Check your table on page 635. ▶

Restricting Domains of Functions

When dealing with fractions and square roots, we learned to make *restrictions* on the domain of the variable to avoid expressions that were undefined in the set of real numbers. Recall that

> The denominator of a fraction *cannot* equal zero.
> The even root of a negative number is *not* real.

We can use these same ideas to restrict domains for functions. Try Example 5.

Example 5. Give the domain for $f(x) = \dfrac{1}{x}$.

Solution: We know that $\dfrac{1}{x}$ is *not* defined when $x = \underline{\quad}$, so

Domain: $x \in R$, $x \neq \underline{\quad}$

Check your work on page 635. ▶

Example 6. Give the domain for
$$h(x) = \frac{x}{x^2 + 2x - 3}.$$

Solution: We rewrite the denominator in factored form.

$$h(x) = \frac{x}{x^2 + 2x - 3}$$

$$= \frac{x}{(x + 3)(x - 1)}$$

The denominator cannot be zero, so

Domain: $x \in R$, $x \neq -3$, $x \neq 1$

Example 7. Give the domain for
$g(x) = \sqrt{x}$.

Solution: We know that $\sqrt{x}$ is *not* a real number when x is *negative*. So

Domain: $x \in R$, $x \geq 0$

Example 8. Give the domain for
$F(x) = \sqrt{2 - x}$.

Solution: We know that $2 - x$ cannot be negative.

$$2 - x \geq 0$$

$$-x \geq -2$$

$$x \leq 2$$

Domain: $x \in R$, $x \leq 2$

⫸ Trial Run

If $f(x) = 3x - 2$, *find*

———— **1.** $f(4)$ ———— **2.** $f(x + h)$

If $g(x) = x^2 - 2x + 3$, *find*

———— **3.** $g(0)$ ———— **4.** $g(-1)$

Give the domain for each function.

———— **5.** $f(x) = \dfrac{1}{x - 1}$ ———— **6.** $h(x) = \dfrac{x}{x^2 - x - 12}$

$x \le 3$ Test

———— **7.** $F(x) = \sqrt{3 - x}$ ———— **8.** $g(x) = \dfrac{1}{\sqrt{x}}$

Answers are on page 635.

▶ Examples You Completed

Example 4. If $f(x) = x^2 - 3x + 1$, find $f(0)$, $f(-2)$, $f(3a)$, and $f(x + h)$.

Solution

$$f(x) = x^2 - 3x + 1$$

$f(0)$	$f(-2)$	$f(3a)$	$f(x + h)$
$= (0)^2 - 3(0) + 1$	$= (-2)^2 - 3(-2) + 1$	$= (3a)^2 - 3(3a) + 1$	$= (x + h)^2 - 3(x + h) + 1$
$= 0 - 0 + 1$	$= 4 + 6 + 1$	$= 9a^2 - 9a + 1$	$= x^2 + 2xh + h^2 - 3x - 3h + 1$
$= 1$	$= 11$		

Example 5. Give the domain for $f(x) = \dfrac{1}{x}$.

Solution: We know that $\dfrac{1}{x}$ is *not* defined when $x = 0$, so

Domain: $x \in R,\ x \neq 0$

Answers to Trial Runs

page 631   **1.** Domain: $\{-1, 0, 1\}$; range: $\{0, 2, 3, 4, 5\}$; not a function
2. Domain: $\{-2, -1, 0\}$; range: $\{0, 4\}$; function   **3.** Domain: $x \in R$; range: $y \in R$; function
4. Function; Do: $x \in R$; Ra: $y \in R$, $y \le 5$
5. Not function; Do: $x \in R$, $0 \le x \le 3$; Ra: $y \in R$, $-3 \le y \le 3$
6. Not function; Do: $x \in R$; Ra: $y \in R$

page 635   **1.** 10   **2.** $3x + 3h - 2$ **3.** 3   **4.** 6   **5.** $x \in R$, $x \neq 1$ **6.** $x \in R$, $x \neq 4$, $x \neq -3$
7. $x \in R$, $x \le 3$ **8.** $x \in R$, $x > 0$

EXERCISE SET 8.5

For each of the following relations, give the domain and the range and state which are functions.

_____ **1.** $\{(1, 0), (2, 3), (3, 5), (2, -3)\}$ _____ **2.** $\{(-5, 0), (-1, 4), (0, 6), (-1, -4)\}$

_____ **3.** $\{(x, y) : y = 2x\}$ _____ **4.** $\{(x, y) : y = x + 2\}$

_____ **5.** $\{(-6, 3), (-2, 3), (0, 3), (3, 3)\}$ _____ **6.** $\{(-1, -4), (0, -4), (2, -4), (5, 4)\}$

_____ **7.** $\{(x, y) : x = 3\}$ _____ **8.** $\{(x, y) : x = 7\}$

_____ **9.** $\{(1, 3), (-1, 3), (2, 5), (-2, 5)\}$

_____ **10.** $\{(2, -3), (-2, -3), (4, -6), (-4, -6)\}$

Use the vertical-line test to tell which of the following graphs represent functions.

_____ **11.**

_____ **12.**

_____ **13.**

_____ **14.**

———— 15.

———— 16.

———— 17.

———— 18.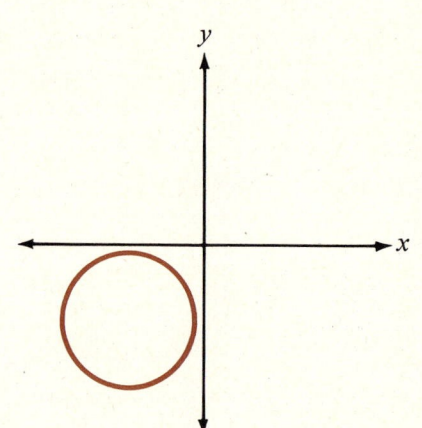

If $f(x) = 3x + 1$ and $g(x) = x^2 + 2x - 8$, find each of the following.

———— 19. $f(2)$

———— 20. $f(4)$

———— 21. $g(-1)$

———— 22. $g(-3)$

———— 23. $f(x + h)$

———— 24. $f(x - h)$

———— 25. $f(-3) + g(0)$

———— 26. $f(-4) - g(0)$

If $f(x) = 2x - 3$ and $g(x) = x^2 - x - 1$, find each of the following.

———— 27. $f(x + 1)$

———— 28. $f(x + 2)$

———— 29. $g(a)$

———— 30. $g(c)$

———— 31. $g(1) - f(2)$

———— 32. $g(3) - f(0)$

———— 33. $g(a + 1)$

———— 34. $g(a + 3)$

Give the domain for each of the following.

_____ **35.** $f(x) = \dfrac{1}{x + 4}$

_____ **36.** $f(x) = \dfrac{1}{x - 6}$

_____ **37.** $f(x) = \dfrac{3}{2x - 1}$

_____ **38.** $f(x) = \dfrac{5}{3x - 2}$

_____ **39.** $g(2x) = \sqrt{x + 1}$

_____ **40.** $g(x) = \sqrt{x + 3}$

_____ **41.** $f(x) = \dfrac{x - 1}{x - 3}$

_____ **42.** $f(x) = \dfrac{x - 4}{x - 5}$

_____ **43.** $h(x) = \sqrt{5 - x}$

_____ **44.** $h(x) = \sqrt{1 - x}$

_____ **45.** $G(x) = \dfrac{x}{x^2 - 16}$

_____ **46.** $G(x) = \dfrac{2x}{x^2 - 25}$

_____ **47.** $f(x) = \sqrt{2x - 6}$

_____ **48.** $f(x) = \sqrt{3x - 9}$

_____ **49.** $F(x) = \dfrac{-1}{2x^2 - 9x + 7}$

_____ **50.** $F(x) = \dfrac{-5}{3x^2 + 14x - 5}$

☆ Stretching the Topics _____

_____ **1.** If $f(x) = 2x^2 - 3x + 4$, find $\dfrac{f(a + h) - f(a)}{h}$, where $h \neq 0$.

_____ **2.** If $g(x) = \dfrac{x^2}{x^2 + 1}$, find $g(\sqrt{a})$.

_____ **3.** Give the domain for $\dfrac{\sqrt{x}}{2x^3 - 11x^2 + 12x}$.

Check your answers in the back of your book.

If you can complete **Checkup 8.5,** you are ready to go on to Section 8.6.

✓ CHECKUP 8.5

For each of the following relations, give the domain and the range and state which are functions.

_____ **1.** $\{(-3, 3), (-2, 2), (-1, 1), (0, 0), (1, 1), (2, 2), (3, 3)\}$

_____ **2.** $\{(x, y) : x = 5\}$

Use the vertical-line test to tell which of the following graphs represent functions.

_____ **3.**

_____ **4.**

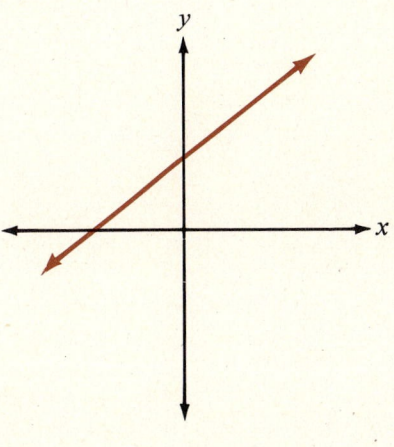

If f(x) = 2x + 1 and g(x) = x² + x − 1, find each of the following.

$-5 = -10 + 1$

_____ $5 = 9$ **5.** $f(3)$

_____ **6.** $g(-2)$

_____ **7.** $f(-5) + g(0)$

Give the domain of each of the following.

_____ **8.** $f(x) = \dfrac{x + 1}{2x - 3}$

_____ **9.** $g(x) = \sqrt{x - 4}$

_____ **10.** $f(x) = \dfrac{x}{x^2 - 3x - 28}$

Check your answers in the back of your book.

If You Missed Problems:	You Should Review Examples:
1	1
2	2
3, 4	3
5–7	4
8–10	5–8

8.6 Switching from Word Statements to Functions

Many relationships between two variables can be expressed as functions or relations.

Working with Variation

If a dealer sells one kind of calculator at $29 each, you should agree that the dealer's revenue from the calculators depends on the number of calculators sold. In other words, the revenue *is a function of* the number of calculators sold. If we let

$$x = \text{number of calculators sold}$$

$$r = \text{revenue from sale of calculators}$$

then we can express the revenue function as

$$r = 29x$$

Of course, for this function to make sense, the domain must be the set of whole numbers. Why?

In the revenue function $r = 29x$, you should agree that the revenue will *increase* as the number of calculators sold *increases*. In such a function, the dependent variable (r) is said to **vary directly** with the independent variable (x). This function is an example of **direct variation.**

Direct Variation. If y varies directly as x, then

$$y = k \cdot x$$

and k is called the constant of variation.

Example 1. Write a function to describe the fact that the number of freshmen at a university varies directly with the total enrollment.

Solution: Let F = number of freshmen

t = total enrollment

then $F = kt$

Example 2. If the number of freshmen is 2500 when total enrollment is 9000, find the constant of variation k. Then find the freshman enrollment when the total enrollment is 13,500.

Solution: If $F = 2500$ when $t = 9000$, we can use substitution to find k.

$$F = kt$$

$$2500 = k(9000) \qquad \text{Substitute.}$$

$$\frac{2500}{9000} = k \qquad \text{Divide by 9000 to isolate } k.$$

$$k = \frac{5}{18} \qquad \text{Reduce the fraction.}$$

Now we know that $F = \dfrac{5}{18} t$ and we can substitute to find F when $t = 13,500$.

$$F = \frac{5}{18} t \qquad \text{Substitute } \frac{5}{18} \text{ for } k.$$

$$= \frac{5}{18} (13,500) \qquad \text{Substitute } 13,500 \text{ for } t.$$

$$= 3750 \qquad \text{Multiply.}$$

When total enrollment is 13,500, freshman enrollment will be 3750.

Now we consider the problem stated at the beginning of this chapter.

Example 3. The distance required to stop a moving car varies directly as the square of the speed of the car. If the distance required to stop a car traveling 20 miles per hour is 30 feet, what distance would be required to stop the same car traveling 40 miles per hour?

Solution: Let d = distance (ft)

s = speed of car (mph)

then $d = k \cdot s^2$

To find k, we substitute 30 feet for d and 20 mph for s.

$$30 = k(20)^2$$

$$30 = k(400)$$

$$\frac{30}{400} = k$$

$$k = \frac{3}{40}$$

Now we know that $d = \frac{3}{40} s^2$ and we can find d when $s = 40$.

$$d = \frac{3}{40} s^2$$

$$= \frac{3}{40} (40)^2$$

$$= 120 \text{ ft}$$

If a car is traveling 40 mph, then 120 feet is the distance required to stop.

If a rectangle with length l and width w has an area of 148 square feet, then we know that

$$l \cdot w = 148$$

We can express the length of the rectangle as a function of the width by writing

$$l = \frac{148}{w}$$

This is a function whose domain is the set of positive real numbers. You should see that the length must *decrease* as the width *increases*. In such a function, the dependent variable (l) is said to **vary inversely** with the independent variable (w). This function is an example of **inverse variation.**

Inverse Variation. If y varies inversely as x, then

$$y = \frac{k}{x}$$

and k is called the constant of variation.

Try Example 4.

Example 4. Write a function to describe the fact that the time (t) for a person to walk a fixed distance varies inversely as the rate (r) at which the person walks.

Solution

$$t = \underline{\quad\quad}$$

Check your function on page 648. ▶

Example 5. If, in Example 4, we know that $t = 3$ hours when $r = 2$ mph, find the constant of variation. Then find t when $r = 1.5$ mph and when $r = \frac{2}{3}$ mph.

Solution

$$t = \frac{k}{r}$$

$$3 = \frac{k}{2} \qquad \text{Substitute 3 for } t \text{ and 2 for } r.$$

$$6 = k \qquad \text{Solve for } k.$$

Therefore $t = \frac{6}{r}$.

When $r = 1.5$ mph

$$t = \frac{6}{1.5}$$

$$= 4 \text{ hr}$$

When $r = \frac{2}{3}$ mph

$$t = \frac{6}{\frac{2}{3}}$$

$$= 6 \cdot \frac{3}{2}$$

$$= 9 \text{ hr}$$

If one variable varies *directly* with the *product* of two other variables, we call such a relationship **joint variation**.

Joint Variation. If y varies jointly with x and z, then

$$y = k \cdot x \cdot z$$

where k is the constant of variation.

Example 6. Write a function that states that the volume (V) of a cone varies jointly with the height (h) and the square of the radius (r) of the base.

Solution

$$V = k \cdot h \cdot r^2$$

You try Example 7.

Example 7. Write a function stating that the simple interest (I) from a certain investment varies jointly with the interest rate (r) and the length of time (t).

Solution

$$I = \underline{\hspace{2cm}}$$

Check your function on page 648. ▶

When a variable varies directly with some variable and inversely with another variable, such a relationship is called **mixed variation.**

Mixed Variation. If y varies directly as x and inversely as z, then

$$y = \frac{kx}{z}$$

where k is the constant of variation.

Example 8. When a customer agrees to pay for a purchase in n installments at an advertised rate of interest r, then the true rate of interest (i) varies jointly as n and r, and inversely as the sum $n + 1$. Write a function to describe this relationship.

Solution

$$i = \frac{knr}{n + 1}$$

Example 9. A certain variable V varies directly with b and inversely with a. Write the function describing this relationship. Then, if $V = -3$ when $b = 12$ and $a = 2$, find k. Finally, find V when $b = 3$ and $a = -1$.

Solution

$$V = \frac{k \cdot b}{a}$$

We know that $V = -3$ when $b = 12$ and $a = 2$. We can substitute to find k.

$$-3 = \frac{k(12)}{2} \qquad \text{Substitute } -3 \text{ for } V, 12 \text{ for } b, \text{ and } 2 \text{ for } a.$$

$$-3 = k \cdot 6 \qquad \text{Simplify.}$$

$$\frac{-3}{6} = k \qquad \text{Divide both sides by 6 to isolate } k.$$

$$k = \frac{-1}{2} \qquad \text{Simplify the fraction.}$$

and the function can be written $V = \dfrac{-1}{2} \cdot \dfrac{b}{a}$. Now find V when $b = 3$ and $a = -1$.

$$V = \frac{-1}{2} \cdot \frac{3}{-1}$$

$$= \frac{3}{2}$$

⟹ Trial Run

Write the function that describes the variation.

_____ **1.** The volume (V) of a sphere varies directly as the cube of its radius (r).

_____ **2.** The average fixed costs ($\bar{c}$) for a company vary inversely as the number of units (n) produced.

_____ **3.** The area (A) of a trapezoid varies jointly as its height (h) and the sum of its bases (a and b).

_____ **4.** The strength (s) of a string varies directly as its thickness (t) and inversely as its length (l).

Answers are on page 649.

Working with Other Functions and Relations

Let's consider some more functions and relations involving two variables that might arise in everyday situations.

Example 10. A car rental agency charges its customers $20 per day plus 32 cents per mile driven. Write an equation to describe one day's total charge (C) as a function of the number of miles driven.

Solution

We Think	*We Write*
Total daily charge is the sum of the flat charge and the mileage charge.	$C = 20 +$ mileage charge
The mileage charge depends on the miles driven.	Let $m =$ number of miles driven
The mileage rate is 32 cents per mile.	$0.32m =$ mileage charge
We write the equation for the total daily charge.	$C = 20 + 0.32m$

You try Example 11.

Example 11. If the length of a rectangular garden is 10 feet more than its width, write an equation for the perimeter (P) of the garden as a function of its width.

Solution

We Think	*We Write*
The perimeter of a rectangle is the sum of 2 lengths and 2 widths.	$P = 2$ lengths $+ 2$ widths
We wish the width to be the independent variable.	Let $w =$ width
The garden is 10 feet longer than it is wide.	_____ + _____ = length
We can write an equation for perimeter.	$P = 2$ lengths $+ 2$ widths $= 2($_____ $+$ _____$) + 2($_____$)$

The perimeter can be expressed as a function of the width by $P = $ _____ .

Check your work on page 649. ▶

Example 12. If a company sells chairs for $90 each and tables for $130 each, write a relation that states that the company wishes its total revenue from chairs and tables to be at least $5850 per week.

Solution

We Think	*We Write*
The company's total revenue comes from selling chairs and tables.	Total revenue = chair revenue + table revenue
Chair revenue depends on how many chairs are sold. Table revenue depends on how many tables are sold.	Let $x =$ number of chairs sold $y =$ number of tables sold
Revenue is found by multiplying price per unit times number of units sold.	$90x =$ chair revenue $130y =$ table revenue
The company wants total revenue to be at least $5850.	$90x + 130y \geq 5850$

The inequality $90x + 130y \geq 5850$ describes the fact that total revenue must be at least $5850. This is a relation whose domain and range are the set of whole numbers.

▶ Examples You Completed _____

Example 3

Solution

$$t = \frac{k}{r}$$

Example 7

Solution

$$I = k \cdot r \cdot t$$

Example 11

Solution

We Write

P = 2 lengths + 2 widths

Let w = width

$w + 10$ = length

P = 2 lengths + 2 widths

$\quad = 2(w + 10) + 2(w)$

$\quad = 2w + 20 + 2w$

$\quad = 4w + 20$

The perimeter can be expressed as a function of the width by $P = 4w + 20$.

Answers to Trial Run

page 647   **1.** $V = kr^3$ **2.** $\bar{c} = \dfrac{k}{n}$ **3.** $A = kh(a + b)$ **4.** $s = \dfrac{kt}{l}$

EXERCISE SET 8.6

_____ **1.** Write an equation of variation and find k if y varies directly as x, and $y = 48$ when $x = 4$.

_____ **2.** Write an equation of variation and find k if y varies directly as x, and $y = 35$ when $x = 7$.

_____ **3.** Write an equation of variation and find k if y varies directly as x, and $y = 16$ when $x = 80$.

_____ **4.** Write an equation of variation and find k if y varies directly as x, and $y = 42$ when $x = 63$.

_____ **5.** Write an equation of variation and find k if y varies inversely as x, and $y = 6$ when $x = 10$.

_____ **6.** Write an equation of variation and find k if y varies inversely as x, and $y = 7$ when $x = 5$.

_____ **7.** Write an equation of variation and find k if y varies inversely as x, and $y = 0.3$ when $x = 15$.

_____ **8.** Write an equation of variation and find k if y varies inversely as x, and $y = 4.5$ when $x = 0.9$.

_____ **9.** Write an equation of variation and find k if z varies jointly as x and y, and $z = 6$ when $x = 4$ and $y = 2$.

_____ **10.** Write an equation of variation and find k if w varies jointly as u and v, and $w = 10$ when $u = 5$ and $v = 3$.

_____ **11.** Write an equation of variation and find k if y varies directly as x and inversely as the square of z, and $y = 12$ when $x = 9$ and $z = 2$.

_____ **12.** Write an equation of variation and find k if y varies directly as the square of z and inversely as x, and $y = 50$ when $x = 2$ and $z = 5$.

_____ **13.** The number of rubber bands produced by a machine varies directly as the amount of time the machine is operating. If 1200 rubber bands are produced in 3 hours, how many can be produced in 8 hours?

_____ **14.** The cost of producing pencils varies directly with the number of pencils produced. If the cost of producing 2000 pencils is $300, how much will it cost to produce 5000 pencils?

_____ **15.** The time required to empty a tank varies inversely as the rate of the pump. If the tank can be emptied in 25 minutes by a pump that is pumping at the rate of 400 gallons per minute, how long will it take to empty the tank with a pump that will pump at the rate of 500 gallons per minute?

_____ **16.** The number of hours required to clean the arena after a ballgame varies inversely as the number of students hired to do the cleaning. If it takes 9 hours for 5 students to do the job, how many hours will it take for 15 students to do the job?

_____ **17.** The length a spring stretches varies directly as the force applied. If the length is 15 inches when the force is 9 pounds, find the force needed to stretch the spring to 20 inches.

_____ **18.** The water pressure at a point under water varies directly as the height of the water above the point. If the pressure is 312 pounds per square foot when the height is 10 feet, find the height of the water if the pressure being exerted at a point is 1248 pounds per square foot.

_____ **19.** The average fixed cost for a manufacturing firm varies inversely as the number of items in daily output. If the average fixed cost is $35 when the daily output is 120 items, how many items must be produced to reduce the average fixed cost to $21?

_____ **20.** The volume of a gas varies inversely as the pressure of the gas on its container. If the volume is 96 cubic feet when the pressure is 25 pounds, what pressure is needed for the volume to be 100 cubic feet?

_____ **21.** The distance a body falls varies directly as the square of the time it has been falling. If an object falls 144 feet in 3 seconds, how long will it take it to fall 400 feet?

_____ **22.** The brightness of light at a given distance from its source varies inversely as the square of this distance. If the brightness of a certain light is 50 (candlepower) at a distance of 20 feet from its source, find the brightness when the distance is 12 feet.

_____ **23.** The weight of a round metal rod varies jointly as the length and the square of the diameter. If the weight of a rod 1 inch in diameter and 8 inches long is 1.5 pounds, find the weight of a rod 2.5 inches in diameter and 16 feet long.

_____ **24.** The weight of a circular disk varies jointly as its thickness and the square of its radius. A disk that is 2 inches thick and has a radius of 5 inches weighs 3.1 pounds. What is the weight of a disk of the same material if it is 3.2 inches thick and has a radius of 8.6 inches?

_____ **25.** A car rental agency charges a fee of $18 per day, plus 30 cents per mile. Write an equation to describe one day's total charge as a function of the number of miles driven.

_____ **26.** A local restaurant charges a fee of $75 plus $7 a plate for serving a banquet. Write an equation to describe the cost of a banquet as a function of how many people attend.

_____ **27.** Mr. Gillenwater sells vacuum cleaners. He earns $100 a week plus $75 commission for each vacuum cleaner he sells. Write an equation to express that his weekly salary is a function of the number of vacuum cleaners he sells.

_____ **28.** Ms. Jolly has a weekly salary of $340 and she earns $10.50 for each hour she works overtime. Write an equation to express that her take-home pay each week is a function of the number of hours she works overtime.

_____ **29.** For traveling on the toll road, Pat needs four times as many quarters as dimes. Write an equation to describe the fact that the total amount needed is a function of the number of dimes.

_____ **30.** Mrs. Abell's recipe for punch calls for twice as many cans of frozen orange juice as frozen lemonade. If a can of orange juice costs 53 cents and a can of lemonade costs 62 cents, write an equation to describe the fact that the total cost is a function of the number of cans of lemonade.

_____ **31.** If the length of a rectangle is 12 feet longer than the width, write an equation that expresses the perimeter as a function of the width.

_____ **32.** If the width of a rectangle is 3 feet less than the length, write an equation that expresses the perimeter as a function of the length.

_____ **33.** If the width of a rectangle is $\frac{1}{3}$ of the length, write an equation that expresses the area as a function of the length.

_____ **34.** If the length of a rectangle is five times the width, write an equation that expresses the area as a function of the width.

_____ **35.** The base of a triangle is 4 more than the altitude. Write an equation that expresses the area as a function of the altitude.

_____ **36.** The altitude of a triangle is 3 less than the base. Write an equation that expresses the area as a function of the base.

_____ **37.** Beth has decided that each day she will walk twice as long as she jogs. If she walks at the rate of 4 miles per hour and jogs at 6 miles per hour, write an equation that expresses the total distance as a function of the time she jogs.

_____ **38.** Wayne can jog 2 miles an hour faster than he can walk. If he decides to do each for a half-hour each day, write an equation that expresses the total distance as a function of the rate he can walk.

_____ **39.** If Simon has lost 5 pounds, 3 pounds, and 4 pounds, during the first three weeks of his diet, write an equation that expresses his average weight loss as a function of the number of pounds he loses during the fourth week.

_____ **40.** If Amanda scored 85%, 72%, and 90% on the first three tests during the semester, write an equation that expresses her test average for the semester as a function of the score she makes on the fourth test.

_____ **41.** Venton buys a bedroom suite for $150 down and $55 a month. Write an equation that expresses the total cost as a function of the number of months he must make payments.

_____ **42.** A consumer service advertises a stereo system for $33 down and only 10 monthly payments. Write an equation that expresses the total cost as a function of the amount of each monthly payment.

_____ **43.** If the Fountain Players sells an adult's ticket for $5.00 and a child's ticket for $3.00, write a relation to state that the theatre group wishes its total receipts from one night's performance to be at least $2000.

———— **44.** Mr. Shouse invested a certain amount of money at 8.5% and another amount at 9.75%. Write a relation that states Mr. Shouse wishes his income from the interest to be at least $1500.

☆ Stretching the Topics ———————————————

———— **1.** If y varies directly as x^2, what is the effect on y when x is multiplied by 3? What is the effect on x when y is multiplied by 4?

———— **2.** The weight that can be safely supported by a beam with rectangular ends varies jointly as the width and the square of the length of the rectangle and inversely as the length of the beam. If a 4-inch by 6-inch beam that is 15 feet long can support a 1200-pound load, what weight can be supported by a 2-inch by 5-inch beam that is 20 feet long?

———— **3.** The profit for manufacturing a certain item is given by $P(x) = -500 + 0.5x + x^2$, where x is the number of items manufactured. Find $P(20)$ and interpret your answer.

Check your answers in the back of your book.

If you can complete **Checkup 8.6,** you are ready to go on to the Review Exercises.

CHECKUP 8.6

_____ 1. The amount of stretching of a spiral spring varies directly as the force applied. If 6 pounds of force stretched the spring $\frac{2}{3}$ inch, how many inches will 18 pounds of force stretch it?

_____ 2. The rate in traveling a certain distance varies inversely as the time. If a car whose rate is 40 miles per hour travels a certain distance in 7 hours, how long will it take a car traveling 50 miles per hour to travel the same distance?

_____ 3. Write an equation of variation and find k if w varies jointly as x and the square of y, and $w = 12$ when $x = 3$ and $y = 2$.

_____ 4. Exhibitors at the county fair must pay a fee of $6.00 plus $1.75 for each item exhibited. Write an equation to describe an exhibitor's fee as a function of the number of items exhibited.

_____ 5. The length of a rectangle is 5 feet less than five times its width. Write an equation that expresses the perimeter as a function of the width.

Check your answers in the back of your book.

If You Missed Problems:	You Should Review Examples:
1	1, 2
2	4
3	6
4	10
5	11

Summary

In this chapter we discovered that solutions for equations and inequalities containing two variables are **ordered pairs** of real numbers. We called the set of replacements for the independent variable the **domain** and the set of replacements for the dependent variable the **range.** Then we learned to distinguish between a **relation** and a **function.**

Relation: any set of ordered pairs.
Function: a set of ordered pairs in which each element of the domain is paired with *one and only one* element of the range.

Using the Cartesian coordinate plane, we learned to graph ordered pairs of real numbers as points and to find certain information about two points (x_1, y_1) and (x_2, y_2) in the plane.

To Find:	Use the Formula:
Distance between the points	$d = \sqrt{(x_2 - x_1)^2 + (y_2 - y_1)^2}$
Midpoint of the line segment joining the points	$(x_M, y_M) = \left(\dfrac{x_1 + x_2}{2}, \dfrac{y_1 + y_2}{2} \right)$
Slope of the line joining the points	$m = \dfrac{y_2 - y_1}{x_2 - x_1}$

We learned that the graph of a **linear equation** of the form $Ax + By = C$ (where A and B are not both zero) is always a straight line and we developed several techniques for graphing lines.

Method	Procedure
Arbitrary points	Choose at least three x-values and find the corresponding y-values. Plot the three points and draw the line.
Intercepts	Let $y = 0$ to find the x-intercept. Let $x = 0$ to find the y-intercept. Plot the two intercept points and draw the line.
Slope-intercept	Rewrite the equation in the form $y = mx + b$. Then use the y-intercept b, and the slope m, to graph the line.
Inspection	For $y = c$, a constant, the graph is a horizontal line crossing the y-axis at $(0, c)$.
	For $x = k$, a constant, the graph is a vertical line crossing the x-axis at $(k, 0)$.

After learning to graph straight lines from given equations, we discovered how to find the equation for a line when given certain information about that line.

Given	Equation to Use	Name
Slope: m y-intercept: b	$y = mx + b$	Slope-intercept equation
Slope: m Point: (x_1, y_1)	$y - y_1 = m(x - x_1)$	Point-slope equation
Two points: (x_1, y_1) and (x_2, y_2)	$m = \dfrac{y_2 - y_1}{x_2 - x_1}$	Slope definition
	$y - y_1 = m(x - x_1)$	Point-slope equation
Line is horizontal	$y = c$	Constant function
Line is vertical	$x = k$	Constant relation

To graph **linear inequalities,** we agreed to *isolate* the dependent variable y, graph the corresponding equation as a solid or dashed line, and then shade the appropriate region.

> If $y < mx + b$, the line is *dashed* and we shade *below* the line.
> If $y \leq mx + b$, the line is *solid* and we shade *below* the line.
> If $y > mx + b$, the line is *dashed* and we shade *above* the line.
> If $y \geq mx + b$, the line is *solid* and we shade *above* the line.

Finally, we discussed function notation and learned to express verbal relationships between variables using functions and relations containing two variables. In particular, we learned to write equations describing *variation*.

Type of Variation	Statement of Variation	Equation
Direct	y varies directly as x	$y = kx$
Inverse	y varies inversely as x	$y = \dfrac{k}{x}$
Joint	y varies directly as x and z	$y = kxz$
Mixed	y varies directly as x and inversely as z	$y = \dfrac{kx}{z}$

☐ Speaking the Language of Algebra

Complete each sentence with the appropriate word or phrase.

1. Solutions for equations and inequalities containing two variables are _____ _____ .

2. In the ordered pair (x, y) we call x the _____ variable and y the _____ variable.

3. The set of permissible replacement values for the independent variable is called the _____ and the set of replacement values for the dependent variable is called the _____ .

4. The graph of $y = mx + b$ is a _____ _____ , for which m is the _____ and b is the _____ .

5. The graph of $y = c$, where c is a constant, is a _____ _____ . The graph of $x = k$, where k is a constant, is a _____ _____ .

6. To graph the inequality $y > 3x - 1$, we first graph $y = 3x - 1$ as a _____ line and then we shade _____ that line.

7. To find the equation of a line through two points, we must first find the _____ of the line.

8. For a linear equation, $Ax + By = C$ is called _____ form and $y = mx + b$ is called _____ form.

9. If $f(x) = 5x + 2$, then $f(3) = $ _____ .

10. The equation $y = kx$ describes _____ variation, while the equation $y = \dfrac{k}{x}$ describes _____ variation.

$$\text{distance} = \sqrt{(-3-1)^2 + (-2+1)^2}$$

$$= \sqrt{(-4)^2 + (3)^2}$$

$$= \sqrt{16 + 9}$$

$$\sqrt{\tfrac{1}{2}(-3+7)^2 + \tfrac{1}{2}(8+5)^2}$$

$$(16)^2 + (13)^2$$

$$\frac{(-3-7)}{(5-0)}$$

$$\frac{-10}{5} = -2$$

REVIEW EXERCISES for Chapter 8: Graphing First-Degree Equations and Inequalities

Graph using the arbitrary-point method.

1. $2x - y = -5$

$-y = 2x - 5$
$y = 2x + 5$

2. $-3x + y = 0$

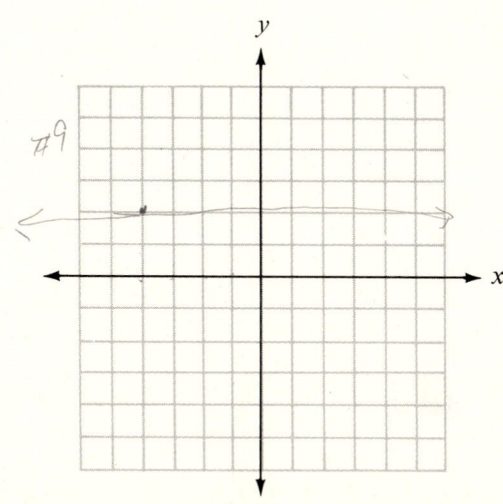

Graph using the intercept method.

3. $2x + 5y = 10$

$5y = -2x + 10$
$y = \frac{-2}{5}x + 2$

4. $-4x + y = 2$

_____ **5.** Find the distance between the points $(-3, 2)$ and $(1, -1)$.

_____ **6.** Find the midpoint of the line segment joining $(-3, 8)$ and $(7, -5)$.

_____ **7.** Find the slope of the line through $(5, -3)$ and $(0, 7)$.

Graph the line through the given point with the given slope.

8. $(1, -2); m = \frac{2}{3}$

9. $(-4, 2); m = 0$

661

Use the slope-intercept method to graph each of the following.

10. $y = 3x + 2$

11. $2x - 4y = 1$

Graph using any method you choose.

12. $3x + 4y = 0$

13. $x - 2y = -6$

14. $2x + 4 = 0$

15. $5 - y = 0$

Find the equation of the line with the given slope and y-intercept.

_____ **16.** $m = -3, b = -1$

_____ **17.** $m = \dfrac{1}{2}, b = 0$

Find the equation of the line through the given point with the given slope.

_____ **18.** $(-2, 2)$, m is undefined

_____ **19.** $(-1, 3)$, $m = \dfrac{3}{4}$

Find the equation of the line through the given points.

_____ **20.** $(-7, 2), (-5, 0)$

_____ **21.** $\left(\dfrac{2}{3}, \dfrac{-1}{2}\right), \left(\dfrac{5}{3}, \dfrac{3}{2}\right)$

_____ **22.** $(3, 2), (3, -5)$

_____ **23.** $(4, -5), (-2, -5)$

_____ **24.** Find the equation of the line through $(-2, 1)$ and parallel to $2x - y = 6$.

_____ **25.** Find the equation of the line through $(6, 0)$ and perpendicular to $x + 2y = -1$.

Graph the inequalities.

26. $x \geq -3$

27. $y < \dfrac{-2}{3} x$

28. $2x - y \leq 6$

29. $3x - 5y < 15$

For each of these relations give the domain and range.

_____ **30.** $\{(-2, 2), (-1, 1), (0, 0), (1, -1), (2, -2)\}$

_____ **31.** $\{(x, y) : y = 3\}$

Use the vertical-line test to tell which of the following graphs represent functions.

———— **32.**

———— **33.**

If $f(x) = 2x - 1$ and $g(x) = x^2 - 9$, find each of the following.

———— **34.** $f(-3)$

———— **35.** $f(2) - g(-1)$

Give the domain of each of the following.

———— **36.** $f(x) = \dfrac{x - 1}{x^2 - 4}$

———— **37.** $f(x) = \sqrt{3 - x}$

———— **38.** Write an equation of variation and find k if y varies directly as x, and $y = 56$ when $x = 8$.

———— **39.** Write an equation of variation and find k if y varies inversely as x, and $y = \dfrac{1}{2}$ when $x = 6$.

———— **40.** Write an equation of variation and find k if z varies jointly as x and the square of y, and $z = \dfrac{1}{4}$ when $x = 9$ and $y = 4$.

———— **41.** A certain variable T varies directly as D and inversely as R. Write a function describing this relationship and find k if $T = 300$ when $D = 20$ and $R = 4$. Then find T when $D = 60$ and $R = 9$.

———— **42.** The cost of copying a pamphlet varies directly with the number of pages in it. If the cost of copying a pamphlet of 50 pages is $3.50, how much will it cost to copy a pamphlet of 80 pages?

———— **43.** The distance a body falls varies directly as the square of the time it has been falling. If an object falls 400 feet in 5 seconds, how far will it fall in 12 seconds?

———— **44.** If the length of a rectangle is 3 feet less than twice the width, write an equation that expresses the area as a function of the width.

Check your answers in the back of your book.

If You Missed Exercises:	You Should Review Examples:	
1, 2	SECTION 8.1	6–8
3, 4		9, 10
5	SECTION 8.2	1, 2
6		3, 4
7		5–8
8, 9		9–12
10, 11		13–15
12–15	SECTIONS 8.1 and 8.2	
16, 17	SECTION 8.3	1–4
18, 19		5–8
20–23		9–12
24		13, 14
25		15
26–29	SECTION 8.4	1–4
30, 31	SECTION 8.5	1, 2
32, 33		3
34, 35		4
36, 37		5–8
38	SECTION 8.6	2
39		4
40		6
41		9
42		2
43		3
44		11

If you have completed the Review Exercises and corrected your errors, you are ready to take the Practice Test.

PRACTICE TEST for Chapter 8: Graphing First-Degree Equations and Inequalities

	Section	Examples

1. Use the arbitrary-point method to graph $2x + y = 5$.   8.1   1, 2

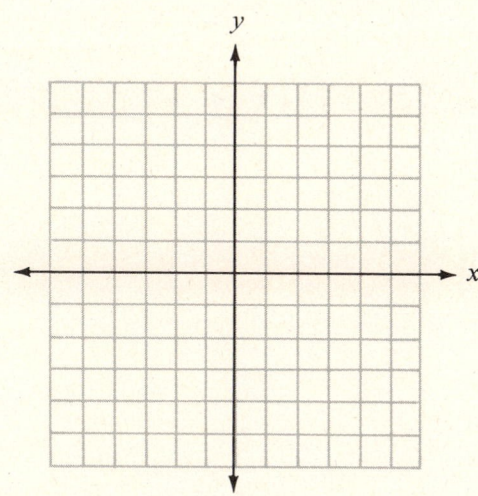

2. Use the intercept method to graph $4x - 3y = 9$.   8.1   9

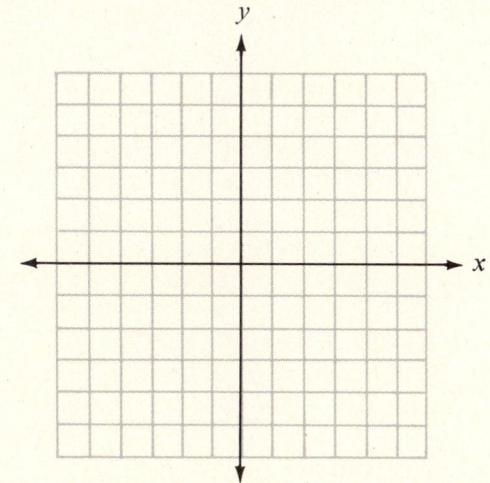

_____ **3.** Find the distance between the points $(-2, 3)$ and $(4, 5)$.   8.2   1, 2

_____ **4.** Find the midpoint of the line segment joining $(0, 5)$ and $(-3, 2)$.   8.2   3, 4

_____ **5.** Find the slope of the line through the points $(3, -1)$ and $(7, -5)$.   8.2   6

6. Graph the line through $(1, -3)$ with slope $m = \dfrac{4}{3}$. 8.2 9

7. Use the slope-intercept method to graph $2x + 5y = 5$. 8.2 13

8. Graph $y - 3x = 0$ using any method you choose. 8.1 7

9. Graph $2y + 6 = 0$ using any method you choose.

8.2　　　15

_____ **10.** Find the equation of the line through the origin with slope $m = -2$.

8.3　　　3

_____ **11.** Find the equation of the line through the point $(-6, 1)$ with slope $m = \dfrac{-2}{3}$.

8.3　　　6

_____ **12.** Find the equation of the line through the points $(5, -3)$ and $(5, 1)$.

8.3　　　12

_____ **13.** Find the equation of the line through the points $(2, -4)$ and $(5, -2)$.

8.3　　　9

_____ **14.** Find the equation of the line through $(3, -5)$ parallel to $x - y = 1$.

8.3　　　13

_____ **15.** Find the equation of the line through $(-1, 4)$ perpendicular to $2x + y = -1$.

8.3　　　15

16. Graph $x \leq 1$.

8.4　　　1

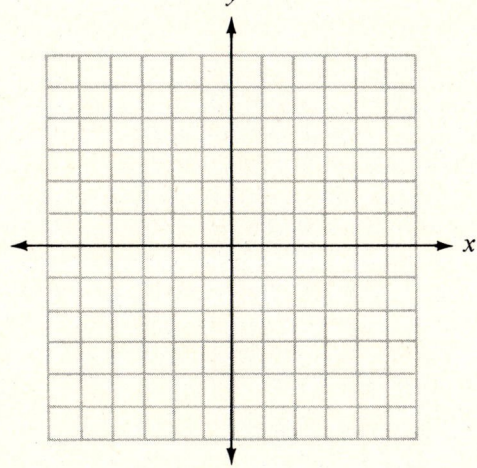

17. Graph $4x - y < -2$.

8.4   2

_____ **18.** For the relation $\{(x, y) : y = 3\}$ give the domain and range. Is this a function?

8.5   1, 2

_____ **19.** If $f(x) = -4x + 1$, find $f\left(\dfrac{-3}{2}\right)$.

8.5   4

_____ **20.** If $g(x) = x^2 - 2x - 3$, find $g(-1)$.

8.5   4

_____ **21.** Give the domain for $f(x) = \sqrt{2x - 1}$.

8.5   8

_____ **22.** Give the domain for $g(x) = \dfrac{x}{4x^2 + 3x}$.

8.5   6

_____ **23.** Write an equation of variation and find k if y varies directly with the square of x, and $y = 18$ when $x = -3$.

8.6   3

_____ **24.** If A varies jointly as b and h, write a function describing this relationship. Find k if $A = 10$ when $b = 5$ and $h = 4$. Then find A when $b = 7$ and $h = 5$.

8.6   5

_____ **25.** If Max owes Charles \$700 and he plans to pay him back \$25 each month, write an equation that expresses the amount Max still owes Charles as a function of the number of months. Find how much Max owes Charles after one year.

8.6   10

SHARPENING YOUR SKILLS after Chapters 1–8

SECTION

———— **1.** Evaluate $(-3)^{-3}$ — 3.2

———— **2.** Write $7 - \sqrt{-12}$ in standard complex form. — 7.5

———— **3.** Evaluate i^{36}. — 7.5

———— **4.** Use the discriminant to describe the solutions for $x^2 - 2x + 5 = 0$. — 5.3

———— **5.** Write $8a^{1/3}$ in radical form. — 7.1

———— **6.** Write $\sqrt[3]{(a-b)^4}$ in exponential form. — 7.1

Simplify.

———— **7.** $5\left[-10 + \dfrac{-18 + 3}{-4 + 9}\right] - 6$ — 1.3

———— **8.** $x - 9[x - 7(2x - 1)]$ — 1.4

———— **9.** $\left(\dfrac{13x^2y^7}{9z^3}\right)^0$ — 3.1

———— **10.** $(a^{-10}b^7c^{-3})(a^{12}b^{-9}c^4)$ — 3.2

———— **11.** $3ab(a^{-1}b^2 - 5a^2b^{-1}) + 4a^7b^{-2}(a^{-4}b^5)$ — 3.3

———— **12.** $(x^2 - 7x + 12)(x^2 - x - 12)$ — 3.4

———— **13.** $\dfrac{2x^2 - 13x - 24}{x^2 - 64} \div \dfrac{4x^2 - 9}{x + 8}$ — 6.2

———— **14.** $\dfrac{x + 8}{x^2 + 3x - 10} + \dfrac{4}{x + 5}$ — 6.3

———— **15.** $\dfrac{1 + x^{-1}}{1 - x^{-2}}$ — 6.4

———— **16.** $\sqrt{48x} - \sqrt{27x} + \sqrt{300x}$ — 7.2

———— **17.** $(\sqrt{3x} + 5)(\sqrt{3x} - 5)$ — 7.3

———— **18.** $\dfrac{2 - 3i}{4 - 5i}$ — 7.5

Factor completely.

———— **19.** $x^3 + 125$ — 4.2

———— **20.** $9x^4 - 30x^2 + 25$ — 4.3

_____ **21.** $63a^2 - 14ab - 45a + 10b$ 4.3

Solve.

_____ **22.** $4(3x - 1) + 1 = 3(9 - x)$ 2.1

_____ **23.** $\left| \dfrac{x}{3} + 5 \right| \geq 10$ 2.4

_____ **24.** $11x^3 - 25x^2 + 6x = 0$ 5.1

_____ **25.** $\dfrac{3}{x - 9} - \dfrac{4}{x + 5} = \dfrac{2}{2x + 10}$ 6.5

_____ **26.** $\sqrt{5x - 21} + \sqrt{2x} = 0$ 7.4

_____ **27.** $x^2 + 20 \leq 12x$ 5.4

_____ **28.** Solve and graph $-2 \leq \dfrac{4a - 6}{2} < 0$. 2.3

_____ **29.** George has doubled the lengths of two opposite sides of his old square garden and has 5.5
increased the other two sides by 4 feet. If the difference between the areas of the old
garden and new garden is 560 square feet, find the dimensions of George's new garden.

Graphing More Relations and Functions

9

If a person throws a ball directly upward at a speed of 96 feet per second, the ball's height y feet above the ground after x seconds is given by the function $y = -16x^2 + 96x$. Use the graph of the function to find the maximum height of the ball.

Although we have glanced briefly at a few graphs that were not straight lines, until now we have concentrated mainly on linear functions and relations. In this chapter we graph sets of ordered pairs satisfying functions and relations that are *not* linear. In particular, we learn to:

1. Graph any function or relation using arbitrary points.
2. Graph quadratic functions.
3. Graph the conic sections.
4. Graph second-degree inequalities.

9.1 Graphing Functions by Arbitrary Points

Recall that our first method for graphing linear functions required that we find several ordered pairs that satisfied the given function. This method of arbitrary points will be even more important to us as we attempt to graph nonlinear functions and relations. Although we used at most three points to graph a straight line, more points are necessary for graphs of most other relations, especially if we have no idea about the appearance of the graph. No matter what relation we are graphing, we shall always be interested in finding the *intercepts* for the graph whenever possible.

Suppose that we try to graph several sets of ordered pairs, using tables to organize our work and labeling the points we find.

Example 1. Graph $y = x^2$.

Solution: Choose several x-values.

x	$y = x^2$	(x, y)
-3	$(-3)^2 = 9$	$(-3, 9)$
-2	$(-2)^2 = 4$	$(-2, 4)$
-1	$(-1)^2 = 1$	$(-1, 1)$
0	$0^2 = 0$	$(0, 0)$
1	$1^2 = 1$	$(1, 1)$
2	$2^2 = 4$	$(2, 4)$
3	$3^2 = 9$	$(3, 9)$

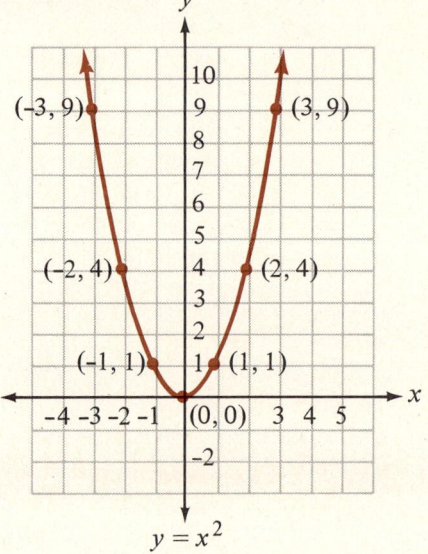

$$y = x^2$$

Plotting these points and connecting them with a smooth curve, we have our graph.

A graph with this shape is called a **parabola.** Is this a function? Since each x-value corresponds to one and only one y-value (remember the vertical-line test), we agree that $y = x^2$ describes a function. In particular, it is a second-degree or **quadratic function.**

From the graph of $y = x^2$ we can determine the domain and range for the function.

$$\text{Domain:} \quad x \in R$$

$$\text{Range:} \quad y \in R \text{ and } y \geq 0$$

Example 2. Graph $y = 2x - x^2$. Find x- and y-intercepts, construct a table of ordered pairs, and decide whether this is a function. Then state its domain and range.

Solution

$$y = 2x - x^2$$

y-INTERCEPT: Let $x = 0$. x-INTERCEPTS: Let $y = 0$.

$$y = 2(0) - 0^2$$

$$y = 0$$

$$(0, 0)$$

$$0 = 2x - x^2$$

$$0 = x(2 - x)$$

$$x = 0 \qquad 2 - x = 0$$

$$x = 2$$

$$(0, 0) \qquad\qquad (2, 0)$$

x	$y = 2x - x^2$	(x, y)
-1	$y = 2(-1) - (-1)^2$ $= -2 - 1 = -3$	$(-1, -3)$
0	$y = 0$	$(0, 0)$
1	$y = 2(1) - (1)^2$ $= 2 - 1 = 1$	$(1, 1)$
2	$y = 0$	$(2, 0)$
3	$y = 2(3) - (3)^2$ $= 6 - 9 = -3$	$(3, -3)$

Once again our graph is a parabola (opening *down* in this example). This is certainly a function; in particular, it is another quadratic function. From the graph we see

$$\text{Domain:} \quad x \in R$$

$$\text{Range:} \quad y \in R \text{ and } y \leq 1$$

Example 3. Graph $y = x^3$ using a table of ordered pairs. Tell whether it is a function, and state its domain and range.

Solution

$$y = x^3$$

y-INTERCEPT: Let $x = 0$.

$$y = 0^3 = 0$$

$$(0, 0)$$

x-INTERCEPT: Let $y = 0$.

$$0 = x^3$$

$$0 = x$$

$$(0, 0)$$

$$y = x^3$$

x	$y = x^3$	(x, y)
-2	$y = (-2)^3 = -8$	$(-2, -8)$
-1	$y = (-1)^3 = -1$	$(-1, -1)$
0	$y = 0$	$(0, 0)$
1	$y = 1^3 = 1$	$(1, 1)$
2	$y = 2^3 = 8$	$(2, 8)$

Is $y = x^3$ a function? Yes; it is a third-degree or **cubic function** and its graph is sort of S-shaped. From the graph we state

$$\text{Domain:} \quad x \in R$$

$$\text{Range:} \quad y \in R$$

Now you try Example 4.

Example 4. Graph $y = x^3 - x$. Find the intercepts, construct a table of ordered pairs and decide whether this is a function. Then state its domain and range.

Solution

$$y = x^3 - x$$

y-INTERCEPT: Let $x = 0$. x-INTERCEPT: Let $y = 0$.

$$y = 0^3 - 0 \qquad\qquad\qquad 0 = x^3 - x$$

$$y = \underline{0} \qquad\qquad\qquad\quad 0 = x(x^2 - 1)$$

$$(0, \underline{0}) \qquad\qquad\qquad 0 = x(x + 1)(x - 1)$$

$$x = 0 \quad\text{or}\quad x + 1 = 0 \quad\text{or}\quad x - 1 = 0$$

$$x = \underline{-1} \qquad x = \underline{1}$$

$$(0, 0) \qquad (\underline{-1}, 0) \qquad (\underline{1}, 0)$$

x	$y = x^3 - x$	(x, y)
-2	$y = (-2)^3 - (-2)$ $= -8 + 2 = \underline{}$	$(-2, \underline{-6})$
-1	$y = 0$	$(-1, 0)$
$\dfrac{-1}{2}$	$y = \left(\dfrac{-1}{2}\right)^3 - \left(\dfrac{-1}{2}\right)$ $= \dfrac{-1}{8} + \dfrac{4}{8} = \underline{}$	$\left(\dfrac{-1}{2}, \underline{\dfrac{3}{8}}\right)$
0	$y = 0$	$(0, 0)$
$\dfrac{1}{2}$	$y = \left(\dfrac{1}{2}\right)^3 - \dfrac{1}{2}$ $= \dfrac{1}{8} - \dfrac{4}{8} = \underline{}$	$\left(\dfrac{1}{2}, \underline{\dfrac{3}{8}}\right)$
1	$y = 0$	$(1, 0)$
2	$y = 2^3 - 2$ $= 8 - 2 = \underline{}$	$(2, \underline{6})$

Is $y = x^3 - x$ a function? _____YES_____ . From the graph we state

Domain: _____

Range: _____

Check your work on page 677. ▶

Example 5. Graph $y = \dfrac{1}{x}$.

Solution: We immediately notice the restriction on the domain and agree that $x \neq 0$. Our graph will *not* cross the y-axis.

y-INTERCEPT: None.

x-INTERCEPT: Let $y = 0$.

$$0 = \frac{1}{x}$$

$$0 \cdot x = 1$$

$$0 = 1$$

This false statement tells us there is *no* x-intercept. Our graph will *not* cross either axis.

x	$y = \dfrac{1}{x}$	(x, y)
-3	$y = \dfrac{-1}{3}$	$\left(-3, \dfrac{-1}{3}\right)$
-1	$y = \dfrac{1}{-1} = -1$	$(-1, -1)$
$\dfrac{-1}{2}$	$y = \dfrac{1}{\dfrac{-1}{2}}$	
	$= 1 \cdot \dfrac{-2}{1} = -2$	$\left(\dfrac{-1}{2}, -2\right)$
$\dfrac{1}{2}$	$y = \dfrac{1}{\dfrac{1}{2}}$	
	$= 1 \cdot \dfrac{2}{1} = 2$	$\left(\dfrac{1}{2}, 2\right)$
1	$y = \dfrac{1}{1} = 1$	$(1, 1)$
3	$y = \dfrac{1}{3}$	$\left(3, \dfrac{1}{3}\right)$

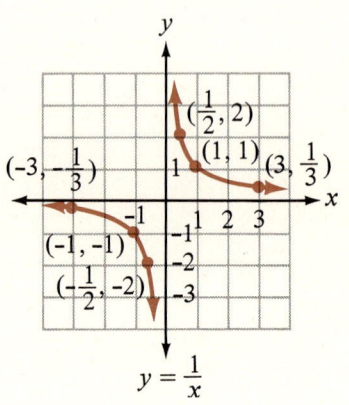

We see that $y = \dfrac{1}{x}$ is a function; in particular, it is called a **rational function.** From the graph, we can state

Domain: $x \in R$ and $x \neq 0$

Range: $y \in R$ and $y \neq 0$

⫸ Trial Run

Graph each function. Then state its domain and range.

Do: _____ **1.** $y = x^2 - 4$
Ra: _____

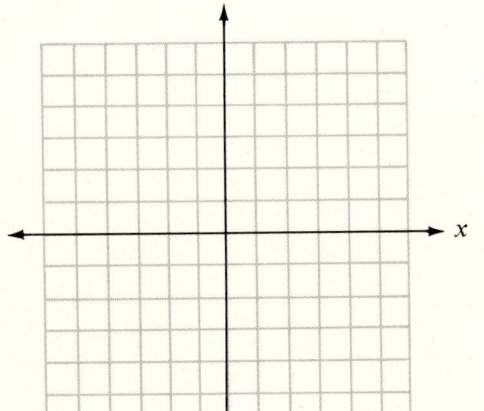

Do: _____ **2.** $y = 1 - x^2$
Ra: _____

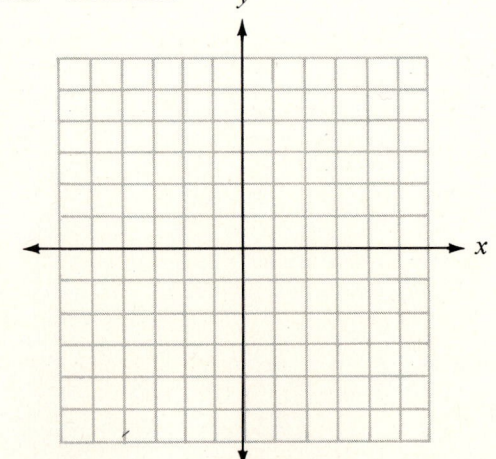

Do: _____ **3.** $y = x^3 - 1$
Ra: _____

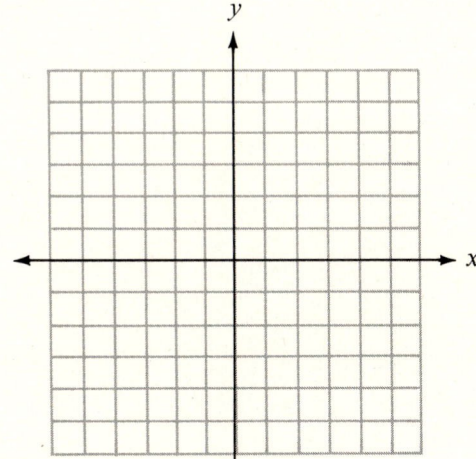

Do: _____ **4.** $y = \dfrac{2}{x}$
Ra: _____

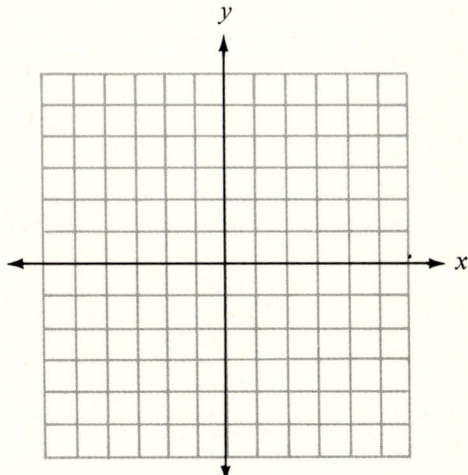

Answers are on page 678.

▶ Example You Completed

Example 4. Graph $y = x^3 - x$. Find the intercepts, construct a table of ordered pairs, and decide whether this is a function. Then state its domain and range.

Solution

$$y = x^3 - x$$

y-INTERCEPT: Let $x = 0$. x-INTERCEPT: Let $y = 0$.

$y = 0^3 - 0$ $0 = x^3 - x$

$y = 0$ $0 = x(x^2 - 1)$

$(0, 0)$ $0 = x(x + 1)(x - 1)$

$x = 0$ or $x + 1 = 0$ or $x - 1 = 0$

$x = -1$ $x = 1$

$(0, 0)$ $(-1, 0)$ $(1, 0)$

x	$y = x^3 - x$	(x, y)
-2	$y = (-2)^3 - (-2)$ $= -8 + 2 = -6$	$(-2, -6)$
-1	$y = 0$	$(-1, 0)$
$\dfrac{-1}{2}$	$y = \left(\dfrac{-1}{2}\right)^3 - \left(\dfrac{-1}{2}\right)$ $= \dfrac{-1}{8} + \dfrac{4}{8} = \dfrac{3}{8}$	$\left(\dfrac{-1}{2}, \dfrac{3}{8}\right)$
0	$y = 0$	$(0, 0)$
$\dfrac{1}{2}$	$y = \left(\dfrac{1}{2}\right)^3 - \dfrac{1}{2}$ $= \dfrac{1}{8} - \dfrac{4}{8} = \dfrac{-3}{8}$	$\left(\dfrac{1}{2}, \dfrac{-3}{8}\right)$
1	$y = 0$	$(1, 0)$
2	$y = 2^3 - 2$ $= 8 - 2 = 6$	$(2, 6)$

Is $y = x^3 - x$ a function? yes. From the graph we state

$$\text{Domain:} \quad x \in R$$
$$\text{Range:} \quad y \in R$$

Answers to Trial Run

page 676 **1.**

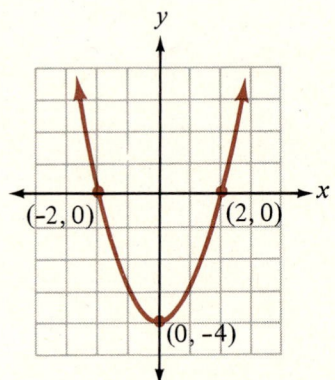

Do: $x \in R$
Ra: $y \in R$ and $y \geq -4$

2.

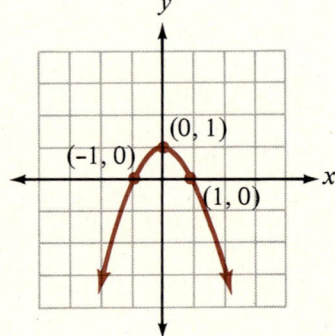

Do: $x \in R$
Ra: $y \in R$ and $y \leq 1$

3.

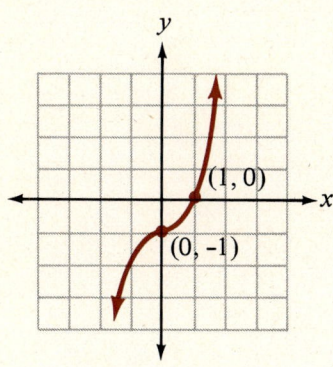

Do: $x \in R$
Ra: $y \in R$

4.

Do: $x \in R$ and $x \neq 0$
Ra: $y \in R$ and $y \neq 0$

EXERCISE SET 9.1

Graph each of the following functions. State the domain and range of each.

Do: _____ **1.** $y = x^2 - 9$ Do: _____ **2.** $y = x^2 - 16$
Ra: _____   Ra: _____

Do: _____ **3.** $y = -x^2 + 1$ Do: _____ **4.** $y = -x^2 + 4$
Ra: _____   Ra: _____

Do: _____ **5.** $y = x^2 - 3$ Do: _____ **6.** $y = x^2 - 5$
Ra: _____   Ra: _____

Do: _____ **7.** $y = x^2 - 2x$ Do: _____ **8.** $y = x^2 + 4x$
Ra: _____   Ra: _____

Do: _____ **9.** $y = x^3 + 1$ Do: _____ **10.** $y = x^3 - 8$
Ra: _____   Ra: _____

Do: _____ **11.** $y = -x^3 - 1$ Do: _____ **12.** $y = -x^3 + 8$
Ra: _____   Ra: _____

Do: _____ **13.** $y = x^3 - 16x$ Do: _____ **14.** $y = x^3 - 4x$
Ra: _____   Ra: _____

Do: _____ **15.** $y = \dfrac{3}{x}$ Do: _____ **16.** $y = \dfrac{4}{x}$
Ra: _____   Ra: _____

Do: _____ **17.** $y = \dfrac{-2}{x}$ Do: _____ **18.** $y = \dfrac{-1}{x}$
Ra: _____   Ra: _____

☆ Stretching the Topics _____

_____ **1.** Graph $y = \dfrac{2}{x - 2}$.

_____ **2.** Graph $y = x^3 - x^2 - 6x$.

Check your answers in the back of your book.

If you can graph the functions in **Checkup 9.1,** you are ready to go on to Section 9.2.

CHECKUP 9.1

Graph each of the following functions. State the domain and range of each.

Do: _____ **1.** $y = x^2 - 5$
Ra: _____

Do: _____ **2.** $y = -2x^2 + 8$
Ra: _____

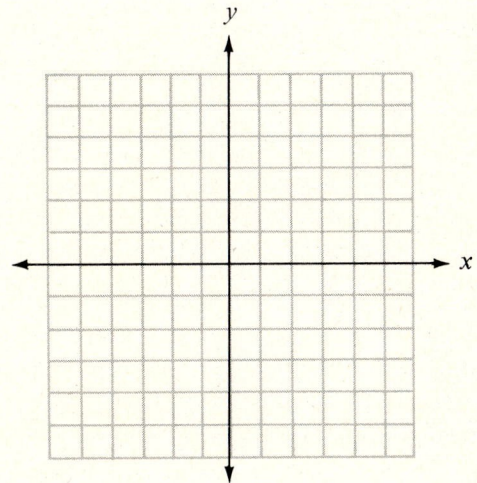

Do: _____ **3.** $y = -x^3$
Ra: _____

Do: _____ **4.** $y = x^3 - 16x$
Ra: _____

Do: _____ 5. $y = \dfrac{-4}{x}$

Ra: _____

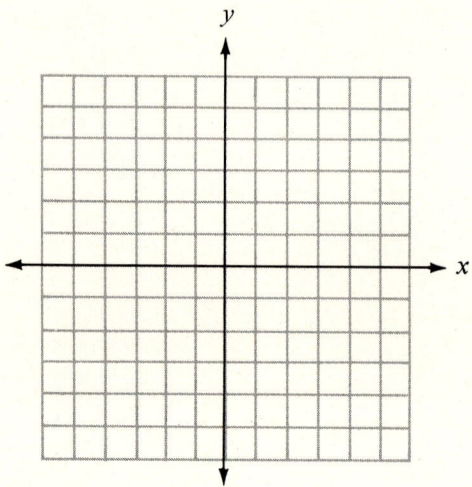

Check your answers in the back of your book.

If You Missed Problems:	You Should Review Examples:
1, 2	1, 2
3	3
4	4
5	5

9.2 Graphing Quadratic Functions

In earlier chapters we learned to solve quadratic equations,

$$ax^2 + bx + c = 0$$

Now we must turn our attention to graphing sets of ordered pairs satisfying equations of the form

$$y = ax^2 + bx + c$$

Because each *x*-value substituted into such an equation will yield one and only one *y*-value, you should agree that the resulting ordered pairs represent a *function*.

Quadratic Function

$$y = ax^2 + bx + c \qquad (a \neq 0)$$

We note that $a \neq 0$ in such a function or else the polynomial would contain no x^2-term and the function would be *linear* rather than quadratic.

Graphing Quadratic Functions: $y = ax^2 + c$

In Section 9.1 we discovered that the graph of a quadratic function is a *parabola*. Let's make some general observations about parabolas that are graphs of quadratic functions.

Characteristics of Graphs of Quadratic Functions (Parabolas)

1. Every parabola opens either upward or downward.
2. There is a *vertical* line, called the **axis of symmetry,** that passes down the middle of a parabola. If the Cartesian plane were folded along this axis of symmetry, the left-hand portion of the parabola would lie exactly on top of the right-hand portion.
3. Every parabola has either a low point (if it opens upward) or a high point (if it opens downward). This point is called the **vertex** of the parabola, and it *always* lies on the axis of symmetry.

$$y = x^2$$

(a)

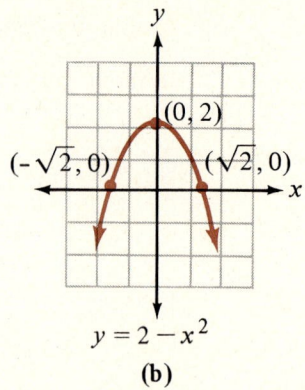

$$y = 2 - x^2$$

(b)

For $y = x^2$

1. The parabola opens upward.
2. The axis of symmetry is the vertical line $x = 0$ (the y-axis).
3. The vertex is the point $(0, 0)$. It is the function's **minimum** (low) point.

For $y = 2 - x^2$

1. The parabola opens downward.
2. The axis of symmetry is the vertical line $x = 0$ (the y-axis).
3. The vertex is the point $(0, 2)$. It is the function's **maximum** (high) point.

Having graphed several parabolas, we observe that whether a parabola opens upward or downward can be determined by the *sign* of a, the coefficient of the x^2-term in $y = ax^2 + bx + c$.

> If a is *positive*, the parabola opens *upward* (see: $y = x^2$).
> If a is *negative*, the parabola opens *downward* (see: $y = 2 - x^2$).

We may also note from the graphs of $y = x^2$ and $y = 2 - x^2$ that

> If $b = 0$ (the x-term is missing), the axis of symmetry will be the y-axis.
>
> If $b = 0$ (the x-term is missing), the vertex must lie on the y-axis (the axis of symmetry). So the vertex must be the graph's y-intercept point.

Armed with these useful pieces of information, we will find that it is not difficult to graph quadratic functions in which the x-term is missing, that is, quadratic functions of the form

$$y = ax^2 + c$$

Example 1. Graph $y = x^2 - 4$. Then state its domain and range.

Solution: First we observe that for $y = x^2 - 4$

1. The parabola opens upward (because a is positive 1).
2. The axis of symmetry is the y-axis (because the x-term is missing).
3. The vertex is the graph's y-intercept point.
4. The vertex is a minimum point (because the parabola opens upward).

The missing pieces of information are the intercepts.

y-INTERCEPT: Let $x = 0$.

$$y = 0^2 - 4$$

$$y = -4$$

$$(0, -4)$$

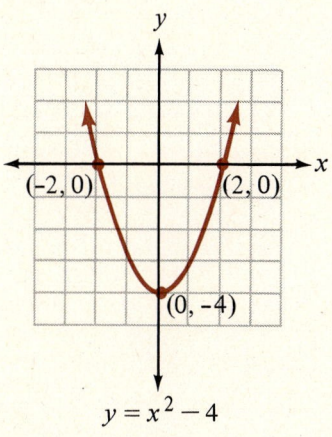

$$y = x^2 - 4$$

x-INTERCEPTS: Let $y = 0$.

$$0 = x^2 - 4$$

$$0 = (x + 2)(x - 2)$$

$$x + 2 = 0 \quad \text{or} \quad x - 2 = 0$$

$$x = -2 \qquad\qquad x = 2$$

$$(-2, 0) \qquad\qquad (2, 0)$$

Domain: $x \in R$

Range: $y \in R$ and $y \geq -4$

Plotting these points and recalling our observations for this parabola, we have our graph.

Example 2. Graph $y = -x^2 - 1$. Then state its domain and range.

Solution: First we observe that for $y = -x^2 - 1$

1. The parabola opens downward.
2. The axis of symmetry is the y-axis.
3. The vertex is the y-intercept point.
4. The vertex is a maximum point.

Now we would like to find the intercepts.

y-INTERCEPT: Let $x = 0$.

$$y = -0^2 - 1$$

$$y = -1$$

$$(0, -1)$$

x-INTERCEPTS: Let $y = 0$.

$$0 = -x^2 - 1$$

$$x^2 = -1$$

Since the square of a real number cannot be negative, this equation has no real number solutions. The graph has *no* x-intercepts.

We conclude that our parabola has its maximum point at its vertex $(0, -1)$ but does *not* cross the x-axis. To determine the shape of this parabola, we can choose an x-value to the left of zero and an x-value to the right of zero, and substitute those x-values into $y = -x^2 - 1$.

Let $x = -1$.

$$y = -(-1)^2 - 1$$
$$= -1 - 1$$
$$= -2$$
$$(-1, -2)$$

Let $x = 1$.

$$y = -(1)^2 - 1$$
$$= -1 - 1$$
$$= -2$$
$$(1, -2)$$

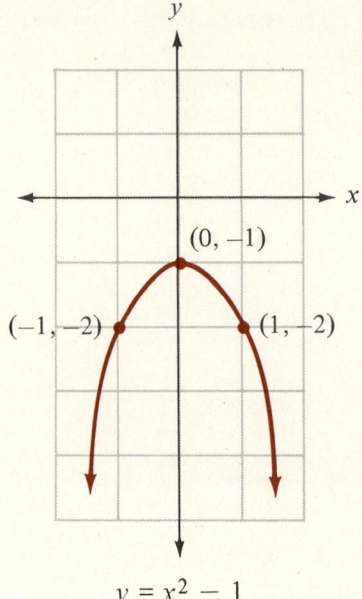

$$y = x^2 - 1$$

Domain: $x \in R$

Range: $y \in R$ and $y \leq -1$

Now you try Example 3.

Example 3. Graph $y = 2x^2$. Then state its domain and range.

Solution: First we observe that for $y = 2x^2$

1. The parabola opens _____ .
2. The axis of symmetry is _____ .
3. The vertex is the _____ point.
4. The vertex is a _____ point.

Now we would like to find the intercepts.

y-INTERCEPT: Let $x = 0$.　　　　x-INTERCEPTS: Let $y = 0$.

$$y = 2(\underline{\quad})^2 \qquad\qquad 0 = 2x^2$$

$$y = \underline{\quad} \qquad\qquad\qquad \underline{\quad} = x^2$$

$$(\underline{\quad}, \underline{\quad}) \qquad\qquad\qquad x = \underline{\quad}$$

$$(\underline{\quad}, \underline{\quad})$$

Since the x-intercept and y-intercept both occur at $(0, 0)$, we must choose an x-value on either side of $(0, 0)$ to substitute into $y = 2x^2$.

Let $x = 1$

$$y = 2(\underline{1})^2$$

$$= \underline{2}$$

$$(1, \underline{2})$$

Let $x = -1$

$$y = 2(\underline{})^2$$

$$= \underline{}$$

$$(-1, \underline{2})$$

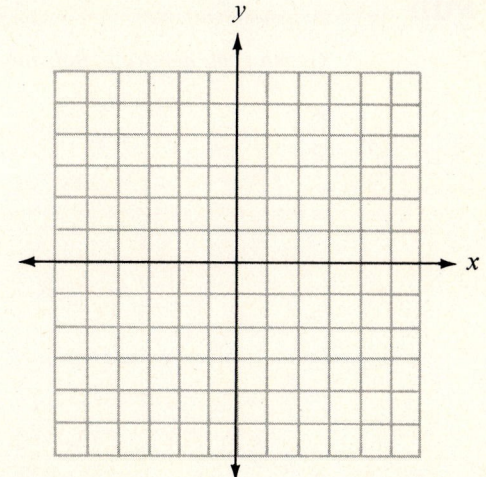

Domain: _____

Range: _____ and _____

Check your work on page 700. ▶

In general, we have discovered that we may graph quadratic functions of the type $y = ax^2 + c$ using the following procedure.

Graphing Quadratic Functions: $y = ax^2 + c$

1. Decide whether the parabola opens upward (if $a > 0$) or downward (if $a < 0$).
2. Realize that the axis of symmetry is the y-axis.
3. Realize that the vertex is the y-intercept point.
4. Decide whether the vertex is a minimum (for an upward parabola) or a maximum (for a downward parabola).
5. Find x- and y-intercepts. If there are not two x-intercepts, find two other points on either side of the vertex.
6. Sketch the parabola and note its domain and range.

Trial Run

Graph each quadratic function. Then state its domain and range.

Do: _____ **1.** $y = 5x^2$
Ra: _____

Do: _____ **2.** $y = -x^2 + 8$
Ra: _____

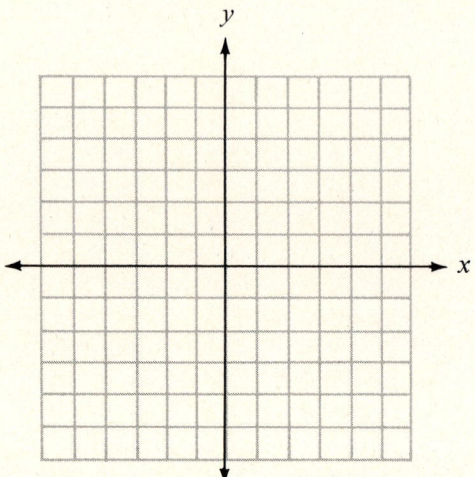

Do: _____ **3.** $y = -3x^2 - 6$
Ra: _____

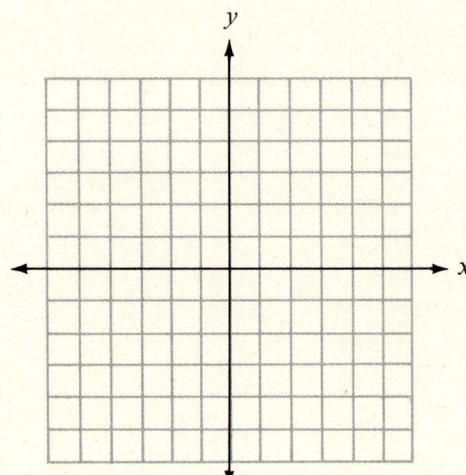

Answers are on page 703.

Graphing Quadratic Functions: $y = ax^2 + bx + c$

Realizing that there are many quadratic functions in which the x-term is *not* missing, we need a procedure for graphing any quadratic function of the form

$$y = ax^2 + bx + c$$

Our approach will be very visual and will rely heavily on common sense and our earlier observations about the nature of parabolas.

Let's try to graph the function

$$y = x^2 - 4x$$

We immediately note that this parabola opens *upward* because the coefficient of the x^2-term is positive. We also note that the axis of symmetry is *not* the y-axis because the x-term is *not* missing.

Let's find the intercepts.

y-INTERCEPT: Let $x = 0$.

$$y = 0^2 - 4(0)$$

$$y = 0$$

$$(0, 0)$$

x-INTERCEPTS: Let $y = 0$.

$$0 = x^2 - 4x$$

$$0 = x(x - 4)$$

$$x = 0 \quad \text{or} \quad x - 4 = 0$$

$$x = 4$$

$$(0, 0) \qquad (4, 0)$$

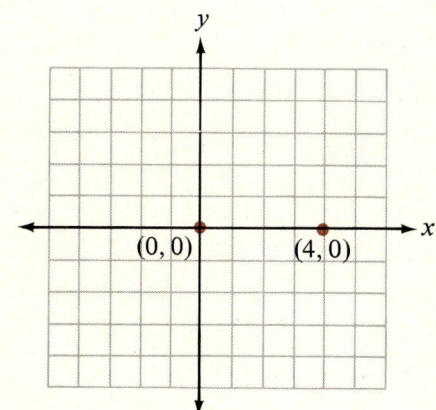

From our understanding of the *axis of symmetry*, we know it is a vertical line half-way between the two x-intercepts. The x-value of the point halfway between $(0, 0)$ and $(4, 0)$ is 2, so the axis of symmetry must be the vertical line $x = 2$.

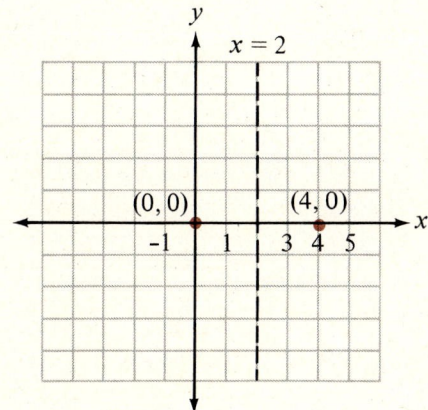

Moreover, we know that the *vertex* of the parabola must lie on the axis of symmetry, $x = 2$. Therefore, the x-coordinate for the vertex must be 2 and we can find the y-coordinate by substituting 2 for x in the equation for the parabola: $y = x^2 - 4x$.

$$y = 2^2 - 4(2)$$
$$= 4 - 8$$
$$= -4$$

The vertex is the point $(2, -4)$.

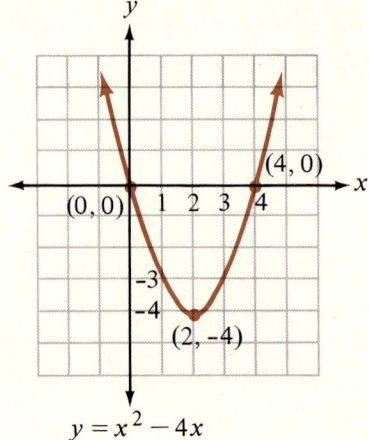

$$y = x^2 - 4x$$

Domain: $x \in R$

Range: $y \in R$ and $y \geq -4$

It is worth mentioning here that we can use part of the *midpoint formula* to find the x-coordinate of the point halfway between two known points on a parabola.

Midpoint Formula

$$x_M = \frac{x_1 + x_2}{2}$$

If you can locate the halfway point by looking, then by all means do so. We shall reserve the formula for situations in which the halfway point is difficult to locate on sight.

As you can see, we need only a few key pieces of information to graph any parabola.

Graphing Parabolas

1. Find the y-intercept.
2. Find the x-intercepts.
3. Find the axis of symmetry.
4. Find the vertex.

Example 4. Graph $y = 2 + x - x^2$. State its domain and range.

Solution: $y = 2 + x - x^2$. This parabola will open downward.

y-INTERCEPT: Let $x = 0$.

$$y = 2 + 0 - 0^2$$
$$= 2$$
$$(0, 2)$$

x-INTERCEPTS: Let $y = 0$.

$$0 = 2 + x - x^2$$
$$x^2 - x - 2 = 0$$
$$(x - 2)(x + 1) = 0$$
$$x - 2 = 0 \quad x + 1 = 0$$
$$x = 2 \quad\quad x = -1$$
$$(2, 0) \quad\quad\quad (-1, 0)$$

AXIS OF SYMMETRY

$$x_M = \frac{2 + (-1)}{2}$$

$$= \frac{1}{2}$$

$x = \dfrac{1}{2}$ is the axis of symmetry

VERTEX: Let $x = \dfrac{1}{2}$.

$$y = 2 + x - x^2$$

$$y = 2 + \frac{1}{2} - \left(\frac{1}{2}\right)^2$$

$$= 2 + \frac{1}{2} - \frac{1}{4}$$

$$= \frac{8}{4} + \frac{2}{4} - \frac{1}{4}$$

$$= \frac{9}{4}$$

$\left(\dfrac{1}{2}, \dfrac{9}{4}\right)$ is the vertex.

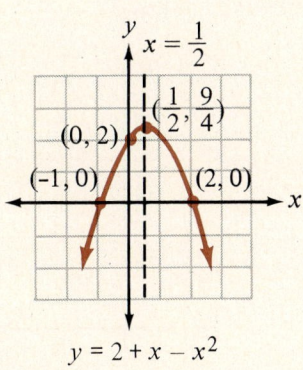

$$y = 2 + x - x^2$$

Domain: $x \in R$

Range: $y \in R$ and $y \le \dfrac{9}{4}$

Now you try the problem stated at the beginning of the chapter.

Example 5. If a person throws a ball directly upward at a speed of 96 feet per second, the ball's height y above the ground after x seconds is given by the function $y = -16x^2 + 96x$. Use the graph of this function to find the maximum height of the ball.

Solution: The graph of $y = -16x^2 + 96x$ is a _____ that opens _____ .

y-INTERCEPT: Let $x = 0$.

$$y = -16(\underline{\quad})^2 + 96(\underline{\quad})$$

$$y = \underline{\quad}$$

$$(0, \underline{\quad})$$

x-INTERCEPTS: Let $y = 0$.

$$0 = -16x^2 + 96x$$

$$0 = -16x(x - \underline{\quad})$$

$$-16x = 0 \quad \text{or} \quad x - \underline{\quad} = 0$$

$$x = 0 \qquad\qquad x = \underline{\quad}$$

$$(0, 0) \qquad\qquad (\underline{\quad}, 0)$$

Axis of Symmetry

$$x_M = \frac{0 + 6}{2}$$

$$= \frac{6}{2}$$

$$= \underline{\hspace{1cm}}$$

$x = \underline{\hspace{1cm}}$ is the axis of symmetry.

Vertex: Let $x = 3$.

$$y = -16x^2 + 96x$$

$$y = -16(3)^2 + 96(3)$$

$$= -16(9) + 288$$

$$= -144 + 288$$

$$y = \underline{\hspace{1cm}}$$

$(3, \underline{\hspace{1cm}})$ is the vertex.

[*Note:* Here we chose to use different scales on the two axes. As long as you clearly label the units, this is perfectly acceptable and is useful when values for either (or both) of the variables are very large or very small.]

Since this parabola opens downward, its vertex is its _____ point. The maximum height of the ball is _____ feet.

Check your work on page 701. ▶

Example 6. Graph $y = x^2 + x - 1$. State its domain and range.

Solution

$y = x^2 + x - 1$. This parabola will open upward.

y-INTERCEPT: Let $x = 0$.

$$y = 0^2 + 0 - 1$$

$$= -1$$

$$(0, -1)$$

x-INTERCEPTS: Let $y = 0$.

$$0 = x^2 + x - 1$$

Since the quadratic expression does not factor, we must use the quadratic formula with $a = 1$, $b = 1$, and $c = -1$.

$$x = \frac{-b \pm \sqrt{b^2 - 4ac}}{2a}$$

$$x = \frac{-1 \pm \sqrt{1^2 - 4(1)(-1)}}{2}$$

$$= \frac{-1 \pm \sqrt{1 + 4}}{2}$$

$$x = \frac{-1 \pm \sqrt{5}}{2}$$

$$\left(\frac{-1 + \sqrt{5}}{2}, 0\right) \quad \text{and} \quad \left(\frac{-1 - \sqrt{5}}{2}, 0\right)$$

AXIS OF SYMMETRY

$$x_M = \frac{\dfrac{-1 + \sqrt{5}}{2} + \dfrac{-1 - \sqrt{5}}{2}}{2}$$

$$= \frac{\dfrac{-1 + \sqrt{5} - 1 - \sqrt{5}}{2}}{2}$$

$$= \frac{\dfrac{-2}{2}}{2}$$

$$x_M = \frac{-1}{2}$$

$x = \dfrac{-1}{2}$ is the axis of symmetry.

VERTEX: Let $x = \dfrac{-1}{2}$.

$$y = x^2 + x - 1$$

$$y = \left(\frac{-1}{2}\right)^2 + \left(\frac{-1}{2}\right) - 1$$

$$= \frac{1}{4} - \frac{1}{2} - 1$$

$$= \frac{1}{4} - \frac{2}{4} - \frac{4}{4}$$

$$y = \frac{-5}{4}$$

$\left(\dfrac{-1}{2}, \dfrac{-5}{4}\right)$ is the vertex.

$$y = x^2 + x - 1$$

Domain: $x \in R$

Range: $y \in R$ and $y \geq \dfrac{-5}{4}$

From Example 6 we notice that our four-step procedure will work even when the x-intercepts are irrational numbers.

⫸ Trial Run

Graph each quadratic function. State the domain and range.

Do: _____ **1.** $y = x^2 - 6x$ Do: _____ **2.** $y = x^2 - 4x + 4$
Ra: _____   Ra: _____

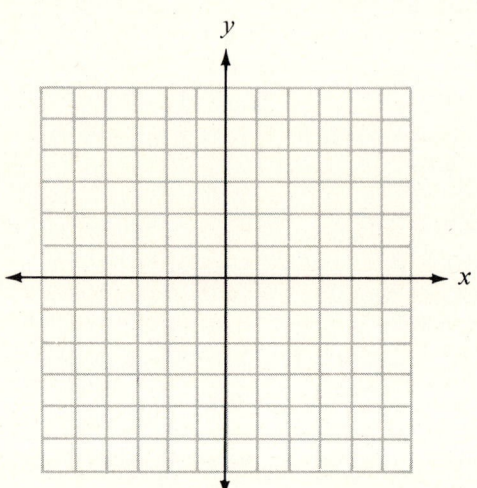

Do: _____ **3.** $y = -x^2 - x + 6$
Ra: _____

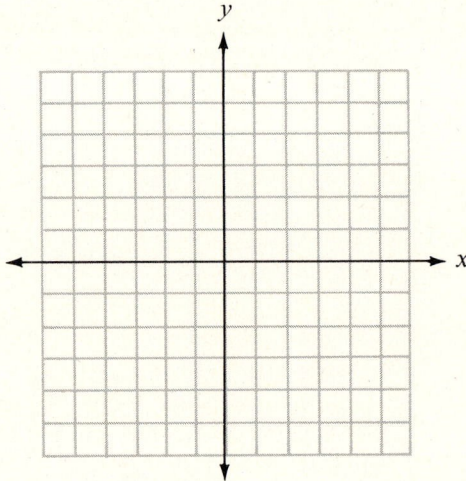

Answers are on page 704.

Graphing Quadratic Functions with No x-Intercepts

Let's graph the quadratic function

$$y = x^2 + x + 1$$

We note that the graph will be a parabola opening upward.

y-INTERCEPT: Let $x = 0$. x-INTERCEPTS: Let $y = 0$.

$$y = 0^2 + 0 + 1$$

$$y = 1$$

$$(0, 1)$$

$$x = \frac{-1\sqrt{1^2 - 4(1)(1)}}{2}$$

$$= \frac{-1 \pm \sqrt{-3}}{2} \quad \text{not real numbers}$$

No x-intercepts.

With no x-intercepts, how do we find the axis of symmetry? We know our graph is a parabola opening upward but never crossing the x-axis. The graph must look something like

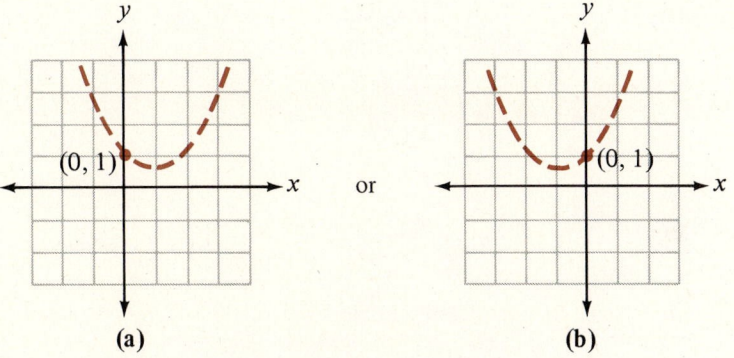

(a)   or   (b)

In either case, there must be another point on the parabola that is symmetric to the point $(0, 1)$. In other words, there must be another point with a y-coordinate of 1. Let's find the other point by letting $y = 1$ in the original equation.

$$y = x^2 + x + 1$$

$$1 = x^2 + x + 1$$

$$0 = x^2 + x$$

$$0 = x(x + 1)$$

$$x = 0 \quad \text{or} \quad x + 1 = 0$$

$$x = -1$$

$$(0, 1) \qquad\qquad (-1, 1)$$

The point $(0, 1)$ is the y-intercept found earlier and the point $(-1, 1)$ is the point on the parabola directly across from $(0, 1)$. Now we can use these two points to locate the axis of symmetry halfway between them.

AXIS OF SYMMETRY:

$$x_M = \frac{-1 + 0}{2}$$

$$= \frac{-1}{2}$$

$x = \dfrac{-1}{2}$ is the axis of symmetry.

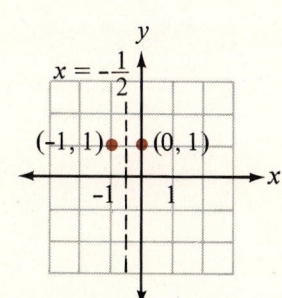

VERTEX: Let $x = \dfrac{-1}{2}$.

$y = x^2 + x + 1$

$y = \left(\dfrac{-1}{2}\right)^2 + \left(\dfrac{-1}{2}\right) + 1$

$= \dfrac{1}{4} - \dfrac{1}{2} + 1$

$= \dfrac{1}{4} - \dfrac{2}{4} + \dfrac{4}{4}$

$= \dfrac{3}{4}$

The vertex is $\left(\dfrac{-1}{2}, \dfrac{3}{4}\right)$.

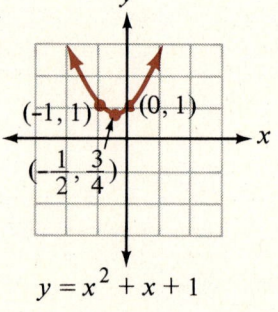

$y = x^2 + x + 1$

Domain: $x \in R$

Range: $y \in R$ and $y \geq \dfrac{3}{4}$

You complete Example 7.

Example 7. Graph $y = -x^2 + 2x - 3$. State its domain and range.

Solution

$y = -x^2 + 2x - 3$. The parabola will open _____ .

y-INTERCEPT: Let $x = 0$.

$y = -0^2 + 2(0) - 3$

$= -3$

$(0, -3)$

x-INTERCEPTS: Let $y = 0$.

$0 = -x^2 + 2x - 3$

$x = \dfrac{-2 \pm \sqrt{2^2 - 4(-1)(-3)}}{2(1)}$

$= \dfrac{-2 \pm \sqrt{4 - 12}}{-2}$

$= \dfrac{-2 \pm \sqrt{-8}}{-2}$

No x-intercepts.

POINT SYMMETRIC TO $(0, -3)$: Let $y = -3$.

$-3 = -x^2 + 2x - 3$

$0 = -x^2 + 2x$

$0 = -x(x - 2)$

$-x = 0$ or $x - 2 = 0$

$x = \underline{\quad}$ $x = \underline{\quad}$

$(\underline{\quad}, -3)$ $(\underline{\quad}, -3)$

Axis of Symmetry

$$x_M = \frac{0 + 2}{2}$$

$$x_M = \underline{\hspace{1cm}}$$

The axis of symmetry is $x = \underline{\hspace{1cm}}$.

Vertex: Let $x = 1$.

$$y = -x^2 + 2x - 3$$

$$y = -(1)^2 + 2(1) - 3$$

$$y = \underline{\hspace{1cm}}$$

The vertex is $(1, \underline{\hspace{1cm}})$.

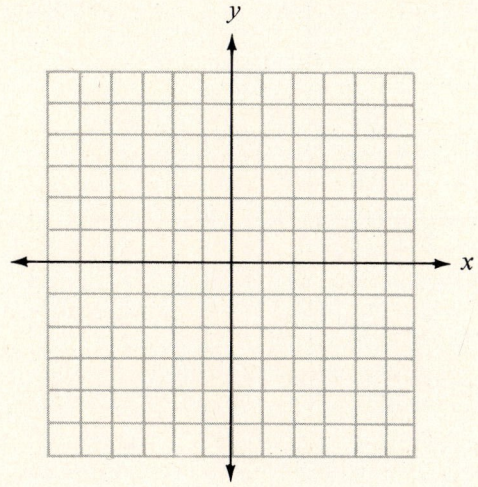

$$y = -x^2 + 2x - 3$$

Domain: _____

Range: _____ and _____

Check your work on page 702. ▶

We summarize our steps for graphing quadratic functions of the form $y = ax^2 + bx + c$.

Graphing Quadratic Functions: $y = ax^2 + bx + c$

1. Find the y-intercept.
2. Find the x-intercepts.
3. Find the axis of symmetry halfway between the x-intercepts.
4. Find the vertex by substituting the x-value from the axis of symmetry into the original equation.
5. Sketch the graph.

If there are *no* x-intercepts, we proceed as follows.

1. Find the y-intercept.
2. Find the point symmetric to the y-intercept.
3. Find the axis of symmetry halfway between the y-intercept and its symmetric point.
4. Find the vertex.
5. Sketch the graph.

⫸ Trial Run

Graph each quadratic function. State the domain and range.

 Do: _____ **1.** $y = x^2 + 2x + 4$

Ra: _____

Do: _____ **2.** $y = -x^2 + 3x - 3$

Ra: _____

Answers are on page 705.

▶ Examples You Completed

Example 3. Graph $y = 2x^2$. Then state its domain and range.

Solution: First we observe that for $y = 2x^2$

1. The parabola opens upward.
2. The axis of symmetry is the y-axis.
3. The vertex is the y-intercept point.
4. The vertex is a minimum point.

Now we would like to find the intercepts.

y-INTERCEPT: Let $x = 0$. x-INTERCEPTS: Let $y = 0$.

$$y = 2(0)^2 \qquad\qquad 0 = 2x^2$$

$$y = 0 \qquad\qquad\quad 0 = x^2$$

$$(0, 0) \qquad\qquad\quad x = 0$$

$$(0, 0)$$

Since the x-intercept and y-intercept both occur at $(0, 0)$, we must choose an x-value on either side of $(0, 0)$ to substitute into $y = 2x^2$.

Let $x = 1$

$y = 2(1)^2$

$= 2$

$(1, 2)$

Let $x = -1$

$y = 2(-1)^2$

$= 2$

$(-1, 2)$

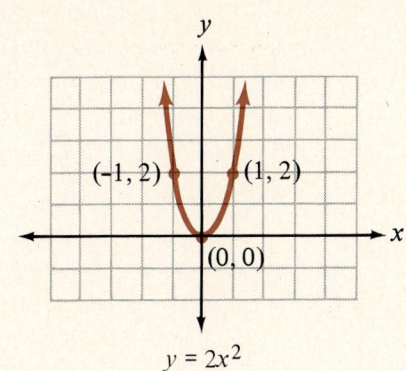

$y = 2x^2$

Domain: $x \in R$

Range: $y \in R$ and $y \geq 0$

Example 5. If a person throws a ball directly upward at a speed of 96 feet per second, then the ball's height y above the ground after x seconds is given by the function $y = -16x^2 + 96x$. Use the graph of this function to find the maximum height of the ball.

Solution: The graph of $y = -16x^2 + 96x$ is a parabola that opens downward.

y-INTERCEPT: Let $x = 0$.

$y = -16(0)^2 + 96(0)$

$y = 0$

$(0, 0)$

x-INTERCEPTS: Let $y = 0$.

$0 = -16x^2 + 96x$

$0 = -16x(x - 6)$

$-16x = 0$ or $x - 6 = 0$

$x = 0$ $\qquad x = 6$

$(0, 0)$ $\qquad (6, 0)$

AXIS OF SYMMETRY

$$x_M = \frac{0 + 6}{2}$$

$$= \frac{6}{2}$$

$$= 3$$

$x = 3$ is the axis of symmetry.

VERTEX: Let $x = 3$.

$y = -16x^2 + 96x$

$y = -16(3)^2 + 96(3)$

$= -16(9) + 288$

$= -144 + 288$

$y = 144$

$(3, 144)$ is the vertex.

Since this parabola opens downward, its vertex is its maximum point. The maximum height of the ball is 144 feet.

Example 7. Graph $y = -x^2 + 2x - 3$. State its domain and range.

Solution: $y = -x^2 + 2x - 3$. The parabola will open downward.

y-INTERCEPT: Let $x = 0$.

$$y = -0^2 + 2(0) - 3$$
$$= -3$$
$$(0, -3)$$

x-INTERCEPTS: Let $y - 0$.

$$0 = -x^2 + 2x - 3$$
$$x = \frac{-2 \pm \sqrt{2^2 - 4(-1)(-3)}}{2(-1)}$$
$$= \frac{-2 \pm \sqrt{4 - 12}}{-2}$$
$$= \frac{-2 \pm \sqrt{-8}}{-2}$$

No real x-intercepts.

POINT SYMMETRIC TO $(0, -3)$: Let $y = -3$.

$$-3 = -x^2 + 2x - 3$$
$$0 = -x^2 + 2x$$
$$0 = -x(x - 2)$$
$$-x = 0 \quad \text{or} \quad x - 2 = 0$$
$$x = 0 \qquad\qquad x = 2$$
$$(0, -3) \qquad\qquad (2, -3)$$

AXIS OF SYMMETRY

$$x_M = \frac{0+2}{2}$$

$$= \frac{2}{2}$$

$$x_M = 1$$

The axis of symmetry is $x = 1$.

VERTEX: Let $x = 1$.

$$y = -x^2 + 2x - 3$$

$$= -(1)^2 + 2(1) - 3$$

$$= -1 + 2 - 3$$

$$y = -2$$

The vertex is $(1, -2)$.

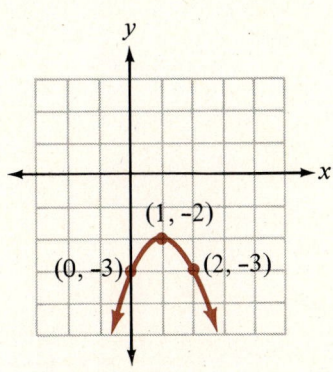

$$y = -x^2 + 2x - 3$$

Domain: $x \in R$

Range: $y \in R$ and $y \le -2$

Answers to Trial Runs

page 690 **1.**

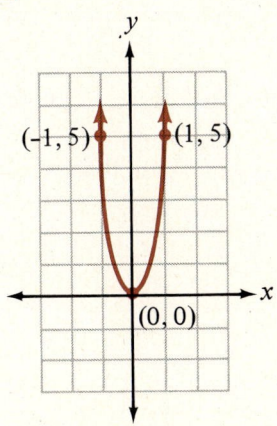

Do: $x \in R$
Ra: $y \in R$ and $y \ge 0$

2.

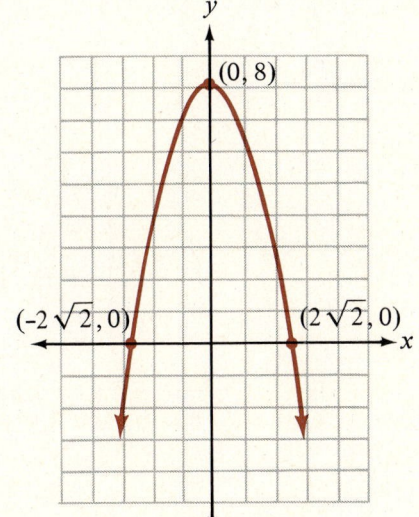

Do: $x \in R$
Ra: $y \in R$ and $y \le 8$

3.

Do: $x \in R$
Ra: $y \in R$ and $y \leq -6$

page 696 **1.**

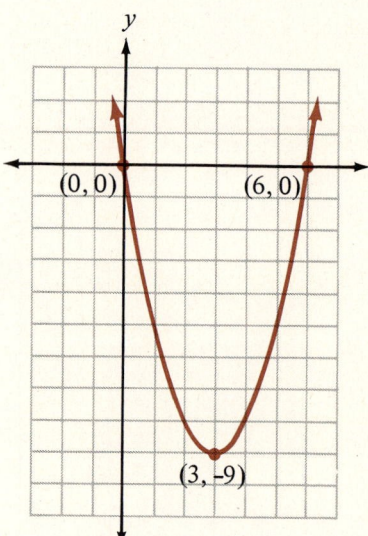

Do: $x \in R$
Ra: $y \in R$ and $y \geq -9$

2.

Do: $x \in R$
Ra: $y \in R$ and $y \geq 0$

3.

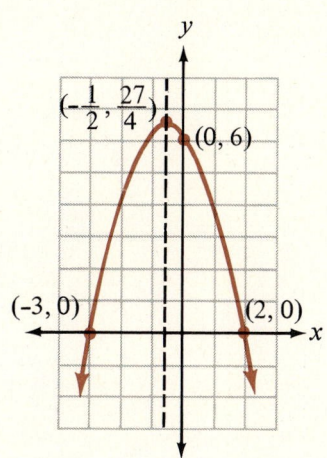

Do: $x \in R$

Ra: $y \in R$ and $y \leq \dfrac{27}{4}$

page 700   **1.**

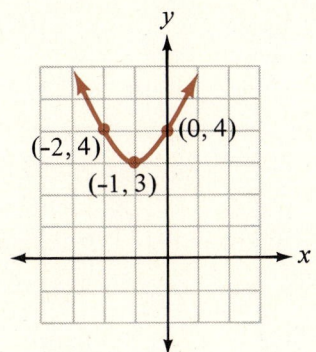

Do: $x \in R$
Ra: $y \in R$ and $y \geq 3$

2.

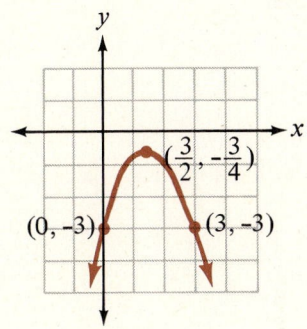

Do: $x \in R$

Ra: $y \in R$ and $y \leq \dfrac{-3}{4}$

EXERCISE SET 9.2

Graph each of the following functions. State the domain and range.

Do: _____   **1.** $y = 3x^2$ Do: _____   **2.** $y = -4x^2$
Ra: _____   Ra: _____

Do: _____   **3.** $y = -x^2 + 6$ Do: _____   **4.** $y = x^2 + 12$
Ra: _____   Ra: _____

Do: _____   **5.** $y = x^2 - 9$ Do: _____   **6.** $y = x^2 - 16$
Ra: _____   Ra: _____

Do: _____   **7.** $y = -5x^2 - 10$ Do: _____   **8.** $y = -2x^2 - 8$
Ra: _____   Ra: _____

Do: _____   **9.** $y = x^2 + 4x$ Do: _____   **10.** $y = x^2 - 2x$
Ra: _____   Ra: _____

Do: _____   **11.** $y = x^2 - 10x + 25$ Do: _____   **12.** $y = x^2 - 6x + 9$
Ra: _____   Ra: _____

Do: _____   **13.** $y = -x^2 + 2x + 15$ Do: _____   **14.** $y = -x^2 - 3x + 4$
Ra: _____   Ra: _____

Do: _____   **15.** $y = -2x^2 - x + 6$ Do: _____   **16.** $y = -3x^2 + 2x + 5$
Ra: _____   Ra: _____

Do: _____   **17.** $y = 3x^2 - 3x + 1$ Do: _____   **18.** $y = 5x^2 - 2x + 1$
Ra: _____   Ra: _____

Do: _____   **19.** $y = x^2 + x + 2$ Do: _____   **20.** $y = x^2 + 3x + 5$
Ra: _____   Ra: _____

Do: _____   **21.** $y = -x^2 - 2x - 4$ Do: _____   **22.** $y = -x^2 + 4x - 6$
Ra: _____   Ra: _____

Do: _____   **23.** $y = -2x^2 + 5x - 5$ Do: _____   **24.** $y = -3x^2 + 8x - 6$
Ra: _____   Ra: _____

_____   **25.** Estelle runs a television repair shop. The formula she uses to approximate her profits is $y = -3x^2 + 108x$, where y is the profit from repairing x television sets. Use a graph to determine how many sets she must repair to make a maximum profit. What is the maximum profit?

_____   **26.** A lamp manufacturer uses the formula $y = -5x^2 + 600x$ to determine the profit y when x lamps are sold. Use a graph to determine how many lamps must be sold to make a maximum profit. What is the maximum profit?

_____   **27.** The function for the cost y of building x picnic tables is $y = x^2 - 12x + 72$. Use a graph to determine how many picnic tables Mr. Denison should build to have a minimum cost. What is the minimum cost?

_____ **28.** The cost y of running a steak house is related to the number of cartons of steaks x that are sold by the function $y = x^2 - 18x + 100$. Use a graph to determine how many cartons of steaks must be sold to keep costs at a minimum. What is the minimum cost?

☆ Stretching the Topics _____

_____ **1.** Estimate to the nearest tenth the x-intercepts of the parabola determined by $y = x^2 - 3x - 5$.

_____ **2.** The sum of the width and length of a rectangle is 18 feet. Write an equation that expresses the fact that the area is a function of the width. Use a graph to determine what the width should be so that the rectangle will have a maximum area. What is the maximum area?

Check your answers in the back of your book.

If you can graph the function in **Checkup 9.2,** you are ready to go on to Section 9.3.

✓ CHECKUP 9.2

Graph each of the following functions. State the domain and range.

Do: _____ **1.** $y = -2x^2$
Ra: _____

Do: _____
Ra: _____

Do: _____ **3.** $y = x^2 - x - 2$
Ra: _____

4 - 2 2
1 - 3

opens up

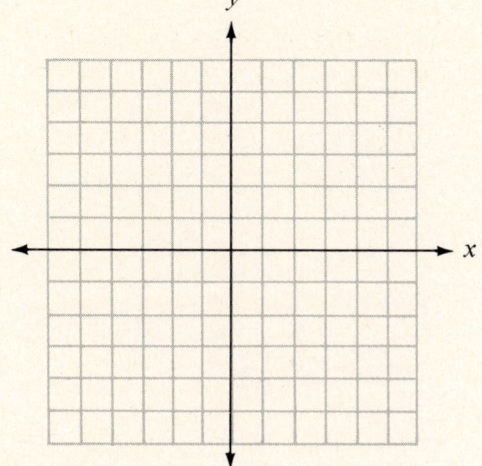

Do: _____ **4.** $y = -x^2 + 5x$
Ra. _____

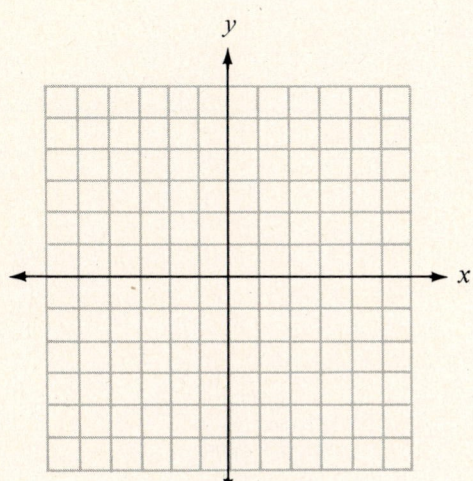

+ 2x − 2

−4 + 4 − 2
−10

opns up

−2

−2 −.0

0 −2

Do: _____ **6.** $y = -x^2 + 3x - 5$

Ra: _____

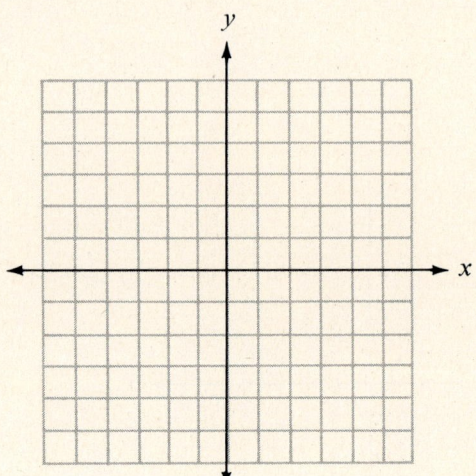

Check your answers in the back of your book.

If You Missed Problems:	You Should Review Examples:
1	2, 3
2	1
3	4
4	5
5	6
6	7

9.3 **Graphing the Conic Sections**

An interesting family of curves results when the surface of a cone is cut by a plane.

Parabola   Circle   Ellipse   Hyperbola

Because of their relationship to the cone, these curves are called the **conic sections.** The first conic section, the **parabola,** was discussed in detail in Section 9.2. Now we must see what kinds of equations would result in the remaining conic sections.

Graphing Circles

Suppose we look at the second conic section—the **circle.** Consider the graph of a circle with center at the origin and radius, r. We know the distance between any point on our circle and the circle's center must be r units.

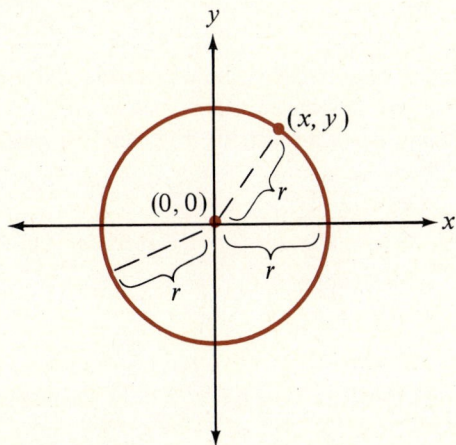

Letting (x, y) represent the coordinates of any point on the circle and $(0, 0)$ represent the origin, we can use the distance formula from Chapter 8 and state that

$$r = \sqrt{(x - 0)^2 + (y - 0)^2}$$
$$= \sqrt{x^2 + y^2}$$
$$r^2 = x^2 + y^2$$

> The graph of the equation
>
> $$x^2 + y^2 = r^2$$
>
> is a *circle* with center (0, 0) and radius r (where $r > 0$).

Suppose that we use this fact to graph the most fundamental circle:

$$x^2 + y^2 = 1$$

Center: (0, 0)   Radius $r^2 = 1$

$$r = 1$$

We may now plot the most easily located points lying $r = 1$ unit from the origin. Those points are, of course, the intercepts and we join them with a smooth curve.

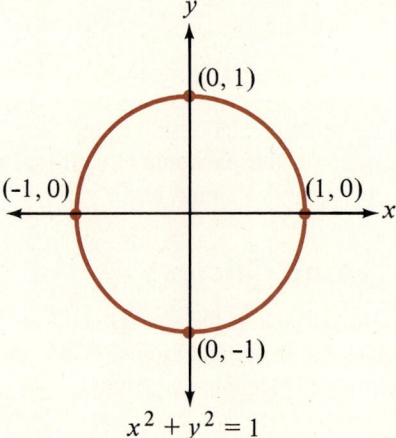

$$x^2 + y^2 = 1$$

This circle is called the **unit circle** and is of great importance in the study of the branch of mathematics called trigonometry. Is $x^2 + y^2 = 1$ a function? The vertical line test tells us that it is *not* a function. It is merely a relation for which

Domain: $x \in R$ and $-1 \le x \le 1$

Range: $y \in R$ and $-1 \le y \le 1$

You complete Example 1.

Example 1. Graph $2x^2 + 2y^2 = 4$. State its domain and range.

Solution

$$2x^2 + 2y^2 = 4$$

Dividing both sides by 2 yields

$$x^2 + y^2 = 2$$

The graph will be a _____ .

Center: (____, ____)

Radius: $r^2 = 2$

$$r = \sqrt{\underline{}}$$

Check your work on page 723. ▶

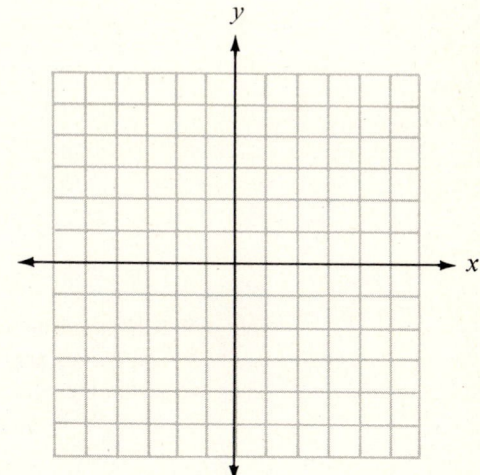

Domain: $x \in R$ and ____ $\leq x \leq$ ____

Range: $y \in R$ and ____ $\leq y \leq$ ____

Graphing circles is quite systematic, but we cannot expect circles to be always centered at the origin. Let's investigate what happens when a circle is centered at some other point, say (h, k). The definition of a circle still demands that the distance from the center (h, k) to any point (x, y) on the circle must be r.

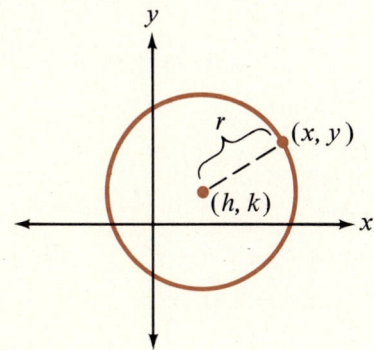

$$r = \sqrt{(x - h)^2 + (y - k)^2}$$
$$r^2 = (x - h)^2 + (y - k)^2$$

The graph of the equation

$$(x - h)^2 + (y - k)^2 = r^2$$

is a *circle* with center (h, k) and radius r (where h, k, and r are constants, and $r > 0$).

Example 2. Graph $(x - 3)^2 + (y - 5)^2 = 4$. State its domain and range.

Solution

$$(x - 3)^2 + (y - 5)^2 = 4$$

Center: $x - h = x - 3 \qquad y - k = y - 5$

$$h = 3 \qquad\qquad k = 5$$

$$(h, k) = (3, 5)$$

Radius: $r^2 = 4$

$$r = 2$$

We locate the center at (3, 5) and then locate four points that are 2 units directly to the left of, right of, above, and below the center. Then we sketch the circle.

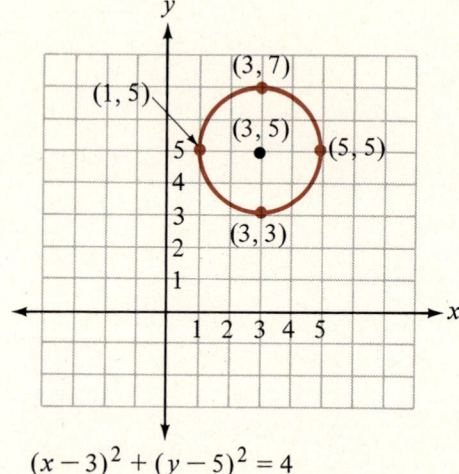

$$(x - 3)^2 + (y - 5)^2 = 4$$

Domain: $x \in R$ and $1 \le x \le 5$

Range: $y \in R$ and $3 \le y \le 7$

Example 3. Graph $(x + 2)^2 + y^2 = 5$. State its domain and range.

Solution

$$(x + 2)^2 + y^2 = 5$$

$$[x - (-2)]^2 + (y - 0)^2 = 5$$

Center: $x - h = x - (-2) \qquad y - k = y - 0$

$$h = -2 \qquad\qquad k = 0$$

$$(h, k) = (-2, 0)$$

Radius: $r^2 = 5$

$$r = \sqrt{5}$$

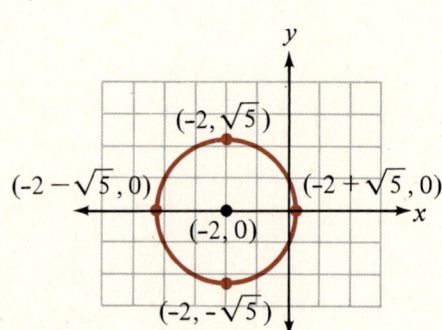

Domain: $x \in R$ and $-2 - \sqrt{5} \le x \le -2 + \sqrt{5}$

Range: $y \in R$ and $-\sqrt{5} \le y \le \sqrt{5}$

When a coordinate of a point is an irrational number, we may find its approximate value using a calculator or the table inside the back cover of this book. Then we approximate the point's location in the plane and label it with its exact irrational coordinates.

⟫ Trial Run

Graph each relation. State the domain and range.

Do: _____ **1.** $x^2 + y^2 = 9$ Do: _____ **2.** $3x^2 + 3y^2 = 6$
Ra: _____   Ra: _____

        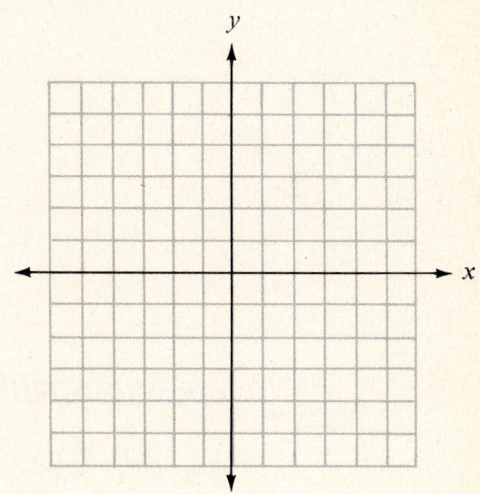

Do: _____ **3.** $(x - 3)^2 + (y - 1)^2 = 1$
Ra: _____

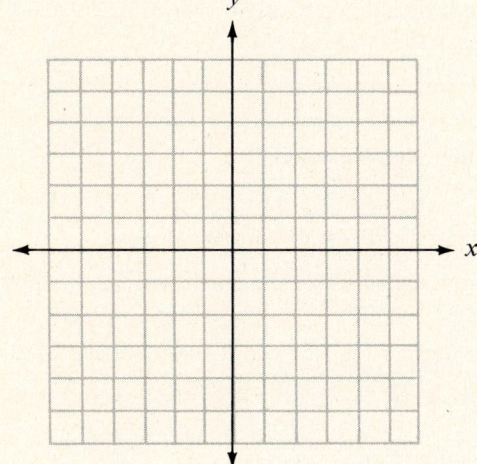

Do: _____   **4.** $(x + 1)^2 + (y + 2)^2 = 25$
Ra: _____

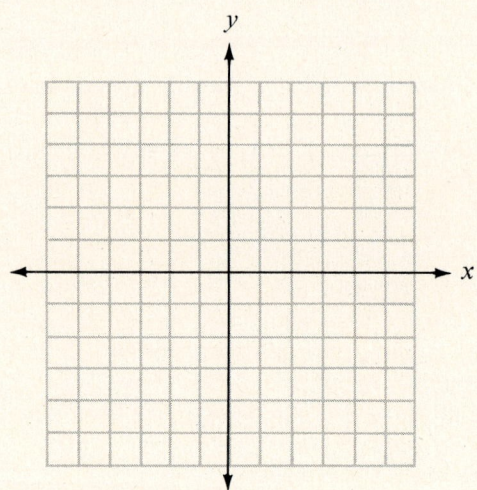

Answers are on page 724.

Graphing Ellipses

The third kind of conic section is the **ellipse.** We concentrate here only on ellipses centered at the origin. Since an ellipse is sort of an elongated circle, we expect its equation to be similar to that of a circle.

> The graph of the equation
> $$Ax^2 + By^2 = C$$
> is an *ellipse* centered at the origin (where A, B, and C are *positive* constants and $A \neq B$).

Note that if $A = B$, this equation represents a circle.

To graph an equation such as

$$9x^2 + 4y^2 = 36$$

we rely heavily on knowing from the start that the graph will be an ellipse. First we find the *intercepts*.

x-INTERCEPTS: Let $y = 0$. y-INTERCEPTS: Let $x = 0$.

$9x^2 + 4(0)^2 = 36$ $9(0)^2 + 4y^2 = 36$

$9x^2 = 36$ $4y^2 = 36$

$x^2 = 4$ $y^2 = 9$

$x = \pm 2$ $y = \pm 3$

$(2, 0)$ and $(-2, 0)$ $(0, 3)$ and $(0, -3)$

Since this is the first ellipse we have graphed, let's plot a few more points.

Let $x = 1$.

$$9(1)^2 + 4y^2 = 36$$

$$9 + 4y^2 = 36$$

$$4y^2 = 27$$

$$y^2 = \frac{27}{4}$$

$$y = \frac{\pm\sqrt{27}}{2}$$

$$y = \frac{\pm 3\sqrt{3}}{2}$$

$$\left(1, \frac{3\sqrt{3}}{2}\right) \quad \text{and} \quad \left(1, \frac{-3\sqrt{3}}{2}\right)$$

Let $x = -1$.

$$9(-1)^2 + 4y^2 = 36$$

$$9 + 4y^2 = 36$$

$$4y^2 = 27$$

$$y^2 = \frac{27}{4}$$

$$y = \frac{\pm\sqrt{27}}{2}$$

$$y = \frac{\pm 3\sqrt{3}}{2}$$

$$\left(-1, \frac{3\sqrt{3}}{2}\right) \quad \text{and} \quad \left(-1, \frac{-3\sqrt{3}}{2}\right)$$

Notice the symmetry of the four points. In order to plot these points, we recall that $\sqrt{3} \doteq 1.732$, so we may *approximate* the location of the points by

$$(1, 2.6) \quad \text{and} \quad (1, -2.6) \qquad (-1, 2.6) \quad \text{and} \quad (-1, -2.6)$$

but we label the points with their *exact* coordinates.

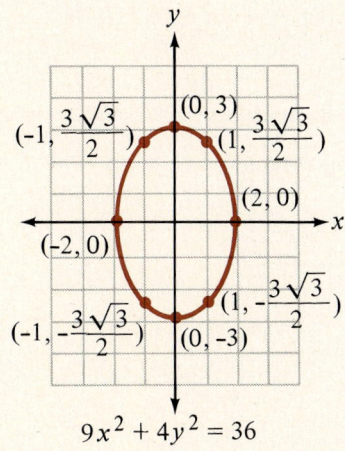

$$9x^2 + 4y^2 = 36$$

Does this graph represent a function? No, it is merely a relation for which

Domain: $x \in R$ and $-2 \le x \le 2$

Range: $y \in R$ and $-3 \le y \le 3$

Now that we have seen that the graph of $Ax^2 + By^2 + C$ (where A, B, $C > 0$, and $A \ne B$) is indeed an ellipse centered at the origin, we shall plot only the four intercept points and connect them with a smooth curve.

You complete Example 4.

Example 4. Graph $x^2 + 2y^2 = 4$. State its domain and range.

Solution

   $x^2 + 2y^2 = 4$

The graph will be an _____ .

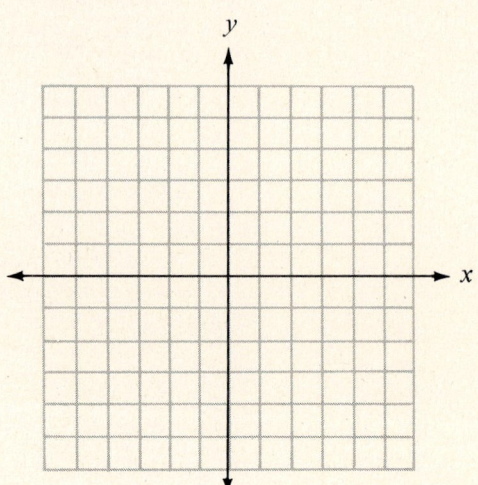

x-INTERCEPTS: Let $y = 0$.

$x^2 + 2(0)^2 = 4$

$x^2 = 4$

$x = \pm$ _____

(_____, 0) and (_____, 0)

y-INTERCEPTS: Let $x = 0$.

$0^2 + 2y^2 = 4$

$2y^2 = 4$

$y^2 = 2$

$y = \pm\sqrt{\underline{}}$

(0, _____) and (0, _____)

Domain: $x \in R$ and _____ $\leq x \leq$ _____

Range: $y \in R$ and _____ $\leq y \leq$ _____

Check your work on page 723. ▶

IIII➡ **Trial Run**

Graph each relation. State the domain and range.

  Do: _____ **1.** $9x^2 + 4y^2 = 36$ Do: _____ **2.** $2x^2 + 3y^2 = 6$
Ra: _____   Ra: _____

Answers are on page 725.

Graphing Hyperbolas

The final conic section that we must investigate is the **hyperbola.** Once again we restrict our discussion to hyperbolas that are centered at the origin. Hyperbolas centered at the origin will look like one of these figures.

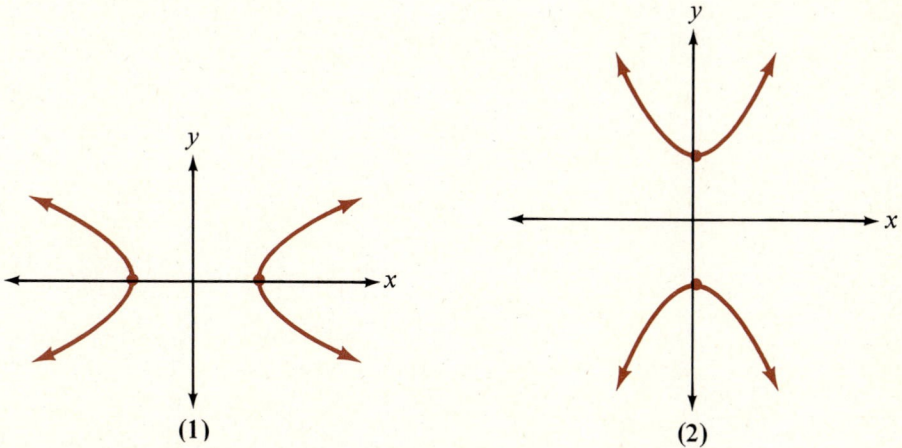

(1)   (2)

The general equation for a hyperbola is similar to the equation for an ellipse, except that *A and B must have opposite signs*.

The graph of the equation

$$Ax^2 + By^2 = C$$

is a *hyperbola* (where A, B, and C are constants, and A and B have *opposite* signs).

Example 5. Graph $x^2 - 4y^2 = 16$. State its domain and range.

Solution

$$x^2 - 4y^2 = 16.$$ Let's find intercepts.

x-INTERCEPTS: Let $y = 0$. y-INTERCEPTS: Let $x = 0$.

$$x^2 - 4(0)^2 = 16 \qquad\qquad 0^2 - 4y^2 = 16$$
$$x^2 = 16 \qquad\qquad\qquad -4y^2 = 16$$
$$x = \pm 4 \qquad\qquad\qquad y^2 = -4$$

$(4, 0)$ and $(-4, 0)$ There are no y-intercepts.

Since our graph does *not* cross the y-axis, it must resemble Figure 1. We need a few more points to be sure of its shape. From Figure 1 we know that only x-values to the left of $(-4, 0)$ or to the right of $(4, 0)$ will correspond to real y-values. To take advantage of the symmetry of the hyperbola, let's substitute two x-values equally spaced to the left of -4 and to the right of 4.

Let $x = -5$.   Let $x = 5$.

$$x^2 - 4y^2 = 16$$ $$x^2 - 4y^2 = 16$$

$$(-5)^2 - 4y^2 = 16$$ $$(5)^2 - 4y^2 = 16$$

$$25 - 4y^2 = 16$$ $$25 - 4y^2 = 16$$

$$-4y^2 = -9$$ $$-4y^2 = -9$$

$$y^2 = \frac{9}{4}$$ $$y^2 = \frac{9}{4}$$

$$y = \pm\frac{3}{2}$$ $$y = \pm\frac{3}{2}$$

$$\left(-5, \frac{3}{2}\right) \quad \text{and} \quad \left(-5, \frac{-3}{2}\right) \qquad \left(5, \frac{3}{2}\right) \quad \text{and} \quad \left(5, \frac{-3}{2}\right)$$

Notice that our symmetrically chosen x-values yielded symmetric y-values. This will always occur if you choose x-values that are opposites. Now we sketch the graph.

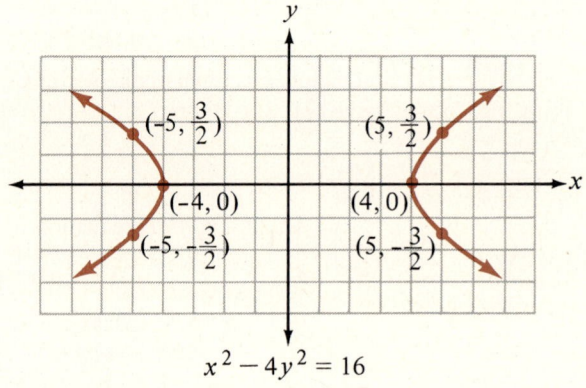

$$x^2 - 4y^2 = 16$$

Does this hyperbola represent a function? No, it is merely a relation for which

Domain: $x \in R$ and $x \geq 4$ or $x \leq -4$

Range: $y \in R$

Example 6. Graph $5y^2 - 2x^2 = 5$. State its domain and range.

Solution

$$5y^2 - 2x^2 = 5.$$ The graph will be a hyperbola.

x-INTERCEPTS: Let $y = 0$. y-INTERCEPTS: Let $x = 0$.

$$5(0)^2 - 2x^2 = 5$$ $$5y^2 - 2(0)^2 = 5$$

$$-2x^2 = 5$$ $$5y^2 = 5$$

$$x^2 = \frac{-5}{2}$$ $$y^2 = 1$$

                                   $$y = \pm 1$$

There are no x-intercepts. $(0, 1)$ and $(0, -1)$

This graph will resemble Figure 2, so we choose x-values equally spaced to the left and right of $(0, 1)$ and $(0, -1)$. Let's substitute $x = -1$ and $x = 1$ into the original equation.

Let $x = -1$.

$$5y^2 - 2x^2 = 5$$
$$5y^2 - 2(-1)^2 = 5$$
$$5y^2 - 2 = 5$$
$$5y^2 = 7$$
$$y^2 = \frac{7}{5}$$
$$y = \pm\sqrt{\frac{7}{5}}$$

Let $x = 1$.

$$5y^2 - 2x^2 = 5$$
$$5y^2 - 2(1)^2 = 5$$
$$5y^2 - 2 = 5$$
$$5y^2 = 7$$
$$y^2 = \frac{7}{5}$$
$$y = \pm\sqrt{\frac{7}{5}}$$

We should rationalize the denominators.

$$y = \frac{\pm\sqrt{7}}{\sqrt{5}} \cdot \frac{\sqrt{5}}{\sqrt{5}}$$
$$= \frac{\pm\sqrt{35}}{5}$$

The points on the hyperbola are

$$\left(-1, \frac{\sqrt{35}}{5}\right), \quad \left(-1, \frac{-\sqrt{35}}{5}\right), \quad \left(1, \frac{\sqrt{35}}{5}\right), \quad \left(1, \frac{-\sqrt{35}}{5}\right)$$

Using the fact that $\sqrt{35} \doteq 5.916$, we can approximate our points by

$$(-1, 1.2), \quad (-1, -1.2), \quad (1, 1.2), \quad (1, -1.2)$$

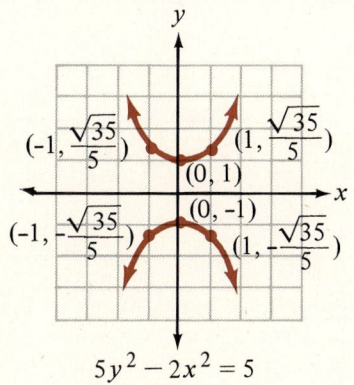

$$5y^2 - 2x^2 = 5$$

Domain: $x \in R$

Range: $y \in R$ and $y \geq 1$ or $y \leq -1$

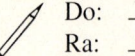 **Trial Run** _____

Graph each relation. State the domain and range.

Do: _____ **1.** $4x^2 - 2y^2 = 16$ Do: _____ **2.** $4y^2 - x^2 = 8$
Ra: _____   Ra: _____

          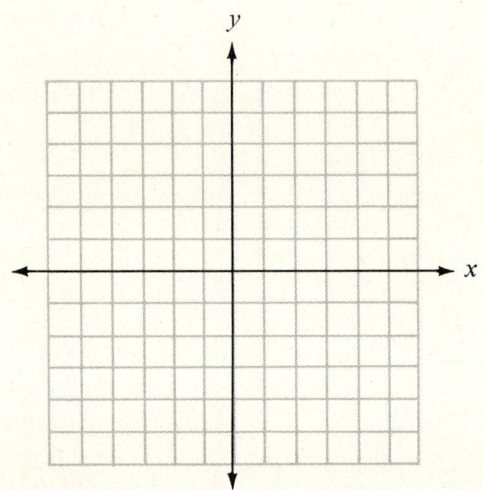

Answers are on page 725.

In discussing the conic sections we have learned to identify each graph from its equation.

Conic Section	Equation	Graphing Method
Parabola	$y = ax^2 + bx + c$ $(a \neq 0)$	Find intercepts Find axis of symmetry Find vertex
Circle	$x^2 + y^2 = r^2$ $(r > 0)$	Center at $(0, 0)$ Radius r
	$(x - h)^2 + (y - k)^2 = r^2$ $(r > 0)$	Center at (h, k) Radius r
Ellipse	$Ax^2 + By^2 = C$ $(A, B, C > 0, A \neq B)$	Find x-intercepts Find y-intercepts
Hyperbola	$Ax^2 + By^2 = C$ $(A$ and B opposite in sign$)$	Find intercepts Find four symmetric points

We discovered that the parabola is the only conic section that represents a function.

▶ Examples You Completed

Example 1. Graph $2x^2 + 2y^2 = 4$. State its domain and range.

Solution

$$2x^2 + 2y^2 = 4$$

Dividing both sides by 2 yields

$$x^2 + y^2 = 2$$

The graph will be a circle.

Center: $(0, 0)$

Radius: $r^2 = 2$

$$r = \sqrt{2}$$

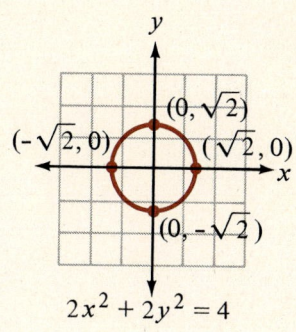

$$2x^2 + 2y^2 = 4$$

Domain: $x \in R$ and $-\sqrt{2} \le x \le \sqrt{2}$

Range: $y \in R$ and $-\sqrt{2} \le y \le \sqrt{2}$

Example 4. Graph $x^2 + 2y^2 = 4$. State its domain and range.

Solution

$$x^2 + 2y^2 = 4$$

The graph will be an ellipse.

x-INTERCEPTS: Let $y = 0$.

$$x^2 + 2(0)^2 = 4$$

$$x^2 = 4$$

$$x = \pm 2$$

$(2, 0)$ and $(-2, 0)$

y-INTERCEPTS: Let $x = 0$.

$$0^2 + 2y^2 = 4$$

$$2y^2 = 4$$

$$y^2 = 2$$

$$y = \pm\sqrt{2}$$

$(0, \sqrt{2})$ and $(0, -\sqrt{2})$

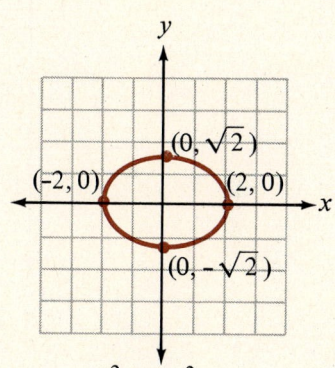

$$x^2 + 2y^2 = 4$$

Domain: $x \in R$ and $-2 \le x \le 2$

Range: $y \in R$ and $-\sqrt{2} \le y \le \sqrt{2}$

Answers to Trial Runs

page 715 **1.**

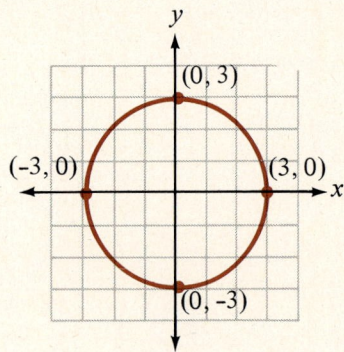

Do: $x \in R$ and $-3 \le x \le 3$
Ra: $y \in R$ and $-3 \le y \le 3$

2.

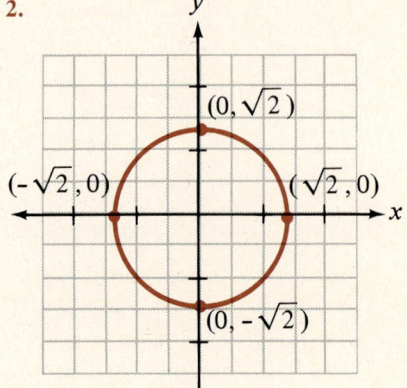

Do: $x \in R$ and $-\sqrt{2} \le x \le \sqrt{2}$
Ra: $y \in R$ and $-\sqrt{2} \le y \le \sqrt{2}$

3.

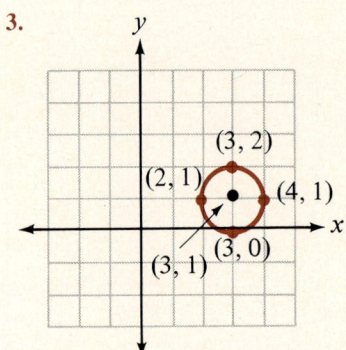

Do: $x \in R$ and $2 \le x \le 4$
Ra: $y \in R$ and $0 \le y \le 2$

4.

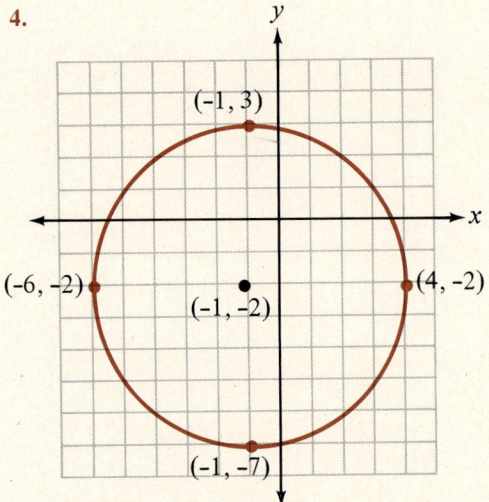

Do: $x \in R$ and $-6 \le x \le 4$
Ra: $y \in R$ and $-7 \le y \le 3$

page 718   **1.**

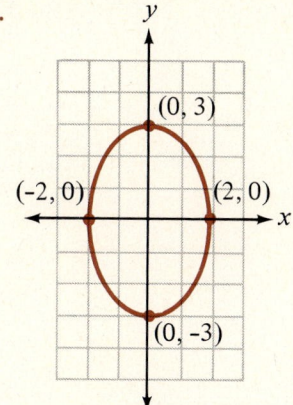

Do: $x \in R$ and $-2 \le x \le 2$
Ra: $y \in R$ and $-3 \le y \le 3$

2.

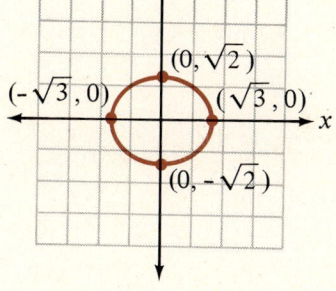

Do: $x \in R$ and $-\sqrt{3} \le x \le \sqrt{3}$
Ra: $y \in R$ and $-\sqrt{2} \le y \le \sqrt{2}$

page 722   **1.**

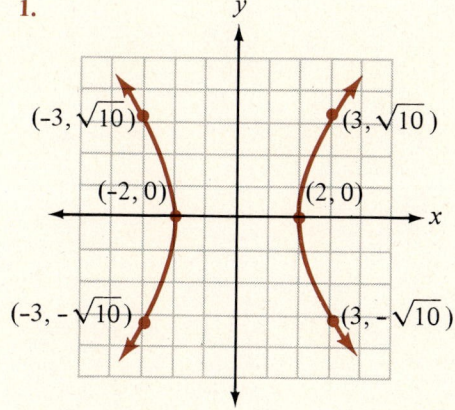

Do: $x \in R$ and $x \le -2$ or $x \ge 2$
Ra: $y \in R$

2.

Do: $x \in R$
Ra: $y \in R$ and $y \le -\sqrt{2}$ or $y \ge \sqrt{2}$

EXERCISE SET 9.3

Graph each of the following relations. State the domain and range.

Do: _____ **1.** $x^2 + y^2 = 16$
Ra: _____

Do: _____ **2.** $x^2 + y^2 = 25$
Ra: _____

Do: _____ **3.** $x^2 + y^2 = 6$
Ra: _____

Do: _____ **4.** $x^2 + y^2 = 10$
Ra: _____

Do: _____ **5.** $5x^2 + 5y^2 = 45$
Ra: _____

Do: _____ **6.** $4x^2 + 4y^2 = 36$
Ra: _____

Do: _____ **7.** $y^2 = 16 - x^2$
Ra: _____

Do: _____ **8.** $y^2 = 36 - x^2$
Ra: _____

Do: _____ **9.** $(x - 1)^2 + (y - 3)^2 = 25$
Ra: _____

Do: _____ **10.** $(x - 2)^2 + (y - 4)^2 = 9$
Ra: _____

Do: _____ **11.** $(x + 3)^2 + (y - 1)^2 = 1$
Ra: _____

Do: _____ **12.** $(x + 9)^2 + (y - 3)^2 = 81$
Ra: _____

Do: _____ **13.** $x^2 + (y - 6)^2 = 4$
Ra: _____

Do: _____ **14.** $x^2 + (y - 8)^2 = 49$
Ra: _____

Do: _____ **15.** $(x + 2)^2 + (y + 7)^2 = 64$
Ra: _____

Do: _____ **16.** $(x + 3)^2 + (y + 5)^2 = 144$
Ra: _____

Do: _____ **17.** $9x^2 + 25y^2 = 225$
Ra: _____

Do: _____ **18.** $4x^2 + 9y^2 = 36$
Ra: _____

Do: _____ **19.** $16x^2 + y^2 = 16$
Ra: _____

Do: _____ **20.** $9x^2 + y^2 = 9$
Ra: _____

Do: _____ **21.** $12x^2 + y^2 = 36$
Ra: _____

Do: _____ **22.** $x^2 + 8y^2 = 16$
Ra: _____

Do: _____ **23.** $3x^2 + 5y^2 = 15$
Ra: _____

Do: _____ **24.** $7x^2 + 5y^2 = 35$
Ra: _____

Do: _____ **25.** $x^2 - 4y^2 = 4$
Ra: _____

Do: _____ **26.** $x^2 - 16y^2 = 16$
Ra: _____

Do: _____ **27.** $4y^2 - 25x^2 = 100$
Ra: _____

Do: _____ **28.** $16y^2 - 9x^2 = 144$
Ra: _____

Do: _____ **29.** $3x^2 - y^2 = 6$
Ra: _____

Do: _____ **30.** $5x^2 - y^2 = 15$
Ra: _____

Do: _____ **31.** $y^2 - x^2 = 100$
Ra: _____

Do: _____ **32.** $y^2 - x^2 = 25$
Ra: _____

☆ Stretching the Topics

_____ **1.** Write the equation of a circle with the center at $(-3, 4)$ and a radius of 5 units.

_____ **2.** Estimate to the nearest tenth the x- and y-intercepts of $6x^2 + 9y^2 = 27$.

_____ **3.** Graph $y = \frac{1}{3}\sqrt{81 - x^2}$. State its domain and range.

Check your answers in the back of your book.

If you can complete Checkup 9.3, you are ready to go on to Section 9.4.

✔ **CHECKUP 9.3**

Graph each of the following relations. State the domain and range.

Do: _____ **1.** $x^2 + y^2 = 64$
Ra: _____

Do: _____ **2.** $3x^2 + 3y^2 = 27$
Ra: _____

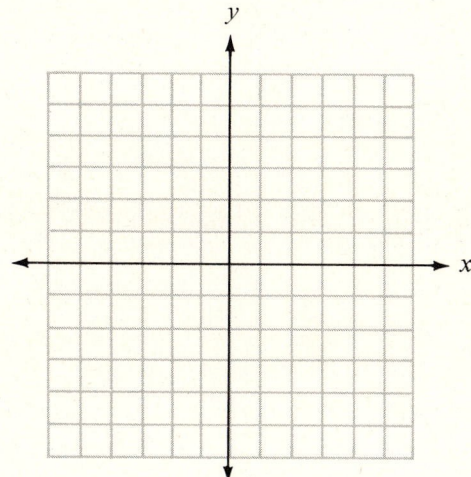

Do: _____ **3.** $(x + 4)^2 + (y - 9)^2 = 81$
Ra: _____

Do: _____ **4.** $9x^2 + 16y^2 = 144$
Ra: _____

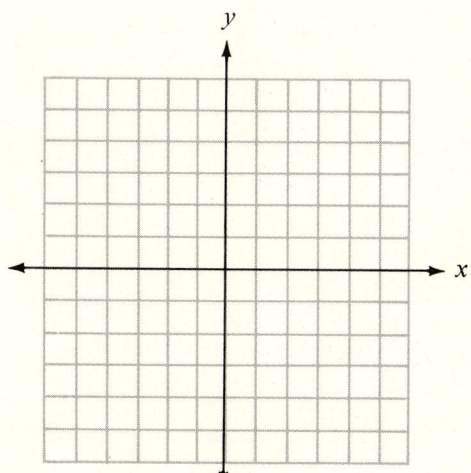

Do: _____ **5.** $16y^2 - 4x^2 = 64$
Ra: _____

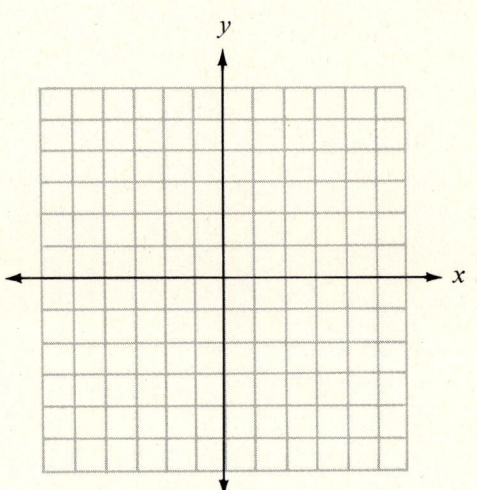

Check your answers in the back of your book.

If You Missed Problems:	You Should Review Examples:
1, 2	1
3	2
4	4
5	5, 6

9.4 Graphing Second-Degree Inequalities

Now that we have learned to graph second-degree equations, we turn our attention to graphing **second-degree inequalities.** Recall from Chapter 8 that we learned to graph a *first-degree inequality* by graphing the corresponding first-degree equation (straight line) and then shading the region containing the points that satisfied the inequality. We shall use the same approach with second-degree inequalities.

Example 1. Graph $y < x^2 - 2x - 8$.

Solution: First we must graph the corresponding boundary equation

$$y = x^2 - 2x - 8$$

which we recognize as a parabola.

y-INTERCEPT: Let $x = 0$.

$$y = -8$$

$$(0, -8)$$

x-INTERCEPTS: Let $y = 0$.

$$x^2 - 2x - 8 = 0$$

$$(x - 4)(x + 2) = 0$$

$$x = 4 \qquad x = -2$$

$$(4, 0) \quad \text{and} \quad (-2, 0)$$

AXIS OF SYMMETRY

$$x_M = \frac{4 - 2}{2}$$

$$x = 1$$

VERTEX: Let $x = 1$.

$$y = 1^2 - 2(1) - 8$$

$$y = -9$$

$$(1, -9)$$

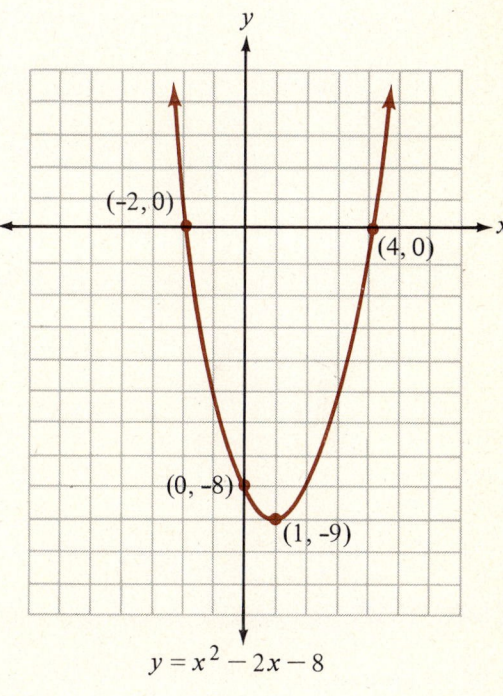

$$y = x^2 - 2x - 8$$

Notice that the parabola divides the plane into two regions: the region above the parabola and the region below the parabola. To find the region in which $y < x^2 - 2x - 8$, we may choose any point in the plane that does not satisfy the equation for the parabola. The origin $(0, 0)$ is an easy point to use, so let's test it.

$$y < x^2 - 2x - 8 \qquad \text{Original inequality.}$$

$$0 \overset{?}{<} 0^2 - 2(0) - 8 \qquad \text{Substitute 0 for } x \text{ and 0 for } y.$$

$$0 \overset{?}{<} -8 \qquad \text{See if the inequality is true.}$$

No; therefore, the region containing $(0, 0)$ is *not* the correct region. The points above the parabola do *not* satisfy the inequality, so the points *below* the parabola must be correct.

We might try a point in that region, say $(0, -10)$.

$$y \overset{?}{<} x^2 - 2x - 8$$

$$-10 \overset{?}{<} 0^2 - 2(0) - 8$$

$$-10 \overset{?}{<} -8$$

The inequality is satisfied, so we graph our inequality using a *dashed* curve (to indicate that the parabola itself is not in the solution) and shading *below* the dashed parabola.

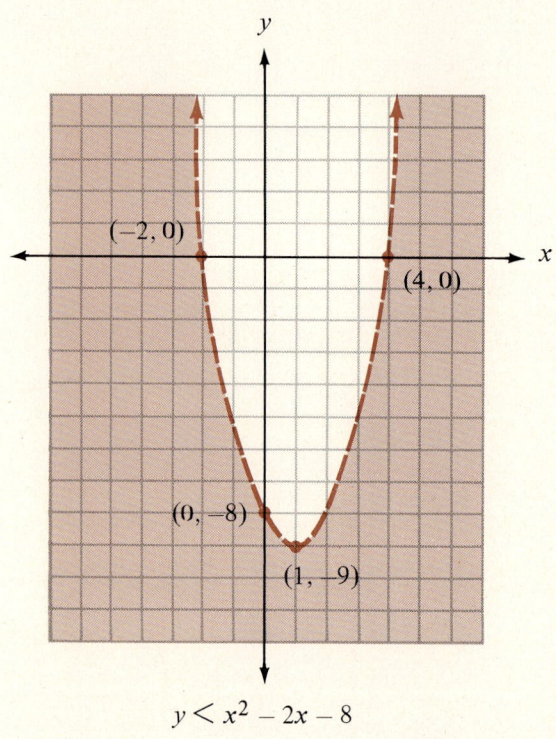

$$y < x^2 - 2x - 8$$

⫸ Trial Run

Graph each inequality.

1. $y \geq 4x^2$

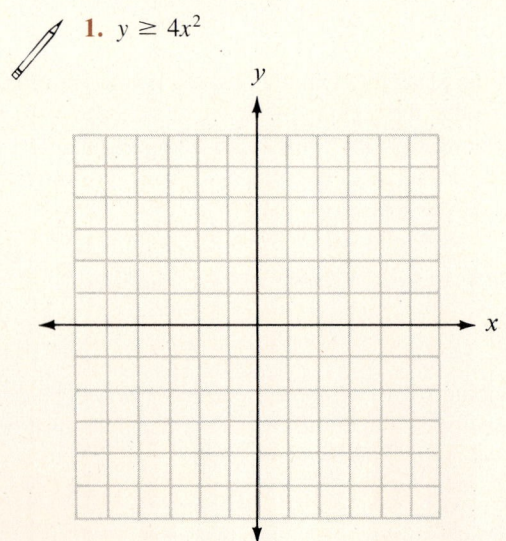

2. $y \geq x^2 - 5x$

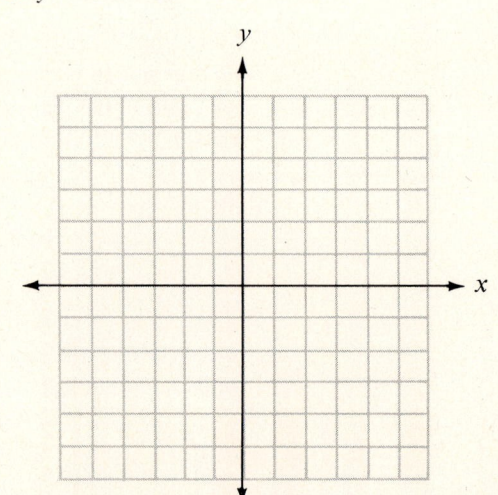

3. $y \leq x^2 + 2x + 3$ **4.** $y < -x^2 + 4x + 12$

          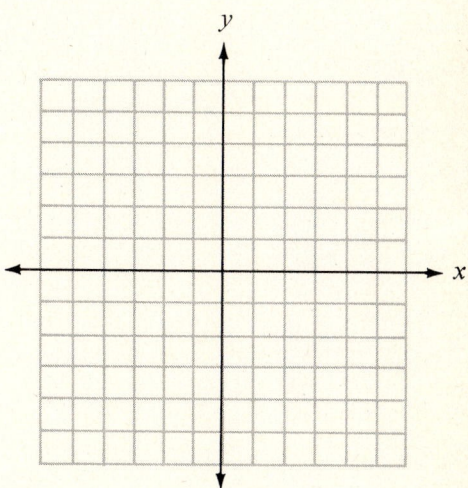

Answers are on page 736.

Similar techniques can be used to graph inequalities whose corresponding equations yield graphs that are circles, ellipses, or hyperbolas. We first graph the ordered pairs satisfying the boundary equation and then use a test point (often the origin) to determine the correct region in the plane.

Example 2. Graph the inequality $x^2 + y^2 \leq 4$.

Solution: First we graph the points satisfying the boundary equation

$$x^2 + y^2 = 4$$

which we recognize as a circle.

$$\text{Center: } (0, 0) \qquad \text{Radius: } r = 2$$

Now we test the point $(0, 0)$ that lies *inside* the circle.

$$x^2 + y^2 \leq 4$$

$$0^2 + 0^2 \overset{?}{\leq} 4$$

$$0 \overset{?}{\leq} 4$$

Yes, points *inside or on* the circle satisfy the inequality.

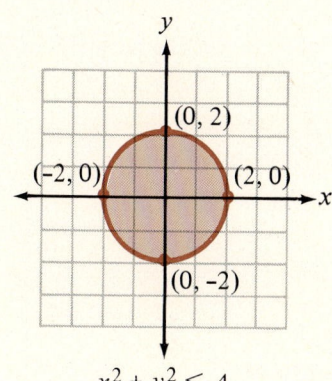

$x^2 + y^2 \leqslant 4$

Example 3. Graph the inequality $3x^2 - y^2 \leq 12$.

Solution: First we graph the points satisfying the boundary equation

$$3x^2 - y^2 = 12$$

which we recognize as a hyperbola. Graphing by the usual methods, we have

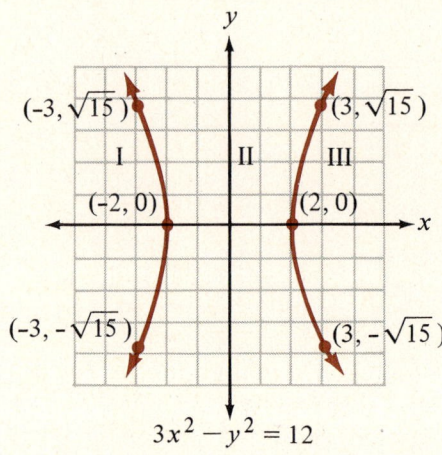

$$3x^2 - y^2 = 12$$

The hyperbola seems to divide the plane into *three* regions, each of which must be tested to see whether the inequality $3x^2 - y^2 \leq 12$ is satisfied.

Region I	Region II	Region III
Test point: $(-3, 0)$	Test point: $(0, 0)$	Test Point: $(3, 0)$
$3x^2 - y^2 \leq 12$	$3x^2 - y^2 \leq 12$	$3x^2 - y^2 \leq 12$
$3(-3)^2 - 0^2 \overset{?}{\leq} 12$	$3(0)^2 - 0^2 \overset{?}{\leq} 12$	$3(3)^2 - 0^2 \overset{?}{\leq} 12$
$3(9) - 0 \overset{?}{\leq} 12$	$0 - 0 \overset{?}{\leq} 12$	$3(9) - 0 \overset{?}{\leq} 12$
$27 \overset{?}{\leq} 12$	$0 \overset{?}{\leq} 12$	$27 \overset{?}{\leq} 12$
No	Yes	No

The inequality is satisfied only in region II *and* on the hyperbola itself. The graph becomes

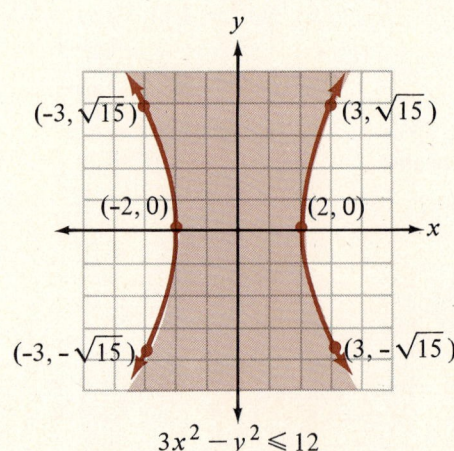

$$3x^2 - y^2 \leq 12$$

⫸ Trial Run

Graph each inequality.

1. $x^2 + y^2 \leq 9$

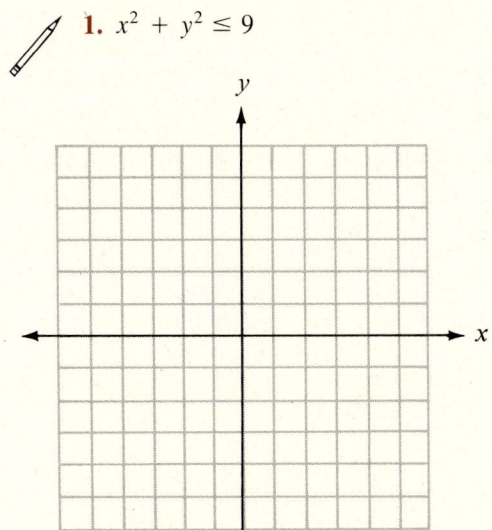

2. $x^2 + 4y^2 > 16$

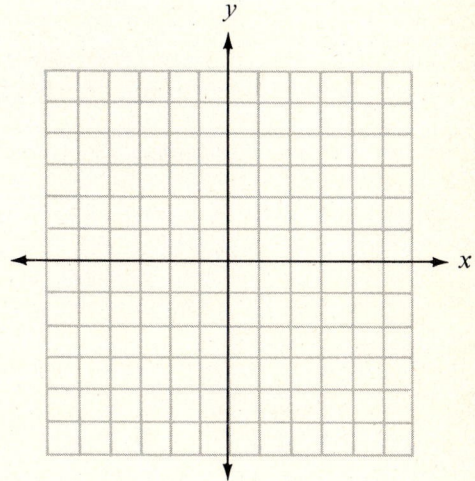

Answers are on page 736.

We have discovered that graphing second-degree inequalities requires a few short steps.

Graphing Second-Degree Inequalities

1. Graph the boundary equation corresponding to the given inequality using a solid or dashed curve.
2. Test a point (often the origin) not on the graph of the boundary equation to determine whether it satisfies the inequality.
3. Decide on the appropriate region and shade it.

Notice that all inequalities represent relations that are *not* functions.

Answers to Trial Runs

page 732 **1.**

2.

3.

4.

page 735 **1.**

2.

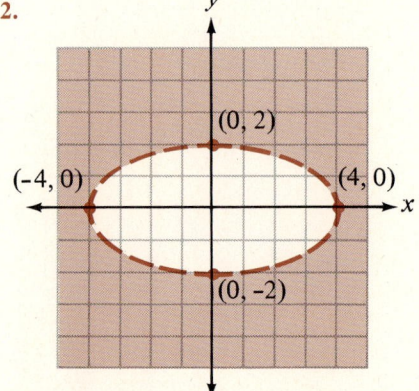

EXERCISE SET 9.4

Graph each of the following relations.

1. $y \geq 4x^2$

2. $y \geq x^2 - 2$

3. $y > 6x - x^2$

4. $y > 4x + x^2$

5. $y \leq x^2 - x - 20$

6. $y \leq x^2 + 5x - 14$

7. $y < -x^2 + 4x - 1$

8. $y < -x^2 + 2x + 5$

9. $x^2 + y^2 \geq 16$

10. $x^2 + y^2 \geq 25$

11. $3x^2 + 3y^2 < 9$

12. $5x^2 + 5y^2 < 10$

13. $(x - 1)^2 + (y + 3)^2 \leq 49$

14. $(x + 3)^2 + (y - 5)^2 \leq 36$

15. $4x^2 + 9y^2 < 36$

16. $25x^2 + y^2 < 25$

17. $9x^2 - y^2 \leq 36$

18. $4x^2 - y^2 \leq 16$

19. $y^2 - 4x^2 > 36$

20. $y^2 - x^2 > 25$

☆ Stretching the Topics _____

_____ **1.** Graph $y > x^3 - x$.

_____ **2.** Graph $xy < 4$.

_____ **3.** Graph $\dfrac{x^2}{9} + \dfrac{y^2}{4} \leq 1$.

Check your answers in the back of your book.

If you can complete **Checkup 9.4,** you are ready to do the Review Exercises.

CHECKUP 9.4

Graph each of the following relations.

1. $y \leq -x^2 + 4$

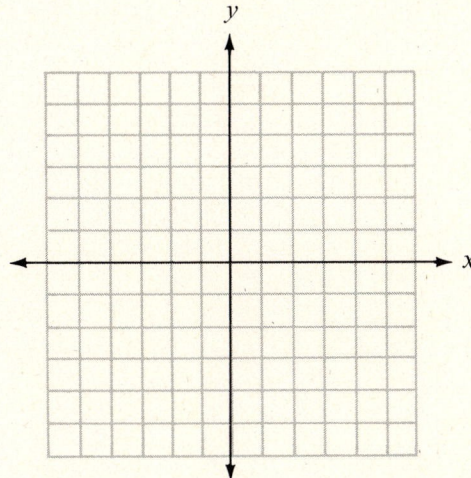

2. $x^2 + y^2 > 25$

3. $(x - 3)^2 + (y + 1)^2 < 25$

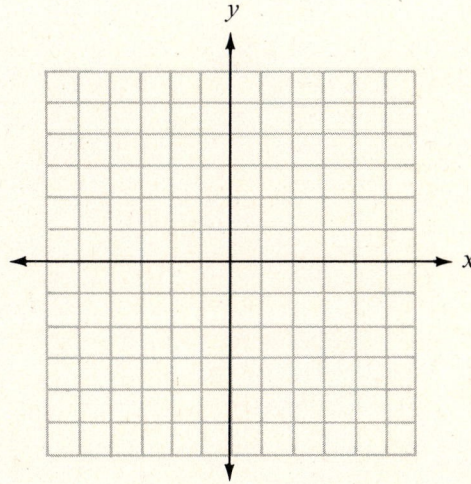

4. $4x^2 + 9y^2 \leq 36$

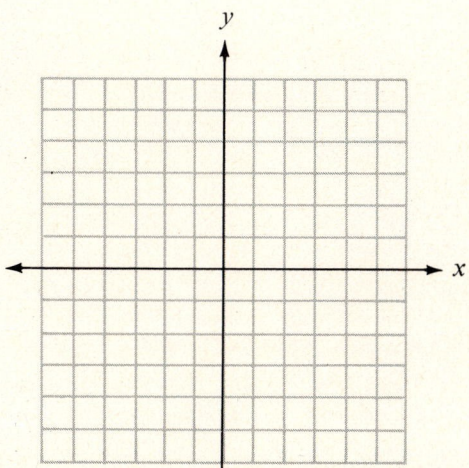

5. $9x^2 - y^2 \le 81$

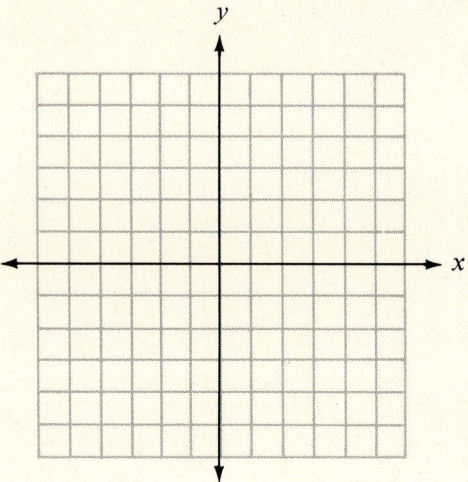

Check your answers in the back of your book.

If You Missed Problems:	You Should Review Example:
1	1
2, 3	2
4	2
5	3

Summary

In this chapter we have extended our understanding of graphing solution sets to include ordered pairs satisfying nonlinear equations and inequalities. After using arbitrary points to determine several unfamiliar graphs, we made some observations about the **conic sections** and generalized the techniques for graphing them.

Conic Section	Equation	Graphing Method
Parabola	$y = ax^2 + bx + c$ $(a \neq 0)$	Find x and y intercepts Find axis of symmetry Find vertex
Circle	$(x - h)^2 + (y - k)^2 = r^2$ $(r > 0)$	Center at (h, k) Radius r
Ellipse	$Ax^2 + By^2 + C$ $(A, B, C > 0; A \neq B)$	Find x-intercepts Find y-intercepts
Hyperbola	$Ax^2 + By^2 = C$ (A and B opposite in sign)	Find x- and y-intercepts Find four symmetric points

To graph **second-degree inequalities,** we learned to graph the corresponding boundary equation and then choose the appropriate region using test points.

Throughout this chapter we agreed that the graph of ordered pairs satisfying an equation or inequality helped us determine whether those ordered pairs represented a function or a relation. We also found that it was possible to determine domain and range from the appearance of the graph.

❑ Speaking the Language of Algebra

Complete each sentence with the appropriate word or phrase.

1. The graph of ordered pairs satisfying the equation $y = ax^2 + bx + c$ is a _____ . If a is positive, we know that the graph opens _____ .

2. For the graph of $y = ax^2 + c$, we know the axis of symmetry is the _____ and the vertex is the _____ point.

3. The graph of $3x^2 + 3y^2 = 9$ is a(n) _____ .
   The graph of $3x^2 + 2y^2 = 9$ is a(n) _____ .
   The graph of $3x^2 - 2y^2 = 9$ is a(n) _____ .

4. The graph of $x^2 + y^2 = 1$ is called the _____ circle. The graph of $x^2 + y^2 < 1$ is the region _____ the circle, while the graph of $x^2 + y^2 > 1$ is the region _____ the circle.

REVIEW EXERCISES for Chapter 9: Graphing More Relations and Functions

Graph each of the following relations. State the domain and range.

Do: _____ **1.** $y = x^2 - 9$ Do: _____ **2.** $y = -3x^2 + 10$
Ra: _____   Ra: _____

Do: _____ **3.** $y = x^3 - 2$ Do: _____ **4.** $y = x^3 - 4x$
Ra: _____   Ra: _____

Do: _____ **5.** $y = \dfrac{5}{x}$ Do: _____ **6.** $y = -x^2 + 6x$
Ra: _____   Ra: _____

Do: _____ **7.** $y = 9x^2 - 12x + 14$ Do: _____ **8.** $y = x^2 - 4x - 5$
Ra: _____   Ra: _____

Do: _____ **9.** $x^2 + y^2 = 121$ Do: _____ **10.** $x^2 + y^2 = 5$
Ra: _____   Ra: _____

Do: _____ **11.** $(x - 3)^2 + y^2 = 49$ Do: _____ **12.** $4x^2 + 25y^2 = 100$
Ra: _____   Ra: _____

Do: _____ **13.** $x^2 - y^2 = 9$ Do: _____ **14.** $9y^2 - 16x^2 = 144$
Ra: _____   Ra: _____

Graph each of the following.

15. $y \geq x^2 - 4x$ **16.** $(x - 5)^2 + (y + 1)^2 > 25$

_____ **17.** Deal on Wheels Tire Company finds its profit y is directly related to the number of sets of tires sold x by the function $y = -x^2 + 200x$. Use the graph to determine the maximum profit.

Check your answers in the back of your book.

If You Missed Problems:	You Should Review Examples:	
1, 2	Section 9.2	1, 2
3, 4	Section 9.1	3, 4
5		5
6–8	Section 9.2	4, 6, 7
9, 10	Section 9.3	1, 3
11		2
12		4
13, 14		5, 6
15	Section 9.4	1
16		2
17	Section 9.2	5

If you have completed the Review Exercises and corrected your errors, you are ready to take the Practice Test.

PRACTICE TEST for Chapter 9: Graphing More Relations and Functions

Graph each of the following relations. State the domain and range.

	SECTION	EXAMPLE

1. $y = 4x - 2x^2$

SECTION 9.1   EXAMPLE 2

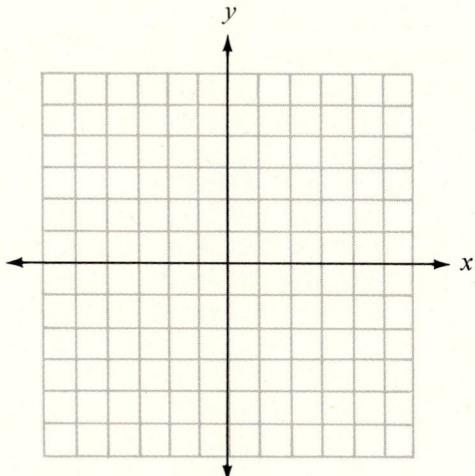

Domain: _____
Range: _____

2. $y = x^3 - 1$

SECTION 9.1   EXAMPLE 3, 4

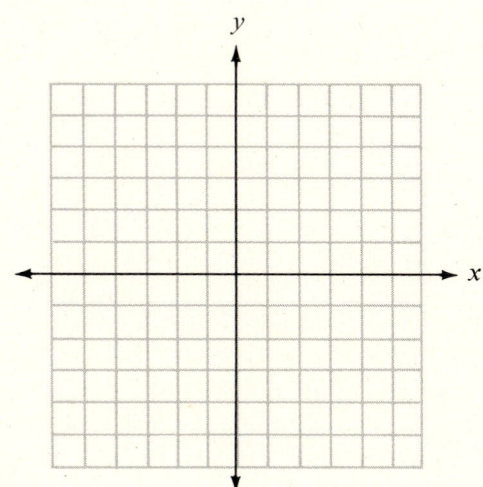

Domain: _____
Range: _____

3. $y = \dfrac{-4}{x}$

9.1   5

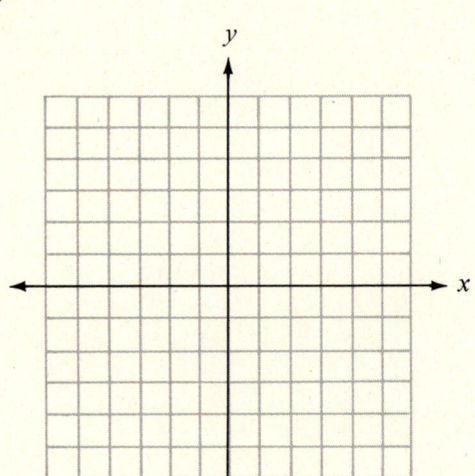

Domain: _____
Range: _____

4. $y = \dfrac{1}{2} x^2$

9.2   3

Domain: _____
Range: _____

5. $y = x^2 - x - 6$ 9.2   4, 6

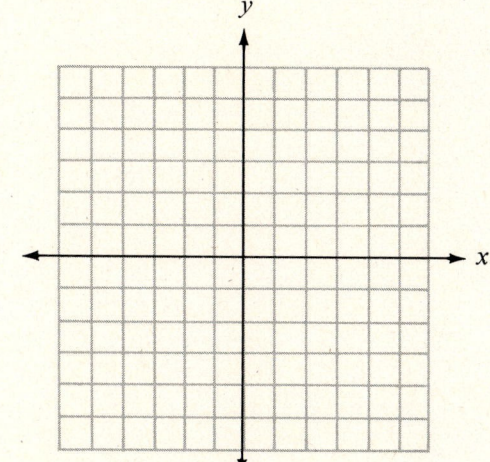

Domain: _____
Range: _____

6. $(x - 2)^2 + (y + 1)^2 = 4$ 9.3   2

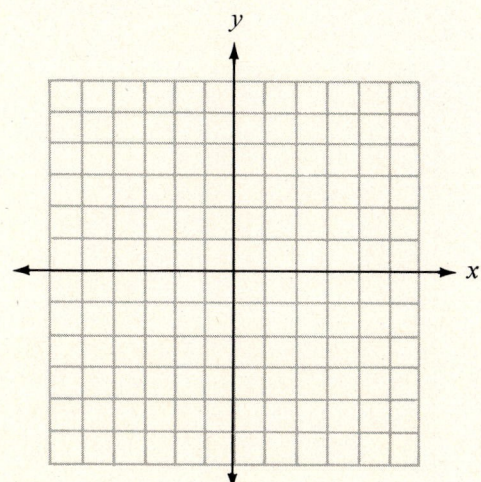

Domain: _____
Range: _____

7. $x^2 + 5y^2 = 25$

9.3   4

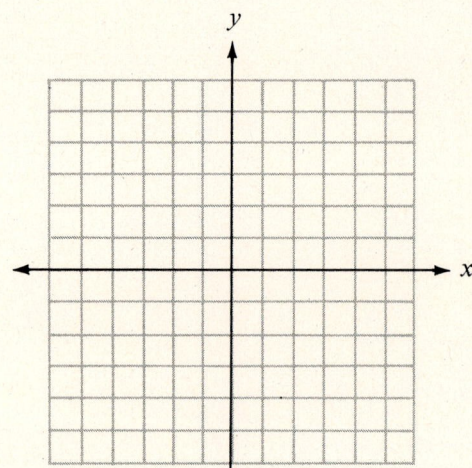

Domain: _____
Range: _____

8. $3y^2 - x^2 = 12$

9.3   6

Domain: _____
Range: _____

Graph each of the following relations.

9. $y \geq x^2 - 2$

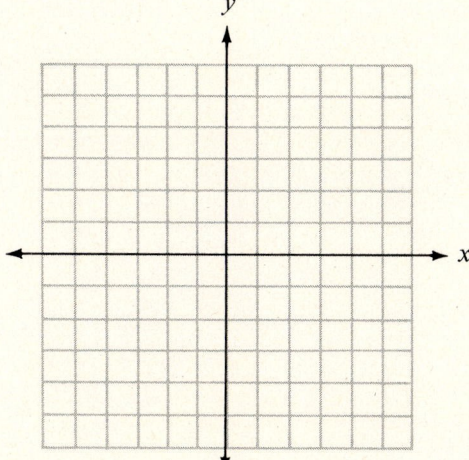

9.4   1

10. $x^2 + y^2 > 1$

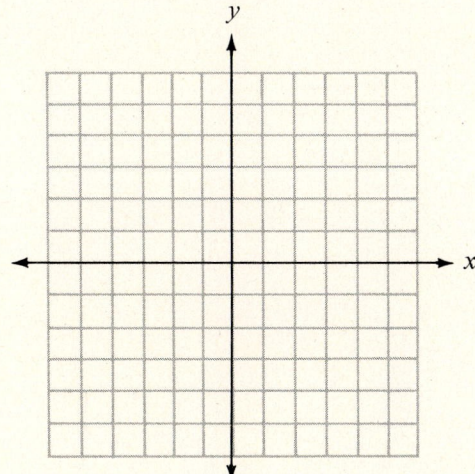

9.4   2

SHARPENING YOUR SKILLS after Chapters 1–9

SECTIONS

_____ 1. Evaluate $3a^2 - 2ab + b^2$ when $a = -2$ and $b = 3$.   3.1

_____ 2. Write $7 - \sqrt{-72}$ in standard complex number form.   7.5

_____ 3. Find the equation of the line through the points $\left(\dfrac{-2}{3}, \dfrac{7}{4}\right)$ and $\left(\dfrac{13}{3}, \dfrac{-1}{4}\right)$.   8.3

_____ 4. Find the distance between $(-3, -1)$ and $(5, -16)$ 8.2

_____ 5. Write a quadratic equation with the solutions $x = -1$ and $x = 7$.   5.1

Simplify.

_____ 6. $-[3(a + 7b) + 7(2a - 3b)]$ 1.4

_____ 7. $10a(-3a^7b^4)^2(ab^3)^3$ 3.1

_____ 8. $(x^2y^{-9}z^{-3})(x^{-5}y^9z^5)$ 3.2

_____ 9. $\dfrac{-7x^4y^{-3}}{x^{-2}y^{-2}}$ 3.2

_____ 10. $a(a - 9)(a + 9) - 12a(a - 3)$ 3.4

_____ 11. $(3a^3 - 8a^2 - 20a + 16) \div (3a - 2)$ 3.5

_____ 12. $\dfrac{4x^2 - 11x + 6}{x^3 - 125} \div \dfrac{4x^2 + 13x - 12}{4x^3 + 20x^2 + 100x}$ 6.3

_____ 13. $\dfrac{5}{x - 4} - \dfrac{3}{x + 3} - \dfrac{35}{x^2 - x - 12}$ 6.3

_____ 14. $\dfrac{x^{-1} + 2y^{-1}}{x^{-2} - 4y^{-2}}$ 6.4

_____ 15. $\left(\dfrac{x^{3/4}y^{-5/12}}{z^{1/4}}\right)^{4/9}$ 7.1

_____ 16. $5\sqrt[4]{x^5y^8} - x\sqrt[4]{81xy^8}$ 7.2

_____ 17. $\dfrac{1}{\sqrt{2} + \sqrt{5}}$ 7.3

Solve.

_____ 18. $|7x - 2| > -8$ 2.4

_____ 19. $2x(x + 1) + 5(x^2 - 4) = 5(2x - 4)$ 5.1

_____ **20.** $5x^2 - 6x - 2 = 0$ $\qquad$ 5.3

_____ **21.** $5(x + 2) \geq 6(x + 1)$ $\qquad$ 5.4

_____ **22.** $\dfrac{3}{3x + 2} - \dfrac{4}{3x - 2} = \dfrac{1}{9x^2 - 4}$ $\qquad$ 6.5

_____ **23.** $\dfrac{x - 1}{2x - 3} < 1$ $\qquad$ 6.5

_____ **24.** $\sqrt[3]{x^2 + 2x} = \sqrt[3]{x + 12}$ $\qquad$ 7.4

_____ **25.** $125x^3 - 27 = 0$ $\qquad$ 7.6

Graph using any method you choose.

26. $5x - y = 6$ $\qquad$ 8.1, 8.2

27. $y = \dfrac{2}{3}x$ $\qquad$ 8.1, 8.2

28. $x + 5y > -10$ $\qquad$ 8.4

_____ **29.** The velocity of a falling body varies directly as the time t during which it falls. Write an equation of variation and find k if at the end of 6 seconds a falling body has a velocity of 192 feet per second. Under the same conditions, what will be the velocity at the end of 8 seconds? $\qquad$ 8.6

_____ **30.** Solve $h = \dfrac{2v^2}{g}$ for v. $\qquad$ 8.5, 5.2

Solving Systems of Equations and Inequalities

10

During the Fourth of July Festival, Marcia sold hot dogs and hamburgers at her refreshment stand. On Friday she made a profit of $130 dollars by selling 200 hot dogs and 150 hamburgers. On Saturday she made a profit of $225 by selling 300 hot dogs and 300 hamburgers. What profit did Marcia make on each hot dog and each hamburger during the two days?

In Chapter 2 we learned how to find all values of *one* variable that satisfied two inequalities at the same time. Now that we have discussed solutions for equations and inequalities containing *two* variables, we must investigate ways to find ordered pairs that satisfy more than one condition at the same time. A set of equations or inequalities that must be satisfied at the same time is called a system of simultaneous equations or inequalities, or simply a **system of equations** or inequalities. Finding the solutions that satisfy all equations in a given system is called **solving a system of equations.** In this chapter we learn how to:

1. Solve systems of linear equations by graphing.
2. Solve systems of linear equations by substitution.
3. Solve systems of linear equations by elimination.
4. Solve systems of linear equations in three variables.
5. Solve nonlinear systems of equations.
6. Solve systems of inequalities.
7. Switch from words to systems of equations.

10.1 Solving Systems of Linear Equations by Graphing and Substitution

Solving Systems by Graphing

If you were asked to find all the ordered pairs that satisfy *both* equations in the system

$$\text{(1)} \quad x + y = 7$$

$$\text{(2)} \quad 2x - y = -10$$

you might be able to come up with an ordered pair by trial and error, but that would be a tedious process.

Considered separately, the set of ordered pairs satisfying each of these equations can be graphed as a straight line. Let's graph those straight lines and look for any points that the two lines have in common. After all, any *common* points would represent ordered pairs satisfying *both* equations, and that is what we are looking for.

(1) $x + y = 7$ (2) $2x - y = -10$

x-INTERCEPT:	Let $y = 0$	*x*-INTERCEPT:	Let $y = 0$
$x + 0 = 7$		$2x - 0 = -10$	
$x = 7$ (7, 0)		$2x = -10$	
y-INTERCEPT:	Let $x = 0$	$x = -5$ (−5, 0)	
$0 + y = 7$		*y*-INTERCEPT:	Let $x = 0$
$y = 7$ (0, 7)		$2(0) - y = -10$	
		$-y = -10$	
		$y = 10$ (0, 10)	

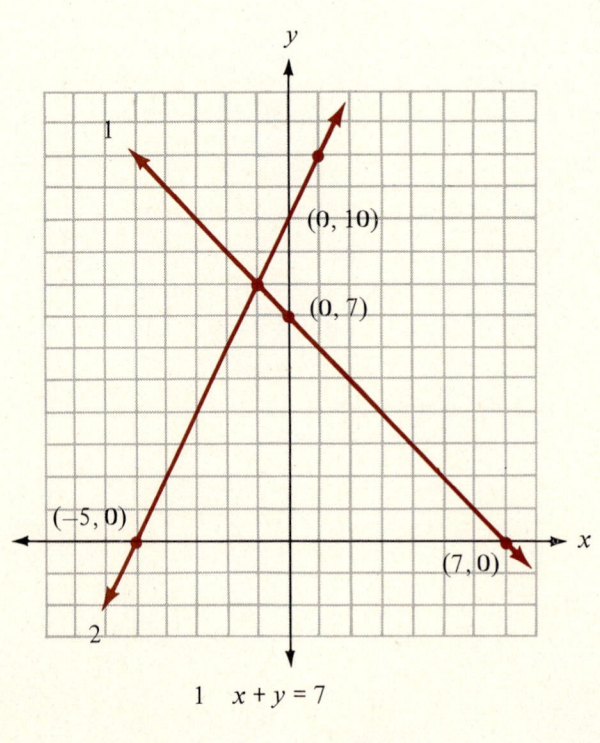

1    $x + y = 7$

2    $2x - y = -10$

You should agree that there is just one point common to both lines, and it appears that the coordinates of that point are $(-1, 8)$. But we must check to be sure that point satisfies both equations in the system.

$$(1) \ x + y = 7 \qquad\qquad (2) \ 2x - y = -10$$

$$-1 + 8 \stackrel{?}{=} 7 \qquad\qquad 2(-1) - 8 \stackrel{?}{=} -10$$

$$7 = 7 \qquad\qquad\qquad -10 = -10$$

Indeed, the ordered pair $(-1, 8)$ is a solution for our system. Are there any other possible solutions? From the graph we see that these lines have exactly one point in common. Since we have found that point, we know that $(-1, 8)$ is the *only* solution for this system. You try to complete Example 1.

Example 1. Use graphing to solve the system.

$$(1) \ y = 3$$

$$(2) \ y = 5x - 2$$

Solution: First we graph both equations.

$$(1) \ y = 3 \qquad\qquad (2) \ y = 5x - 2$$

Horizontal line   Slope: $m =$ _____

y-intercept: 3 y-intercept: $b =$ _____

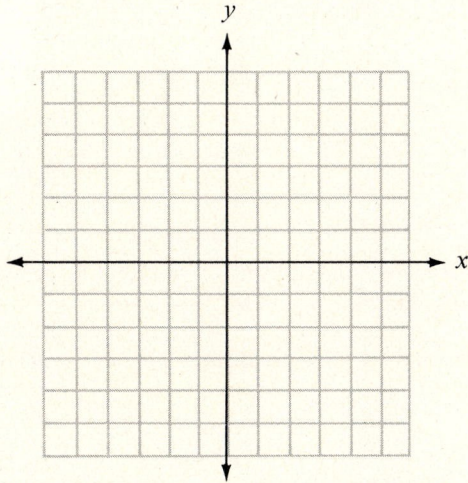

It appears that (_____ , _____) is the common point for the lines. We can check this ordered pair in both equations.

$$(1) \qquad y = 3 \qquad (2) \qquad y = 5x - 2$$

$$\text{\_\_\_\_} = 3 \qquad\qquad \text{\_\_\_\_} \stackrel{?}{=} 5(\text{\_\_\_\_}) - 2$$

$$\text{\_\_\_\_} = \text{\_\_\_\_}$$

Check your work on page 763. ▶

Example 2. Using graphing to solve the system.

$$(1)\ y = -2x$$

$$(2)\ 6x + 3y = 9$$

Solution: First we graph each equation.

(1) $y = -2x$ (2) $6x + 3y = 9$

    Slope: $m = -2$ $3y = -6x + 9$

    y-intercept: $b = 0$ $y = -2x + 3$

                                  Slope: $m = -2$

                                  y-intercept: $b = 3$

    ① $y = -2x$

    ② $6x + 3y = 9$

The fact that these lines are parallel should not be surprising, since they have the same slope and different y-intercepts. They share *no* common point, so there is *no solution* for this system of equations.

Example 3. Use graphing to solve the system.

$$(1)\ 2x - 3y = 9$$

$$(2)\ 6y = 4x - 18$$

Solution: First we graph each equation.

(1) $2x - 3y = 9$ (2) $6y = 4x - 18$

    $-3y = -2x + 9$ $y = \dfrac{4}{6}x - 3$

    $y = \dfrac{2}{3}x - 3$ $y = \dfrac{2}{3}x - 3$

    Slope: $m = \dfrac{2}{3}$ Slope: $m = \dfrac{2}{3}$

    y-intercept: $b = -3$ y-intercept: $b = -3$

Since both lines have the same slope and y-intercept, their graphs must be the same straight line.

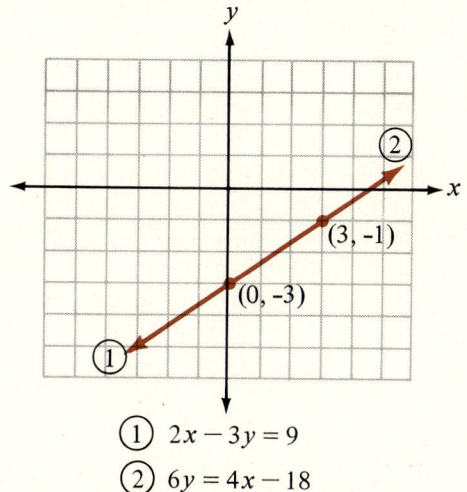

① $2x - 3y = 9$

② $6y = 4x - 18$

What solutions do the two equations share? Any ordered pair that satisfies one equation will satisfy the other. We may write the solution to the system as:

all ordered pairs satisfying $2x - 3y = 9$

or

all ordered pairs satisfying $6y = 4x - 18$

Systems of linear equations can be classified according to the number of solutions they have.

> *Independent and consistent system.* The system has *one* solution (intersecting lines).
> *Inconsistent system.* The system has *no* solutions (parallel lines).
> *Dependent system.* The system has an *infinite number* of solutions (coinciding lines).

Now try Example 4.

Example 4. Solve and classify the system.

(1) $3x + y = 5$

(2) $y = \dfrac{-1}{2} x$

Solution: First we graph each equation.

(1) $3x + y = 5$ (2) $y = \dfrac{-1}{2} x$

$\quad\quad y = -3x + 5$

Slope: $m = $ _____   Slope: $m = $ _____

y-intercept: $b = $ _____ y-intercept: $b = $ _____

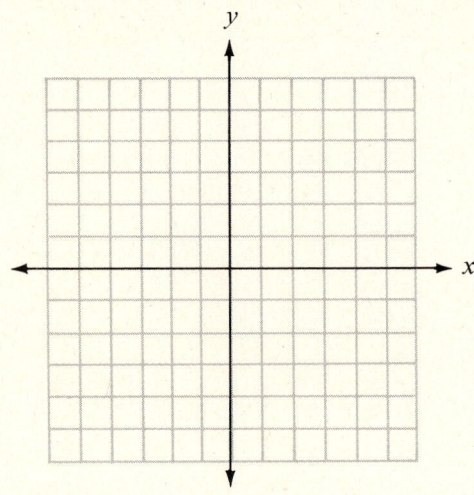

(1) $3x + y = 5$

(2) $y = \dfrac{-1}{2} x$

The common point seems to be (_____ , _____). Let's check in each equation.

(1) $3x + y = 5$ (2) $y = \dfrac{-1}{2} x$

$3(\text{\_\_\_\_}) + (\text{\_\_\_\_}) \overset{?}{=} 5$

$\text{\_\_\_\_} - \text{\_\_\_\_} \overset{?}{=} 5$ $(\text{\_\_\_\_}) \overset{?}{=} \dfrac{-1}{2} (\text{\_\_\_\_})$

$\text{\_\_\_\_} = 5$ $\text{\_\_\_\_} = \text{\_\_\_\_}$

The solution for the system is (_____ , _____) and the system is _____ and _____ .

Check your work on page 763. ▶

⇛ Trial Run

Solve by graphing and classify the system.

 1. $3x - y = 4$
$2x + y = 11$

 2. $y = 5x - 2$
$10x - 2y = 4$

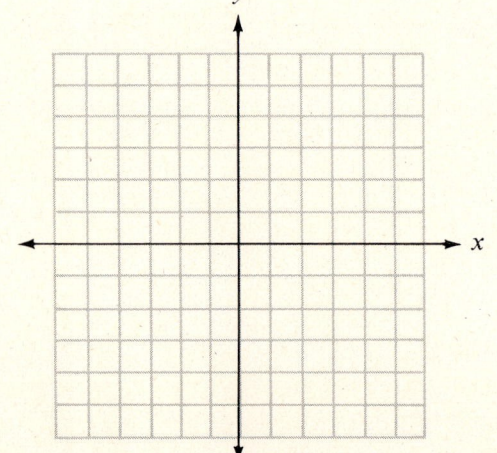

_____ **3.** $2x - 3y = 3$
$y = 3$

_____ **4.** $x + 2y = 6$
$y = \dfrac{-1}{2}x + 1$

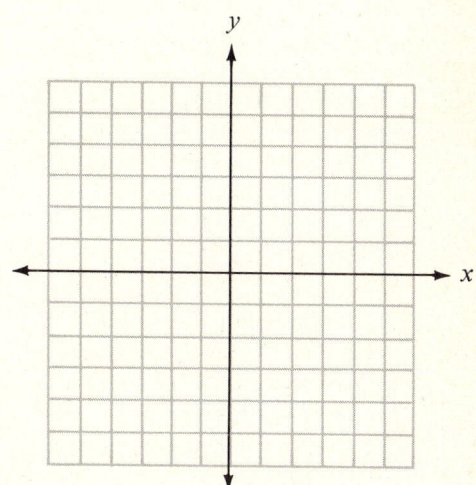

Answers are on page 765.

Solving Systems by Substitution

It has probably occurred to you that the method of graphing is useful in solving linear systems only when it is easy to determine the point where two lines intersect. If two lines intersected at a point such as $\left(\dfrac{2}{17}, \dfrac{-4}{17} \right)$, it would be almost impossible to locate that point with any accuracy using only a graph. Or if the lines intersected at a point such as $(124, -67)$, graphing would be a very awkward way of locating that point.

Therefore, we must find an algebraic method for solving systems of equations without graphing. In fact, there are two algebraic methods that can be used; we first discuss the **method of substitution.**

To solve a system such as

$$(1) \quad x + y = 7$$

$$(2) \quad y = 2x + 10$$

remember that we are looking for the ordered pair that will satisfy equation (1) and equation (2) at the same time. Because we are looking for a *common* (x, y), we may "borrow" the expression for y from equation (2) and use it in equation (1).

$x + y = 7$	Equation (1).
$y = 2x + 10$	Equation (2).
$x + (2x + 10) = 7$	Substitute $2x + 10$ for y in equation (1).
$3x + 10 = 7$	Simplify.
$3x = -3$	Isolate x-term.
$x = -1$	Isolate x.

We have found the x-coordinate for the common point. To find the corresponding y-coordinate, we can substitute -1 for x in either of the original equations.

$$(1)\ x + y = 7 \qquad \text{or} \qquad (2)\ y = 3x + 10$$
$$-1 + y = 7 \qquad\qquad\qquad y = 2(-1) + 10$$
$$y = 8 \qquad\qquad\qquad\qquad y = 8$$

The solution for the system is $(-1, 8)$, which matches the solution obtained earlier by graphing.

The method of substitution is especially useful if in either of the original equations, one of the variables is already isolated or has a coefficient of 1 and therefore can be easily isolated.

Example 5. Solve the system by substitution.

$$(1)\ 3y - 2x = 10$$
$$(1)\ 2y + \ \ x = \ \ 9$$

Solution: In equation (2) we can isolate x.

$$2y + x = 9$$
$$x = 9 - 2y$$

Now we substitute $9 - 2y$ for x in equation (1) and solve for y.

$$3y - 2(9 - 2y) = 10$$
$$3y - 18 + 4y = 10$$
$$7y - 18 = 10$$
$$7y = 28$$
$$y = 4$$

We find the corresponding x-value by substituting 4 for y in either equation. The easiest way to find x is by using equation (2) with x isolated.

$$x = 9 - 2y$$
$$x = 9 - 2(4)$$
$$x = 1$$

The ordered pair $(1, 4)$ is the solution for the system. You should check to be sure that the solution satisfies equation (1).

You try Example 6.

Example 6. Solve the system by substitution.

$$(1)\ \ y - \ \ 5 = 0$$
$$(2)\ 3x + 2y = 8$$

Solution: We can isolate _____ in equation (1).

$$y - 5 = 0$$
$$y = \underline{\quad\quad}$$

Now we know the y-value for our solution, so we can substitute _____ for y in equation (2) and solve for x.

$$3x + 2y = 8$$

$$3x + 2(\underline{\qquad}) = 8$$

$$3x + \underline{\qquad} = 8$$

$$3x = \underline{\qquad}$$

$$x = \underline{\qquad}$$

The ordered pair ($\underline{\qquad}$, $\underline{\qquad}$) is the solution for the system.

Check your work on page 764. ▶

Example 7. Solve the system by substitution.

$$(1) \ 4x - y = 3$$

$$(2) \ 2y - 8x = 1$$

Solution: We can isolate y in equation (1).

$$4x - y = 3$$

$$4x - 3 = y$$

Now substitute $4x - 3$ for y in equation (2) and solve for x.

$$2y - 8x = 1$$

$$2(4x - 3) - 8x = 1$$

$$8x - 6 - 8x = 1$$

$$-6 = 1$$

What does this false statement imply? We know that $-6 \neq 1$.

Let's return to our original equations and write each one in slope-intercept form.

$$(1) \ 4x - y = 3 \qquad\qquad (2) \ 2y - 8x = 1$$

$$4x - 3 = y \qquad\qquad\qquad 2y = 8x + 1$$

$$\text{Slope: } m = 4 \qquad\qquad\qquad y = 4x + \frac{1}{2}$$

$$\text{Slope: } m = 4$$

$$y\text{-intercept: } b = -3 \qquad y\text{-intercept: } b = \frac{1}{2}$$

The graphs of these lines have the same slope but different y-intercepts, so they must be *parallel lines*. The system is inconsistent and has *no solution*.

> If, in solving a system of equations algebraically, we arrive at a statement that is *false*, we know that the system is *inconsistent* and has *no solution*.

Example 8. Solve the system by substitution.

$$(1) \ y = 2x - 3$$

$$(2) \ 5y + 15 = 10x$$

Solution: Since y is isolated in equation (1) we can substitute $2x - 3$ for y in equation (2) and solve for x.

$$5y + 15 = 10x$$
$$5(2x - 3) + 15 = 10x$$
$$10x - 15 + 15 = 10x$$
$$10x - 10x = 0$$
$$0 = 0$$

Once again we seem to have lost our variable, but this time we are left with the *true* statement $0 = 0$. To interpret this result, let's discuss the graphs of the original equations.

(1) $y = 2x - 3$ (2) $5y + 15 = 10x$

$$5y = 10x - 15$$
$$y = 2x - 3$$

Slope: $m = 2$ Slope: $m = 2$

y-intercept: $b = -3$ y-intercept: $b = -3$

Since the graphs have the same slope and the same y-intercept, we know they represent the *same straight line*. The system is dependent and the solutions are all ordered pairs satisfying $y = 2x - 3$ (or all ordered pairs satisfying $5y + 15 = 10x$).

> If, in solving a system of equations algebraically, we arrive at a statement that is *always true*, we know that the system is *dependent*, and any ordered pair that satisfies one equation will satisfy the other.

⫸ Trial Run

Solve each system by substitution.

_____ **1.** $x + y = 7$
        $2x - y = 5$

_____ **2.** $5x - y = 2$
        $y = 3$

_____ **3.** $3x - y = 1$
        $y = 3x + 2$

_____ **4.** $5x + 2y = 9$
        $x + 3y = 4$

_____ **5.** $y = 3 - x$
        $2x + 2y = 6$

Answers are on page 765.

▶ **Examples You Completed** _____

Example 1. Use graphing to solve the system

$$\text{(1) } y = 3$$

$$\text{(2) } y = 5x - 2$$

Solution: First we graph both equations.

$$\text{(1) } y = 3 \qquad\qquad \text{(2) } y = 5x - 2$$

Horizontal line $\qquad$ Slope: $m = \dfrac{5}{1}$

y-intercept: 3 $\qquad$ y-intercept: $b = -2$

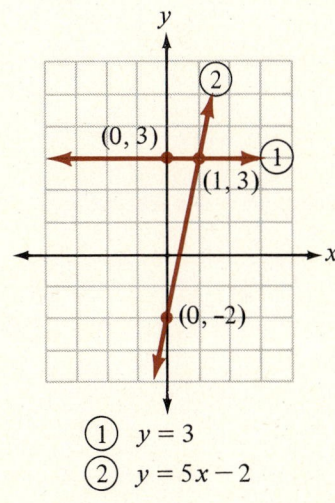

① $y = 3$
② $y = 5x - 2$

It appears that $(1, 3)$ is the common point for the lines. We can check this ordered pair in both equations.

$$\text{(1) } y = 3 \qquad\qquad \text{(2) } y = 5x - 2$$

$$3 = 3 \qquad\qquad\quad 3 \overset{?}{=} 5(1) - 2$$

$$ 3 = 3$$

Example 4. Solve and classify the system.

$$\text{(1) } 3x + y = 5$$

$$\text{(2) } y = \dfrac{-1}{2}x$$

Solution: First we graph each equation.

$$\text{(1) } 3x + y = 5 \qquad\qquad \text{(2) } y = \dfrac{-1}{2}x$$

$$y = -3x + 5$$

Slope: $m = \dfrac{-3}{1}$ $\qquad\qquad$ Slope: $m = \dfrac{-1}{2}$

y-intercept: $b = 5$ $\qquad\qquad$ y-intercept: $b = 0$

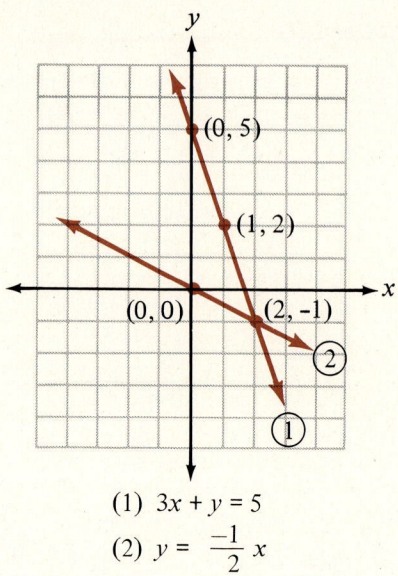

$$(1) \;\; 3x + y = 5$$

$$(2) \;\; y = \frac{-1}{2}x$$

The common point seems to be $(2, -1)$. Let's check in each equation.

$$(1) \qquad 3x + y = 5 \qquad (2) \qquad y = \frac{-1}{2}x$$

$$3(2) + (-1) \stackrel{?}{=} 5$$

$$6 - 1 \stackrel{?}{=} 5 \qquad\qquad (-1) \stackrel{?}{=} \frac{-1}{2}(2)$$

$$5 = 5 \qquad\qquad -1 = -1$$

The solution for the system is $(2, -1)$ and the system is independent and consistent.

Example 6. Solve the system by substitution.

$$(1) \;\; y - \;\; 5 = 0$$

$$(2) \;\; 3x + 2y = 8$$

Solution: We can isolate y in equation (1).

$$y - 5 = 0$$

$$y = 5$$

Now we know the y-value for our solution, so we can substitute 5 for y in equation (2) and solve for x.

$$3x + 2y = 8$$

$$3x + 2(5) = 8$$

$$3x + 10 = 8$$

$$3x = -2$$

$$x = \frac{-2}{3}$$

The ordered pair $\left(\dfrac{-2}{3}, 5 \right)$ is the solution for the system.

Answers to Trial Runs

page 758   **1.**

Independent, consistent

2.

Dependent

3.

Independent, consistent

4.

Inconsistent

page 762   **1.** (4, 3)   **2.** (1, 3)   **3.** No solution (inconsistent)   **4.** $\left(\dfrac{19}{13}, \dfrac{11}{13}\right)$

5. All ordered pairs satisfying $y = 3 - x$ (dependent)

EXERCISE SET 10.1

Solve by graphing and classify the system.

_____ **1.** $2x - y = 5$
$\quad\quad x + y = 7$

_____ **2.** $3x - y = 4$
$\quad\quad x - y = 2$

_____ **3.** $y = 3x + 7$
$\quad\quad 2y - 6x = 5$

_____ **4.** $y = -2x + 3$
$\quad\quad 3y + 6x = 6$

_____ **5.** $3x - y = 6$
$\quad\quad\quad y = 6x$

_____ **6.** $-2x + y = 8$
$\quad\quad\quad\quad y = 4x$

_____ **7.** $2x - y = 10$
$\quad\quad\quad y = 4$

_____ **8.** $3x + y = 9$
$\quad\quad\quad y = -3$

_____ **9.** $x + 4y = 1$
$\quad\quad 8y = 2 - 2x$

_____ **10.** $x - 3y = 2$
$\quad\quad 6y = 2x - 4$

_____ **11.** $3x - 2y = 0$
$\quad\quad\quad x - y = -1$

_____ **12.** $4x + 3y = 0$
$\quad\quad\quad x + y = -1$

_____ **13.** $3x + 2y = 6$
$\quad\quad 2x + y = 2$

_____ **14.** $5x + 4y = 20$
$\quad\quad\quad x + y = 6$

_____ **15.** $4x - 2y = 16$
$\quad\quad -2x + y = -4$

_____ **16.** $-3x + 2y = -6$
$\quad\quad 6x - 4y = 8$

Solve the systems by substitution.

_____ **17.** $3x - 2y = 12$
$\quad\quad\quad x - y = 5$

_____ **18.** $2x - 3y = 11$
$\quad\quad\quad x + y = 3$

_____ **19.** $3x + y = 2$
$\quad\quad 4x - 2y = -3$

_____ **20.** $5x + y = -1$
$\quad\quad 13x + 3y = 2$

_____ **21.** $8x - 3y = 7$
$\quad\quad\quad\quad x = 2$

_____ **22.** $7x - 11y = -1$
$\quad\quad\quad\quad\quad x = 3$

_____ **23.** $3x - 8y = 13$
$\quad\quad\quad\quad y = 3x$

_____ **24.** $5x - 3y = 28$
$\quad\quad\quad\quad y = -4x$

_____ **25.** $14x - 2y = 10$
$\quad\quad 7x - y = 2$

_____ **26.** $10x + 2y = 3$
$\quad\quad 5x + y = -1$

_____ **27.** $4x - y = 8$
$\quad\quad 2x + 3y = 1$

_____ **28.** $5x + y = 7$
$\quad\quad 4x - 2y = -3$

_____ **29.** $10x - 5y = 20$
$\quad\quad\quad y = 2x - 4$

_____ **30.** $3x - 15y = 15$
$\quad\quad\quad x = 5y + 5$

_____ **31.** $x - y = -2$
$\quad\quad 11x - 3y = 2$

_____ **32.** $x - y = -4$
$\quad\quad 7x + 5y = -4$

☆ Stretching the Topics

Solve by substitution.

_____ **1.** $0.3x - 0.2y = 0.7$
$0.4x - 0.1y = 0.6$

_____ **2.** Find the value(s) of k for which the system

$$y = kx + 5$$
$$y = 3x - 2$$

will be inconsistent.

Check your answers in the back of your book.

If you can complete **Checkup 10.1,** you are ready to go on to Section 10.2.

✓ CHECKUP 10.1

Solve by graphing and classify the system.

_____ **1.** $x - 2y = 8$
$3x + 2y = 0$

_____ **2.** $-x + 2y = 4$
$5x - 10y = -20$

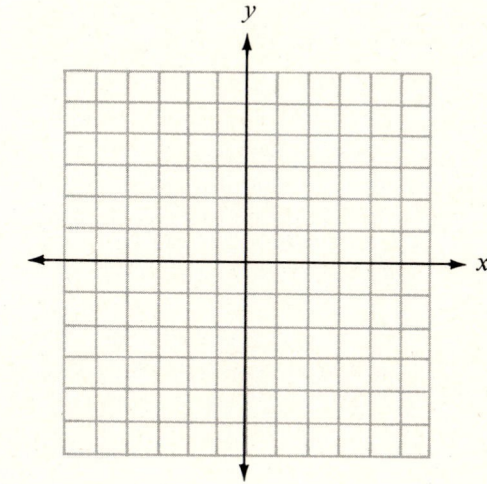

Solve the systems by substitution.

_____ **3.** $9x - 4y = 10$
$2x + y = 6$

_____ **4.** $x - y = -3$
$-4x + 4y = 1$

_____ **5.** $4x + 7y = 16$
$x + 4y = 7$

Check your answers in the back of your book.

If You Missed Problems:	You Should Review Examples:
1	1, 4
2	3
3	5
4	7
5	5

10.2 Solving Systems of Linear Equations by Addition

What happens when we want to solve a system of equations algebraically but neither variable can easily be isolated in either equation? To solve a system such as

$$(1) \quad 5x - 4y = 8$$

$$(2) \quad 3x + 2y = 7$$

by the method of substitution would be a tedious process because neither equation contains a variable with a coefficient of 1.

In this section we make use of two important facts from earlier chapters.

> We may multiply both sides of an equation by any nonzero constant.
> We may add equal quantities to both sides of an equation.

Using the Method of Addition

In the system of equations above,

$$(1) \quad 5x - 4y = 8$$

$$(2) \quad 3x + 2y = 7$$

we may *multiply* both sides of equation (2) by the constant 2 and the system becomes

$$5x - 4y = 8$$

$$6x + 4y = 14$$

Since the quantities on the left and right sides of these equations are equal, we may *add* the left-hand quantities and *add* the right-hand quantities.

$$\begin{array}{r} 5x - 4y = 8 \\ 6x + 4y = 14 \\ \hline 11x + 0y = 22 \end{array}$$

We have eliminated the y-terms, and now we can solve for x.

$$11x = 22$$

$$x = 22$$

We have found the x-value of the common solution. To find the corresponding y-value, we can substitute 2 for x in either of the original equations.

$$(1) \quad 3x + 2y = 7 \qquad \text{or} \qquad (2) \quad 5x - 4y = 8$$

$$3(2) + 2y = 7 \qquad\qquad\qquad 5(2) - 4y = 8$$

$$6 + 2y = 7 \qquad\qquad\qquad 10 - 4y = 8$$

$$2y = 1 \qquad\qquad\qquad -4y = -2$$

$$y = \frac{1}{2} \qquad\qquad\qquad y = \frac{-2}{-4}$$

$$y = \frac{1}{2}$$

The ordered pair $\left(2, \dfrac{1}{2}\right)$ is the solution for the system.

Now let's discuss how we arrived at the solution. When we chose to multiply equation (2) by the constant 2, we were trying to make the coefficients of the y-term *opposites* of each other. Then, when we added the equations together, the y-terms were eliminated. For obvious reasons, this method is called the **method of addition** (or the **method of elimination**).

Example 1. Solve the system by addition.

$$(1)\ 6x - 10y = 5$$

$$(2)\ 2x - 3y = 1$$

Solution: We can make the x-coefficients opposites by multiplying equation (2) by -3.

$6x - 10y = 5$	Equation (1).
$-6x + 9y = -3$	Multiply equation (2) by -3.
$0x - y = 2$	Add the equations.
$-y = 2$	Simplify.
$y = -2$	Solve for y.

We find the corresponding x-value by substituting -2 for y in either of the original equations. Let's use equation (2).

$$2x - 3y = 1$$

$$2x - 3(-2) = 1$$

$$2x + 6 = 1$$

$$2x = -5$$

$$x = \frac{-5}{2}$$

The solution for the system is $\left(\dfrac{-5}{2}, -2\right)$ and you should check this ordered pair in equation (1).

Now you try Example 2.

Example 2. Solve the system by addition.

$$(1)\ x + 5y = 3$$

$$(2)\ 10y = 4 - 2x$$

Solution: First we rewrite equation (2) so that like variables are lined up beneath each other in the system.

$$(1)\ \ x + \ 5y = 3$$

$$(2)\ 2x + 10y = 4$$

To eliminate the x-terms, we can multiply equation (1) by -2.

$\underline{\hspace{1cm}}x - \underline{\hspace{1cm}}y = \underline{\hspace{1cm}}$	Multiply equation (1) by -2.
$2x + 10y = 4$	Equation (2).
$\underline{\hspace{1cm}}x + \underline{\hspace{1cm}}y = \underline{\hspace{1cm}}$	Add the equations.
$0 = -2$	Simplify.

In attempting to eliminate the *x*-terms, we have also eliminated the *y*-terms! Since $0 = -2$ is a *false* statement, this system must be _____ ; it has _____ solution. Graphing the original equation would yield a pair of _____ lines.

Check your work on page 777. ▶

Example 3. Solve the system by addition.

$$(1)\ 5x + 2y = 1$$

$$(2)\ 3x - 3y = 2$$

Solution: The coefficients of the *y*-terms are opposite in sign, so let's try to eliminate the *y*-terms. Unfortunately (unless we use fractions), we cannot make the coefficients match by multiplying just one equation by a constant. But if we multiply equation (1) by 3 and multiply equation (2) by 2, the coefficients of the *y*-terms will be opposites.

$5x + 2y = 1$	Equation (1).
$3x - 3y = 2$	Equation (2).
$15x + 6y = 3$	Multiply equation (1) by 3.
$6x - 6y = 4$	Multiply equation (2) by 2.
$\overline{21x + 0y = 7}$	Add the equations.
$21x = 7$	Simplify.
$x = \dfrac{7}{21}$	Solve for *x*.
$x = \dfrac{1}{3}$	

Now we substitute $\dfrac{1}{3}$ for *x* in either equation. Let's use equation (2).

$$3x - 3y = 2$$

$$3\left(\frac{1}{3}\right) - 3y = 2$$

$$1 - 3y = 2$$

$$-3y = 1$$

$$y = \frac{-1}{3}$$

The ordered pair $\left(\dfrac{1}{3}, \dfrac{-1}{3}\right)$ is the solution for the system.

Now you try Example 4.

Example 4. Solve the system by addition.

$$(1)\ 3x - 3y = 6$$

$$(2)\ 5x - 10 = 5y$$

Solution: Line up variables beneath each other and try to eliminate the *x*-terms.

$$3x - 3y = 6 \qquad \text{Equation (1).}$$
$$5x - 5y = 10 \qquad \text{Equation (2).}$$
$$15x - 15y = 30 \qquad \text{Multiply equation (1) by 5.}$$
$$\underline{-15x + 15y = -30} \qquad \text{Multiply equation (2) by } -3.$$
$$\_x + \_y = \underline{\quad\quad} \qquad \text{Add the equations.}$$
$$\underline{\quad\quad} = \underline{\quad\quad} \qquad \text{Simplify.}$$

Since this a statement that is *always* true, the system is _____ . The solutions are all ordered pairs satisfying _____ . Graphing the original equations yields the _____ line.

Check your work on page 778. ▶

⫸ Trial Run

Solve each system by addition.

_____ **1.** $5x - y = 7$
$\qquad\quad\;\; 3x + 4y = -5$

_____ **2.** $2x + 3y = 7$
$\qquad\quad\;\; x + 5y = 4$

_____ **3.** $4x - 3y = -5$
$\qquad\quad\;\; 2x - 5y = 1$

_____ **4.** $\;\;\; 3x - 2y = 2$
$\qquad\quad\;\; -6x + 4y = -1$

_____ **5.** $7x - 3y = 9$
$\qquad\quad\;\; 2x + 5y = -4$

_____ **6.** $4x - 5y = 2$
$\qquad\quad\;\; 12x = 15y + 6$

Answers are on page 778.

Solving Systems of Linear Equations in Three Variables

The solutions for an equation containing three variables *x*, *y*, and *z* are **ordered triples** rather than ordered pairs. For instance, the ordered triple $(x, y, z) = (1, 3, 2)$ is one solution for the equation

$$x + 2y - 3z = 1$$

There are infinitely many other ordered triples that satisfy this equation.

Graphing equations in three variables requires a three-dimensional coordinate system. We shall not investigate such graphs in this book. Instead, we shall find algebraic methods for finding ordered triples that satisfy three equations at the same time. We shall use the methods of *addition* and/or *substitution* to solve systems of three equations in three variables.

Our method is basically the same method used for solving systems of two equations in two variables. Of course, more steps will be required and we must keep careful track of what we are doing.

Example 5. Solve the system by addition.

$$(1) \quad x - y + z = 4$$
$$(2) \quad 2x + y + 2z = 5$$
$$(3) \quad x + 3y - z = -4$$

Solution

We can eliminate y-terms from equation (1) and equation (2).

$$
\begin{array}{l}
(1) \quad x - y + z = 4 \\
(2) \ \underline{2x + y + 2z = 5} \\
\quad \ \ 3x + 0y + 3z = 9 \\
\quad \ \ 3x + 3z = 9
\end{array}
$$

Let's label this new equation (a).

(a) $3x + 3z = 9$

We must bring equation (3) into the discussion and use another equation with it to eliminate y-terms again. Multiply equation (1) by 3.

$$
\begin{array}{l}
3x - 3y + 3z = 12 \\
\underline{x + 3y - z = -4} \\
4x + 0y + 2z = 8 \\
 4x + 2z = 8
\end{array}
$$

Label this new equation (b).

(b) $4x + 2z = 8$

Now equations (a) and (b) represent a system of two equations in two variables, x and z.

(a) $3x + 3z = 9$

(b) $4x + 2z = 8$

We proceed as before to eliminate x.

$$
\begin{array}{ll}
12x + 12z = 36 & \text{Multiply equation (a) by 4.} \\
\underline{-12x - 6z = -24} & \text{Multiply equation (b) by } -3. \\
 6z = 12 & \text{Add the equations.} \\
 z = 2 & \text{Solve for } z.
\end{array}
$$

Substitute 2 for z in equation (a) or equation (b) and solve for x.

(a) $3x + 3z = 9$

$3x + 3(2) = 9$

$3x = 3$

$x = 1$

Substitute 2 for z and 1 for x in equation (1) or (2) or (3) and solve for y.

(1) $x - y + z = 4$

$1 - y + 2 = 4$

$-y = 1$

$y = -1$

The ordered triple is $(1, -1, 2)$. You should verify that this ordered triple also satisfies equations (2) and (3).

Example 6. Solve the system by elimination.

(1) $3x + z = 7$

(2) $x + 5y = 2$

(3) $3y - z = -1$

Solution: We line up like terms beneath each other.

(1) $3x + z = 7$

(2) $x + 5y = 2$

(3) $ 3y - z = -1$

First use equations (1) and (3) to eliminate z.

$$3x \qquad + z = 7$$
$$\underline{\qquad 3y - z = -1}$$
$$\text{(a)} \quad 3x + 3y \qquad = 6$$

Now use equation (a) with equation (2) to eliminate x. Multiply equation (2) by -3.

$$\text{(a)} \quad 3x + 3y = 6$$
$$\text{(2)} \quad \underline{-3x - 15y = -6}$$
$$-12y = 0$$
$$y = 0$$

Substitute 0 for y in equation (a) or equation (2) to find x.

$$\text{(2)} \quad x + 5y = 2$$
$$x + 5(0) = 2$$
$$x = 2$$

Substitute 2 for x in equation (1) to find z.

$$\text{(1)} \quad 3x + z = 7$$
$$3(2) + z = 7$$
$$z = 1$$

The solution is $(2, 0, 1)$. Verify that this solution satisfies each of the original equations.

Example 7. Solve the system by elimination.

$$\text{(1)} \quad x - \frac{1}{2}y - \frac{1}{2}z = 4$$

$$\text{(2)} \quad x - \frac{3}{2}y - 2z = 3$$

$$\text{(3)} \quad \frac{1}{4}x + \frac{1}{4}y - \frac{1}{4}z = 9$$

Solution: First we clear the fractions from each equation.

(1) $2x - y - z = 8$	Multiply equation (1) by LCD: 2.
(2) $2x - 3y - 4z = 6$	Multiply equation (2) by LCD: 2.
(3) $x + y - z = 0$	Multiply equation (3) by LCD: 4.

Eliminate y from equations (1) and (3).

$$\text{(1)} \quad 2x - y - z = 8$$
$$\text{(3)} \quad \underline{x + y - z = 0}$$
$$\text{(a)} \quad 3x \qquad - 2z = 8$$

Eliminate y from equations (2) and (3). Multiply equation (3) by 3.

$$2x - 3y - 4z = 6$$
$$\underline{3x + 3y - 3z = 0}$$
$$\text{(b)} \quad 5x \qquad - 7z = 6$$

Eliminate x from equations (a) and (b).

$$\text{(a)} \quad 3x - 2z = 8$$
$$\text{(b)} \quad 5x - 7z = 6$$

$15x - 10z = 40$	Multiply equation (a) by 5.
$\underline{-15x + 21z = -18}$	Multiply equation (b) by -3.
$11z = 22$	Add the equations.
$z = 2$	Solve for z.

Substitute 2 for z in equation (a) to find x.

$$\text{(a)} \quad 3x - 2z = 8$$
$$3x - 2(2) = 8$$
$$3x = 12$$
$$x = 4$$

Substitute 4 for x and 2 for z in equation (3) to find y.

$$\text{(3)} \quad x + y - z = 0$$
$$4 + y - 2 = 0$$
$$y + 2 = 0$$
$$y = -2$$

The solution for the system is $(4, -2, 2)$.

⟫ Trial Run

Solve each system by addition.

1.
$$x + y - z = 4$$
$$x - 2y + z = -1$$
$$3x - y + z = 4$$

2.
$$2x + z = 4$$
$$x - 5y = 3$$
$$4y - z = 2$$

3.
$$x + \frac{1}{2}y - \frac{1}{4}z = 0$$

$$\frac{1}{2}x + y + \frac{1}{2}z = 3$$

$$\frac{3}{2}x + \frac{1}{4}y + z = -2$$

Answers are on page 778.

As was the case with a system of two equations in two variables, a system of three equations in three variables may not yield exactly one ordered triple. When solving such a system algebraically, we must watch for the same signals as before.

> If we arrive at a *false* statement, the system is *inconsistent*. There is no solution for the system.
>
> If we arrive at a statement that is *always true*, the system is *dependent*. There is not just one solution for the system.

▶ Examples You Completed

Example 2. Solve the system by elimination.

$$\text{(1)} \quad x + 5y = 3$$
$$\text{(2)} \quad 10y = 4 - 2x$$

Solution: First we rewrite equation (2) so that like variables are lined up beneath each other in the system.

$$\text{(1)} \quad x + 5y = 3$$
$$2x + 10y = 4$$

To eliminate the x terms, we can multiply equation (1) by -2.

$$\begin{array}{ll} -2x - 10y = -6 & \text{Multiply equation (1) by } -2. \\ \underline{2x + 10y = 4} & \text{Equation (2).} \\ 0x + 0y = -2 & \text{Add the equations.} \\ 0 = -2 & \text{Simplify.} \end{array}$$

In attempting to eliminate the x-terms, we have also eliminated the y-terms! Since $0 = -2$ is a *false* statement, this system must be inconsistent; it has no solution. Graphing the original equations would yield a pair of parallel lines.

Example 4. Solve the system by elimination.

$$(1) \quad 3x - 3y = 6$$

$$(2) \quad 5x - 10 = 5y$$

Solution: Line up variables beneath each other and try to eliminate the x-terms.

$$\begin{array}{ll} 3x - 3y = 6 & \text{Equation (1).} \\ 5x - 5y = 10 & \text{Equation (2).} \end{array}$$

$$\begin{array}{ll} 15x - 15y = 30 & \text{Multiply equation (1) by 5.} \\ \underline{-15x + 15y = -30} & \text{Multiply equation (2) by } -3. \\ 0x + 0y = 0 & \text{Add the equations.} \\ 0 = 0 & \text{Simplify.} \end{array}$$

Since this is a statement that is *always true*, the system is dependent. The solutions are all ordered pairs satisfying $3x - 3y = 6$. Graphing the original equations yields the same line.

Answers to Trial Runs

page 774   **1.** $(1, -2)$ **2.** $\left(\dfrac{23}{7}, \dfrac{1}{7}\right)$ **3.** $(-2, -1)$ **4.** No solution   **5.** $\left(\dfrac{33}{41}, \dfrac{-46}{41}\right)$

6. All ordered pairs satisfying $4x - 5y = 2$

page 777   **1.** $(2, 1, -1)$ **2.** $(3, 0, -2)$ **3.** $(-2, 4, 0)$

EXERCISE SET 10.2

Solve the systems by addition.

_____ **1.** $3x - y = 4$
$2x + y = 1$

_____ **2.** $4x - y = 5$
$3x + y = 9$

_____ **3.** $3x - 2y = 0$
$5x + 2y = 6$

_____ **4.** $-4x + 5y = 0$
$4x + y = 2$

_____ **5.** $7x - 10y = 4$
$x + 5y = 7$

_____ **6.** $8x - 4y = 6$
$3x + 2y = 4$

_____ **7.** $2x - 8y = 9$
$-x + 4y = -6$

_____ **8.** $-3x + 3y = 1$
$x - y = 8$

_____ **9.** $5x - 4y = 1$
$2x + 3y = 5$

_____ **10.** $7x - 6y = 20$
$3x + 5y = 1$

_____ **11.** $7x - 5y = 2$
$2x - 3y = -1$

_____ **12.** $2x - 3y = 5$
$5x - 2y = -4$

_____ **13.** $3x - 6y = 9$
$-2x + 4y = -6$

_____ **14.** $8x - 4y = -4$
$-10x + 5y = 5$

_____ **15.** $y = \dfrac{5}{2}x - \dfrac{5}{2}$
$4x - 6y = 15$

_____ **16.** $10x - 9y = 12$
$y = \dfrac{2}{3}x - \dfrac{4}{3}$

_____ **17.** $2x - 7y = 3$
$3x + 2y = -1$

_____ **18.** $5x - 2y = -2$
$4x + 3y = 5$

_____ **19.** $\dfrac{1}{2}x - \dfrac{1}{9}y = 0$
$\dfrac{1}{5}x - \dfrac{1}{4}y = \dfrac{3}{20}$

_____ **20.** $\dfrac{1}{3}x - \dfrac{1}{7}y = 0$
$\dfrac{1}{2}x - \dfrac{1}{4}y = \dfrac{-3}{8}$

_____ **21.** $\dfrac{1}{3}x - \dfrac{2}{3}y = -2$
$\dfrac{1}{2}x + \dfrac{1}{6}y = 2$

_____ **22.** $\dfrac{2}{3}x + \dfrac{1}{2}y = -1$
$x - \dfrac{1}{4}y = \dfrac{-7}{2}$

_____ **23.** $0.5x - y = -1$
$2.5x + 6y = 28$

_____ **24.** $0.25x - 2y = -4$
$2x + 7y = -9$

_____ **25.** $0.2x + 0.7y = 15$
$0.8x + 2.1y = 11$

_____ **26.** $0.3x + 0.8y = 3$
$0.9x + 3.4y = 12$

_____ **27.** $x - 2y + z = 3$
$2x + 3y - z = -4$
$-x + y + z = 0$

_____ **28.** $2x - y + z = 3$
$x + 3y - 2z = 5$
$3x - 4y + z = 2$

_____ 29. $2x - y + 4z = 17$
$3x + 4y - z = -1$
$4x + 3y + 2z = 11$

_____ 30. $4x - 3y - z = 16$
$2x + 2y + 3z = 8$
$x + 4y + 5z = 3$

_____ 31. $2x - y - z = 6$
$2x - 2y + z = 10$
$x + y - 3z = -2$

_____ 32. $x + y + z = 6$
$3x - 2y - z = 13$
$2x - y + 3z = 26$

_____ 33. $x + y = 0$
$3y + z = 1$
$2x - 3z = -10$

_____ 34. $2x - y = 0$
$3y - z = 5$
$x + z = 2$

_____ 35. $\frac{1}{3}x + \frac{1}{6}y + z = -2$

$x + \frac{1}{2}y + z = 0$

$\frac{2}{3}x + y - \frac{1}{3}z = 5$

_____ 36. $\frac{1}{3}x + y + \frac{1}{2}z = 7$

$\frac{1}{2}x + \frac{3}{2}y + z = 11$

$\frac{1}{3}x + \frac{1}{3}y - z = 0$

☆ Stretching the Topics _____

Solve.

_____ 1. $3(2x - 3) + 4y - 20 = 0$
$4(y + 3) - 3x - 10 = 0$

_____ 2. $2(x + 1) - 3(y - 4) = 2$
$4(x - 1) + 2(y + 1) = 6$

_____ 3. Find the values of m and b that will make the line $y = mx + b$ pass through the points $(3, 2)$ and $(5, 4)$.

Check your answers in the back of your book.

If you can complete **Checkup 10.2,** you are ready to go on to Section 10.3.

✓ CHECKUP 10.2

Solve the systems by addition.

_____ **1.** $x - 3y = 2$
$2x + y = -3$

2. $2x + 3y = 5$
$x + y = 2$

_____ **3.** $2x - 3y = 4$
$5x + 11y = 47$

_____ **4.** $3x - 6y = 3$
$5x - 10y = 5$

_____ **5.** $x + y - z = 4$
$2x - 3y + z = -4$
$3x - y - z = 2$

Check your answers in the back of your book.

If You Missed Problems:	You Should Review Examples:
1–3	1, 3
4	4
5	5

10.3 Using Determinants to Solve Systems (Optional)

Mathematicians invented the **matrix** as a convenient way to organize a pattern of numbers.

> **Definition of a Matrix.** A matrix is a rectangular array of numbers.

A matrix has horizontal rows and vertical columns, enclosed in brackets.

$$\begin{bmatrix} 1 & 2 \\ 3 & 0 \\ 4 & -1 \end{bmatrix}$$ This matrix has 3 rows and 2 columns. It is called a 3×2 ("three by two") matrix.

$$\begin{bmatrix} 5 & -2 \\ 3 & 1 \end{bmatrix}$$ This matrix has 2 rows and 2 columns. It is a 2×2 matrix.

If a matrix has m rows and n columns, it is called an $m \times n$ matrix. If a matrix has the same number of rows as columns, ($m = n$), it is called a **square matrix.**

Finding Determinants

Associated with every *square* matrix is a unique real number called its **determinant.** Although every $n \times n$ matrix has a determinant, we shall discuss here the method for finding the determinant of a 2×2 matrix only. We denote the determinant of a matrix by enclosing the array with vertical bars rather than brackets.

For the matrix

$$\begin{bmatrix} a_1 & b_1 \\ a_2 & b_2 \end{bmatrix}$$ the determinant is written $\begin{vmatrix} a_1 & b_1 \\ a_2 & b_2 \end{vmatrix}$

Now we must find the real number value of this determinant using the definition of a 2×2 determinant.

> **A 2 × 2 Determinant**
> $$\begin{vmatrix} a_1 & b_1 \\ a_2 & b_2 \end{vmatrix} = a_1 b_2 - a_2 b_1$$

Example 1. Find $\begin{vmatrix} 3 & 5 \\ -1 & 2 \end{vmatrix}$.

Solution

$$\begin{vmatrix} 3 & 5 \\ -1 & 2 \end{vmatrix} = 3(2) - (-1)(5)$$

$$= 6 + 5$$

$$= 11$$

You complete Example 2.

Example 2. Find $\begin{vmatrix} 2 & -3 \\ -4 & 6 \end{vmatrix}$.

Solution

$$\begin{vmatrix} 2 & -3 \\ -4 & 6 \end{vmatrix} =$$

Check your work on page 787. ▶

▶ **Trial Run**

Find each determinant.

_____ **1.** $\begin{vmatrix} 2 & 3 \\ 1 & -4 \end{vmatrix}$ 　　_____ **2.** $\begin{vmatrix} -\dfrac{1}{2} & 0 \\[2ex] \dfrac{7}{3} & 4 \end{vmatrix}$ 　　_____ **3.** $\begin{vmatrix} 0.7 & 0.6 \\ 0.1 & 0.2 \end{vmatrix}$

Answers are on page 788.

Using Cramer's Rule

Determinants can be used to solve a system of linear equations according to a theorem called Cramer's rule. We shall discuss the use of that rule in solving systems of *two* equations in *two* unknowns only. Consider such a system.

$$(1) \quad a_1x + b_1y = c_1$$

$$(2) \quad a_2x + b_2y = c_2$$

Here we are using subscripts on a, b, and c to indicate *any* constants.

To solve this system by addition, we can eliminate the y-terms by multiplying equation (1) by b_2 and multiplying equation (2) by $-b_1$.

$b_2[a_1x + b_1y = c_1]$	Multiply equation (1) by b_2.
$-b_1[a_2x + b_2y = c_2]$	Multiply equation (2) by $-b_1$.
$a_1b_2x + b_1b_2y = c_1b_2$	Remove parentheses.
$\underline{-a_2b_1x - b_1b_2y = -c_2b_1}$	
$a_1b_2x - a_2b_1x = c_1b_2 - c_2b_1$	Add equations.
$x(a_1b_2 - a_2b_1) = c_1b_2 - c_2b_1$	Factor out x on left side.
$x = \dfrac{c_1b_2 - c_2b_1}{a_1b_2 - a_2b_1}$	Isolate x.

But notice that the numerator and denominator on the right can both be rewritten as determinants.

$$x = \frac{\begin{vmatrix} c_1 & b_1 \\ c_2 & b_2 \end{vmatrix}}{\begin{vmatrix} a_1 & b_1 \\ a_2 & b_2 \end{vmatrix}}$$

Observe that the denominator is the determinant of the matrix made up of the coefficients of the variables in the original system. The determinant of that coefficient matrix is often represented by D.

Observe that the numerator is a similar determinant in which the column of x-coefficients in D has been replaced by the column of constants from the original system. This determinant is represented by D_x. Another determinant, D_y, is formed when the column of y-coefficients in D is replaced by the column of constants.

For the system

$$a_1 x + b_1 y = c_1$$

$$a_2 x + b_2 y = c_2$$

we define

$$D = \begin{vmatrix} a_1 & b_1 \\ a_2 & b_2 \end{vmatrix} \qquad D_x = \begin{vmatrix} c_1 & b_1 \\ c_2 & b_2 \end{vmatrix} \qquad D_y = \begin{vmatrix} a_1 & c_1 \\ a_2 & c_2 \end{vmatrix}$$

In our system we discovered that $x = \dfrac{D_x}{D}$. A similar procedure (eliminating x-terms instead) would result in the statement that $y = \dfrac{D_y}{D}$. These facts are summarized in Cramer's rule.

Cramer's Rule. For the system

$$a_1 x + b_1 y = c_1$$

$$a_2 x + b_2 y = c_2$$

the solutions are found by

$$x = \frac{D_x}{D} \qquad \text{and} \qquad y = \frac{D_y}{D}$$

where $D \neq 0$ and

$$D = \begin{vmatrix} a_1 & b_1 \\ a_2 & b_2 \end{vmatrix} \qquad D_x = \begin{vmatrix} c_1 & b_1 \\ c_2 & b_2 \end{vmatrix} \qquad D_y = \begin{vmatrix} a_1 & c_1 \\ a_2 & c_2 \end{vmatrix}$$

Example 3. Use Cramer's rule to solve the system.

$$3x + 4y = 2$$

$$7x - 2y = 9$$

Solution: Here

$$a_1 x + b_1 y = c_1 \qquad \text{is} \qquad 3x + 4y = 2$$

$$a_2 x + b_2 y = c_2 \qquad \text{is} \qquad 7x - 2y = 9$$

We must find each determinant.

$$D = \begin{vmatrix} a_1 & b_1 \\ a_2 & b_2 \end{vmatrix} \qquad D_x = \begin{vmatrix} c_1 & b_1 \\ c_2 & b_2 \end{vmatrix} \qquad D_y = \begin{vmatrix} a_1 & c_1 \\ a_2 & c_2 \end{vmatrix}$$

$$= \begin{vmatrix} 3 & 4 \\ 7 & -2 \end{vmatrix} \qquad = \begin{vmatrix} 2 & 4 \\ 9 & -2 \end{vmatrix} \qquad = \begin{vmatrix} 3 & 2 \\ 7 & 9 \end{vmatrix}$$

$$= 3(-2) - 7(4) \qquad = 2(-2) - 9(4) \qquad = 3(9) - 7(2)$$

$$= -6 - 28 \qquad = -4 - 36 \qquad = 27 - 14$$

$$D = -34 \qquad\qquad D_x = -40 \qquad\qquad D_y = 13$$

Now we can find x and y.

$$x = \frac{D_x}{D} \qquad\qquad y = \frac{D_y}{D}$$

$$= \frac{-40}{-34} \qquad\qquad = \frac{13}{-34}$$

$$x = \frac{20}{17} \qquad\qquad y = \frac{-13}{34}$$

The solution is $\left(\dfrac{20}{17}, \dfrac{-13}{34}\right)$.

Example 4. Use Cramer's rule to solve the system.

$$x + y = 1$$
$$3x + 3y = 4$$

Solution: Here

$$a_1 x + b_1 y = c_1$$
$$a_2 x + b_2 y = c_2$$

is

$$1x + 1y = 1$$
$$3x + 3y = 4$$

$$D = \begin{vmatrix} a_1 & b_1 \\ a_2 & b_2 \end{vmatrix} = \begin{vmatrix} 1 & 1 \\ 3 & 3 \end{vmatrix} = 1(3) - 3(1) = 3 - 3 = 0$$

Since $D = 0$, we cannot use Cramer's rule. The system does not have one unique solution; it is either inconsistent or dependent. We must solve by some other method to decide.

Using the addition method, we multiply the first equation by -3 to eliminate x.

$$
\begin{array}{ll}
-3x - 3y = -3 & \text{Multiply equation (1) by } -3. \\
\underline{3x + 3y = 4} & \text{Equation (2).} \\
\, 0 = 1 & \text{Add equations.}
\end{array}
$$

This system is inconsistent; there is *no* solution.

As this example indicates, Cramer's rule provides a quick procedure for determining whether a system has a unique solution. If $D = 0$, we know that the system is inconsistent or dependent.

Example 5. Use Cramer's rule to solve the system.

$$y = 3x + 2$$
$$5x + 4y = 8$$

Solution: Our system is

$$-3x + y = 2$$
$$5x + 4y = 8$$

$$D = \begin{vmatrix} -3 & 1 \\ 5 & 4 \end{vmatrix} \qquad D_x = \begin{vmatrix} 2 & 1 \\ 8 & 4 \end{vmatrix} \qquad D_y = \begin{vmatrix} -3 & 2 \\ 5 & 8 \end{vmatrix}$$

$$= -3(4) - 5(1) \qquad\quad = 2(4) - 8(1) \qquad\quad = -3(8) - 5(2)$$

$$= -12 - 5 \qquad\qquad\quad = 8 - 8 \qquad\qquad\quad = -24 - 10$$

$$= -17 \qquad\qquad\qquad = 0 \qquad\qquad\qquad = -34$$

Now we can find x and y.

$$x = \frac{D_x}{D} \qquad \text{and} \qquad y = \frac{D_y}{D}$$

$$= \frac{0}{-17} \qquad\qquad\qquad = \frac{-34}{-17}$$

$$x = 0 \qquad\qquad\qquad\quad y = 2$$

The solution is $(0, 2)$.

Cramer's rule can also be extended to solving systems of more equations in more variables, but that is a topic for your next algebra course.

⫸ Trial Run

Use Cramer's rule to solve each system.

_____ **1.** $3x - y = 1$
            $x + 3y = 2$

_____ **2.** $4x + 3y = -8$
            $-x + y = 2$

_____ **3.** $y = 5 - 2x$
            $4x + 5y = 7$

_____ **4.** $-2x + 3y = 5$
            $5x + 7y = 1$

Answers are on page 788.

▶ Example You Completed

Example 2. Find $\begin{vmatrix} 2 & -3 \\ -4 & 6 \end{vmatrix}$.

Solution

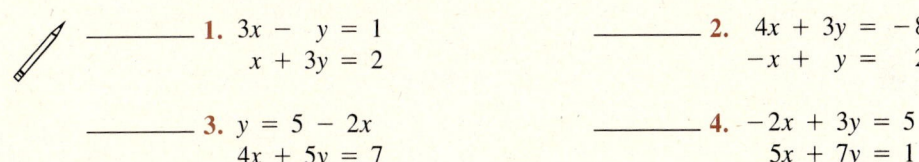

$$\begin{vmatrix} 2 & -3 \\ -4 & 6 \end{vmatrix} = 2(6) - (-4)(-3) = 12 - 12 = 0$$

Answers to Trial Runs

page 784   **1.** -11 **2.** -2 **3.** 0.08

page 787   **1.** $\left(\dfrac{1}{2}, \dfrac{1}{2}\right)$ **2.** $(-2, 0)$ **3.** $(3, -1)$ **4.** $\left(\dfrac{-32}{29}, \dfrac{27}{29}\right)$

EXERCISE SET 10.3

Find the value of each determinant.

_____ 1. $\begin{vmatrix} 2 & 3 \\ -2 & 5 \end{vmatrix}$

_____ 2. $\begin{vmatrix} -3 & -2 \\ 5 & 3 \end{vmatrix}$

_____ 3. $\begin{vmatrix} 0 & -3 \\ 4 & 0 \end{vmatrix}$

_____ 4. $\begin{vmatrix} -7 & 0 \\ 0 & 9 \end{vmatrix}$

_____ 5. $\begin{vmatrix} \frac{3}{4} & \frac{1}{6} \\ -\frac{1}{2} & \frac{1}{3} \end{vmatrix}$

_____ 6. $\begin{vmatrix} \frac{-1}{3} & \frac{5}{6} \\ \frac{-3}{5} & \frac{7}{10} \end{vmatrix}$

Use Cramer's rule to solve each system.

_____ 7. $2x + 3y = 7$
$x + 2y = 3$

_____ 8. $5x - 12y = 4$
$4x - 7y = -2$

_____ 9. $3x + 2y = 4$
$x + \frac{2}{3}y = -2$

_____ 10. $4x - y = 5$
$-x + \frac{1}{4}y = -2$

_____ 11. $-x + y = 1$
$3x - 4y = -3$

_____ 12. $2x + 3y = 4$
$4x + y = -2$

_____ 13. $3x + 2y = 1$
$3x - 2y = -5$

_____ 14. $2x + 3y = 5$
$x + 5y = -2$

_____ 15. $\frac{1}{2}x + \frac{1}{3}y = \frac{3}{2}$
$\frac{2}{3}x - \frac{1}{4}y = \frac{5}{3}$

_____ 16. $\frac{1}{4}x + \frac{1}{2}y = \frac{5}{4}$
$\frac{1}{2}x - \frac{3}{4}y = \frac{3}{2}$

_____ 17. $0.3x + 1.5y = 6$
$1.5x - 0.5y = 6$

_____ 18. $0.8x + 0.3y = 3$
$0.3x + 0.5y = 1.9$

☆ Stretching the Topics

_____ 1. If $\begin{vmatrix} a + 2 & -3 \\ a + 10 & 4 \end{vmatrix} = 0$, find the value of a.

_____ 2. Use Cramer's rule to find the value of x and y in terms of a and b.

$$ax - 2y = 3$$
$$2x + by = -2$$

Check your answers in the back of your book.

If you can complete **Checkup 10.3,** you are ready to go on to Section 10.4.

✓ **CHECKUP 10.3**

Find the value of each determinant.

_____ **1.** $\begin{vmatrix} -4 & 5 \\ 3 & 0 \end{vmatrix}$

_____ **2.** $\begin{vmatrix} -12 & -3 \\ 9 & 2 \end{vmatrix}$

Use Cramer's rule to solve each system.

_____ **3.** $2x - y = 5$
$-x + y = -3$

_____ **4.** $\frac{2}{3}x - y = -2$
$2x - 3y = 3$

_____ **5.** $3x + 2y = 5$
$4x - 5y = 22$

_____ **6.** $4x + 9y = 3$
$9x - 4y = 2$

Check your answers in the back of your book.

If You Missed Problems:	You Should Review Examples:
1, 2	1, 2
3	3
4	4
5, 6	3–5

10.4 Solving Nonlinear Systems of Equations and Systems of Inequalities

Solving Nonlinear Systems of Equations

As you might imagine, a *nonlinear system* of equations is a system in which at least one of the equations is *not* linear. The graph of a nonlinear system might contain:

1. A line and a parabola, circle, ellipse, or hyperbola.

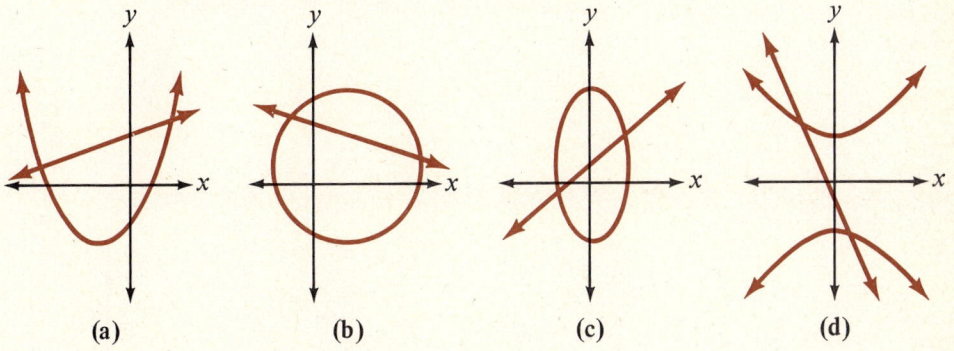

(a)   (b)   (c)   (d)

2. A parabola and a circle, ellipse, or hyperbola.

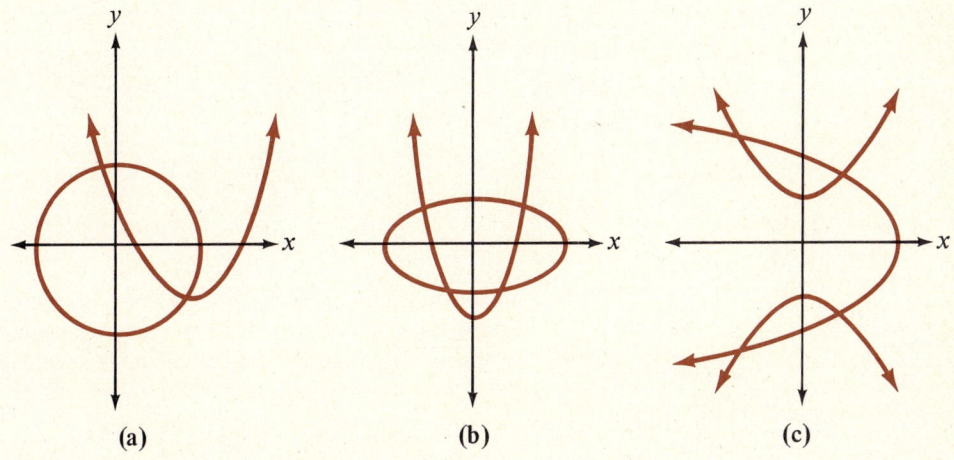

(a)   (b)   (c)

3. A circle and an ellipse or hyperbola.

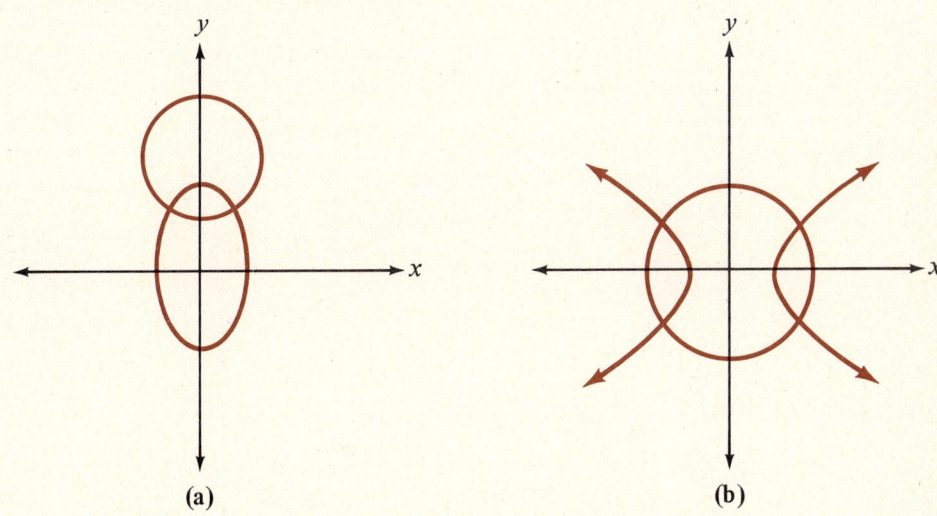

(a)   (b)

4. An ellipse and a hyperbola.

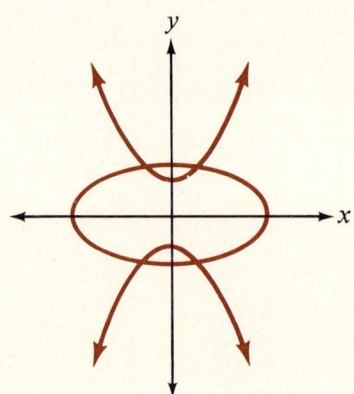

5. A pair of parabola, circles, ellipses, or hyperbolas.

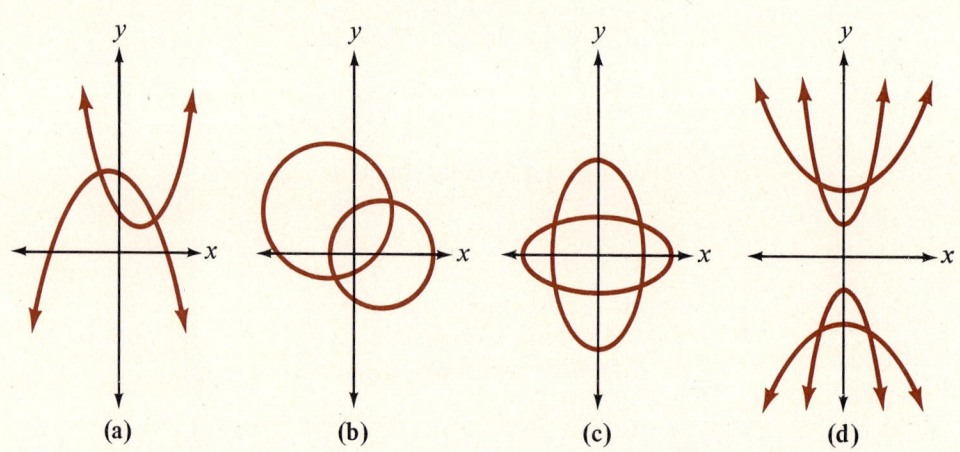

(a)   (b)   (c)   (d)

It is a good idea to have a mental picture of the graph of a system before attempting to solve it so that you will know the maximum number of possible solutions. In solving nonlinear systems we shall use whichever of our established methods seems most appropriate.

Example 1. Solve the system.

$$(1) \quad y = -3x - 1$$

$$(2) \quad y = x^2 - 2x - 3$$

Solution: The graph of equation (1) is a straight line, and the graph of equation (2) is a parabola. We seek the points of intersection of the line and parabola; there can be at most *two* such points.

Since y is isolated in at least one equation, the method of *substitution* seems appropriate here. We can substitute $-3x - 1$ for y in equation (2) and solve for x.

$$-3x - 1 = x^2 - 2x - 3 \qquad \text{Substitute } -3x - 1 \text{ for } y \text{ in equation (2).}$$

$$0 = x^2 + x - 2 \qquad \text{Set the quadratic expression equal to zero.}$$

$$0 = (x + 2)(x - 1) \qquad \text{Factor the trinomial.}$$

$$x + 2 = 0 \qquad x - 1 = 0$$
$$\text{or}$$
$$x = -2 \qquad x = 1$$

The graphs intersect at the points with x-coordinates of -2 and 1. To find the corresponding y-coordinates, we substitute the x-value into equation (1) or (2). Let's use equation (1) $y = -3x - 1$.

Let $x = -2$	Let $x = 1$
$y = -3(-2) - 1$	$y = -3(1) - 1$
$= 6 - 1$	$= -3 - 1$
$y = 5$	$y = -4$
$(-2, 5)$	$(1, -4)$

The solutions for the system are $(-2, 5)$ and $(1, -4)$. Look at the graph of the system and check both solutions in equation (2) $y = x^2 - 2x - 3$.

CHECK: $x = -2$ and $y = 5$

$$5 \stackrel{?}{=} (-2)^2 - 2(-2) - 3$$

$$5 \stackrel{?}{=} 4 + 4 - 3$$

$$5 = 5$$

CHECK: $x = 1$ and $y = -4$

$$-4 \stackrel{?}{=} (1)^2 - 2(1) - 3$$

$$-4 \stackrel{?}{=} 1 - 2 - 3$$

$$-4 = -4$$

① $y = -3x - 1$
② $y = x^2 - 2x - 3$

Example 2. Solve the system.

$$(1) \quad x^2 + y^2 = 4$$

$$(2) \quad x^2 + 2y^2 = 6$$

Solution: Equation (1) represents a circle and equation (2) represents an ellipse. There can be at most *four* points of intersection.

Since neither variable is easily isolated in either equation, we should try *elimination* rather than substitution here. We can eliminate the x^2-terms by multiplying equation (1) by -1.

$-x^2 - y^2 = -4$	Multiply equation (1) by -1.
$\underline{x^2 + 2y^2 = 6}$	Equation 2.
$y^2 = 2$	Add the equations.
$y = \pm\sqrt{2}$	Solve for y.
$y = \sqrt{2} \quad \text{or} \quad y = -\sqrt{2}$	

To find corresponding x-values, substitute these y-values into equation (1).

Let $y = \sqrt{2}$	Let $y = -\sqrt{2}$
$x^2 + y^2 = 4$	$x^2 + y^2 = 4$
$x^2 + (\sqrt{2})^2 = 4$	$x^2 + (-\sqrt{2})^2 = 4$
$x^2 + 2 = 4$	$x^2 + 2 = 4$
$x^2 = 2$	$x^2 = 2$
$x = \pm\sqrt{2}$	$x = \pm\sqrt{2}$
$(\sqrt{2}, \sqrt{2})$ and $(-\sqrt{2}, \sqrt{2})$	$(\sqrt{2}, -\sqrt{2})$ and $(-\sqrt{2}, -\sqrt{2})$

This system has four solutions. Let's sketch the graph.

① $x^2 + y^2 = 4$
② $x^2 + 2y^2 = 6$

Example 3. Solve the system.

$$(1) \quad x^2 + y^2 = 1$$
$$(2) \quad 2x^2 - 3y^2 = 8$$

Solution: The graph of equation (1) is a circle and equation (2) is a hyperbola. There can be at most *four* points of intersection. Let's solve the system by eliminating x^2-terms.

$-2x^2 - 2y^2 = -2$	Multiply equation (1) by -2.
$\underline{2x^2 - 3y^2 = 8}$	Equation (2).
$-5y^2 = 6$	Add the equations.
$y^2 = \dfrac{-6}{5}$	Isolate y^2.
$y = \pm\sqrt{\dfrac{-6}{5}}$	Solve for y.

There are no real y-values. The system has *no real number solutions;* the circle and hyperbola do *not* intersect. Look at the graph.

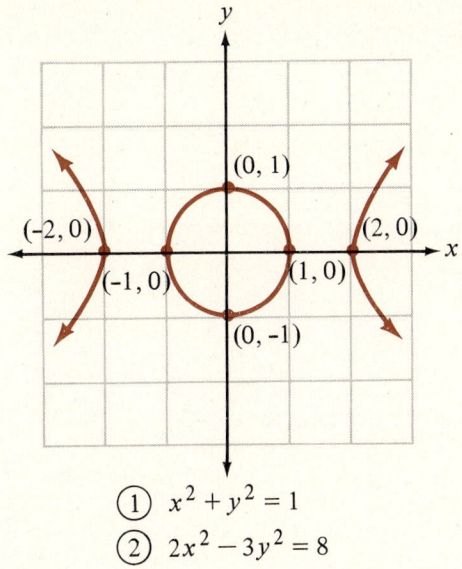

$$\textcircled{1} \quad x^2 + y^2 = 1$$
$$\textcircled{2} \quad 2x^2 - 3y^2 = 8$$

Now you complete Example 4.

Example 4. Solve the system.

$$(1) \quad y = x^2 + 2$$
$$(2) \quad x^2 + y^2 = 4$$

Solution: The graph of equation (1) is a _____ and the graph of equation (2) is a _____ . There can be at most _____ points of intersection.

We could solve this equation by substitution, since y is isolated in equation (1), but such a substitution would change equation (2) into a fourth-degree equation. Instead, let's line up the variables and eliminate the x^2-terms.

$$(1) \quad -x^2 \qquad + y = 2$$
$$(2) \quad \underline{\quad x^2 + y^2 \qquad = 4}$$
$$y^2 + y = 6$$
$$y^2 + y - 6 = 0$$

$$(y + \underline{\quad})(y - \underline{\quad}) = 0$$

$$y + \underline{\quad} = 0 \qquad or \quad y - \underline{\quad} = 0$$

$$y = \underline{\quad} \qquad\qquad y = \underline{\quad}$$

We substitute these y-values into equation (1).

Let $y = -3$	Let $y = 2$
$y = x^2 + 2$	$y = x^2 + 2$
$-3 = x^2 + 2$	$2 = x^2 + 2$
$-5 = x^2$	$0 = x^2$
$\pm\sqrt{-5} = x$	$x = \underline{\quad}$
No real solutions	$(\underline{\quad} , 2)$

The only solution is $(\underline{\quad} , 2)$. The parabola and circle intersect in just one point. Let's see if this result is verified by the graph.

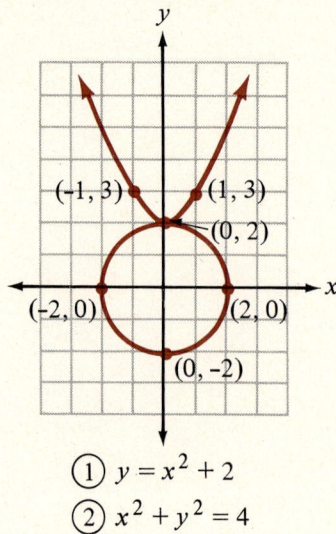

$$\textcircled{1} \; y = x^2 + 2$$
$$\textcircled{2} \; x^2 + y^2 = 4$$

Check your work on page 801. ▶

⫸ Trial Run

Solve each system.

1. $y = x^2 - 2$
   $2x + y = 1$

2. $x^2 + y^2 = 9$
   $y = 2$

3. $x^2 + y^2 = 13$
   $2x^2 + 3y^2 = 30$

4. $x^2 + y^2 = 12$
   $y = x^2 - 6$

Answers are on page 802.

Solving Systems of Inequalities

In Chapter 8 we learned to graph inequalities as shaded regions in the plane. If we wish to find all points satisfying more than one inequality, we must graph each inequality and look for shaded regions *common* to all the given inequalities.

Example 5. Solve the system of inequalities.

$$(1) \; y > 2x - 1$$

$$(2) \; y \le 3 - x$$

Solution: We graph each inequality by the usual methods of Chapter 8.

(1) $y > 2x - 1$

Consider the boundary equation

$$y = 2x - 1$$

Slope: $m = 2$

y-intercept: $b = -1$

The line is *dashed* and we use vertical shading for the region *above* the line.

(2) $y \le 3 - x$

Consider the boundary equation

$$y = -x + 3$$

Slope: $m = -1$

y-intercept: $b = 3$

The line is *solid* and we use horizontal shading for the region *below* the line.

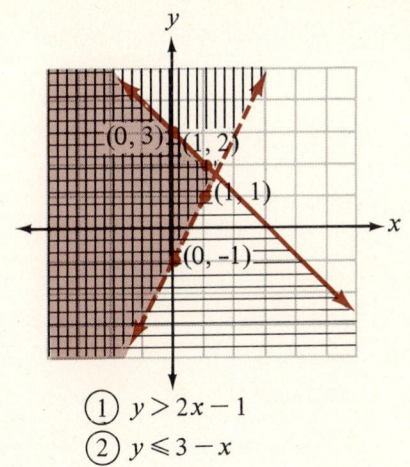

① $y > 2x - 1$

② $y \leqslant 3 - x$

The region in which both inequalities are satisfied is the region containing both kinds of shading. It is the region shaded in red in the figure.

You try Example 6.

Example 6. Solve the system of inequalities.

$$(1)\ (x - 2)^2 + (y + 1)^2 < 4$$

$$(2)\ y + x \geq 0$$

Solution: We graph each inequality.

(1) $(x - 2)^2 + (y + 1)^2 < 4$

Consider the boundary equation

$$(x - 2)^2 + (y + 1)^2 = 4$$

Center: (___, ___)

Radius: $r =$ ___

The graph is a dashed _____ and we shade the region _____ the circle.

(2) $y + x \geq 0$

$$y \geq -x$$

Consider the boundary equation

$$y = -x$$

Slope: $m =$ ___

y-intercept: $b =$ ___

The graph is a _____ line and we shade the region _____ the line.

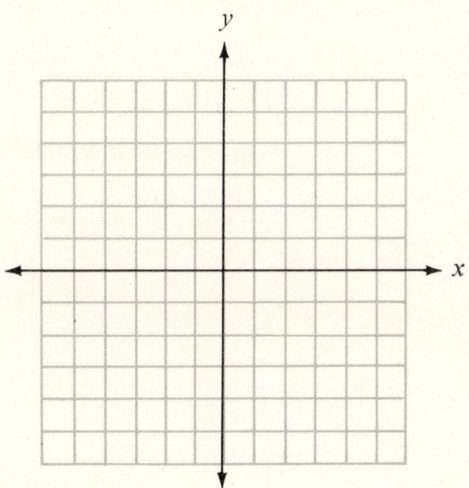

Check your work on page 801.

Example 7. Solve the system of inequalities.

$$(1)\ y \leq 5 - x^2$$

$$(2)\ x < 1$$

$$(3)\ y \geq 2$$

Solution: We graph each inequality.

(1) $y \leq 5 - x^2$

Consider the boundary equation

$$y = 5 - x^2$$

This is a *solid* parabola that opens down and is symmetric to the *y*-axis. The vertex is (0, 5). We shade *below* the parabola.

(2) $x < 1$

Consider the boundary equation

$$x = 1$$

This is a *dashed* vertical line and we shade to the *left* of the line.

(3) $y \geq 2$

Consider the boundary equation

$$y = 2$$

This is a *solid* horizontal line and we shade *above* the line.

1 $y \leq 5 - x^2$

2 $x < 1$

3 $y \geq 2$

The region in which all *three* kinds of shading appear is the region shaded in red in the graph.

⫸ Trial Run

Solve each system of inequalities.

1. $5x - 2y \leq 6$
   $y < x$

2. $x^2 + y^2 < 25$
   $y \geq -x + 5$

3. $y < -x^2 + 9$
   $y > x + 1$
   $x \geq 0$

Answers are on page 802.

▶ **Examples You Completed** ────────────────────────────

Example 4. Solve the system.

$$(1) \ y = x^2 + 2$$

$$(2) \ x^2 + y^2 = 4$$

Solution: The graph of equation (1) is a parabola and the graph of equation (2) is a circle. There can be at most four points of intersection.

$$
\begin{array}{rl}
(1) & -x^2 \quad\quad + y \ = 2 \\
(2) & \underline{x^2 + y^2 \quad\quad = 4} \\
& y^2 + y \ = 6 \\
& y^2 + y - 6 \ = 0 \\
& (y + 3)(y - 2) \ = 0
\end{array}
$$

$$y + 3 = 0 \quad \text{or} \quad y - 2 = 0$$

$$y = -3 \quad\quad\quad y = 2$$

We substitute these y-values into equation (1).

Let $y = -3$	Let $y = 2$
$y = x^2 + 2$	$y = x^2 + 2$
$-3 = x^2 + 2$	$2 = x^2 + 2$
$-5 = x^2$	$0 = x^2$
$\pm\sqrt{-5} = x$	$x = 0$
No real solutions	$(0, 2)$

The solution is $(0, 2)$.

Example 6. Solve the system of inequalities.

$$(1) \ (x - 2)^2 + (y + 1)^2 < 4$$

$$(2) \ y + x \geq 0$$

Solution: We graph each inequality.

$$(1) \ (x - 2)^2 + (y + 1)^2 < 4$$

Consider the boundary equation

$$(x - 2)^2 + (y + 1)^2 = 4$$

Center: $(2, -1)$

Radius: $r = 2$

The graph is a dashed circle and we shade the region inside the circle.

(2) $y + x \geq 0$

$y \geq -x$

Consider the boundary equation

$y = -x$

Slope: $m = \dfrac{-1}{1}$

y-intercept: $b = 0$

The graph is a solid line and we shade the region above the line.

① $(x - 2)^2 + (y + 1)^2 < 4$
② $y + x \geq 0$

Answers to Trial Runs

page 798   **1.** $(1, -1)$ and $(-3, 7)$ **2.** $(\sqrt{5}, 2)$ and $(-\sqrt{5}, 2)$
3. $(-3, 2), (3, 2), (-3, -2),$ and $(3, -2)$
4. $(\sqrt{3}, -3), (-\sqrt{3}, -3), (2\sqrt{2}, 2),$ and $(-2\sqrt{2}, 2)$

page 800   **1.**

2.

3.

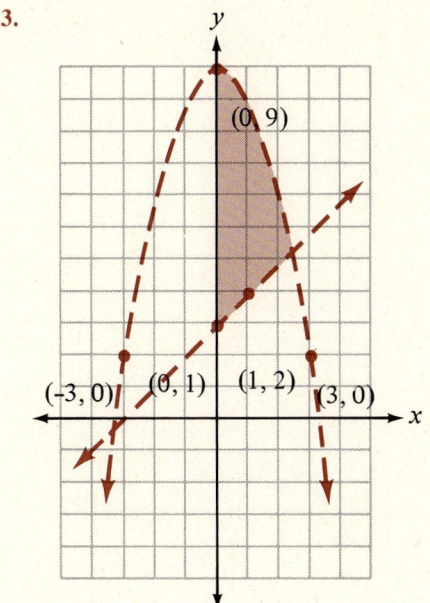

EXERCISE SET 10.4

Solve the following systems.

_____ **1.** $y = x^2 - 5$
$x + y = -3$

_____ **2.** $y = x^2 + 3$
$x + y = 9$

_____ **3.** $x^2 + y^2 = 20$
$x - y = 2$

_____ **4.** $x^2 + y^2 = 25$
$x + y = -1$

_____ **5.** $x^2 + y^2 = 25$
$y = -x^2 + 13$

_____ **6.** $x^2 + y^2 = 16$
$y = x^2 - 10$

_____ **7.** $4x^2 - 9y^2 = 16$
$y = \dfrac{2}{5}x$

_____ **8.** $9x^2 - 27y^2 = 9$
$y = \dfrac{1}{2}x$

_____ **9.** $x^2 + y^2 = 8$
$3x^2 + 4y^2 = 28$

_____ **10.** $x^2 + y^2 = 25$
$2x^2 + 5y^2 = 98$

_____ **11.** $4x^2 + 9y^2 = 36$
$x^2 - 4y^2 = 49$

_____ **12.** $4x^2 + 25y^2 = 100$
$2x^2 - 9y^2 = 72$

_____ **13.** $y = x + 3$
$xy = 10$

_____ **14.** $y = 2x - 3$
$xy = -1$

_____ **15.** $x = y^2 - 3$
$x = 1$

_____ **16.** $x = y^2 - 5$
$x = 4$

Solve the following systems.

17. $2x + 4y \le 20$
$3x - 2y < 6$

18. $2x - y < 7$
$3x + 4y \ge -6$

19. $x^2 + y^2 \le 16$
$3x - y \ge 1$

20. $x^2 + y^2 \ge 4$
$x - y \ge 0$

21. $-3x + 2y \le -6$
$6x - 4y \ge 8$

22. $4x - 2y \ge 16$
$-2x + y \le -4$

23. $y > x^2 - 4x$
$3y < 2x$
$x > 2$

24. $y > x^2 + 6x$
$y < -x$
$y > 1$

☆ Stretching the Topics _____

_____ **1.** Solve the system for y in terms of a.

$$x^2 + y^2 = 16$$
$$x + y = a$$

_____ **2.** In Problem 1, for what values of a does the line intersect the circle in two points? For what values of a does the line fail to meet the circle? For what values of a is the line tangent to a?

————— **3.** Solve graphically

$$x^2 + y^2 \geq 4$$
$$4x^2 + 9y^2 \leq 36$$

Check your answers in the back of your book.

If you can complete **Checkup 10.4,** you are ready to go on to Section 10.5.

CHECKUP 10.4

Solve the systems.

_____ **1.** $y = x^2 - 8$
$3x + y = -4$

_____ **2.** $x^2 + y^2 = 9$
$y = -x^2 + 3$

_____ **3.** $x^2 - y^2 = 16$
$4x^2 + y^2 = 9$

4. Graph the system

$y > 5x - 2$

$2x - 3y > -3$

$y \geq 1$

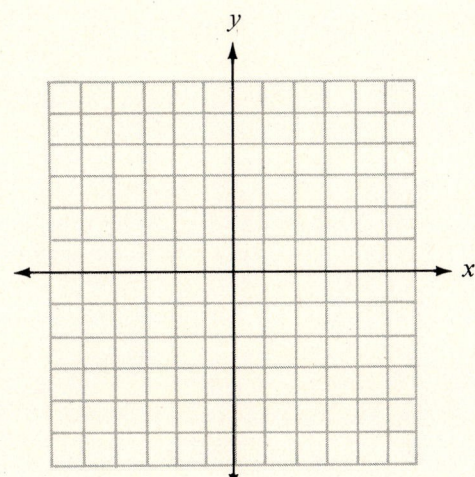

Check your answers in the back of your book.

If You Missed Problem:	You Should Review Example:
1	1
2	4
3	2
4	7

10.5 Switching from Words to Systems of Equations

Systems of equations are useful in solving word problems involving more than one unknown quantity that must meet more than one condition at the same time. Let's look at the problem stated at the beginning of this chapter.

Example 1. During the Fourth of July Festival, Marcia sold hot dogs and hamburgers at her refreshment stand. On Friday she made a profit of $130 by selling 200 hot dogs and 150 hamburgers. On Saturday she made a profit of $225 by selling 300 hot dogs and 300 hamburgers. What profit did Marcia make on each hot dog and each hamburger during the two days?

Solution

We Think	*We Write*
We do not know the profit on each hot dog and each hamburger.	Let x = profit on each hot dog (dollars) y = profit on each hamburger (dollars)

We can find the profit for each kind of sandwich by *multiplying* the number sold times the profit per sandwich.

	Hot Dog Profit	Hamburger Profit
Friday	$200x$	$150y$
Saturday	$300x$	$300y$

Each day's profit is the *sum* of the hot dog profit and the hamburger profit.

Friday: $200x + 150y = 130$
Saturday: $300x + 300y = 225$

Now we have a system of two equations to solve by elimination.

$$(1) \quad 200x + 150y = 130$$

$$(2) \quad 300x + 300y = 225$$

$-400x - 300y = -260$	Multiply equation (1) by -2.
$300x + 300y = 225$	Equation (2).
$-100x = -35$	Add the equations.
$x = \dfrac{-35}{-100}$	Isolate x.
$x = 0.35$	Simplify.

We find the corresponding y-value by substituting 0.35 for x in either of the original equations. Let's use equation (1).

$$200x + 150y = 130$$

$$200(0.35) + 150y = 130$$

$$70 + 150y = 130$$

$$150y = 60$$

$$y = \frac{60}{150}$$

$$y = \frac{2}{5}$$

$$y = 0.40$$

Marcia made a profit of $0.35 on each hot dog and $0.40 on each hamburger.

Example 2. A sail in the shape of a right triangle has a perimeter of 30 feet. If the longest side (hypotenuse) of the sail measures 13 feet, what are the lengths of the other two sides (legs)?

Solution

We Think	*We Write*
We do not know the lengths of the two legs.	Let x = length of one leg (feet) y = length of other leg (feet)
An illustration is always useful in a geometry problem.	
The perimeter of a triangle is the *sum* of the lengths of the three sides.	Perimeter $x + y + 13 = 30$ $x + y = 17$
We need another equation. Since the sail is a *right* triangle, we can use the Pythagorean theorem. We now have a system of two equations to solve.	Pythagorean theorem: $x^2 + y^2 = 13^2$ $x^2 + y^2 = 169$

(1) $x + y = 17$

(1) $x^2 + y^2 = 169$

Elimination will not work here, so we use the method of substitution. Let's solve equation (1) for y.

(1) $x + y = 17$	
$y = 17 - x$	Isolate y.
(2) $\qquad x^2 + y^2 = 169$	Equation (2).
$x^2 + (17 - x)^2 = 169$	Substitute $17 - x$ for y in equation (2).
$x^2 + 289 - 34x + x^2 = 169$	Remove parentheses.
$2x^2 - 34x + 289 = 169$	Combine like terms.
$2x^2 - 34x + 120 = 0$	Set the quadratic expression equal to zero.
$2(x^2 - 17x + 60) = 0$	Factor out the common 2.
$2(x - 5)(x - 12) = 0$	Factor the trinomial.

$$x - 5 = 0 \quad \text{or} \quad x - 12 = 0$$

$$x = 5 \qquad\qquad x = 12$$

Substitute 5 for x in equation (1).   Substitute 12 for x in equation (1).

(1) $x + y = 17$	(1) $x + y = 17$
$5 + y = 17$	$12 + y = 17$
$y = 12$	$y = 5$
$(5, 12)$	$(12, 5)$

The sides of the sail must measure 5 feet and 12 feet.

Example 3. Anna and Beth are planning to drive to Canada, sharing the time at the wheel. Anna always drives at one constant rate and Beth always drives at another constant rate. If Anna drives for 7 hours and Beth for 4 hours, they can travel 570 miles. If Anna drives for 4 hours and Beth for 7 hours, they can travel 585 miles. How fast does each one drive?

Solution

We Think	*We Write*
We do not know either driver's rate.	Let a = Anna's rate (mph)
	b = Beth's rate (mph)

Distance is found by *multiplying* rate × time.

	Anna's Distance	Beth's Distance
First plan	$7a$	$4b$
Second plan	$4a$	$7b$

The total distance for either plan is the *sum* of the drivers' distances.

First plan: $7a + 4b = 570$
Second plan: $4a + 7b = 585$

Now we have our system of two equations to solve by elimination. Let's eliminate b.

$$(1) \quad 7a + 4b = 570$$

$$(2) \quad 4a + 7b = 585$$

$$
\begin{array}{ll}
49a + 28b = 3990 & \text{Multiply equation (1) by 7.} \\
-16a - 28b = -2340 & \text{Multiply equation (2) by } -4. \\
\hline
33a \qquad\quad = 1650 & \text{Add the equations.}
\end{array}
$$

$$a = \frac{1650}{33} \qquad \text{Isolate } a.$$

$$a = 50 \qquad \text{Simplify.}$$

We can find the corresponding b-value by substituting 50 for a in either equation.

$$(1) \qquad 7a + 4b = 570$$

$$7(50) + 4b = 570$$

$$350 + 4b = 570$$

$$4b = 220$$

$$b = \frac{220}{4}$$

$$= 55$$

Anna's rate is 50 mph and Beth's rate is 55 mph.

Example 4. Beside his house, Carlo wishes to plant a rectangular garden which is to be enclosed on the two shorter sides and one of the longer sides by 38 feet of fencing. If he wishes the area of the garden to be 176 square feet, what should be its dimensions?

Solution

We Think	*We Write*
We do not know the dimensions of the garden.	Let x = length of shorter side (feet)
	y = length of longer side (feet)

An illustration is helpful in a geometry problem.

The fencing will be used on three sides.

Fencing: $x + x + y = 38$
$2x + y = 38$

The area of a rectangle is found by *multiplying* length times width.

Area: $x \cdot y = 176$

We now have a nonlinear system that can be solved by substitution.

(1) $2x + y = 38$

(2) $\quad xy = 176$

(1) $\quad y = 38 - 2x$ Solve equation (1) for y.

(2) $x(38 - 2x) = 176$ Substitute $38 - 2x$ for y in equation (2).

$\quad 38x - 2x^2 = 176$ Remove parentheses.

$\quad 0 = 2x^2 - 38x + 176$ Set the quadratic expression equal to zero.

$\quad 0 = 2(x^2 - 19x + 88)$ Factor out the common 2.

$\quad 0 = 2(x - 8)(x - 11)$ Factor the trinomial.

$x - 8 = 0 \quad$ or $\quad x - 11 = 0$

$x = 8 \qquad\qquad x = 11$

Substitute 8 for x in equation (2).   Substitute 11 for x in equation (2).

(2) $\quad xy = 176$ (2) $\quad xy = 176$

$\quad 8y = 176$ $11y = 176$

$\quad y = \dfrac{176}{8}$ $y = \dfrac{176}{11}$

$\quad y = 22$ $y = 16$

$(8, 22)$ $(11, 16)$

In each case the y-value is larger than the x-value, so either pair of dimensions satisfies the condition of our problem. Carlo's garden can measure 8 feet by 22 feet *or* 11 feet by 16 feet.

EXERCISE SET 10.5

_____ 1. For a United Way fund-raising dinner, a total of 245 tickets were sold. Adults' tickets sold for $6.50 each and children's tickets sold for $3.00. If $1410.50 worth of tickets were sold, how many adults' tickets and how many children's tickets were sold?

_____ 2. A total of 2825 tickets were sold for the concert for total ticket receipts of $25,898. If advance tickets sold for $8 and tickets at the door sold for $10, how many tickets of each kind were sold?

_____ 3. Harry works part-time mowing lawns and lifeguarding at the pool. When he mows for 5 hours and guards for 9 hours, he earns $61. When he mows 3 hours and guards 12 hours, he earns $63. How much does he earn per hour at each job?

_____ 4. Five pounds of cashews and 8 pounds of peanuts cost $26.27, and 7 pounds of cashews and 4 pounds of peanuts cost $23.89. What is the price of each per pound?

_____ 5. Art sold a box of books at a yard sale for $56.00. He sold some of the books for $2.00 each and some for $1.75 each. If there were 30 books in the box, how many did he sell at each price?

_____ 6. At Garvin's Department Store, Martha and Laura found skirts on sale at one price and shoes on sale at another price. Martha bought 3 skirts and 4 pairs of shoes for $85. Laura bought 5 skirts and 2 pairs of shoes for $81. Find the sale price for the skirts and shoes.

_____ 7. J. K. invested part of his money at 10% and the rest at 12%. The income from both investments was $2700. If the amount invested at 10% is $5000 less than twice the amount invested at 12%, find the amount that he has invested at each rate.

_____ 8. Mr. Van Hooser has $20,000 to invest. He invests part of it at 9% and the rest at 13%. If he earns $2120 interest after 1 year, how much does he have invested at each rate?

_____ 9. Sam and his roommate are planning to drive to Florida during spring break. Sam averages one rate of speed when he drives and his rommate averages another rate. If Sam drives for 5 hours and his roommate for 5 hours, they can travel 600 miles, but if Sam drives for only 3 hours and his roommate drives for 7 hours, they can travel 620 miles. How fast does each one drive?

_____ 10. When Craig went to interview for a job, he made part of the trip by bus and the other part by taxi. The time he traveled by bus was 10 minutes longer than the amount of time by taxi. He averaged 40 miles per hour by bus and 30 miles per hour by taxi. If the total trip was 30 miles, how long did he spend traveling by bus and how long by taxi?

_____ 11. The length of a carpet is 20 feet less than three times the width. The difference between the length and the width is 4 feet. Find the dimensions of the carpet.

_____ **12.** The width of a patio is 5 feet less than twice its length. The difference between the length and width is 1 foot. Find the dimensions of the patio.

_____ **13.** A community organization distributed 500 pounds of mixed candy in its Christmas cheer boxes. Part of the candy was bought at $1.10 a pound and part at $0.95 a pound. How many pounds were bought at each price if the total cost was $501.25?

_____ **14.** A horse breeder used a mixture of shelled corn and oats to feed his horses. He found a 100 bushels of this mixture weighed 4160 pounds. How many bushels of each kind did he use if a bushel of shelled corn weighs 56 pounds and a bushel of oats weighs 32 pounds?

_____ **15.** A pharmacist wishes to put 500 grains of a certain drug into 3-grain and 2-grain capsules. If he fills 200 capsules, how many capsules of each kind does he fill?

_____ **16.** Find the capacity of each of Mr. Ratley's two trucks if 5 trips of the smaller truck and 2 trips of the larger will carry 10 tons, while 3 trips of the smaller truck and 6 trips of the larger will carry 24 tons.

_____ **17.** A piece of wire 60 inches long is bent in the form of a right triangle whose hypotenuse is 25 inches. Find the other two sides of the triangle.

_____ **18.** Nelson has a flower bed in the corner of his yard that is a right triangle. If the flower bed requires 30 feet of fencing and the hypotenuse is 13 feet, find the lengths of the other two sides.

_____ **19.** Agnes is redecorating her sewing room. She will need 34 feet of a decorative ceiling border and 72 square feet of new carpeting. Find the dimensions of the sewing room.

_____ **20.** Find the dimensions of Bridget's garden if she needs 260 feet of fencing to enclose an area of 4200 square feet.

_____ **21.** The area of a rectangular piece of material is 12 square feet. The material is cut along the diagonal forming two triangles. If the length of the diagonal is 5 feet, find the dimensions of the rectangle.

_____ **22.** The area of a rectangular piece of wooden paneling is 120 square feet. The panel is sawed along the diagonal forming two triangles. If the length of the diagonal is 17 feet, find the dimensions of the rectangle.

☆ Stretching the Topics _____

_____ **1.** The sum of the perimeters of two squares is 52 inches and the sum of the areas is 97 square inches. Find the length of each side of both squares.

_____ **2.** A farmer pressed 3000 pounds of tobacco into bales. If he had made each bale 15 pounds heavier, he would have had 10 fewer bales. Find the number of bales that he pressed and also the weight of each bale.

_____ **3.** Find the dimensions of a rectangle whose area remains unchanged if the length is decreased 2 feet and the width is increased 1 foot, and whose area is decreased by 44 square feet if the length is decreased by 4 and the width is decreased by 3 feet.

Check your answers in the back of your book.

If you can solve the problems in **Checkup 10.5,** you are ready to do the Review Exercises.

CHECKUP 10.5

_____ **1.** A grocer bought 4 crates of lettuce and 3 crates of cabbage. Later he bought 2 crates of lettuce and 5 crates of cabbage. His first bill was $34.20 and the second bill was $31.80. Find the cost of each crate of lettuce and each crate of cabbage.

_____ **2.** The color flags hanging in the stadium at State University are right triangles. The perimeter of each flag is 36 feet and the longest side measures 15 feet. Find the lengths of the other two sides.

_____ **3.** Willie drove his car 289 miles in 6 hours. Part of the time he averaged 45 miles per hour and the rest of the time he averaged 53 miles per hour. How much time did he drive at each speed?

_____ **4.** Acme Trucking Company has a parking lot which is enclosed on the two shorter sides and one of the longer sides by 260 feet of fencing. If the area of the parking lot is 8000 square feet, what are its dimensions?

Check your answers in the back of your book.

If You Missed Problem:	You Should Review Example:
1	1
2	2
3	3
4	4

Summary

In this chapter we learned to find ordered pairs of numbers that satisfy two equations at the same time. We discussed several methods for solving such systems of equations and established ways of determining which method to use in different circumstances.

Method	Graphing	Substitution	Addition
When to use:	When common points are easily located	When either variable is easily isolated in either equation	When neither variable is easily isolated
How to use:	1. Graph the original equations. 2. Locate the point of intersection. 3. Write the solution as an ordered pair.	1. Isolate one variable in one equation. 2. Substitute the found expression into the other equation, and solve for the value of the remaining variable. 3. Substitute that value into either original equation and solve for the other variable. 4. Write the solution as an ordered pair.	1. Write both equations in form $ax + by = c$. 2. Multiply one or both equations by constants to make the coefficients of one variable *opposites* in the two equations. 3. Add equations to eliminate one variable, and solve for the value of the remaining variable. 4. Substitute that value into either original equation and solve for the other variable. 5. Write the solution as an ordered pair.

We became alerted to the possibility that a system of linear equations might have *no* solutions in common or *all* solutions in common.

Label	Common Solutions	Signal when Solving	Graph
Consistent and independent	One ordered pair	Statement is sometimes true	Intersecting lines
Inconsistent	No ordered pairs	Statement is never true	Parallel lines 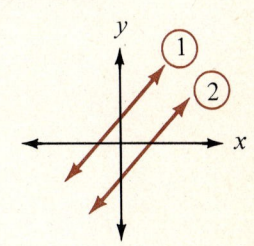

Label	Common Solutions	Signal when Solving	Graph
Dependent	All ordered pairs	Statement is always true	Coinciding lines

We discovered that we could use the same methods of substitution and/or elimination to solve *systems of three equations in three variables* and to solve *nonlinear systems of equations.*

To solve *systems of inequalities in two variables,* we used the graphing method, looking for the shaded regions where all the given inequalities were satisfied.

Finally, we discussed the usefulness of systems of equations in finding solutions to problems stated in words.

❑ Speaking the Language of Algebra

1. In solving a system of linear equations, we must find the _____ _____ that satisfies both equations.

2. If it is easy to isolate one variable in one of the equations, we may solve the system by the method of _____ .

3. In using the method of elimination, we wish the coefficient of one variable in one equation to be the _____ of the coefficient of the same variable in the other equation.

4. If we arrive at a statement such as "0 = 3" in solving a linear system algebraically, we know the system has _____ solution and the graph of the system is a pair of _____ lines.

5. If the graph of one equation is a circle and the graph of another equation is a parabola, the system of these two equations will have *at most* _____ solutions.

Name _____ **Date** _____

REVIEW EXERCISES for Chapter 10: Solving Systems of Equations and Inequalities

Solve by graphing and classify the systems.

_____ 1. $3x - y = 5$
$2x + y = 0$

_____ 2. $2x - 3y = -1$
$-4x + 6y = 12$

Solve the systems by substitution.

_____ 3. $2x + y = -5$
$3x + 2y = -6$

_____ 4. $5x + 2y = 4$
$x + 2y = -4$

Solve the systems by addition.

_____ 5. $2x + 3y = -1$
$-x + y = 13$

_____ 6. $4x - 4y = 9$
$-3x + 3y = 2$

_____ 7. $4x + 5y = 25$
$3x + 7y = 9$

_____ 8. $x + y + z = 1$
$5x - 2y - z = -3$
$2x + y - 3z = 5$

Solve the following systems.

_____ 9. $y = x^2 - 15$
$5x - y = -9$

_____ 10. $x^2 + y^2 = 20$
$2x^2 - y^2 = 28$

Solve the following systems.

11. $x^2 + y^2 < 25$
$y > 3x - 1$

12. $y > 3x + 4$
$x \leq 3$
$x + y > -1$

_____ 13. During the month of May, the Union County Sheriff's Department issued 117 traffic tickets for speeding or failing to obey traffic signals. The fine for each speeding ticket was $50 and the fine for failure to obey traffic signals was $35. If the Sheriff's Department collected $5340 in fines from these two offenses, how many of each kind of ticket were issued?

_____ 14. From Henshaw, Henry bicycled due east to Grove Center. He then bicycled due north to Boxville. Finally, he returned 35 miles to Henshaw by completing the triangle. Henry traveled a total of 84 miles. If the shortest stretch was between Henshaw and Grove Center, how far apart are those two towns?

_____ 15. The area of a rectangle is 60 square meters and the diagonal is 13 meters. Find the dimensions of the rectangle.

Check your answers in the back of your book.

If You Missed Exercises:	You Should Review Examples:	
1, 2	Section 10.1	1–4
3, 4		5–8
5–7	Section 10.2	1–4
8		5
9	Section 10.4	1
10		3
11		6
12		7
13	Section 10.5	1
14		2
15		4

If you have completed the Review Exercises and corrected your errors, you are ready to take the Practice Test.

PRACTICE TEST for Chapter 10: Solving Systems of Equations and Inequalities

		SECTION	EXAMPLE

1. Solve the system by graphing.

$$-3x + 2y = -4$$
$$x - y = 3$$

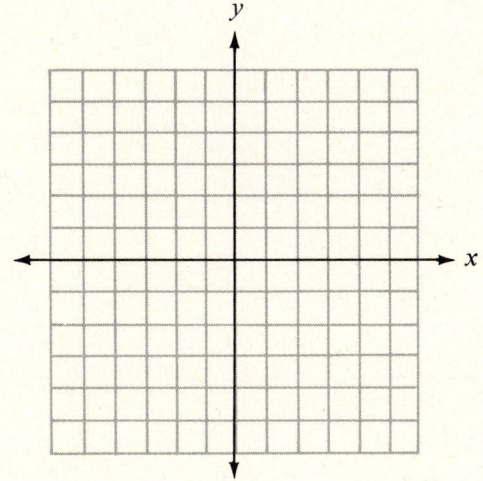

SECTION 10.1   EXAMPLE 4

_____ **2.** Solve the system by substitution.

$$6x - 2y = -9$$
$$4x + y = 1$$

10.1   5

_____ **3.** Solve the system by addition.

$$2x + 5y = 4$$
$$3x + 6y = 1$$

10.2   1

Solve each system.

_____ **4.** $6x - 4y = -20$
$2y = 3x + 10$

10.2   4

_____ **5.** $7 + y = 0$
$x - 4y = 13$

10.1   6

_____ **6.** $x + 2y = 3$
$3x + 6y = 6$

10.2   2

_____ **7.** $x + 2y + z = -3$
$-x + y + 3z = -3$
$2x - 3y + 4z = 8$

10.2   5

_____ **8.** $y = x^2 - 2x$
$2x - y = 3$

10.4   1

_____ **9.** $x + y < 0$
$y \geq 2x + 3$

10.4   5

_____ **10.** During one week, Barbara ate 5 apples and 7 peaches for a total of 820 calories. Sam ate 10 apples and 3 peaches for a total of 980 calories. Find the number of calories in one apple and in one peach.

10.5   1

SHARPENING YOUR SKILLS after Chapters 1–10

SECTION

_____ 1. Give the restrictions on the variable and reduce the fraction $\dfrac{a^2 - 6a - 55}{5a^2 + 23a - 10}$.   6.1

_____ 2. Find the real values for x and y so that $(3x - 5) - (5y + 6)i = 2x + i$.   7.5

_____ 3. Find the equation of a horizontal line through $(-2, 1)$.   8.3

_____ 4. Find the midpoint of the line segment joining $(3, -2)$ and $(5, 9)$.   8.2

_____ 5. Find the equation of a line through $(-4, 1)$ parallel to $2x - y = 3$.   8.3

_____ 6. If $g(x) = -x^2 + 2x - 1$, find $g(-2)$.   8.5

_____ 7. Give the domain for $f(x) = \sqrt{5 - 2x}$.   8.5

_____ 8. Use synthetic division to find the remainder when $x^4 + x^3 + 5x - 2$ is divided by $x + 2$.   3.5

_____ 9. Solve by graphing: $\dfrac{3x - 1}{7} < 1 - x$ or $\dfrac{x}{6} - 1 < -2$ 2.3

Simplify.

_____ 10. $\left(\dfrac{3a}{2b}\right)^{-1}\left(\dfrac{a^2 b}{a^{-3} b^4}\right)^2\left(\dfrac{7a^{-11}b^7}{9a^{-1}b^{-9}}\right)^0$ 3.2

_____ 11. $7x - x[3x^2 - 4x(x + 2)]$ 3.3

_____ 12. $(x - 1)^4$ 3.4

_____ 13. $\dfrac{6x^2 + 19x - 20}{5x^2 - 17x + 6} \div \dfrac{20 - 9x - 18x^2}{3x^2 - 5x - 12}$ 6.2

_____ 14. $\dfrac{3x - 1}{x^2 - 3x + 2} + \dfrac{2x + 1}{x^2 + 2x - 8} - \dfrac{x - 3}{x^2 + 3x - 4}$ 6.3

_____ 15. $\dfrac{x - \dfrac{15}{x + 2}}{x + 10 + \dfrac{40}{x - 3}}$ 6.4

_____ 16. $\left(\dfrac{8}{27}\right)^{-2/3}$ 7.1

_____ 17. $(x^{4/5}y^{-2/3}z^{8/3})^{1/16}$ 7.2

_____ 18. $\sqrt[3]{\dfrac{14x^2}{5}} \cdot \sqrt[3]{\dfrac{-40x}{7}}$ 7.3

_____ **19.** $(\sqrt{3} - 2\sqrt{y})(4\sqrt{3} + \sqrt{y})$ 　　　　7.3

_____ **20.** $\dfrac{1 - \sqrt{-6}}{1 + \sqrt{-6}}$ 　　　　7.5

Solve.

_____ **21.** $(5x - 1)(x + 2) = -5(x^2 + 1) - 8x$ 　　　　5.1

_____ **22.** $16x^3 - 48x^2 - x + 3 = 0$ 　　　　5.1

_____ **23.** $8x^2 + 10x + 1 = 0$ 　　　　5.3

_____ **24.** $144 - x^2 > 0$ 　　　　5.4

_____ **25.** $\dfrac{x}{x - 6} + \dfrac{1}{2x - 1} = \dfrac{2x^2 - 18}{2x^2 - 13x + 6}$ 　　　　6.5

_____ **26.** $\dfrac{x - 10}{x} \geq \dfrac{1}{3}$ 　　　　6.5

_____ **27.** $\sqrt{x - 11} = 11 - \sqrt{x}$ 　　　　7.4

_____ **28.** $\dfrac{3}{x^2} + 7 = \dfrac{8}{x}$ 　　　　7.6

Graph using any method you choose.

29. $8x - 3y = -15$ 　　　　8.1

30. $3x - 8 = 0$ 　　　　8.1

31. $x + y < -4$ 　　　　8.4

Graph the following relations. State the domain and range.

32. $y = x^3 - 9x$ 　　　　9.1

33. $y = -x^2 - x + 12$ 　　　　9.2

34. $(x + 5)^2 + (y - 8)^2 = 64$ 　　　　9.3

35. $y > 8x - x^2$ 　　　　9.4

_____ **36.** Lela can rake all the leaves from the Harper's yard in 5 hours. If her sister Linda helps her, they can finish the yard in 3 hours. How long would it take Linda by herself to rake the yard? 　　　　6.6

Working with Exponential and Logarithmic Functions

11

When $100 is deposited in a savings account paying interest at the rate of 2% per quarter (every 3 months), the total amount, A, in the account after n quarters is given by the function $A = 100(1.02)^n$. Use a calculator to find the amount in the account after 7 quarters; after 5 years.

We have worked with many functions in the preceding chapters. Most functions we have looked at have been polynomial functions. Now it is time to investigate two functions of different types. In this chapter we:

1. Work with exponential functions.

2. Work with logarithmic functions.

3. Use the properties of logarithms.

4. Use common logarithms.

5. Solve exponential and logarithmic equations.

11.1 Working with Exponential Functions

In earlier discussions of polynomial functions such as

$$y = 2x + 3 \qquad y = x^2 + 7x + 1 \qquad y = 1 - x^3$$

we were dealing with expressions containing constant powers of a variable base. Now we turn our attention to expressions in which the base is a constant and the variable appears in the exponent. We wish to investigate expressions such as

$$2^x \qquad \left(\frac{1}{4}\right)^x \qquad (-3)^x$$

Evaluating Exponential Expressions

To evaluate exponential expressions for different values of the variable, we substitute the given x-value wherever x appears in the exponent.

x	2^x	$\left(\frac{1}{4}\right)^x$	$(-3)^x$
3	$2^3 = 8$	$\left(\frac{1}{4}\right)^3 = \frac{1}{64}$	$(-3)^3 = -27$
0	$2^0 = 1$	$\left(\frac{1}{4}\right)^0 = 1$	$(-3)^0 = 1$
-1	$2^{-1} = \frac{1}{2}$	$\left(\frac{1}{4}\right)^{-1} = 4$	$(-3)^{-1} = \frac{-1}{3}$
$\frac{1}{2}$	$2^{1/2} = \sqrt{2}$	$\left(\frac{1}{4}\right)^{1/2} = \sqrt{\frac{1}{4}} = \frac{1}{2}$	$(-3)^{1/2} = \sqrt{-3} = \sqrt{3}\,i$ (not a real number)

As you can see from the last entry in our table, an exponential expression with a *negative* base will have a *real* value only if we avoid x-values in the exponent that cause us to take an *even* root. Therefore, we must avoid substituting x-values that are fractions with *even* denominators. Such a restriction on x-values can become a real headache, because there are infinitely many fractions with even denominators that must be avoided. We shall discuss this dilemma in more detail shortly.

For now let's practice evaluating some more exponential expressions.

Example 1. Complete the table of values for 10^x and 5^{x+1}.

x	10^x	5^{x+1}
-2	$10^{-2} = \frac{1}{10^2} = $ ____	$5^{-2+1} = 5^{-1} = $ ____
-1	$10^{-1} = $ ____	$5^{-1+1} = 5^0 = $ ____
0	$10^0 = $ ____	$5^{0+1} = 5^1 = $ ____
1	$10^1 = $ ____	$5^{1+1} = 5^2 = $ ____
2	$10^2 = $ ____	$5^{2+1} = 5^3 = $ ____

Check your table on page 837. ▶

x	$y = \left(\dfrac{1}{2}\right)^x$	(x, y)
-2	$\left(\dfrac{1}{2}\right)^{-2} = \dfrac{1}{2^{-2}} = 4$	$(-2, 4)$
-1	$\left(\dfrac{1}{2}\right)^{-1} = 2$	$(-1, 2)$
0	$\left(\dfrac{1}{2}\right)^{0} = 1$	$(0, 1)$
$\dfrac{1}{2}$	$\left(\dfrac{1}{2}\right)^{1/2} = \dfrac{1}{\sqrt{2}} = \dfrac{\sqrt{2}}{2}$	$\left(\dfrac{1}{2}, \dfrac{\sqrt{2}}{2}\right)$
1	$\left(\dfrac{1}{2}\right)^{1} = \dfrac{1}{2}$	$\left(1, \dfrac{1}{2}\right)$
2	$\left(\dfrac{1}{2}\right)^{2} = \dfrac{1}{4}$	$\left(2, \dfrac{1}{4}\right)$

Notice that graph of $y = \left(\dfrac{1}{2}\right)^x$ looks like the graph of $y = 2^x$, flipped across the y-axis.

Before summarizing our observations about the exponential function, let's consider the graph of $y = 1^x$.

Example 5. Graph $y = 1^x$.

Solution

x	$y = 1^x$	(x, y)
-2	$1^{-2} = \dfrac{1}{1^2} = 1$	$(-2, 1)$
-1	$1^{-1} = 1$	$(-1, 1)$
0	$1^0 = 1$	$(1, 1)$
$\dfrac{1}{2}$	$1^{1/2} = \sqrt{1} = 1$	$\left(\dfrac{1}{2}, 1\right)$
1	$1^1 = 1$	$(1, 1)$
2	$1^2 = 1$	$(2, 1)$

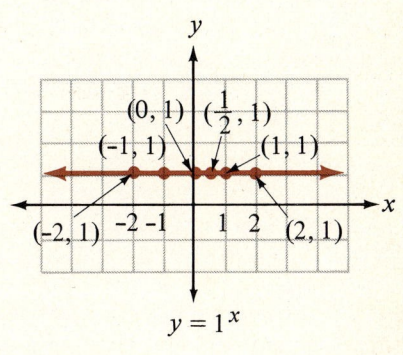

These ordered pairs are indeed boring and their graph does not resemble the other examples of exponential relations. In fact, this graph looks more like the graph of a constant function. For this reason, we exclude the base 1 from our discussion of exponential relations.

From our examples we see that the exponential relation is indeed a function because each x-value corresponds to one and only one y-value. We also note that the exponent can be *any* real number, but the resulting y-value will always be a *positive* real number. Now we are ready to summarize our findings regarding **exponential functions**.

> **Definition of Exponential Function**
>
> $$y = a^x$$
>
> where $a > 0$ and $a \neq 1$ is called an exponential function.
>
> Domain: $x \in R$
> Range: $y \in R$, $y > 0$

For our own convenience, we make note of the fact that all exponential functions of the form $y = a^x$ contain the point $(0, 1)$. Moreover, we can determine whether $y = a^x$ is increasing or decreasing from left to right by looking at the base a.

> If $0 < a < 1$, then $y = a^x$ is decreasing (Example 4).
> If $a > 1$, then $y = a^x$ is increasing (Example 3).

To graph exponential functions that are variations on the basic function $y = a^x$, we continue to use arbitrary points.

Example 6. Graph $y = 5^{x+1}$.

Solution

x	$y = 5^{x+1}$	(x, y)
-2	$5^{-2+1} = 5^{-1} = \dfrac{1}{5}$	$\left(-2, \dfrac{1}{5}\right)$
-1	$5^{-1+1} = 5^0 = 1$	$(-1, 1)$
$\dfrac{1}{2}$	$5^{-1/2+1} = 5^{1/2} = \sqrt{5}$	$\left(\dfrac{-1}{2}, \sqrt{5}\right)$
0	$5^{0+1} = 5^1 = 5$	$(0, 5)$
1	$5^{1+1} = 5^2 = 25$	$(1, 25)$

$y = 5^{x + 1}$

⫸ Trial Run

Graph each function.

1. $y = 3^x$

2. $y = \left(\dfrac{1}{3}\right)^x$

3. $y = 2^{x+1}$

4. $y = \left(\dfrac{1}{2}\right)^{2x}$

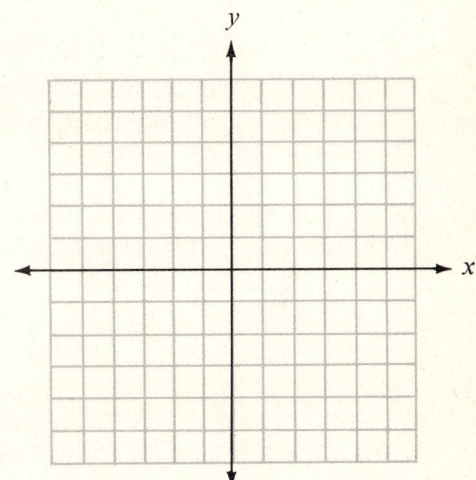

Answers are on page 837.

Exponential functions have many applications in the real world. Population growth, chemical decay, and compound interest problems can all be solved using exponential functions. Consider the problem stated at the beginning of this chapter.

Example 7. When \$100 is deposited in a savings account paying compound interest at the rate of 2% per quarter (every 3 months), the total amount, A, in the account after n quarters is given by the function $A = 100(1.02)^n$. Use a calculator to find the amount in the account after 7 quarters; after 5 years.

Solution: We can use a calculator to evaluate the function

$$A = 100(1.02)^n$$

After 7 quarters:

$$n = 7 \text{ quarters}$$
$$A = 100(1.02)^7$$
$$\doteq 100(1.1486)$$
$$\doteq 114.86$$

There will be approximately \$114.86 in the account after 7 quarters.

After 5 years:

$$n = 5(4) = 20 \text{ quarters}$$
$$A = 100(1.02)^{20}$$
$$\doteq 100(1.4859)$$
$$\doteq \$148.59$$

There will be approximately \$148.59 in the account after 5 years (or 20 quarters).

Solving Exponential Equations

Suppose that we wish to find x-values corresponding to specific y-values for an exponential function. For instance, suppose that we wish to find the value of x for which

$$2^x = 8$$

Looking back at the graph of $y = 2^x$, you should agree that there will be only one x-value corresponding to the y-value of 8. Since we can rewrite 8 as a power of 2, our equation becomes

$$2^x = 8$$
$$2^x = 2^3$$

Since each side of this equation has been expressed as a power of the same base and we have agreed that each power of 2 corresponds to one and only one exponent, we are safe in concluding that

$$\text{if} \quad 2^x = 2^3$$
$$\text{then } x = 3$$

To solve an exponential equation, we should first *try to rewrite each side of the equation as a power of the same base*. If this is possible, we may equate the exponents and solve the equation.

Exponential Equations

If $a^m = a^n$

then $m = n$

Example 8. Solve $3^x = \dfrac{1}{3}$.

Solution

$$3^x = \dfrac{1}{3}$$

$$3^x = 3^{-1}$$

$$x = -1$$

You try Example 9.

Example 10. Solve $2^{x+1} = \dfrac{1}{8}$.

Solution

$$2^{x+1} = \dfrac{1}{8}$$

$$2^{x+1} = \dfrac{1}{2^3}$$

$$2^{x+1} = 2^{-3}$$

$$x+1 = -3$$

Check your work on page 837. ▶

Example 12. Solve $\left(\dfrac{1}{2}\right)^{3x} = 16^{x+1}$.

Solution

$$\left(\dfrac{1}{2}\right)^{3x} = 16^{x+1}$$

$(2^{-1})^{3x} = (2^4)^{x+1}$	Rewrite each base as a power of 2.
$2^{-3x} = 2^{4(x+1)}$	Raise each power to its power.
$2^{-3x} = 2^{4x+4}$	Remove parentheses.
$-3x = 4x + 4$	Equate exponents.
$-7x = 4$	Isolate x-term.
$x = \dfrac{-4}{7}$	Isolate x.

Example 9. Solve $5^{-x} = 25$.

Solution

$$5^{-x} = 25$$

$$5^{-x} = 5^2$$

Check your work on page 837. ▶
Then try Example 10.

Example 11. Solve $9^x = 3^{x-5}$.

Solution

$$9^x = 3^{x-5}$$

$$(3^2)^x = 3^{x-5}$$

$$3^{2x} = 3^{x-5}$$

$$2x = x - 5$$

$$x = -5$$

Example 13. Solve $2 \cdot 2^x = 8^{1-x}$.

Solution

$$2 \cdot 2^x = 8^{1-x}$$

$$2^1 \cdot 2^x = 8^{1-x} \qquad \text{Rewrite 2 as } 2^1.$$

$$2^{1+x} = 8^{1-x} \qquad \text{Multiply powers of 2 on left.}$$

$$2^{1+x} = (2^3)^{1-x} \qquad \text{Rewrite 8 as power of 2.}$$

$$2^{1+x} = 2^{3(1-x)} \qquad \text{Raise the power to a power.}$$

$$2^{1+x} = 2^{3-3x} \qquad \text{Remove parentheses.}$$

$$1 + x = 3 - 3x \qquad \text{Equate exponents.}$$

$$4x = 2 \qquad \text{Isolate } x\text{-term.}$$

$$x = \frac{1}{2} \qquad \text{Isolate } x.$$

If it is not possible to rewrite both sides of an exponential equation as powers of the same base, the equation cannot be solved using our techniques. Equations such as

$$5^x = 7 \qquad \text{and} \qquad 2^{x+1} = 3^x$$

will be discussed in Section 11.5.

⫸ Trial Run

Solve.

_____ 1. $10^x = \dfrac{1}{10}$

_____ 2. $2^{-x} = \dfrac{1}{4}$

_____ 3. $3^{x+1} = \dfrac{1}{27}$

_____ 4. $4^x = 2^{x+3}$

_____ 5. $\left(\dfrac{1}{5}\right)^{2x} = 25^{x+2}$

_____ 6. $3^2 \cdot 3^x = 9^{3-x}$

Answers are on page 838.

▶ **Examples You Completed** ──────────────────────────────

Example 1. Complete the table of values for 10^x and 5^{x+1}.

x	10^x	5^{x+1}
-2	$10^{-2} = \dfrac{1}{10^2} = \dfrac{1}{100}$	$5^{-2+1} = 5^{-1} = \dfrac{1}{5}$
-1	$10^{-1} = \dfrac{1}{10}$	$5^{-1+1} = 5^0 = 1$
0	$10^0 = 1$	$5^{0+1} = 5^1 = 5$
1	$10^1 = 10$	$5^{1+1} = 5^2 = 25$
2	$10^2 = 100$	$5^{2+1} = 5^3 = 125$

Example 9. Solve $5^{-x} = 25$.

Solution

$$5^{-x} = 25$$
$$5^{-x} = 5^2$$
$$-x = 2$$
$$x = -2$$

Example 10. Solve $2^{x+1} = \dfrac{1}{8}$.

Solution

$$2^{x+1} = \frac{1}{8}$$
$$2^{x+1} = \frac{1}{2^3}$$
$$2^{x+1} = 2^{-3}$$
$$x + 1 = -3$$
$$x = -4$$

Answers to Trial Runs ──────────────────────────────

page 829   **1.** $\dfrac{1}{25}$; 1; 25   **2.** $\dfrac{1}{4}$; $\dfrac{1}{2}$; 2   **3.** $\dfrac{-1}{9}$; -1; $-\sqrt{3}$; -9 **4.** $\dfrac{1}{9}$; 1; $\sqrt{3}i$ (not real); 9

page 833   **1.**

2.

3.

4.

page 836   **1.** $x = -1$ **2.** $x = 2$ **3.** $x = -4$ **4.** $x = 3$ **5.** $x = -1$ **6.** $x = \dfrac{4}{3}$

EXERCISE SET 11.1

Evaluate each of the following for the given values of x.

_____ 1. 4^x; $x = \dfrac{-1}{2}$, $x = 0$, $x = \dfrac{1}{2}$, $x = 2$

_____ 2. 9^x; $x = \dfrac{-1}{2}$, $x = 0$, $x = \dfrac{1}{2}$, $x = 2$

_____ 3. 5^{x-1}; $x = -1$, $x = \dfrac{1}{2}$, $x = \dfrac{3}{2}$, $x = 3$

_____ 4. 6^{x-2}; $x = 0$, $x = \dfrac{3}{2}$, $x = 2$, $x = 4$

_____ 5. -2^{3x}; $x = -1$, $x = \dfrac{-1}{3}$, $x = 0$, $x = \dfrac{2}{3}$

_____ 6. -3^{5x}; $x = \dfrac{-3}{5}$, $x = 0$, $x = \dfrac{2}{5}$, $x = 1$

_____ 7. $(-2)^{4x}$; $x = \dfrac{-1}{2}$, $x = 0$, $x = \dfrac{1}{8}$, $x = 1$

_____ 8. $(-3)^{2x}$; $x = \dfrac{-1}{4}$, $x = \dfrac{1}{2}$, $x = 0$, $x = 2$

_____ 9. $\left(\dfrac{1}{2}\right)^{2-x}$; $x = 1$, $x = 0$, $x = \dfrac{3}{2}$, $x = 4$

_____ 10. $\left(\dfrac{1}{3}\right)^{1-x}$; $x = 1$, $x = 0$, $x = \dfrac{1}{2}$, $x = 3$

Graph each of the following.

11. $y = 6^x$

12. $y = 4^x$

13. $y = \left(\dfrac{1}{4}\right)^x$

14. $y = \left(\dfrac{1}{6}\right)^x$

15. $y = 3^{x+1}$

16. $y = 2^{x-3}$

17. $y = \left(\dfrac{1}{3}\right)^{2x}$

18. $y = \left(\dfrac{1}{2}\right)^{3x}$

19. $y = 9^{x/2}$

20. $y = 10^{x/2}$

21. $y = 5^{3-x}$

22. $y = 6^{2-x}$

23. $y = \left(\dfrac{1}{2}\right)^{2x-1}$

24. $y = \left(\dfrac{1}{3}\right)^{2x+1}$

Solve.

———— **25.** $5^x = \dfrac{1}{5}$

———— **26.** $3^x = \dfrac{1}{3}$

———— **27.** $4^{-x} = \dfrac{1}{16}$

———— **28.** $5^{-x} = \dfrac{1}{25}$

———— **29.** $10^{x+1} = \dfrac{1}{1000}$

———— **30.** $8^{x-1} = \dfrac{1}{64}$

———— **31.** $\left(\dfrac{1}{9}\right)^{2x} = 81^{x+2}$

———— **32.** $\left(\dfrac{1}{4}\right)^{3x} = 64^{x+1}$

———— **33.** $16^x = 2^{2x-1}$

———— **34.** $25^x = 5^{3x+1}$

———— **35.** $\left(\dfrac{1}{25}\right)^{1/x} = 5$

———— **36.** $\left(\dfrac{1}{81}\right)^{2/x} = 9$

———— **37.** $2^3 \cdot 2^x = 8^{-x}$

———— **38.** $5^2 \cdot 5^x = 25^{x-1}$

☆ Stretching the Topics _____

———— **1.** Graph $y = \left(\dfrac{1}{2}\right)^{x-1} + 2$

———— **2.** Solve $(2^x)^{x+1} = 16^{x+1}$ for x.

———— **3.** A bond paying 8% yearly interest is bought for $4000. How much will the
bond be worth 10 years from now if the interest is compounded quarterly?

Check your answers in the back of your book.

If you can complete **Checkup 11.1,** you are ready to go on to Section 11.2.

✓ **CHECKUP 11.1**

_____ **1.** Evaluate 3^{x-1} for $x = -1$, $x = 0$, $x = \frac{1}{2}$, $x = 2$, $x = \frac{3}{2}$.

Graph.

2. $y = 4^{x-1}$

3. $y = 2^{-2x}$

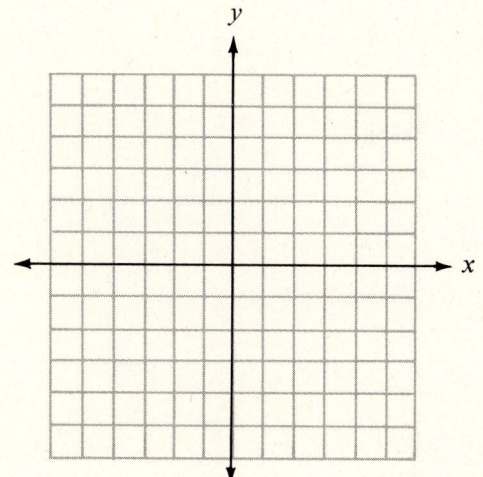

Solve.

_____ **4.** $3^{2x-1} = \dfrac{1}{243}$

_____ **5.** $\left(\dfrac{1}{32}\right)^x = 4^{x-1}$

Check your answers in the back of your book.

If You Missed Problem:	You Should Review Examples:
1	1, 2
2	6
3	4
4	10
5	12

11.2 Working with Logarithmic Functions

In an exponential statement such as $a^x = b$, we would like to find a way to express the exponent x in terms of a and b.

Writing Logarithmic Expressions

When solving equations such as $2^x = 8$, we were looking for the exponent to which we could raise the base 2 to arrive at the number 8. In other words, we could say

x = the exponent to which we must raise the base 2 to get 8

As is typical, mathematicians have invented a way of symbolizing such statements, and their invention is called a **logarithm.** We may rewrite our statement as

x = the logarithm, base 2, of 8

or simply

$$x = \log_2 8$$

The subscript 2 identifies the base, the x represents the exponent on the 2, and the number 8 tells the result of raising 2 to the x power. In other words,

$$x = \log_2 8 \quad \text{means} \quad 2^x = 8$$

These two statements say *exactly the same thing*. The logarithmic form of the statement merely allows us to isolate the exponent in the statement. Every logarithmic statement corresponds to an exponential statement, and vice versa. The sooner you realize this fact, the better off you will be in your understanding of logarithms.

Logarithmic Notation

$$x = \log_a b \quad \text{means} \quad a^x = b$$

Because these statements are equivalent, you should see that *a logarithm is just an exponent*.

Because of the close relationship between logarithms and the exponential functions discussed in Section 11.1, we shall continue to restrict ourselves to *positive bases* different from 1. Let's practice switching back and forth between exponential statements and logarithmic statements.

Example 1. Write each exponential statement as a logarithmic statement.

$$3^2 = 9$$

$$2^{-2} = \frac{1}{4}$$

$$5^{1/2} = \sqrt{5}$$

Solution

$$3^2 = 9 \quad \text{means} \quad \log_3 9 = 2$$

$$2^{-2} = \frac{1}{4} \quad \text{means} \quad \log_2 \frac{1}{4} = -2$$

$$5^{1/2} = \sqrt{5} \quad \text{means} \quad \log_5 \sqrt{5} = \frac{1}{2}$$

You complete Example 2.

Example 2. Write each logarithmic statement as an exponential statement.

$$\log_5 125 = 3$$

$$\log_7 1 = 0$$

$$\log_8 2 = \frac{1}{3}$$

Solution

$$\log_5 125 = 3 \quad \text{means} \quad \underline{} = \underline{}$$

$$\log_7 1 = 0 \quad \text{means} \quad \underline{} = \underline{}$$

$$\log_8 2 = \frac{1}{3} \quad \text{means} \quad \underline{} = \underline{}$$

Check your work on page 850. ▶

⫸ Trial Run

Write each exponential statement as a logarithmic statement.

_____ **1.** $2^2 = 4$

_____ **2.** $5^{-2} = \dfrac{1}{25}$

_____ **3.** $9^{1/2} = 3$

Write each logarithmic statement as an exponential statement.

_____ **4.** $\log_3 81 = 4$

_____ **5.** $\log_2 \dfrac{1}{8} = -3$

_____ **6.** $\log_4 2 = \dfrac{1}{2}$

Answers are on page 851.

Simplifying Logarithmic Expressions

Most of us feel more comfortable working with exponential statements than with logarithmic statements because we have been dealing with exponents throughout algebra. To simplify logarithmic expressions, it is probably wise to assign a variable name to the logarithm and then switch to exponential form. For example, let's find $\log_4 16$.

$\log_4 16 = x$	Call the logarithm ''x.''
$4^x = 16$	Switch to exponential form.
$4^x = 4^2$	Rewrite both sides as powers of 4.
$x = 2$	Solve the exponential equation.

Therefore, $\log_4 16 = 2$.

You try Example 3.

Example 3. Find $\log_2 32$.

Solution: Let

$$\log_2 32 = x$$

$$2^x = 32$$

$$2^x = 2^{\underline{}}$$

$$x = \underline{}$$

Therefore, $\log_2 32 = \underline{}$.

Check your work on page 850. ▶

Then complete Example 4.

Example 4. Find $\log_3 \frac{1}{3}$.

Solution: Let

$$\log_3 \frac{1}{3} = x$$

$$3^x = \frac{1}{3}$$

$$3^x = 3^{\underline{}}$$

$$x = \underline{}$$

Therefore, $\log_3 \frac{1}{3} = \underline{}$.

Check your work on page 850. ▶

Example 5. Find $\log_9 1$.

Solution: Let

$$\log_9 1 = x$$

$$9^x = 1$$

$$9^x = 9^0$$

$$x = 0$$

Therefore, $\log_9 1 = 0$.

Example 6. Find $\log_6 6^5$.

Solution: Let

$$\log_6 6^5 = x$$

$$6^x = 6^5$$

$$x = 5$$

Therefore, $\log_6 6^5 = 5$.

Example 7. Find $\log_3 3 + \log_2 \frac{1}{8}$.

Solution: Let

$$\log_3 3 = x \qquad \text{and} \qquad \log_2 \frac{1}{8} = y$$

$$3^x = 3 \qquad\qquad\qquad 2^y = \frac{1}{8}$$

$$3^x = 3^1 \qquad\qquad\qquad 2^y = 2^{-3}$$

$$x = 1 \qquad\qquad\qquad y = -3$$

So

$$\log_3 3 = 1 \qquad \text{and} \qquad \log_2 \frac{1}{8} = -3$$

Therefore,

$$\log_3 3 + \log_2 \frac{1}{8}$$

$$= 1 + (-3)$$

$$= -2$$

Example 8. Find $\log_5 0$.

 Solution: Let

$$\log_5 0 = x$$

$$5^x = 0$$

There is *no* exponent to which we can raise 5 to arrive at the answer 0. Therefore, $\log_5 0$ does *not* exist.

Example 9. Find $\log_2 (-2)$.

 Solution: Let

$$\log_2 (-2) = x$$

$$2^x = -2$$

There is *no* exponent to which we can raise 2 to arrive at the answer -2. Therefore, $\log_2 (-2)$ does *not* exist.

From these examples we can make some useful observations.

For $a > 0$, $a \neq 1$:

$\log_a 1 \;\; = 0$ (because $a^0 = 1$)

$\log_a a \;\; = 1$ (because $a^1 = a$)

$\log_a a^n = n$ (because $a^n = a^n$)

$\log_a b$ exists only when $b > 0$.

Our last observation deserves some emphasis.

> The logarithm of a negative number *or* zero does not exist.

⫸ Trial Run

Find the value of each expression.

_____ **1.** $\log_5 \dfrac{1}{5}$

_____ **2.** $\log_4 1$

_____ **3.** $\log_3 27$

_____ **4.** $\log_7 7^3$

_____ **5.** $\log_2 32 + \log_3 \dfrac{1}{81}$

_____ **6.** $\log_3 (-9)$

Answers are on page 851.

Solving Logarithmic Equations

Now that we have learned to treat logarithmic statements as exponential statements, we should be able to find any unknown quantity in a logarithmic statement. Let's solve each equation for the variable.

 You try Example 10.

Example 10. Solve $\log_4 \frac{1}{4} = x$.

Solution

$$\log_4 \frac{1}{4} = x$$

$$4^x = \frac{1}{4}$$

$$4^x = 4^{\underline{}}$$

$$x = \underline{}$$

Check your work on page 850. ▶

Example 11. Solve $\log_a 25 = 2$.

Solution

$$\log_a 25 = 2$$

$$a^2 = 25$$

$$a = \pm 5$$

But remember that the base, a, must be positive. So

$$a = 5$$

Example 12. Solve $\log_6 b = -2$.

Solution

$$\log_6 b = -2$$

$$6^{-2} = b$$

$$\frac{1}{36} = b$$

Now try Example 13.

Example 13. Solve $\log_2 (x + 1) = 3$.

Solution

$$\log_2 (x + 1) = 3$$

$$2^3 = x + 1$$

$$\underline{} = x + 1$$

$$\underline{} = x$$

Check your work on page 851. ▶

Example 14. Solve $\log_4 64 = 2 - 3x$.

Solution

$$\log_4 64 = 2 - 3x$$

$4^{2-3x} = 64$	Switch to exponential form.
$4^{2-3x} = 4^3$	Write 64 as a power of base 4.
$2 - 3x = 3$	Equate exponents.
$-3x = 1$	Isolate x-term.
$x = \dfrac{-1}{3}$	Isolate x.

Example 15. Solve $\log_x (x + 2) = 2$.

Solution

$$\log_x (x + 2) = 2$$

$x^2 = x + 2$	Switch to exponential form.
$x^2 - x - 2 = 0$	Set quadratic expression equal to zero.
$(x - 2)(x + 1) = 0$	Factor the trinomial.
$x - 2 = 0 \quad \text{or} \quad x + 1 = 0$	Set each factor equal to zero.
$x = 2 \qquad\qquad x = -1$	Solve for x.

Since x appears as the base in the original logarithmic statement, x cannot be negative. The solution is $x = 2$.

Trial Run

Solve.

_____ **1.** $\log_{10} \dfrac{1}{10} = x$

_____ **2.** $\log_a 36 = 2$

_____ **3.** $\log_3 b = -4$

_____ **4.** $\log_5 (x - 1) = 2$

_____ **5.** $\log_2 32 = 3 - 2x$

_____ **6.** $\log_x (2x + 3) = 2$

Answers are on page 851.

Graphing the Logarithmic Function

Suppose that we look at ordered pairs (x, y) satisfying the equation

$$y = \log_2 x$$

We could come up with some ordered pairs by choosing x and finding y, but we must be careful to choose x-values whose logarithms are easily found. It is easier to rewrite this statement in exponential form before finding ordered pairs.

$$y = \log_2 x \qquad \text{means} \qquad 2^y = x$$

Now we can find ordered pairs more easily by substituting values for y and finding corresponding x-values. Let's try that; but remember that our ordered pairs should be written (x, y).

Example 16. Graph $y = \log_2 x$.

Solution

$$y = \log_2 x \qquad \text{means} \qquad 2^y = x$$

y	$x = 2^y$	(x, y)
-2	$2^{-2} = \dfrac{1}{4}$	$\left(\dfrac{1}{4}, -2\right)$
-1	$2^{-1} = \dfrac{1}{2}$	$\left(\dfrac{1}{2}, -1\right)$
0	$2^0 = 1$	$(1, 0)$
$\dfrac{1}{2}$	$2^{1/2} = \sqrt{2}$	$\left(\sqrt{2}, \dfrac{1}{2}\right)$
1	$2^1 = 2$	$(2, 1)$
2	$2^2 = 4$	$(4, 2)$

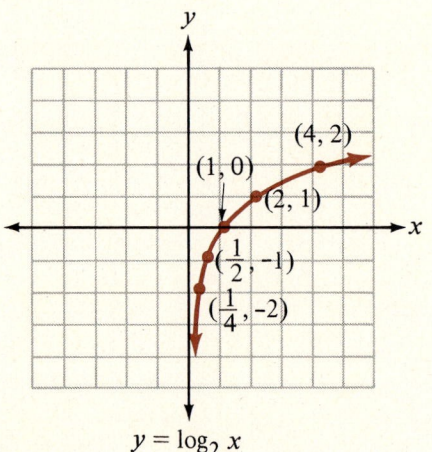

$$y = \log_2 x$$

Notice that our y-values could be positive, negative, or zero, but the x-values always turned out to be *positive*.

Example 17. Graph $y = \log_{1/3} x$.

Solution

$$y = \log_{1/3} x \qquad \text{means} \qquad \left(\frac{1}{3}\right)^y = x$$

y	$x = \left(\frac{1}{3}\right)^y$	(x, y)
-1	$\left(\frac{1}{3}\right)^{-1} = 3$	$(3, -1)$
0	$\left(\frac{1}{3}\right)^{0} = 1$	$(1, 0)$
$\dfrac{1}{2}$	$\left(\frac{1}{3}\right)^{1/2} = \sqrt{\frac{1}{3}} = \frac{\sqrt{3}}{3}$	$\left(\frac{\sqrt{3}}{3}, \frac{1}{2}\right)$
1	$\left(\frac{1}{3}\right)^{1} = \frac{1}{3}$	$\left(\frac{1}{3}, 1\right)$

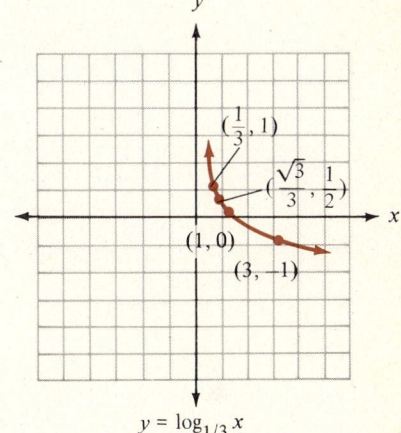

$$y = \log_{1/3} x$$

From the graphs of $y = \log_2 x$ and $y = \log_{1/3} x$, you should agree that the sets of ordered pairs satisfying such equations are indeed *functions*. Each x in the domain corresponds to one and only one y in the range. We notice that x is always *positive*, even though y can be any real number.

Let's summarize our findings for **logarithmic functions.**

Definition of Logarithmic Function

$$y = \log_a x$$

where $a > 0$, $a \neq 1$, is called a logarithmic function.

Domain: $x \in R$ and $x > 0$

Range: $y \in R$

To graph a logarithmic function more easily, we have agreed to follow these steps.

Graphing a Logarithmic Function

1. Rewrite the logarithmic statement as an exponential statement.
2. Solve for x.
3. Substitute y-values and find the corresponding x-values.
4. Write the ordered pairs (x, y).
5. Use the ordered pairs to sketch the graph.

⫸ Trial Run

Graph.

1. $y = \log_{10} x$

2. $y = \log_{1/2} x$

Answers are on page 851.

▶ Examples You Completed

Example 2

Solution

$$\log_5 125 = 3 \text{ means } 5^3 = 125$$

$$\log_7 1 = 0 \text{ means } 7^0 = 1$$

$$\log_8 2 = \frac{1}{3} \text{ means } 8^{1/3} = 2$$

Example 3. Find $\log_2 32$.

Solution: Let

$$\log_2 32 = x$$

$$2^x = 32$$

$$2^x = 2^5$$

$$x = 5$$

Example 4. Find $\log_3 \frac{1}{3}$.

Solution: Let

$$\log_3 \frac{1}{3} = x$$

$$3^x = \frac{1}{3}$$

$$3^x = 3^{-1}$$

$$x = -1$$

Example 10. Solve $\log_4 \frac{1}{4} = x$.

Solution

$$\log_4 \frac{1}{4} = x$$

$$4^x = \frac{1}{4}$$

$$4^x = 4^{-1}$$

$$x = -1$$

Example 13. Solve $\log_2 (x + 1) = 3$.

Solution

$$\log_2 (x + 1) = 3$$
$$2^3 = x + 1$$
$$8 = x + 1$$
$$7 = x$$

Answers to Trial Runs

page 844   **1.** $\log_2 4 = 2$ **2.** $\log_5 \dfrac{1}{25} = -2$ **3.** $\log_9 3 = \dfrac{1}{2}$ **4.** $3^4 = 81$ **5.** $2^{-3} = \dfrac{1}{8}$
6. $4^{1/2} = 2$

page 846   **1.** -1 **2.** 0 **3.** 3 **4.** 3 **5.** 1 **6.** Does not exist

page 848   **1.** -1 **2.** 6 **3.** $\dfrac{1}{81}$ **4.** 26 **5.** -1 **6.** 3

page 850   **1.**   **2.**

EXERCISE SET 11.2

Write each of the following as a logarithmic statement.

_____ **1.** $3^2 = 9$

_____ **2.** $2^4 = 16$

_____ **3.** $4^{-2} = \dfrac{1}{16}$

_____ **4.** $6^{-2} = \dfrac{1}{36}$

_____ **5.** $16^{1/2} = 4$

_____ **6.** $27^{1/3} = 3$

_____ **7.** $9^{-1/2} = \dfrac{1}{3}$

_____ **8.** $25^{-1/2} = \dfrac{1}{5}$

Write each of the following as an exponential statement.

_____ **9.** $\log_5 125 = 3$

_____ **10.** $\log_4 64 = 3$

_____ **11.** $\log_2 \dfrac{1}{32} = -5$

_____ **12.** $\log_3 \dfrac{1}{27} = -3$

_____ **13.** $\log_{16} 4 = \dfrac{1}{2}$

_____ **14.** $\log_{27} 3 = \dfrac{1}{3}$

_____ **15.** $\log_{36} \dfrac{1}{6} = \dfrac{-1}{2}$

_____ **16.** $\log_{64} \dfrac{1}{8} = \dfrac{-1}{2}$

Find each of the following.

_____ **17.** $\log_8 \dfrac{1}{8}$

_____ **18.** $\log_7 \dfrac{1}{7}$

_____ **19.** $\log_{10} 1$

_____ **20.** $\log_{12} 1$

_____ **21.** $\log_5 125$

_____ **22.** $\log_6 36$

_____ **23.** $\log_2 2^7$

_____ **24.** $\log_3 3^5$

_____ **25.** $\log_5 0$

_____ **26.** $\log_7 (-49)$

_____ **27.** $\log_5 25 - \log_2 \dfrac{1}{8}$

_____ **28.** $\log_3 \dfrac{1}{27} + \log_2 2$

_____ **29.** $\log_{25} \dfrac{1}{5} + \log_9 3$

_____ **30.** $\log_{49} \dfrac{1}{7} + \log_4 2$

Solve for the variable.

_____ **31.** $\log_{12} \dfrac{1}{12} = x$

_____ **32.** $\log_2 \dfrac{1}{2} = x$

_____ **33.** $\log_a 49 = 2$ _____ **34.** $\log_a 121 = 2$

_____ **35.** $\log_2 b = -5$ _____ **36.** $\log_3 b = -2$

_____ **37.** $\log_4 (x + 1) = 3$ _____ **38.** $\log_3 (2x + 1) = 3$

_____ **39.** $\log_3 81 = 5 + 2x$ _____ **40.** $\log_2 64 = 4 - 3x$

_____ **41.** $\log_x (x + 6) = 2$ _____ **42.** $\log_x (2x + 8) = 2$

Graph each of the following.

43. $y = \log_3 x$ **44.** $y = \log_5 x$

45. $y = \log_{1/3} x$ **46.** $y = \log_{1/10} x$

☆ Stretching the Topics _____

_____ **1.** Solve $\log_2 (x^2 - 5x + 4) = 3$ for x.

_____ **2.** Graph $y = \log_3 (2 + x)$.

_____ **3.** Solve $\dfrac{x + 5}{\log_3 9} = \dfrac{\log_2 8}{x}$ for x.

Check your answers in the back of your book.

If you can complete **Checkup 11.2,** you are ready to do Section 11.3.

✓ CHECKUP 11.2

_____ **1.** Write $5^2 = 25$ as a logarithmic statement.

_____ **2.** Write $64^{-1/2} = \dfrac{1}{8}$ as a logarithmic statement.

_____ **3.** Write $\log_3 81 = 4$ as an exponential statement.

_____ **4.** Write $\log_5 \dfrac{1}{25} = -2$ as an exponential statement.

_____ **5.** Find $\log_2 \dfrac{1}{32}$.   _____ **6.** Find $\log_8 2$.

_____ **7.** Solve $\log_a 125 = 3$.   _____ **8.** Solve $\log_4 (x - 3) = \dfrac{1}{2}$.

9. Graph $y = \log_6 x$.   **10.** Graph $y = \log_{1/5} x$.

 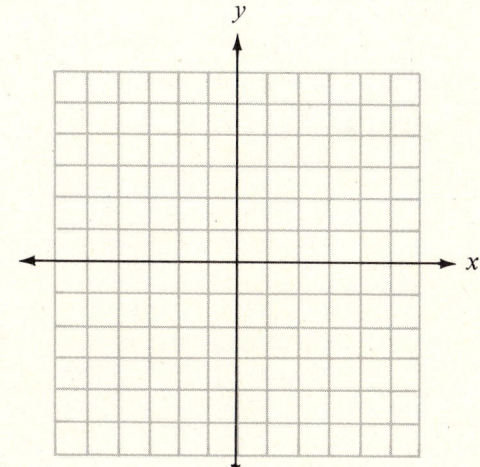

Check your answers in the back of your book.

If You Missed Problems:	You Should Review Examples:
1, 2	1
3, 4	2
5, 6	3–7
7	11, 15
8	13
9, 10	16, 17

11.3 Using the Properties of Logarithms

Since *logarithms are exponents*, you should suspect that the rules for working with logarithms will be similar to the laws of exponents. Let's recall a few laws of exponents.

First Law of Exponents: $a^m \cdot a^n = a^{m+n}$

Second Law of Exponents: $(a^m)^n = a^{m \cdot n}$

Fifth Law of Exponents: $\dfrac{a^m}{a^n} = a^{m-n}$

Using the Laws of Logarithms

First let's consider the logarithm of a *product*. The logarithm of a product of factors is the *sum* of the logarithms of each of those factors.

First Law of Logarithms

$$\log_a (M \cdot N) = \log_a M + \log_a N$$

Let's verify this law with a numerical example.

$$\log_2 (16 \cdot 8) \stackrel{?}{=} \log_2 16 + \log_2 8$$

$$\log_2 128 \stackrel{?}{=} \log_2 16 + \log_2 8$$

$$7 \stackrel{?}{=} 4 + 3 \qquad \text{Remember: } 2^7 = 128, 2^4 = 16, \text{ and } 2^3 = 8.$$

$$7 = 7$$

We have seen that this law "works" for our numerical example, but let's prove it for any base a. Let

$$\log_a M = m \quad \text{and} \quad \log_a N = n$$

$$a^m = M \quad \text{and} \quad a^n = N \qquad \text{Switch to exponential form.}$$

$$M \cdot N = a^m \cdot a^n \qquad \text{Consider the product } M \cdot N.$$

$$M \cdot N = a^{m+n} \qquad \text{Use first law of Exponents.}$$

$$\log_a (M \cdot N) = m + n \qquad \text{Switch to logarithmic form.}$$

$$\log_a (M \cdot N) = \log_a M + \log_a N \qquad \text{Substitute: } m = \log_a M$$

$$n = \log_a N$$

We have proved the first law of logarithms.

Example 1. Find $\log_3 (27 \cdot 81)$.

Solution

$$\log_3 (27 \cdot 81) = \log_3 27 + \log_3 81 \qquad \text{First law of logarithms.}$$

$$= 3 + 4 \qquad \text{Remember: } 3^3 = 27 \text{ and } 3^4 = 81$$

$$= 7$$

Let's consider a rule for finding the logarithm of a *quotient*. The logarithm of a quotient is the *difference* between the logarithm of the numerator and the logarithm of the denominator.

Second Law of Logarithms

$$\log_a \frac{M}{N} = \log_a M - \log_a N$$

Once again, we shall verify this law with a numerical example before proving it in general.

$$\log_3 \frac{243}{27} \overset{?}{=} \log_3 243 - \log_3 27$$

$$\log_3 9 \overset{?}{=} \log_3 243 - \log_3 27$$

$$2 \overset{?}{=} 5 - 3 \qquad \text{Remember: } 3^2 = 9, \ 3^5 = 243, \text{ and } 3^3 = 27.$$

$$2 = 2$$

Now we can prove the second law of logarithms for any base *a*. Let

$$\log_a M = m \quad \text{and} \quad \log_a N = n$$

$$a^m = M \quad \text{and} \qquad a^n = N \qquad \text{Switch to exponential form.}$$

$$\frac{M}{N} = \frac{a^m}{a^n} \qquad \text{Consider the quotient } \frac{M}{N}.$$

$$\frac{M}{N} = a^{m-n} \qquad \text{Use fifth law of exponents.}$$

$$\log_a \frac{M}{N} = m - n \qquad \text{Switch to logarithmic form.}$$

$$\log_a \frac{M}{N} = \log_a M - \log_a N \qquad \text{Substitute: } m = \log_a M$$
$$n = \log_a N$$

We have proved the second law of logarithms.

Example 2. Find $\log_2 \dfrac{32}{256}$.

Solution

$$\log_2 \frac{32}{256} = \log_2 32 - \log_2 256 \qquad \text{Second law of logarithms.}$$

$$= 5 - 8 \qquad \text{Remember: } 2^5 = 32 \text{ and } 2^8 = 256.$$

$$= -3$$

Finally, we seek a way to simplify the logarithm of a *power*. The logarithm of a power is the *product* of the exponent and the logarithm of the base.

> **Third Law of Logarithms**
>
> $$\log_a M^c = c \log_a M$$

We again verify this law with a numerical example before proving it in general.

$$\log_2 4^3 \overset{?}{=} 3 \log_2 4$$

$$\log_2 64 \overset{?}{=} 3 \log_2 4$$

$$6 \overset{?}{=} 3 \cdot 2 \qquad \text{Remember: } 2^6 = 64 \text{ and } 2^2 = 4.$$

$$6 = 6$$

Now let's prove the third law of logarithms. Let

$$\log_a M = m$$

$a^m = M$	Switch to exponential form.
$M^c = (a^m)^c$	Consider the power M^c.
$M^c = a^{cm}$	Use second law of exponents.
$\log_a M^c = cm$	Switch to logarithmic form.
$\log_a M^c = c \log_a M$	Substitute: $m = \log_a M$

Our proof is complete.
You finish Example 3.

Example 3. Find $\log_4 16^7$.

Solution

$$\log_4 16^7$$

$$= 7 \log_4 16$$

$$= 7(\underline{\quad\quad})$$

$$= \underline{\quad\quad}$$

Check your work on page 864. ▶

Example 4. Find $\log_5 \sqrt{125}$.

Solution

$$\log_5 \sqrt{125}$$

$$= \log_5 (125)^{1/2}$$

$$= \frac{1}{2} \log_5 125$$

$$= \frac{1}{2} (3)$$

$$= \frac{3}{2}$$

The three laws of logarithms are very useful in changing logarithms of complicated expressions into sums, differences, or constant multiples of simpler logarithmic expressions. Often it will be necessary to use more than one law in the simplifying process.
You complete Example 5.

Example 5. Find $\log_2 (4^7 \cdot 2^5)$.

Solution

$$\log_2 (4^7 \cdot 2^5)$$

$= \log_2 4^7 + \log_2 2^5$ Use first law of logarithms.

$= 7 \log_2 4 + 5 \log_2 2$ Use third law of logarithms.

$= 7(\underline{2}) + 5(\underline{1})$ Simplify the logarithms.

$= \underline{14} + \underline{5}$ Find each product.

$= \underline{19}$ Find the sum.

Check your work on page 864. ▶

Example 6. Find $\log_3 \dfrac{3^5 \cdot 81^2}{27^{10}}$.

Solution

$$\log_3 \frac{3^5 \cdot 81^2}{27^{10}}$$

$= \log_3 3^5 + \log_3 81^2 - \log_3 27^{10}$ Use first and second laws of logarithms.

$= 5 \log_3 3 + 2 \log_3 81 - 10 \log_3 27$ Use third law of logarithms.

$= 5(1) + 2(4) - 10(3)$ Remember: $3^1 = 3$, $3^4 = 81$, and $3^3 = 27$.

$= 5 + 8 - 30$ Find each product.

$= -17$ Simplify.

In Examples 7 through 10 assume that we know that

$$\log_{10} 2 \doteq 0.3010$$

$$\log_{10} 3 \doteq 0.4771$$

and find approximate values for each logarithm.

Example 7. Find $\log_{10} 6$.

Solution

$\log_{10} 6 = \log_{10} (3 \cdot 2)$ Rewrite 6 as $3 \cdot 2$.

$= \log_{10} 3 + \log_{10} 2$ Use first law of logarithms.

$\doteq 0.3010 + 0.4771$ Substitute given values.

$\log_{10} 6 \doteq 0.7781$ Find the sum.

You complete Example 8.

Example 8 Find $\log_{10} \dfrac{8}{9}$.

Solution

$$\log_{10} \frac{8}{9} = \log_{10} 8 - \log_{10} 9 \qquad \text{Use second law of logarithms.}$$

$$= \log_{10} 2^3 - \log_{10} 3^2 \qquad \text{Rewrite 8 as } 2^3 \text{ and 9 as } 3^2.$$

$$= 3 \log_{10} 2 - 2 \log_{10} 3 \qquad \text{Use third law of logarithms.}$$

$$\doteq 3(\underline{.30103}) - 2(\underline{.47712^{(3)}}) \qquad \text{Substitute given values.}$$

$$\doteq \underline{.90309} - \underline{} \qquad \text{Find each product.}$$

$$\log_{10} \frac{8}{9} \doteq \underline{-.0515} \qquad \text{Simplify.}$$

Check your work on page 864. ▶

Example 9. Find $\log_{10} 600$.

Solution

$$\log_{10} 600 = \log_{10} (2 \cdot 3 \cdot 100) \qquad \text{Rewrite 600 as } 2 \cdot 3 \cdot 100.$$

$$= \log_{10} 2 + \log_{10} 3 + \log_{10} 100$$

$$\doteq 0.3010 + 0.4771 + 2 \qquad \text{Remember: } 10^2 = 100.$$

$$\log_{10} 600 \doteq 2.7781$$

Example 10. Find $\log_{10} 1500$.

Solution

$$\log_{10} 1500 = \log_{10} \frac{3000}{2} \qquad \text{Rewrite 1500 as } \frac{3000}{2}.$$

$$= \log_{10} \frac{3 \cdot 1000}{2} \qquad \text{Rewrite 3000 as } 3 \cdot 1000.$$

$$= \log_{10} 3 + \log_{10} 1000 - \log_{10} 2$$

$$\doteq 0.4771 + 3 - 0.3010 \qquad \text{Remember: } 10^3 = 1000.$$

$$\doteq 3.4771 - 0.3010$$

$$\log_{10} 1500 \doteq 3.1761$$

⫸ Trial Run

Find the logarithms.

_____ **1.** $\log_3 (9 \cdot 27)$ \qquad\qquad _____ **2.** $\log_2 \dfrac{8}{128}$

_____ **3.** $\log_5 25^3$ \qquad\qquad _____ **4.** $\log_6 \sqrt{36}$

_____ **5.** $\log_3 (9^4 \cdot 3^2)$

If $\log_{10} 2 \doteq 0.3010$ and $\log_{10} 3 \doteq 0.4771$, approximate each logarithm.

———— **6.** $\log_{10} 18$ ———— **7.** $\log_{10} 400$

———— **8.** $\log_{10} \dfrac{16}{27}$

Answers are on page 864.

Rewriting Logarithmic Expressions Containing Variables

In the examples above we used the laws of logarithms to simplify logarithmic expressions to a single number. Sometimes it is not possible or even necessary to reduce a logarithmic expression to a numerical value. In such cases we may merely wish to rewrite the original expression in another form, using the laws of logarithms.

Example 11. Use the laws of logarithms to rewrite $\log_a \dfrac{x^2(x + 2)}{x - 3}$.

Solution

$$\log_a \frac{x^2(x + 2)}{x - 3}$$

$$= \log_a x^2(x + 2) - \log_a(x - 3) \qquad \text{Use second law of logarithms.}$$

$$= \log_a x^2 + \log_a (x + 2) - \log_a (x - 3) \qquad \text{Use first law of logarithms.}$$

$$= 2 \log_a x + \log_a (x + 2) - \log_a (x - 3) \qquad \text{Use third law of logarithms.}$$

You complete Example 12.

Example 12. Use the laws of logarithms to rewrite $\log_a \dfrac{(x + 5)^3}{\sqrt{x - 4}}$.

Solution

$$\log_a \frac{(x + 5)^3}{\sqrt{x - 4}}$$

$$= \log_a \frac{(x + 5)^3}{(x - 4)^{1/2}} \qquad \text{Rewrite } \sqrt{x - 4} \text{ as } (x - 4)^{1/2}.$$

$$= \log_a (x + 5)^3 - \log_a (x - 4)^{1/2} \qquad \text{Use second law of logarithms.}$$

$$= \text{\_\_\_\_} \log_a (\text{_____}) - \text{\_\_\_\_} \log_a (\text{_____}) \qquad \text{Use third law of logarithms.}$$

Check your work on page 864. ▶

On the other hand, sometimes we may wish to rewrite a sum or difference of several logarithmic expressions as a *single* logarithmic expression. To accomplish this, we must reverse the procedure used above.

Example 13. Rewrite as a single logarithm $\log_a (2x + 3) - 5 \log_a x$.

Solution

$$\log_a (2x + 3) - 5 \log_a x$$

$$= \log_a (2x + 3) - \log_a x^5 \qquad \text{Use third law of logarithms.}$$

$$= \log_a \frac{2x + 3}{x^5} \qquad \text{Use second law of logarithms.}$$

Example 14. Rewrite as a single logarithm.

$$3 \log_a x + \frac{1}{2} \log_a (x + 7) - \frac{1}{3} \log_a (x^2 + 5)$$

Solution

$$3 \log_a x + \frac{1}{2} \log_a (x + 7) - \frac{1}{3} \log_a (x^2 + 5)$$

$$= \log_a x^3 + \log_a (x + 7)^{1/2} - \log_a (x^2 + 5)^{1/3}$$

$$= \log_a x^3 + \log_a \sqrt{x + 7} - \log_a \sqrt[3]{x^2 + 5}$$

$$= \log_a x^3 \sqrt{x + 7} - \log_a \sqrt[3]{x^2 + 5}$$

$$= \log_a \frac{x^3 \sqrt{x + 7}}{\sqrt[3]{x^2 + 5}}$$

▐▶ Trial Run

Rewrite, using the laws of logarithms.

_____ **1.** $\log_a (x + 1)(2x - 5)^2$ _____ **2.** $\log_a \dfrac{x^2 + 1}{\sqrt[3]{x + 9}}$

Rewrite as a single logarithm.

_____ **3.** $3 \log_a x + \dfrac{1}{2} \log_a (x^2 + 2)$

_____ **4.** $2 \log_a 5x + 4 \log_a (x^2 + 9) - \dfrac{1}{5} \log_a 7x - 3 \log_a (x - 1)$

Answers are on page 864.

As you may have noticed from our work with logarithmic expressions, there is no rule for simplifying the logarithm of a sum or difference.

$$\log_a (M + N) \neq \log_a M + \log_a N$$

$$\log_a (M - N) \neq \log_a M - \log_a N$$

▶ **Examples You Completed** _____

Example 3. Find $\log_4 16^7$.

Solution

$$\log_4 16^7$$
$$= 7 \log_4 16$$
$$= 7(2)$$
$$= 14$$

Example 5. Find $\log_2 (4^7 \cdot 2^5)$.

Solution

$$\log_2 (4^7 \cdot 2^5)$$
$$= \log_2 4^7 + \log_2 2^5$$
$$= 7 \log_2 4 + 5 \log_2 2$$
$$= 7(2) + 5(1)$$
$$= 14 + 5$$
$$= 19$$

Example 8. Find $\log_{10} \dfrac{8}{9}$.

Solution

$$\log_{10} \frac{8}{9}$$
$$= \log_{10} 8 - \log_{10} 9$$
$$= \log_{10} 2^3 - \log_{10} 3^2$$
$$= 3 \log_{10} 2 - 2 \log_{10} 3$$
$$\doteq 3(0.3010) - 2(0.4771)$$
$$\doteq 0.9030 - 0.9542$$
$$\doteq -0.0512$$

Example 12. Use the laws of logarithms to re-write $\log_a \dfrac{(x + 5)^3}{\sqrt{x - 4}}$.

Solution

$$\log_a \frac{(x + 5)^3}{\sqrt{x - 4}}$$
$$= \log_a \frac{(x + 5)^3}{(x - 4)^{1/2}}$$
$$= \log_a (x + 5)^3 - \log_a (x - 4)^{1/2}$$
$$= 3 \log_a (x + 5) - \frac{1}{2} \log_a (x - 4)$$

Answers to Trial Runs _____

page 861   **1.** 5   **2.** -4 **3.** 6   **4.** 1   **5.** 10   **6.** 1.2552   **7.** 2.6020   **8.** -0.2273

page 863   **1.** $\log_a (x + 1) + 2 \log_a (2x - 5)$ **2.** $\log_a (x^2 + 1) - \dfrac{1}{3} \log_a (x + 9)$ **3.** $\log_a x^3 \sqrt{x^2 + 2}$

4. $\log_a \dfrac{(5x)^2 (x^2 + 9)^4}{\sqrt[5]{7x}\,(x - 1)^3}$

EXERCISE SET 11.3

Find the following logarithms.

_____ **1.** $\log_2 (16 \cdot 8)$ _____ **2.** $\log_2 (4 \cdot 32)$

_____ **3.** $\log_3 \dfrac{27}{81}$ _____ **4.** $\log_3 \dfrac{9}{243}$

_____ **5.** $\log_7 49^2$ _____ **6.** $\log_6 36^3$

_____ **7.** $\log_3 \sqrt{27}$ _____ **8.** $\log_2 \sqrt[3]{16}$

_____ **9.** $\log_5 (5^3 \cdot 25^2)$ _____ **10.** $\log_2 (8^3 \cdot 16^2)$

_____ **11.** $\log_3 \dfrac{27^2}{9^3}$ _____ **12.** $\log_5 \dfrac{25^3}{125^2}$

_____ **13.** $\log_2 \dfrac{2^5 \cdot 8^2}{16^3}$ _____ **14.** $\log_3 \dfrac{9^2 \cdot 27^3}{81^2}$

Find each of the following logarithms, given that $\log_{10} 2 \doteq 0.3010$, $\log_{10} 3 \doteq 0.4771$, $\log_{10} 5 \doteq 0.6990$, and $\log_{10} 7 \doteq 0.8451$.

_____ **15.** $\log_{10} 15$ _____ **16.** $\log_{10} 14$

_____ **17.** $\log_{10} \sqrt{200}$ _____ **18.** $\log_{10} \sqrt{300}$

_____ **19.** $\log_{10} 350$ _____ **20.** $\log_{10} 210$

_____ **21.** $\log_{10} \dfrac{49}{6}$ _____ **22.** $\log_{10} \dfrac{25}{14}$

_____ **23.** $\log_{10} 45$ _____ **24.** $\log_{10} 63$

Rewrite using the laws of logarithms.

_____ **25.** $\log_a x^3(x + 5)^2$ _____ **26.** $\log_a x^2(x + 1)^5$

_____ **27.** $\log_a \dfrac{(x - 3)^5}{\sqrt{3x}}$ _____ **28.** $\log_a \dfrac{7x - 5}{\sqrt[3]{2x}}$

_____ **29.** $\log_a \dfrac{\sqrt[3]{x^2}}{x^2(x - 9)^5}$ _____ **30.** $\log_a \dfrac{\sqrt{x^5}}{x^4(x^2 + 1)}$

_____ **31.** $\log_a \dfrac{(x - 3)\sqrt{2 - x}}{\sqrt{x}(x + 1)^5}$ _____ **32.** $\log_a \dfrac{(x - 2)\sqrt{3 - x}}{\sqrt[3]{x}(x + 2)^4}$

Rewrite as a single logarithm.

_____ **33.** $2 \log_a (x - 1) + \dfrac{1}{3} \log_a x$ _____ **34.** $3 \log_a x + \dfrac{1}{5} \log_a (x + 1)$

_____ 35. $\frac{1}{2} \log_a x - 4 \log_a (x + 2)$ _____ 36. $\frac{1}{3} \log_a x - 2 \log_a (x^2 + 2)$

_____ 37. $\frac{1}{2} \log_a 3x + 3 \log_a x - \log_a (x^2 + 1)$

_____ 38. $\frac{1}{3} \log_a 2x + 4 \log_a x - \log_a (2x - 1)$

_____ 39. $3 \log_a (x + 1) + \frac{2}{3} \log_a (x + 1) - \log_a x - 5 \log_a (x^2 + 1)$

_____ 40. $2 \log_a (x - 1) + \frac{3}{2} \log_a (x - 1) - \log_a 2x - 3 \log_a (x - 2)$

☆ Stretching the Topics _____

_____ 1. For the formula $V = \frac{1}{3} \pi r^2 h$, write $\log_a r$ in expanded form.

_____ 2. Show that $5 \log_a \sqrt[3]{a} - 7 \log_a \sqrt{a} + 10 \log_a \sqrt[3]{a} - \frac{3}{2} \log_a a = 0$.

Check your answers in the back of your book.

If you can complete **Checkup 11.3,** you are ready to go on to Section 11.4.

CHECKUP 11.3

Find the following logarithms.

_____ **1.** $\log_2 (16 \cdot 8)$

_____ **2.** $\log_3 \dfrac{9}{\sqrt{27}}$

Find the following logarithms given that $\log_{10} 2 \doteq 0.3010$ and $\log_{10} 3 \doteq 0.4771$.

_____ **3.** $\log_{10} 72$

_____ **4.** $\log_{10} \sqrt{80}$

_____ **5.** $\log_{10} 360$

_____ **6.** $\log_{10} \dfrac{90}{8}$

Rewrite using the laws of logarithms.

_____ **7.** $\log_a (x^2 \sqrt{x^2 + 1})$

_____ **8.** $\log_a \dfrac{\sqrt[3]{x^2}\,(x^2 + 1)}{(x^2 + 4)^5}$

Rewrite as a single logarithm.

_____ **9.** $2 \log_a (x + 1) - \dfrac{2}{3} \log_a x$

_____ **10.** $\dfrac{1}{2} \log_a (3x) + \log_a (x + 5) - 4 \log_a x - \log_a (2x + 1)$

Check your answers in the back of your book.

If You Missed Problems:	You Should Review Examples:
1, 2	1–6
3–6	7–10
7, 8	11, 12
9, 10	13, 14

11.4 Using Common Logarithms

Before the advent of the inexpensive hand-held calculator, logarithms provided the tool for scientific calculations. Scientists did not relish lengthy computation with decimal numbers any more than college students do, and they used logarithms to make the arithmetic of such numbers less tedious.

Using Scientific Notation

The key to using logarithms for computation is our ability to write every decimal number as a product of some number between 1 and 10, and an appropriate power of 10. Let us spend a moment reviewing some integer powers of 10.

$10^0 = 1$

$10^1 = 10$ $\qquad\qquad\qquad\qquad 10^{-1} = \dfrac{1}{10^1} = \dfrac{1}{10} = 0.1$

$10^2 = 100$ $\qquad\qquad\qquad\quad 10^{-2} = \dfrac{1}{10^2} = \dfrac{1}{100} = 0.01$

$10^3 = 1000$ $\qquad\qquad\qquad\quad 10^{-3} = \dfrac{1}{10^3} = \dfrac{1}{1000} = 0.001$

$10^4 = 10,000$ $\qquad\qquad\qquad 10^{-4} = \dfrac{1}{10^4} = \dfrac{1}{10,000} = 0.0001$

$10^5 = 100,000$ $\qquad\qquad\qquad 10^{-5} = \dfrac{1}{10^5} = \dfrac{1}{100,000} = 0.00001$

Suppose we assume that n represents a natural number. To write a power of 10 without using exponents, we first write down the numeral 1. Then if the power is n, we add zeros to the *right* of that 1 and stop and place a decimal point when we have moved n places right. If the power is $-n$, we add zeros to the left of that 1 and stop and place a decimal point when we have moved n places left. If the power is 0, we leave the 1 as is.

Example 1. Write 10^9.

Solution

$$10^9 = 1,\underbrace{000,000,000}_{\text{9 places to the right of 1}}$$

Example 2. Write 10^{-7}.

Solution

$$10^{-7} = 0.\underbrace{0000001}_{\text{7 places to the left of 1}}$$

You should agree that each of the numbers below has been rewritten in an equivalent form.

$$325 = 3.25 \times 100 \quad = 3.25 \times 10^2$$
$$69,000 = 6.9 \times 10,000 = 6.9 \times 10^4$$
$$87 = 8.7 \times 10 \quad\quad = 8.7 \times 10^1$$
$$0.325 = 3.25 \times 0.1 \quad = 3.25 \times 10^{-1}$$
$$0.0069 = 6.9 \times 0.001 \quad = 6.9 \times 10^{-3}$$
$$1.3 = 1.3 \times 1 \quad\quad\;\; = 1.3 \times 10^0$$

The third version of each number is called its **scientific notation** form, and it is not difficult to rewrite a number using scientific notation if you have understood our discussion of powers of 10.

> **Scientific Notation.** To write a number using scientific notation, we must express it as a *product* of some factor between 1 and 10, and an appropriate power of 10.

We summarize here the steps needed to write a number using scientific notation.

> 1. Insert a caret ($\wedge$) in the given number to obtain the necessary factor between 1 and 10.
> 2. Count the places from your caret to the actual position of the decimal point in the given number.
> 3. If the actual decimal point is n places to the *right* of your caret, the appropriate power of 10 is 10^n. If the actual decimal point is n places to the *left* of your caret, the appropriate power of 10 is 10^{-n}.
> 4. Write the given number as the product of your factor between 1 and 10, and the appropriate power of 10.

For instance, to write 578 using scientific notation, we proceed as follows:

$$578 = 5{\wedge}78{\wedge}$$

2 places right

$$= 5.78 \times 10^2$$

Similarly, we may write 0.00578 using scientific notation by reasoning as follows:

$$0.00578 = 0.005{\wedge}78$$

3 places left

$$= 5.78 \times 10^{-3}$$

Now you complete Examples 3 and 4.

Example 3. Use scientific notation to write 76,300.

Solution

$$76,300 = 7{\wedge}6300$$

4 places right

$$= 7.63 \times 10^{4}$$

Check your work on page 877. ▶

Example 4. Use scientific notation to write 0.000012.

Solution

$$0.000012 = 0.00001{\wedge}2$$

5 places left

$$= 1.2 \times 10^{-5}$$

Check your work on page 877. ▶

Example 5. Use scientific notation to write

123

123,000

0.123

0.000123.

Solution

$$123 = 1_\wedge 23 = 1.23 \times 10^2$$

$$123{,}000 = 1_\wedge 23000 = 1.23 \times 10^5$$

$$0.123 = 0.1_\wedge 23 = 1.23 \times 10^{-1}$$

$$0.000123 = 0.0001_\wedge 23 = 1.23 \times 10^{-4}$$

Notice in Example 5 that all the numbers shared the same factor between 1 and 10. The numbers differed only in their powers of 10. We could think of these numbers as belonging to the same "family," perhaps called the 1.23 family. We shall make use of this observation shortly.

We can reverse our steps to change a number from scientific notation to standard form.

Example 6. Write 6.4×10^5 in standard form.

Solution

$$6.4 \times 10^5 = 6_\wedge 40000_\wedge$$

5 places right

$$= 640{,}000$$

Example 7. Write 3×10^{-8} in standard form.

Solution

$$3 \times 10^{-8} = 3.0 \times 10^{-8}$$

$$= {}_\wedge 00000003_\wedge$$

8 places left

$$= 0.00000003$$

⟫ Trial Run ━━━━━━━━━━━━━━━━━━━━━━━━━━━━━━━━━━━━━━━

Use scientific notation to write each number.

_____ **1.** 899 _____ **2.** 0.00054 _____ **3.** 2,010,000

Write each number in standard form.

_____ **4.** 10^{-5} _____ **5.** 7.1×10^3 _____ **6.** 1.05×10^{-3}

Answers are on page 877.

Finding Common Logarithms

Suppose that we wish to find numerical values for *logarithms* of numbers written in scientific notation. Since every such number is expressed as a product, we use the first law of logarithms to write

$$\log_a (1.23 \times 10^2) = \log_a 1.23 + \log_a 10^2$$

$$\log_a (1.23 \times 10^5) = \log_a 1.23 + \log_a 10^5$$

$$\log_a (1.23 \times 10^{-1}) = \log_a 1.23 + \log_a 10^{-1}$$

$$\log_a (1.23 \times 10^{-4}) = \log_a 1.23 + \log_a 10^{-4}$$

What would be a good choice for the base, a, in computing these logarithms? Since the second term on the right of each logarithmic statement requires that we find the logarithm of a power of 10, it seems that a base of 10 might be a wise choice. Why? Because from earlier work with logarithms, we recall that $\log_a a^n = n$, so $\log_{10} 10^n = n$.

Therefore, using a base of 10 in our logarithmic statements we may say that

$$\log_{10} (1.23 \times 10^2) = \log_{10} 1.23 + \log_{10} 10^2 = \log_{10} (1.23) + 2$$

$$\log_{10} (1.23 \times 10^5) = \log_{10} 1.23 + \log_{10} 10^5 = \log_{10} (1.23) + 5$$

$$\log_{10} (1.23 \times 10^{-1}) = \log_{10} 1.23 + \log_{10} 10^{-1} = \log_{10} (1.23) + (-1)$$

$$\log_{10} (1.23 \times 10^{-4}) = \log_{10} 1.23 + \log_{10} 10^{-4} = \log_{10} (1.23) + (-4)$$

To finish finding the logarithm of each of these numbers, we only need to know $\log_{10} 1.23$, and we shall return to that matter in a minute. First, however, let's notice how the logarithms of the given numbers differ from each other.

$$\log_{10} (1.23 \times 10^2) = 2 + \log_{10} 1.23$$

$$\log_{10} (1.23 \times 10^5) = 5 + \log_{10} 1.23$$

$$\log_{10} (1.23 \times 10^{-1}) = -1 + \log_{10} 1.23$$

$$\log_{10} (1.23 \times 10^{-4}) = -4 + \log_{10} 1.23$$

The part of each logarithm that we have already computed *exactly matches the power of 10* in the scientific notation form of each of the given numbers. This part is called the **characteristic** of the logarithm because it characterizes the given number by indicating its decimal-point location.

For each of our logarithms, the $\log_{10} 1.23$ is called the **mantissa** and it is the same for *all* numbers in the 1.23 family. For example,

$$\log_{10} (1230) = \log_{10} (1.23 \times 10^3) = 3 + \log_{10} 1.23$$

characteristic — mantissa

Let's finish finding each of these logarithms by finding $\log_{10} 1.23$. Fortunately, the actual computation of base 10 logarithms of numbers between 1 and 10 has already been done by mathematicians. Also called **common logarithms,** they appear in the Table of Common Logarithms inside the back cover.

To locate $\log_{10} 1.23$ in the table, we first locate 1.2 in the leftmost column. Then we move across the top row to find the 3 and look for the entry in the body of the table where the 1.2 row meets the 3 column. That entry is 0.0899 and we conclude that $\log_{10} 1.23 = 0.0899$.

We may now write the common logarithms of all our original numbers.

$$\log_{10} (1.23 \times 10^2) = 2 + 0.0899$$

$$\log_{10} (1.23 \times 10^4) = 4 + 0.0899$$

$$\log_{10} (1.23 \times 10^{-1}) = -1 + 0.0899$$

$$\log_{10} (1.23 \times 10^{-5}) = -5 + 0.0899$$

To find the common logarithm of any number, then, we must use the following steps.

Finding the Common Logarithm of a Number

1. Rewrite the given number in scientific notation.
2. Use the power of 10 as the characteristic of the common logarithm of the given number.
3. Use the common logarithm of the factor between 1 and 10 as the mantissa of the common logarithm of the given number.
4. Find the mantissa in the table of common logarithms.
5. Write the common logarithm of the given number as the sum of its characteristic and its mantissa.

If the characteristic of a logarithm is positive, we are permitted to write the characteristic and mantissa as a single number. For example,

$$\log_{10} (1.23 \times 10^2) = 2 + 0.0899 = 2.0899$$

$$\log_{10} (1.23 \times 10^4) = 4 + 0.899 = 4.0899$$

It is understood that the number preceding the decimal point in each of these logarithms represents the characteristic and the number after the decimal point is the mantissa.

However, when the characteristic is negative, we are not permitted to write the characteristic and mantissa together as a single number. Learning the method for combining characteristic and mantissa in such cases serves no useful purpose here. For our work with logarithms, we shall agree to write a negative characteristic separate from its mantissa.

By popular agreement, mathematicians decided to use the notation **log N**, with no written base, to represent $\log_{10} N$.

$$\log N = \log_{10} N$$

Example 8. Find log 639,000.

Solution

$$\log 639{,}000 = \log (6.39 \times 10^5)$$
$$= 5 + \log 6.39$$
$$= 5 + 0.8055$$
$$= 5.8055$$

Example 9. Find log 0.0048.

Solution

$$\log 0.0048 = \log (4.80 \times 10^{-3})$$
$$= -3 + \log 4.80$$
$$= -3 + 0.6812$$

If we are given the logarithm of a number, we may find that number merely by reversing our steps. For example, suppose we know that

$$\log N = 2.5514$$
$$= 2 + 0.5514$$

We know that the characteristic is 2 and the mantissa is 0.5514. Locating 0.5514 in the *body* of our table, we see that it is the logarithm of 3.56. Of course, we know that 2 is the logarithm of 10^2. We may write

$$\log N = 2 + 0.5514$$
$$= \log 10^2 + \log 3.56$$
$$= \log (3.56 \times 10^2)$$

We conclude that

$$N = 3.56 \times 10^2$$
$$= 356$$

Finding the number with a given logarithm is sometimes called finding the **antilogarithm** of the given logarithm. In our last example,

$$356 = \text{antilog}_{10} (2.5514)$$

Definition of Antilogarithm

$N = \text{antilog}_a L$

    means

"N is the number whose logarithm is L"

Therefore,

If $N = \text{antilog } L$

then $L = \log N$

Example 10. Find antilog $(-2 + 0.9557)$.

Solution: We use the table to find the number whose logarithm is 0.9557.

$$\text{antilog } (-2 + 0.9557) = 9.03 \times 10^{-2}$$
$$= 0.0903$$

Example 11. Find antilog$_{10}$ (7.7619).

Solution: We use the table to find the number whose logarithm is 0.7619.

$$\text{antilog } (7 + 0.7619) = 5.78 \times 10^7$$
$$= 57,800,000$$

⟫ **Trial Run**

Find the common logarithms.

_____ **1.** log 6.45   _____ **2.** log 180

_____ **3.** log 478,000   _____ **4.** log 0.00759

Find the antilogarithms.

_____ **5.** antilog 3.9504   _____ **6.** antilog $(-2 + 0.5416)$

Answers are on page 877.

Computing with Common Logarithms

Let's turn our attention to using common logarithms as a computational tool. Keep in mind that logarithms were available to students of mathematics long before calculators came into existence. The primary reason for the inclusion of this section is to reinforce your understanding of the laws of logarithms, *not* to suggest that computation using logarithms is more efficient than computation using a calculator.

Since $y = \log x$ is a function, we recall that if $A = B$, then $\log A = \log B$. This fact allows us to "take the logarithm of both sides of an equation."

Suppose that you wish to calculate a power such as $(2.9)^5$ without tedious multiplication.

$$N = (2.9)^5$$

$\log N = \log (2.9)^5$	Take logarithms of both sides.
$= 5 \log 2.9$	Use third law of logarithms.
$= 5 \log (2.9 \times 10^0)$	Rewrite 2.9 in scientific notation.
$= 5(0.4624)$	Find the logarithm in the table.
$\log N = 2.3120$	Remove parentheses.
$N = \text{antilog } 2.3120$	Isolate N.
$\doteq 2.05 \times 10^2$	
$\doteq 205$	
$(2.9)^5 \doteq 205$	

Let's see whether this answer is reasonable. We know that

$$2 < 2.9 < 3$$

so

$$2^5 < (2.9)^5 < 3^5$$
$$32 < (2.9)^5 < 243$$

Since $32 < 205 < 243$ our answer seems appropriate, especially since 2.9 is much closer to 3 than to 2. It is not a bad idea to roughly check your answers to see whether they make sense.

Example 12. Use common logarithms to compute $(6280)(0.000132)$.

Solution

$$N = (6280)(0.000132)$$

$\log N = \log (6280)(0.000132)$	Take logarithms of both sides.
$= \log (6280) + \log (0.000132)$	Use first law of logarithms.
$= \log (6.28 \times 10^3) + \log (1.32 \times 10^{-4})$	Use scientific notation.
$= (3 + 0.7980) + (-4 + 0.1206)$	Find logarithms in the table.
$= 3 + (-4) + 0.7980 + 0.1206$	Combine characteristics and mantissas.
$\log N = -1 + 0.9186$	
$N = \text{antilog } (-1 + 0.9186)$	Isolate N.
$\doteq 8.29 \times 10^{-1}$	Approximate N from the table.
$\doteq 0.829$	Write N in standard form.

Therefore,

$$(6280)(0.000132) \doteq 0.829$$

Example 13. Use common logarithms to find $\dfrac{(753)(0.054)^3}{2200}$.

Solution

$$N = \frac{(753)(0.054)^3}{2200}$$

$$\log N = \log \frac{(753)(0.054)^3}{2200}$$

$$= \log 753 + \log (0.054)^3 - \log 2200 \qquad \text{Use first and second laws of logarithms.}$$

$$= \log 753 + 3 \log 0.054 - \log 2200 \qquad \text{Use third law of logarithms.}$$

$$= \log (7.53 \times 10^2) + 3 \log (5.4 \times 10^{-2})$$
$$\qquad - \log (2.2 \times 10^3)$$

$$= (2 + 0.8768) + 3(-2 + 0.7324) - (3 + 0.3424) \qquad \text{Find each logarithm.}$$

$$= 2 + 0.8768 - 6 + 2.1972 - 3 - 0.3424 \qquad \text{Remove parentheses.}$$

$$= 2 + 0.8768 - 6 + 2 + 0.1972 - 3 - 0.3424 \qquad \text{Separate mantissas and characteristics.}$$

$$= (2 - 6 + 2 - 3) + (0.8768 + 0.1972 - 0.3424)$$

$$= -5 + (1.0740 - 0.3424) \qquad \text{Combine characteristics.}$$

$$\log N = -5 + 0.9316 \qquad \text{Combine mantissas.}$$

$$N = \text{antilog } (-5 + 0.9316) \qquad \text{Isolate } N.$$

$$\doteq 8.54 \times 10^{-5} \qquad \text{Approximate } N \text{ from the table.}$$

$$\doteq 0.0000854 \qquad \text{Write } N \text{ in standard form.}$$

Therefore,

$$\frac{(753)(0.054)^3}{2200} \doteq 0.0000854$$

This may seem like a tedious process, but think of the arithmetic necessary in computing such a quotient without logarithms. The use of logarithms reduces multiplication and/or division of very large or very small numbers to addition and/or subtraction of decimal numbers. It reduces finding powers and/or roots of very large or very small numbers to multiplication by a rational number.

Example 14. Use logarithms to find $N = \sqrt{0.0753}$.

Solution

$$N = \sqrt{0.0753}$$

$$\log N = \log \sqrt{0.0753}$$

$$= \log (0.0753)^{1/2}$$

$$= \frac{1}{2} \log (0.0753)$$

$$= \frac{1}{2} \log (7.53 \times 10^{-2})$$

$$= \frac{1}{2} (-2 + 0.8768)$$

$$= \frac{1}{2} (-2) + \frac{1}{2} (0.8768)$$

$$\log N = -1 + 0.4384$$

$$N = \text{antilog} (-1 + 0.4384)$$

$$\doteq 2.74 \times 10^{-1}$$

$$\doteq 0.274$$

Therefore

$$\sqrt{0.0753} \doteq 0.274$$

To those of us who are accustomed to doing calculations with calculators, the benefits of using logarithms as a tool in computations may seem questionable. However, such practice does reinforce understanding of the laws of logarithms and the process for finding antilogarithms.

⫸ Trial Run

Use logarithms to approximate each N.

_____ **1.** $N = (271)(0.0136)$ _____ **2.** $N = \dfrac{5730}{28}$

_____ **3.** $N = (0.0619)^4$ _____ **4.** $N = \sqrt[3]{32}$

Answers are on page 877.

▶ Examples You Completed

Example 3. Use scientific notation to write $76{,}300$.

Solution

$$76{,}300 = 7{\underset{\curvearrowright}{\wedge}}6300$$

4 places right

$$= 7.63 \times 10^4$$

Example 4. Use scientific notation to write 0.000012.

Solution

$$0.000012 = 0.00001{\underset{\curvearrowleft}{\wedge}}2$$

5 places left

$$= 1.2 \times 10^{-5}$$

Answers to Trial Runs

page 871 **1.** 8.99×10^2 **2.** 5.4×10^{-4} **3.** 2.01×10^6 **4.** 0.00001 **5.** 7100 **6.** 0.00105

page 874 **1.** 0.8096 **2.** 2.2553 **3.** 5.6794 **4.** $-3 + 0.8802$ **5.** 8920 **6.** 0.0348

page 877 **1.** $N \doteq 3.69$ **2.** $N \doteq 205$ **3.** $N \doteq 1.47 \times 10^{-5} = 0.0000147$ **4.** $N \doteq 3.17$

EXERCISE SET 11.4

Use scientific notation to write each number.

_____ **1.** 0.000542

_____ **2.** 0.0000361

_____ **3.** 23.5

_____ **4.** 46.7

_____ **5.** 89,000,000

_____ **6.** 73,000,000

_____ **7.** 0.00071

_____ **8.** 0.00523

_____ **9.** 78,300

_____ **10.** 67,400

Write each number in standard form.

_____ **11.** 1.62×10^{-4}

_____ **12.** 2.31×10^{-5}

_____ **13.** 3.82×10^{5}

_____ **14.** 7.16×10^{4}

_____ **15.** 3.01×10^{1}

_____ **16.** 9.23×10^{2}

_____ **17.** 4.21×10^{-2}

_____ **18.** 8.37×10^{-1}

_____ **19.** 1.50×10^{7}

_____ **20.** 2.40×10^{6}

Find the common logarithms.

_____ **21.** log 87,500

_____ **22.** log 293,000

_____ **23.** log 0.000139

_____ **24.** log 0.00245

_____ **25.** log 18.7

_____ **26.** log 23.6

_____ **27.** log 85,000,000

_____ **28.** log 9,310,000

_____ **29.** log 0.00000000621

_____ **30.** log 0.000000000725

Find the antilogarithms.

_____ **31.** antilog $(-2 + 0.8791)$

_____ **32.** antilog $(-3 + 0.7818)$

_____ **33.** antilog 5.1931

_____ **34.** antilog 7.5551

_____ **35.** antilog $(-8 + 0.6474)$

_____ **36.** antilog $(-9 + 0.7782)$

_____ **37.** antilog 0.3560

_____ **38.** antilog 0.9355

_____ **39.** antilog 4.9474

_____ **40.** antilog 3.8228

Use common logarithms to approximate each N.

_____ 41. $N = (0.316)(9.72)$

_____ 42. $N = (0.813)(7.46)$

_____ 43. $N = (0.871)^4$

_____ 44. $N = (0.0316)^3$

_____ 45. $N = \dfrac{83.6}{28.7}$

_____ 46. $N = \dfrac{42.8}{17.5}$

_____ 47. $N = \dfrac{(34.7)^2}{80.3}$

_____ 48. $N = \dfrac{(7.31)^3}{62.3}$

_____ 49. $N = \dfrac{(86)(12.3)}{217}$

_____ 50. $N = \dfrac{(59)(15.7)}{328}$

_____ 51. $N = \sqrt[3]{(48.5)(3.13)}$

_____ 52. $N = \sqrt[3]{(56.7)(9.36)}$

_____ 53. $N = \dfrac{\sqrt{87.2}\,(96)}{8.09}$

_____ 54. $N = \dfrac{\sqrt{96.3}\,(28)}{7.63}$

☆ Stretching the Topics

_____ 1. Use logarithms to approximate $N = \dfrac{\sqrt{451}\,\sqrt[3]{73.4}}{(1.65)^5(2.01)^2}$.

_____ 2. The time T for one period of a simple pendulum is given by the formula

$T = 2\pi \sqrt{\dfrac{l}{g}}$, where l is the length of the pendulum and g is the acceleration

due to gravity. Find l if $T = 6.5$ seconds, $g = 98$ meters per second, and $\pi = 3.14$. Use logarithms.

Check your answers in the back of your book.

If you can complete **Checkup 11.4,** you are ready to go on to Section 11.5.

✔ CHECKUP 11.4

_____ **1.** Use scientific notation to write 0.00231.

_____ **2.** Write 1.21×10^4 in standard form.

Find the common logarithms.

_____ **3.** log 726

_____ **4.** log 0.00983

Find the antilogarithms.

_____ **5.** antilog $(-3 + 0.0899)$

_____ **6.** antilog (4.9106)

Use common logarithms to find N.

_____ **7.** $N = (5.37)(0.00325)$

_____ **8.** $N = \dfrac{75.9}{36.3}$

_____ **9.** $N = \sqrt[3]{8360}$

_____ **10.** $N = (0.0725)^4$

Check your answers in the back of your book.

If You Missed Problems:	You Should Review Examples:
1	4
2	6
3, 4	8, 9
5, 6	10, 11
7–10	12–14

11.5 Solving More Logarithmic and Exponential Equations

In Section 11.2 we learned to solve logarithmic equations using the definition of a logarithm. Recall that for $a > 0$, $a \neq 1$, $b > 0$ and $n \in R$,

$$\log_a b = n \qquad \text{means} \qquad a^n = b$$

If a logarithmic equation contains a single logarithmic expression, we shall continue to use this definition to solve the equation. Recalling that *the logarithm of zero or a negative number does not exist*, we shall always *check solutions* for logarithmic equations.

Example 1. Solve $\log_2 (3x + 1) = 4$.

Solution

$$\log_2 (3x + 1) = 4$$

$$2^4 = 3x + 1 \qquad \text{Switch to exponential form.}$$

$$16 = 3x + 1 \qquad \text{Simplify the power.}$$

$$15 = 3x \qquad \text{Isolate the } x\text{-term.}$$

$$5 = x \qquad \text{Isolate } x.$$

CHECK

$$\log_2 (3 \cdot 5 + 1) \overset{?}{=} 4$$

$$\log_2 16 \overset{?}{=} 4$$

$$4 = 4$$

The solution is $x = 5$.

Example 2. Solve $\log_5 (x + 1)(x - 3) = 1$.

Solution

$$\log_5 (x + 1)(x - 3) = 1$$

$$5^1 = (x + 1)(x - 3) \qquad \text{Switch to exponential form.}$$

$$5 = x^2 - 2x - 3 \qquad \text{Remove parentheses.}$$

$$0 = x^2 - 2x - 8 \qquad \text{Set quadratic expression equal to zero.}$$

$$0 = (x - 4)(x + 2) \qquad \text{Factor the trinomial.}$$

$$x - 4 = 0 \quad \text{or} \quad x + 2 = 0 \qquad \text{Use the zero product rule.}$$

$$x = 4 \qquad\qquad x = -2 \qquad \text{Solve for } x.$$

CHECK

$$x = 4$$

$$\log_5 (4 + 1)(4 - 3) \overset{?}{=} 1$$

$$\log_5 (5)(1) \overset{?}{=} 1$$

$$\log_5 5 \overset{?}{=} 1$$

$$1 = 1$$

CHECK

$$x = -2$$

$$\log_5 (-2 + 1)(-2 - 3) \overset{?}{=} 1$$

$$\log_5 (-1)(-5) \overset{?}{=} 1$$

$$\log_5 5 \overset{?}{=} 1$$

$$1 = 1$$

The solutions are $x = 4$ or $x = -2$.

Solving Equations Containing Only Logarithmic Expressions

From the graph of the logarithmic function, you should agree that each y in the range corresponds to one and only one x in the domain. This observation is very important because it implies that if we know that the logarithm of one quantity equals the logarithm of another quantity, we may conclude that those two quantities are equal.

$$\text{If } \log_a M = \log_a N$$
$$\text{then} \quad M = N$$

With this fact in mind, we can solve logarithmic equations that contain several logarithmic expressions and no other terms.

Example 3. Solve $\log_2 (x + 5) = \log_2 (4x - 1)$.

Solution

$$\log_2 (x + 5) = \log_2 (4x - 1)$$
$$x + 5 = 4x - 1$$
$$6 = 3x$$
$$2 = x$$

CHECK

$$x = 2$$
$$\log_2 (2 + 5) \stackrel{?}{=} \log_2 (4 \cdot 2 - 1)$$
$$\log_2 7 = \log_2 7$$

You complete Example 4.

Example 4. Solve $\log_3 (5 - x) = \log_3 (x^2 - 1)$.

Solution

$$\log_3 (5 - x) = \log_3 (x^2 - 1)$$
$$5 - x = x^2 - 1$$
$$0 = x^2 + x - 6$$

$$x = \underline{\quad} \qquad \text{or} \qquad x = \underline{\quad}$$

CHECK

$$x = -3$$
$$\log_3 [5 - (-3) \stackrel{?}{=} \log_3 [(-3)^2 - 1]$$
$$\log_3 [5 + \underline{\quad}] \stackrel{?}{=} \log_3 [\underline{\quad} - 1]$$
$$\log_3 \underline{\quad} = \log_3 \underline{\quad}$$

CHECK

$$x = 2$$
$$\log_3 (5 - 2) \stackrel{?}{=} \log_3 (2^2 - 1)$$
$$\log_3 \underline{\quad} = \log_3 \underline{\quad}$$

The solutions are $x = \underline{\quad}$ or $x = \underline{\quad}$.

Check your work on page 894. ▶

When an equation contains several logarithmic expressions, it is best to combine those expressions so that there is just one logarithmic expression on each side of the equation. The laws of logarithms provide us with the key for such manipulation.

Solving Equations Containing Only Logarithmic Expressions

1. Use the laws of logarithms to rewrite each side of the equation as the logarithm of a single quantity.
2. Set those quantities equal to each other and solve for the variable.
3. Check each solution in the original equation to be sure that the logarithm of a negative number does not appear.

Example 5. Solve $\log_a x + \log_a (x + 2) = \log_a (3x + 2)$.

Solution

$$\log_a x + \log_a (x + 2) = \log_a (3x + 2)$$

$$\log_a x(x + 2) = \log_a (3x + 2) \qquad \text{Use first law of logarithms.}$$

$$x(x + 2) = 3x + 2 \qquad \text{If } \log_a M = \log_a N, \text{ then } M = N.$$

$$x^2 + 2x = 3x + 2 \qquad \text{Remove parentheses.}$$

$$x^2 - x - 2 = 0 \qquad \text{Set quadratic expression equal to zero.}$$

$$(x - 2)(x + 1) = 0 \qquad \text{Factor the trinomial.}$$

$$x - 2 = 0 \quad \text{or} \quad x + 1 = 0 \qquad \text{Set each factor equal to 0.}$$

$$x = 2 \qquad\qquad x = -1 \qquad \text{Solve.}$$

We must check our solutions. Since our base, a, is not a specific number, our check will only inform us whether either solution puts us in the position of writing the logarithm of a *negative* number. If so, we must discard that solution.

$$\log_a x + \log_a (x + 2) = \log_a (3x + 2)$$

CHECK   CHECK

$$x = 2 \qquad\qquad\qquad\qquad x = -1$$

$$\log_a 2 + \log_a (2 + 2) \stackrel{?}{=} \log_a (3 \cdot 2 + 2) \qquad \log_a (-1) + \log_a (-1 + 2) \stackrel{?}{=}$$

$$\log_a 2 + \log_a 4 \stackrel{?}{=} \log_a 8 \qquad\qquad \log_a [3(-1) + 2]$$

$$\log_a (-1) + \log_a (1) \stackrel{?}{=} \log_a (-1)$$

In checking -1 in the original equation, logarithms of negative numbers occurred. Therefore, we must reject that solution. Our only solution is $x = 2$.

Example 6. Solve $\log_a (x + 1) - \log_a 3 = \log_a (1 - 2x)$.

Solution

$$\log_a (x + 1) - \log_a 3 = \log_a (1 - 2x)$$

$$\log_a \frac{x + 1}{3} = \log_a (1 - 2x) \qquad \text{Use second law of logarithms.}$$

$$\frac{x + 1}{3} = 1 - 2x \qquad \text{If } \log_a M = \log_a N, \text{ then } M = N.$$

$$\frac{3}{1}\left(\frac{x + 1}{3}\right) = 3(1 - 2x) \qquad \text{Multiply by LCD: 3.}$$

$$x + 1 = 3 - 6x \qquad \text{Remove parentheses.}$$

$$7x = 2 \qquad \text{Isolate } x\text{-term.}$$

$$x = \frac{2}{7} \qquad \text{Isolate } x.$$

CHECK

$$\log_a\left(\frac{2}{7} + 1\right) - \log_a 3 \stackrel{?}{=} \log_a\left[1 - 2\left(\frac{2}{7}\right)\right]$$

$$\log_a\left(\frac{2}{7} + \frac{7}{7}\right) - \log_a 3 \stackrel{?}{=} \log_a\left(\frac{7}{7} - \frac{4}{7}\right)$$

$$\log_a \frac{9}{7} - \log_a 3 \stackrel{?}{=} \log_a \frac{3}{7}$$

Since no logarithms of negative numbers appear, we accept our solution, $x = \frac{2}{7}$.

Example 7. Solve $\log_a (x + 5) + \log_a (x - 3) = \log_a (2x + 1)$.

Solution

$$\log_a (x + 5) + \log_a (x - 3) = \log_a (2x + 1)$$

$$\log_a (x + 5)(x - 3) = \log_a (2x + 1)$$

$$(x + 5)(x - 3) = 2x + 1$$

$$x^2 + 2x - 15 = 2x + 1$$

$$x^2 - 16 = 0$$

$$(x - 4)(x + 4) = 0$$

$$x - 4 = 0 \quad \text{or} \quad x + 4 = 0$$

$$x = 4 \qquad\qquad x = -4$$

CHECK

$$x = 4$$

$$\log_a (4 + 5) + \log_a (4 - 3) \stackrel{?}{=} \log_a [2(4) + 1]$$

$$\log_a 9 + \log_a 1 \stackrel{?}{=} \log_a 9$$

CHECK

$$x = -4$$

$$\log_a (-4 + 5) + \log_a (-4 - 3) \stackrel{?}{=} \log_a [2(-4) + 1]$$

$$\log_a 1 + \log_a (-7) \stackrel{?}{=} \log_a (-7)$$

Since logarithms of negative numbers appear in the check for $x = -4$, we must reject that solution. The only solution is $x = 4$.

⫸ **Trial Run**

Solve.

_____ **1.** $\log_3 (2x - 1) = 2$ _____ **2.** $\log_2 (x - 2)(x + 5) = 3$

_____ **3.** $\log_5 (8 - 5x) = \log_5 (x^2 - 6)$

_____ **4.** $\log_a (2x - 1) - \log_a 5 = \log_a (4 - x)$

_____ **5.** $\log_3 x + \log_3 (x + 4) = \log_3 (x + 10)$

_____ **6.** $\log_{10} \left(\dfrac{x + 3}{2} \right) + \log_{10} (x + 1) = \log_{10} (x + 9)$

Answers are on page 895.

Solving Equations Containing Logarithmic Expressions and Constants

If a logarithmic equation contains several logarithmic expressions together with one or more constants, we must use the laws of logarithms and the definition of a logarithm to solve it. In general, we shall use the following procedure.

Solving Equations Containing Logarithmic Expressions and Constants

1. Write all logarithmic expressions on one side of the equation and combine all constants on the other side.
2. Use the laws of logarithms to rewrite the logarithmic expressions as a single logarithmic expression.
3. Use the definition of a logarithm to rewrite the equation in exponential form.
4. Simplify the exponential equation and solve for the variable.
5. Check each solution in the original equation to be sure that the logarithm of a negative number does not appear.

Example 8. Solve $\log_3 (2x - 1) = 2 + \log_3 4$.

   Solution

$$\log_3 (2x - 1) = 2 + \log_3 4$$

$\log_3 (2x - 1) - \log_3 4 = 2$ Write logarithmic expressions on one side.

$\log_3 \left(\dfrac{2x - 1}{4} \right) = 2$ Rewrite logarithms as a single logarithm.

$3^2 = \dfrac{2x - 1}{4}$ Switch to exponential form.

$9 = \dfrac{2x - 1}{4}$ Simplify power.

$$36 = 2x - 1 \qquad \text{Multiply both sides by 4.}$$

$$37 = 2x \qquad \text{Isolate } x\text{-term.}$$

$$\frac{37}{2} = x \qquad \text{Isolate } x.$$

CHECK

$$\log_3 \left[2\left(\frac{37}{2}\right) - 1 \right] \overset{?}{=} 2 + \log_3 4$$

$$\log_3 (37 - 1) \overset{?}{=} 2 + \log_3 4$$

$$\log_3 36 \overset{?}{=} 2 + \log_3 4$$

Since no logarithms of negative numbers appear, we accept the solution $x = \dfrac{37}{2}$.

You complete Example 9.

Example 9. Solve $\log_2 5 + \log_2 (x + 3) = 3$.

Solution

$$\log_2 5 + \log_2 (x + 3) = 3$$

$$\log_2 5(x + 3) = 3 \qquad \text{Rewrite logarithms as a single logarithm.}$$

$$\text{Switch to exponential form.}$$

$$\text{Simplify both sides.}$$

$$\text{Isolate } x\text{-term.}$$

$$\underline{} = x \qquad \text{Isolate } x.$$

CHECK

$$x = \frac{-7}{5}$$

$$\log_2 5 + \log_2 \left(\frac{-7}{5} + 3 \right) \overset{?}{=} 3$$

$$\log_2 5 + \log_2 \left(\frac{-7}{5} + \frac{15}{5} \right) \overset{?}{=} 3$$

$$\log_2 5 + \log_2 \left(\frac{8}{5} \right) \overset{?}{=} 3$$

Since no logarithms of negative numbers appear, we accept the solution $x = \dfrac{-7}{5}$.

Check your work on page 895. ▶

Example 10. Solve $\log_{10} (x - 2) + 4 = 5 - \log_{10} (x + 1)$.

Solution

$$\log_{10} (x - 2) + 4 = 5 - \log_{10} (x + 1)$$

$$\log_{10} (x - 2) + \log_{10} (x + 1) = 5 - 4 \qquad$$ Write logarithms on one side, constants on other side.

$$\log_{10} (x - 2)(x + 1) = 1 \qquad$$ Rewrite logarithms as a single logarithm. Combine constants.

$$10^1 = (x - 2)(x + 1) \qquad$$ Switch to exponential form.

$$10 = x^2 - x - 2 \qquad$$ Remove parentheses.

$$0 = x^2 - x - 12 \qquad$$ Set quadratic expression equal to zero.

$$0 = (x - 4)(x + 3) \qquad$$ Factor the trinomial.

$$x - 4 = 0 \quad \text{or} \quad x + 3 = 0 \qquad$$ Set each factor equal to zero.

$$x = 4 \quad \text{or} \qquad x = -3 \qquad$$ Solve for x.

CHECK   CHECK

$$x = 4 \qquad\qquad\qquad x = -3$$

$$\log_{10} (4 - 2) + 4 \overset{?}{=} 5 - \log_{10} (4 + 1) \qquad \log_{10} (-3 - 2) + 4 \overset{?}{=} 5 - \log_{10} (-3 + 1)$$

$$\log_{10} 2 + 4 \overset{?}{=} 5 - \log_{10} (5) \qquad\qquad \log_{10} (-5) + 4 \overset{?}{=} 5 - \log_{10} (-2)$$

Since logarithms of negative numbers appear in the check of $x = -3$, we must reject that solution. The only solution is $x = 4$.

Example 11. Solve $\log_3 (2x - 1) + \log_3 (2x + 7) = 2 + \log_3 x$.

Solution

$$\log_3 (2x - 1) + \log_3 (2x + 7) = 2 + \log_3 x$$

$$\log_3 (2x - 1) + \log_3 (2x + 7) - \log_3 x = 2 \qquad$$ Write logarithms on one side.

$$\log_3 \frac{(2x - 1)(2x + 7)}{x} = 2 \qquad$$ Rewrite logarithms as a single logarithm.

$$3^2 = \frac{(2x - 1)(2x + 7)}{x} \qquad$$ Switch to exponential form.

$$9 = \frac{(2x - 1)(2x + 7)}{x} \qquad$$ Simplify power.

$$9x = (2x - 1)(2x + 7) \qquad$$ Multiply by LCD: x.

$$9x = 4x^2 + 12x - 7 \qquad$$ Remove parentheses.

$$0 = 4x^2 + 3x - 7 \qquad$$ Set quadratic expression equal to zero.

$$0 = (4x + 7)(x - 1) \qquad$$ Factor trinomial.

$$4x + 7 = 0 \qquad \text{or} \qquad x - 1 = 0$$

$$4x = -7$$

$$x = \frac{-7}{4} \qquad\qquad\qquad x = 1$$

CHECK

$$x = \frac{-7}{4}$$

$$\log_3\left[2\left(\frac{-7}{4}\right) - 1\right] + \log_3\left[2\left(\frac{-7}{4}\right) + 7\right] = 2 + \log_3\left(\frac{-7}{4}\right)$$

We immediately see the logarithm of a negative number on the right side and *reject* the solution $x = \frac{-7}{4}$.

CHECK

$$x = 1$$

$$\log_3(2 \cdot 1 - 1) + \log_3(2 \cdot 1 + 7) \stackrel{?}{=} 2 + \log_3 1$$

$$\log_3 1 + \log_3(9) \stackrel{?}{=} 2 + \log_3 1$$

No logarithms of negative numbers appear, so we accept the solution $x = 1$.

⫸ Trial Run

Solve.

————— **1.** $\log_2 x + \log_2 6 = 1$ ————— **2.** $\log_3(x + 2) - \log_3 5 = 2$

————— **3.** $\log_6 x + \log_6(x + 5) = 2$

————— **4.** $\log_2(x - 1) = \log_2(2x + 3) + 1$

————— **5.** $\log_{10}(x + 3) + 3 = 4 - \log_{10} x$

Answers are on page 895.

Solving More Exponential Equations

In Section 11.1 we learned to solve exponential equations by rewriting both sides of the equation as powers of the same base. At that time we agreed that equations such as $5^x = 7$ and $2^{x+1} = 3^x$ could not be solved using that technique. Instead, we can now solve them by taking the common logarithm of each side.

$$5^x = 7$$

$$\log 5^x = \log 7 \qquad \text{If } M = N, \text{ then } \log M = \log N.$$

$$x \log 5 = \log 7 \qquad \text{Use third law of logarithms.}$$

$$x = \frac{\log 7}{\log 5} \qquad \text{Isolate } x.$$

Keep in mind that log 7 and log 5 are constants; their value can be *approximated* using a table of common logarithms.

$$x = \frac{\log 7}{\log 5} \doteq \frac{0.8451}{0.6990}$$

$$x \doteq 1.209$$

Remember, the *exact* value for x is $\dfrac{\log 7}{\log 5}$, but the *approximate* value is 1.209.

Example 12. Solve $3^{x+2} = 4$. Then approximate the solution to the nearest thousandth.

Solution

$3^{x+2} = 4$	
$\log 3^{x+2} = \log 4$	If $M = N$, then $\log M = \log N$.
$(x + 2) \log 3 = \log 4$	Use third law of logarithms.
$x \log 3 + 2 \log 3 = \log 4$	Use distributive property.
$x \log 3 = \log 4 - 2 \log 3$	Isolate x-term.
$x = \dfrac{\log 4 - 2 \log 3}{\log 3}$	Isolate x.
$x \doteq \dfrac{0.6021 - 2(0.4771)}{0.4771}$	Substitute values from table.
$= \dfrac{0.6021 - 0.9542}{0.4771}$	Remove parentheses.
$x \doteq \dfrac{-0.3521}{0.4771}$	Simplify the numerator.
$x \doteq -0.738$	Find the quotient.

Example 13. Solve $2^{x+1} = 3^x$. Then approximate the solution to the nearest thousandth.

Solution

$2^{x+1} = 3^x$	
$\log 2^{x+1} = \log 3^x$	If $M = N$, then $\log M = \log N$.
$(x + 1) \log 2 = x \log 3$	Use third law of logarithms.
$x \log 2 + \log 2 = x \log 3$	Use distributive property.
$\log 2 = x \log 3 - x \log 2$	Isolate x-terms.
$\log 2 = x(\log 3 - \log 2)$	Factor out x on right.
$\dfrac{\log 2}{\log 3 - \log 2} = x$	Isolate x.
$\dfrac{0.3010}{0.4771 - 0.3010} \doteq x$	Substitute values from table.
$\dfrac{0.3010}{0.1761} \doteq x$	Simplify the denominator.
$1.709 \doteq x$	Find the quotient.

We used logarithms to solve these equations because we could not rewrite both sides as powers of the same base. However, do not forget to check that possibility before beginning to solve an exponential equation. An equation such as $2^{x+1} = 4^x$ can be solved more efficiently by writing each side as a power of 2 than by using common logarithms (although either method can be used).

⟫ Trial Run

Solve for the variable. Approximate your solution to the nearest thousandth.

_____ **1.** $6^x = 10$ _____ **2.** $4^{x-1} = 5$

_____ **3.** $(1.03)^r = 2$ _____ **4.** $7^x = 9^{x+2}$

Answers are on page 895.

When money is deposited in an account that pays **compound interest,** the depositor earns interest on the original amount (called the **principal,** P) and also on previously earned interest. If the interest rate is r per payment period, the total amount, A, in the account after n payment periods is found by the formula

$$A = P(1 + r)^n$$

To evaluate the right side without a calculator we may use common logarithms.

Example 14. If \$2000 is deposited in an account paying 2% compound interest per quarter, find the total amount after 10 years.

Solution Since there are 4 quarters in one year, there are 40 quarters in 10 years. So $n = 40$, $P = 2000$, and $r = 2\% = 0.02$.

$A = P(1 + r)^n$	Formula for compound amount.
$= 2000(1 + 0.02)^{40}$	Substitute into formula.
$= 2000(1.02)^{40}$	
$\log A = \log[2000\,(1.02)^{40}]$	If $M = N$, then $\log M = \log N$.
$= \log 2000 + \log (1.02)^{40}$	Use first law of logarithms.
$= \log (2000) + 40 \log (1.02)$	Use third law of logarithms.
$\doteq 3.3010 + 40(0.0086)$	Substitute logarithms from table.
$\doteq 3.3010 + 0.344$	Remove parentheses.
$\log A \doteq 3.6450$	Simplify.
$A = \text{antilog } 3.6450$	Isolate A.
$\doteq 4.42 \times 10^3$	Approximate A from the table.
$\doteq 4420$	Write A in standard form.

After 10 years in an account paying 2% per quarter, a principal of \$2000 will have increased to \$4420.

Example 15. If Hannah deposits $1500 today in an account paying 7% compound interest per year, in how many years will the amount in her account have doubled?

Solution: We want the final amount A to be twice the original principal, so $A = 2(1500) = 3000$. Here we know that $P = 1500$, $A = 3000$, and $r = 7\% = 0.07$, but we must find n.

$A = P(1 + r)^n$	Formula for compound amount.
$3000 = 1500(1 + 0.07)^n$	Substitute into formula.
$\dfrac{3000}{1500} = (1.07)^n$	Divide by 1500.
$2 = (1.07)^n$	Simplify left side.

We must solve an exponential equation.

$\log 2 = \log (1.07)^n$	If $M = N$, then $\log M = \log N$.
$\log 2 = n \log 1.07$	Use third law of logarithms.
$\dfrac{\log 2}{\log 1.07} = n$	Isolate n.
$\dfrac{0.3010}{0.0294} \doteq n$	Substitute values from table.
$10.2 \doteq n$	Find the quotient.

The amount in the account paying 7% per year will double in about 10.2 years.

In biology and chemistry, patterns of growth and decay are often described by exponential functions.

Example 16. If the number of bacteria in a certain culture triples every hour and there are 100 bacteria present initially, then the number of bacterial present after t hours is found by the formula

$$B = 100 \cdot 3^t$$

Find the approximate number of bacteria present after 30 minutes.

Solution: Here $t = 30$ minutes or $\dfrac{1}{2}$ hour.

$B = 100 \cdot 3^t$	
$= 100 \cdot 3^{1/2}$	Substitute $\dfrac{1}{2}$ for t.
$= 100\sqrt{3}$	Interpret fractional exponent.
$\doteq 100(1.732)$	Substitute value for $\sqrt{3}$.
$\doteq 173.2$	Simplify.

After 30 minutes, there will be approximately 173 bacteria.

You complete Example 17.

Example 17. Under the conditions of Example 16, in how many hours would the number of bacteria reach 5000?

Solution: Here we know $B = 5000$, but we must find t.

$$B = 100 \cdot 3^t$$

$$5000 = 100 \cdot 3^t$$

$$\frac{5000}{100} = 3^t$$

$$50 = 3^t$$

$$\log 50 = \log 3^t$$

$$\log 50 = t \log 3$$

$$= t$$

The number of bacteria will reach 5000 after approximately _____ hours.

Check your work on page 895. ▶

▶ Examples You Completed _____

Example 4. Solve $\log_3 (5 - x) = \log_3 (x^2 - 1)$.

Solution

$$\log_3 (5 - x) = \log_3 (x^2 - 1)$$

$$5 - x = x^2 - 1$$

$$0 = x^2 + x - 6$$

$$0 = (x + 3)(x - 2)$$

$$x + 3 = 0 \qquad\qquad x - 2 = 0$$

$$x = -3 \qquad\qquad x = 2$$

CHECK

$$x = -3$$

$$\log_3 [5 - (-3)] = \log_3 [(-3)^2 - 1]$$

$$\log_3 [5 + 3] \stackrel{?}{=} \log_3 (9 - 1)$$

$$\log_3 8 = \log_3 8$$

CHECK

$$x = 2$$

$$\log_3 (5 - 2) \stackrel{?}{=} \log_3 (2^2 - 1)$$

$$\log_3 3 = \log_3 3$$

The solutions are $x = -3$ or $x = 2$.

Example 9. Solve $\log_2 5 + \log_2 (x + 3) = 3$.

Solution

$$\log_2 5 + \log_2 (x + 3) = 3$$

$$\log_2 5(x + 3) = 3$$

$$2^3 = 5(x + 3)$$

$$8 = 5x + 15$$

$$-7 = 5x$$

$$\frac{-7}{5} = x$$

Example 17. Under the conditions of Example 16, in how many hours would the number of bacteria reach 5000?

Solution: Here we know $B = 5000$, but we must find t.

$$B = 100 \cdot 3^t$$

$$5000 = 100 \cdot 3^t$$

$$\frac{5000}{100} = 3^t$$

$$50 = 3^t$$

$$\log 50 = \log 3^t$$

$$\log 50 = t \log 3$$

$$\frac{\log 50}{\log 3} = t$$

$$\frac{1.6990}{0.4771} \doteq t$$

$$3.56 \doteq t$$

The number of bacteria will reach 5000 after approximately 3.56 hours (or about 3 hours 34 minutes).

Answers to Trial Runs

page 887   **1.** $x = 5$ **2.** $x = -6, x = 3$ **3.** $x = -7$ **4.** $x = 3$ **5.** $x = 2$ **6.** $x = 3$

page 890   **1.** $x = \dfrac{1}{3}$ **2.** $x = 43$ **3.** $x = 4$ **4.** No solution   **5.** $x = 5$

page 892   **1.** $x 2 \doteq 1.285$ **2.** $x 2 \doteq 2.161$ **3.** $r 2 \doteq 23.516$ **4.** $x 2 \doteq 17.492$

EXERCISE SET 11.5

Solve.

_____ **1.** $\log_5 (3x + 4) = 2$ _____ **2.** $\log_3 (5x + 2) = 3$

_____ **3.** $\log_4 x(3x - 8) = 2$ _____ **4.** $\log_6 2x(x + 2) = 1$

_____ **5.** $\log_3 (x + 7)(x - 1) = 2$ _____ **6.** $\log_2 (x - 5)(x - 1) = 5$

_____ **7.** $\log_2 (5x - 17) = \log_2 (x + 3)$

_____ **8.** $\log_4 (3x - 2) = \log_4 (12 + x)$

_____ **9.** $\log_5 (9 - 2x) = \log_5 (x - 6)$

_____ **10.** $\log_3 (8 - 5x) = \log_3 (2x - 13)$

_____ **11.** $\log_a 3 + \log_a (x - 2) = \log_a (x + 1)$

_____ **12.** $\log_a 5 + \log_a (2x - 1) = \log_a (4x + 7)$

_____ **13.** $\log_3 x + \log_3 (x + 9) = \log_3 (8x + 12)$

_____ **14.** $\log_a x + \log_a (2x - 1) = \log_a (4x - 3)$

_____ **15.** $\log_{10} (3x + 2) - \log_{10} 4 = \log_{10} (x - 9)$

_____ **16.** $\log_{10} (2x - 5) - \log_{10} 3 = \log_{10} (x - 5)$

_____ **17.** $\log_2 (x - 3) = \log_2 (3x - 8) - \log_2 x$

_____ **18.** $\log_2 (2x + 1) = \log_2 (2x + 3) - \log_2 x$

_____ **19.** $\log_a x + \log_a (x - 7) = \log_a 2 + \log_a (x + 11)$

_____ **20.** $\log_a (x - 3) + \log_a (x + 1) = \log_a (7x - 23)$

_____ **21.** $\log_3 x + \log_3 4 = 2$ _____ **22.** $\log_3 x + \log_3 2 = 2$

_____ **23.** $\log_4 x - \log_4 3 = 1$ _____ **24.** $\log_4 x - \log_4 5 = 1$

_____ **25.** $\log_2 (x - 1) - \log_2 4 = 3$ _____ **26.** $\log_2 (2x - 3) - \log_2 5 = 2$

_____ **27.** $\log_4 x + \log_4 (x - 3) = 1$ _____ **28.** $\log_2 x + \log_2 (x - 4) = 5$

_____ **29.** $\log_5 (2x + 7) - \log_5 x = 1$ _____ **30.** $\log_3 (2x + 3) - \log_3 x = 2$

_____ **31.** $\log_2 (x - 3) = \log_2 (x - 2) + 2$

_____ **32.** $\log_3 (x - 5) = \log_3 (x + 2) + 1$

_____ **33.** $\log_5 (2x - 9) + 2 = 3 - \log_5 x$

_____ **34.** $\log_6 (x - 1) + 3 = 4 - \log_6 x$

_____ **35.** $\log_2 x + \log_2 (x - 3) = 2 + \log_2 (x + 2)$

_____ **36.** $\log_2 (x - 4) + \log_2 (x - 6) = 2 + \log_2 x$

Solve each equation and approximate each solution to the nearest thousandth.

_____ **37.** $4^x = 25$

_____ **38.** $3^x = 28$

_____ **39.** $3^{2x} = 16$

_____ **40.** $2^{3x} = 12$

_____ **41.** $5^{x-1} = 17$

_____ **42.** $6^{x+2} = 37$

_____ **43.** $9^{2x+1} = 78.3$

_____ **44.** $8^{3x-1} = 94.5$

_____ **45.** $1.76^x = 23.4$

_____ **46.** $1.37^x = 42.3$

_____ **47.** $8 = 14^{x-1}$

_____ **48.** $9 = 15^{1-x}$

_____ **49.** $5^{x-1} = 6^x$

_____ **50.** $7^{x+1} = 10^{2x}$

_____ **51.** $3^{2x+1} = 7^{x-2}$

_____ **52.** $4^{2x+3} = 5^{x-2}$

_____ **53.** If \$3000 is deposited in an account paying 3% per quarter, find the total amount after 6 years.

_____ **54.** If \$4000 is deposited in an account paying 2% per quarter, find the total amount after 5 years.

_____ **55.** Find the total amount in an account after 8 years if \$10,000 is invested at 10% per year compounded semiannually.

_____ **56.** Find the total amount in an account after 6 years if \$8000 is invested at 12% per year compounded semiannually.

_____ **57.** If \$2000 is invested today in an account paying 8% per year compounded yearly, in how many years will the amount have tripled?

_____ **58.** If \$5000 is invested today in an account paying 10% per year compounded semiannually, in how many years will the amount have doubled?

_____ **59.** If the number of bacteria in a culture triples every hour and there are 200 bacteria present initially, find the number of bacteria present after 4 hours. In how many hours would the number of bacteria reach 10,000?

_____ **60.** If the number of bacteria in a culture doubles every hour and there are 50 present initially, find the number of bacteria present after 30 minutes. In how many hours would the number of bacteria reach 400?

☆ Stretching the Topics _____

_____ **1.** Solve $3 \cdot 7^{x+1} = 7 \cdot 3^{x+2}$ for x.

_____ **2.** Find the total amount on deposit at the end of 4 years if \$68,900 is invested at a yearly rate of 13.2% and the interest is compounded monthly.

_____ **3.** Under certain conditions the number of bacteria in a culture is given by $B = n \cdot 2^{t/20}$, where n is the original number of bacteria and t is the number of minutes after the experiment begins. If the original number of bacteria is 250,000, find how many minutes it would take for the number of bacteria to reach 1,000,000.

Check your answers in the back of your book.

If you can complete **Checkup 11.5,** you are ready to do the Review Exercises.

CHECKUP 11.5

Solve.

_____ **1.** $\log_a x + \log_a (x + 10) = \log_a (13x + 28)$

_____ **2.** $\log_{10} (3x - 1) - \log_{10} 3 = \log_{10} (2x + 1)$

_____ **3.** $\log_3 (x + 12) - \log_3 2 = 2$

_____ **4.** $\log_2 (3x + 2) - 4 = \log_2 (x - 2) - 2$

_____ **5.** $5^{x+1} = 13$

_____ **6.** $3^{x+1} = 7^x$

Check your answers in the back of your book.

If You Missed Problems:	You Should Review Examples:
1, 2	5–7
3	8
4	10
5	12
6	13

Summary

In this chapter we studied two new functions that are very closely related.

Exponential Function	Logarithmic Function
$y = a^x$	$y = \log_a x$
$(a > 0, a \neq 1)$	$(a > 0, a \neq 1)$
Domain: $x \in R$	Domain: $x \in R, x > 0$
Range: $y \in R, y > 0$	Range: $y \in R$

We agreed that

$$y = \log_a x \qquad \text{means} \qquad a^y = x$$

and discussed the three laws of logarithms that can be used to rewrite logarithmic expressions.

Name	Symbols	Example
First law of logarithms	$\log_a M \cdot N = \log_a M + \log_a N$	$\log_2 x(x + 3) = \log_2 x + \log_2 (x + 3)$
Second law of logarithms	$\log_a \dfrac{M}{N} = \log_a M - \log_a N$	$\log_3 \left(\dfrac{x + 5}{8} \right) = \log_3 (x + 5) - \log_3 8$
Third law of logarithms	$\log_a M^c = c \log_a M$	$\log_{10} (x + 3)^2 = 2 \log_{10} (x + 3)$

Common logarithms are logarithms for which the base is 10. Because every positive number can be rewritten in scientific notation as a product of some number between 1 and 10, and an appropriate power of 10, we can readily approximate the common logarithm of such numbers using the table inside the back cover.

In solving exponential and logarithmic equations, we used four important facts about exponential and logarithmic statements.

Fact	Example
If $a^m = a^n$ then $m = n$	*Solve:* $2^x = 4^{x+1}$ $2^x = 2^{2(x+1)}$ $x = 2(x + 1)$ $x = 2x + 2$ $-2 = x$
If $\log_a M = \log_a N$ then $M = N$	*Solve:* $\log_3 (2x - 3) = \log_3 (x + 1)$ $2x - 3 = x + 1$ $x = 4$
$y = \log_a x$ means $a^y = x$	*Solve:* $\log_3 (5x - 1) = 2$ $3^2 = 5x - 1$ $9 = 5x - 1$ $10 = 5x$ $2 = x$
If $M = N$ then $\log M = \log N$	*Solve:* $3^x = 8$ $\log 3^x = \log 8$ $x \log 3 = \log 8$ $x = \dfrac{\log 8}{\log 3} \doteq \dfrac{0.9031}{0.4771}$ $x \doteq 1.89$

When solving logarithmic equations, we agreed to always check our solutions. If the logarithm of a negative number appeared in checking a solution, we agreed to reject that solution because *the logarithm of a negative number or zero does not exist.*

❏ Speaking the Language of Algebra

Complete each sentence with the appropriate word or phrase.

1. The function $y = a^x$ (where $a > 0$, $a \neq 1$) is called the _____ function. Its domain is _____ and its range is _____ , _____ .

2. The function $y = \log_a x$ (where $a > 0$, $a \neq 1$) is called the _____ function. Its domain is _____ , _____ and its range is _____ .

3. The statement $y = \log_a x$ means _____ .

4. A logarithm for which the base is 10 is called a _____ logarithm.

5. To solve an exponential equation, we first try to rewrite both sides as powers of the _____ _____ .

6. To solve an equation such as $\log_2 (x + 5) = 4$, we should rewrite it in _____ form.

7. After solving a logarithmic equation, we must check the solutions to be sure that the logarithm of a _____ number or zero does not appear.

REVIEW EXERCISES for Chapter 11: Working with Exponential and Logarithmic Functions

_____ **1.** Evaluate 5^{2x-1} when $x = 1$, $x = 0$, $x = -1$, $x = \frac{1}{4}$, $x = \frac{1}{2}$.

_____ **2.** Graph $y = 3^{x+1}$.

_____ **3.** Solve $\left(\frac{1}{81}\right)^x = 3^{x+5}$.

_____ **4.** Write $3^{-2} = \frac{1}{9}$ as a logarithmic statement.

_____ **5.** Write $125^{-1/3} = \frac{1}{5}$ as a logarithmic statement.

_____ **6.** Write $\log_2 64 = 6$ as an exponential statement.

_____ **7.** Write $\log_8 \frac{1}{2} = \frac{-1}{3}$ as an exponential statement.

_____ **8.** Find $\log_3 \frac{1}{27}$.

_____ **9.** Find $\log_4 \frac{1}{2} - \log_2 1$.

_____ **10.** Solve $\log_a 216 = 3$.

_____ **11.** Solve $\log_4 b = -3$.

_____ **12.** Solve $\log_8 (x + 2) = \frac{1}{3}$.

_____ **13.** Graph $y = \log_5 x$.

_____ **14.** Find $\log_3 \sqrt[4]{243}$.

_____ **15.** Find $\log_2 \dfrac{\sqrt{16} \cdot 32^2}{128}$

Given that $\log_{10} 2 = 0.3010$ _and_ $\log_{10} 3 = 0.4771$.

_____ **16.** Find $\log_{10} \frac{32}{27}$.

_____ **17.** Find $\log_{10} \sqrt{540}$.

_____ **18.** Rewrite $\log_a x^5(x + 5)$ using the laws of logarithms.

_____ **19.** Rewrite $\log_a \dfrac{x^3 \sqrt{x^2 - 1}}{(x + 1)^5}$ using the laws of logarithms.

_____ **20.** Rewrite $2 \log_a x + \dfrac{1}{2} \log_a (x - 1)$ as a single logarithm.

_____ **21.** Rewrite $3 \log_a (2x + 1) - \dfrac{4}{5} \log_a x - \log_a (x + 1)$ as a single logarithm.

_____ **22.** Use scientific notation to write 0.00729.

_____ **23.** Write 9.34×10^5 in standard form.

_____ **24.** Use common logarithms to approximate $N = \dfrac{(34.7)^2(123)}{\sqrt{80.3}}$.

Solve.

_____ **25.** $\log_{10} (7x - 2) = \log_{10} (3x + 2)$

_____ **26.** $\log_a x + \log_a (x + 14) = \log_a (16x + 63)$

_____ **27.** $\log_3 (x - 9) - \log_3 5 = 1$

_____ **28.** $\log_{10} x + \log_{10} (3x - 5) = 2$

_____ **29.** $4^{2x} = 27.3$

_____ **30.** $9^{x - 1} = 12^{2x}$

Check your answers in the back of your book.

If You Missed Exercises:	You Should Review Examples:	
1	Section 11.1	1, 2
2		6
3		12
4, 5	Section 11.2	1
6, 7		2
8, 9		3–7
10–12		10–15
13		16
14, 15	Section 11.3	1–6
16, 17		7–10
18, 19		11, 12
20, 21		13, 14
22	Section 11.4	4
23		6
24		12–14
25	Section 11.5	3
26		5–7
27, 28		8–11
29, 30		12, 13

If you have completed the Review Exercises and corrected your errors, you are ready to take the Practice Test.

PRACTICE TEST for Chapter 11: Working with Exponential and Logarithmic Functions

		SECTION	EXAMPLES
_____ **1.** Evaluate 9^{2-2x} for $x = 0$, $x = 1$, $x = \dfrac{3}{4}$.		11.1	1
_____ **2.** Write $36^{-1/2} = \dfrac{1}{6}$ as a logarithmic statement.		11.2	2
_____ **3.** Write $\log_5 125 = 3$ as an exponential statement.		11.2	4
_____ **4.** Find $\log_7 \dfrac{1}{49}$.		11.3	4, 5
_____ **5.** Find $\log_2 \dfrac{32\sqrt{2}}{(8)^4}$.		11.3	4, 5
_____ **6.** Given that $\log_{10} 3 = 0.4771$ and $\log_{10} 7 = 0.8451$, find $\log_{10} 6300$.		11.3	9, 10
_____ **7.** Rewrite $\log_a \dfrac{x^5(x^2 - 3)^2}{\sqrt[3]{x + 1}}$ using the laws of logarithms.		11.3	11, 12
_____ **8.** Rewrite $\dfrac{1}{5} \log_a (x + 1) - 2 \log_a (x - 6) - \dfrac{2}{3} \log_a x$ as a single logarithm.		11.3	13, 14
_____ **9.** Use scientific notation to write 2,350,000.		11.4	4
_____ **10.** Write 1.72×10^{-3} in standard form.		11.4	6
_____ **11.** Use common logarithms to approximate $N = (68.5)\sqrt[3]{3.24}$		11.4	12–14
12. Graph $y = 3^{x+2}$.		11.1	6

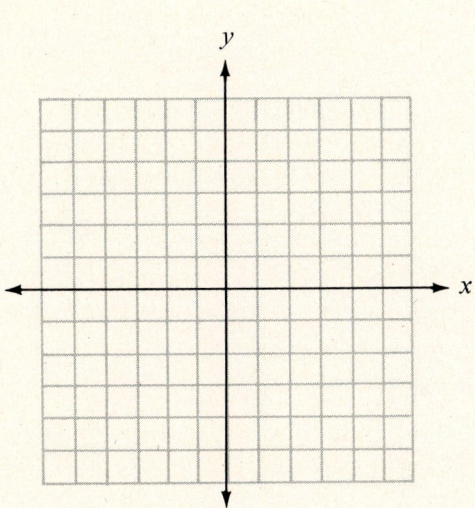

13. Graph $y = \log_2 x$.   11.2   16

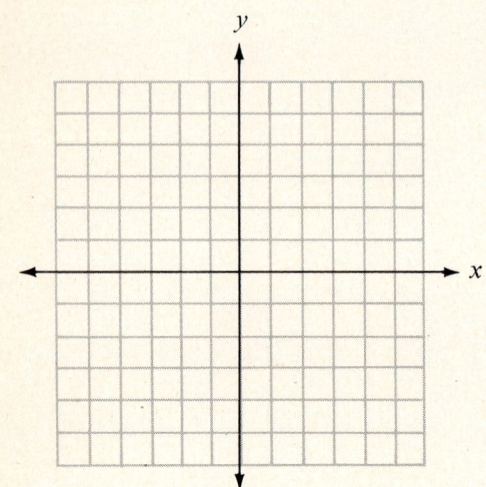

Solve for the variable.

_____ **14.** $8^{2x} = 2^{x-3}$ 11.1   11

_____ **15.** $\log_a \dfrac{1}{16} = -4$ 11.2   11

_____ **16.** $\log_{25}(2x + 1) = \dfrac{1}{2}$ 11.5   1

$$25^{\frac{1}{2}} = (2x + 1)$$
$$5 = 2x + 1$$

_____ **17.** $\log_4(2x - 1) = \log_4(8 - x)$ 11.5   3

_____ **18.** $\log_7(x + 4) - \log_7 x = \log_7 5$ 11.5   8

_____ **19.** $\log_2(x + 6) + \log_2 x = 4$ 11.5   9

_____ **20.** $8^x = 3^{x-1}$ 11.5   13

SHARPENING YOUR SKILLS after Chapters 1–11

_____ **1.** Evaluate $8(x - 2)^2 - 11x$ when $x = 0$. 1.4

_____ **2.** Evaluate $|(-2)(4)| + |-3|$. 1.4

_____ **3.** Give the restrictions on the variable and reduce the fraction $\dfrac{9y^2 - 144}{3y^2 - 24y + 48}$. 6.1

_____ **4.** Write $\left(\dfrac{27}{64}\right)^{-2/3}$ in radical form and simplify as much as possible. 7.1

_____ **5.** Evaluate $-i^{11}$. 7.5

_____ **6.** Find the slope of the line joining $\left(\dfrac{-1}{2}, 2\right)$ and $\left(\dfrac{1}{4}, \dfrac{2}{3}\right)$. 8.3

_____ **7.** Find the equation of the line with x-intercept of $\dfrac{1}{2}$ and y-intercept of $\dfrac{-1}{4}$. 8.3

_____ **8.** If $f(x) = 4x - 3$ and $g(x) = x^2 - 5x + 6$, find $g(1) - f(-3)$. 8.5

Simplify.

_____ **9.** $\dfrac{-6(3 - 8) - 5(-8)}{-4[(-12) + 7]}$ 1.3

_____ **10.** $x^2y^3(3xy^2)^2 - 2xy^4(xy)^3$ 3.1

_____ **11.** $\left(\dfrac{-7x^4y^{-3}}{x^{-2}y^{-2}}\right)^{-2}$ 3.2

_____ **12.** $7x^2 - 9[2(3x^2 + x) - 5x] + 2$ 3.3

_____ **13.** $(2a - 3)^3$ 3.4

_____ **14.** $(3a^3 - 8a^2 - 20a + 16) \div (3a - 2)$ 3.5

_____ **15.** $\dfrac{9x^2 - 25}{x^2 - 3x - 40} \cdot \left(\dfrac{x^2 - 64}{3x^2 + x - 10} \div \dfrac{3x^2 + 29x + 40}{2x^2 + 9x - 5}\right)$ 6.2

_____ **16.** $\dfrac{2}{x + 5} - \dfrac{3}{x + 3} - \dfrac{x - 3}{x^2 + 8x + 15}$ 6.3

_____ **17.** $\dfrac{8 - 10x^{-1} + 3x^{-2}}{2 - 13x^{-1} + 6x^{-2}}$ 6.4

_____ **18.** $7\sqrt[3]{a^4b^3} - 2b\sqrt[3]{27a^4}$ 7.2

——————— **19.** $(\sqrt{3} - 2\sqrt{y})(4\sqrt{3} + \sqrt{y})$ 7.3

——————— **20.** $\dfrac{-10}{2 - \sqrt{-36}}$ 7.5

Factor completely.

——————— **21.** $(x + 3)^3 + 64$ 4.2

——————— **22.** $5h^2 + 22hk + 24k^2$ 4.3

——————— **23.** $27 + 42(2x - y) - 5(2x - y)^2$ 4.3

Solve.

——————— **24.** $4[x - (2x + 1)] = -2[3(x - 1) + 7]$ 2.1

——————— **25.** $2x - 9 < 5x - 3$ or $x - 1 > 7x + 17$ 2.3

——————— **26.** $|5m - 9| \le -3$ 2.4

——————— **27.** $5x(x - 2) = 10(9 - x)$ 5.3

——————— **28.** $(x - 3)(x + 1) < x(5 - x) - 6$ 5.4

——————— **29.** $\dfrac{x + 5}{x - 4} = \dfrac{64}{x^2 - 16} + \dfrac{8}{x + 4}$ 6.5

——————— **30.** $\dfrac{6}{x(x + 2)} < 0$ 6.5

——————— **31.** $\sqrt{x - 4} = 5 + \sqrt{x + 1}$ 7.4

——————— **32.** $\dfrac{y}{y - 7} = \dfrac{1}{y}$ 7.6

——————— **33.** $\begin{aligned} 2x - 3y &= 7 \\ x + 4y &= -2 \end{aligned}$ 10.2

——————— **34.** $\begin{aligned} x &= y^2 - 4 \\ x &= 3y \end{aligned}$ 10.4

Graph.

35. $2x - 3y > -9$

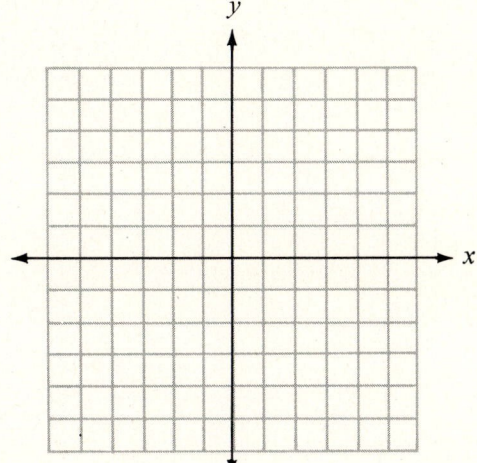

8.4

36. $y = x^2 + 3x + 7$

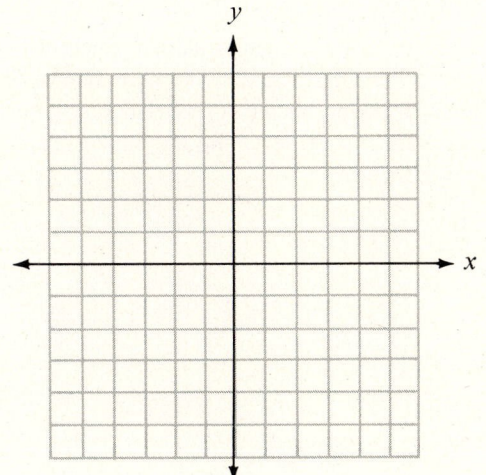

9.2

37. $(x - 2)^2 + (y - 2)^2 \leq 81$

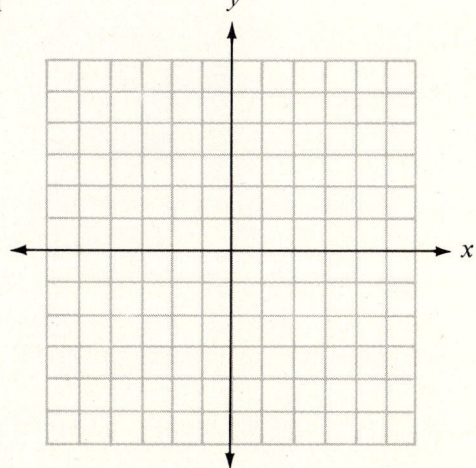

9.4

38. $x^2 + y^2 \leq 25$ 10.4
  $2x - y \geq 1$

_____ **39.** Solve $t = 2\pi \sqrt{\dfrac{l}{g}}$ for g.   7.4

_____ **40.** Katrina has a rectangular garden and a triangular garden. Each of two sides of the   10.5
triangle is as long as the length of the rectangle. The third side of the triangle is as long
as the width of the rectangle. If the perimeter of the triangular garden is 24 feet and the
perimeter of the rectangular garden is 30 feet, find the dimensions of the rectangular
garden.

Answers

CHAPTER 1

Exercise Set 1.1 (page 11)
1. Q, R **3.** W, J, Q, R **5.** H, R **7.** Q, R **9.** $0, 9$

11. $-8, -5.2, \dfrac{-1}{2}, 0, \dfrac{7}{2}, 5.666\ldots, 9$ **13.** -8 **15.** True   **17.** True   **19.** True

21.

23. $3 > \sqrt{7}$

25. $\dfrac{-9}{2} < -3$ **27.** $\dfrac{10}{2} = \sqrt{25}$ **29.** $-1.6 < -0.7$ **31.** $\dfrac{-8}{3} > -\sqrt{16}$ **33.** $0.29 > 0$

35. $-\sqrt{6} < 0$ **37.** $\dfrac{3}{4} > 0$ **39.** $-9 < 0$ **41.** -5 **43.** $\dfrac{7}{8}$ **45.** 3.2 **47.** $-\sqrt{5}$

49. 6 **51.** 6.8 **53.** $\sqrt{3}$ **55.** 5 **57.** 8

Stretching the Topics
1. 7 **2.** $>$ **3.** 0

Checkup 1.1 (page 13)
1. J, Q, R **2.** $-3, 0, 4$ **3.** $\dfrac{8}{2} = \sqrt{16}$ **4.** $\dfrac{-7}{2} < -2.5$ **5.** $\dfrac{-9}{5} < 0$ **6.** $\sqrt{36} > 0$

7. $\dfrac{5}{12}$ **8.** -9 **9.** 3 **10.** 9

Exercise Set 1.2 (page 25)
1. 14 **3.** -10 **5.** 4 **7.** -9 **9.** $\dfrac{-7}{9}$ **11.** -7 **13.** -2 **15.** 1.5

17. -40 **19.** 74 **21.** -4 **23.** 72 **25.** -105 **27.** 0 **29.** 16 **31.** -4

33. 7 **35.** $\dfrac{-6}{7}$ **37.** 12 **39.** 0 **41.** 23 **43.** 0 **45** Not uniquely defined

47. -13 **49.** Undefined

Stretching the Topics
1. -5 **2.** Undefined   **3.** $1°, 22°$

Checkup 1.2 (page 27)
1. -52 **2.** -38 **3.** 33 **4.** 7 **5.** 7 **6.** 28 **7.** -27 **8.** -9 **9.** 15
10. Undefined

Exercise Set 1.3 (page 39)
1. Associative property for multiplication   **3.** Closure property for addition
5. Multiplicative identity   **7.** Distributive property for multiplication over addition

9. Multiplicative inverse   **11.** $8 + (-4)$ **13.** $-9 + (3 + 4)$ **15.** 0   **17.** -19
19. $2(3 - 5)$ **21.** 23   **23.** 290   **25.** -3 **27.** 168   **29.** 0   **31.** 18   **33.** -87
35. -20 **37.** -65 **39.** -8 **41.** 0   **43.** -13 **45.** -3 **47.** Undefined
49. $\dfrac{-3}{4}$ **51.** 11   **53.** 2   **55.** -47 **57.** $\dfrac{-5}{12}$

Stretching the Topics
1. -6.7 **2.** 0   **3.** 496,800; 360 gain

Checkup 1.3 (page 41)
1. Commutative property for multiplication   **2.** Multiplicative inverse   **3.** 17   **4.** 28   **5.** 5
6. -180 **7.** -92 **8.** 15   **9.** $\dfrac{-1}{2}$ **10.** $\dfrac{10}{3}$

Exercise Set 1.4 (page 55)
1. 9   **3.** 10   **5.** 0   **7.** 45   **9.** 30   **11.** 38   **13.** 459.2   **15.** 13   **17.** 100
19. $-4a - 7$ **21.** $2x^2 - 11x$ **23.** $-9y^2 - 1$ **25.** $11x - 6y - 2$
27. $3x^2 - 5xy + y^2$ **29.** $-12a^2 - 11b^2$ **31.** a **33.** $-7x$ **35.** $12y^2$ **37.** $6x^2 - 7$
39. $13y^2 - 16y + 14$ **41.** $8x + 15y - 52$ **43.** $-14x$ **45.** $19m - 4$
47. $-7x + 10y$ **49.** $-67x + 12$ **51.** $-20x + 37$ **53.** $70x + 120$ **55.** $-2x + 38y$
57. $-38a + 6b$ **59.** $9x + 5y$ **61.** 11   **63.** $3x$ if $3x \geq 0$; $-3x$ if $3x < 0$ **65.** $5 - \pi$
67. $x - 2$ if $x - 2 \geq 0$; $-(x - 2)$ if $x - 2 < 0$ **69.** $3 - \sqrt{7}$ **71.** -5 **73.** 13

Stretching the Topics
1. $5x - 83y$; 43   **2.** $11x - 5$ **3.** 29

Checkup 1.4 (page 59)
1. 18   **2.** 784   **3.** $-4x^2 + x - 4$ **4.** $3x^2 - 9y^2$ **5.** $-26a$ **6.** $17y^2 + 6y - 8$
7. $-14n + 45$ **8.** $-28x + 51y$ **9.** $4x$ if $4x \geq 0$; $-4x$ if $4x < 0$ **10.** $9 - \sqrt{2}$

Exercise Set 1.5 (page 67)
1. $27.50x$ dollars   **3.** $(18 + 0.3n)$ dollars   **5.** $6w$ ft or $3l$ ft   **7.** $2x + 7$
9. $(940 + 10n)$ dollars   **11.** $110t$ mi or $55b$ mi   **13.** $9x$ dollars   **15.** $24s$ or $8w$
17. $(0.8c + 27.8)$ dollars or $(-0.8p + 43.8)$ dollars
19. $(15f + 70)$ dollars or $(15t - 35)$ dollars   **21.** $(2500 + 3.85x)$ dollars
23. $(41.3x - 2475)$ dollars   **25.** $(1.75x - 10{,}250)$ dollars

Stretching the Topics
1. $2n - 4$ **2.** $58h$ **3.** $(0.0125x + 900)$ dollars or $(-0.0125y + 1050)$ dollars

Checkup 1.5 (page 71)
1. $(420 + 12x)$ dollars   **2.** $(4l - 10)$ ft or $(4w + 10)$ ft   **3.** $(50 + 3n)$ dollars
4. $(-25t + 220)$ mi or $(25h + 120)$ mi   **5.** $(4x + 50)$ dollars

Speaking the Language of Algebra (page 74)
1. Constants; variables   **2.** Real   **3.** Terms; sum; difference   **4.** Add (or subtract) like terms
5. Factors; product   **6.** Dividend; divisor; quotient   **7.** 0; undefined   **8.** Negative

Review Exercises (page 75)
1. $-\sqrt{2}, \sqrt{3}$ **2.** $-4, 0, 5$ **3.** $\dfrac{15}{3} = \sqrt{25}$ **4.** $\dfrac{-9}{4} < 13$ **5.** $-19 < \dfrac{-3}{4}$
6. $5 > \sqrt{7}$ **7.** $-\sqrt{6} < 0$ **8.** $8.75 > 0$ **9.** -55 **10.** -14 **11.** -4.29
12. -4 **13.** 17   **14.** 45   **15.** -5.6 **16.** -26 **17.** 5   **18.** Undefined   **19.** 4
20. 0   **21.** 840   **22.** 3360   **23.** 48   **24.** -72 **25.** 96   **26.** $\dfrac{1}{2}$ **27.** -8
28. 0   **29.** -3 **30.** 14   **31.** $2720   **32.** 22   **33.** $-2x^2 + x - 7$ **34.** $3x^2 - 3y^2$
35. $-37a$ **36.** $4y^2 - 21y - 15$ **37.** $2n + 53$ **38.** $3x - 36$ **39.** $23 - 48x$
40. $-2x - 9y$ **41.** $-84x - 8y$ **42.** $48x + 53y$ **43.** 21   **44.** a if $a \geq 0$; $-a$ if $a < 0$

45. $6 - \sqrt{10}$ **46.** 4   **47.** -4 **48.** 22   **49.** $6l - 14$
50. $(2.7p - 4.75)$ dollars or $(2.7c + 8.75)$ dollars   **51.** $6i - 15$ **52.** $500 - 12x$

Practice Test (page 79)
1. $\sqrt{10} > 3$ **2.** $-\sqrt{5} < 0$ **3.** -47 **4.** -3 **5.** 15   **6.** 44   **7.** -4 **8.** 0
9. -32 **10.** 28   **11.** -3 **12.** 142   **13.** $-3x^2 - 2x - 11$ **14.** $-ab - b^2$
15. $9m$ **16.** $-9x^2 - 2x - 20$ **17.** $12a - 24$ **18.** $16y - 19$ **19.** $-21x + 12$
20. $6x - y$ **21.** -7 **22.** -6 **23.** $3 - \sqrt{5}$ **24.** $(10i + 135)$ mi or $(-10t + 165)$ mi
25. $(8w + 32)$ in. or $\left(\dfrac{8}{3}l + 32\right)$ in.

Chapter 2

Exercise Set 2.1 (page 93)

1. $x = 5$ **3.** $y = -2$ **5.** $m = -6$ **7.** $y = \dfrac{2}{3}$ **9.** $a = 0$ **11.** $x = \dfrac{-5}{9}$

13. $x = 12$ **15.** $x = -26$ **17.** $x = 7$ **19.** $x = \dfrac{-1}{2}$ **21.** $y = 3$ **23.** $x = 12$

25. $x = -4$ **27.** $a = 0$ **29.** $x = 4$ **31.** $y = 3$ **33.** $x = -4$ **35.** $m = \dfrac{5}{3}$

37. $a = 1$ **39.** $x = -10$ **41.** $x = 0$ **43.** Any real number   **45.** $y = 21$

47. No solution   **49.** $x = \dfrac{95}{4}$ **51.** $x = -3$ **53.** $x = 3$ **55.** $h = \dfrac{2A}{b}$ **57.** $t = \dfrac{d}{r}$

59. $t = \dfrac{A - P}{Pr}$ **61.** $x = \dfrac{2y + 1}{3}$ **63.** $x = \dfrac{5 - by}{a}$ **65.** $x = 6a$ **67.** $n = \dfrac{l - a + d}{d}$

69. $r = \dfrac{V}{2d}$

Stretching the Topics

1. $x = \dfrac{3b}{2a}$ **2.** $a = d + l - nd$ **3.** $x = 16$

Checkup 2.1 (page 95)

1. $x = 27$ **2.** $x = \dfrac{-36}{11}$ **3.** $a = -24$ **4.** $x = 3$ **5.** $x = \dfrac{1}{3}$ **6.** $y = 4$

7. No solution   **8.** $x = \dfrac{-16}{11}$ **9.** $x = \dfrac{15}{2}$ **10.** $r = \dfrac{A - P}{Pt}$

Exercise Set 2.2 (page 105)

1.

3.

5.

7.

9.

11. $x < -3$ **13.** $x \geq 3$ **15.** $\varnothing$
17. $2 < x \leq 5$ **19.** $-1 \leq x \leq 4$

21. $\varnothing$ **23.** $x = -3$ **25.** $x \geq 2$ **27.** $x < 4$ **29.** $x \in R, x \neq 3$ **31.** $x \in R$

33. $x < \dfrac{-1}{2}$ or $x \geq 0$ **35.** $x \in R$ **37.** $x < -3$ **39.** $-4 < x < \dfrac{7}{2}$

41. $x < -1$ or $x > 0$ **43.** $x < 5$ **45.** $x \geq \dfrac{3}{5}$ **47.** $x \in R$ **49.** $\varnothing$

A4   ANSWERS

Stretching the Topics

1. $x = 3.5$ **2.** $x \geq \dfrac{-3}{4}$ **3.** $\pi \leq x \leq \dfrac{22}{7}$

Checkup 2.2 (page 107)

1.

2.

3.

4. $0 < x \leq 2$ **5.** $x < 3$ **6.** $\varnothing$

7. $x \in R$ **8.** $0 \leq x < 5$

9. $x < -3$ or $x > 2$ **10.** $x \in R,\ x \neq \dfrac{1}{2}$

Exercise Set 2.3 (page 121)

1. $x \geq 2$

3. $y < 1$

5. $a \leq \dfrac{-1}{2}$

7. $x > -2$

9. $a \geq 2$

11. $x > \dfrac{3}{2}$

13. $m \leq 4$

15. $x > 12$

17. $x \leq \dfrac{7}{3}$

19. $y \geq -4$

21. $\varnothing$ **23.** $\varnothing$

25. $0 \leq x < 2$

27. $x < 16$

29. $x < 5$ or $x \geq 6$

31. $x \in R$

33. $7 \le m \le 12$

35. $-8 < a \le -6$

37. $-7 \le y < 5$

39. $0 < x \le 2$

Stretching the Topics

1. $b \ge 13$

2. $x \ge 4$

3. $-2 \le x < 3$

Checkup 2.3 (page 123)

1. $x \ge 3$

2. $a > 4$

3. $y < -10$

4. $x \le -2$

5. $\dfrac{1}{2} < x < 2$

6. $x > 3$

7. $x > \dfrac{-7}{2}$

8. $x \le 2$ or $x > 3$

9. $-4 \le a \le -2$

10. $-4 < y \le 2$

Exercise Set 2.4 (page 135)

1. $x = 9$ or $x = -9$ **3.** $a = 21$ or $a = -21$ **5.** $x = 0.7$ or $x = -0.7$ **7.** $y = 5$

9. No solution   **11.** $x = 1$ or $x = \dfrac{4}{3}$ **13.** No solution   **15.** $a = -24$ or $a = -36$

17. $x = \dfrac{8}{3}$ or $x = 6$ **19.** $x = -13$ or $x = 27$ **21.** $\dfrac{-5}{3} < y < \dfrac{5}{3}$ **23.** $-35 < a < 35$

25. $-2 < x < 2$ **27.** $\dfrac{-2}{3} \le x \le 2$ **29.** $12 \le x \le 18$ **31.** No solution

33. $-9 < x < \dfrac{29}{3}$ **35.** $a > 2.5$ or $a < -2$.   **37.** All real numbers   **39.** $x \le -2$ or $x \ge 2$

41. $x > 10$ or $x < -10$ **43.** $x \ge 1$ or $x \le \dfrac{-3}{5}$ **45.** $m < -15$ or $m > 21$

47. All real numbers   **49.** $x \ge \dfrac{13}{7}$ or $x \le -1$

Stretching the Topics

1. $m = 1$ or $m = \dfrac{3}{2}$ **2.** $\dfrac{1}{2} < x < \dfrac{3}{2}$ **3.** $1 < x < 8$

Checkup 2.4 (page 137)

1. $x = 5.2$ or $x = -5.2$ **2.** $x = -1$ or $x = 17$ **3.** No solution   **4.** $-3 < x < 3$

5. $4 \le x \le 24$ **6.** $\dfrac{-5}{3} \le a \le 3$ **7.** $x > 2$ or $x < -2$ **8.** $m \le 6$ or $m \ge 12$

9. $x > 6$ or $x < -5$ **10.** $x > 14$ or $x < -2$

Exercise Set 2.5 (page 145)

1. 6   **3.** 400 boxes of chocolate chip, 250 boxes of sugar   **5.** 12 lb   **7.** 9 quarters, 14 dimes
9. $10,000   **11.** 12 lb of ground beef, 8 lb of sausage   **13.** 50 and 55 mph   **15.** 247.5 mi
17. 12 books for $2.50, 18 books for $4   **19.** 12   **21.** 51 $10 bills, 43 $5 bills, 86 $1 bills
23. At least 19   **25.** No more than 14   **27.** No more than 60 by 120 ft   **29.** At least 150
31. Less than 40   **33.** At least 4000

Stretching the Topics
1. $12,500   **2.** 7 by 16 ft   **3.** No more than 11 cm

Checkup 2.5 (page 149)

1. 3 by 5 ft   **2.** $3\dfrac{11}{24}$ hr   **3.** No more than 9 hr   **4.** At least 35

Speaking the Language of Algebra (page 151)
1. 1   **2.** Isolate the variable   **3.** Remove the parentheses and combine like terms
4. Open; left   **5.** $a < x < b$ **6.** Intersection; union   **7.** Preserved; reversed
8. $x = a$ or $x = -a$; $-a \le x \le a$; $x \ge a$ or $x \le -a$

Review Exercises (page 153)

1. $x = 7$ **2.** $x = -6$ **3.** $x = -42$ **4.** $x = -45$ **5.** $x = -3$ **6.** $x = \dfrac{-1}{2}$

7. $x = \dfrac{8}{15}$ **8.** $x = 5$ **9.** $x = \dfrac{-19}{11}$ **10.** $y = \dfrac{5}{7}$ **11.** $x = 0$ **12.** $C = \dfrac{5}{9}(F - 32)$

13.   **14.**

15. $-5 \le x < 2$ **16.** $\varnothing$ **17.** $x > -1$ **18.** $x \in R$

19. $x \leq -10$

20. $x < 1$

21. $x > -5$

22. $y \geq \frac{1}{2}$

23. $x \leq -2$

24. $3 < x < 4$

25. $x \geq -5$ or $x < -7$

26. $x < 3$

27. $7 \leq x \leq 13$

28. $4 < x \leq 10$

29. $x = 5$ or $x = -2$ **30.** $x = 5$ or $x = 13$ **31.** No solution   **32.** $\dfrac{-11}{3} \leq x \leq 5$

33. No solution   **34.** $x > 7$ or $x < -4$ **35.** All real numbers   **36.** $x \leq 2$ or $x \geq 10$
37. At least 5   **38.** \$12,000 at 14%, \$17,000 at 12%   **39.** 45 and 55 mph   **40.** 15 by 45 ft

Practice Test (page 157)

1. $x = 2$ **2.** $y = 1$ **3.** $x = 4$ **4.** $a = \dfrac{-2}{5}$ **5.** $x = \dfrac{1}{9}$ **6.** $b = \dfrac{2A - ha}{h}$

7. $x \geq 2$ **8.** $y > 2$ **9.** $\dfrac{-1}{8} \leq x \leq 3$ **10.** $x = \dfrac{9}{4}$ or $x = \dfrac{5}{4}$ **11.** $x = -5$ or $x = 10$

12. $-5 < x < 6$ **13.** $x \geq 1$ or $x \leq \dfrac{-7}{3}$ **14.** $-6 < x \leq -2$ **15.** $x < 7$

16. $-3 < x < 4$ **17.** $x < -1$ or $x \geq 3$ **18.** 8 mph   **19.** 435   **20.** $1\frac{1}{2}$ in.

CHAPTER 3

Exercise Set 3.1 (page 171)
1. 186   **3.** 2   **5.** 1   **7.** 2   **9.** $-20x^8$ **11.** $-12x^7y^5$ **13.** $6x^7y^4$ **15.** $42x^4y^4$
17. $35a^3b^4c^5$ **19.** $14x^2y^8z^7$ **21.** $64y^4$ **23.** $-32x^{10}y^{15}$ **25.** $20x^7$ **27.** $9x$
29. $-50x^8y^5$ **31.** $288a^7b^{11}$ **33.** $12a^{24}b^8$ **35.** $-448x^{17}y^7z^8$ **37.** $11x^2y^6$
39. $-a^6b^{12} + 3a^{12}b^6$ **41.** $\dfrac{1}{9y^2}$ **43.** $\dfrac{125a^6}{27b^3}$ **45.** $\dfrac{a^4b^{20}}{16c^8}$ **47.** $\dfrac{16y^{12}}{81z^{20}}$ **49.** 1   **51.** x^4

53. $-3a^4$ **55.** x^6y^2 **57.** $-x^3y^3z$ **59.** $\dfrac{a^4}{7}$ **61.** $\dfrac{3a^6b^2c}{2}$ **63.** $8a^3b^3c^3$ **65.** $49a^5b^5$

Stretching the Topics
1. $-64a^5b^{12}c^5$; 262,144   **2.** $x^{3n}y^{2n+11}$ **3.** $a = -7$

Checkup 3.1 (page 173)

1. -55 **2.** 13 **3.** $42x^8y^3$ **4.** $-45a^5b^6c^3$ **5.** $121y^6$ **6.** $-75x^{18}$ **7.** $\dfrac{-27a^6}{b^9}$

8. $\dfrac{16a^6}{9c^4}$ **9.** $\dfrac{27a^5}{8}$ **10.** $-3a^4b^4c$

Exercise Set 3.2 (page 185)

1. $\dfrac{1}{16}$ **3.** $\dfrac{-1}{243}$ **5.** $\dfrac{1}{81}$ **7.** $\dfrac{-1}{25}$ **9.** $\dfrac{7}{36}$ **11.** $\dfrac{16}{9}$ **13.** $\dfrac{1}{x^6}$ **15.** $\dfrac{a^2}{b^6}$ **17.** $\dfrac{1}{xz^6}$

19. $\dfrac{ac^3}{b^4}$ **21.** $\dfrac{1}{x^6}$ **23.** $\dfrac{1}{32x^5}$ **25.** $\dfrac{9}{a^2}$ **27.** $x^{15}y^3$ **29.** $\dfrac{-x^3}{27y^6}$ **31.** $\dfrac{49y^4}{x^{10}}$ **33.** a^9b^7

35. xy^9z^3 **37.** $\dfrac{a^{12}}{125}$ **39.** $\dfrac{y^3}{x^6z^3}$ **41.** $\dfrac{y^6}{x^5}$ **43.** $\dfrac{c^6}{a^9b^{12}}$ **45.** $\dfrac{1}{x^{10}y^{12}}$ **47.** $\dfrac{1}{9x^{16}y^{10}}$

49. $\dfrac{-64c^9}{a^{18}b^{12}}$ **51.** $\dfrac{4b^{20}}{27a^{16}}$ **53.** $\dfrac{2}{b^7}$ **55.** y^{10}

Stretching the Topics

1. $\dfrac{1}{x^{3b-2a}}$ **2.** $\dfrac{-32b^{13}e^8}{a^5c^9d^{12}f^2}$ **3.** $x^{4a+6}y^{2b-5}$

Checkup 3.2 (page 187)

1. $\dfrac{1}{64}$ **2.** $\dfrac{-1}{64}$ **3.** xz^5 **4.** $\dfrac{1}{x^9}$ **5.** $\dfrac{1}{36x^2}$ **6.** $\dfrac{x^6}{y^{14}z^6}$ **7.** $\dfrac{y^{14}}{x^5}$ **8.** $\dfrac{y^4}{x^5z^3}$ **9.** $\dfrac{y^{10}}{9x^{18}}$

10. $\dfrac{1}{3x^4y^5}$

Exercise Set 3.3 (page 195)

1. Fifth-degree binomial in x **3.** Third-degree binomial in x **5.** Constant monomial
7. Second-degree trinomial in y **9.** Fourth-degree, four-term polynomial in x
11. $x^3 - 4x^2 + 3x + 7$ **13.** $x^4 + 8x^2 + 12$ **15.** $-a^3 + a^2 - a + 1$ **17.** $3x + 5$
19. $x^3 + 4x^2 - 3x + 2$ **21.** $-x^2 + 4x - 6$ **23.** $3a^2 - 8ab + 16b^2$
25. $-a^2 + 10a - 7$ **27.** $-a^2 - 7a + 6$ **29.** $-3b$ **31.** $-x^2 - y^2$
33. $-3a^2 - 18a - 35$ **35.** $-y^2 + 3$ **37.** $9x^2 - 2x + 7$ **39.** $-3x^2 - 6x - 4$
41. $7a + 23b + 3$ **43.** $3x^2 + 3x - 4$ **45.** $2m^3 - 10m^2 + 12m$
47. $-15x^3y^3 + 5x^2y^2 + 20xy$ **49.** $14a^4b^2 - 2a^2b^4$
51. $32m^4n - 8m^3n^2 + 4m^2n^3 - 12mn^4$ **53.** $7x^2y - 10xy^2$ **55.** $3x^3 - 6x^2 + 9x$
57. $-14a^3 - 6a^2 + 35a$ **59.** $-5x^3y^3 - 13x^3y^7$

Stretching the Topics

1. $9x + 1$ **2.** $-9a^3b - a^2b^2 + 28ab^3$; -20 **3.** $2y^2 + 2xy - \dfrac{2x^2}{y^2} - 4x^2$

Checkup 3.3 (page 197)

1. Fourth-degree binomial in x **2.** Second-degree trinomial in y **3.** $-x^4 - 2x^3 + x^2 + 2$
4. $-7a^2 + 8a - 9$ **5.** $5a^2 + 2ab - 15b^2$ **6.** $6x^2 + 17x - 7$ **7.** $15a - 65b$
8. $5x^3 - 15x^2 + 25x$ **9.** $13m^2n + 45mn^2$ **10.** $15a^3 - a^2 - 24a$

Exercise Set 3.4 (page 209)

1. $12y^2 - 23y + 5$ **3.** $12x^2 + 8x - 15$ **5.** $18 - 13y + 2y^2$ **7.** $28x^2 + 29xy + 6y^2$

9. $a^2 + \dfrac{2}{3}ab + \dfrac{1}{9}b^2$ **11.** $0.04x^2 - 0.09y^2$ **13.** $x^2y^2 - 5xy - 50$

15. $9x^2y^2 - 12xyz + 4z^2$ **17.** $0.36a^2 - 25b^2$ **19.** $3x^4 - x^2y^2 - 24y^4$

21. $a^4 + a^2b^2 + \dfrac{1}{4}b^4$ **23.** $x^3 - 5x^2 + 7x - 3$ **25.** $-56y^2 - y^3 + y^4$

27. $3x^2 + x - 15$ **29.** $4a^2 + 24a + 13$ **31.** $-4x^2 + 32x - 44$ **33.** $79x^2 - 70x + 18$
35. $a^3 - 11a^2 + 6a$ **37.** $a^3 - 4a^2b - 7ab^2 + 10b^3$ **39.** $x^3 - 4x^2 - 3x + 18$
41. $a^3 + 6a^2 + 12a + 8$ **43.** $x^5 - 2x^4 - 3x^2 - x + 5$ **45.** $3x^4 + 5x^3 - 10x^2 - 9x + 5$

47. $a^4 - 10a^2b^2 + 25b^4$ **49.** $x^4 - 4y^4$ **51.** $x^2 - 2xy + y^2 - 10x + 10y + 25$
53. $a^2 + 8a + 16 - 9b^2$ **55.** $16 - 32x + 24x^2 - 8x^3 + x^4$

Stretching the Topics
1. $4a^2 - 4ab + b^2 - 12ac + 6bc + 9c^2$ **2.** $8x^{6a} - 18x^{3a}y^b + 7y^{2b}$
3. $x^{-3} + x^{-2}y + 2x^{-1}y^{-2} - 4y^{-3}$

Checkup 3.4 (page 211)
1. $2x^2 + 11x - 21$ **2.** $a^2 - 20ab + 100b^2$ **3.** $9x^2 - 25y^2$ **4.** $x^3 - 8x^2 + 9x + 18$
5. $13x^2 - 2xy - 4y^2$ **6.** $2a^2 - 19a + 9$ **7.** $x^3 - 2x^2y - xy^2 + 2y^3$
8. $a^3 - 15a^2 + 75a - 125$ **9.** $x^4 - 4x^3 + 6x^2 - 5x + 2$
10. $a^2 + 2ab + b^2 - 18a - 18b + 81$

Exercise Set 3.5 (page 221)
1. $x^3 - 7x^2 + 3x + 8$ **3.** $3a^2 - 2ab + 5b^2$ **5.** $x - 2 + \dfrac{5}{3x}$ **7.** $-1 + 4z - 5xy$

9. $x^3 - 3x^2 + 2x - 1 + \dfrac{3}{x}$ **11.** $x - 8$ **13.** $4x + 3$ **15.** $x - 2 - \dfrac{2}{2x + 5}$

17. $x^2 - 4x + 9 - \dfrac{2}{x + 1}$ **19.** $a^2 + 3a - 4$ **21.** $x^2 - xy + 3y^2$ **23.** $x^2 - 4x + 2$

25. $x^2 + 2x + 4$ **27.** $x^3 + 2x^2 + 3x + 6 + \dfrac{25}{2x - 4}$ **29.** $2x^2 - x + 4 + \dfrac{x + 3}{4x^2 - x - 3}$

31. $x^3 - 3x^2 + 2x - 2 + \dfrac{1}{x + 1}$ **33.** $3x^3 - x^2 + x$ **35.** $x^3 - 2x^2 + 4x - 8 + \dfrac{32}{x + 2}$

37. $x^4 - 2x^3 + x - 2 - \dfrac{2}{x + 5}$ **39.** $x^5 - x^4 + x^3 - x^2 + x + 1 - \dfrac{1}{x - 1}$

Stretching the Topics

1. $4x^2 - 5xy + 3xy^2 - \dfrac{2xy^3}{2x + y}$ **2.** $x^{2a} + 2x^a - 3$ **3.** No; 205

Checkup 3.5 (page 223)
1. $2x^2 - xy + 3y^2$ **2.** $\dfrac{-x^2}{2} + x - 2 + \dfrac{4}{x}$ **3.** $2x + 3$ **4.** $x^2 - 4x - 5 - \dfrac{2}{x + 3}$

5. $x^2 + 2x + 1 + \dfrac{1}{2x^2 - 5x - 3}$ **6.** $a^4 + 2a^3 + 4a^2 + 8a + 16$

7. $x^2 - 2x - 4 + \dfrac{2}{x + 3}$ **8.** $2x^3 - x^2 + x - 2$

Exercise Set 3.6 (page 231)
1. $2x^2 + 12x$ **3.** $6x^2 - 35x$ or $6y^2 + 49y + 49$ **5.** $18x^2 + 18x + 9$
7. $7x^3 + 21x^2 + 14x$ **9.** $2l^2 - 8l$ **11.** $4l - 16$ or $4w + 16$ **13.** $20s + 100$

15. $5x^2 + 20x + 25$ **17.** $\dfrac{x^2 - 7x}{2}$ or $\dfrac{a^2 + 7a}{2}$ **19.** $2n^2 - 12n + 9$

21. $w^2 + 50w$ or $l^2 - 50l$ **23.** $\dfrac{x^2 + 9x}{2}$ or $\dfrac{y^2 - 9y}{2}$ **25.** $x^2 + 9x + 20$ **27.** $10w + 65$
29. $w^3 + 4w^2$

Stretching the Topics

1. $300x - 2x^2$ **2.** $18w^2 - 288w + 864$ **3.** $\dfrac{35x - x^2}{2}$

Checkup 3.6 (page 235)
1. $4w^2 - 3w$ **2.** $\dfrac{2x^2 + 5x}{2}$ **3.** $2x^2 + 12x + 36$ or $2y^2 - 12y + 36$ **4.** $2x^2 + 2x$
5. $30s + 225$

Speaking the Language of Algebra (page 238)
1. Base; exponent; a; n **2.** Add   **3.** Multiply   **4.** Factor   **5.** Numerator; denominator
6. Subtract; denominator; numerator   **7.** First; binomial   **8.** Second; trinomial
9. Difference of the squares   **10.** Square of the hypotenuse

Review Exercises (page 239)
1. $-50x^6y^5$ **2.** $49y^{10}$ **3.** $-288x^{17}$ **4.** $\dfrac{-27x^{12}}{y^6}$ **5.** $\dfrac{8a^3}{27c^3}$ **6.** $\dfrac{14a^3}{3}$ **7.** $\dfrac{1}{16}$

8. $\dfrac{-1}{8}$ **9.** $\dfrac{z^3}{x^2}$ **10.** $\dfrac{5}{x^3}$ **11.** $\dfrac{x^6z^4}{y^{10}}$ **12.** $\dfrac{y^{12}}{x^3}$ **13.** $\dfrac{y^{12}}{8x^{21}}$ **14.** $\dfrac{9y^8}{x^4}$

15. Monomial in x and y **16.** Third-degree trinomial in x **17.** $2y^4 + y^3 - y^2 + 6$
18. $-2x^2 + 3x - 5$ **19.** $5x^2 + 23$ **20.** $13a - b$ **21.** $3x^3 - 15x^2 + 18x$
22. $17m^2n + mn^2$ **23.** $6a^3 - 2a^2 + 10a$ **24.** $-2x^2y^3 + 7x^3y^5$ **25.** $3x^2 + 14x - 5$
26. $a^2 - 14ab + 49b^2$ **27.** $16x^2 - 9y^2$ **28.** $x^3 - 6x^2 + 5x + 6$ **29.** $7x^2 + 2xy - 9y^2$
30. $a^2 - 25a + 27$ **31.** $2a^3 - 3a^2b - 3ab^2 + 2b^3$ **32.** $a^3 - 6a^2 + 12a - 8$
33. $x^4 + x^3 - 6x^2 + x + 3$ **34.** $a^2 + 4ab + 4b^2 + 6a + 12b + 9$ **35.** $a^2 - 3ab + 2b^2$

36. $3x - 1 - \dfrac{3}{x + 6}$ **37.** $x^2 - 4x + 16$ **38.** $x^2 + 3xy - 10y^2$ **39.** $x^2 - 3x + 9$

40. $2x^2 + 3x + 7 + \dfrac{12}{x - 2}$ **41.** $3x^3 - 4x^2 + 2x - 2 + \dfrac{4}{x + 1}$ **42.** $\dfrac{5x^2}{2}$ **43.** $5x^2 + 5x$

44. $x^2 + 2x - 15$ **45.** $-6x^2 + 2$

Practice Test (page 243)
1. -96 **2.** $\dfrac{-1}{32}$ **3.** $8a^{12}b^3$ **4.** $-2x^7y^4$ **5.** $-125x^9y^3$ **6.** $-a^{12}b^6c^9$ **7.** $\dfrac{81x^6}{y^4z^6}$

8. $-2ab^3$ **9.** $\dfrac{x^9z^3}{8}$ **10.** $\dfrac{8x^2}{y}$ **11.** $\dfrac{a^3c^6}{b^9}$ **12.** $-x^9y^3z^3$ **13.** $\dfrac{-2y^4}{x^6}$

14. $-6x^2 + 69x - 6$ **15.** $12a^2 + 41a - 36$ **16.** $12x^2 - x - 35$
17. $9c^2 - 42cd + 49d^2$ **18.** $81y^2 - 4z^2$ **19.** $3x^3 - 17x^2 + 9x + 5$
20. $a^3 - 15a^2 + 75a - 125$ **21.** $2x^2 - 5x + 14$ **22.** $-4x^2y^2 - xy + 2$

23. $7x + 24 + \dfrac{70}{x - 3}$ **24.** $y^2 - 3y + 9$ **25.** $2x^2 + 16x + 64$ or $2y^2 - 16y + 64$

Sharpening Your Skills after Chapters 1–3 (page 245)
1. $\dfrac{-11}{2} < 4.5$ **2.** $8.2 > 0$ **3.** -9 **4.** 7 **5.** Commutative property of addition   **6.** 0

7. 18 **8.** $-x + 9y$ **9.** $-2x^2 - 2$ **10.** $x = \dfrac{1}{2}$ **11.** $y = 8$ **12.** $x < -7$

13. $-1 < y \le 5$ **14.** $x = -4, x = 22$ **15.** $\dfrac{-4}{5} \le a \le 2$ **16.** $\varnothing$

17. $-3 \le x \le -2$ **18.** $x \le 3$ or $x > 6$ **19.** $500 - 15n$ **20.** $\dfrac{1}{2}$ hr

CHAPTER 4

Exercise Set 4.1 (page 253)
1. $10(5x^2 - xy + 4y^2)$ **3.** $2b(b^2 - 3b - 2)$ **5.** $-y(y^2 - y + 2)$
7. $3x(x^2 - 2xy + 4y^2)$ **9.** $-a^2b(a^2 - 8ab + 5b^2)$ **11.** $5x^2y^2(x - 16y)$
13. $3xy(3x^2 - 4xy + 8y^2)$ **15.** $8a^2(9a^2 - 5a + 1)$ **17.** $4x^2y^2(3x^2 + 2xy + y^2)$
19. $9xyz(x^3 + 2x^2y - 3xy^2 - 5y^3)$ **21.** $(y - 1)(y - 3)$ **23.** $(x + 3)(2x - 1)$
25. $(3x - y)(2a - b)$ **27.** $(x - y)(x - y - 3)$ **29.** $(x - y)(a + b - 2c)$
31. $2x(2a + b)$ **33.** $(x - 2)[5(x - 2)^2 - 4(x - 2) + 1]$
35. $2(x + 2y)[2(x + 2y)^2 - 3(x + 2y) - 1]$ **37.** $(6 + c)(a + b)$ **39.** $(x - 7)(x^2 + 6)$
41. $(2x + 3y)(7a - 5)$ **43.** $(7a - 4b)(10a + 3)$ **45.** $3(2x + y)(a^2 + b^2)$
47. $(x + 1)(x^2 + 1)$ **49.** $(b - c)(9a - 7)$ **51.** $(2a - 3b)(x^2 + 1)$

Stretching the Topics
1. $x^{4a}(x^a + 4)$ **2.** $x^2(x^{3a} + 3x^{2a} - 5)$ **3.** $x^{-3}(1 + 5x + x^2)$

Checkup 4.1 (page 255)
1. $7(5x^2 - 2x + 1)$ **2.** $7b(b^2 - 2b + 4)$ **3.** $5y(x^2 - 5y^2)$ **4.** $-a^4b^2(a^2 - 3ab^2 - 6b^4)$
5. $9x^3yz(x^2y^2 - 3xy - 2)$ **6.** $(x - 9)(2x - 3)$ **7.** $(a + b)(a + b - 12)$
8. $(x - y)^2(6 - x + y)$ **9.** $(a + b)(2x - 3y)$ **10.** $(x - 7)(x^2 + 1)$

Exercise Set 4.2 (page 263)
1. $(3y + 1)(3y - 1)$ **3.** $\dfrac{1}{4}\left(a + \dfrac{5}{3}\right)\left(a - \dfrac{5}{3}\right)$ **5.** $(mn + 15)(mn - 15)$
7. $(5xy + 4z)(5xy - 4z)$ **9.** $3(x + 0.5)(x - 0.5)$ **11.** $6m(m + 5)(m - 5)$
13. $2y^2(x + 6y)(x - 6y)$ **15.** $(5x^2 + 11)(5x^2 - 11)$ **17.** $(m^2 + 4)(m + 2)(m - 2)$
19. $3(m^2 + 9)(m + 3)(m - 3)$ **21.** $(x + y + 7)(x + y - 7)$
23. $(3a + 2b + 2c)(3a + 2b - 2c)$ **25.** $2xy(x + 2y)(x - 2y)$
27. $(16 + x + 2y)(16 - x - 2y)$ **29.** $(x + y + 1)(x - y - 5)$
31. $(x + 9)(x + 3)(x - 3)$ **33.** $(x - 2)(x + 5)(x - 5)$ **35.** $(x - y)(x^2 + xy + y^2)$
37. $(a + 2)(a^2 - 2a + 4)$ **39.** $(4 - m)(16 + 4m + m^2)$ **41.** $(xy + 1)(x^2y^2 - xy + 1)$
43. $(5x - 3y)(25x^2 + 15xy + 9y^2)$ **45.** $2x(x + 2)(x^2 - 2x + 4)$
47. $(ab - 4c)(a^2b^2 + 4abc + 16c^2)$ **49.** $4x^2y^2(x + 2y)(x^2 - 2xy + 4y^2)$
51. $7xy^2(x - 2y)(x^2 + 2xy + 4y^2)$ **53.** $(x - y + 5)(x^2 - 2xy + y^2 - 5x + 5y + 25)$
55. $(5 - x)(7 - 4x + x^2)$ **57.** $3(x + y)(x - y)(x^2 + xy + y^2)(x^2 - xy + y^2)$
59. $(x - 1)(x + 1)(x^2 - x + 1)$

Stretching the Topics
1. $4(x^a + 3y^a)(x^a - 3y^a)$ **2.** $(3x^2 + 1)(3x^2 - 1)(2x^2 + 1)(2x^2 - 1)$
3. $x^a(x^a - 1)(x^{2a} + x^a + 1)$

Checkup 4.2 (page 265)
1. $(5a + 2)(5a - 2)$ **2.** $(10xy + 11z)(10xy - 11z)$ **3.** $3(x + 12)(x - 12)$
4. $5m(2m + 3n)(2m - 3n)$ **5.** $(x^2 + 16)(x + 4)(x - 4)$ **6.** $(x + 4y)(x^2 - 4xy + 16y^2)$
7. $3(a - 2b)(a^2 + 2ab + 4b^2)$ **8.** $(x - 2 + 3y)(x - 2 - 3y)$
9. $(x - y + 6)(x^2 - 2xy + y^2 - 6x + 6y + 36)$
10. $(a + b)(a - b)(a^2 + ab + b^2)(a^2 - ab + b^2)$

Exercise Set 4.3 (page 279)
1. $(x - 12)^2$ **3.** $(y - 6)(y + 5)$ **5.** $(b + 11)(b - 3)$ **7.** $(3a - 4b)^2$ **9.** $(x - 5y)^2$
11. $(13 - m)(3 - m)$ **13.** $(xy + 5)(xy - 3)$ **15.** $3(x - 3)(x + 2)$ **17.** $5a(a - 4)^2$
19. $2y(5 - y)(2 + y)$ **21.** $-10a(b^2 - 3)^2$ **23.** $xy(x - 10y)(x - 8y)$
25. $abc(a - 9b)(a + 7b)$ **27.** $(a - 5)^2$ **29.** $(x + y - 4)(x + y - 3)$
31. $(x + 13)(x + 1)$ **33.** $(2y + 1)(y + 3)$ **35.** $(5x - 2)(2x - 3)$
37. $(3a + 8b)(2a - 7b)$ **39.** $5(3c - d)(2c - 9d)$ **41.** $-x^2(3x - 7)(3x - 4)$
43. $(4m + 11n)(3m - 5n)$ **45.** $2xy(4x - 5)(3x + 7)$ **47.** $(5a - 5b + 2)(a - b + 1)$
49. $(5x + 5y - 2)^2$ **51.** $(4 - 13x + 26y)(2 - 3x + 6y)$ **53.** $(x^2 + 8)(x + 3)(x - 3)$
55. $3(x^2 - 5)(x + 2)(x - 2)$ **57.** $3x^2(x^2 - 5y^2)^2$ **59.** $2(x^5 + 11)(x^5 - 3)$
61. $A = \dfrac{1}{2}h(a + b)$; 28 sq in.   **63.** $s = 2(5 + 8t)(3 - t)$; 36.16 ft
65. $C = (2x - 3)(x + 2)$; \$5044

Stretching the Topics
1. $(2x^a + 11)^2$ **2.** $[3(x + y)^m + 5][2(x + y)^m + 3]$ **3.** $(x^{a+3} + 6y^{a-2})^2$

Checkup 4.3 (page 281)
1. $(x - 6y)(x - 3y)$ **2.** $2(y + 9)(y - 6)$ **3.** $(5x - 9y)(2x + 3y)$ **4.** $(7a - 3b)^2$
5. $-6y(y - 2)(y + 1)$ **6.** $(x^2 + 7)(x + 3)(x - 3)$ **7.** $(2x + 1)(2x - 1)(x + 3)(x - 3)$
8. $(x + y + 10)(x + y - 2)$ **9.** $(2x + 2y - 7)(x + y - 3)$
10. $S = 16t(50 - t)$; 10,000 ft

Exercise Set 4.4 (page 285)
1. $(x - y)(4x - 4y + 1)(x - y - 1)$ **3.** $2m(m + 5)(m - 5)$ **5.** $6a(a - 8)(a + 3)$
7. $abc(a - 9b)(a + 6b)$ **9.** $(3x^3 - 1)(x - 2)(x^2 + 2x + 4)$
11. $(7a - 7b - 3)(a - b + 1)$ **13.** $(x - 5)(x^2 + 1)$ **15.** $3(x - 10)(x + 9)$
17. $-2x(x - 4)(x^2 + 4x + 16)$ **19.** $8x^2(3x^2 - 5xy^3 + 7y^4)$ **21.** $(9x - 2y)^2$

23. $(11x + 11y - 2)(x + y - 3)$ **25.** $4x^2(x^2 - 5y^2)(x + y)(x - y)$ **27.** $(x - 15y)^2$
29. $3(x^2 - 2x - 90)$ **31.** $(xy + 6)(x^2y^2 - 6xy + 36)$ **33.** Prime
35. $4(x - 2y)(a + b)(a - b)$ **37.** $(x^2 + 1)(3x + 4)(3x - 4)$
39. $3(2x - 5)(4x^2 + 10x + 25)$ **41.** $4ab(3a + 5b)^2$ **43.** $(0.5xy + 1.1z)(0.5xy - 1.1z)$
45. $(2a^3b^2 + 7)^2$ **47.** $-2(3x + 4)(3x - 4)$ **49.** $4x^3yz^2(11z + 8y)(z - 3y)$

Stretching the Topics
1. $2x^a(x^{2a-4} + 3x^{a-2} + 9)$ **2.** $ab(a^x - 9b^x)(a^{2x} + 9a^xb^x + 81b^{2x})$
3. $(-x - 7y)(37x^2 + 14xy + 13y^2)$

Checkup 4.4 (page 287)
1. $(5x - 9y)^2$ **2.** $3y^2(x + 8y)(x - 8y)$ **3.** $4x^2y^2(x^3 - 3)$ **4.** $(3x - 2y)(5a - 3)$
5. $(3x + 8)^2$ **6.** $(x + 6)(x - 6)(x^2 + 1)$ **7.** $(7a - 7b - 3)(a - b + 1)$
8. $(x + 2y)(x - 2y)(x^2 - 2xy + 4y^2)(x^2 + 2xy + 4y^2)$ **9.** $(6x - 7y)(4x - 3y)$
10. $(x + 1)(x - 1)^2(x^2 + x + 1)$

Speaking the Language of Algebra (page 289)
1. Product   **2.** Common factor   **3.** Grouping   **4.** Trial and error   **5.** Prime

Review Exercises (page 291)
1. $5a(a - 2)(a - 1)$ **2.** $-x^3y^2(x^2 - 5xy + 2y^2)$ **3.** $3x^2yz(xy - 1)^2$ **4.** $(x - 2)(3x + 4)$
5. $(a - 2b)(a - 2b - 3)$ **6.** $(b + 2)(a - 2)$ **7.** $(a + b)(5 + c)$ **8.** $(x + y)(3a - 2b)$
9. $(x - 9)(x^2 + 2)$ **10.** $2(y - z)(3x - 5)$ **11.** $(3a + 4)(3a - 4)$
12. $(2ab + 3c)(2ab - 3c)$ **13.** $3(x + 4)(x - 4)$ **14.** $6m(2m + 5n)(2m - 5n)$
15. $(x^2 + 9)(x + 3)(x - 3)$ **16.** $(x + 3y)(x^2 - 3xy + 9y^2)$
17. $(3a - 2b)(9a^2 + 6ab + 4b^2)$ **18.** $3z(xy - 2)(x^2y^2 + 2xy + 4)$
19. $(x + y + 3)(x^2 + 2xy + y^2 - 3x - 3y + 9)$
20. $(a + 4)(a - 1)(a^2 - 4a + 16)(a^2 + a + 1)$ **21.** $(x - 7y)(x - 5y)$
22. $3(a - 8)(a + 7)$ **23.** $-2y(y + 5)(y - 3)$ **24.** $(x^2 + 1)(x + 4)(x - 4)$
25. $(x + y + 9)(x + y - 7)$ **26.** $(9x - 2)(2x + 3)$ **27.** $(3x - 5y)^2$
28. $3x^2(6x - 7)(2x + 3)$ **29.** $(3x - 7)(x - 7)$ **30.** $(2a + 1)(2a - 1)(a + 5)(a - 5)$

Practice Test (page 293)
1. $4w(2w^3 - w + 3)$ **2.** $-5ab(ab^2 - 3a + 5b)$ **3.** $(3x - 4)(7 - 6x)$
4. $(3y - 2z)(5 + 3y - 2z)$ **5.** $(a + b)(2 + c)$ **6.** $(5 - x)(a - 1)$
7. $(11x + 2y)(11x - 2y)$ **8.** $4(x + 9)(x - 9)$ **9.** $(a^2 + 1)(a + 1)(a - 1)$
10. $(2a - 5b + 10)(2a - 5b - 10)$ **11.** $(2x + 1)(x + 5)(x - 5)$
12. $(5x + y)(25x^2 - 5xy + y^2)$ **13.** $5(a - 2)(a^2 + 2a + 4)$ **14.** $(5x - 3)(6x + 1)$
15. $7(x - 3y)(x + 2y)$ **16.** $(5a - 3b)^2$ **17.** $x^3(2x + 3)(x - 4)$ **18.** $(10x + 9)(4x - 1)$
19. $(3x + 3y - 4)(2x + 2y + 1)$ **20.** $(x + 3)(x - 3)(x + 1)(x - 1)$

Sharpening Your Skills after Chapters 1–4 (page 295)
1. $-9, -3.2, \frac{1}{3}, 0, \frac{5}{2}, 4, 5.1313\ldots$ **2.** $-3x^4 + 3x^3 + 4x^2 - 5x$ **3.** -20 **4.** -21

5. $\pi - 2$ **6.** -20 **7.** $\frac{1}{16}$ **8.** 1 **9.** -1 **10.** $-15n + 19$ **11.** $14a - 50b$

12. $8x^2 + 22xy - 21y^2$ **13.** $x^3 + 6x^2 + 12x + 8$ **14.** $a^3 - 6a^2 + 12a - 8$

15. $x^2 - 5x + 25$ **16.** $8x^9y^5$ **17.** $9a^2b^4c^6$ **18.** x^2y^5 **19.** $\dfrac{x^{12}}{8y^{18}z^6}$ **20.** $\dfrac{7x^{11}}{y^5}$

21. $x = -7$ **22.** $x \leq \frac{4}{3}$ **23.** $m \leq 11$ or $m \geq 15$

24. $x > -6$

25. $12s + 36$

CHAPTER 5

Exercise Set 5.1 (page 309)

1. $x = 9, x = -11$　**3.** $x = \dfrac{3}{5}$　**5.** $x = \dfrac{1}{8}, x = \dfrac{-5}{3}$　**7.** $x = \dfrac{9}{2}, x = -7$

9. $x = \dfrac{1}{3}, x = \dfrac{-1}{3}$　**11.** $x = 0, x = 2, x = \dfrac{7}{3}$　**13.** $x = 6, x = 5$　**15.** $x = -10, x = 2$

17. $x = -5$　**19.** $m = 4, m = -11$　**21.** $y = -1, y = \dfrac{-2}{5}$　**23.** $a = 5, a = \dfrac{2}{3}$

25. $a = 8$　**27.** $x = 3, x = -3$　**29.** $y = 0, y = 13$　**31.** $m = \dfrac{3}{4}, m = \dfrac{-5}{2}$

33. $y = 3, y = \dfrac{7}{2}$　**35.** $m = \dfrac{8}{3}, m = \dfrac{-2}{3}$　**37.** $x = 5, x = -3$　**39.** $x = 0, x = -9$

41. $x = \dfrac{23}{14}$　**43.** $x = \dfrac{9}{2}, x = -1$　**45.** $x = 0, x = 2$　**47.** $x = \dfrac{-10}{3}, x = -1$

49. $x = 4, x = -2, x = \dfrac{3}{2}$　**51.** $x = 0, x = \dfrac{8}{5}, x = \dfrac{2}{5}$

53. $x = 1, x = -1, x = 4, x = -4$　**55.** $x = 3, x = -3, x = 1, x = -1$

57. $x = 0, x = 2, x = -2$　**59.** $x = 0, x = 8, x = -8$　**61.** $x = -3, x = 4, x = -4$

63. $x = \dfrac{1}{2}, x = \dfrac{3}{2}, x = \dfrac{-3}{2}$　**65.** $x^2 + 11x + 18 = 0$　**67.** $x^3 - 9x = 0$

69. $3x^2 - 17x + 10 = 0$

Stretching the Topics

1. $x = 4, x = \dfrac{-17}{2}$　**2.** $x = 2, x = \dfrac{-6}{23}$　**3.** $x = 3$

Checkup 5.1 (page 311)

1. $x = \dfrac{10}{3}, x = \dfrac{-1}{5}$　**2.** $x = 9, x = -7$　**3.** $y = \dfrac{7}{3}, y = \dfrac{-1}{2}$　**4.** $a = \dfrac{5}{2}, a = \dfrac{-2}{3}$

5. $x = 3, x = \dfrac{1}{3}$　**6.** $x = \dfrac{5}{4}, x = -3$　**7.** $x = 0, x = \dfrac{2}{13}, x = \dfrac{3}{2}$

8. $x = 6, x = -6, x = 2, x = -2$　**9.** $x = 0, x = 1, x = 64$　**10.** $x^2 - 3x - 4 = 0$

Exercise Set 5.2 (page 323)

1. $3\sqrt{6}$　**3.** $11\sqrt{2}$　**5.** $6\sqrt{5}$　**7.** $13\sqrt{3}$　**9.** 15　**11.** $\dfrac{3}{4}$　**13.** $\dfrac{11}{2}$　**15.** $\dfrac{\sqrt{7}}{5}$

17. $\dfrac{2\sqrt{2}}{3}$　**19.** $\dfrac{7\sqrt{2}}{17}$　**21.** $x = 2\sqrt{2}, x = -2\sqrt{2}$; irrational　**23.** No real number solutions

25. $x = 7, x = -7$; rational　**27.** $y = \dfrac{1}{3}, y = \dfrac{-1}{3}$; rational　**29.** $x = \dfrac{7}{15}, x = \dfrac{-7}{15}$; rational

31. $a = \dfrac{1}{4}, a = \dfrac{-1}{4}$; rational　**33.** No real number solutions

35. $x = \sqrt{10}, x = -\sqrt{10}$; irrational　**37.** $x = \dfrac{10}{3}, x = \dfrac{-10}{3}$; rational

39. $x = \sqrt{15}, x = -\sqrt{15}$; irrational　**41.** $a = \pm\sqrt{\dfrac{x+y}{3}}$　**43.** $s = \pm\sqrt{\dfrac{A}{6}}$

45. $t = \pm\sqrt{\dfrac{2S}{g}}$　**47.** $r = \pm\sqrt{\dfrac{V}{\pi h}}$

Stretching the Topics

1. $x = \dfrac{\sqrt{5}}{2}, x = \dfrac{-\sqrt{5}}{2}, x = \sqrt{3}, x = -\sqrt{3}$　**2.** $x = 2, x = -2$

3. $x = a, x = -a, x = 3a, x = -3a$

Checkup 5.2 (page 325)

1. $2\sqrt{3}$　　**2.** $5\sqrt{7}$　　**3.** $\dfrac{5}{9}$　　**4.** $\dfrac{5\sqrt{5}}{12}$　　**5.** $\dfrac{4\sqrt{2}}{5}$　　**6.** $x = 2\sqrt{6},\ x = -2\sqrt{6}$; irrational

7. No real number solutions　　**8.** $x = \dfrac{9}{5},\ x = \dfrac{-9}{5}$; rational

9. $x = \dfrac{2\sqrt{5}}{3},\ x = \dfrac{-2\sqrt{5}}{3}$, irrational　　**10.** $b = \pm\sqrt{c^2 - a^2}$

Exercise Set 5.3 (page 339)

1. $x = -5,\ x = 3$　　**3.** $x = 0,\ x = -4$　　**5.** $x = \dfrac{-1 + \sqrt{37}}{6},\ x = \dfrac{-1 - \sqrt{37}}{6}$

7. $x = \dfrac{-9 + \sqrt{17}}{8},\ x = \dfrac{-9 - \sqrt{17}}{8}$　　**9.** No real number solutions　　**11.** $x = 3,\ x = 1$

13. $x = 0,\ x = 8$　　**15.** $x = \dfrac{-1 + 3\sqrt{5}}{2},\ x = \dfrac{-1 - 3\sqrt{5}}{2}$

17. $x = \dfrac{-1 + \sqrt{17}}{4},\ x = \dfrac{-1 - \sqrt{17}}{4}$　　**19.** $x = \dfrac{1 + \sqrt{3}}{5},\ x = \dfrac{1 - \sqrt{3}}{5}$

21. $x = 4,\ x = 5$　　**23.** $x = 3,\ x = \dfrac{-1}{2}$　　**25.** No real number solutions

27. $x = \dfrac{-3 + \sqrt{77}}{2},\ x = \dfrac{-3 - \sqrt{77}}{2}$　　**29.** $x = \dfrac{-4 + \sqrt{10}}{6},\ x = \dfrac{-4 - \sqrt{10}}{6}$

31. $x = -3$　　**33.** $x = \sqrt{2},\ x = -\sqrt{2}$　　**35.** $x = 1 + \sqrt{5},\ x = 1 - \sqrt{5}$

37. $x = \dfrac{-7 + \sqrt{37}}{6},\ x = \dfrac{-7 - \sqrt{37}}{6}$　　**39.** No real number solutions

41. $x = 2 + \sqrt{2},\ x = 2 - \sqrt{2}$　　**43.** $x = \dfrac{5}{4},\ x = -1$　　**45.** $x = \dfrac{3 + \sqrt{2}}{2},\ x = \dfrac{3 - \sqrt{2}}{2}$

47. $x = \dfrac{-c + \sqrt{c^2 - 4aM}}{2a}\quad x = \dfrac{-c - \sqrt{c^2 - 4aM}}{2a}$　　**49.** $r = \dfrac{-\pi h \pm \sqrt{\pi^2 h^2 + A\pi}}{\pi}$

51. One rational solution　　**53.** Two different rational solutions
55. Two different irrational solutions　　**57.** Two different rational solutions
59. Two different rational solutions

Stretching the Topics
1. $x = 2\sqrt{3},\ x = -4\sqrt{3}$　　**2.** $x = -3\sqrt{5}$　　**3.** Two different rational solutions

Checkup 5.3 (page 341)
1. $y = 1 + 2\sqrt{3},\ y = 1 - 2\sqrt{3}$　　**2.** No real number solutions

3. $x = 1 + \sqrt{5},\ x = 1 - \sqrt{5}$　　**4.** $x = \dfrac{-3 + \sqrt{13}}{2},\ x = \dfrac{-3 - \sqrt{13}}{2}$

5. $a = \dfrac{4 + \sqrt{6}}{5},\ a = \dfrac{4 - \sqrt{6}}{5}$　　**6.** $x = \dfrac{-1 + \sqrt{7}}{2},\ x = \dfrac{-1 - \sqrt{7}}{2}$

7. $r = \dfrac{-\pi h \pm \sqrt{\pi^2 h^2 + 2\pi A}}{2\pi}$　　**8.** Two different rational solutions

Exercise Set 5.4 (page 349)

1. $-7 < x < 7$　　**3.** $x < \dfrac{-5}{2}$ or $x > \dfrac{5}{2}$　　**5.** $x \le -3$ or $x \ge 0$　　**7.** $0 \le x \le \dfrac{1}{2}$

9. $-2 < x < -1$　　**11.** $-4 \le x \le 1$　　**13.** $x < -5$ or $x > 1$　　**15.** $3 \le x \le 7$

17. $-5 < x < 5$　　**19.** $x < -6$ or $x > \dfrac{1}{3}$　　**21.** $\dfrac{-1}{2} \le x \le \dfrac{8}{3}$　　**23.** $x \in R$

25. $x \le -5$ or $x \ge \dfrac{1}{3}$　　**27.** $-3 < x < \dfrac{2}{7}$

Stretching the Topics
1. $x \le -\sqrt{2}$ or $x \ge \sqrt{2}$　　**2.** $x < -2$ or $0 < x < 2$　　**3.** $-3 < y < 1$ or $y > 3$

Checkup 5.4 (page 351)

1. $x \leq -5$ or $x \geq -3$ **2.** $0 < x < 4$ **3.** $x < -12$ or $x > 3$ **4.** $\frac{1}{2} < x < 9$

5. $x \leq \frac{-3}{4}$ or $x \geq 7$ **6.** $3 < x < 6$

Exercise Set 5.5 (page 357)

1. 8 by 11 ft   **3.** 150 by 200 ft   **5.** 38 by 50 ft; 300 sq. ft   **7.** 5 by 5 cm and 10 by 10 cm
9. 12 by 18 ft   **11.** 12 by 12 by 4 in.; 20 by 20 in.   **13.** 10 rods
15. 11 and 12 or -3 and -2 **17.** 8 cm and 15 cm   **19.** 14 ft   **21.** 25 ft   **23.** 10
25. 35   **27.** 3 sec

Stretching the Topics

1. 18 by 24 ft.   **2.** 6, 7, 8 or -8, -7, -6

Checkup 5.5 (page 361)

1. 12 by 16 ft   **2.** $\frac{1}{2}$ ft   **3.** 17 cm

Speaking the Language of Algebra (page 364)

1. Quadratic   **2.** At least one of the factors must be zero   **3.** $x - r$; factor

4. $\frac{-b \pm \sqrt{b^2 - 4ac}}{2a}$ **5.** Discriminant; no real   **6.** Positive; negative

Review Exercises (page 365)

1. $x = -7, x = \frac{5}{3}$ **2.** $x = 8, x = -3$ **3.** $x = -7, x = \frac{5}{2}$ **4.** $x = -6$

5. $x = 3, x = -3$ **6.** $x = 0, x = 4$ **7.** $a = 4, a = \frac{2}{3}$ **8.** $a = 3, a = \frac{1}{5}$

9. $x = 4, x = -4$ **10.** $y = 0, y = \frac{7}{5}$ **11.** $x = 7, x = \frac{4}{3}$ **12.** $x = 6, x = \frac{1}{5}$.

13. $x = 0, x = -3, x = \frac{5}{12}$ **14.** $x = 2, x = -2, x = 5, x = -5$

15. $x = 0, x = 7, x = -8$ **16.** $x^2 - 5x - 24 = 0$ **17.** $5\sqrt{3}$ **18.** $9\sqrt{2}$ **19.** $\frac{\sqrt{5}}{3}$

20. $\frac{3\sqrt{2}}{7}$ **21.** $x = 4\sqrt{2}, x = -4\sqrt{2}$; irrational   **22.** No real number solutions

23. $x = \frac{5}{3}, x = \frac{-5}{3}$; rational   **24.** $x = \frac{1}{4}, x = \frac{-1}{4}$; rational   **25.** $x = 8, x = -6$

26. $x = \frac{3 + \sqrt{33}}{2}, x = \frac{3 - \sqrt{33}}{2}$ **27.** $x = \frac{5}{2}$ **28.** $x = -2 + \sqrt{5}, x = -2 - \sqrt{5}$

29. $x = \frac{1 + \sqrt{37}}{18}, x = \frac{1 - \sqrt{37}}{18}$ **30.** $y = -1, y = \frac{5}{3}$

31. $x = -1 + \sqrt{7}, x = -1 - \sqrt{7}$ **32.** $x = \frac{3 + \sqrt{41}}{8}, x = \frac{3 - \sqrt{41}}{8}$

33. $n = \frac{3 \pm \sqrt{9 + 8d}}{2}$ **34.** $t = \frac{-g \pm \sqrt{g^2 + 64S}}{32}$ **35.** Two different irrational solutions

36. Two different rational solutions   **37.** No real number solutions   **38.** One rational solution

39. $-12 < x < 12$ **40.** $x \leq 0$ or $x \geq 3$ **41.** $\frac{-1}{2} \leq x \leq 5$ **42.** $-2 < x < 0$ or $x > 2$

43. 60 by 84 in.   **44.** 6 by 6 cm, 12 by 12 cm   **45.** 3 in.   **46.** 7 ft

Practice Test (page 367)

1. $x = 4, x = -9$ **2.** $x = \frac{5}{3}$ **3.** $x = 0, x = 3$ **4.** $x = 0, x = 2, x = \frac{-3}{5}$

5. $x = 2, x = -7$ **6.** $x = 2, x = -2, x = 4, x = -4$ **7.** $x = \dfrac{1}{2}, x = 1, x = -1$

8. $x^2 - 16 = 0$ **9.** $a = \dfrac{2}{9}, a = \dfrac{-2}{9}$ **10.** $x = \dfrac{2\sqrt{5}}{3}, x = \dfrac{-2\sqrt{5}}{3}$

11. $x = \dfrac{-7 + \sqrt{37}}{2}, x = \dfrac{-7 - \sqrt{37}}{2}$ **12.** $x = \dfrac{1 + \sqrt{7}}{3}, x = \dfrac{1 - \sqrt{7}}{3}$

13. No real number solutions   **14.** $y = \dfrac{1 \pm \sqrt{1 + 12aK}}{6a}$ **15.** One rational solution

16. $0 \le x \le \dfrac{7}{2}$ **17.** $x < \dfrac{-4}{5}$ or $x > 2$ **18.** 12 in., 13 in.

Sharpening Your Skills after Chapters 1–5 (page 369)

1. $x - 2 < 0$ **2.** $\dfrac{-16}{3} > -\sqrt{36}$ **3.** Undefined   **4.** 2   **5.** $\dfrac{-1}{16}$ **6.** 1

7. $-2x - 2y$ **8.** $-63x^6y^4z^6$ **9.** $\dfrac{9x^{10}}{25z^6}$ **10.** $\dfrac{y^9}{x^{17}}$ **11.** $-21x^6y^6 - 7x^5y^5$

12. $a^3 - 4a^2b - 31ab^2 + 70b^3$ **13.** $x^2 + 2xy + y^2 - 12x - 12y + 36$
14. $2a^2 - 4ab + 3b^2$ **15.** $x^2 + 2x - 5$ **16.** $(a + 9b)(a + 3b)$
17. $5(x^2 + 4)(x + 2)(x - 2)$ **18.** $(3x - 4)(9x^2 + 12x + 16)$ **19.** $(x - 5)(x^2 + 9)$
20. $-9x^2(2x - 3)(x + 1)$ **21.** $(x - 9)^2$ **22.** \$6900   **23.** $x = 1$ **24.** $x < 3$
25. No solution   **26.** $-1 < x \le 3$ **27.** $\dfrac{-10}{3} < x < 4$ **28.** $a = \dfrac{by}{b - x}$ **29.** At least 10
30. $(s + 3)(s - 2) = s^2 + s - 6$

CHAPTER 6

Exercise Set 6.1 (page 381)

1. $a \neq 0, b \neq 0; \dfrac{-2a^2}{b^3}$ **3.** $a \neq 0, b \neq 0, c \neq 0; \dfrac{c}{3a^2b^2}$ **5.** $x \neq 0, y \neq 0, z \neq 0; \dfrac{1}{x^2y^2}$

7. $x \neq 0; 2(2y - 1)$ **9.** $y \neq 0, y \neq \dfrac{4}{5}; \dfrac{1}{5y - 4}$ **11.** $a \neq 0, b \neq 0; \dfrac{-(4a + 5c)}{6}$

13. $x \neq -y; \dfrac{1}{2}$ **15.** $x \neq -y; \dfrac{x - y}{2(x + y)}$ **17.** $x \neq -9; \dfrac{x - 9}{3}$ **19.** $x \neq 4, x \neq -4; \dfrac{x + 5}{x - 4}$

21. $x \neq -2, x \neq -5; \dfrac{x + 4}{x + 5}$ **23.** $y \neq 7, y \neq 3; \dfrac{5(y + 3)}{y - 7}$ **25.** $b \neq \dfrac{1}{2}; \dfrac{1}{2b - 1}$

27. $x \neq \dfrac{-2}{3}, x \neq \dfrac{7}{2}; \dfrac{x}{2x - 7}$ **29.** $a \neq 0, a \neq \dfrac{1}{2}, a \neq 5; \dfrac{a + 7}{a(2a - 1)}$

31. $x \neq -3, x \neq \dfrac{5}{4}; \dfrac{x^2 - 3x + 9}{4x - 5}$ **33.** $\dfrac{3}{a - 5}$ **35.** $\dfrac{(2a + 1)(2a - 1)}{5}$ **37.** $\dfrac{1}{x + 1}$

39. $\dfrac{x + 5y}{x + y}$ **41.** $\dfrac{-1}{b - 2a}$ **43.** $\dfrac{-(x + 5)}{4 - 3x}$ or $\dfrac{x + 5}{3x - 4}$ **45.** $\dfrac{-(x + y)}{4y - 5x}$ or $\dfrac{x + y}{5x - 4y}$

47. $\dfrac{a - b}{3(a + 5)}$ **49.** $2x - 3$ **51.** $\dfrac{4(3x - 2)}{(x - 2)^3}$ **53.** $(a + 5)(a - 2)$

Stretching the Topics

1. $\dfrac{-6(x - 2)}{x^2(x + 6)}$ **2.** $x^n + 2$ **3.** $x \neq 0, x \neq 5y, x \neq -y$

Checkup 6.1 (page 383)

1. $x \neq 0, y \neq 0, z \neq 0; \dfrac{-2x}{z^2}$ **2.** $y \neq 0, y \neq \dfrac{3}{2}; \dfrac{1}{2y - 3}$ **3.** $x \neq y; \dfrac{x + y}{2(x - y)}$

4. $a \neq 0, b \neq 0, \dfrac{-(2a + 3)}{4}$ **5.** $x \neq 9, x \neq -9; \dfrac{x + 6}{x + 9}$ **6.** $x \neq 6, x \neq -6; \dfrac{3x}{x + 6}$

7. $\dfrac{9x + y}{x + y}$ **8.** $\dfrac{-1}{2b - 3a}$ **9.** $a + 5$ **10.** 8

Exercise Set 6.2 (page 393)

1. $\dfrac{6}{x}$ **3.** $\dfrac{a^2 + b^2}{a + b}$ **5.** $\dfrac{2(x - 5)}{3(x + 5)}$ **7.** $3(x + 9y)$ **9.** $\dfrac{5y(2x + 5)}{8z^2(2x - 5)}$ **11.** 1

13. $\dfrac{x + 5y}{5}$ **15.** $\dfrac{2(x - 3)}{x - 2}$ **17.** $\dfrac{-3}{2(3x - 8)}$ **19.** $\dfrac{-(x + 3)}{2x - 1}$ **21.** $\dfrac{-1}{x(1 - x)}$

23. $\dfrac{x - 3}{x + 4}$ **25.** $3x^3$ **27.** $\dfrac{y^2}{9(x + 9)}$ **29.** $\dfrac{x + 5}{49}$ **31.** $\dfrac{x(2x - 3)}{3(x - 8)}$ **33.** $\dfrac{3}{5(x - 12)}$

35. $\dfrac{1}{x + 3}$ **37.** $\dfrac{3x}{x - 3}$ **39.** 1 **41.** -1 **43.** $\dfrac{-(x + 5)}{6x + 1}$ **45.** $\dfrac{5}{2}$

47. $\dfrac{(2x + 1)(3x + 2)}{4(3x - 2)(2x - 1)}$ **49.** $\dfrac{2}{x}$

Stretching the Topics

1. $\dfrac{1}{2}$ **2.** $\dfrac{3x^a + 1}{3x^a(x^a + 3)}$ **3.** $2x^{2a - 2c}$

Checkup 6.2 (page 397)

1. $\dfrac{12}{x}$ **2.** $\dfrac{x(x - 7)}{8(x - 4)}$ **3.** $\dfrac{5(2x - 7)}{6y(2x + 7)}$ **4.** $\dfrac{x + 9y}{2y}$ **5.** $\dfrac{y^3(x - 15)}{3x^5}$ **6.** $\dfrac{y^3}{4(x + 4)}$

7. $\dfrac{1}{5x + 4}$ **8.** $\dfrac{7x(x - 10)}{(x + 2)(x - 7)}$ **9.** $\dfrac{-2}{5x}$ **10.** 1

Exercise Set 6.3 (page 411)

1. 2 **3.** $\dfrac{2}{x}$ **5.** $\dfrac{y + 3}{x}$ **7.** $\dfrac{-2x}{2x - 1}$ **9.** 3 **11.** $\dfrac{1}{x - 9}$ **13.** 2 **15.** $\dfrac{1}{x + 4}$

17. $\dfrac{2a - 3}{a}$ **19.** $\dfrac{x + 1}{x - 4}$ **21.** $x - 4$ **23.** $\dfrac{15x - 27}{5x^2}$ **25.** $\dfrac{36y - 50x}{15xy}$ **27.** $\dfrac{x + 3}{x(x + 1)}$

29. $\dfrac{2x^2 - 5x - 15}{2x^2(x - 3)}$ **31.** $\dfrac{7x + 9}{(x + 2)^2}$ **33.** $\dfrac{-2x - 32}{(x + 4)(x - 4)}$ **35.** $\dfrac{-11}{3(x + 1)}$

37. $\dfrac{x^2 - x - 5}{x(x + 5)(x - 5)}$ **39.** $\dfrac{x + 4}{x^4}$ **41.** $\dfrac{4x}{(x + 4)(x - 2)}$ **43.** $\dfrac{2x - 5}{x(2x - 1)}$

45. $\dfrac{3x - 7}{(x + 3)(x - 2)}$ **47.** $\dfrac{1}{x + 5}$ **49.** $\dfrac{2}{x + 5}$ **51.** $\dfrac{x - 2}{x^2 + 2x + 4}$

Stretching the Topics

1. $\dfrac{6x^{3a} - 4x^a y^b + 5y^{3b}}{10x^{2a}y^{2b}}$ **2.** $\dfrac{x^a - 1}{x^a}$ **3.** $\dfrac{x - 1}{x + 1}$

Checkup 6.3 (page 413)

1. $\dfrac{1}{x^2}$ **2.** $\dfrac{10x}{5x - 3}$ **3.** $\dfrac{-1}{x + 2}$ **4.** $\dfrac{x + 5}{4x - 3}$ **5.** $\dfrac{9x + 10y}{6x^2y^2}$ **6.** $\dfrac{11x - 21}{x(x - 3)}$

7. $\dfrac{7x - 2}{(x + 2)(x - 2)}$ **8.** $\dfrac{x - 3}{2x^2(x - 1)^2}$ **9.** $\dfrac{17}{(x - 9)(x + 7)}$ **10.** $\dfrac{x + 10}{(x - 6)(x + 2)}$

Exercise Set 6.4 (page 423)

1. $\dfrac{1}{10}$ **3.** $\dfrac{x^2}{3y}$ **5.** $\dfrac{5x^2}{x - y}$ **7.** $\dfrac{3}{5(x + 4)}$ **9.** $\dfrac{x^2}{x + 2}$ **11.** $\dfrac{x}{2(x^2 + 3x + 9)}$

13. $2x^3(3x + 4)$ **15.** $\dfrac{28}{5}$ **17.** $\dfrac{x - 3}{x}$ **19.** $\dfrac{1}{x + 1}$ **21.** $\dfrac{x}{4x + 1}$ **23.** $\dfrac{4a^2 + 2a + 1}{a^2}$

25. $\dfrac{2}{15}$ **27.** $\dfrac{9 - x}{9x}$ **29.** $\dfrac{15(2x + 1)}{x^2}$ **31.** $\dfrac{5(x + 6)}{x + 5}$ **33.** $\dfrac{(x + 7)^2}{(x + 10)^2}$ **35.** $\dfrac{5x + 6}{9x - 2}$

37. $a - 4$ **39.** $2b$ **41.** $\dfrac{(x + 1)(x + 2)}{(x - 2)(x + 14)}$ **43.** $\dfrac{x}{x - 4}$ **45.** $\dfrac{1 - 2x + 4x^2}{x^2}$

47. $\dfrac{2x - 1}{x - 2}$ **49.** $\dfrac{y - 2x}{y + 2x}$ **51.** $\dfrac{4x^2}{1 + 2x^2y^3}$ **53.** $\dfrac{x^2}{(x + 2)(x - 2)}$

Stretching the Topics

1. $\dfrac{(2x^2 - 3x + 25)(x + 3)}{(x - 4)(7 - x)}$ 2. $\dfrac{1}{x(x + d)}$ 3. $3(x^n + 1)$

Checkup 6.4 (page 427)

1. $\dfrac{x^7 y^3}{5}$ 2. $\dfrac{5x^3}{2x + 3}$ 3. $\dfrac{x^3}{x - 11}$ 4. $\dfrac{x - 9}{x}$ 5. $\dfrac{4a^2 + 6a + 9}{a^2}$ 6. $\dfrac{7 + x}{7x}$

7. $\dfrac{7x(x - 5)}{(x + 2)(x - 7)}$ 8. $\dfrac{2(a - 18)}{7(a - 2)}$ 9. $\dfrac{x(4x + 1)}{x - 5}$ 10. $\dfrac{xy}{y - 3x}$

Exercise Set 6.5 (page 437)

1. $x = \dfrac{-14}{5}$ 3. $x = 3$ 5. $x = 3$ 7. $x = \dfrac{44}{9}$ 9. $x = \dfrac{33}{4}$ 11. $x = \dfrac{-1}{10}$

13. $x = \dfrac{-2}{5}$ 15. $x = 39$ 17. $x = \dfrac{-17}{3}$ 19. $a = \dfrac{3}{5}$ 21. $x = 0$ 23. No solution

25. $x = -7$ 27. $x = \dfrac{5}{3}, x = -2$ 29. $y = 2\sqrt{2}, y = -2\sqrt{2}$ 31. $x = -6, x = 5$

33. $y = \dfrac{x}{1 - x^2}$ 35. $y = \dfrac{4xz}{2x - z}$ 37. $t = \dfrac{A - P}{Pr}$ 39. $x < 0$ 41. $x > 3$

43. $x \le -3$ 45. $x > \dfrac{5}{2}$ 47. $x \le 3$ 49. $x > 5$ 51. $x < 0$ or $x > 4$ 53. $0 < x < 1$

55. $x < 0$ or $x \ge \dfrac{9}{2}$ 57. $x < -1$ or $x > 1$

Stretching the Topics

1. $x = 0, x = -4$ 2. $x = \dfrac{5 \pm \sqrt{41}}{2}$ 3. $x = \dfrac{-3k}{4}$

Checkup 6.5 (page 441)

1. $x = -23$ 2. $a = \dfrac{23}{2}$ 3. $b = \dfrac{1}{13}$ 4. $a = \dfrac{-12}{7}$ 5. $x = -5$ 6. $y = \dfrac{x}{x - 5}$

7. $x > 5$ 8. $0 \le x < 10$ 9. $-3 \le x < 0$ 10. $x < -1$ or $x > 3$

Exercise Set 6.6 (page 451)

1. $12\dfrac{1}{3}$ 3. 4 5. Ann \$680, Meredith \$850 7. Burt 22, Johnny 33

9. Gretchen 10 mph, George 12 mph 11. Michelle 24 min, Carolyn 30 min

13. Bus 50 mph, train 80 mph 15. $7\dfrac{1}{2}$ days 17. $17\dfrac{1}{7}$ hr 19. $17\dfrac{1}{2}$ min

Stretching the Topics

1. $22\dfrac{1}{2}$ days 2. 14.8 mph

Checkup 6.6 (page 453)

1. 280 2. $26\dfrac{2}{3}$ hr 3. 55 mph

Speaking the Language of Algebra (page 456)

1. Rational algebraic expression 2. Division by zero 3. Divide; common factor 4. Factor
5. LCD 6. Negative 7. Proportion 8. Rate of work, time worked

Review Exercises (page 457)

1. $x \ne 0, y \ne 0, z \ne 0$; $\dfrac{-3x^3}{2z^3}$ 2. $x \ne 0, x \ne 2$; $\dfrac{1}{x - 2}$ 3. $x \ne y$; $\dfrac{x + y}{4(x - y)}$

4. $x \ne 7, x \ne -7$; $\dfrac{x - 5}{x - 7}$ 5. $x \ne \dfrac{1}{7}y, x \ne -2y$; $\dfrac{3x - 2y}{x + 2y}$

6. $a \neq \frac{5}{3}b, a \neq 2b; \frac{a^2 + 2ab + 4b^2}{3a - 5b}$ **7.** $x \neq \frac{1}{4}y, x \neq y; 1$ **8.** $x \neq 8, x \neq \frac{3}{2}; \frac{-(x + 8)}{2x - 3}$

9. $\frac{2}{x}$ **10.** $\frac{x(x + 6)}{5(x + 2)}$ **11.** $\frac{3x - 5y}{2x + y}$ **12.** $\frac{y(x + 11y)}{5x^4}$ **13.** $\frac{-(9x - y)}{x(3x - y)}$ **14.** $\frac{8}{3}$

15. $\frac{8x}{2x - 3}$ **16.** $\frac{1}{x - 3}$ **17.** $\frac{8x + 30}{x(x + 10)}$ **18.** $\frac{3x^2 + 5x}{(x + 5)(x - 5)}$ **19.** $\frac{1}{x + 7}$

20. $\frac{2}{x + 2}$ **21.** $\frac{x^3}{3y}$ **22.** $\frac{3x}{3x + 1}$ **23.** $\frac{x + 8}{x}$ **24.** $\frac{5 + x}{5x}$ **25.** $\frac{a^2 + 2a + 4}{a^2}$

26. $\frac{2x + 1}{x}$ **27.** $x = 29$ **28.** $a = 7$ **29.** $x = -6$ **30.** $x > 9$ **31.** $x < 0$ or $x \geq 1$

32. $-1 < x < 5$ **33.** 20 ft **34.** $4\frac{2}{7}$ hr **35.** Mac 45 mph, Becky 55 mph

Practice Test (page 459)

1. $x \neq 0, y \neq 0, c \neq 0; \frac{3x}{2c^2}$ **2.** $x \neq -5, x \neq 2; \frac{x - 3}{x - 2}$ **3.** $y \neq 1, y \neq -1; \frac{y^2 + y + 1}{y + 1}$

4. $m \neq -2n, m \neq 3n; \frac{m - 1}{m - 3n}$ **5.** $\frac{2(x + 8)}{x(x + 1)}$ **6.** $4x^2$ **7.** $\frac{3b(a + 5b)}{4a^3}$ **8.** $\frac{7}{x^2}$

9. $\frac{1}{x - 7}$ **10.** $\frac{2(5x - 1)}{5x(5x + 2)}$ **11.** $\frac{-(y^2 + 36)}{(y + 6)(6 - y)}$ or $\frac{y^2 + 36}{(y + 6)(y - 6)}$ **12.** $\frac{3}{x + 4}$

13. $\frac{-3x^2 + 15x + 62}{(3x - 4)(x + 6)}$ **14.** $\frac{2}{7x(1 + 2x)}$ **15.** $\frac{5(2x - 3)}{x - 5}$ **16.** $\frac{2x - 5}{x}$ **17.** $x = \frac{8}{21}$

18. No solution **19.** $-4 < x \leq \frac{1}{5}$ **20.** $23\frac{1}{3}$ hr

Sharpening Your Skills after Chapters 1–6 (page 461)

1. $-\sqrt{13} < 0$ **2.** 0 **3.** 27 **4.** $\frac{-1}{16}$ **5.** $16\sqrt{3}$ **6.** Two different irrational solutions

7. $x^2 - x - 6 = 0$ **8.** $14x + 9y$ **9.** $a^{15}b^7 + 3a^7b^{15}$ **10.** $\frac{27b^{24}}{16a^{12}}$ **11.** $a^3 - 9a + 6$

12. $-63y^2 + 16y^3 - y^4$ **13.** $x^2 - 9x + 5 - \frac{11}{x}$ **14.** $x^2 - 5xy - 6y^2$

15. $-3x^2y(x^4y^2 - 7x^2y + 9)$ **16.** $(a - b)(a - b - 5)$ **17.** $(3xy + 11)(3xy - 11)$

18. $(5y + 4)(2y - 3)$ **19.** $3(8 + y)(5 - y)$ **20.** $y = -2$ **21.** No solution

22. $m < -5$ or $m > 25$ **23.** $x = 0, x = -8, x = \frac{5}{9}$ **24.** $m = \frac{2}{5}, m = \frac{-1}{6}$

25. $x = \frac{8}{11}, x = \frac{-8}{11}$ **26.** $x = 1 + \sqrt{5}, x = 1 - \sqrt{5}$ **27.** $y \geq \frac{20}{3}$ **28.** $-3 < a \leq 6$

29. $-4 \leq x \leq \frac{2}{5}$ **30.** $x = \frac{-1 \pm \sqrt{1 + 8aM}}{4a}$

CHAPTER 7

Exercise Set 7.1 (page 473)

1. x^2 **3.** $a^{5/3}$ **5.** x^4 **7.** $\frac{x}{y^4}$ **9.** $\frac{b^{2/3}}{a^{3/2}}$ **11.** $x^{16/15}$ **13.** $\frac{x^{1/2}}{y^{1/4}z^{1/3}}$ **15.** $x^{1/2}y^{1/3}$

17. $\frac{b^{5/2}}{a^{3/8}}$ **19.** $\frac{x^{1/9}z^{1/14}}{y^{1/20}}$ **21.** -3 **23.** 3 **25.** Not a real number **27.** $\frac{-4}{5}$ **29.** 2

31. $\frac{5}{9}$ **33.** $(2x)^{1/4}$ **35.** $5a^{1/2}$ **37.** $(x^2 + 16)^{1/2}$ **39.** $-y^{1/5}$ **41.** $(-a)^{1/3}$

43. $x^{1/3} + y^{1/3}$ **45.** $x^{2/5}y^{1/5}$ **47.** $(x - y)^{2/5}$ **49.** $\frac{1}{x^{4/3}y^{2/3}}$ or $x^{-4/3}y^{-2/3}$ **51.** $\sqrt[5]{3y^3}$

53. $4\sqrt{x}$ **55.** $\sqrt[3]{a + b}$ **57.** $\sqrt[5]{x} + \sqrt[5]{y}$ **59.** 5 **61.** $\sqrt{13}$ **63.** $\sqrt[3]{y^2}$

65. $\sqrt[3]{(x - 3)^2}$ **67.** $\frac{1}{\sqrt[5]{(a - b)^3}}$ **69.** 4 **71.** $\frac{1}{27}$ **73.** 5 **75.** $\frac{6}{5}$

Stretching the Topics

1. $\dfrac{1}{250}$ **2.** $\dfrac{n^4}{36,864}$ **3.** $9xy^{5/12}$

Checkup 7.1 (page 475)

1. a **2.** $\dfrac{y^2}{x^{1/2}}$ **3.** $\dfrac{x^{1/4}}{y^{31/10}}$ **4.** $\dfrac{x^{1/15}z^{1/2}}{y^{9/4}}$ **5.** Not a real number **6.** $\dfrac{-1}{2}$

7. $x^{2/7}y^{3/7}$ **8.** $(x + y)^{3/2}$ **9.** $\sqrt[3]{(x - y)^2}$ **10.** 32

Exercise Set 7.2 (page 485)

1. $2b\sqrt{3}$ **3.** $2x^2y^3\sqrt{6xy}$ **5.** $36a^7$ **7.** $2c^5d\sqrt{6d}$ **9.** $-2ab\sqrt[3]{5a^2b}$ **11.** $\dfrac{4a^2\sqrt{3a}}{3b}$

13. $\dfrac{2x\sqrt[3]{2x}}{3y^2}$ **15.** $\dfrac{12x^3\sqrt{x}}{5y^2}$ **17.** $2x\sqrt[5]{x^2}$ **19.** $7x^2$ **21.** $2x\sqrt[3]{2x}$ **23.** $2x^2y^2\sqrt[4]{2xy^2}$

25. $(x - 3)\sqrt[5]{x - 3}$ **27.** $x - 3$ **29.** $2x + 1$ **31.** $7\sqrt{5}$ **33.** $6\sqrt{3} + 6\sqrt{7}$ **35.** 0

37. $-4\sqrt[3]{5}$ **39.** $15 - 7\sqrt{x}$ **41.** $-3\sqrt{5} - 11y$ **43.** $-2\sqrt{a} - 23\sqrt[3]{b}$

45. $-6\sqrt[5]{2} - 3\sqrt[3]{2}$ **47.** $3\sqrt{3}$ **49.** $\sqrt{5}$ **51.** $4x\sqrt{3}$ **53.** $22\sqrt[3]{2}$ **55.** 0 **57.** $\dfrac{4\sqrt{2}}{15}$

59. $13\sqrt{2} - 5\sqrt{3}$ **61.** $11ab\sqrt[3]{3a^2b}$

Stretching the Topics

1. $x^{2m}\sqrt{x}$ **2.** $x^{2m}y^{3n}$ **3.** $\dfrac{3y^2\sqrt{2y}}{z^2} - \dfrac{10y^2\sqrt{2}}{z^2}$

Checkup 7.2 (page 487)

1. $2x^2y\sqrt{11x}$ **2.** $\dfrac{-3a\sqrt[3]{a}}{2b}$ **3.** $(x + 1)\sqrt[4]{x + 1}$ **4.** $-9x\sqrt{2}$ **5.** $5\sqrt{7} - 21x$ **6.** 0

7. $-\sqrt[3]{10}$ **8.** $4\sqrt[3]{3} + 4\sqrt{3}$ **9.** $\dfrac{\sqrt{2}}{21}$ **10.** $3x^2y\sqrt[3]{xy}$

Exercise Set 7.3 (page 501)

1. $7\sqrt{2}$ **3.** $3a^2\sqrt{2}$ **5.** $-3x\sqrt[3]{2}$ **7.** $3x^2$ **9.** $-2x\sqrt[3]{3}$ **11.** $2\sqrt[5]{x^3}$

13. $8x\sqrt{xy}$ **15.** $5x$ **17.** $3x\sqrt{10y}$ **19.** $12x^2y^2z\sqrt{y}$ **21.** $2\sqrt{3} + 2$ **23.** $x\sqrt{3} - 3\sqrt{x}$

25. $3x\sqrt{2y} + 2y\sqrt{3x}$ **27.** $3\sqrt[3]{a} - a$ **29.** $2 - 3\sqrt{x} + x$

31. $10 - 14\sqrt{2y} - 3y$ **33.** $x - 6\sqrt{x} + 9$ **35.** $2x - 9$ **37.** $16x - 20$ **39.** $x^2 - 2y$

41. $9x - 30\sqrt{xy} + 25y$ **43.** $\dfrac{2\sqrt{6}}{3}$ **45.** $\dfrac{\sqrt{2}}{2}$ **47.** $\dfrac{\sqrt{6}}{3}$ **49.** $\dfrac{3x\sqrt{10xy}}{5y}$

51. $\dfrac{3xy\sqrt{15xz}}{5z}$ **53.** $\dfrac{4\sqrt{a + 5}}{a + 5}$ **55.** $\dfrac{\sqrt{x^2 - 9}}{x + 3}$ **57.** $\dfrac{3\sqrt[3]{25}}{5}$ **59.** $\dfrac{64\sqrt[4]{27}}{3}$ **61.** $\dfrac{2\sqrt[3]{9x^2}}{x}$

63. $\dfrac{\sqrt[4]{12y^3}}{2y}$ **65.** $\dfrac{\sqrt[3]{10y}}{2y}$ **67.** $3(2 + \sqrt{3})$ **69.** $\dfrac{\sqrt{7} - \sqrt{2}}{5}$ **71.** $\dfrac{x + 2\sqrt{x}}{x - 4}$

73. $\dfrac{a + 3\sqrt{a}}{a - 9}$ **75.** $3\sqrt{2} - 2\sqrt{3}$ **77.** $\dfrac{18 - 7\sqrt{5}}{79}$

Stretching the Topics

1. $x - 2$ **2.** $(3\sqrt{x} + 2\sqrt{y})(3\sqrt{x} - 2\sqrt{y})$ **3.** $\dfrac{x + 2 + 5\sqrt{x + 2}}{x - 23}$

Checkup 7.3 (page 505)

1. $2x\sqrt{15x}$ **2.** $4x\sqrt{y}$ **3.** $5\sqrt{2} + 5$ **4.** 10 **5.** $x - 12\sqrt{x} + 36$ **6.** $\dfrac{2\sqrt{15}}{5}$

7. $\dfrac{\sqrt{15}}{5}$ **8.** $-5\sqrt[3]{9x^2}$ **9.** $\dfrac{3x\sqrt{10xy}}{2y}$ **10.** $-4 + 2\sqrt{6}$

Exercise Set 7.4 (page 515)

1. $x = 25$ **3.** $x = 50$ **5.** $x = 6$ **7.** $x = 3$ **9.** No solution **11.** $x = 4$
13. $x = 6$ **15.** $x = -7$ **17.** $x = 3$ **19.** $x = 4$ **21.** $x = 9, x = 2$ **23.** $x = 2$
25. No solution **27.** $x = 1$ **29.** $x = 9$ **31.** No solution **33.** $x = -10, x = 2$
35. $x = 2$ **37.** $y = -1$ **39.** 3 cm, 4 cm, 5 cm **41.** 972π cu in.
43. $w = \sqrt{d^2 - l^2}$

Stretching the Topics

1. $x = 3, x = 2$ **2.** $x = 6$ **3.** $l = \dfrac{T^2 g}{4\pi^2}$

Checkup 7.4 (page 517)

1. $x = 7$ **2.** $x = \dfrac{4}{3}$ **3.** No solution **4.** $x = 9$ **5.** No solution **6.** $x = -9$

7. $x = 9$ **8.** $x = 6, x = 3$ **9.** $x = 2$ **10.** $x = 2$

Exercise Set 7.5 (page 531)

1. $0 + 4i$ **3.** $0 + \sqrt{21}i$ **5.** $0 + 5\sqrt{2}i$ **7.** $0 - 5\sqrt{3}i$ **9.** $4 - 3\sqrt{6}i$

11. $18 + 7i$ **13.** $x = 1, y = -4$ **15.** $x = -4, y = \dfrac{1}{3}$ **17.** $x = \dfrac{-1}{2}, y = \dfrac{-5}{3}$

19. $x = 3, y = \dfrac{3}{2}$ **21.** $x = 2, y = 0$ **23.** $x = 4, y = \dfrac{-4}{3}$ **25.** i **27.** 1 **29.** i

31. i **33.** $5 + 9i$ **35.** $0 - 9i$ **37.** $9 - 5\sqrt{2}i$ **39.** $5 + 0i$ **41.** $-2 + i$
43. $12 + 2i$ **45.** $-18 - 12i$ **47.** $-3 + 21i$ **49.** $-10 + 10i$ **51.** $31 + 29i$
53. $4 - 6\sqrt{5}i$ **55.** $-2 - 21\sqrt{6}i$ **57.** $19 + 0i$ **59.** $-26 + 7i$ **61.** $2 - 5i$

63. $\dfrac{4}{5} - 2i$ **65.** $3 - \dfrac{\sqrt{3}}{4}i$ **67.** $0 - 5i$ **69.** $0 + 4\sqrt{5}i$ **71.** $\dfrac{-1}{2} + \dfrac{\sqrt{3}}{2}i$

73. $6 + 9i$ **75.** $\dfrac{7}{25} + \dfrac{1}{25}i$ **77.** $\dfrac{-\sqrt{3}}{4} + \dfrac{7}{4}i$ **79.** $\dfrac{-2}{29} - \dfrac{5}{29}i$ **81.** $\dfrac{13}{15} - \dfrac{2}{5}i$

83. $\dfrac{-1}{9} - \dfrac{4\sqrt{5}}{9}i$

Stretching the Topics

1. -8 **2.** $(x + 3i)(x - 3i)$ **3.** $x = -4, y = 5; -4 + 5i$

Checkup 7.5 (page 535)

1. $0 + 11i$ **2.** $9 - 2\sqrt{7}i$ **3.** $x = 4, y = -4$ **4.** -1 **5.** $6 - 5i$ **6.** $3 - 9\sqrt{3}i$
7. $17 - 6i$ **8.** $21 - 20i$ **9.** $-2 - 3i$ **10.** $-1 - i$

Exercise Set 7.6 (page 539)

1. $x = 3i, x = -3i$ **3.** $x = \dfrac{5}{2}i, x = \dfrac{-5}{2}i$ **5.** $x = \sqrt{3}i, x = -\sqrt{3}i$

7. $x = \dfrac{-1}{2} \pm \dfrac{\sqrt{3}}{2}i$ **9.** $x = \dfrac{4}{3} \pm \dfrac{\sqrt{5}}{3}i$ **11.** $x = \dfrac{3}{4} \pm \dfrac{\sqrt{23}}{4}i$ **13.** $x = \dfrac{-3}{2} \pm \dfrac{5\sqrt{3}}{2}i$

15. $x = \dfrac{-3}{5} \pm \dfrac{\sqrt{11}}{5}i$ **17.** $x = -2, x = 1 \pm \sqrt{3}i$ **19.** $x = 5, x = \dfrac{-5}{2} \pm \dfrac{5\sqrt{3}}{2}i$

21. $x = \dfrac{1}{2}, x = \dfrac{-1}{4} \pm \dfrac{\sqrt{3}}{4}i$ **23.** $x = 3, x = -3, x = 3i, x = -3i$
25. $x = \sqrt{11}, x = -\sqrt{11}, x = \sqrt{11}i, x = -\sqrt{11}i$ **27.** $x = 3i, x = -3i, x = 2i, x = -2i$
29. $x = \sqrt{5}i, x = -\sqrt{5}i, x = \sqrt{2}, x = -\sqrt{2}$ **31.** $x = \dfrac{-2}{3} \pm \dfrac{\sqrt{2}}{3}i$ **33.** $x = \dfrac{1}{3} \pm \dfrac{\sqrt{2}}{3}i$

35. $x = \dfrac{1}{6} \pm \dfrac{\sqrt{23}}{6}i$

Stretching the Topics

1. $x = -8$ **2.** $x = -2i, x = \dfrac{-1}{2}i$ **3.** $x^2 + ix + 6 = 0$

Checkup 7.6 (page 541)

1. $x = 5i, x = -5i$ **2.** $x = 2\sqrt{6}i, x = -2\sqrt{6}i$ **3.** $x = \dfrac{1}{2} \pm \dfrac{3\sqrt{11}}{2}i$ **4.** $x = \dfrac{6}{5} \pm \dfrac{3}{5}i$

5. $x = -5, x = \dfrac{5}{2} \pm \dfrac{5\sqrt{3}}{2}i$ **6.** $x = \dfrac{3}{4}, x = \dfrac{-3}{8} \pm \dfrac{3\sqrt{3}}{8}i$

7. $x = \sqrt{10}, x = -\sqrt{10}, x = \sqrt{10}i, x = -\sqrt{10}i$ **8.** $x = 4i, x = -4i, x = 3i, x = -3i$

9. $x = 1 \pm \sqrt{6}i$ **10.** $x = \dfrac{3}{4} \pm \dfrac{\sqrt{15}}{4}i$

Speaking the Language of Algebra (page 545)

1. Real   **2.** Index; radicand   **3.** Like radicals   **4.** Rationalizing the denominator
5. Extraneous   **6.** Complex number; real; real; $\sqrt{-1}$ **7.** Real; imaginary; imaginary unit
8. Real parts; imaginary parts   **9.** Conjugate   **10.** Conjugate; denominator

Review Exercises (page 547)

1. x **2.** $\dfrac{x^2}{y^8}$ **3.** $\dfrac{x^{3/4}y^{7/4}}{z}$ **4.** $x^{1/6}y^{1/10}$ **5.** -5 **6.** $\dfrac{1}{3}$ **7.** $a^{2/5}b^{3/5}$

8. $(x - y)^{-2/5}$ or $\dfrac{1}{(x - y)^{2/5}}$ **9.** $(\sqrt[3]{27})^2 = 9$ **10.** $\sqrt[4]{(a + b)^3}$ **11.** $3xy\sqrt{2x}$

12. $\dfrac{-2x\sqrt[3]{3x^2}}{5y}$ **13.** $2\sqrt{7} - 11x$ **14.** $x\sqrt{5}$ **15.** $5\sqrt[3]{2} - \sqrt{3}$ **16.** $\dfrac{-16\sqrt{2}}{9}$

17. $5x^2\sqrt{6}$ **18.** $3a\sqrt{b}$ **19.** $3\sqrt{2} + 3$ **20.** 78 **21.** $2x - 6\sqrt{2x} + 9$ **22.** $\dfrac{2\sqrt{6a}}{3a}$

23. $\dfrac{3\sqrt[3]{4x}}{2x}$ **24.** $\dfrac{5x\sqrt{6xy}}{3y}$ **25.** $\dfrac{\sqrt{xy}}{2y^2}$ **26.** $-(2\sqrt{3} + 3\sqrt{2})$ **27.** $x = 20$

28. $v = \pi r^2 h$ **29.** $x = 3$ **30.** $x = -13$ **31.** $x = 5$ **32.** $x = 2$ **33.** $0 + 12i$

34. $2 + 5i$ **35.** $6 - 2\sqrt{3}i$ **36.** $0 + 9i$ **37.** $x = 3, y = -4$ **38.** $-i$

39. $1 - 12\sqrt{2}i$ **40.** $23 + 14i$ **41.** $\dfrac{-2}{3} - \dfrac{1}{3}i$ **42.** $\dfrac{-1}{2} - \dfrac{5}{2}i$

43. $x = 3\sqrt{3}i, x = -3\sqrt{3}i$ **44.** $x = \dfrac{1}{3} \pm \dfrac{\sqrt{2}}{3}i$ **45.** $x = 1 \pm \sqrt{3}i$

46. $x = 4, x = -2 \pm 2\sqrt{3}i$ **47.** $x = \dfrac{-1}{2} \pm \dfrac{\sqrt{3}}{2}i$ **48.** $x = \dfrac{-1}{5} \pm \dfrac{3}{5}i$

Practice Test (page 551)

1. $x^{3/4}y^{1/4}$ **2.** $3\sqrt[5]{(x - y)^2}$ **3.** $\dfrac{1}{\sqrt[3]{a^2 - b^2}}$ **4.** $0 + 7i$ **5.** $5 - 5\sqrt{3}$ **6.** 6 **7.** $\dfrac{4}{9}$

8. $x^{7/5}$ **9.** $\dfrac{b^{10}}{a^9}$ **10.** $x^{3/8}y^{1/5}$ **11.** $\dfrac{y^{1/2}}{x^{1/4}z^{5/4}}$ **12.** $\dfrac{3x\sqrt{7x}}{7y}$ **13.** $13\sqrt{2x}$

14. $5\sqrt[3]{3} + 4\sqrt{3}$ **15.** $\dfrac{4x\sqrt{5}}{3}$ **16.** $6x^2\sqrt{6x}$ **17.** $-2 + 5\sqrt{7}$ **18.** $3x - 10\sqrt{3x} + 25$

19. $3x^2$ **20.** $\dfrac{10x\sqrt{14xy}}{7y}$ **21.** $\dfrac{\sqrt[3]{15x}}{5x}$ **22.** $\dfrac{5(\sqrt{7} - \sqrt{3})}{2}$ **23.** $-18 - 24i$

24. $\dfrac{-1}{4} - i$ **25.** $\dfrac{13}{10} + \dfrac{11}{10}i$ **26.** $x = 2, y = \dfrac{-3}{7}$ **27.** $x = 13$ **28.** No solution

29. $x = \dfrac{1}{3} \pm \dfrac{\sqrt{2}}{3}i$ **30.** $S = 4\pi r^2$

Sharpening Your Skills after Chapters 1–7 (page 553)

1. -24 **2.** $5 - \sqrt{10}$ **3.** 8 **4.** $\dfrac{1}{64}$ **5.** $\dfrac{3\sqrt{2}}{5}$ **6.** 7 **7.** $-118x + 16$

8. $5x^5y^{11}z^5$ **9.** $\dfrac{-54}{y^8}$ **10.** $-27a^3 - 8a^2$ **11.** $x^3 + 3x^2 - 9x + 5$

12. $7x - 2 + \dfrac{1}{3x - 2}$ **13.** $\dfrac{a - 11}{5a - 2}$ **14.** $\dfrac{x^2(x - 13)}{2y^5}$ **15.** $\dfrac{x + 3}{x(x + 6)}$ **16.** $\dfrac{5(2x - 5)}{x}$

17. $2xyz(3x^3 - 6x^2y - 15xy^2 - 2y^3)$ **18.** $3x^2y^2(x + 3y)(x^2 - 3xy + 9y^2)$

19. $(3a - 2)(a - 5)$ **20.** $2(x + 3)(x - 3)(x^2 + 2)$ **21.** $(x - 2)(x + 2)(x^2 - 2x + 4)$

22. $x = -1$ **23.** $x \geq 1$ or $x \leq \dfrac{-3}{7}$ **24.** $x = \dfrac{7}{20} \pm \dfrac{\sqrt{71}}{20}i$

25. $x = \dfrac{1 + \sqrt{2}}{3},\ x = \dfrac{1 - \sqrt{2}}{3}$ **26.** $x \in R$ **27.** $x = 0$ **28.** $x < 0$ or $x > 9$

29. $2250 at 12%, $2750 at 9%   **30.** 4 ft

CHAPTER 8

Exercise Set 8.1 (page 577)

1.

3.

5.

7.

9.

11.

13.

15.

17.

19.

21.

23.

25.

27.

29.

31.

33.

35.

37.

39.

Stretching the Topics

1.

2.

3.

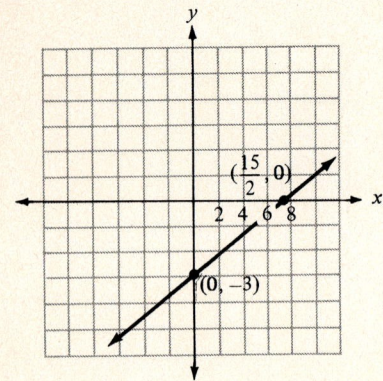

Checkup 8.1 (page 579)

1.

2.

3.

4.

5.

6.

7.

8.

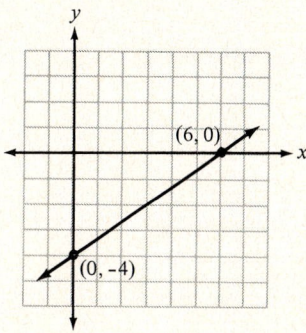

Exercise Set 8.2 (page 599)

1. 5   **3.** $3\sqrt{5}$ **5.** $\sqrt{5}$ **7.** 13   **9.** $\dfrac{\sqrt{5}}{2}$ **11.** $\sqrt{29}$ **13.** $(6, 11)$ **15.** $(1, -2)$

17. $\left(\dfrac{-5}{2}, 3\right)$ **19.** $\left(-6, \dfrac{11}{2}\right)$ **21.** $(4, 0)$ **23.** $\left(\dfrac{-1}{4}, \dfrac{7}{4}\right)$ **25.** $\dfrac{5}{3}$ **27.** Undefined

29. -1 **31.** 0   **33.** $\dfrac{-39}{2}$ **35.** 3

37.

39.

41.

43.

45.

47.

49.

51.

53.

55.

57.

59.

Stretching the Topics

1. 24   **2.** $\left(\dfrac{-5}{3}, \dfrac{-19}{6}\right)$ **3.** $m = \dfrac{13}{42}, b = \dfrac{37}{28}$

Checkup 8.2 (page 601)

1. 5   **2.** $(7, -8)$ **3.** -1 **4.** $\dfrac{-3}{4}$

5.

6.

7.

8.

9.

10.

Exercise Set 8.3 (page 615)

1. $y = -5$ **3.** $y = \dfrac{1}{2}$ **5.** $x = 2$ **7.** $x = \dfrac{-2}{3}$ **9.** $y = 4x + 2$ **11.** $y = -2x + 1$

13. $y = \dfrac{1}{3}x - \dfrac{2}{3}$ **15.** $y = 10$ **17.** $y = \dfrac{1}{2}x - \dfrac{5}{4}$ **19.** $y = 7x - 2$ **21.** $y = 8x + 13$

23. $y = -x + 9$ **25.** $y = -3$ **27.** $y = \dfrac{3}{4}x + \dfrac{19}{4}$ **29.** $x = -2$

31. $y = \dfrac{-2}{3}x + \dfrac{7}{12}$ **33.** $y = -2x + 2$ **35.** $y = \dfrac{-5}{3}x + \dfrac{16}{3}$ **37.** $y = -8$

39. $y = \dfrac{-5}{2}x - \dfrac{1}{2}$ **41.** $x = 6$ **43.** $y = x - \dfrac{11}{10}$ **45.** $y = -2x - 4$

47. $y = 2x + 3$ **49.** $y = \dfrac{-1}{3}x + 4$ **51.** $y = \dfrac{-1}{3}x + \dfrac{7}{3}$ **53.** $y = 2x + 4$

55. $y = \dfrac{-1}{2}x - \dfrac{1}{2}$ **57.** $y = \dfrac{5}{3}x - 5$ **59.** $y = -2x + 4$

Stretching the Topics

1. $y = \dfrac{13}{8}x + \dfrac{33}{2}$ **2.** $y = 4x - 4$ **3.** $y = \dfrac{10}{3}x - \dfrac{34}{3}; y = \dfrac{-3}{7}x + \dfrac{26}{7}; y = \dfrac{-13}{4}x - \dfrac{19}{4}$

Checkup 8.3 (page 617)

1. $y = 7$ **2.** $x = \dfrac{1}{2}$ **3.** $y = -4x + 2$ **4.** $y = \dfrac{-2}{3}x$ **5.** $y = -4x - 3$

6. $x = -3$ **7.** $y = 5x - 6$ **8.** $y = -x + \dfrac{3}{10}$ **9.** $y = 3x + 5$ **10.** $y = 4x - 5$

Exercise Set 8.4 (page 625)

1.

3.

5.

7.

9.

11.

13.

15.

17.

19.

Stretching the Topics

1.

2.

3.

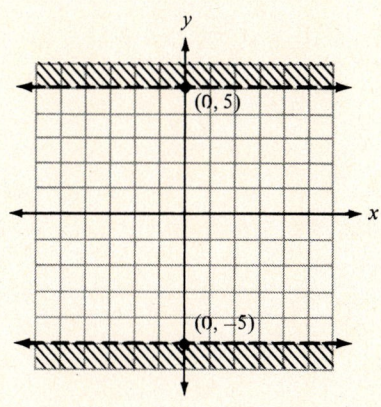

Checkup 8.4 (page 627)

1.

2.

3.

4.

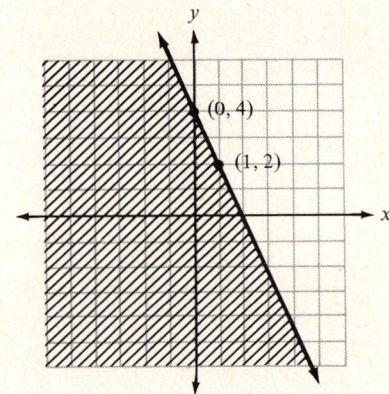

Exercise Set 8.5 (page 637)

1. Domain $\{1, 2, 3\}$; range: $\{-3, 0, 3, 5\}$, no   **3.** Domain: $x \in R$; range; $y \in R$, yes

5. Domain: $\{-6, -2, 0, 3\}$; range: $\{3\}$; yes   **7.** Domain: $\{3\}$; range: $y \in R$, no

9. Domain: $\{-2, -1, 1, 2\}$; range: $\{3, 5\}$, yes   **11.** Yes   **13.** Yes   **15.** No   **17.** No

19. 7   **21.** -9 **23.** $3x + 3h + 1$ **25.** -16 **27.** $2x - 1$ **29.** $a^2 - a - 1$

31. -2 **33.** $a^2 + a - 1$ **35.** $x \in R, x \neq -4$ **37.** $x \in R, x \neq \frac{1}{2}$

39. $x \in R, x \geq -1$ **41.** $x \in R, x \neq 3$ **43.** $x \in R, x \leq 5$ **45.** $x \in R, x \neq 4, x \neq -4$

47. $x \in R, x \geq 3$ **49.** $x \in R, x \neq 1, x \neq \frac{7}{2}$

Stretching the Topics

1. $4a + 2h - 3$ **2.** $\dfrac{a}{a + 1}$ **3.** $x \in R, x > 0, x \neq 4, x \neq \dfrac{3}{2}$

Checkup 8.5 (page 641)

1. Domain: $\{-3, -2, -1, 0, 1, 2, 3\}$; range $\{0, 1, 2, 3\}$, yes   **2.** Domain: $\{5\}$; range: $y \in R$, no

3. No   **4.** Yes   **5.** 7   **6.** 1   **7.** -10 **8.** $x \in R, x \neq \dfrac{3}{2}$ **9.** $x \in R, x \geq 4$

10. $x \in R, x \neq 7, x \neq -4$

Exercise Set 8.6 (page 651)

1. $y = kx, k = 12$ **3.** $y = kx, k = \dfrac{1}{5}$ **5.** $y = \dfrac{k}{x}, k = 60$ **7.** $y = \dfrac{k}{x}, k = 4.5$

9. $z = kxy, k = \dfrac{3}{4}$ **11.** $y = \dfrac{kx}{z^2}, k = \dfrac{16}{3}$ **13.** 3200 rubber bands   **15.** 20 min

17. 12 lb   **19.** 200   **21.** 5 sec   **23.** 225 lb   **25.** $C = 18 + 0.30m$

27. $s = 100 + 75n$ **29.** $A = 5d$ **31.** $P = 4w + 24$ **33.** $A = \frac{1}{3}l^2$

35. $A = \frac{1}{2}h^2 + 2h$ **37.** $d = 14j$ **39.** $a = \frac{12 + x}{4}$ **41.** $C = 150 + 55n$

43. $5x + 3y \geq 2000$

Stretching the Topics
1. Multiplied by 9; multiplied by 16   **2.** 312.5 lb
3. -90; The loss is \$90 when 20 items are manufactured.

Checkup 8.6 (page 655)

1. 2   **2.** $5\frac{3}{5}$ hr   **3.** $w = kxy^2, k = 1$ **4.** $F = 6.00 + 1.75n$ **5.** $P = 12w - 10$

Speaking the Language of Algebra (page 658)
1. Ordered pairs   **2.** Independent; dependent   **3.** Domain; range
4. Straight line; slope; y-intercept   **5.** Horizontal line; vertical line   **6.** Dashed; above
7. Slope   **8.** Standard; slope-intercept   **9.** 17   **10.** Direct; inverse

Review Exercises (page 661)
1.

2.

3.

4.

5. 5   **6.** $\left(2, \frac{3}{2}\right)$ **7.** -2

8.

9.

10.

11.

12.

13.

14.

15.

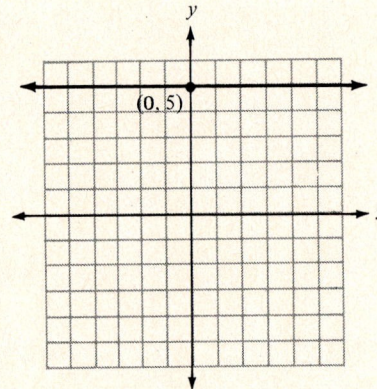

16. $y = -3x - 1$ **17.** $y = \frac{1}{2}x$ **18.** $x = -2$ **19.** $y = \frac{3}{4}x + \frac{15}{4}$ **20.** $y = -x - 5$

21. $y = 2x - \frac{11}{6}$ **22.** $x = 3$ **23.** $y = -5$ **24.** $y = 2x + 5$ **25.** $y = 2x - 12$

26.

27.

28.

29.

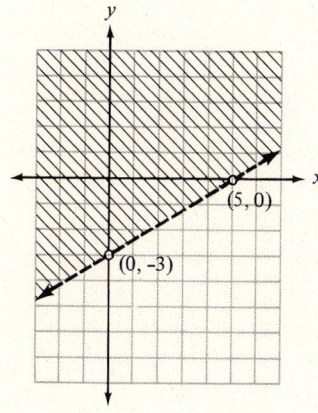

30. Domain: $\{-2, -1, 0, 1, 2\}$; range: $\{-2, -1, 0, 1, 2\}$ **31.** Domain: $x \in R$; range: $\{3\}$

32. yes   **33.** No   **34.** -7 **35.** 11   **36.** $x \in R$, $x \neq 2$, $x \neq -2$ **37.** $x \in R$, $x \leq 3$

38. $y = kx$, $k = 7$ **39.** $y = \dfrac{k}{x}$, $k = 3$ **40.** $z = kxy^2$, $k = \dfrac{1}{576}$

41. $T = \dfrac{kD}{R}$, $k = 60$, $T = 400$ **42.** \$5.60   **43.** 2304 ft   **44.** $A = 2w^2 - 3w$

Practice Test (page 665)

1.

2.

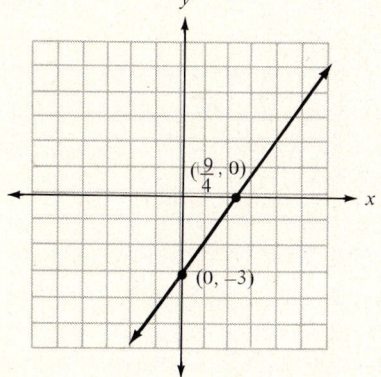

3. $2\sqrt{10}$ **4.** $\left(\dfrac{-3}{2}, \dfrac{7}{2}\right)$ **5.** $m = -1$

6.

7.

8.

9.

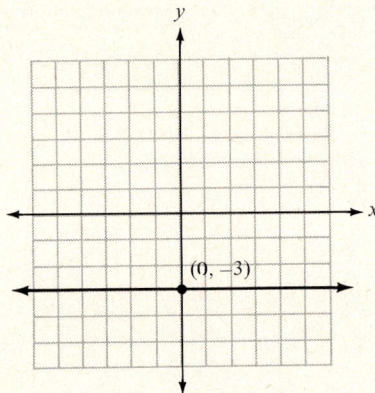

10. $y = -2x$ **11.** $y = \dfrac{-2}{3}x - 3$ **12.** $x = 5$ **13.** $y = \dfrac{2}{3}x - \dfrac{16}{3}$ **14.** $y = x - 8$

15. $y = \dfrac{1}{2}x + \dfrac{9}{2}$

16.

17.

18. Domain: $x \in R$; range: $\{3\}$, yes   **19.** 7   **20.** 0   **21.** $x \in R$, $x \geq \dfrac{1}{2}$

22. $x \in R$, $x \neq 0$, $x \neq \dfrac{-3}{4}$ **23.** $y = kx^2$, $k = 2$ **24.** $A = kbh$, $k = \dfrac{1}{2}$, $A = \dfrac{35}{2}$

25. $A = 700 - 25m$, \$400

Sharpening Your Skills after Chapters 1–8 (page 669)

1. $\dfrac{-1}{27}$ **2.** $7 - 2\sqrt{3}i$ **3.** 1   **4.** No real number solutions   **5.** $8\sqrt[3]{a}$ **6.** $(a - b)^{4/3}$

7. -71 **8.** $118x - 63$ **9.** 1 **10.** $\dfrac{a^2c}{b^2}$ **11.** $3b^3 - 15a^3 + 4a^3b^3$

12. $x^4 - 8x^3 + 7x^2 + 72x - 144$ **13.** $\dfrac{1}{2x - 3}$ **14.** $\dfrac{5x}{(x + 5)(x - 2)}$ **15.** $\dfrac{x}{x - 1}$

16. $11\sqrt{3x}$ **17.** $3x - 25$ **18.** $\dfrac{23}{41} - \dfrac{2}{41}i$ **19.** $(x + 5)(x^2 - 5x + 25)$

20. $(3x^2 - 5)^2$ **21.** $(9a - 2b)(7a - 5)$ **22.** $x = 2$ **23.** $x \geq 15$ or $x \leq -45$

24. $x = 0, x = \dfrac{3}{11}, x = 2$ **25.** $x = 30$ **26.** No solution   **27.** $2 \leq x \leq 10$

28. $\dfrac{1}{2} \leq a < \dfrac{3}{2}$ **29.** 24 by 40 ft

CHAPTER 9

Exercise Set 9.1 (page 681)

1.

(-3, 0)   (3, 0)
(0, -9)

Do: $x \in R$
Ra: $y \in R, y \geqslant -9$

3.

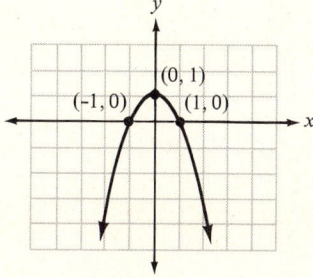

(0, 1)
(-1, 0)   (1, 0)

Do: $x \in R$
Ra: $y \in R, y \leqslant 1$

5.

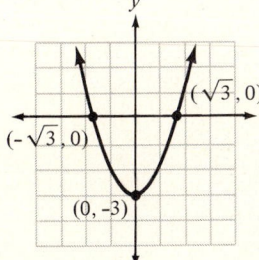

$(\sqrt{3}, 0)$
$(-\sqrt{3}, 0)$
(0, -3)

Do: $x \in R$
Ra: $y \in R, y \geqslant -3$

7.

(2, 0)
(0, 0)
(1, -1)

Do: $x \in R$
Ra: $y \in R, y \geqslant -1$

9.

(1, 2)
(-1, 0)   (0, 1)

Do: $x \in R$
Ra: $y \in R$

11.

(-1, 0)   (0, -1)
(1, -2)

Do: $x \in R$
Ra: $y \in R$

13.

Do: $x \in R$
Ra: $y \in R$

15.

Do: $x \in R, x \neq 0$
Ra: $y \in R, y \neq 0$

17.

Do: $x \in R, x \neq 0$
Ra: $y \in R, y \neq 0$

Stretching the Topics

1.

2.

Checkup 9.1 (page 683)

1.

Do: $x \in R$
Ra: $y \in R, y \geqslant -5$

2.

Do: $x \in R$
Ra: $y \in R, y \leqslant 8$

|

3.

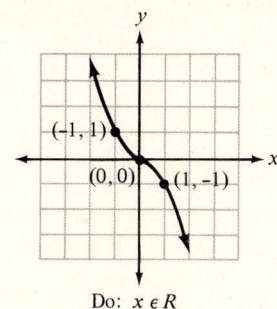

Do: $x \in R$
Ra: $y \in R$

4.

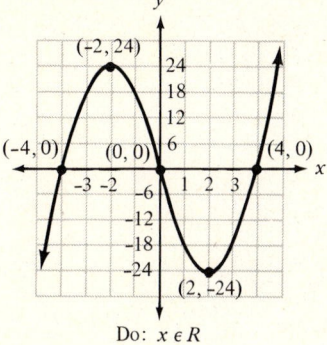

Do: $x \in R$
Ra: $y \in R$

5.

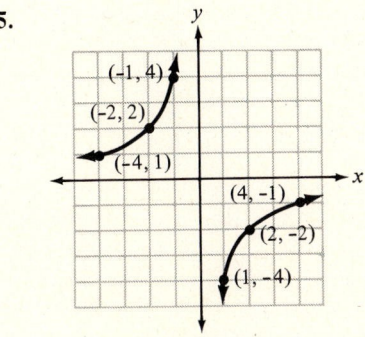

Do: $x \in R, x \neq 0$
Ra: $y \in R, y \neq 0$

Exercise Set 9.2 (page 707)

1.

Do: $x \in R$
Ra: $y \in R, y \geqslant 0$

3.

Do: $x \in R$
Ra: $y \in R, y \leqslant 6$

5.

Do: $x \in R$
Ra: $y \in R, y \geqslant -9$

7.

Do: $x \in R$
Ra: $y \in R, y \leqslant -10$

9.

Do: $x \in R$
Ra: $y \in R, y \geqslant -4$

11.

Do: $x \in R$
Ra: $y \in R, y \geqslant 0$

13.

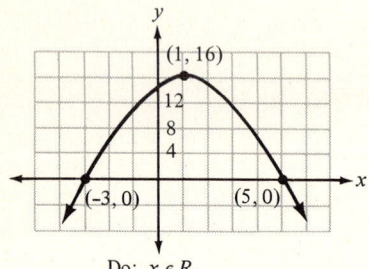

Do: $x \in R$
Ra: $y \in R, y \leqslant 16$

15.

Do: $x \in R$
Ra: $y \in R, y \leqslant \frac{49}{8}$

17.

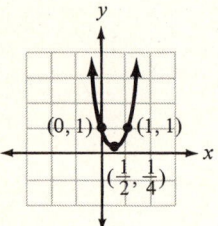

Do: $x \in R$
Ra: $y \in R, y \geqslant \frac{1}{4}$

19.

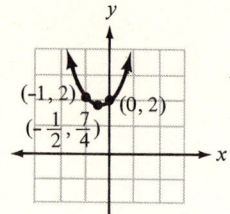

Do: $x \in R$
Ra: $y \in R, y \geqslant \frac{7}{4}$

21.

Do: $x \in R$
Ra: $y \in R, y \leqslant -3$

23.

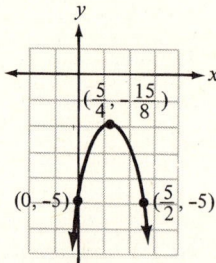

Do: $x \in R$
Ra: $y \in R, y \leqslant -\frac{15}{8}$

25.

18, $972

27.

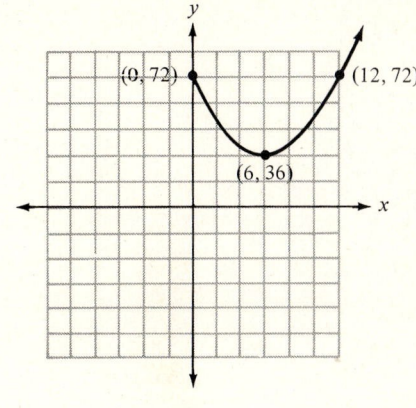

6, $36

Stretching the Topics
1. $x = 4.2$, $x = -1.2$

2.

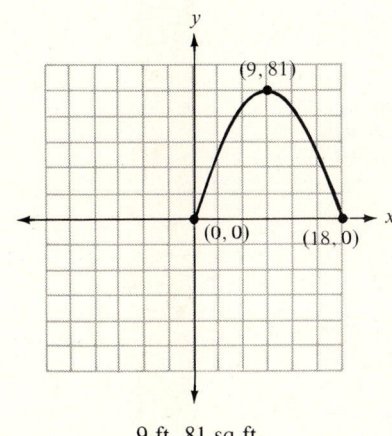

9 ft, 81 sq ft

Checkup 9.2 (page 709)
1.

Do: $x \in R$
Ra: $y \in R$, $y \leq 0$

2.

3.

4.

Do: $x \in R$

Ra: $y \in R, y \leqslant \dfrac{25}{4}$

5.

6.

Do: $x \in R$

Ra: $y \in R, y \leqslant -\dfrac{11}{4}$

Exercise Set 9.3 (page 727)

1.

Do: $-4 \leqslant x \leqslant 4$

Ra: $-4 \leqslant y \leqslant 4$

3.

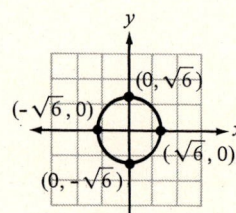

Do: $-\sqrt{6} \leqslant x \leqslant \sqrt{6}$

Ra: $-\sqrt{6} \leqslant y \leqslant \sqrt{6}$

5.

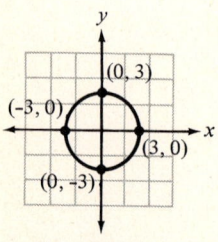

Do: $-3 \leqslant x \leqslant 3$

Ra: $-3 \leqslant y \leqslant 3$

7.

Do: $-4 \leqslant x \leqslant 4$

Ra: $-4 \leqslant y \leqslant 4$

9.

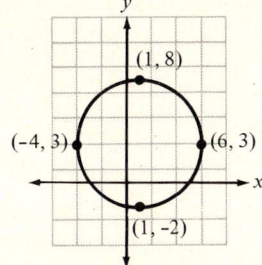

Do: $-4 \leqslant x \leqslant 6$
Ra: $-2 \leqslant y \leqslant 8$

11.

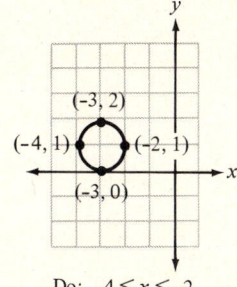

Do: $-4 \leqslant x \leqslant -2$
Ra: $0 \leqslant y \leqslant 2$

13.

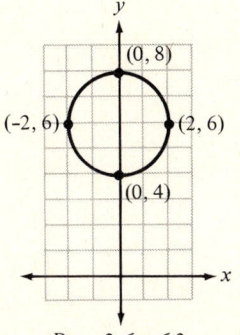

Do: $-2 \leqslant x \leqslant 2$
Ra: $4 \leqslant y \leqslant 8$

15.

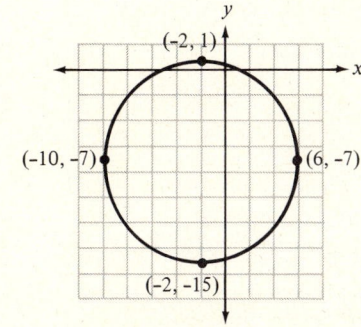

Do: $-10 \leqslant x \leqslant 6$
Ra: $-15 \leqslant y \leqslant 1$

17.

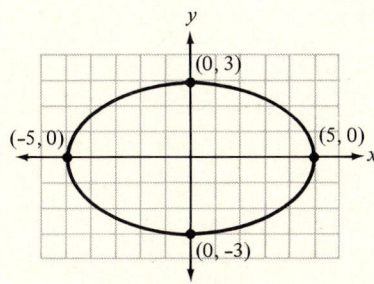

Do: $-5 \leqslant x \leqslant 5$
Ra: $-3 \leqslant y \leqslant 3$

19.

Do: $-1 \leqslant x \leqslant 1$
Ra: $-4 \leqslant y \leqslant 4$

21.

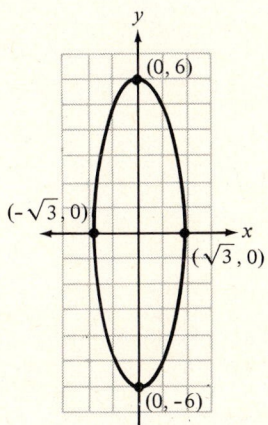

Do: $-\sqrt{3} \leqslant x \leqslant \sqrt{3}$
Ra: $-6 \leqslant y \leqslant 6$

23.

Do: $-\sqrt{5} \leqslant x \leqslant \sqrt{5}$
Ra: $-\sqrt{3} \leqslant y \leqslant \sqrt{3}$

25.

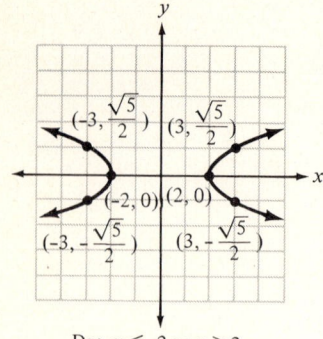

Do: $x \leqslant -2$ or $x \geqslant 2$
Ra: $y \in R$

27.

Do: $x \in R$
Ra: $y \geqslant 5$ or $y \leqslant -5$

29.

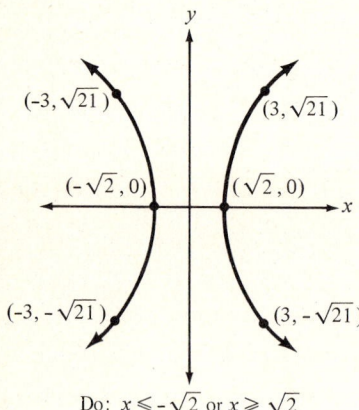

Do: $x \leqslant -\sqrt{2}$ or $x \geqslant \sqrt{2}$
Ra: $y \in R$

31.

Do: $x \in R$
Ra: $y \leqslant -10$ or $y \geqslant 10$

Stretching the Topics
1. $(x + 3)^2 + (y - 4)^2 = 25$ **2.** $(0, 1.7), (0, -1.7), (2.1, 0), (-2.1, 0)$

3.

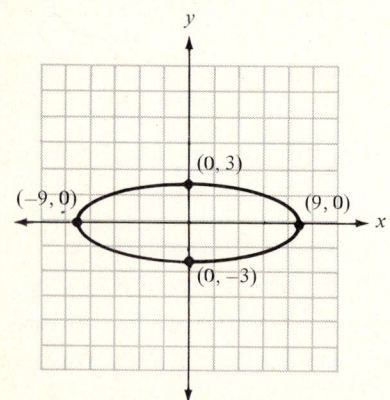

Do: $-9 \leqslant x \leqslant 9$
Ra: $-3 \leqslant y \leqslant 3$

Checkup 9.3 (page 729)

1.

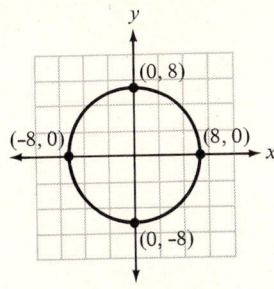

Do: $-8 \leqslant x \leqslant 8$
Ra: $-8 \leqslant y \leqslant 8$

2.

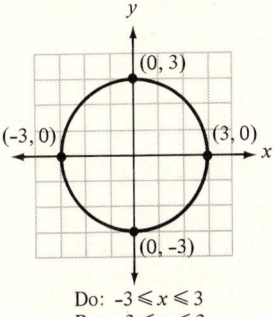

Do: $-3 \leqslant x \leqslant 3$
Ra: $-3 \leqslant y \leqslant 3$

3.

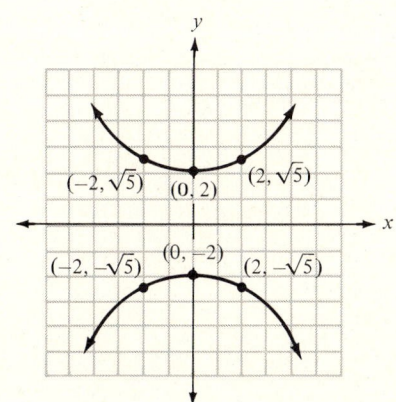

Do: $-13 \leqslant x \leqslant 5$
Ra: $0 \leqslant y \leqslant 18$

4.

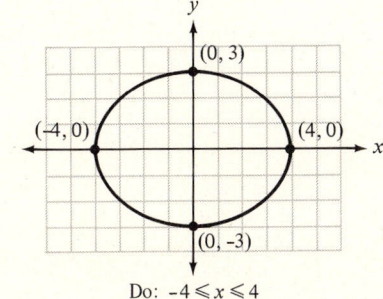

Do: $-4 \leqslant x \leqslant 4$
Ra: $-3 \leqslant y \leqslant 3$

5.

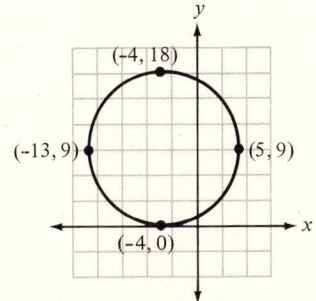

Do: $x \in \mathbb{R}$
Ra: $y \geqslant 2$ or $y \leqslant -2$

Exercise Set 9.4 (page 737)

1.

3.

5.

7.

9.

11.

13.

15.

17.

19.

Stretching the Topics

1.

2.

3.

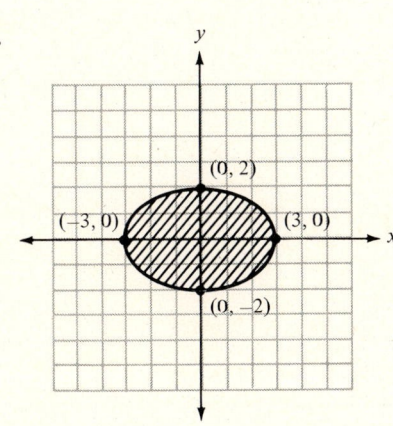

Checkup 9.4 (page 739)

1.

2.

3.

4.

5.

Speaking the Language of Algebra (page 741)
1. Parabola; upward **2.** *y*-axis; *y*-intercept **3.** Circle; ellipse; hyperbola
4. Unit; inside; outside

Review Exercises (page 743)

1.

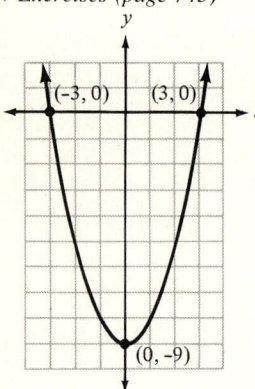

Do: $x \in R$
Ra: $y \in R, y \geq -9$

2.

Do: $x \in R$
Ra: $y \in R, y \leq 10$

3.

Do: $x \in R$
Ra: $y \in R$

4.

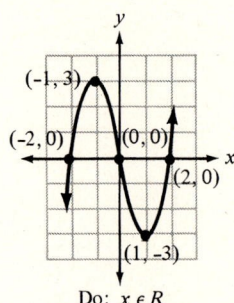

Do: $x \in R$
Ra: $y \in R$

5.

$(1, 5)$
$(3, \frac{5}{3})$
$(5, 1)$
$(-5, -1)$
$(-3, -\frac{5}{3})$
$(-1, -5)$

Do: $x \in R, x \neq 0$
Ra: $y \in R, y \neq 0$

6.

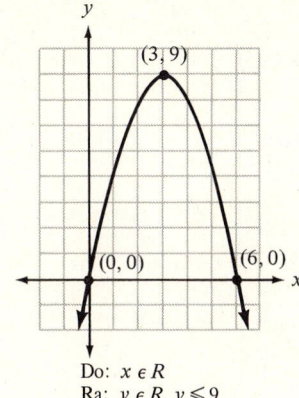

$(3, 9)$
$(0, 0)$
$(6, 0)$

Do: $x \in R$
Ra: $y \in R, y \leqslant 9$

7.

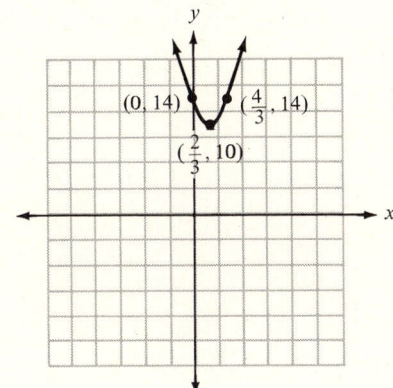

$(0, 14)$
$(\frac{4}{3}, 14)$
$(\frac{2}{3}, 10)$

Do: $x \in R$
Ra: $y \in R, y \geqslant 10$

8.

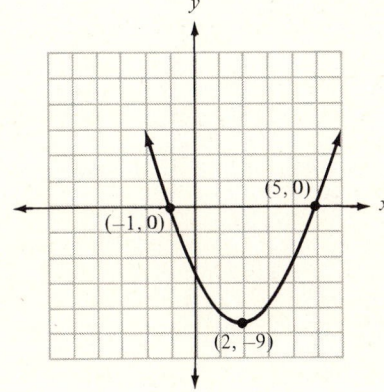

$(5, 0)$
$(-1, 0)$
$(2, -9)$

Do: $x \in R$
Ra: $y \in R, y \geqslant -9$

9.

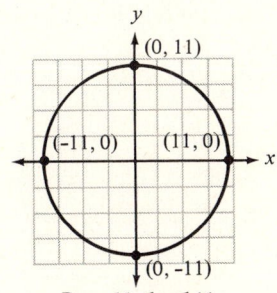

$(0, 11)$
$(-11, 0)$
$(11, 0)$
$(0, -11)$

Do: $-11 \leqslant x \leqslant 11$
Ra: $-11 \leqslant y \leqslant 11$

10.

$(0, \sqrt{5})$
$(-\sqrt{5}, 0)$
$(\sqrt{5}, 0)$
$(0, -\sqrt{5})$

Do: $-\sqrt{5} \leqslant x \leqslant \sqrt{5}$
Ra: $-\sqrt{5} \leqslant y \leqslant \sqrt{5}$

11.

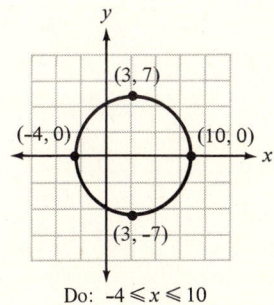

$(3, 7)$
$(-4, 0)$
$(10, 0)$
$(3, -7)$

Do: $-4 \leqslant x \leqslant 10$
Ra: $-7 \leqslant y \leqslant 7$

12.

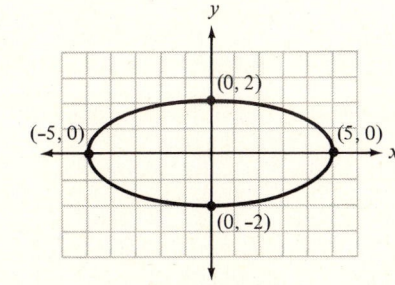

$(0, 2)$
$(-5, 0)$
$(5, 0)$
$(0, -2)$

Do: $-5 \leqslant x \leqslant 5$
Ra: $-2 \leqslant y \leqslant 2$

13.

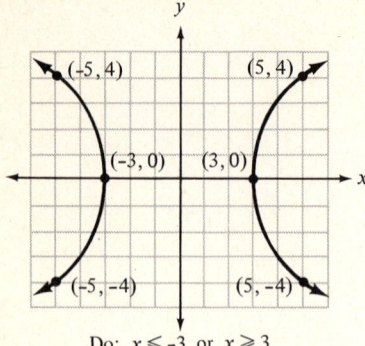

Do: $x \leqslant -3$ or $x \geqslant 3$
Ra: $y \in R$

14.

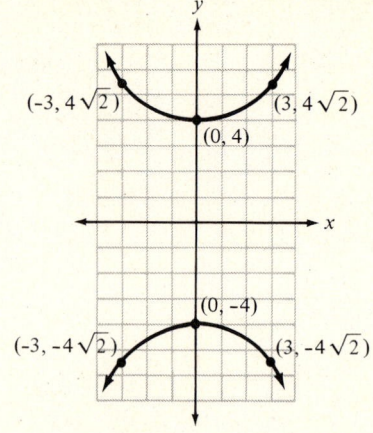

Do: $x \in R$
Ra: $y \leqslant -4$ or $y \geqslant 4$

15.

16.

17.

$10,000

Practice Test (page 745)

1.

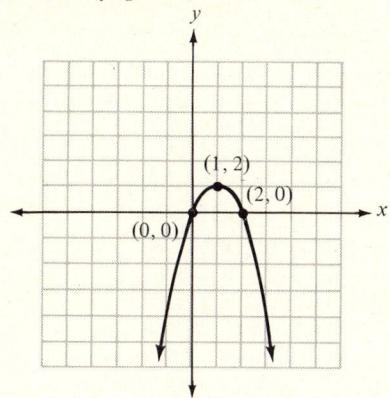

Do : $x \in R$
Ra : $y \in R, y \leqslant 2$

2.

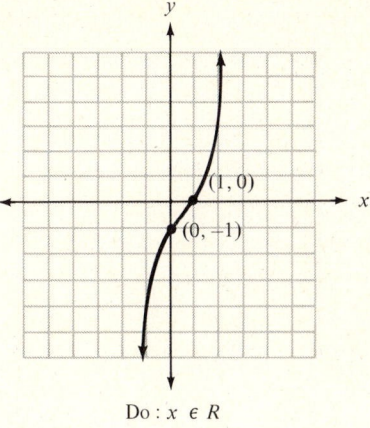

Do : $x \in R$
Ra : $y \in R$

3.

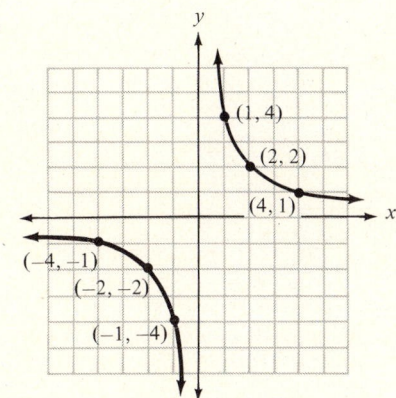

Do : $x \in R, x \neq 0$
Ra : $y \in R, y \neq 0$

4.

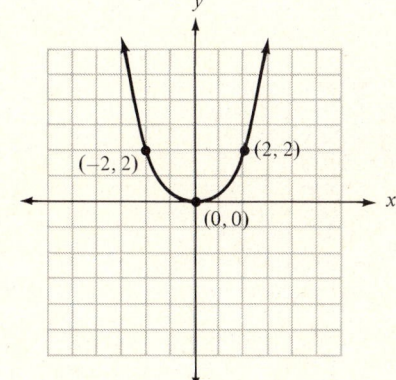

Do : $x \in R$
Ra : $y \in R, y \geqslant 0$

5.

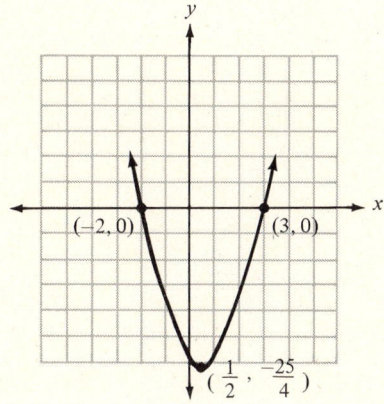

Do : $x \in R$
Ra ; $y \in R, y \geqslant \dfrac{-25}{4}$

6.

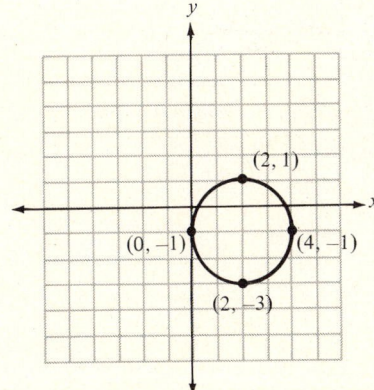

Do : $0 \leqslant x \leqslant 4$
Ra : $-3 \leqslant y \leqslant 1$

7.

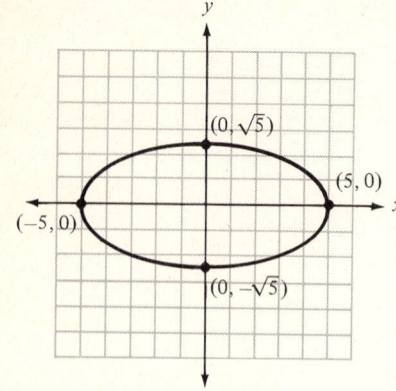

Do : $-5 \leqslant x \leqslant 5$
Ra : $-\sqrt{5} \leqslant y \leqslant \sqrt{5}$

8.

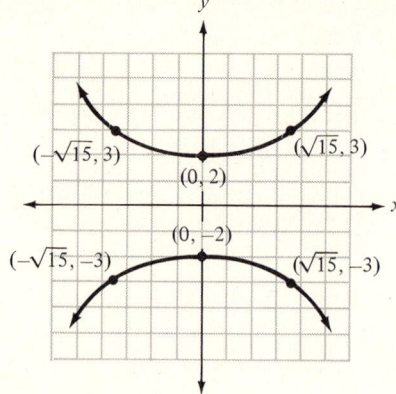

Do : $x \in R$
Ra : $y \in R, y \geqslant 2$ or $y \leqslant -2$

9.

10.

Sharpening Your Skills after Chapters 1–9 (page 751)

1. 33   **2.** $7 - 6\sqrt{2}i$ **3.** $y = \dfrac{-2}{5}x + \dfrac{89}{60}$ **4.** 17   **5.** $x^2 - 6x - 7 = 0$ **6.** $-17a$

7. $90a^{18}b^{17}$ **8.** $\dfrac{z^2}{x^3}$ **9.** $\dfrac{-7x^6}{y}$ **10.** $a^3 - 12a^2 - 45a$ **11.** $a^2 - 2a - 8$

12. $\dfrac{4x(x - 2)}{(x + 4)(x - 5)}$ **13.** $\dfrac{2}{x + 3}$ **14.** $\dfrac{xy}{y - 2x}$ **15.** $\dfrac{x^{1/3}}{y^{5/27}z^{1/9}}$ **16.** $2xy^2\sqrt[4]{x}$

17. $\dfrac{-(\sqrt{2} - \sqrt{5})}{3}$ **18.** All real numbers   **19.** $x = 0, x = \dfrac{8}{7}$ **20.** $x = \dfrac{3 \pm \sqrt{19}}{5}$

21. $x \leq 4$ **22.** $x = -5$ **23.** $x < \dfrac{3}{2}$ or $x > 2$ **24.** $x = -4, x = 3$

25. $x = \dfrac{3}{5}, x = \dfrac{-3}{10} \pm \dfrac{3\sqrt{3}}{10}i$ **26.**

27.

28.

29. $V = kt$; $k = 32$; 256 ft per sec   **30.** $V = \pm \sqrt{\dfrac{2hg}{2}}$

CHAPTER 10

Exercise Set 10.1 (page 767)

1.

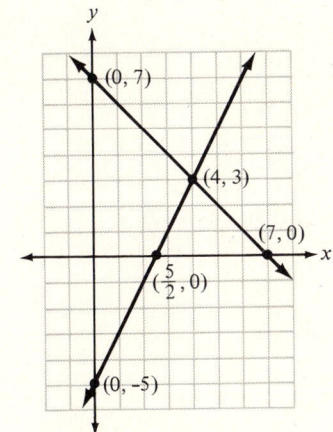

Solution: $(4, 3)$
Independent and consistent

3.

Solution: None
Inconsistent

5.

Solution: $(-2, -12)$
Independent and consistent

7.

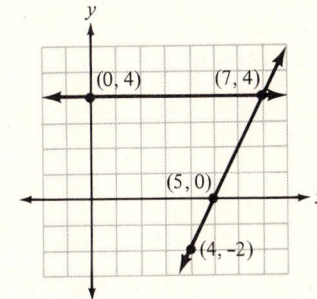

Solution: $(7, 4)$
Independent and consistent

9.

Solution: All ordered pairs satisfying
$x + 4y = 1$
Dependent

11.

Solution: (2, 3)
Independent and consistent

13.

Solution: (-2, 6)
Independent and consistent

15.

Solution: None
Inconsistent

17. $(2, -3)$ **19.** $\left(\dfrac{1}{10}, \dfrac{17}{10}\right)$ **21.** $(2, 3)$ **23.** $\left(\dfrac{-13}{21}, \dfrac{-13}{7}\right)$ **25.** No solution

27. $\left(\dfrac{25}{14}, \dfrac{-6}{7}\right)$ **29.** All ordered pairs satisfying $y = 2x - 4$ **31.** $(1, 3)$

Stretching the Topics
1. $(1, -2)$ **2.** $k = 3$

Checkup 10.1 (page 769)
1.

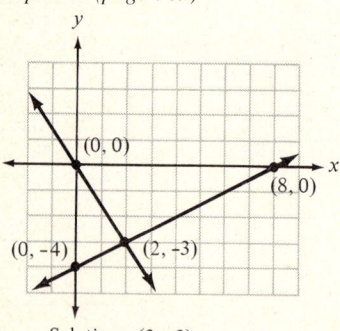

Solution: (2, -3)
Independent and consistent

2.

Solution: All ordered pairs
satisfying $-x + 2y = 4$
Dependent

3. $(2, 2)$ **4.** No solution   **5.** $\left(\dfrac{5}{3}, \dfrac{4}{3}\right)$

Exercise Set 10.2 (page 779)
1. $(1, -1)$ **3.** $\left(\dfrac{3}{4}, \dfrac{9}{8}\right)$ **5.** $(2, 1)$ **7.** No solution   **9.** $(1, 1)$ **11.** $(1, 1)$

13. All ordered pairs satisfying $3x - 6y = 9$ **15.** $\left(0, \dfrac{-5}{2}\right)$ **17.** $\left(\dfrac{-1}{25}, \dfrac{-11}{25}\right)$

19. $\left(\dfrac{-6}{37}, \dfrac{-27}{37}\right)$ **21.** $\left(\dfrac{18}{7}, \dfrac{30}{7}\right)$ **23.** $(4, 3)$ **25.** $(-170, 70)$ **27.** $(0, -1, 1)$

29. $(2, -1, 3)$ **31.** $(4, 0, 2)$ **33.** $(1, -1, 4)$ **35.** $\left(\dfrac{3}{2}, 3, -3\right)$

Stretching the Topics
1. $\left(\dfrac{31}{9}, \dfrac{25}{12}\right)$ **2.** $(0, 4)$ **3.** $m = 1, b = -1$

Checkup 10.2 (page 781)
1. $(-1, -1)$ **2.** $(1, 1)$ **3.** $(5, 2)$ **4.** All ordered pairs satisfying $3x - 6y = 3$
5. $(2, 3, 1)$

Exercise Set 10.3 (page 789)
1. 16 **3.** 12 **5.** $\dfrac{1}{3}$ **7.** $(5, -1)$ **9.** No solution **11.** $(-1, 0)$ **13.** $\left(\dfrac{-2}{3}, \dfrac{3}{2}\right)$

15. $\left(\dfrac{67}{25}, \dfrac{12}{25}\right)$ **17.** $(5, 3)$

Stretching the Topics
1. $a = \dfrac{-38}{7}$ **2.** $\left(\dfrac{3b - 4}{ab + 4}, \dfrac{-2(a + 3)}{ab + 4}\right)$

Checkup 10.3 (page 791)
1. -15 **2.** 3 **3.** $(2, -1)$ **4.** No solution **5.** $(3, -2)$ **6.** $\left(\dfrac{30}{97}, \dfrac{19}{97}\right)$

Exercise Set 10.4 (page 805)
1. $(1, -4), (-2, -1)$ **3.** $(-2, -4), (4, 2)$ **5.** $(4, -3), (-4, -3), (3, 4), (-3, 4)$
7. $\left(\dfrac{5}{2}, 1\right), \left(\dfrac{-5}{2}, -1\right)$ **9.** $(2, 2), (2, -2), (-2, 2), (-2, -2)$ **11.** No solution
13. $(-5, -2), (2, 5)$ **15.** $(1, 2), (1, -2)$

17.

19.

21.

23.

Stretching the Topics

1. $y = \dfrac{a \pm \sqrt{32 - a^2}}{2}$ **2.** $-4\sqrt{2} < a < 4\sqrt{2};\ a > 4\sqrt{2}$ or $a < -4\sqrt{2};\ a = \pm 4\sqrt{2}$

Checkup 10.4 (page 807)

1. $(-4, 8), (1, -7)$ **2.** $(0, 3), (\sqrt{5}, -2), (-\sqrt{5}, -2)$ **3.** No solution

4.

Exercise Set 10.5 (page 813)

1. 193 adults, 52 children   **3.** $5 mowing, $4 guarding   **5.** 14 at $2, 16 at $1.75
7. $15,000 at 10%, $10,000 at 12%   **9.** Sam 55 mph, roommate 65 mph   **11.** 12 by 16 ft
13. 175 at $1.10, 325 at $0.95   **15.** 100 of each   **17.** 15 and 20 in.   **19.** 8 by 9 ft
21. 3 by 4 ft

Stretching the Topics

1. 4 and 9 in.   **2.** 50 bales weighing 60 lb each   **3.** 5 by 12 ft

Checkup 10.5 (page 817)

1. Lettuce $5.40, cabbage $4.20   **2.** 9 and 12 ft   **3.** $3\frac{5}{8}$ hr at 45 mph, $2\frac{3}{8}$ hr at 53 mph

4. 50 by 160 ft or 80 by 100 ft

Speaking the Language of Algebra (page 820)

1. Ordered pair   **2.** Substitution   **3.** Opposite   **4.** No; parallel   **5.** 4

Review Exercises (page 821)

1.

Solution: $(1, -2)$
Independent and consistent

2.

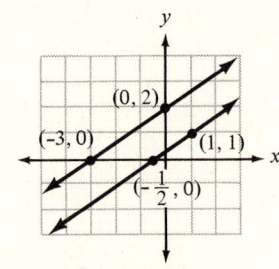

Solution: None
Inconsistent

3. $(-4, 3)$ **4.** $(2, -3)$ **5.** $(-8, 5)$ **6.** No solution   **7.** $(10, -3)$ **8.** $(0, 2, -1)$
9. $(8, 49), (-3, -6)$ **10.** $(4, 2), (4, -2), (-4, 2), (-4, -2)$

11.

12.

13. 83 speeding, 34 failure to obey traffic signals   **14.** 21 mi   **15.** 5 by 12 m

Practice Test (page 823)

1.

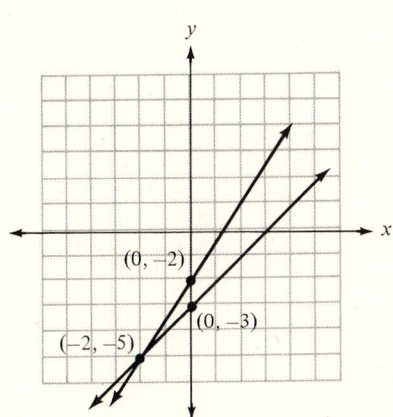

2. $\left(\dfrac{-1}{2}, 3\right)$

3. $\left(\dfrac{-19}{3}, \dfrac{10}{3}\right)$ **4.** All ordered pairs satisfying $6x - 4y = -20$ **5.** $(-5, -7)$

6. No solution   **7.** $(1, -2, 0)$ **8.** $(3, 3), (1, -1)$

9.

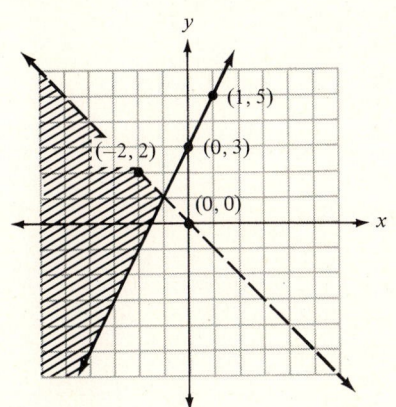

10. Apple 80 calories, peach 60 calories

Sharpening Your Skills after Chapters 1–10 (page 825)

1. $a \neq \frac{2}{5}, a \neq -5, \frac{a-11}{5a-2}$ **2.** $x = 5, y = \frac{-7}{5}$ **3.** $y = 1$ **4.** $\left(4, \frac{7}{2}\right)$

5. $y = 2x + 9$ **6.** -9 **7.** $x \leq \frac{5}{2}$ **8.** -4

9. $x < \frac{4}{5}$

10. $\frac{2a^9}{3b^5}$ **11.** $x^3 + 8x^2 + 7x$ **12.** $x^4 - 4x^3 + 6x^2 - 4x + 1$ **13.** $\frac{-(x+4)}{5x-2}$

14. $\frac{4x^2 + 15x - 11}{(x-2)(x-1)(x+4)}$ **15.** $\frac{(x-3)^2}{(x+2)^2}$ **16.** $\frac{9}{4}$ **17.** $\frac{x^{1/20}z^{1/6}}{y^{1/24}}$ **18.** $-2x\sqrt[3]{2}$

19. $12 - 7\sqrt{3y} - 2y$ **20.** $\frac{-5}{7} - \frac{2\sqrt{6}}{7}i$ **21.** $x = \frac{-1}{5}, x = \frac{-3}{2}$

22. $x = 3, x = \frac{1}{4}, x = \frac{-1}{4}$ **23.** $x = \frac{-5 + \sqrt{17}}{8}, x = \frac{-5 - \sqrt{17}}{8}$ **24.** $-12 < x < 12$

25. No solution   **26.** $x < 0$ or $x \geq 15$ **27.** $x = 36$ **28.** $x = \frac{4}{7} \pm \frac{\sqrt{5}}{7}i$

29.

30.

31.

32.

Do: $x \in R$
Ra: $y \in R$

33.

Do: $x \in R$

Ra: $y \in R, y \leq 12\frac{1}{4}$

34.

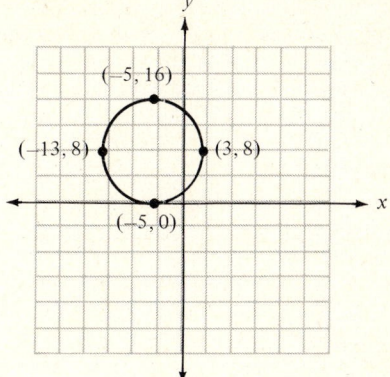

Do: $-13 \leq x \leq 3$

Ra: $0 \leq y \leq 16$

35.

36. $7\frac{1}{2}$ hr

CHAPTER 11

Exercise Set 11.1 (page 839)

1. $\frac{1}{2}$, 1, 2, 16   **3.** $\frac{1}{25}$, $\frac{\sqrt{5}}{5}$, $\sqrt{5}$, 25   **5.** $\frac{-1}{8}$, $\frac{-1}{2}$, -1, -4 **7.** $\frac{1}{4}$, 1, $\sqrt{2}i$, 16

9. $\frac{1}{2}$, $\frac{1}{4}$, $\frac{\sqrt{2}}{2}$, 4

11.

13.

(0, 3)

(-2, $\frac{1}{3}$)

(-1, 1)

17.

(-1, 9)

(0, 1) (1, $\frac{1}{9}$)

19.

(1, 3)

(-1, $\frac{1}{3}$)

(0, 1)

21.

(2, 5)

(3, 1) (4, $\frac{1}{5}$)

23.

(0, 2) ($\frac{1}{2}$, 1)

(1, $\frac{1}{2}$)

25. $x = -1$ **27.** $x = 2$ **29.** $x = -4$ **31.** $x = -1$ **33.** $x = \dfrac{-1}{2}$ **35.** $x = -2$

37. $x = \dfrac{-3}{4}$

Stretching the Topics

1.

(-1, 6)

(0, 4) (1, 3)

(2, $\frac{5}{2}$)

2. $x = 4, x = -1$ **3.** 8830

Checkup 11.1 (page 841)

1. $\dfrac{1}{9}, \dfrac{1}{3}, \dfrac{\sqrt{3}}{3}, 3, \sqrt{3}$

2.

3.

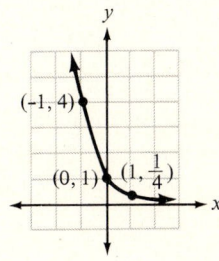

4. $x = -2$ **5.** $x = \dfrac{2}{7}$

Exercise Set 11.2 (page 853)

1. $\log_3 9 = 2$ **3.** $\log_4 \dfrac{1}{16} = -2$ **5.** $\log_{16} 4 = \dfrac{1}{2}$ **7.** $\log_9 \dfrac{1}{3} = \dfrac{-1}{2}$ **9.** $5^3 = 125$

11. $2^{-5} = \dfrac{1}{32}$ **13.** $16^{1/2} = 4$ **15.** $36^{-1/2} = \dfrac{1}{6}$ **17.** -1 **19.** 0 **21.** 3 **23.** 7

25. Does not exist   **27.** 5 **29.** 0 **31.** $x = -1$ **33.** $a = 7$ **35.** $b = \dfrac{1}{32}$

37. $x = 63$ **39.** $x = \dfrac{-1}{2}$ **41.** $x = 3$

43.

45.

Stretching the Topics

1. $x = \dfrac{5 + \sqrt{41}}{2}$

3. $x = -6, x = 1$

2.

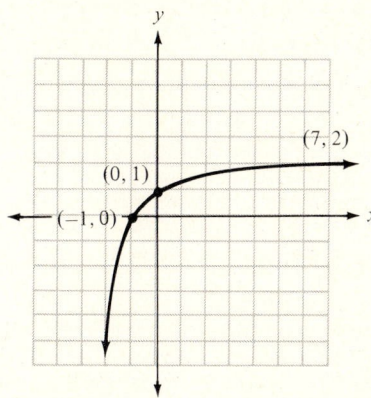

Checkup 11.2 (page 855)

1. $\log_5 25 = 2$ **2.** $\log_{64} \dfrac{1}{8} = \dfrac{-1}{2}$ **3.** $3^4 = 81$ **4.** $5^{-2} = \dfrac{1}{25}$ **5.** -5 **6.** $\dfrac{1}{3}$

7. $a = 5$ **8.** $x = 5$

9.

10.

Exercise Set 11.3 (page 865)

1. 7 **3.** −1 **5.** 4 **7.** $\frac{3}{2}$ **9.** 7 **11.** 0 **13.** −1 **15.** 1.1761 **17.** 1.1505

19. 2.5441 **21.** 0.9121 **23.** 1.6532 **25.** $3 \log_a x + 2 \log_a (x + 5)$

27. $5 \log_a (x - 3) - \frac{1}{2} \log_a 3x$ **29.** $\frac{2}{3} \log_a x - 2 \log_a x - 5 \log_a (x - 9)$

31. $\log_a (x - 3) + \frac{1}{2} \log_a (2 - x) - \frac{1}{2} \log_a x - 5 \log_a (x + 1)$ **33.** $\log_a \sqrt[3]{x} \, (x - 1)^2$

35. $\log_a \dfrac{\sqrt{x}}{(x + 2)^4}$ **37.** $\log_a \dfrac{x^3 \sqrt{3x}}{x^2 + 1}$ **39.** $\log_a \dfrac{\sqrt[3]{(x + 1)^2} \, (x + 1)^3}{x(x^2 + 1)^5}$

Stretching the Topics

1. $\log_a r = \frac{1}{2} (\log_a 3 + \log_a V - \log_a \pi - \log_a h)$

2. $5 \log_a \sqrt[3]{a} - 7 \log_a \sqrt{a} + 10 \log_a \sqrt[3]{a} - \frac{3}{2} \log_a a$

$= \log_a \dfrac{a^{5/3} \cdot a^{10/3}}{a^{7/2} \cdot a^{3/2}}$

$= \log_a \dfrac{a^{15/3}}{a^{10/2}}$

$= \log_a \dfrac{a^5}{a^5}$

$= \log_a 1$

$= 0$

Checkup 11.3 (page 867)

1. 7 **2.** $\frac{1}{2}$ **3.** 1.8572 **4.** 0.9515 **5.** 2.5562 **6.** 1.0512

7. $2 \log_a x + \frac{1}{2} \log_a (x^2 + 1)$ **8.** $\frac{2}{3} \log_a x + \log_a (x^2 + 1) - 5 \log_a (x^2 + 4)$

9. $\log_a \dfrac{(x + 1)^2}{\sqrt[3]{x^2}}$ **10.** $\log_a \dfrac{\sqrt{3x} \, (x + 5)}{x^4(2x + 1)}$

Exercise Set 11.4 (page 879)

1. 5.42×10^{-4} **3.** 2.35×10^1 **5.** 8.9×10^7 **7.** 7.1×10^{-4} **9.** 7.83×10^4
11. 0.000162 **13.** 382,000 **15.** 30.1 **17.** 0.0421 **19.** 15,000,000 **21.** 4.9420
23. $-4 + 0.1430$ **25.** 1.2718 **27.** 7.9294 **29.** $-9 + 0.7931$ **31.** 0.0757
33. 156,000 **35.** 0.0000000443 **37.** 2.27 **39.** 88,600 **41.** $N \doteq 3.07$
43. $N \doteq 0.575$ **45.** $N \doteq 2.91$ **47.** $N \doteq 15.0$ **49.** $N \doteq 4.87$ **51.** $N \doteq 5.34$
53. $N \doteq 111$

Stretching the Topics
1. $N \doteq 1.80$ 2. 105 m

Checkup 11.4 (page 881)
1. 2.31×10^{-3} 2. 12,100   3. 2.8609   4. $-3 + 0.9926$ 5. 0.00123   6. 81,400
7. $N \doteq 0.0175$ 8. $N \doteq 2.09$ 9. $N \doteq 20.3$ 10. $N \doteq 0.0000276$

Exercise Set 11.5 (page 897)
1. $x = 7$ 3. $x = 4, x = \dfrac{-4}{3}$ 5. $x = -8, x = 2$ 7. $x = 5$ 9. No solution

11. $x = \dfrac{7}{2}$ 13. $x = 3$ 15. $x = 38$ 17. $x = 4$ 19. $x = 11$ 21. $x = \dfrac{9}{4}$

23. $x = 12$ 25. $x = 33$ 27. $x = 4$ 29. $x = \dfrac{7}{3}$ 31. No solution   33. $x = 5$

35. $x = 8$ 37. $x \doteq 2.322$ 39. $x \doteq 1.262$ 41. $x \doteq 2.760$ 43. $x \doteq 0.492$
45. $x \doteq 5.577$ 47. $x \doteq 1.788$ 49. $x \doteq -8.826$ 51. $x \doteq -19.865$ 53. $6090
55. $21,800   57. 15 years (14.28)   59. 16200, 3.56 hr

Stretching the Topics
1. $x = 1.296$ 2. $116,000   3. 40 min

Checkup 11.5 (page 901)
1. $x = 7$ 2. No solution   3. $x = 6$ 4. $x = 10$ 5. $x \doteq 0.594$ 6. $x \doteq 1.297$

Speaking the Language of Algebra (page 904)
1. Exponential function; $x \in R$; $y \in R, y > 0$ 2. Logarithmic function; $x \in R, x > 0, y \in R$
3. $a^y = x$ 4. Common   5. Same base   6. Exponential   7. Negative

Review Exercises (page 905)
1. $5, \dfrac{1}{5}, \dfrac{1}{125}, \dfrac{\sqrt{5}}{5}, 1$

2.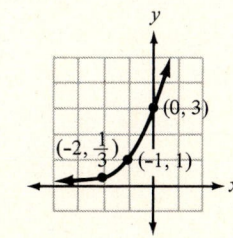

3. $x = -1$ 4. $\log_3 \dfrac{1}{9} = -2$ 5. $\log_{125} \dfrac{1}{5} = \dfrac{-1}{3}$ 6. $2^6 = 64$ 7. $8^{-1/3} = \dfrac{1}{2}$

8. -3 9. $\dfrac{-1}{2}$ 10. $a = 6$ 11. $b = \dfrac{1}{64}$ 12. $x = 0$

13.

14. $\dfrac{5}{4}$ 15. 5   16. 0.0737   17. 1.3662
18. $5 \log_a x + \log_a (x + 5)$

19. $3 \log_a x + \dfrac{1}{2} \log_a (x^2 - 1) - 5 \log_a (x + 1)$ 20. $\log_a x^2 \sqrt{x - 1}$

21. $\log_a \dfrac{(2x + 1)^3}{\sqrt[5]{x^4} (x + 1)}$ 22. 7.29×10^{-3} 23. 934,000   24. $N \doteq 16,500$ 25. $x = 1$

26. $x = 9$ 27. $x = 24$ 28. $x = \dfrac{20}{3}$ 29. $x \doteq 1.193$ 30. $x \doteq -0.792$

Practice Test (page 907)

1. 81, 1, 3   **2.** $\log_{36} \frac{1}{6} = \frac{-1}{2}$ **3.** $5^3 = 125$ **4.** -2 **5.** $\frac{-13}{2}$ **6.** 3.7993

7. $5 \log_a x + 2 \log_a (x^2 - 3) - \frac{1}{3} \log_a (x + 1)$ **8.** $\log_a \dfrac{\sqrt[5]{x + 1}}{\sqrt[3]{x^2} (x - 6)}$ **9.** 2.35×10^6

10. 0.00172   **11.** $N \doteq 101$

12.

13.

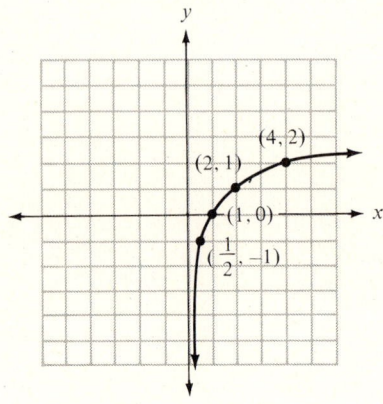

14. $x = \dfrac{-3}{5}$ **15.** $a = 2$ **16.** $x = 2$ **17.** $x = 3$ **18.** $x = 1$ **19.** $x = 2$

20. $x = \dfrac{\log 3}{\log 3 - \log 8}$

Sharpening Your Skills after Chapters 1–11 (page 909)

1. 32   **2.** 11   **3.** $y \neq 4, \dfrac{3(y + 4)}{y - 4}$ **4.** $\dfrac{16}{9}$ **5.** i **6.** $\dfrac{-16}{9}$ **7.** $y = \dfrac{1}{2}x - \dfrac{1}{4}$

8. 17   **9.** $\dfrac{7}{2}$ **10.** $7x^4y^7$ **11.** $\dfrac{y^2}{49x^{12}}$ **12.** $-47x^2 + 27x + 2$

13. $8a^3 - 36a^2 + 54a - 27$ **14.** $a^2 - 2a - 8$ **15.** $\dfrac{2x - 1}{x + 2}$ **16.** $\dfrac{-2}{x + 5}$ **17.** $\dfrac{4x - 3}{x - 6}$

18. $ab\sqrt[3]{a}$ **19.** $12 - 7\sqrt{3y} - 2y$ **20.** $\dfrac{-1}{2} - \dfrac{3}{2}i$ **21.** $(x + 7)(x^2 + 2x + 13)$

22. Prime   **23.** $(9 - 2x + y)(3 + 10x - 5y)$ **24.** $x = -2$ **25.** $x > -2$ or $x < -3$

26. No solution   **27.** $x = 3\sqrt{2}, x = -3\sqrt{2}$ **28.** $\dfrac{1}{2} < x < 3$ **29.** $x = 3$

30. $-2 < x < 0$ **31.** No solution   **32.** $y = \dfrac{1}{2} \pm \dfrac{3\sqrt{3}}{2}i$ **33.** $(2, -1)$

34. $(12, 4), (-3, -1)$

35.

36.

37.

38.

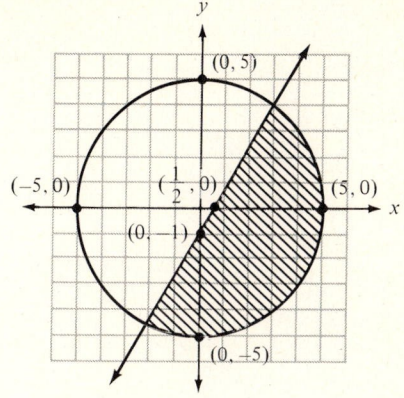

39. $g = \dfrac{4\pi^2 l}{t^2}$ **40.** 6 by 9 ft

Index